D1709619

THE BOOK OF THE FARM

DETAILING THE LABOURS OF THE STEWARD, PLOWMAN, HEDGER, CATTLE-MAN, SHEPHERD, FIELD-WORKER, AND DAIRYMAID.

WITH FOUR HUNDRED AND FIFTY ILLUSTRATIONS

Volume I

by

Henry Stephens

THE

BOOK OF THE FARM.

1. OF THE DIFFICULTIES WHICH THE YOUNG FARMER HAS TO ENCOUNTER AT THE OUTSET OF LEARNING PRACTICAL HUSBANDRY.

"One, but painted thus,
Would be interpreted a thing perplex'd
Beyond self-explication." CYMBELINE.

THE young farmer, left to his own guidance, when beginning to learn his profession, encounters many perplexing difficulties. The difficulty which at first most prominently obtrudes itself on his notice consists in the *distribution* of the labor of the farm; and it presents itself in this way:— He observes the teams employed one day in one field, at one kind of work, and perhaps the next day in another field, at a different sort of work. He observes the persons employed as field-workers assisting the teams one day, and in the next, perhaps, working by themselves in another field or elsewhere. He observes those changes with attention, considers of their utility, but cannot discover the reasons for making so very varied arrangements; not because he entertains the least doubt of their propriety, but, being as yet uninitiated in the art of farming, he cannot foresee the purpose for which those labors are performed. The reason why he cannot at once foresee this is, that in all cases, excepting at the finishing operations, the end is unattained at the time of his observation.

The next difficulty the young farmer encounters is in the *variety* of the labors performed. He not only sees various arrangements made to do the same sort of work, but various kinds of work. He discovers this difference on examining more closely into the nature of the work he sees performing. He observes one day the horses at work in the plow in one field, moving in a direction quite opposite, in regard to the ridges, to what they were in the plow in another field. On another day he observes the horses at work with quite a different implement from the plow. The field-workers, he perceives, have laid aside the implement with which they were working, and are performing the labor engaged in with the hand. He cannot comprehend why one sort of work should be prosecuted one day, and quite a different sort of work the next. This difficulty is inexplicable for the same reason why he could not overcome the former one: because he cannot foresee the end for which those varieties of work are performed. No doubt he is aware that every kind and variety of work which are performed on a farm, are preparatives to the attainment of certain crops; but what portion of any work is intended as a certain part of the preparation for a particular crop, is a knowledge which he cannot acquire by intuition. Every preparatory work is thus perplexing to the young farmer.

Field work being thus chiefly *anticipatory*, is the circumstance which renders its object so perplexing to the learner. He cannot possibly perceive the connection between preparatory labors and their ultimate ends; and yet, until he learn to appreciate their necessary connection, he will remain incapable of managing a farm. It is in the exercise of this faculty of anticipation or foresight that the experienced and careful farmer is contradistinguished from the ignorant and careless. Indeed, let the experience of farming be ever so extensive, or, in other words, let the knowledge of minutiæ be ever so intimate, unless the farmer use his experience by foresight, he will never be enabled to conduct a farm aright. Both foresight and experience are acquired by observation, though the former is matured by reflection. Observation is open to all farmers, but all do not profit by it. Every farmer may acquire, in time, sufficient *experience* to conduct a farm in a passable manner; but many farmers never acquire *foresight*, because they never reflect, and therefore cannot make their experience tell to the most advantage. Conducting a farm by foresight is thus a higher acquirement than the most intimate knowledge of the minutiæ of labor. Foresight cannot be exercised without the assistance of experience; though the latter may exist independently of the former. As the elements of every art must first be acquired by observation, a knowledge of the minutiæ of labor should be the first subject for acquirement by the young farmer. By carefully tracing the connection betwixt combined operations and their ultimate ends, he will acquire foresight.

The necessity of possessing foresight in arranging the minutiæ of labor, before the young farmer can with confidence undertake the direction of a farm, renders *farming* more difficult of acquirement, and a longer time of being acquired, than most other arts. This statement may appear incredible to those who have been accustomed to hear of farming being easily and soon learned by the meanest capacity. No doubt it may be acquired in time, to a certain degree, by all who are capable of improvement by observation and experience; but, nevertheless, the ultimate ends for which the various kinds of field-work are prosecuted, are involved in obscurity to every learner. In most other arts no great space usually elapses between the commencement and completion of the piece of work, and the piece is worked at until finished. The beginner can thus soon perceive the connection between the minutest portion of the work in which he is engaged, and the object for which it is intended. There is in this no obscurity to perplex his mind. He is purposely led, by degrees, from the simplest to the most complicated parts of his art, so that his mind is not bewildered at the outset by participating in a multiplicity of works at one time. He thus begins to acquire true experience from the outset.

The young farmer has no such advantages in his apprenticeship. There is no simple, easy work, or one object only to engage his attention at first. On the contrary, many minutiæ connected with the various works in progress, claim his attention at one and the same time, and if the requisite attention to any one of them be neglected for the time, no other opportunity for observing it can occur for a twelvemonth. It is a misfortune to the young farmer, in such circumstances, to be thrown back in his progress by a trifling neglect. He cannot make up his lee-way until after the revolution of a year. And though ever so attentive, he cannot possibly learn to anticipate operations in a shorter time, and therefore cannot possibly understand the drift of a single operation in the first year of his apprenticeship. The first year is generally spent almost unprofitably, and certainly unsatisfactorily to an inquisitive mind. But attentive observation during the first year will enable him, in the second, to anticipate the successive operations ere they arrive, and arrange every minutia of labor as it is required. Many of the events of the first year, which had left no adequate impression of their importance on his memory, crowd upon his observation in the second, as essential components of recognized operations. A familiar recognition of events tends, in a rapid degree, to enlarge the sphere of experience and to inspire confidence in one's own judgment: and this quality greatly facilitates the acquisition of foresight.

Let it not be imagined by those who have never passed through the perplexing ordeals incident to the first year of farming, that I have described them in strong colors, in order to induce to the belief that farming is an art more difficult of attainment than it really is. So far is this from being the case, I may safely appeal to the experience of every person who had attained manhood before beginning to learn farming, whether I have not truly depicted his own condition at the outset of his professional career. So that every young man learning farming must expect to meet with those difficulties.

---

2. OF THE MEANS OF OVERCOMING THOSE DIFFICULTIES.

"We can clear these ambiguities."
ROMEO AND JULIET.

EXPERIENCE undoubtedly dissipates doubt and removes perplexity; but experience, though a sure and a safe, is a slow teacher. A whole year must revolve ere the entire labors of a farm can be exhibited in the field. and the young farmer satisfactorily understand what he is about; and a whole year is too much time for most young men to *sacrifice*. Could the young farmer find a monitor to explain to him, during the first year of his apprenticeship, the purpose for which every operation on a farm is performed,—foretell to him the results which every operation is intended to effect,—and indicate to him the relative progress which all the operations should make, from time to time, toward the attainment of their various ends, he would thereby acquire a far greater quantity of professional information, and have greater confidence in its accuracy, than he could possibly obtain for himself in that anxious period of his novitiate. Such a monitor would best be an experienced and intelligent farmer, were he duly attentive to his pupil. Farmers, however, can scarcely bestow so much attention as would be desired by pupils at all times; because the lapses of time occasioned by necessary engagements, in the fulfilment of which farmers are sometimes obliged to leave home, produce inattention on the part of the farmer; and inattention and absence combined constitute sad interruptions to tuition, and cannot always be avoided by the most painstaking farmer. But a *book* might be made an efficient assistant-monitor. If expressly written for the purpose, it might not only corroborate what the farmer inculcated, but serve as a substitute in his temporary absence. In this way tuition might proceed uninterruptedly, and the pupil never want a monitor upon whom he could confidently rely. Were a book, purposely so arranged, put into the hands of young farmers so circumstanced, the usual deprecations against recommending the acquirement of practical farming from books alone would not here apply. I would give no such counsel to any young farmer; because books on farming, to be really serviceable to the learner, ought not to constitute the arena on which to study farming—the field being the best place for perceiving the fitness of labor to the purposes it is designed to attain—but as monitors for indicating the best modes of management, and showing the way of learning those modes most easily. *By these, the practice of experienced farmers might be communicated and recommended to beginners. By consulting those which had been purposely written for their guidance, while they themselves were carefully observing the daily operations of the farm, the import of labors—which are often intricate, always protracted over considerable portions of time, and necessarily separated from each other—would be acquired in a much shorter time than if left to be discovered by the sagacity of beginners.*

It is requisite to explain that, by the phrase "*young farmer*," I mean the young man who, having finished his scholastic and academical education, directs his attention, for the first time, to the acquirement of practical farming; or who, though born on a farm, having spent the greater part of his life at school, determines, at length, on following his father's profession. For the latter class of young men, tuition in farming, and information from books, are as requisite as for the former. Those who have constantly seen farming from infancy can never be said to have been young farmers; for, by the time they are fit to act for themselves, they are proficients in farming. Having myself, for a time, been placed precisely in the position of the first description of young men, I can bear sincere testimony to the truth of the difficulties I have described as having to be encountered in the first year of apprenticeship. I felt that a guide-book would have been an invaluable monitor to me, but none such existed at the time. No doubt it is quite reasonable to expect of the farmer ability to instruct the pupils committed to his charge in a competent manner. This is certainly his duty; which, if rightly performed, no guide-book would be required by pupils; but very few farmers who receive pupils undertake the onerous task of instruction. Practical farming they leave the pupils to acquire for themselves in the fields, by imperfect observation and slow experience, as they themselves had previously done; theoretical knowledge, very few, if any, are competent to impart. The pupils, being thus very much left to their own application, can scarcely avoid being beset with difficulties, and losing much time. At the same time it must be acknowledged that the practice gained by slow experience is, in the end, the most valuable and enduring. Still, a book on farming, expressly written to suit his circumstances, might be a valuable instructor to the young farmer; it might guard him against the difficulties which learners are apt to encounter; and it would recompense him for loss of time, by imparting sound professional information.

Such a book, to be really a useful instructor and correct guide, should, in my estimation, possess these necessary qualifications. Its principal matter should consist of a clear narrative of all the labors of the farm, as they occur in succession; and it should give the reasons fully for which each piece of work is undertaken. While the principal operations are narrated in this way, the precise method of executing every species of work, whether manual or implemental, should be minutely detailed. The construction of the various implements by which work is performed—the mode of using them—the accidents to which each is liable—should be circumstantially described. A seasonable narrative of the principal operations will show the young farmer that farming is really a systematic business, having a definite object in view, and possessing the means of attaining it. The reasons for doing every piece of work in one way, rather than another, will convince him that farming is an art founded on rational and known principles. A description of the implements, and of the method of using them, will give him a closer insight into the nature and fitness of field-work for attaining its end, than by any other means. A perusal of these narratives, all having a common object, will impart a more comprehensive and clearer view of the management of a farm in a given time, than he could acquire by himself from witnessing ever so many isolated operations. The influence of the seasons on all the labors of the field is another consideration which should be attended to in such a book. In preparing the ground, and during the growth of the crops, the labor appropriated to each kind of crop terminates for a time, and is not resumed until a fit season arrive. These periodical cessations from labor form natural epochs in the progress of the crops toward maturity, and afford convenient opportunities for performing the work peculiarly appropriate to each epoch; and, since every operation of the farm is made to conform with its season, these epochs correspond exactly with the *natural* seasons of the year. I say with the *natural* seasons, in contradistinction to the common yearly seasons, which are entirely conventional. This necessary and opportune agreement between labor and the natural seasons induces a corresponding division of the labors of the farm into four great portions, or *seasons*, as they are usually termed. Labor should, therefore, be described with particular reference to its appropriate season.

[Reflecting on the preceding chapters with a view to give them practical bearing on our own country, one is led to remark that the struggles which ensued immediately after the establishment of our National Independence, and which had for their object the settlement of the great working principles of the Constitution, produced intense political excitement throughout the country. Unfortunately, this rage for politics, dignified in its commencement by great national aims, settled down into a sort of political *monomania;* and hence these struggles have become perennial, but with sad degeneracy as to motive. Anxiety about the great administrative principles of the Government has been superseded by an abiding and unquenchable thirst for office, for the sake of official emolument and power. $25,000,000 constitute the annual premiums to be contended for and distributed among *party* competitors, sufficiently numerous and active, and so widely dispersed among the people as to draw off their minds from the practical bearing of public legislation and its indissoluble connection with the landed interest of the country. Thus has it happened that, while the People of the United States—landholders in very large proportion—pay annually, and, as it would seem, willingly, so many millions to provide for military instruction and to maintain military institutions, little or nothing is done by their Representatives for the diffusion of agricultural knowledge, or for the construction of highways or conveniences to develop and improve the *industrial resources of the country!*

If the people—the cultivators of the soil—would force their legislators to appropriate, for the dissemination of agricultural knowledge through all the common schools of the country, one-tenth of that which the landholders now pay for prolonged debates and useless legislation, and for the pay and maintenance of the military machinery of the Government, our young men destined for farming might enter on their career in full possession of the knowledge recommended in the preceding chapters; and such knowledge could not fail to be followed by an immense increase in all the fruits of productive industry, as well as in that additional security for the public peace and prosperity which is the natural fruit of superior intelligence. The young men of the country would then enter upon life with a well-founded conviction that farming, truly, "*is* an art, founded on rational and known principles."

With these views it was that Wadsworth and Van Rensselaer, of New-York, exerted their influence for the establishment of Common Schools, in all of which the Sciences *connected with Agriculture* ought now to have preference and prominence over all others. The spontaneous and unassisted growth of agricultural schools, which may now be seen springing up in various sections of the country, clearly indicates that public sentiment is, fortunately, beginning to gather force and to take a right direction on this great subject of popular agricultural education—giving reason to hope that, eventually, even those who deny that the Government possesses any constitutional control over it, will at least agree that if public treasure is to be collected and appropriated to the diffusion of any sort of knowledge, or the construction of any sort of road, or survey or map, or the publication of any sort of book, it were far better for the interests of the people of the United States, and of humanity, that such expenditures should be applied to the dissemination of that sort of knowledge, and to the construction of that sort of road, or map, or book, which shall have a tendency to increase the "staff of life"—to make bread abundant—and provide, in a word, the materials of manufactures—the elements of commerce—and the basis, not so much of military glory, as of *national prosperity*.

There seems, in truth—and every Christian will hail all such omens with delight—to be a growing conviction that, as the field of Science enlarges, the practical man cannot fulfil his calling, whatever that calling may be, without *some acquaintance with those branches of Science which bear upon it*. The Divine, the Lawyer, the Physician, the Merchant—and he, still honored and rewarded above the best, whose art it is to destroy his fellow-man—are all of them acting upon this principle. The advance of Science in all other pursuits, except farming, is making empiricism in them degrading and unprofitable. Yet the follower of each of them was once an empiric. The Farmer alone is so still. Does not, then, the advance of Science—may we not say, his own character and self-respect—require him, too, to be a man of certainty—independent on, or rather a controller of, circumstances? *Ed. Farm. Lib.*]

3. OF THE KIND OF INFORMATION TO BE FOUND IN EXISTENT WORKS ON AGRICULTURE.

"Tire the bearer with a book of words."
MUCH ADO ABOUT NOTHING.

UNLESS the business of a farm be treated in books somewhat in the manner thus described, I consider it impossible for a young farmer to derive from them the requisite information for conducting a farm, even though he should be constantly resident upon it. By even the most careful perusal of books, which relate methods of cultivating crops and treating live stock in the most general terms and in detached sections having no relative connection with each other, the young farmer will never, in my opinion, understand how to apportion labor and modify its application to the raising of crops and rearing of live-stock, in accordance with the nature of the season. He will never learn to know by perusing a narrative couched in the most general terms, when an operation is really well performed; because, to be able to judge of the quality of work, all its minutiæ ought previously to have been fully and carefully detailed to him. Narratives couched in general terms, to the exclusion of essential minutiæ, will never impart that precision of ideas which the mind should possess in conducting any piece of field work; and without precision of ideas in regard to labor, no man will ever be able to conduct a farm aright. But to be told how to conduct a farm aright, is the chief motive of the young farmer for consulting a book at all.

Now, on examining works of any pretensions which have, for years past, been written on practical Agriculture, none will be found to have been written and arranged on the principles I have recommended, and much less for the special benefit of beginners in farming. All are so arranged as to constitute books of reference for experienced, rather than as guides for young farmers. Yet, how few of the former will condescend to *consult* agricultural works! The aversion of experienced farmers to to consult books on Agriculture has long been proverbial.* No doubt this

[*This aversion doubtless arises in a great measure from the neglect of parents to have Agriculture and studies nearly akin to it made *a part of the education* of their sons. Amusements followed, courses of reading indulged in, and habits contracted when we are young, continue to have their influence over us in after life, and to possess attraction, if only by force of early association, while he who grows to man's estate in ignorance of that great perennial source of enjoy-

aversion may be explained; but whether the explanation is to be found in a general indifference to book-farming, or in the quality of the books themselves, or in both circumstances combined, it is not easy to determine. The aversion, however, appears to be felt more toward systematic than periodical works on Agriculture. The latter class receives favor because, possibly, they may contain something that is not generally known, and their information bears the character of freshness. As to young farmers, if they cannot find books suited to their particular state of knowledge, they have no alternative but to peruse those that are extant.

For the sake of the young farmer, the usual contents of agricultural books require farther consideration. Let any systematic work on practical Agriculture be examined, and it will be found to contain an arrangement of the various particulars of farming, somewhat in this order. The soil and the various methods of working it are first described. The implements are then most probably particularized, or their description deferred to a later portion of the work. The methods of raising and securing the different kinds of crops are then detailed; and the treatment of live-stock is delayed to the last. We suppose that no satisfactory reasons can be given for adopting this particular arrangement of subjects. It is, perhaps, considered a *simple* arrangement, because it proceeds from what is considered the elementary process of preparing the soil, to the more complicated process of cultivating the plants for which the soil has been prepared. But the simplicity of the arrangement, I apprehend, is to be found rather in what is assumed than what is apparent; for plowing land is not a more simple process, or more elementary than sowing seed. Indeed, some sorts of plowing require far greater dexterity and ingenuity in the performance than any process connected with the production of crops. Perhaps it is considered a *natural* arrangement, because the ground is first prepared, and the crop is then sown. The ground, it is true, must be partially, if not wholly, prepared before the crop be put into it; but, in the cultivation of the summer crops, much of the labor bestowed on the land is performed while the crops are in a rapid progress toward maturity.

Although the seasons visibly influence the operations and products of the farm, systematic works on Agriculture scarcely disclose the subdivision of the year into seasons, much less the very different operations performed in different seasons, and still less the difference of character of the same season in different years. For all that is given in them by way of advice, every operation may as well be performed in one season as in another. No doubt, reference is made, and cannot altogether be avoided being made, to the season in which the piece of work described should be performed; but the reference seems to allude to the season more as an accidental concomitant, than as constituting the sole influential power that

ment, the love of books, very rarely contracts a fondness for them in after life. To put them in his way, when his habits are already fixed, and expect to force them on his attention, is like attempting to espalierise trees already half grown: hence the force of the maxim that you should "bring up the child in the way he should go." The boy should be led, while yet a boy, to take pleasure in books, and especially in such books as are best calculated to store his mind with the practice, and to embellish it with the literature of *his particular calling*. All that can enlighten Agriculture and instruct in the natural history and properties of whatever belongs to the country, should, at country schools, take the place of the antiquated trash which continues to form the bulk of Common School education, with the exception, in a great and honorable measure, of the schools in New-York, Massachusetts, and some other Northern States, where the subject has attracted the earnest regards of able, benevolent and conscientious men, who so well deserve and so rarely receive the tribute due to the real benefactors of mankind. *Ed. Farm. Lib.*]

regulates the order of time in which the work should be performed. The allusion to the season, in short, only forms an isolated hint, which, being singly repeated in a number of places, it is impossible for the reader to keep in mind the particular operation that should be performed in its own season. This apparent neglect of the great influential power which regulates all farm business, constitutes an insuperable objection to describing, in an uninterrupted narrative, a piece of work which is performed at intervals. Such a dissertation might bewilder the reader on its perusal, but could not satisfy the mind of the young inquiring farmer.

But the minuter arrangements in the books I am remarking on, are fully more objectionable than the general. The entire process usually adopted for working the land for a particular crop, is described in an uninterrupted narrative, before a description of the nature of the crop is given for which the soil is preparing; and, in consequence, before the connection between the preparation and the crop can be understood by the young farmer. This is not the usual procedure on farms, and cannot therefore be accounted natural; and it certainly tends to mislead the beginner. The usual practice is, that the land destined for any particular crop is prepared to a certain degree, at stated times, in accordance with the natural seasons, and between those times many operations intervene which bear no relation to that particular crop. Every operation thus occupies a portion of time, intermittent in its season, and cannot truly be described in a continued dissertation. The finishing operation of every crop is always deferred until the appropriate season.

The descriptions of implements are very unsatisfactory, and their construction, for the most part, is very imperfectly represented. None trace their action from the first start to the entire completion of the work. Implements of husbandry having, only a few years ago, been made in the rudest manner, their actions were necessarily imperfect, and their absolute weight a serious drag on the draught. They are now constructed on true principles of mechanical science—are light in motion, perfect in action, and elegant in form. It is remarkable that a correct description of improved implements has not ere this been undertaken by some skillful machinist.*

Some works treat first of the science of Agriculture, and then of the practice, as if the science of the art had been ascertained by studying abundance of facts derived from practice; or, as if its science already possesses such a superiority as to be allowed the precedence of practice. Others make science follow practice, as if the science had been derived from the practice described; whereas what is offered as science is generally presented in isolated speculations, volunteered chiefly by theorists unacquainted with the practice of Agriculture. Some authors theorize on agricultural subjects from as slight a foundation of facts as in the experimental sciences, although they profess to give no preference to science over practice. Theorizing writers, however, sometimes throw out hints which, when improved by more practical experimenters, really lead to useful results; but whatever may be the origin of the hints of theorists, the ability to give a convincing and philosophical reason for every operation in husbandry, is an accomplishment which every young farmer should endeavor to attain. Efforts to discover reasons for practice derived from

[* It is to be presumed that when this was written, the author had not seen "RANSOME'S BOOK OF THE IMPLEMENTS OF AGRICULTURE." It contains drawings and descriptions of all the implements of Agriculture employed in England. Such of them as are adapted to our country will be given in the FARMERS' LIBRARY AND MONTHLY JOURNAL OF AGRICULTURE, as two of the harrows were in Vol. 1. pp. 591-2. *Ed. Farm. Lib.*]

principles applicable alike to science and good husbandry, is a healthful exercise of the mind, and tend to render it capable of accommodating practice to existing circumstances. Conformity of practice with the season exhibits in the farmer superior ability for conducting farming operations: like the experienced mariner, who renders every change in the gale subservient to the safety of his ship, navigation itself not being more dependent on weather than is farming. By pursuing a course of observation and investigation such as this, the mind of the young farmer will soon become scientifically enlightened; but books on farming usually afford no assistance in pursuing such a course of study.

The treatment of live-stock is usually deferred to the conclusion in works on Agriculture, as if it were either the most important, or the most complicated, occupation of the farm. Breeding for the improvement of a particular race of animals, and judicious crossing betwixt two fixed races, are indeed occupations which tax the judgment severely; but the ordinary treatment of live-stock is as easily managed as most of the operations of the field. The complete separation, moreover, made in books betwixt live-stock and field-operations, is apt to impress the mind of the inexperienced reader that no necessary connection subsists betwixt stock and crop, whereas neither can be treated with advantage either to the farmer or themselves, unless both are attended to simultaneously.

From what I have stated regarding the arrangement of the subjects in systematic works on Agriculture, it will be observed that they are better adapted for reference than tuition. They form a sort of dictionary or cyclopedia, in which the different subjects are treated independently of each other, under different heads, though they may not be placed in alphabetical order. Being strictly works of reference, they may be consulted at any time; and are only valuable as such, in proportion to the accuracy of the information they contain; and being such, they are unfitted to impart agricultural knowledge suited to *beginners;* because, 1st, operations are not described in the order in which they occur on the farm; 2d, the descriptions omit many of the minutiæ of management, and yet constant attention to these constitutes an essential characteristic of a good farmer; 3d, they contain no precautionary warnings against the probability of failure in operations from various incidental causes, which ought to be anticipated, and attempted to be shunned; and, 4th, they afford no idea of the mode of carrying on various operations simultaneously in the different departments of management. Such works, therefore, impart no notion of *how* to set about to conduct a farm; and yet, without this essential information, to obtain which the earnest young farmer toils incessantly, they can render him no assistance as guides. Indeed, the authors of such works do not profess to be teachers of young farmers.

Experience has made me well acquainted with the nature of the difficulties tyros in Agriculture have to contend with; and I clearly see that the books on farming extant are incompetent to assist them in overcoming those difficulties. I consider it, therefore, very desirable that a work should be written for the express purpose of presenting facilities to young farmers in the acquirement of their profession. This opinion I have entertained for many years, and see no cause to change it for all the works on Agriculture that have been published of late years. To me it is matter of surprise that such a work has never been written by any of the prominent writers on Agriculture in this prolific age of books, when assistance in the acquirement of learning is proffered in so many shapes to the youths of all classes. In most other branches of art, there is no want of facilities in books for acquiring their elementary principles and practice.

On the kindred art of gardening, in particular, every possible variety of publication exists, from the ponderous folio to the tiny duodecimo, containing all the minutiæ of practice and the elucidation of principles. It is difficult to account for the want of solicitude shown by agricultural writers, for the early advancement of the young farmer. Perhaps many of them have never experienced the irksome difficulties of acquiring a practical knowledge of Agriculture, and therefore cannot extend their sympathies to those who have; perhaps the exhibition of an intimate acquaintance with the minutiæ of farming appears too trivial an accomplishment to arrest the attention of general writers; perhaps they think when a young man begins to farm, it is sufficient for him to have a steward in whose skill he can confide; perhaps the tuition of young farmers is beneath their dignity, and they would rather aspire to the higher object of instructing experienced men; or perhaps they have never condescended to trouble themselves with practical farming, which, to judge of their lucubrations by the sterlingness of their practical worth, many of them, I dare say, never have.

---

4. OF THE CONSTRUCTION OF "THE BOOK OF THE FARM."

> "A book! O rare one!
> Be not as is our fangled world, a garment
> Nobler than that it covers: let thy effects
> So follow, to be most unlike our courtiers,
> As good as promise."
>
> CYMBELINE.

A BOOK for the special purpose of instructing young farmers, such as it should be, and such as they are entitled to expect from the hands of experienced agriculturists, is yet a desideratum in the agricultural literature of this country. I am disposed to question the ability of any one man to write such a work, as its accomplishment would require a rare combination of qualities. The writer would require, as a primary qualification, to be a highly experienced agriculturist, able to indite lucid instructions for conducting a farm. He should also be a clear-headed mechanician, to describe with minute distinctness the principles and construction of agricultural implements. He should, moreover, be an accomplished man of science, to explain to conviction the rationale of every operation. Onerous as the task thus appears, I shall, nevertheless, attempt to write such a book. With adequate assistance, I trust I shall be able to overcome, at least, the practical difficulties of the undertaking; and, as to the scientific part, men of science have not yet brought Science to bear upon Agriculture in so satisfactory a manner as to justify them in contemning the rational explanations given of the various operations by practical men.—Could I but succeed in *arranging* the various operations as they successively and actually occur on a farm, in so lucid a manner as that any young farmer might comprehend the exact purport of each piece of work, as it developed itself in the field, I should certainly do him essential service.—In accomplishing this, it is scarcely possible to invest with sufficiently attractive interest the descriptions of the minute details of the various operations, so that their aptitude to the purpose intended may be appreciated. Careful attention to these details—in themselves, I own, irksome—will the sooner enable the young farmer to understand thoroughly the connection of successive operations; and by the understanding of which he will be forewarned of the approach, and be able to ascertain the import, of the particular end for which they are preparatory. Besides showing by anticipation the successive operations as they arrive, could I also give clear descriptions of the labor performed for each crop, as it is carried on *simultaneously* on a farm, I should achieve a still greater service for the young farmer. He would then clearly comprehend a difficult department of his art.

To accomplish these ends, I purpose to arrange the matter in the following manner, and for the following reasons. The entire business of a farm necessarily occupies a year; but that year embraces in some years more, and in others less, than twelve months. The agricultural year, moreover, both in its commencement and termination, does not correspond with that of the calendar; and those periods are determined in this way. The beginning and ending of every agricultural year are entirely dependent on the duration of the life of cultivated vegetables, which constitute the chief product of the farm. In the temperate regions of the globe, vegetable life becomes dormant, or extinct, according as the vegetable is perennial or annual, at the beginning of winter. The beginning of winter is therefore chosen, in the temperate zones, to commence the agricultural year, and, of course, the labors of the farm; and, when winter again approaches, the labors of the field have performed their annual revolution. The same sort of work is performed year after year. To understand those labors throughout the year is the chief aim of the young farmer; and to describe them to him satisfactorily is the principal object of this book.

Two modes of describing farm-business may be adopted. One is to arrange it under different heads, and describe all similar operations under the same head, as has hitherto been done in systematic works on Agriculture. The other mode is to describe the operations as they *actually occur, singly, in succession*, as is to be done in this work. Both methods describe the general farm business, and both may be consulted for any particular part of the business. But how the relative position of any particular part of the business stands in regard to, and influences any other, can only be shown by the latter method, and it does so at a glance of the eye. Moreover, as some parts of farm business commence, and others terminate, at one or other period of the year, the latter method can clearly indicate, what the other cannot so well do, in which period any particular operation is commenced, continued, or terminated; and it gives the details of each operation much more minutely than the other method.

The agricultural year, like the common year, is distinctly and conveniently divided into seasons, which regulate all farm work. I have given the seasons as full an influence over the arrangements of the matter in this book as they really possess over the business-matter of the farm. The whole business-matter is divided into four parts, each bearing the name of the season that influences the operations that are performed in it. By this arrangement every operation, whether requiring longer or shorter time for completion, is described as it takes its turn in the fields. The work that occupies only a short time to begin and complete, in any one season, is described in a single narrative. Very few of the operations of a farm, however, are begun and completed in one of the seasons; some extending over the whole four, and most into two or three. Any piece of work that extends over almost all the seasons can, nevertheless, be described with great accuracy; for although, in its progress toward completion, it may altogether occupy an extended range of time, each season imposes a peculiar kind of operation toward the advancement of the work; which peculiar operation ceases, and a different kind is entered upon, at the season which concludes the work. These cessations of labor, connected with the same work which extends over several seasons, are thus not mere conveniences, but necessary and temporary finishings of work, which it would be improper to resume but at a subsequent and appropriate season. In this way all the more extensive pieces of work are gradually advanced, in progressive steps, season after season, until their completion: while the smaller are concurrently brought onward and completed, each in its proper season.

Before proceeding farther, let me guard the young farmer against imbibing a misconception regarding the *length* of the seasons. In the year of the calendar, each season extends over a period of three calendar months; and the same three months every year compose the same season, whatsoever may be the nature of the weather. Every season of the calendar is thus of the same length. The seasons of the agricultural year, though bearing the same names as those of the calendar, are, on the other hand, not of the same length every year, but their duration is regulated by the state of the weather. The agricultural seasons have characteristic signs to distinguish them. The spring revives the dormant powers of vegetables; the summer enlarges their growth; the autumn develops the means of reproduction; and the winter puts a stop to vegetable energy. In the year of the calendar these characteristics are assumed to last just three months in each season; but in the agricultural year, notwithstanding that the characteristics of one season extend over or are contracted within three months, still that season bears its proper name, whether it encroaches on or is encroached upon by another season. The spring, for example, may be encroached on by the protraction of winter on the one hand, and the earliness of summer on the other; a case in which results both a late and short spring—a state of spring which creates very bustling work to the farmer. So with the rest of the seasons. This elastic property in the agricultural seasons contradistinguishes them from the seasons of the calendar which possess no elasticity. The commencement, continuance, and termination of field work being, therefore, entirely dependent on the seasons of the agricultural year—and those seasons, in their turn, being as dependent on the weather—it follows that field operations are entirely dependent on the state of the weather, and not on the conventional seasons of the calendar. Whether an agricultural season be long or short, the work that properly belongs to it *must* be finished in it while it lasts. If it be of sufficient length, the work to be performed, admitting of a considerable latitude of time, may be well finished; and, if not so finished, the crop runs the risk of failure. Should any season happen to be shortened by the weather, by the preceding season encroaching upon it, the work should be so far advanced during the preceding prolonged season that, when the proper season for its completion arrives—as arrive it will—the finishing may be accomplished before its expiring. Should any season be curtailed by the earliness of the succeeding one, and the weather improve, as in the case of summer appearing before its time, no apprehension need be entertained of accomplishing the finishing work in a satisfactory manner; but should the weather prove worse, as in the premature approach of winter upon autumn, then extraordinary exertions are required to avert the disastrous consequences of winter weather upon the crops. The unusual protraction of any of the seasons in which a work should be completed is attended with no risk, except that too frequently, from the consciousness of having plenty of time to complete the work, unnecessary delay is permitted, until the succeeding season unexpectedly makes its appearance. In such cases, procrastination is truly the thief of time. During the protraction of a season, much time is often wasted in waiting for the arrival of the succeeding one, in which a particular work is most properly finished; but, in a contracted season, a great part of the work is hurriedly gone through, and of course slovenly performed. The most perfect field-work is performed when the agricultural and conventional seasons happen to coincide in duration.*

The greatest difficulty which the farmer experiences, when first assuming the management of a farm, is in distributing and adjusting labor. To accomplish this distribution and adjustment correctly, in reference to the work, and with ease as regards the laborer, a thorough knowledge is requisite of the quantity of work that can be performed in a given time by all the instruments of labor, animal and mechanical, usually employed. It is the duty of the young farmer to acquire this knowledge with all diligence and dispatch; for a correct *distribution* of the instruments of labor enables the work to be performed in the most perfect manner in regard to the soil—with the smallest exertion as regards physical force—and with the greatest celerity in regard to time; and, in the *adjustment* of those instruments, every one should just perform its own share of work. These essential particulars I shall point out, in their connection with the work in hand. In descanting on the distribution of labor, I shall incur the hazard of being prolix rather than superficial. The general reader may dislike the perusal of minute details; but the ardent student will recieve with thankfulness the minutest portion of instruction, especially as he can only otherwise acquire this kind of instruction by long experience. The distribution and adjustment of labor is a branch of farm management that has been entirely overlooked by every writer on systematic Agriculture.

Constant attention on the part of the young farmer to the minutiæ of labor evinces in him that sort of acuteness which perceives the quickest mode of acquiring his profession. The distribution of the larger pieces of work may proceed satisfactorily enough under the skill of ordinary work-people; but the minuter can *best* be adjusted by the master or steward. The larger operations would always be left in a coarse state, were the smaller not to follow, and finish them off neatly. There are many minor operations, unconnected with greater, which should be skillfully performed for the sake of their own results; and they should be so arranged as to be performed with neatness and dispatch. Many of them are frequently performed concurrently with the larger operations; and, to avoid confusion, both their concurrent labors should harmonize. Many of the minuter operations are confined to the tending of live-stock, and the various works performed about the farmstead. Attention to minutiæ constituting the chief difference betwixt the neat and careless farmer, I have be-

[* Every young farmer may lay it down as a good rule to endeavor in all his work to be a *little before* the best manager in his neighborhood, "Drive your work or your work will drive you," says Dr. Franklin, and "Time enough always proves little enough." The best way, as a general rule, and which is a cardinal one with all successful farmers, is to *begin early*—get your land in the most perfect tilth—plant and sow as early as the season will admit, but let nothing tempt you to plow stiff or clay land when wet enough to bake into clods—endeavor to be the first to harvest and house your crops, and then prepare and send them at once to market, and sell them for the best price you can get. By that means you keep your work before you, avoid much mortification, and save your crop from waste and depredation by rats that gnaw, and thieves that "break in and steal."

N. B.—It is doubtful if any farmer ever yet lost anything by cutting his *wheat* too early and too green—not that it *could not* be done, but probably never is done—while millions have been lost by leaving it too long in the field *Ed. Farm. Lib.*]

stowed due consideration on them. They form another particular which has been too much overlooked by systematic writers on Agriculture.

Implements of husbandry may be considered the right hand of the farmer; because, without their aid, he could not display the *skill* of his art. Modern mechanical skill has effected much by the improvement of old, and the invention of new implements. Modifications of construction and unusual combinations of parts are frequently attempted by mechanics; and, though many such attempts issue in failure, they nevertheless tend to divulge new combinations of mechanical action. It is desirable that all mechanists of implements should understand practical Agriculture, and all farmers study the principles of mechanics and the construction of machines, so that their conjoined judgment and skill might be exercised in testing the practical utility of implements. When unacquainted with farming, mechanists are apt to construct implements that are obviously unsuited to the work they are intended to execute; but having been put together after repeated alterations, and, probably, at considerable expense, the makers endeavor to induce those farmers who are no adepts at mechanics to give them a trial. After some unsatisfactory trials they are thrown aside.—Were farmers acquainted with the principles of mechanics, the discrimination which such knowledge would impart would, through them, form a barrier against the spread of implements of questionable utility, and only those find circulation which had been proved to be simple, strong, and efficient.* It may be no easy matter to contrive implements possessing all those desirable qualities; but, as they are much exposed to the weather, and the ground upon which they have to act being ponderous and uncouth, it is necessary they should be of simple construction. Simplicity of con struction, however, has its useful limits. Most farm operations being of themselves simple, should be performed with simple implements; and all the *primary* operations, which are simple, requiring considerable power, the implements executing them should also be *strong;* but operations that are complicated, though stationary, require to be performed with comparatively complicated machinery, which, being stationary, may be used without derangement. Operations that are both complicated and locomotive should be performed with implements producing complicated action by simple means, in order to avoid derangement of their constituent parts.—This last is a difficult, if not impossible problem, to solve in practical mechanics. The common plow approaches more nearly to its practical solution than any other implement; yet that truly wonderful implement, executing difficult work by simple means, should yet be so modified in construction as to permit the plowman to wield it with greater ease. These considerations tend to show that the form and construction of implements of husbandry, and the circumstances in which they may be used, are still subjects affording ample scope upon which mechanical skill can exercise itself.

Implements have not received in works on Agriculture that consideration which their importance demands. The figures of them have been made by draftsmen who have evidently had no accurate conception of the

[* Who can deny that the principles of Mechanics, as far as all agricultural machinery is concerned, ought to take the place of some other things on which so much time is bestowed in all our country schools? Should any boy, who is to be a farmer, come to the possession of his estate without having been made to understand the principles of action—for example, of the *wedge*, the *screw*, the *inclined plane*, and the *lever?* How plainly such principles are illustrated by men of science, and how easily they may be comprehended by the commonest capacity, may be seen in the December and January (1845–6) Numbers of the MONTHLY JOURNAL OF AGRICULTURE.

Ed. Farm. Lib.]

functions of their constituent parts. The descriptions given of those constituent parts are generally meager, and not unfrequently erroneous; and as to the best mode of using implements, and the accidents to which they are liable, one would never discover that there was any peculiarity in the one, or liability to the other. In order to avoid both these classes of errors, much care has been bestowed in this work in delineating the figures, and giving descriptions of all the implements requisite for conducting a farm.

To ensure accuracy in these respects, I consider myself fortunate in having acquired the assistance of Mr. James Slight, Curator of the Machines and Models in the Museum of the Highland and Agricultural Society of Scotland, whose high qualifications as a describer and maker of machines are duly appreciated in Scotland. His son George, yet a very young man, is a beautiful delineator of them, as the drawings of the cuts and engravings in the work amply testify. And having myself paid close attention to the applicability of most of the implements used in farm operations, I have undertaken to describe the mode of using them—to state the quantity of work which each should perform, the accidents to which each is liable, and the precautions which should be used to avoid accidents. With our united efforts, I have confidence of giving such an *exposé* of farm implements as will surpass every other work of the kind. We have the advantage of having the field to ourselves. To assist the right understanding of the implements, they are represented by figures.

So much for the practical, and now for the scientific portion of the work. Agriculture may, perhaps, truly be considered one of the experimental sciences, as its principles are, no doubt, demonstrable by the test of experiment, although farmers have not yet been able to deduce principles from practice. It is remarkable that very few scientific men have, as yet, been induced to subject agricultural practice to scientific research; and those of them who have devoted a portion of their time to the investigation of its principles have imparted little or no satisfactory information on the subject. This unfortunate result may probably have arisen from the circumstance that Agriculture has so intimate a relation to every physical science that, until all those relations are first investigated, no sufficient data can be offered for a satisfactory scientific explanation of its practice The difficulty of the investigation is, no doubt, much enhanced by husbandry being usually pursued as a purely practical art, because the facility of thus pursuing it successfully renders practical men indifferent to Science. They consider it unnecessary to burden their minds with scientific research, while practice is sufficient for their purpose. Could the man of practice, however, supply the man of science with a series of accurate observations on the leading operations of the farm, the principles of those operations might be much elucidated; but I conceive the greatest obstacle to the advancement of scientific Agriculture is to be sought for in the unacquaintance of men of science with practical Agriculture. Would the man of science become acquainted with practice, much greater advancement in scientific Agriculture might be expected than if the practical man were to become a man of science, because men of science are best capable of conducting scientific research, and, being so qualified, could best understand the relation which their investigations bore to practice; and, until the relation betwixt principles and practice is well understood, scientific researches, though perhaps important in themselves, and interesting in their results, tend to no practical utility in Agriculture. In short, until the facts of husbandry be acquired by practice, men of science will in vain endeavor to construct a satisfactory theory of Agriculture on the principles of the inductive philosophy.

If this view of the present position of the science of Agriculture be correct, it may be expected to remain in a state of quiescence until men of science become practical agriculturists, or, what would still prolong its state of dormancy, until farmers acquire scientific knowledge. It is a pity to damp the ardor of scientific pursuit where it is found to exist; but, from what I have observed of the scanty services science has hitherto conferred on Agriculture, and knowing the almost helpless dependency of farming on the seasons, I am reluctantly impelled to the belief that it is less in the power of science to benefit Agriculture, than the sanguine expectations of many of its true friends would lead farmers to believe. It is wrong to doubt the power of science to assist Agriculture materially; and it is possible, in this age of successful art, that an unexpected discovery in science may yet throw a flood of light on the path of the husbandman; but I am pretty sure, unless the man of science become also the practical husbandman, it will be difficult, if not impossible, for him to discover which department of the complicated art of husbandry is most accessible to the research of science.

Hitherto, as it appears to me, Agriculture has derived little benefit from the sciences, notwithstanding its obvious connection with many of them. A short review of the relation which the physical sciences bear to Agriculture will render this opinion more reasonable. In the first place, the action of the electric agency in the atmosphere and on vegetation is yet as little understood in a practical sense as in the days of Franklin and of Ellis. No doubt, the magnetic and electric influences are now nearly identified; but the mode of action of either, or of both, in producing and regulating atmospherical phenomena, is still ill understood; and, so long as obscurity exists in regard to the influence of their elementary principles, the history of atmospherical phenomena cannot advance, and the anticipations of atmospherical changes cannot be trusted.

Geologists, at first engaged in ascertaining the relative positions of the harder rocks composing the crust of the earth, have only of late years directed their attention to the investigation of the more recent deposits; but, even with these, they have afforded no assistance in the classification of natural soils and subsoils. They have never yet explained the origin of a surface-soil, almost always thin, though differing in thickness, over subsoils composed of different kinds of deposits. They have never yet ascertained the position and structure of subsoil deposits, so as to inform the farmer whether land would be most effectually drained with drains running parallel with, or at right angles to, the courses of valleys and rivers.

Systematic botany can only be useful to Agriculture in describing the natural plants which are indigenous to different soils. Botanists have successfully shown the intimate relation subsisting betwixt plants and the soils on which they grow; but much yet remains to be ascertained of the relation betwixt different soils and trees, and the effects of different subsoils on the same kind of tree. Planting cannot be pursued on fixed principles, if planters are unacquainted with this knowledge; and, until a fixed and generally received classification of soils and subsoils is determined on, it is impossible to comprehend, by description, what particular soil or soils the plants referred to affect.

Botanical physiology has developed many remarkable phenomena, and explained most of the important functions of plants—investigations which tend to give a clearer insight into the growth of crops. In this department of science, too much discussion to be of benefit to Agriculture has, as I conceive, been expended on what really constitutes the food of plants. Whether the food is taken up by the plant in a gaseous, a solid, or a liquid state, may in itself be a very interesting inquiry, but it tends to no utility in Agriculture so long as no manures are supplied to crops in a gaseous or liquid state. All that can practically be done in supplying food to plants, is to observe the increased quantity of their secretions in a given condition from an increased given quantity of manure. Thus may the increased quantities of mucilage, farina, gluten, in the various cultivated plants, be observed. It is of little moment to the farmer whether the manure administered is taken up by the crops in a gaseous, liquid, or solid state, since all these secretions are elaborated from the same manure. The anatomical structure of plants, the situations, soils, and manures which crops affect, the secretions which they elaborate, and the prolificacy and value of their products, are the results that most interest the farmer; and, if botanical physiologists desire to benefit Agriculture they must direct their attention to the emendation and increase of products. Again, the results from the cross impregnation of plants of the same kind, so as to produce valuable permanent varieties, may confer as valuable a boon on Agriculture as the successful crossings of different breeds of live-stock have already conferred by increasing their value. Many varieties of plants having their origin in this way have been brought into notice, and some are now established and extensively cultivated; but most of the varieties in use have been obtained from casual impregnations effected by Nature herself, and not by the efforts of man to obtain varieties possessing superior properties, as in the case of the domesticated animals. Thus botanical physiology might confer great benefit on Agriculture, if its views were directed to increasing the prolificacy of valuable plants already in cultivation, and introducing others that would withstand the modes of culture and changes of climate incidental to this country.

But there is one view in which botanical physiology may be of use to Agriculture, and that is, in ascertaining correctly the nature, properties, and relative values of plants. To show the importance of such an investigation, a case may here be specified. A variety of rye-grass, called Italian, has been lately introduced into this country. It is found to be a very free grower in this climate; and it is highly acceptable to all kinds of live-stock, whether in a green or dried state. Could this grass be rendered certainly perennial, it would be an invaluable acquisition to the pastures of this country. Its character, however, is rather capricious, for in some places it disappears after two years' cultivation, while in others it displays undiminished vigor of growth for four or five years, and may perhaps continue so to do for an indefinite period of time. Judging by these various results, it is probable that there is more than one variety of the plant, and distinguishing varieties seem to be known to foreigners. Keeping in view the existence of varieties, if different varieties were affected differently by the same locality, there would be nothing in the phenomenon to excite surprise; but when the same variety, derived from the same stock, and placed in similar circumstances, exhibits different instances of longevity, there must be characteristics of the plant still unknown to cultivators. In this dilemma, the assistance of the botanical physiologist would be desirable to discover those latent characteristics. It would be desirable to know the conditions that regulate the existence of plants into permanent and temporary varieties—a property of plants at present involved in mystery. Hitherto, no practical explanation of the subject has been proffered to the farmer; and so long as he shall be permitted to discover the true properties of plants for himself, botanical physiology cannot be regarded by him as of much use to Agriculture.

The Italian rye-grass exhibits in its nature an anomaly that no other variety of rye-grass does. The *annual* rye-grass, as it is commonly called, is seldom seen in the ground, even to the extent of a few plants, in any kind of soil, and under any treatment, after the second year; and the *perennial* is as seldom observed to fail in any circumstances, except when it may have been too closely cropped by sheep to the ground too late in autumn, when it generally dies off in the following spring. But the Italian may be annual or perennial in the same circumstances. Farmers cannot account for such an anomaly. High condition of good soil may tend to prolong, while the opposite state of poor soil may tend to shorten, its existence. But *why* those circumstances should not produce the same effects on all varieties of rye-grass, it is for science to explain.

Entomology might be made to serve Agriculture more than it has yet done. In this department of science farmers might greatly assist the entomologist, by observing the minute, but varied and interesting, habits of insects. The difficulty of comprehending the true impulses of insects, as well as of identifying species in the different states of transformation, render the observations of farmers less exact than those of entomologists who have successfully studied the technicalities of the science. The field of observation in the insect creation being very wide, and there being comparatively but few explorers in it, a large portion of a man's life would be occupied in merely observing species and their habits, and a much larger in forming general deductions from repeated observation. The result would be, were farmers to study entomology, that a long period must elapse ere the habits of even the most common destructive insects, and the marks of their identity, would become familiarized to them. In consequence of this obstacle to the study of the farmer, the obligation ought to be the greater to those entomologists who daily observe the habits of insects in the fields and woods, and simplify their individual characteristics; and at the same time devise plans to evade their extensive ravages, and recommend simple and effective means for their destruction. The English farmer, living in a climate congenial to the development of insect life, painfully experiences their destructive powers on crops and woods; and, although in England entomologists are ever vigilant and active, yet their efforts easily to overcome the tenacity of insect life, with a regard to the safety of the plant, have hitherto proved unavailing.*

Chemistry is somehow imagined to be the science that can confer the

[*Writers on Entomology, though they have described the structure and habits of insects, have probably done little to prevent their ravages. The mischief done by the Hessian Fly has been mitigated by the labors of those whose studies have enabled them to indicate the period of their nidification, or egg-laying; and thus teaching the farmer to delay his time of sowing—whereby, however, his crop is necessarily diminished. But it does not follow that because a knowledge of the physiology and habits of insects may lead to no practical remedy against their ravages, that their natural history should not be studied. All such studies form a part, and an elegant part too, of agricultural literature, and deserve, therefore to be cultivated by every country gentleman. KOLLAR on the Insects injurious to Farmers and Gardeners, and HARRIS on the Insects injurious to Vegetation in Massachusetts, ought assuredly to form a part of every farmer's library. There are books enough of the most entertaining character, *closely allied to his own profession*, sufficient to beguile and improve every leisure hour the country gentleman can command, and it behooves every farmer to supply them to his sons. The time is coming, may we not hope, when to be a good farmer will carry with it the presumption of being a man of various and elegant, as well as of practical knowledge—when the agriculturist, ceasing to be a mere empiric, will know as well the why and the wherefore as the how and the when. The foundation for all this, it cannot be too often repeated, must be laid in our schools. *Ed. Farm. Lib.*]

greatest benefits on Agriculture. This opinion seems confirmed in the minds of most writers and agriculturists, and especially the English, most probably from the circumstance of an eminent chemist having been the first to undertake the explanation of agricultural practice on strictly scientific principles. Sir Humphry Davy has, no doubt, been the cause of bestowing on that science the character, whose influence was imagined to be more capable of benefiting Agriculture than its eulogists have since been able to establish. He endeavored to explain with great acuteness many of the most familiar phenomena of Agriculture, when in possession of very limited acquaintance with practical facts; and the result has been, that while his own chemical researches have conferred no practical benefit on Agriculture—his conclusions being in collision with practice—the field of observation and experiment which he explored and traversed has since been carefully avoided by succeeding chemists, in the conviction, no doubt, that wherein he failed they were not likely to succeed. The idea seemed never to have struck them that Sir Humphry had attempted to enforce a connection betwixt Chemistry and Agriculture which both were incapable of maintaining. Viewing the relation betwixt them merely in a practical point of view, I can see no very obvious connection betwixt tilling the soil and forcing crops by manure for the support of man and beast—which is the chief end of Agriculture—and ascertaining the constituent parts of material bodies, organic and inorganic—which is the principal business of Chemistry. A knowledge of the constituent parts of soils, or plants, or manures, now forms a necessary branch of general chemical education, but *how* that knowledge *can improve agricultural practice*, has never yet been practically demonstrated. No doubt, Chemistry informs us that plants will not vegetate in pure earths, and that those earths constitute the principal basis of all soils; but as pure earths are never found in soils in their ordinary state, farmers can have no chance of raising crops on them. It may be true, as Chemistry intimates, that plants imbibe their food only when in a state of solution; but what avails this fact to Agriculture, if fact it be, when manures are only applied in a solid state? It may be quite true, as Chemistry declares, that plants cannot supply, from their composition, any substance they have not previously derived from the air, earth, or decomposed organic matter; but of what practical use to Agriculture is this declaration, as long as farmers successfully raise every variety of crop from the same manure? Chemistry may be quite correct in its views with regard to all these particulars, but so is practice, and yet both are very far from agreeing; and as long as this constitutes the only sort of information that Chemistry affords, it is unimportant to the farmer. He wishes to be shown *how* to render the soil more fertile, manures more effective, and crops more prolific, by the practical application of chemical principles.

There are many writers, I am convinced, who recommend the study of Chemistry to farmers little acquainted with the true objects of chemical research, and not much more with practical Agriculture. At all events, they expatiate, only in vague generalities, on the advantages of analyzing soils, manures, &c. but do not attempt to demonstrate *how* any practice of husbandry may *certainly* be improved by the suggestions of Chemistry. The truth is, until chemists become thoroughly acquainted with agricultural facts, they cannot see the bearings of chemical principles on agricultural practice, any more than the most uncouth farmer; and until they prove the farmer's practice in any one instance wrong, and are *certain* of its being put right by their suggestions, there is no use of lauding Chemistry as a paramount science for Agriculture.

In this view of the science, I would rather underrate the ability of Chemistry to benefit Agriculture than excite the fallacious hopes of the farmer by extolling it with undue praise. At the same time, were a chemist to recommend suggestions promising a favorable issue, *that* might tend to excite a well-grounded hope in chemical assistance, and I am sure the suggestions would even be fairly tried by farmers who entertain pretty strong suspicions against science. If, for example, on carefully analyzing a plant in common culture, it was found to contain an ingredient which it could not obviously have derived from the manure or the soil, were a suggestion made to mix a quantity of that particular ingredient with the soil or manure, it would at once be cheerfully put to the test of experiment by farmers. If, on the other hand, were the same chemist to suggest making heavy clay land friable by the mechanical admixture of sand, the physical impracticability of the proposal would at once convince the farmer that the chemist had no adequate notion of farm work. And yet propositions as absurd as this have frequently been suggested to farmers by writers who are continually maintaining the ability of Chemistry to benefit Agriculture. But let me appeal to facts—to ordinary experience.

I am not aware of a single agricultural practice that has been adopted from the suggestions of Chemistry. I am not speaking unadvisedly while making this unqualified statement. In truth, I do not know a single operation of the farm that has not originated in sheer practice. But is it not somewhat unreasonable to expect improvement in agricultural practice, and still more, an entirely improved system of Agriculture, from the suggestions of Chemistry? Some chemical results may appear to bear analogy to certain operations of the farm, such as the preparation of manures; but such analogies, being chiefly accidental, are of themselves insufficient grounds upon which to recommend chemical affinity as the principle which ought to regulate a system of practically mechanical operations. How can the most familiar acquaintance with the chemical constituents of all the substances found on a farm, suggest a different mode of making them into manure, inasmuch as *practice* must first pronounce the treatment to be an improvement, before it can really be an improvement, whatever Chemistry may suggest? Besides, Chemistry, with all its knowledge of the constituent parts of substances, cannot foretell, more confidently than practice, the results of the combinations with the soil, of the substances analyzed among themselves, and the combined effects of these and the soil upon cultivated plants. I am aware that hints may be suggested by science which may prove beneficial to practice; but unless they *accord with the nature of the practice* to which they are proposed to be applied, they are certain of proving unserviceable. Many hints thrown out at random have frequently been put to the test of experiment; but to experimentize on hints is quite a different thing in farming from that sort of farming which is proposed to be entirely based on theoretical suggestions, whether of Chemistry or of any other science.

For these reasons, I conceive, Chemists would be more usefully employed in following than in attempting to lead practical Agriculture. If it were practicable, it would certainly be very desirable for the farmer to be assured that his practice was in accordance with chemical principles; if, for example, it could be explained on chemical principles *why* a certain class of soils is better suited to a certain kind of crop than other classes, and *why* animal manure is better suited than vegetable to a certain kind of crop; when Chemistry shall explain *why* certain results are obtained by practice, it will accomplish much, it will elucidate that which was before obscure in principle. Were chemists to confine the first stage of their investigations of agricultural matters to this extent, farmers would be much gratified with the assurance of their practice being in unison with the principles of chemical science; and this would tend more than any other circumstance to inspire them with confidence in the utility of that science. This is the position which Chemistry, in my opinion, should occupy in relation to Agriculture; for how successful soever it may be in assisting other arts, such as dyeing, soap-making, and ink-making, it assists them both by synthesis and analysis; whereas it can only investigate agricultural subjects by analysis, because every substance employed in Agriculture, especially a manure, is used by farmers in the state it is found in the markets, without reference to its chemical constituent parts; and, when used, should an analytical or synthetical process go on among those parts, or with the soil, with which they are intimately brought into contact, the process going on in the *soil* would change the chemical composition of the whole, and place them beyond the reach of chemical research. The investigation of the soil after the removal of the crop might then be curious, but nothing more.* In this investigation the farmer, the vegetable physiologist, and the chemist, would all disagree as to the extent of the influence exercised by the favorite substance of each in producing the

[* In all these views we cannot agree, but the author would seem to be supported in them by DAVID LAW, Esq. Professor of Agriculture in the University of Edinburgh, who says in reference to the use of chemistry in analyzing soils, "The chemist may draw useful conclusions from a careful analysis of the matter of the soil, and may from time to time be able to communicate results that may be serviceable to the practical farmer; but it is not necessary for the ends of practice that the farmer himself should be a chemist. The farmer cannot arrive at the science of mineral analysis, without a knowledge of chemistry and the business of the laboratory, which he can rarely acquire, and which it is in no degree necessary to his success as a farmer that he should be possessed of."

Mr. COLMAN, too, whose judgment and zeal in the cause of education and science as applied to Agriculture are so well known, seems to think the actual importance of chemistry, in its connection with that pursuit, so far, has been overrated. He says "the application of sulphuric acid to bones seems as yet to be the only case of the application of chemical science to the improvement of Agriculture upon scientific principles, and this affords strong grounds to hope for much more."

In relation to agricultural schools, Mr. COLMAN's Report may be read with profit. One ought to take for granted that it, too, will be in all our country schools. On the principle that the Father who wished to beget in his sons a capacity for labor, told them he had buried his treasure in the garden, without telling the spot, leaving them to dig it all over, we recommend the reader not to adopt, *in extenso*, the opinions of these two distinguished authors, until he shall have read carefully Davy and Boussingault and Liebig and Petzboldt and Johnstone, and *then* form his own judgment. Surely every young Farmer should know enough of chemistry to be able, as he may by a very simple process, to analyze his own soil that he may know in what most important ingredients it is redundant or deficient, to the end that having learned from analysis made by professional chemists the ingredients which make the necessary food for certain crops, he may be able to supply such as are needful in the soil, or to avoid the expense of applying others, in which he finds his land to be redundant. Mr. COLMAN, in his personal observations on European Agriculture, vol. 1, part 3, gives us the best accounts of the practical working and benefit of Agricultural schools in Ireland, and of the one then about to be established at Cirencester, England. Referring to the school at *Larne*, he says:

"It was from this establishment that a detachment of five pupils was sent for examination to the great meeting of the Agricultural Society of Scotland the last autumn, where their attainments created a great sensation, and produced an impression, on the subject of the importance of agricultural education, which is likely to lead to the adoption of some universal system on the subject.

"I shall transcribe the account given of the occasion: 'Five boys from the school at Larne were introduced to the meeting, headed by their teacher. They seemed to belong to the better class of peasantry, being clad in homely garbs; and they appeared to be from twelve to fifteen years of age. They were examined, in the first instance, by the inspector of schools, in grammar, geography, and arithmetic; and scarcely a single question did they fail to answer correctly. They were then examined, by an agricultural professor, in the scientific branches, and by two

crop. In settling the question, however, the farmer would have the same advantage over his rivals, in taking possession of the crop as the reward of *his practical* skill, as the lawyer who, in announcing the judgment of the court to two contending parties, gave a shell to each, and kept the oyster to himself.

Of all the sciences, mechanics have proved the most useful to Agriculture. If implements may be characterized as the right hand of Agriculture, mechanical science, in improving their form and construction, may be said to have given cunning to that right hand; for, mechanical science, testing the strength of materials, both relatively and absolutely, employs no more material in implements than is sufficient to overcome the force of resistance, and it induces to the discovery of that form which overcomes resistance with the least power. Simplicity of construction, beauty of form of the constituent parts, mathematical adjustment, and symmetrical proportion of the whole machine, are now the characteristics of our implements; and it is the fault of the hand that guides them, if field-work is not now dexterously, neatly, and quickly performed. In saying thus much for the science that has improved our implements to the state they now are, when compared with their state some years ago, I am not averring they are quite perfect. They are, however, so far perfect as to be correct in mechanical principle, and light in operation, though not yet simple enough in construction. No doubt many may yet be much simplified in construction; and I consider the machinist who simplifies the action of any useful implement, thereby rendering it less liable to derangement, does a good service to Agriculture as the inventor of a new and useful implement.

These are the principles which determine the arrangement adopted in this book. In applying these principles, as the *seasons* supremely rule the destiny of every farming operation, so to them is given full sway over the whole arrangement. This is accomplished by describing every operation *in the season it should be performed*, and this condition necessarily implies the subdivision of the arrangement into *four seasons*. Authors of Farmers' Calendars divide their subject-matter into calendar or fixed months, being apparently inattentive to the influences of the seasons. Such an arrangement cannot fail to create confusion in the minds of young farmers; as any operation that is directed to be done in any month, may not in every year, be performed in the same month, on account of the fluctuating nature of the seasons.

In adopting the seasons as the great divisor of the labors of the farm, the months which each season occupies are not specified by name, because the same season does not occupy the same number of months, nor even exactly the same months, in every year. The same work, however, is performed in the same season every year, though not, perhaps, in the same month or months.

In arranging the seasons themselves, the one which commences the agricultural year, which is *Winter*, has the precedence. The rest follow in the natural succession of Spring, Summer, and Autumn; in which last all farming operations, having finished their annual circuit, finally terminate. A few remarks, illustrative of its nature, and the work performed in it, are

practical farmers in the practical departments of Agriculture. Their acquaintance with these was alike delightful and astonishing. They detailed the chemical constitution of the soil and the effect of manures, the land best fitted for green crops, the different kinds of grain, the dairy, and the system of rotation of crops. Many of these answers required considerable exercise of reflection; and as previous concert between themselves and the gentlemen who examined them was out of the question, their acquirements seemed to take the meeting by surprise; at the same time they afforded the utmost satisfaction, as evincing how much could be done by a proper system of training.'"

given at the commencement of each season. By comparing these introductory remarks, one with the others, the nature of the principal operations throughout the year may be discovered; and, by perusing them in succession as they follow, an epitome of the entire farm operations for the year may be obtained.

Throughout the four seasons, from the commencement of winter to the end of autumn, the operations of the farm, both great and small, are described in a continued narrative. This narrative is printed in the larger type *(long primer)*. The reader will soon discover that this narrative does not extend uninterruptedly through the whole pages—portions of smaller type *(brevier)* intervening, and apparently interrupting it. On passing over the small type, it will be perceived that it is really written, and may be perused without interruption. The object of this plan is to permit the necessary descriptions of all the operations, performed in succession throughout the year, to be read in the large type, to the exclusion of every other matter that might distract the attention of the reader from the principal subject. A peristrephic view, so to speak, of the entire operations of the farm is thus obtained. The leading operations, forming the principal subjects of the narrative, are distinguished by appropriate titles in CAPITALS placed across the middle of the page. The titles are numbered, and constitute, in the aggregate, a continuous succession, running through all the seasons. The leading operations thus easily attract the eye. Wood-cut figures of implements, and other objects, requiring no detailed descriptions, and representing at once their form and use, are inserted in the paragraph which alludes to them in the narrative.

Implements that require detailed descriptions to explain, and complicated figures to represent them; reasons for preferring one mode to another of doing the same kind of work; and explanations of agricultural practice on scientific principles—together constituting the subsidiary portion of the work—are given in paragraphs in the medium-sized type *(brevier)*, and this matter is that which apparently interrupts the principal narrative. Each paragraph is numbered within parentheses, the same as in the principal narrative, and these paragraphs carry on the numbers arithmetically with the paragraphs of the principal narrative. When references are made from the large to the small type, they are made in corresponding numerals. The words most expressively characteristic of the illustration contained in the paragraph are placed in *italics* at or near the beginning of it.

Marking *all* the paragraphs with numerals greatly facilitates the finding out of any subject alluded to—saves repetition of descriptions when the same operation is performed in different seasons—and furnishes easy reference to subjects in the index.

Wood-cut figures of the intricate implements and other objects requiring detailed descriptions, are placed among the descriptions of them in the brevier type. The portraits of the animals given are intended to illustrate the *points* required to be attended to in the breeding of the domesticated animals. The portraits are taken from life by eminent artists. The wood-cuts are enumerated as they occur in the order of succession, whether they belong to the large or the small type, and each wood-cut is designated by its distinctive appellation—both the numeral and appellative being requisite for quick and easy reference.

The matter in the small type appears somewhat like foot-notes in ordinary books; but, in this instance, it differs in character from foot-notes, inasmuch as it occurs in unbroken pages at the end of the description of every leading operation. By this plan the principal narrative is not interfered with, and both it and its illustrations may be perused before the succeeding leading operation and its illustrations are taken into consideration. This plan has the advantage of relieving the principal narrative of heavy foot-notes—the perusal of which, when long, not only seriously interrupts the thread of the narrative, but causes the leaves gone over to be turned back again; both interferences being serious drawbacks to the pleasant perusal of any book.

Foot-notes required either for the principal narrative or illustrations are distinguished by the usual marks, and printed at the bottom of the page in the smallest type used in this work.

The paragraphs containing the matter supplied by Mr. Slight are enclosed within brackets (thus, [ ]), and attested by his initials, J. S.

[The additions by John S. Skinner are designated by his title of "*Editor of the Farmers' Library.*"]

---

5. OF THE EXISTING METHODS OF LEARNING PRACTICAL HUSBANDRY.

"I have vowed to hold the plow for her sweet love three year."
LOVE'S LABOR LOST.

I HAVE hinted that there are three states, in one of which the young farmer will be found when beginning to learn his profession. One is when he himself is born and brought up on a farm, on which, of course, he may acquire a knowledge of farming intuitively, as he would his mother tongue. Another is when he goes to school in boyhood, and remains there until ready to embark in the active business of farming; the impressions of his younger years will become much effaced, and he will require to renew his acquaintance with farming as he would of a language that he had forgotten. Young men thus early grounded generally make the best farmers, because the great secret of knowing practical farming consists in bestowing particular attention on minor operations, which naturally present themselves to the youthful mind before it can perceive the use of general principles. Farmers so brought up seldom fail to increase their capital; and, if their education has been superior to their rank in life, frequently succeed in improving their status in society. It is to the skillful conduct and economical management of farmers so situated, that Scotland owes the high station she occupies among the agricultural nations of the world.

The third state in which the learning of farming is requisite is when a young man who has been educated and entirely brought up in a town, or perhaps passed his boyhood in the country, but may have bestowed little attention on farming, wishes to learn it as his profession. In either of these cases, it is absolutely necessary for him to learn it practically on a farm; for total ignorance of his business, and entire dependence on the skill and integrity of his servants, will soon involve him in pecuniary difficulties. To meet the wishes of seekers of agricultural knowledge, there are farmers who receive pupils as boarders, and undertake to teach them practical husbandry.

The chief inducement, as I conceive, which at first prompts young men who have been nurtured in towns to adopt farming as a profession, is an undefined desire to lead a country life. The desire often originates in this way. Most boys spend a few weeks in the country during the school vacation in summer, on a visit to relations, friends, or school companions. To them the period of vacation is a season of true enjoyment. Free of the task—in the possession of unbounded liberty—untrammeled by the restraints of time, and partaking of sports new to them and solely appertaining to the country, they receive impressions of a state of happiness which are ever after identified with a country life. They regret the period of return to school—leave the scene of those enjoyments with reluctance—and conceive that their happiness would be perpetual, were their hearts wedded to the objects that captivated them. Hence the desire to return to those scenes.

It is conducive to the promotion of Agriculture that young birds of fortune are thus occasionally ensnared by the love of rural life. They bring capital into the profession; or, at all events, it will be forthcoming when the scion of his father's house has made up his mind to become a farmer. Besides, these immigrations into farms are requisite to supply the places of farmers who retire or die out. Various motives operate to bring farms into the market. Sons do not always follow their father's profession, or there may not be a son to succeed, or he may die, or choose another kind of life, or may have experienced ill treatment at home, or been guilty of errors which impel him to quit the paternal roof. For these drains, a supply must flow from other quarters to maintain the equilibrium of agricultural industry. This young race of men, converted into practical farmers, being generally highly born and well educated, assume at once a superior status in, and improve the tone of, rural society. Though they may amass no large fortunes, they live in good style. In the succeeding generation, another change takes place. Unless he is well provided with a patrimony, the son seldom succeeds his father in the farm. The father finds he cannot give the farm free of burdens to one son in justice to the rest of the family. Rather than undertake to liquidate *such* a burden by means of a *farm*—that is, from land that is not to be his own—the son wisely relinquishes farming, which, in these circumstances, would be to him a life of pecuniary thralldom.

The young man who wishes to learn farming practically on a farm, should enter upon his task at the end of harvest, as immediately after that the preparatory operations commence for raising the next year's crop; and that is the season, therefore, which begins the new-year of farming. He should provide himself with an ample stock of *stout* clothing and shoes, capable of repelling cold and rain, and so made as to answer at once for walking and riding. From the outset, he must make up his mind to encounter all the difficulties I have described under the first head. Formidable as they may seem, I encourage him with the assurance that it is in his power to overcome them all. The most satisfactory way of overcoming them is to resolve to learn his business in a truly practical manner. Merely being domiciled on a farm is not of itself a sufficient means of overcoming them, for the advantages of residence may be squandered away in idleness, by frequent absence, by spending the hours of work in the house in light reading, or by casual and capricious attendance on field operations. Such habits must be eschewed, before there can be a true desire to become a *practical* farmer. *Every* operation, whether important or trifling, should be *personally* attended to, as there is none but what tends to produce an anticipated result. Attention alone can render them familiar; and, without a familiar acquaintance with every operation, the management of a farm need never be undertaken.

Much assistance in promoting this attention should not be *expected* from the farmer. No doubt it is his *duty* to communicate all he knows to his pupils; and, as I believe, most are willing to do so; but, as efficient tuition implies constant attendance on work, the farmer himself cannot constantly attend to every operation, or even explain any, unless his attention is directed to it; and much less will he deliver extempore lectures at appointed times. Reservedness in him does not necessarily imply *unwillingness* to communicate his skill; because, being himself familiar with every operation that can arrest the attention of his pupils, any explanation of minutiæ at any other time than when the work is in the act of being performed, and when only it could be *understood* by the pupils, would only serve to render the subject more perplexing. In these circumstances the best plan for the pupil to follow is to *attend constantly*, and *personally observe* every change that takes place in every piece of work. Should the farmer happen to be present, and be appealed to, he will, as a matter of course, immediately clear up every difficulty in the most satisfactory way; but should he be absent, being otherwise engaged, then the steward or grieve, or any of the plowmen, or shepherd, as the nature of the work may be, will, on inquiry, afford as much information on the spot as will serve to enlighten his mind until he associates with the farmer at the fireside.

To be enabled to discover that particular point in every operation which, when explained, renders the whole intelligible, the pupil should put his hand to every kind of work, be it easy or difficult, irksome or pleasant.—Experiences acquired by himself, however slightly affecting his mind—desirous of becoming acquainted with every professional incident—will solve difficulties much more *satisfactorily* than the most elaborate explanations given by others. The larger the stock of these personal experiences he can accumulate, the sooner will the pupil understand the purport of every thing that occurs in his sight. Daily opportunities occur on a farm for joining in work, and acquiring those experiences. For example, when the plows are employed, the pupil should walk from the one to the other, and observe which plowman or pair of horses perform the work with the greatest apparent difficulty or ease. He should also mark the different styles of work executed by each plow. A considerate comparison of these particulars will enable him to ascertain the best and worst specimens of work. He should then endeavor to discover the cause why different styles of work are produced by apparently very similar means, in order to enable himself to rectify the worst and practice the best. The surest way of detecting error and discovering the best method is to take hold of each plow successively, and he will find, in the endeavor to maintain each in a steady position, and perform the work evenly, that all require *considerable* labor—every muscle being awakened into energetic action, and the brow most probably moistened. As these symptoms of fatigue subside with repetitions of the exercise, he will eventually find one of the plows more easily guided than any of the rest. The reasons for this difference he must himself endeavor to find out by comparison, for its holder cannot inform him, because he professes to have—indeed, can have—no knowledge of any other plow but his own. In prosecuting this system of trials with the plows, he will find himself becoming a plowman, as the mysteries of the art divulge themselves to his apprehension; but the reason why the plow of one of the men moves more easily, does better work, and oppresses the horses less than any of the rest, is not so obvious; for the land is in the same state to them all—there cannot be much difference in the strength of the pairs of horses, as each pair are generally pretty well matched—and, in all probability, the construction of the plows is the same, if they have been made by the same plow-wright, yet one plowman evidently exhibits a decided superiority in his work over the rest. The inevitable conclusion is, that plowman understands his business better than the others. He shows this by trimming the irons of his plow to the state of the land, and the nature of the work he is about to perform, and by training his horses more in accordance with their natural temperament, whereby they are guided more tractably. Having the shrewdness to acquire these essential accomplishments to a superior degree, the execution of superior work is an easier task to him than inferior work to the other plowmen. This case, which I have selected for an example, is not altogether a supposititious one; for, however dexterous all the plowmen on a farm may be, one will always be found to show a superiority over the rest.

Having advanced thus far in the knowledge and practice and capability of judging of work, the pupil begins to feel the importance of his acquisition; and this success will fan the flame of his enthusiasm, and prompt him to greater acquirements. But even in regard to the plow, the pupil has much to learn. Though he has picked out the best plowman, and knows why he is so, he is himself still ignorant of how practically to trim a plow, and to drive the horses with discretion. The plowman will be able to afford him ocular proof *how* he places *(tempers)* all the irons of the plow in relation to the state of the land, and *why* he yokes and drives the horses as he does in preference to any other plan. Illiterate and unmechanical as he is, and his language full of *technicalities*, his explanations will nevertheless give the pupil a clearer sight into the *minutiæ* of plowing than he could acquire by himself as a spectator in an indefinite length of time.

I have selected the plow as being the most useful implement to illustrate the method which the pupil should follow, in all cases, to learn a *practical* knowledge of every operation in farming. In like manner, he may become acquainted with the particular mode of managing all the larger implements which require the combined agency of man and horse to put into action; as well as become accustomed to wield the simpler implements used by the hand easily and ambidexterously, a great part of farm-work being executed with simple but very efficient tools. Frequent *personal* attendance at the farm-stead, during the winter months, to view and conduct the threshing-machine, while threshing corn, and afterward to superintend the winnowing-machine, in cleaning it for the market, will be amply repaid by the acquisition of essential knowledge regarding the nature and value of the cereal and leguminous grains. There is, moreover, no better method of acquiring an extensive knowledge of *all* the minor operations of the farm, than for the pupil personally to superintend the labors of the field-workers. Their labors are essential, methodical, almost always in requisition, and mostly consisting of minutiæ; and their general utility is shown, not only in their intrinsic worth, but in relation to the labors of the teams.

The general introduction of sowing-machines, particularly those which sow broadcast, has nearly superseded the beautiful art of sowing corn by hand.* Still a great deal of corn is sown by the hand, especially on small

* It were to be wished that this remark were more applicable to the U. S. Since the introduction of an English sowing-machine near Wilmington, Delaware, and the improvement on it, as it is alleged, by PENNOCK, it may be expected that this operation, recommended as well by its neatness as by its economy, will be extended through the country. It will be seen on referring to the June, 1846, Monthly Journal of Agriculture (which, be it always understood, is published along with the Farmers' Library) that Mr. Jones, a very observing and diligent practical Farmer, gives it as his opinion that the use of Pennock's drill or sowing-machine effects a saving or increase of 25 per cent.

farms, on which expensive machines have not yet found their way. In the art of hand-sowing, the pupil should endeavor to excel, for, being difficult to perform it in an easy and neat manner, the superior execution of it is regarded as an accomplishment. It is, besides, a manly and healthful exercise, conducive to the establishment of a robust frame and sound constitution.

The feeding of cattle in the farm-stead, or of sheep in the fields on turnips, does not admit of much participation of labor with the cattle-man or shepherd; but nevertheless, either practice will form an interesting subject of study to the pupil, and without strict attention to both, he will never acquire a knowledge of fattening live-stock, and of computing their value.

By steadily pursuing the course of observation which I have thus chalked out, and particularly in the first year of his apprenticeship, the pupil in a short time, will acquire a *considerable* knowledge of the minutiæ of labor; and it is only in this way that the groundwork for a familiar acquaintance with them can be laid. A truly familiar acquaintance with them requires years of experience. Indeed, observant farmers are learning some new, or modifications of some old, practice every day, and such new-like occurrences serve to keep alive in them a regard for the most trivial incident that happens on a farm.

In urging on the pupil the *necessity* of putting his hand to every kind of labor, I do not mean to say he should become a first-rate workman. To become so would require a much longer time than he could spare in a period of pupilage. His personal acquaintance with every implement and operation should, however, enable him by that time to decide quickly whether work is well or ill done, and whether it has been executed in a reasonable time. No doubt this extent of knowledge may be acquired *in time*, without the actual labor of the hands; but, as it is the interest of the pupil to learn his profession not only in the shortest possible time, but in the best manner, and as these can be acquired sooner through the joint coöperation of the head and hands, than by either singly, it would seem imperative on him to begin to acquire his profession by labor.

Other considerations regarding the acquisition of practical knowledge deserve the attention of the pupil. It is most conducive to his interest to learn his profession in youth, and before the meridian of life has set in, when labor of every kind becomes irksome. It is also much better to have a *thorough* knowledge of farming, *before* engaging in it on his own account, than to acquire it in the course of a lease, during which heavy losses may be incurred by the commission of comparatively trivial errors, especially at the early period of its tenure, when farms in all cases are most difficult to conduct. It is an undeniable fact, that the work of a farm never proceeds *so* smoothly and satisfactorily to all parties engaged in its culture, as when the farmer is thoroughly master of his business. His orders are then implicitly obeyed, not because they are pronounced more authoritatively, but because a skillful master's plans and directions inspire that degree of confidence in the laborers as to believe them to be the

The spread of improved agricultural implements is proverbially slow over the world, but less so in this, probably, than in any other part of it; owing to the freedom of our institutions, the frequent and extensive intercourse of our people and their aptness both at invention and imitation. But even here the adoption of the costly machinery of improved construction in England, is much restrained by the want of capital, which counteracts even our greater necessity for all labor-saving contrivances. All associations or uses of capital, therefore, to supply means to the industrious and frugal agriculturist, on fair terms and for periods corresponding with the nature of his pursuits, would deserve to be rewarded with honor and with profit.

best that could be devised in the circumstances. Shame is often acutely felt by servants, on being detected in error, whether of the head or heart, by so competent and discriminating a judge as a skillful master; because rebuke from such a one implies ignorance or negligence in those against whom it is directed. The fear of having ignorance and idleness imputed to them, by a farmer who has become acquainted with the capabilities of work-people by dint of his own experience, and can estimate their services as they really deserve, urges laborers to do a fair day's work in a workmanlike style.

Let the converse of all these circumstances be imagined; let the losses to which the ignorant farmer is a daily prey, by many ways—by hypocrisy, by negligence, by idleness, and by dishonesty of servants—be calculated, and it must be admitted that it is infinitely safer for the farmer to trust to his own skill for the fulfillment of his engagements, than entirely to depend on that of his servants, which he will be obliged to do if they know his business better than he himself does. No doubt a trustworthy steward may be found to manage well enough for him—and such an assistant is at all times valuable—but, in such a position, the steward himself is placed in a state of temptation, in which he should never be put; and, besides, the inferior servants never regard him as a master, and his orders are never so punctually obeyed, where the master himself is resident. I would, therefore, advise every young farmer to acquire a *competent* knowledge of his profession, before embarking in the complicated undertaking of conducting a farm. I only say a competent knowledge; for the gift to excel is not imparted to all who select farming as their profession; "it is not in man who walketh to direct his *steps*" aright, much less to attune his *mind* to the highest attainments.

Before the pupil fixes on any particular farm for his temporary abode, he should duly consider the objects he wishes to attain. I presume his chief aim is to attain such an intimate knowledge of farming as to enable him to employ his capital safely in the prosecution of the highest department of his profession. This will, probably, be best attained by learning that system which presents the greatest safeguards against unforeseen contingencies. Now, there is little doubt that the kind of farming which cultivates a variety of produce is more likely to be safe, during a lease, in regard both to highness of price and quantity of produce, than that which only raises one kind of produce, whether wholly of animals, or wholly of grain. For, although one kind of produce, when it happens to be prolific or high priced, may, in one year, return a greater profit than a variety of produce in the same year, yet the probabilities are much against the frequent recurrence of such a circumstance. The probability rather is that one of the varieties of produce will succeed, in price or produce, every year; and, therefore, in every year there will be a certain degree of success in that mode of farming which raises a variety of produce. Take, as an example, the experience of late years. All kinds of live-stock have been reared with profit for some years past; but the case is different in regard to grain. Growers of grain have suffered greatly in their capital in that time. And yet, to derive the fullest advantages from even the rearing of stock, it is necessary to cultivate a certain extent of land upon which to raise straw and green crops for them in winter. Hence, that system is the best for the young farmer to learn, which cultivates a relative proportion of stock and crop, and not either singly. This has been characterized as the *mixed* system of husbandry. It avoids, on the one hand, the monotony and inactivity attendant on the raising of grain, and subdues, on the other, the roving disposition engendered in the tending of live-stock in a pastoral district; so it blends both occupations into a happy union of cheerfulness and quiet.

Most farmers in the lowlands of Scotland practice the mixed husbandry, but it is reduced to a perfect system nowhere so fully as in the Border Counties of England and Scotland. There many farmers accept pupils, and thither many of the latter go to prepare themselves to become farmers. The usual fee for pupils, in that part of the country, is one hundred pounds [$500] per annum for bed and board, with the use of a horse to occasional markets and shows.* If the pupil desire to have a horse of his own, about thirty pounds a-year more are demanded. On these moderate terms, pupils are generally very comfortably situated.

I am very doubtful of it being good policy to allow the pupil a horse of his own at first. Constant attention to field-labor is not unattended with irksomeness; and, on the other hand, exercise on horseback is a tempting recreation to young minds. It is enough for a young man to feel the removal of parental restraint, without also having the dangerous incentive of an idle life placed at his disposal. They should consider that, upon young men arrived at the years when they become farming pupils, it is not in the power, and is certainly not the inclination, of farmers to impose ungracious restraints. It is the duty of their parents and guardians to impose these; and the most effectual way that I know of, in the circumstances, to avoid temptations, is the denial of a riding-horse. Attention to business in the first year will, most probably, induce a liking for it in the second; and, after that, the indulgence of a horse may be granted to the pupil with impunity, as the reward of diligence. Until then, the horse occasionally supplied by the farmer to attend particular markets, or pay friendly visits to neighbors, should suffice; and, as that is the farmer's own property, it will be more in his power to curb in his pupils any propensity to wander abroad too frequently, and thereby preserve his own character as a tutor.

Three years of apprenticeship are, in my opinion, requisite to give a pupil an adequate knowledge of farming—such a knowledge, I mean, as would impress him with the confidence of being himself able to manage a farm; and no young man should undertake such a management until he feels this confidence in himself. Three years may be considered by many as too long a time to spend in learning *farming;* but, after all, it is much less time than that given to many other professions, whose period of apprenticeship extends to five and even seven years; and, however highly esteemed those professions may be, none possess a deeper interest, in a national point of view, than that of Agriculture. There is a condition attendant on the art of farming—which is common to it and gardening, but inapplicable to most other arts—that a year must elapse before the same work can again be performed. Whatever may be the ability of the learner

[* A gentleman of ample fortune, residing at Paterson, New-Jersey, has lately, with his son's entire concurrence and desire, sent him to reside with a Scotch farmer, of Shields, near Ayr; to whom he pays one hundred pounds sterling a year—say $500—for board and education as a farmer. Mr. Turnant (that appears to be the name) is Vice President of the Ayrshire Agricultural Society; and is, withal, a gentleman, and lives as such. The young man alluded to keeps no horse, but is fully instructed in all the departments of Agriculture. We mention this particular case of a young gentleman, not urged by necessity, betaking himself to Agriculture as a profession, with the acquiescence of his father, to show that public sentiment is taking a right turn, and that those who have been so assiduously laboring to elevate this pursuit in the public esteem have not entirely lost their labor, but may hope yet to see practical Agriculture followed as an intellectual occupation—one in which success may warrant the presumption of some merit besides the mere faculties of imitation and plodding industry. *Ed. Farm Lib.*]

to acquire farming, *time* must thus necessarily elapse before he can have the opportunity of again witnessing a bygone operation. There is no doubt of his natural capacity to acquire, in two years, the art to manage a farm, but the operations necessarily occupying a year in their performance, prevent that acquisition in less time than three. This circumstance, of itself, will cause him to spend a year in merely observing passing events. This is in his first year. As the operations of farming are all anticipatory, the second year may be fully employed in studying the progress of work in preparation of anticipated results. In the third year, when his mind has been stored with all the modes of doing work, and the purposes for which they are performed, the pupil may attempt to put his knowledge into practice; and his first efforts at management cannot be attempted with so much ease of mind to himself as on the farm of his tutor, under his correcting guidance.

This is the *usual* progress of tuition during the apprenticeship of the pupil; but, could he be brought to anticipate results while watching the progress of passing events, one year might thus be cut off his apprenticeship. Could a *book* enable him to acquire the experience of the second year in the course of the first, a year of probationary trial would be saved him, as he would then acquire in two what requires three years to accomplish. *This* book will accomplish no small achievement—will confer no small benefit on the agricultural pupil—if it accomplish this.

---

6. OF THE ESTABLISHMENT OF SCIENTIFIC INSTITUTIONS OF PRACTICAL AGRICULTURE.

"Here let us breathe, and happily institute
A course of learning and ingenious studies."
TAMING OF THE SHREW.

ALTHOUGH I know of no existing plan so suited to the learning of practical farming as a protracted residence on a farm, yet I feel assured a more efficient one might easily be proposed for the purpose. An evident and serious objection against the present plan is the want of that solicitous superintendence over the progress of the pupils, on the part of the farmer, which is implied in his receiving them under his *charge*. The pupils are left too much to their own discretion to learn farming effectually. They are not sufficiently warned of the obstacles they have to encounter at the outset of their career. Their minds are not sufficiently guarded against receiving a wrong bias in the methods of performing the operations. The advantages of performing them in one way rather than another are not sufficiently indicated. The effects which a change of weather has in altering the arrangements of work fixed upon, and of substituting another more suited to the change, is not sufficiently explained. Instead of receiving explanatory information on these and many more particulars, the pupils are mostly left to find them out by their own diligence. If they express a desire to become acquainted with these things, no doubt it will be cheerfully gratified by the farmer; but how can the uninitiated pupil know the precise subject with which to express his desire to become acquainted?

In such a system of tuition, it is obvious that the diligent pupil may be daily perplexed by doubtful occurrences, and the indifferent pupil permit unexplained occurrences to pass before him, without notice. Reiterated occurrences will, in time, force themselves upon the attention of every class of pupils: but, unless their attention is purposely drawn to, and explications proffered of, the more hidden difficulties in the art of farming, they will spend much time ere they be capable of discovering important occult matters by their own discernment.

It is in this respect that farmers, who profess to be tutors, show, as I conceive, remissness in their duty to their pupils; for all of them can impart the information alluded to, and give, besides, a common-sense explanation of every occurrence that usually happens on a farm, otherwise they should decline pupils.

It is obvious that pupils should not be placed in this disadvantageous position. They ought to be *taught* their profession; because the art of husbandry should be acquired, like every other art, by teaching, and not by intuition. On the other hand, pupils in this, as in every other art, ought to endeavor to acquire the largest portion of the knowledge of external things by their own observation; and they should be made aware, by the farmer, that he can at most only *assist* them in their studies; so that, without much study on their parts, all the attention bestowed on their tuition by the most pains-taking farmer will prove of little avail. Practical experience forms the essential portion of knowledge which farmers have to impart, and it is best imparted on the farm; but they have not always the leisure, by reason of their other avocations, to communicate even this on the spot in its due time. More than mere practical knowledge, however, is requisite to satisfy the mind of the diligent pupil. He wishes to be satisfied that he is learning the *best* method of conducting a farm: he wishes to be informed of the *reasons* why one mode of management is preferable to every other: he wishes to become *familiar* with the explanations of all the phenomena that are observable on a farm.

To afford all the requisite information to the pupil in the highest perfection, and to assist the farmer in affording it to him in the easiest manner, I propose the following plan of tuition for adoption, where circumstances will permit it to be established. The more minutely its details are explained, the better will it be understood by those who may wish to form such establishments.

Let a farmer of good natural abilities, of firm character, fair education, and pleasant manners—leasing a farm of not less than five hundred acres, and pursuing the mixed system of husbandry—occupy a house of such a size as would afford accommodation to from ten to twenty pupils. The farm should contain different varieties and conditions of soil—be well fenced, well watered, and not at an inconvenient distance from a town.

With regard to the internal arrangements of the house, double-bedded rooms would form suitable enough sleeping apartments. Besides a dining-room and drawing-room, for daily use, there should be a large room, fitted up with a library, containing books affording *sound* information on all agricultural subjects, in various languages—forming at one time a lecture-room for the delivery of lectures on the elementary principles of those sciences which have a more immediate reference to Agriculture, and at another a reading or writing room or parlor for conversations on farming subjects.—There should be fixed, at suitable places, a barometer, a sympiesometer, thermometers (one of which should mark the lowest degree of temperature in the night), a rain-gauge, an anemometer, and a weathercock. No very useful information, in my opinion, can be derived by the *farmer* from a bare *register* of the hights and depths of the barometer and thermometer.

A more useful register for him would be that of the directions of the wind, accompanied with remarks on the state of the weather, the heat of the air as indicated by the feelings, and the character of the clouds as expressed by the most approved nomenclature. The dates of the commencement and termination of every leading operation on the farm should be noted down, and appropriate remarks on the state of the weather during its performance recorded. A small chemical laboratory would be useful in affording the means of analyzing substances whose component parts were not well known. Microscopes would be useful in observing the structure of plants and insects, for the better understanding of their respective functions.

The slaughter-house required for the preparation of the meat used by the family should be fitted up to afford facilities for dissecting those animals which have been affected by peculiar disease. Skeletons and preparations for illustrating comparative anatomy could thus be formed with little trouble. A roomy dairy should be fitted up for performing experiments on the productive properties of milk in all its various states. A portion of the farm-offices should be fitted up with apparatus for making experiments on the nutritive properties of different kinds and quantities of food, and the fattening properties of different kinds of animals. A steelyard, for easily ascertaining the live-weight of animals, is a requisite instrument.—The bakery, which supplies the household bread, would be a proper place for trying the relative panary properties of different kinds of flour and meal. Besides these, apparatus for conducting experiments on other subjects, as they were suggested, could be obtained when required.

Another person besides the farmer will be required to put all this apparatus into use. He should be a man of science, engaged for the express purpose of showing the relation betwixt science and Agriculture. There would be no difficulty of obtaining a man of science, quite competent to explain natural phenomena on scientific principles. For that purpose, he would require to have a familiar acquaintance with the following sciences: With meteorology and electricity, in order to explain atmospherical phenomena, upon the mutations of which all the operations of farming are so dependent: with hydrostatics and hydraulics, to explain the action of streams and of dammed-up water on embankments, to suggest plans for the recovery of land from rivers and the sea, and to indicate the states of the weather which increase or diminish the statical power of the sap in vegetables: with botany and vegetable physiology, to show the relations between the natural plants and the soils on which they grow, with a view to establish a closer affinity between the artificial state of the soil and the perfect growth of cultivated plants; to exhibit the structure of the different orders of cultivated plants; and to explain the nature and uses of the healthy, and the injurious effects of the diseased secretions of plants: with geology, to explain the nature and describe the structure of the superficial crust of the earth, in reference to draining the soil; to show the effects of subsoils on the growth of trees; to explain the effects of damp subsoils on trees, and of the variations of the surface of the ground on climate: with mechanics, to explain the principles which regulate the action of all machines, and which acquirement previously implies a pretty familiar acquaintance with the mathematics: with chemistry, to explain the nature of the composition of, and changes in, mineral, vegetable, and animal substances: with anatomy and animal physiology, to explain the structure and functions of the animal economy, with a view to the prevention of disease, incidental to the usual treatment of animals, and to particular localities.—All young men, educated for what are usually termed the learned professions—theology, law, and medicine—are made acquainted with these sciences, and a young man from either faculty would be competent to take charge of such an establishment. Of the three I would give preference to the *medical* man, as possessing professionally a more intimate knowledge of chemistry, and animal and vegetable physiology, than the others. But the most learned graduate of either profession will display his scientific acquirements to little advantage in teaching pupils in Agriculture, unless he has the judgment to select those parts of the various sciences whose principles can most satisfactorily explain the operations of Agriculture.—Ere he can do this successfully, he would, I apprehend, require to know Agriculture practically, by a previous residence of at least two years on a farm. Without such a preparation, he would never become a *useful* teacher of agricultural pupils.

On the supposition that he is so qualified, his duty is to take the direct charge of the pupils. His chief business should be to give demonstrations and explanations of all the phenomena occurring during operations in the farm field. The more popular demonstrations on botany, animal and vegetable physiology, and geology, as also on meteorology, optics, and astronomy, whenever phenomena occur which would call forth the application of the principles of any of those sciences, would be best conducted in the fields. In the library, short lectures on the elementary principles of science could be regularly delivered—conversations on scientific and practical subjects conducted—and portions of the most approved authors on Agriculture, new and old, read. These latter subjects could be most closely prosecuted when bad weather interrupted field-labor. In the laboratory, slaughter-house, farmstead, and dairy, he could command the attendance of the pupils, when any subject in those departments was to be explained.

The duty of the farmer himself, the governor or head of the establishment, is to enforce proper discipline among the pupils, both within and without doors. He should teach them practically how to perform every species of work, explain the nature and object of every operation performing, and foretell the purport of every operation about to be performed. For these important purposes he should remain at home as much as is practicable with his avocations abroad.

The duties of the pupils are easily defined. They should be ready at all times to hear instruction, whether in science or practice, within or without doors. Those pupils who wish to study practice more than science, should not be constrained to act against their inclinations, as science possesses little allurement to some minds; and it should be borne in mind by the tutors, that practical farming is what the pupils have chiefly come to learn, and that practice may prove successful in after life without the assistance of science, whereas science can never be applied without practice.

The duty common to all, is the mutual conducting of experiments, both in the fields and garden; for which purpose both should be of ample dimensions. All new varieties of plants might be first tried in the garden, until their quantity warranted the more profuse and less exact, though more satisfactory culture of the field. On ridges in the fallow-field, with different kinds and quantities of manure, and different modes of working the soil and sowing the seeds, experiments should be continually making with new and old kinds of grains, roots, tubers, bulbs, and herbaceous plants. In course of time, the sorts best suited to the locality will show themselves, and should be retained, and the worthless abandoned. In like manner, experiments should be made in the crossing of animals, whether with the view of maintaining the purity of blood in one, the improvement of the blood of another, or the institution of an entirely new blood. In either class of experiments many new and interesting facts regarding the constitutional differences of animals, could not fail to be elicited.

Any farmer establishing such an institution, which could only be done at considerable expense, in fitting up a house in an adequate manner, and securing the services of a man of science, would deserve to be well remunerated. I before mentioned that one hundred pounds a year as board were cheerfully given by pupils to farmers under ordinary circumstances. In such an institution, less than one hundred and fifty pounds a year would not suffice to remunerate the farmer. Supposing that ten pupils at that fee each, were accommodated on one farm, the board would amount to fifteen hundred pounds a year. In regard to the expense of maintaining such an establishment, with the exception of foreign produce and domestic luxuries, all the ordinary means of good living exist on a farm. The procuring of these necessaries and luxuries and maintaining a retinue of grooms and domestic servants, together with the salary of an accomplished tutor, which should not be less than three hundred pounds a year besides board, would probably incur an annual disbursement of a thousand pounds a year. The farmer would thus receive five hundred pounds a year for risk of the want of the full complement of pupils, and for interest on the outlay of capital. Such a profit may be considered a fair, but not an extravagant remuneration for the comfortable style of living and superior kind of tuition afforded in such an establishment.

Were the particulars pitched at a lower scale, a profit might be derived from ten pupils, of not less a sum than that derived from the usual board of one hundred pounds a year. Were two hundred a year exacted, pupils of the highest class of society might be expected. Were different institutions at different rates of board, established, all the classes of society would be accommodated.

Would farmers who have accommodation for conducting such an institution, but duly consider the probable certainty of obtaining a considerable increase to their income, besides the higher distinction of conducting so useful an institution, I have no doubt many would make the attempt. There are insuperable obstacles to some farmers making the attempt; but there are many who possess the requisite qualifications of accommodation in house and farm locality, personal abilities, influence, and capital, for instituting such an establishment. But even where all these qualifications do not exist, most of the obstacles might be overcome. In the case of the house, it could be enlarged at the farmer's own expense, for the landlord cannot be expected to erect a farm-house beyond the wants of an ordinary family; nor, perhaps, would every small landed proprietor permit the unusual enlargement of a farm-house, in case it should be rendered unsuitable to the succeeding tenant. To avoid this latter difficulty, the farmer who could afford accommodation to the fewest number, could receive the highest class of pupils, were his own education and manners competent for the highest society, while those who had more accommodation could take a more numerous and less elevated rank of pupils. In either way the profit might be equally compensatory.

In regard to other considerations, a tutor entirely competent could not at once be found. It may safely be averred, that a really scientific man, thoroughly acquainted with the practice of Agriculture, is not to be found in this country. But were a demand for the services of scientific men to arise from the increase and steady prosperity of such establishments, no doubt, men of science would qualify themselves for the express purpose. As to pupils, the personal interest of the farmer might not avail him much at first in influencing parents in his favor, but if he possess the reputation of being a good farmer, he would soon acquire fame for his institution. I have no doubt of an eminent farmer entirely succeeding to his wishes, who occupies a commodious house, on a large farm, in an agricultural district of high repute, and possessing sufficient capital, were he to make the experiment by engaging a competent scientific tutor, and teaching the practical department himself. Such a combination of alluring circumstances could not fail of attracting pupils from all parts of the country, who were really desirous of learning Agriculture in a superior manner.

There might still be another, though less attractive and efficacious, mode of accomplishing a similar end. Let a scientific tutor, after having acquired a competent knowledge of practical Agriculture, procure a commodious and comfortable house in any village in the vicinity of some large farms, in a fine agricultural district. Let him receive pupils into his house, on his own account, in such numbers and at such fees as he conceives would remunerate him for his trouble and risk. Every thing relating to science within doors, could be conducted as well in such a house as in any farm-house; and as to a field for practice, let the tutor give a douceur to each of the large farmers in his neighborhood, for liberty for himself and pupils to come at will and inspect all the operations of the farms. In this way a very considerable knowledge of farming might be imparted. Having every article of consumption to purchase at market-price, such an establishment would cost more to maintain than that on a farm; but, on the other hand, the salary of the tutor would in this case be saved, and there would be no farmer and his family to support. To assist in defraying the extra expenses of such an establishment, let the tutor permit, for a moderate fee, the sons of those farmers whose farms he has liberty to inspect, and of those who live at a distance, to attend his lectures and readings in the house, and his prelections in the fields. A pretty extensive knowledge of and liking for the science of Agriculture might thus be diffused throughout the country, among a class of young farmers who might never have another opportunity of acquiring it, because they would never become permanent inmates in any such establishment.

I have known a mode of learning farming adopted by young farmers of limited incomes, from remote and semi-cultivated parts of the country, of lodging themselves in villages in cultivated districts adjacent to large farms, occupied by eminent farmers, and procuring leave from them to give their daily personal labor and superintendence in exchange for the privilege of seeing and participating in all the operations of the farm.

There are still other modes than those described above of learning farming, which deserve attention, and require remark. Among these the only one in this country, apart from the general practice of boarding with practical farmers, is the Class of Agriculture in the University of Edinburgh. This chair was endowed in 1790 by Sir William Pulteney, with a small salary, and placed under the joint patronage of the Judges of the Court of Session, the Senatus Academicus of the University of Edinburgh, and the Town Council of the City of Edinburgh. The first professor elected by the patrons to this chair in 1791, was the late Dr. Coventry, whose name, in connection with the Agriculture of the country, stood prominent at one time. He occupied the chair until his death in 1831. His prelections, at the earlier period of his career as a professor, were successful, when his class numbered upward of seventy students. When I attended it, the number of students was upward of forty. Dr. Coventry was a pleasing lecturer, abounding in anecdote, keeping his hearers always in good humor, courting interrogation, and personally showing great kindness to every student. At the latter period of his incumbency, the class dwindled away, and for some years before his death he delivered a course of lectures only every two years. He delivered, I understand, thirty-four courses in forty years.

The present Professor Low succeeded Dr. Coventry. Since his installation into the chair, he has rekindled the dying embers of the agricultural class, by delivering an annual course of lectures suited to the improved state of British Agriculture, and by forming a museum of models of agricultural implements, and portraits of live-stock illustrative of his lectures, of the most extensive and valuable description. In Dr. Coventry's time there was no museum deserving the name, and seeing this, Professor Low had no doubt been impressed with the important truth, that without models of the most approved implements, and portraits of the domesticated animals, serving to illustrate the principal operations and breeds of animals to be seen on the best cultivated farms, and pastoral districts, a mere course of lectures would prove nugatory. This museum is attached to the University, and to show the zeal and industry by which the present Professor has been actuated in its formation, the objects in it must be worth more than £2,000. The funds which obtained those objects were derived from the revenues under the management of the Board of Trustees for the encouragement of Arts and Manufactures in Scotland. This Board was instituted by the 15th Article of the Treaty of Union between Scotland and England. Besides forming the museum, Professor Low has, during his yet short incumbency in the chair, already contributed much important matter to the agricultural literature of the country, by the publication of his Elements of Practical Agriculture, which contain almost the entire substance of his lectures, and the series of colored portraits of animals taken from the pictures in the museum, now coming out periodically in numbers.

There has lately been appointed a lecturer on Agriculture in Marischal College, Aberdeen, at a salary of £40 a year. Being but an experiment, the appointment, I believe, has only been made for three years.

There is no public institution in England for teaching Agriculture.*— Some stir is making in the establishment of an Agricultural College in Kent, the prospectus of which I have seen; and, some time ago, I saw a statement which said that provision exists for the endowment of a chair of Agriculture in one of the Colleges of Oxford.

An agricultural seminary has existed at Templemoyle, in the county of Londonderry, Ireland, for some years. It originated with the members of the North-west of Ireland Farming Society, and the first intention was to form it on such a scale as to teach children of the higher orders every science and accomplishment, while those of the lower orders, the sons of farmers and tradesmen, were to be taught Agriculture. But the latter arrangement only has been found to be practicable. In a statement circulated by a member of the committee, I find that "the formation of this establishment has caused its founders an expenditure of above £4,000— of which about £3,000 were raised at its commencement by shares of £25 each, taken by the noblemen, gentlemen, and members of the North-west Society. The Grocers' Company of London, on whose estate it is situated, have been most liberal in their assistance, and have earned a just reward in the improvement of their property, by the valuable example the farm of Templemoyle presents to their tenantry.

"In sending a pupil to Templemoyle, it is necessary to have a nomina-

[* The one at Cirencester, mentioned by Mr. Colman, is probably now in operation.

tion from one of the shareholders, or from a subscriber of £2 annually.—The annual payment for pupils is £10 a year, and for this trifling sum they are found in board, lodging, and washing, and are educated so as to fit them for land-stewards, directing agents, practical farmers, schoolmasters, and clerks. From fifteen to seventeen is the age best suited to entrance at Templemoyle, as three years are quite sufficient to qualify a student possessed of ordinary talents, and a knowledge of the rudiments of reading and writing, to occupy any of the above situations.

"Upward of two hundred young men, natives of sixteen different counties in Ireland, have passed or remain in the school. Of these, between forty and fifty have been placed in different situations, such as land-stewards, agents, schoolmasters, and clerks, or employed on the Ordnance Survey. Nearly one hundred are now conducting their own or their fathers' farms, in a manner very superior to that of the olden time.*

"The school and farm of Templemoyle are situated about six miles from Londonderry—about a mile distant from the mailcoach-road leading from Londonderry to Newtonlimavady. The house, placed on an eminence, commands an extensive and beautiful view over a rich and highly cultivated country, terminated by Lough Foyle. The base of the hill is occupied by a kitchen and ornamental garden, cultivated by the youths of the establishment, under an experienced gardener. The house and farm-offices behind contain spacious, lofty, and well ventilated schoolrooms; refectory; dormitories; apartments for the masters, matron, servants, &c. Each pupil occupies a separate bed; the house can accommodate seventy-six, and the number of pupils is sixty. They receive an excellent education in reading,

[* It may be of practical service in the United States to give here the general regulations of the "*Larne*" School, of which Mr. Colman seems to think so favorably. He says they were given to him in printed form by the intelligent principal. *Ed. Farm. Lib.*

"1. As the great object is to make the boys practical farmers, one-half of them will be at all times on the farm, where they will be employed in manual labor, and receive from the head farmer such instructions, reasons, and explanations, as will render the mode of proceeding, in all the various operations performed on the farm, sufficiently intelligible to them. Every pupil is to be made a plowman, and taught, not only how to use, but how to settle the plow-irons for every soil and work, and to be instructed and made acquainted with the purpose and practical management of every other implement generally used. And all are to be kept closely to their work, either by the head farmer or his assistant, or, in their unavoidable absence, by the monitor placed in charge of them.

"2. Their attention is to be drawn to stock of all kinds, and to the particular points which denote them to be good, bad, indifferent, hardy, delicate, good feeders, good milkers, &c.

"3. At the proper season of the year, the attention of the boys is to be directed to the making and repairing of fences, that they may know both how to make a new one, and, what is of great advantage, how to repair and make permanent those of many years' standing.†

"4. The head farmer will deliver evening lectures to the pupils on the theory and practice of Agriculture, explaining his reasons for adopting any crop, or any particular rotation of crops, as well as the most suitable soil and the most approved modes of cultivating for each; the proper management and treatment of working, feeding, and dairy stock; the most approved breeds, and their adaptation to different soils. He will point out the best method of reclaiming, draining, and improving land; and will direct attention to the most recent inventions in agricultural implements, detailing the respective merits of each.

"5. After the boys have been taught to look at stock on a farm with a farmer's eye, the committee propose that they should in rotation attend the head farmer to fairs and markets, in order to learn how to buy and sell stock. At the same time, the committee expect the head farmer will make his visits to fairs as few as possible, as his attention to the pupils of the establishment is always required, and he should therefore be as seldom as possible absent from Templemoyle.

"An annual examination of the school is held before the committee and subscribers, and conducted by examiners totally independent of the school. The examination is attended by the leading gentlemen in the neighborhood, and many of these take a part in the examination, by either asking or suggesting questions—a practice which is deserving of recommendation, as adapted to give additional value and dignity to the examination.

"Such are some of the principal regulations of the school, which I have copied, that its management might be fully understood."]

† This, of course, applies principally to live fences, or hedges. It could at present have little pertinency to the United States, where certainly there • very little mystery in making the fences, and as little labor expended in keeping them in repair.

writing, arithmetic, book-keeping, mathematics, land-surveying, and geography. This department is managed by an excellent head-master, and assistant-master, both resident in the house. The pupils are so classed that one-half are receiving their education in the house, while the remainder are engaged in the cultivation of a farm of 165 Scotch acres, in the management of which they are directed by the head-farmer, an experienced and clever man, a native of Scotland, who has a skillful plowman under him. The pupils who are employed one part of the day on the farm are replaced by those in the school, so that the education always advances in and out of doors *pari passu*."*

In enumerating the means of obtaining agricultural knowledge in this country, I cannot omit mentioning those coässistant institutions, the Veterinary Colleges. Their great object is to form a school of veterinary science, in which the anatomical structure of quadrupeds of all kinds—horses, cattle, sheep, dogs, &c.—the diseases to which they are subject, and the remedies proper to be applied for their removal, might be investigated and regularly taught; in order that, by this means, enlightened practitioners of liberal education, whose sole study has been devoted to the veterinary art in all its branches, may be gradually dispersed over all the kingdom. The Veterinary College of London was instituted in 1791, according to the plan of Mr. Sain Bel, who was appointed the first professor. Parliamentary grants have been afforded at times to aid this institution, when its finances rendered such a supply essential. It is supported by subscription. Every subscriber of the sum of £21 is a member of the society for life.—Subscribers of two guineas annually are members for one year, and are equally entitled to the benefits of the institution. A subscriber has the privilege of having his horses admitted into the infirmary, to be treated, under all circumstances of disease, at 3s. 6d. per night, including keep, medicines, or operations of whatever nature that may be necessary; likewise of bringing his horses to the college for the advice of the professor *gratis*, in cases where he may prefer the treatment of them at home.†—Until last year, care was chiefly bestowed in this institution on the horse, when the Royal Agricultural Society of England, deeming it as important for the promotion of Agriculture to attend to the diseases of the other animals reared on farms as well as the horse, voted £300 a year out of their funds for that purpose.

The Veterinary College of Edinburgh had its origin in the personal exertions of its present professor, Mr. William Dick, in 1818, who, after five years of unrequited labor, fortunately for himself and the progress of the veterinary science in Scotland, obtained the patronage of the Highland and Agricultural Society of Scotland, who have afforded him a small salary since 1823. Since then the success of his exertions has been extraordinary—not fewer than from seventy to one hundred pupils attending the college every session, of whom about twenty every year, after at least two years' study of practical anatomy and medicine, become candidates for the diploma of veterinary surgeon. Their qualifications are judged of after an examination by the most eminent medical practitioners in Edinburgh.—The students enjoy free admission to the lectures on human anatomy and physiology in Queen's College, by the liberality of its professors. Through the influence of the Highland Society, permission has been obtained for the graduates to enter as veterinary surgeons into her Majesty's cavalry regiments, as well as those of the Honorable East India Company.

In recommending farmers to attend lectures on veterinary science, it

* Irish Farmer's Magazine, No. 51.

† Beauties of England and Wales, vol. x. Part IV. p. 181.

must not be imagined that I wish them to become veterinary surgeons.—Let every class of people adhere to their own profession. But there is no doubt that a knowledge of veterinary science is of great use to the farmer, not in enabling him to administer to the diseased necessities of his live-stock—for that requires more professional skill and experience than any farmer can attain to, and is the proper province of the regularly bred veterinary surgeon—but to enable him readily to detect a disease by its symptoms, in order to apply immediate checks to its progress until he can communicate with and inform the veterinary surgeon of the nature of the complaint, whereby he may bring with him materials for treating it correctly on his arrival. The death of a single animal may be a serious loss to the farmer; and if, by his knowledge of the *principles* of the veterinary art, he can stay the progress of any disease, he may not only avert the loss, but prevent his animal being much affected by disease. Disease, even when not fatal to animals, leaves injurious effects on their constitutions for a long time.

With regard to attending lectures on Agriculture, I should say, from my own experience, that more benefit will be derived from attending them after having acquired a practical knowledge of husbandry than before; because many of the details of farming cannot be comprehended, unless the descriptions of them are given where the operations themselves can be referred to.

Abroad are several institutions for the instruction of young men in Agriculture, among which is the far-famed establishment of Hofwyl, in the canton of Berne, in Switzerland, belonging to M. de Fellenberg.* This establishment is not intended so much for a school of Agriculture, as that of education and moral discipline. All the pupils are obliged to remain nine years, at least until they attain the age of twenty-one—during which time they undergo a strict moral discipline, such as the inculcation of habits of industry, frugality, veracity, docility, and mutual kindness, by means of good example rather than precepts, and chiefly by the absence of all bad example. The pupils are divided into the higher and lower orders, among the former of whom may be found members of the richest families in Germany, Russia, and Italy. For these the course of study is divided into three periods of three years each. In the first, they study Greek, Grecian history, and the knowledge of animals, plants, and minerals; in the second, Latin, Roman history, and the geography of the Roman world; and, in the third, modern languages and literature, modern history to the last century, geography, the physical sciences, and chemistry. During the whole nine years they apply themselves to mathematics, drawing, music, and gymnastic exercises. The pupils of the canton of Berne only pay M. de Fellenberg 45 louis each, and do not cost their parents above 100 louis or 120 louis a year. Strangers pay him 125 louis, including board, clothing, washing, and masters.

The pupils of the lower orders are divided into three classes according to their age and strength. The first get a lesson of half an hour in the morning, then breakfast, and afterward go to the farm to work. They return at noon. Dinner takes them half an hour, and after another lesson of one hour, they go again to work on the farm until six in the evening. This is their summer occupation; and in winter they plait straw for chairs, make baskets, saw logs and split them, thresh and winnow corn, grind colors, knit stockings; for all of which different sorts of labor an

[* For a valuable notice of this institution, see the Report of Professor BACHE, on his return from Europe—having been sent by the Girard College, on a tour of observation, and to purchase a library (we believe) for that institution. [*Ed. Farm. Lib.*

adequate salary is credited to each boy's class until they are ready to leave the establishment. Such as have a turn for any of the trades in demand at Hofwyl, wheelwright, carpenter, smith, tailor or shoemaker, are allowed to apply to them. Thus the labor of the field, their various sports, their lessons, their choral songs, and necessary rest, fill the whole circle of the twenty-four hours; and judging from their open, cheerful, contented countenances, nothing seems wanting to their happiness.

It is admitted that, on leaving the establishment, the pupils of the higher classes are emmently moral and amiable in their deportment, that they are very intelligent, and that their ideas have a wide range; and though they may not be so advanced in science as some young men brought up elsewhere, they are as much so as becomes liberal-minded gentlemen, though not professors. The pupils of the lower classes leave at the age of twenty-one, understanding Agriculture better than any peasants ever did before, besides being practically acquainted with a trade, and with a share of learning quite unprecedented among the same class of people; and yet as hard-working and abstemious as any of them, and with the best moral habits and principles. It seems impossible to desire or imagine a better condition of peasantry.

As all the instruction at this establishment is conveyed orally, a great many teachers are required in proportion to the number of the pupils. In 1819, there were thirty professors for eighty pupils. This entails a considerable expense upon M. de Fellenberg, who besides extends the erection of buildings as he finds them necessary. He is, however, upon the whole, no loser by the speculation. Each pupil of the lower orders costs him £56 a year to maintain and educate, which is £3 8s. a year beyond the value of his work, and yet the investment is a profitable one, yielding something more than 8¼ per cent. interest, net of all charges. "The farm is undoubtedly benefited by the institution, which affords a ready market for its produce, and perhaps by the low price at which the labor of the boys is charged. But the farm, on the other hand, affords regular employment to the boys, and also enables M. de Fellenberg to receive his richer pupils at a lower price than he could otherwise do. Hofwyl, in short, is a great whole, where one hundred and twenty or one hundred and thirty pupils, more than fifty masters and professors, as many servants, and a number of day-laborers, six or eight families of artificers and tradesmen, altogether about three hundred persons, find a plentiful, and in many respects a luxurious subsistence, exclusive of education, out of a produce of one hundred and seventy* acres; and a money income of £6,000 or £7,000, reduced more than half by salaries, affords a very considerable surplus to lay out in additional buildings."† It seems that, since 1807, two convents, one in the canton of Fribourg, and the other in that of Thurgovie, have formed establishments analogous to those of M. de Fellenberg.‡

The celebrated German institution for teaching Agriculture is at Möeglin, near Frankfort on the Oder. It is under the direction of M. Von Thaër. There are three professors besides himself—one for mathematics, chemistry, and geology; one for veterinary knowledge; and a third for botany and the use of the different vegetable productions in the materia medica, as well as for entomology. Besides these, an experienced agriculturist is engaged, whose office it is to point out to the pupils the mode

* This is the number of acres in the farm as stated in the Edinburgh Review for October, 1819; but a correspondent in Hull's Philanthropic Repertory for 1832, makes it 250 acres.
† Edinburgh Review, No. 64.
‡ Ebel, Manuel du Voyageur en Suisse, tome ii.

of applying the sciences to the practical business of Husbandry. Such a person would be difficult to be found in this country. The course commences in *September*, the best season, in my opinion, for commencing the learning of Agriculture. During the winter months the time is occupied in mathematics, and in the summer the geometrical knowledge is practically applied to the measurement of land, timber, buildings, and other objects. The first principles of chemistry are unfolded. Much attention is paid to the analyzation of soils. There is a large botanic garden, with a museum containing models of implements of husbandry. The various implements used on the farm are all made by smiths, wheelwrights and carpenters residing round the institution; the workshops are open to the pupils, and they are encouraged by attentive inspection to become masters of the more minute branches of the economy of an estate.

As the sum paid by each pupil, who are from twenty to twenty-four years of age, is 400 rix-dollars annually (equal to about £60 sterling, if the rix-dollar is of Prussian currency), and besides which they provide their own beds and breakfasts, none but youths of good fortune can attend at Möeglin. Each has a separate apartment. They are very well-behaved young men, and their conduct to each other, and to the professors, is polite even to punctilio.

The estate of Möeglin consists of twelve hundred English acres. About thirty years ago it was given in charge by the King of Prussia to M. Von Thaër, who at that time was residing as a physician at Celle, near Luneburg, in Hanover, with the view of diffusing agricultural knowledge in Prussia, which it was known M. Von Thaër possessed in an eminent degree, as evinced by the translations of numerous agricultural works from the English and French, by his management in setting an example to the other great landed proprietors, and stimulating them to adopt similar improvements. His Majesty also wished him to conduct a seminary, in which the knowledge of the sciences might be applied to Husbandry, for the instruction of the young men of the first families.* When M. Von Thaër undertook the management of this estate, its rental was only 2,000 rix-dollars a year (£300), and twenty years ago the rental had increased to 12,000 rix-dollars (£1,800). This increased value, besides the buildings erected, has arisen from the large flocks of sheep which in summer are folded on the land, and in winter make abundant manure in houses constructed for their lodging.

These particulars are taken from Mr. Jacob's travels, who visited Möeglin in 1819, and who, in considering of the utility of such an institution in this country, makes these remarks on the personal accomplishments of M. Von Thaër. "We have already carried the division of labor into our Agriculture, not certainly so far as it is capable of being carried, but much farther than is done in any other country. We have some of the best sheep farmers; of the best cattle and horse breeders; of the best hay, turnip, potato, and corn farmers in the world; but we have perhaps no one individual that unites in his own person so much knowledge of chemistry, of botany, of mathematics, of comparative anatomy, and of the application of these various sciences to *all* the practical purposes of Agriculture as Von Thaër does; nor is the want felt, because we have numbers of individuals who, by applying to each branch separately, have

[* For his biography and a general account of his writings, and the whole of his great work on "THE PRINCIPLES OF AGRICULTURE," see the FARMERS' LIBRARY, commencing with Number 3, Vol. I. The whole has been published, and may be had, bound by itself, and ought undoubtedly to be a standard book in every school in the United States. *Ed. Farm. Lib.*]

reached a hight of knowledge far beyond what any man can attain who divides his attention between several objects. In chemistry we have now most decidedly the lead. In all of botany that is not mere nomenclature, it is the same. In mechanics we have no equals. There are thus abundant resources, from which practical lessons may be drawn, and be drawn to the greatest advantage; and that advantage has excited, and will continue to excite, many individuals to draw their practical lessons for each particular branch of Agriculture, from that particular science on which it depends; and thus the whole nation will become more benefited by such divisions and subdivisions of knowledge, than by a slight tincture of all the sciences united in the possession of some individuals."*

France also possesses institutions for the teaching of Agriculture. The first was that model farm at Roville, near Nancy, founded by M. Mathieu de Dombasle.† Though it is acknowledged that this farm has done service to the Agriculture of France, its situation being so far-removed from the centre of that country, its influence does not extend with sufficient rapidity. Its limited capital does not permit the addition of schools, which are considered necessary for the instruction of young proprietors who wish to manage their own properties with advantage, and of agents capable of following faithfully the rules of good Husbandry.

To obviate the disadvantages apparent in the institution of Roville, "a number of men distinguished for their learning and zeal for the prosperity of France, and convinced of the utility of the project, used means to form an association of the nature of a joint-stock company, with 500 shares of 1,200 francs each, forming a capital of 600,000 francs (£25,000). The first half of this sum was devoted to the advancement of superior culture, and the second half to the establishment of two schools, one for pupils who, having received a good education, wish to learn the theory and the application of Agriculture, and of the various arts to which it is applicable; and the other for children without fortune, destined to become laborers, instructed as good plowmen, gardeners, and shepherds, worthy of confidence being placed in them."‡ This society began its labors in 1826 by purchasing the domain of Grignon, near Versailles, in the valley of Gally, in the commune of Thiverval, and appointing M. Bella, a military officer who had gained much agricultural information from M. Von Thaër during two years' sojourn with his corps at Celle. M. Bella traveled through France, in the summer of 1826, to ascertain the various modes of culture followed in the different communes. Grignon was bought in the name of the king, Charles X. who attached it to his domain, and gave the society the title of the Royal Agricultural Society for a period of forty years. The statutes of the society were approved of by royal ordinance on the 23d May, 1827, and a council of administration was named from the list of shareholders, consisting of a president, two vice-presidents, a secretary, a treasurer, and directors.

The domain, which occupies the bottom and the two sides of the valley, in length 2,254 metres (a metre being equal to 3 feet and 11½ lines), is divided into two principal parts; the one is composed of a park of 290 hectares (387 acres), inclosed with a stone wall, containing the mansion-house and its dependencies, the piece of water, the trees, the gardens, and the land appropriated to the farm; the other, called the outer farm, is composed of 176 hectares (234 acres), of uninclosed land, to the south of the park.

* Jacob's Travels in Germany, &c. pp. 173–186.
† Annales de Roville.
‡ Rapport General sur la ferme de Grignon, Juin, 1828, p. 3.

With regard to the nature of the schools at Grignon, this account has been published: "The council of administration being occupied in the organization of regular schools, has judged that it would be convenient and useful to open, in 1829, a school for work-people, into which to admit boys of from twelve to sixteen years of age, to teach them reading, writing, arithmetic, and the primary elements of the practice of geometry. The classes to meet two hours every day in summer, and four hours in winter, the rest of the time to be employed in manual work. The fee to be 300 francs the first year, 200 francs the second, and 100 francs the third. After three years of tuition the fee to cease, when an account is to be opened to ascertain the value of their work against the cost of their maintenance, and the balance to go to form a sum for them when they ultimately leave the institution."

"Meanwhile, as the director has received several applications for the admission of young men, who, having received a good education, are desirous of being instructed in Agriculture, the council has authorized the conditional admission of six pupils. But as there are yet no professors, the pupils who are at present at Grignon can only actually receive a part of the instruction which it is intended to be given. They every day receive lessons from the director on the theory of Agriculture, besides lessons on the veterinary art, and the elements of botany, from the part of the veterinary school attached to the establishment; also lessons of the art of managing trees and making plantations, given by a forester of the crown forests, and some notions of gardening by the gardener. During the rest of the time, they follow the agricultural labors and other operations of the establishment. They pay 100 francs a month, including bed, board, and washing.

"Several proprietors who occupy farms, having expressed a desire to see young farm-servants taught the use of superior implements, and the regular service on a farm, the director has admitted a few, lessening the fee to the payment of board and lodging. There are two just now. To such are given the name of 'farm pupils.'"*

The course of education proposed to be adopted at Grignon, is divided into theoretical and practical. The course to continue for two years. In the first year to be taught mathematics, topography, physics, chemistry, botany and botanical physiology, veterinary science, the principles of culture, the principles of rural economy applied to the employment of capital, and the interior administration of farms. The second year to comprehend the principles of culture in the special application to the art of producing and using products; the mathematics applied to mechanics, hydraulics, and astronomy; physics and chemistry applied to the analysis of various objects; mineralogy and geology applied to Agriculture; gardening, rural architecture, legislation in reference to rural properties, and the principles of health as applicable both to man and beast.

There are two classes of pupils, free and internal. Any one may be admitted a free pupil that has not attained twenty years of age, and every free pupil to have a private chamber. The pupils of the interior must be at least fifteen years of age.

The fee of the free pupils is 1,500 francs a year; that of the pupils of the interior 1,300 francs. They are lodged in the dormitories in box-beds; those who desire private apartments pay 300 francs more, exclusive of furniture, which is at the cost of the pupils.†

There is an agricultural school at Hohenheim, in the Duchy of Wir-

* Annales de Grignon, 2d livraison, 1829, p. 48.
† Annales de Grignon, 3d livraison, 1830, p. 108.

temberg, and another at Flottbeck in Flanders, belonging to M. Voght. An account of both these institutions is given by M. Bella, in the third number of the Annales of Grignon. There are, I understand, schools of Agriculture, both in St. Petersburg and Moscow, but have not been so fortunate as to meet with any account of them.

It appears to me from the best consideration I can give to the manner in which Agriculture is taught at these schools, that as means of imparting real practical knowledge to pupils, they are inferior to the usual mode adopted in this country, of living with farmers. In reference to the results of the education obtained at Möeglin, Mr. Jacob says: "It appeared to me that there was an attempt to crowd too much instruction into too short a compass, for many of the pupils spend but one year in the institution, and thus only the foundation, and that a very slight one, can be laid in so short a space of time. It is, however, to be presumed, that the young men come here prepared with considerable previous knowledge, as they are mostly between the ages of twenty and twenty-four, and some few appeared to be still older."*

Although the pupils are kept at Hofwyl for nine years, and are fined if they leave it sooner, it is obvious that the higher class of them bestow but little attention on farming, and most on classical literature. And the particulars given in the elaborate programme of the school of Agriculture at Grignon, clearly evince that attention to minute discipline, such as marking down results, and to what are termed *principles*, which just mean vague theorizings, form a more important feature of tuition than the practice of husbandry. The working pupils may acquire some knowledge of practice by dint of participating in work, but the other class can derive very little benefit from all the practice they see.

---

7. OF THE EVILS ATTENDANT ON LANDOWNERS NEGLECTING TO LEARN PRACTICAL AGRICULTURE.

"——leaving me no sign—
Save men's opinions, and my living blood—
To show the world I am a gentleman."
RICHARD II.

THERE would be no want of pupils of the highest class for institutions such as I have recommended for promoting agricultural education, did landed proprietors study their true interests, and learn practical Agriculture. Besides the usual succession of young farmers to fill the places of those who retire, and these of themselves would afford the largest proportion of the pupils, were every son of a landowner, who has the most distant prospect of being a landed proprietor himself, to become an agricultural pupil, in order to qualify himself to fulfill *all* the onerous duties of his station, when required to occupy that important position in the country, that class of pupils would not only be raised in respectability, but the character of landed proprietors, as agriculturists, would also be much elevated. The expectant landlord should therefore undergo that tuition, though he may intend to follow, or may have already followed, any other profession. The camp and the bar seem to be the especially favorite arenas upon which the young scions of the gentry are desirous of displaying their

* Jacob's Travels in Germany, &c. p. 186.

first acquirements.* These professions are highly honorable, none more so, and they are, no doubt, conducive to the formation of the character of the gentleman; but, after all, are seldom followed out by the young squire. The moment he attains rank above a subaltern, or dons his gown and wig, he quits the public service, and assumes the functions of an incipient country gentleman. In the country he becomes at first enamored of field sports, and the social qualities of sportsmen. Should these prove too rough for his taste, he travels abroad peradventure in search of sights, or to penetrate more deeply into the human breast. Now, all the while he is pursuing this course of life, quite unexceptionable in itself, he is neglecting a most important part of his duty—that of learning to become a good landlord. On the other hand, though he devote himself to the profession of arms or the the law, either of which may confer distinction on its votaries; yet if either be preferred by him to Agriculture, he is doing much to unfit himself from being an influential landlord. To become a soldier or a lawyer, he willingly undergoes initiatory drillings and examinations; but, to become a landlord, he considers it quite unnecessary, to judge by his conduct, to undergo any initiatory tuition. That is a business, he conceives, that can be learned at any time, and seems to forget that it is *his* profession, and does not consider that it is one as difficult of thorough attainment as ordinary soldiership or legal lore. No doubt the army is an excellent school for confirming, in the young, principles of honor and habits of discipline; and the bar for giving clear insight into the principles upon which the rights of property are based, and into the true theory of the relation betwixt landlord and tenant; but while these matters may be attained, a knowledge of Agriculture, the weightiest matter to a landlord, should not be neglected. The laws of honor and discipline are now well understood, and no army is required to inculcate their acceptableness on good society. A knowledge of law, to be made applicable to the occurrences of a country life, must be *matured by long experience;* for, perhaps, no sort of knowledge is so apt to render landed proprietors litigious and uncompromising with their tenants as a smattering

[* So is it in the United States; and who can wonder at it, seeing that the bar and the habit of public speaking acquired at it open the broadest and easiest road to public distinction, while the *military* is almost the only *life* commission bestowed by the Government. Having once received that, the officer is placed for the remainder of his days beyond the reach of political vicissitude. Even his education has been at the public expense, and having once received his sword, he has only to keep his head above ground and *he* is sure of promotion and of increased pay. Courage in the line of his profession is, properly, sure of honors and rewards; while the same virtue in civil life offers no immunity against proscription and party despotism. All liberty of thought is deemed to be incompatible with party loyalty. This preference and elevation of the *military* over *civil* virtues is but a servile prejudice derived from despotic Governments, where rulers, cut off from sympathy with and dependence on the people, have to rely on the military arm for support on all occasions of popular commotion and outbreak under irremediable oppression. Nor will civil virtues and the capacity to promote the substantial interests of the people enjoy that eminence in the public esteem, and that encouragement and reward which it should be the care of a republican government to bestow, until the mass of the cultivators of the soil become *more generally and thoroughly instructed,* not only in the practical principles of their calling, but in the preference which they have a right to assert as due from Government to the landed interest—an interest on which all others live, and without which they would all dwindle and perish, as does the misletoe when the oak on which it grows falls under the strokes of the woodman's ax.

When will agriculturists force a system of legislation in which honor shall be rendered to the *men of the country* for talents and civil virtues; and military and other parasitical institutions and classes be reluctantly tolerated and supported as necessary ev'ls, or at least as mere appendages in the great machinery of Government?

of law. Instances have come under my own notice, of the injurious propensities which a slight acquaintance with law engenders in landed proprietors, as exhibited on their own estates, and at county and parochial meetings. No class of persons require Pope's admonition regarding the evil tendency of a "little learning" to be more strongly inculcated on them, than the young barrister who doffs his legal garments, to assume in ignorance the part of the country squire:

> "A little learning is a dangerous thing!
> Drink deep, or taste not the Pierian spring:
> There shallow draughts intoxicate the brain."

I do not assert that a knowledge of military tactics, or of law, is inconsisten with Agriculture. On the contrary, a *competent* knowledge of either, and particularly of the latter, confers a value on the character of a country gentleman versant with Agriculture; but what I do assert most strongly is, that the most intimate acquaintance with either will never serve as a substitute for ignorance of Agriculture in a country gentleman.

One evil arising from studying those exciting professions before Agriculture is, that, however short the time spent in acquiring them, it is sufficiently long to create a distaste to learning Agriculture practically, for such a task can only be undertaken, after the turn of life, by enthusiastic minds. But as farming is necessarily *the profession* of the country gentleman, for all *have* a farm, it should be learned, theoretically and practically, before his education should be considered finished. If he so incline, he can afterward enter the tented field, or exercise his forensic eloquence, when the tendency which I have noticed in these professions will be unable to efface the knowledge of Agriculture previously acquired. This is the proper course for every young man destined to become a landed proprietor to pursue, and who wishes to be otherwise employed as long as he cannot exercise the functions of a landlord. Were this course always pursued, the numerous engaging ties which a country life never fails to form, rendered more interesting by a knowledge of Agriculture, would tend to extinguish the kindling desire for any other profession. Such a result would be most desirable for the country; for only contemplate the effects of the course pursued at present by landowners. Does it not strike every one as an incongruity for a country gentleman to be unacquainted with country affairs? Is it not "passing strange" that he should require inducements to learn his hereditary profession—to know a business which alone can enable him to maintain the value of his estate, and secure his income? Does it not infer a species of infatuation to neglect becoming well acquainted with the true relation he stands to his tenants, and by which, if he did, he might confer happiness on many families; but to violate which, he might entail lasting misery on many more? In this way the moral obligations of the country gentleman are too frequently neglected. And no wonder, for these cannot be perfectly understood, or practiced aright but by tuition in early life, or by very diligent and irksome study in maturer years. And no wonder that great professional mistakes are frequently committed by proprietors of land. Descending from generalities to particulars, it would be no easy task to describe all the evils attendant on the neglect of farming by landowners; for though some are obvious enough, others can only be morally discerned.

1. One of the most obvious of those evils is, when country gentlemen take a prominent share in discussions on public measures connected with Agriculture, and which, from the position they occupy, they are frequently called upon to do, it may be remarked that their speeches are usually introduced with apologies for not having sufficiently attended to agricultural

matters. The avowal is candid, but it is any thing but creditable to the position they hold in the agricultural commonwealth. When, moreover, it is their lot or ambition to be elected members of the legislature, it is deplorable to find so many so little acquainted with the questions which bear directly or indirectly on Agriculture. On these accounts, the tenantry are left to fight their own battles on public questions. Were landowners practically acquainted with Agriculture, such painful avowals would be spared, as a familiar acquaintance with it enables the man of cultivated mind at once to perceive its practical bearing on most public questions.

2. A still greater evil consists in their consigning the management of valuable estates to the care of men as little acquainted as themselves with practical Agriculture. A factor or agent, in such a condition, always affects much zeal for the interest of his employer; but it is "a zeal not according to knowledge." Fired by this zeal, and undirected, as it most probably is, by sound judgment, he soon discovers something at fault among the poorer tenants. The rent, perhaps, is somewhat in arrear—the strict terms of the lease have been deviated from—things appear to him to be going down hill. These are fruitful topics of contention. Instead of being "kindly affectioned," and thereby willing to interpret the terms of the lease in a generous spirit, the factor hints that the rent must be better secured, through the means of another tenant. Explanation of circumstances affecting the condition of the farmer, over which he has perhaps no control—the inapplicability, perhaps, of the peculiar covenants of the lease to the particular circumstances of the farm—the lease having perhaps been drawn up by himself, or some one as ignorant as himself—are excuses unavailingly offered to one who is confessedly unacquainted with country affairs, and the result ensues in interminable disputes betwixt him and the tenants. With these the landlord is *unwilling* to interfere, in order to preserve intact the authority of the factor; or, what is still worse, is *unable* to interfere, because of his own unacquaintance with the actual relations subsisting betwixt himself and his tenants, and, of course, the settlement is left with the originator of the disputes. Hence originate actions at law, criminations and recriminations—much alienation of feeling; and at length a settlement of matters, at best, perhaps, unimportant, is left to the arbitration of practical men, in making which submission the factor acknowledges as much as he himself was unable to settle the dispute. The tenants are glad to submit to arbitration to save their money. In all such disputes they, being the weaker parties, suffer most in purse and character. The landlord, who should have been the natural protector, is thus converted into the unconscious oppressor, of his tenants. This is confessedly an instance of a bad factor; but have such instances of oppression never occurred, and from the same cause, that of ignorance in both landlord and factor?

A factor acquainted with practical Agriculture would conduct himself ve y differently in the same circumstances. He would endeavor to prevent legitimate differences of opinion on points of management terminating into disputes, by skillful investigation and well-timed compromise.— He studies to uphold the honor of both landlord and tenants. He can see whether the terms of the lease are strictly applicable to prevailing circumstances, and judging thereby, checks every improper deviation from appropriate covenants, while he makes ample allowance for unforeseen contingencies. He can discover whether the condition of the tenants is influenced more by their own doings, than by the nature of the farms they occupy. He regulates his conduct toward them accordingly; encouraging the industrious and skillful, admonishing the indolent, and amending the unfavorable circumstances of the farms. Such a man is highly respected, and his opinion and judgment are greatly confided in by the tenantry. Mutual kindliness of intercourse always subsists between them. No landlord, whether himself acquainted or unacquainted with farming, but especially the latter, should confide the management of his estate to any other kind of factor.

3. Another obvious evil is one which affects the landed proprietor's own comfort and interest, and which is the selection of a steward or grieve for conducting the home-farm. In all cases it is necessary for a landowner to have a home-farm, and to have a steward to conduct it. But the steward of a squire, acquainted and unacquainted with farming, is placed in very dissimilar circumstances. The steward of a squire acquainted with farming, enjoying good wages, and holding a respectable and responsible situation, *must* conduct himself as an honest and skillful manager, for he knows he is superintended by one who can criticize his management well. A steward in the other position alluded to, must necessarily have, and will soon take care to have, everything his own way. He soon becomes proud in his new charge, because he is in the service of a squire. He soon displays a haughty bearing, because he knows he is the only person on the farm who knows anything about his business. He becomes overbearing to the rest of the servants, because, in virtue of his office, he is appointed purveyor to the entire establishment; and he knows he can starve the garrison into a surrender whenever he pleases. He domineers over the inferior work-people, because, dispensing weekly wages, he is the custodier of a little cash. Thus advancing in his own estimation step by step, and finding the most implicit reliance placed in him by his master, who considers his services as invaluable, the temptations of office prove too powerful for his virtue, he aggrandizes himself, and conceals his malpractices by deception. At length his peculations are detected, by perhaps some trivial event, the insignificance of which had escaped his watchfulness. Then loss of character and loss of place overtake him at once.—Such flagrant instances of unworthy factors and stewards of country gentlemen, are not supposititious. I could specify instances of both, whose mismanagment has come under my own observation. Both species of pests are engendered from the same cause—the ignorance of landowners in country affairs.*

4. Another injurious effect it produces is absenteeism. When farming possesses no charms to the country gentleman, and field-sports become irksome by monotonous repetition, his taste for a country life declines, and to escape *ennui* at home, he banishes himself abroad. If such lukewarm landed proprietors, when they go abroad, would always confide the management of their estates to unexceptionable factors, their absence would be little felt by the tenants, who would proceed with the substantial improvement of their farms with greater zest under the countenance of a sensible factor, than of a landlord who contemns a knowledge of Agriculture. But it must be admitted that tenants farm with much greater *confidence* under a landlord acquainted with farming, who is always at home, than under the most unexceptionable factor. The disadvantages of absenteeism are only felt by tenants left in charge of a litigious factor, and

[* With obvious allowance for difference of circumstances, some of these remarks may be made to apply to gentlemen, more especially in the South, who devolve their affairs too much on their managers or overseers; and to merchants, and other gentlemen of fortune, who cannot reside through the year on their estates. *Ed. Farm. Lib.*]

it is always severely felt by day-laborers, tradesmen, and shop-keepers in villages and small country towns.

Now, all these evils—for evils they certainly are—and many more I have not touched upon, would be avoided, if landowners would make it a point to acquire a knowledge of practical Agriculture. This can best be done in youth, when it should be studied as a *necessary* branch of education, and learned as the most *useful* business which country gentlemen can know. It will qualify them to appoint competent factors—to determine upon the terms of the lease most suited to the nature of each of the farms on their properties, and to select the fittest tenants for them. This qualification could not fail to inspire in tenants confidence in their landlords, by which they will be encouraged to cultivate their farms in the best manner for the land and for themselves, in even the most trying vicissitudes of seasons; and without which confidence the land, especially on estates on which no leases are granted, would never be cultivated with spirit. It confers on landlords the power to judge for themselves of the proper fulfillment of the onerous and multifarious duties of a factor. It enables them to converse freely in technical terms with their tenants on the usual courses of practice, to criticize work, and to predicate the probability of success or failure of any proposed course of culture. The reproving or approving remarks of such landlords operate powerfully with tenants. How many useful hints is it all times in the power of such landlords to suggest to their tenants or managers, on skillfulness, economy, and neatness of work; and how many salutary precepts may they inculcate on cottagers, on the beneficial effects of parental discipline and domestic cleanliness! The degree of good which the direct moral influence of such landlords among their tenantry can effect, can scarcely be over estimated; its primary effect being to ensure respect, and create regard.—The good opinion, too, of a judicious factor is highly estimated by the tenantry; but the discriminating observations of a practical and well-disposed landlord go much farther in inducing tenants to maintain their farms in the highest order, and to cherish a desire to remain on them from generation to generation. Were all landlords so actuated—and acquaintance with farming would certainly prompt them thus to act—they could at all times command the services of superior factors and skillful tenants.

They would then find there is not a more pleasing, rational, and interesting study than practical Agriculture; and soon discover that to know the minutiæ of farming is just to create an increasing interest in every farm operation. In applying this knowledge to practice, they would soon find it to operate beneficially for their estates, by the removal of objects which offend the eye or taste, and the introduction of others that would afford shelter, promote improvement, and contribute to the beauty of the landscape of the country around.

These maxims of Bacon seem not an inapt conclusion to our present remarks: "He that cannot look into his own estate at all, had need both choose well those whom he employeth, and change them often, for new are more timorous and less subtle. He that can look into his estate but seldom, it behoveth him to turn all to certainties."* [Essays, p. 106.]

[* It has doubtless already occurred to the reader that some of the preceding observations do not apply to our country, but where they are not exactly applicable, they are mixed up with others that may be applied, or from which hints may be drawn. In republishing an author, we should remember the anecdote told by Doctor Franklin when Members of Congress were picking holes in the Declaration of Independence—how a party from the country criticized a hatter's sign with the picture of a hat, and under it, "*Hats sold here for cash, by John Smith.*" One said the *sign* showed it was *hats*, and nothing else; another, that it was useless to say *sold*, be-

8. EXPERIMENTAL FARMS AS PLACES FOR INSTRUCTION IN FARMING.

"Things done without example, in their issue
Are to be feared."
HENRY VIII.

It seems to be a favorite notion with some writers on agricultural subjects, that, of all places for learning farming, experimental farms are the best. They even recommend the formation of experimental farms, with the view of affording to young men the best system of agricultural education. They go the length of confidently asserting that all the field operations and experiments, on experimental farms, could be conducted by pupils. And they are nearly unanimous in conceiving that 200 acres would be a large enough extent for an experimental farm, and that on such a farm 100 pupils could be trained to become farmers, stewards and plowmen.

A very slight consideration of the nature of an experimental farm, will serve to show how unsuitable such a place is for *learning* farming. The sole object of an experimental farm is, to become acquainted with the best properties of plants and animals by experiment, and thereby to ascertain whether those properties are such as would recommend them for introduction to an ordinary farm. It is obvious, from this statement, that it is needless following, on an experimental farm, the *usual modes* of cultivating the ordinary plants and rearing the ordinary animals of a farm. Either *new* plants, and *other* modes than the usual ones of cultivating and rearing the ordinary plants and animals, should be tried on an experimental farm, otherwise it would not be an *experimental* farm, or of more use than an ordinary farm. In witnessing new or unusual modes of culture, the pupil would thus learn nothing of the particulars of *ordinary farming*. Extraordinary modes of cultivating ordinary plants, by changes in the rotation or of manure, and the risk of failure in both—for failure is a necessary condition of experiment—would only serve to impress the minds of *pupils* with experimental schemes, instead of guiding them to the most approved plan of cultivating each sort of plant. To confound the mind of a beginner, by presenting before it various modes of doing the same thing, without the ability to inform it which is the best, is to do him a lasting injury. Were a pupil, who had been trained up on an ordinary farm, to have opportunities of witnessing varieties of experiments conducted on an experimental farm, he might then derive benefit from numerous hints which would be suggested in the course of making the experiments. But if pupils would be unfavorably placed on an experimental farm, by remaining constantly on it, much more would the farm itself be injured, by having its experiments performed by inexperienced pupils. So far from *pupils* being able to conduct experiments to a satisfactory issue, the most

cause it might be inferred they were not to be given away; the third was for striking out *here*, as the place would be presumed to be where the sign was hung out; the fourth said all would know that when the hats were *sold* the money was to be paid; and a fifth insisted that the *name* of the seller was entirely superfluous—until they struck out all except the *picture of the hat*. So we must be cautious in striking out too freely from an author, because what might seem superfluous to most readers may for peculiar reasons be acceptable or useful to others; and these observations will apply to other sections of the work, though they may not be repeated.

Ed. Farm. Lib.]

experienced cultivators are at times baffled by unforeseen difficulties; and so far would *such* experiments inspire confidence in farmers, that they would assuredly have quite an opposite tendency. So far, therefore, would the services of pupils in any degree compensate for the extraordinary outlay occasioned on experimental farms, by unsuccessful or unprofitable experiments, that even those of the most experienced cultivators would most probably produce no such desirable result; for no *experimenter* can command success, and failure necessarily implies extraordinary outlay. So far, therefore, could the services of pupils accomplish what those of experienced cultivators could not command, that their very presence on an *experimental* farm, with the right of coöperating in the experiments, would be a constant source of inconvenience to the experienced experimenters.

But, besides these objections, the mode of conducting experiments on so small farms as those recommended by most writers, would be quite unsuitable to pupils desirous of *learning* farming. Where varieties of culture on various sorts of plants are prosecuted on a small extent of ground, only a very small space can be allotted to each experiment. It is true that, should any of the varieties of plants be new to this country, the seed of which at first being of course only obtainable in small quantities, to procure such being a primary object with the promoters of experimental farms the space required for them at first must be very small. But although each lot of ground should be small, the great varieties of seeds cultivated in so many different ways, will nevertheless require a great number of lots, which altogether will cover a considerable extent of ground. How all these lots are to be apportioned on 200 acres, together with ground for experimenting on different breeds of animals, and different kinds of forest trees, is more than I can imagine. It would require more than double that extent of ground to give mere standing-room to all the objects that should be cultivated on an experimental farm, and over and above which, 100 pupils on such a farm would form a perfect crowd. Besides, the lots being so small, would require to be worked with the spade instead of the plow; and this being the case, let the experiments on such a farm be ever so perfectly performed, they could give pupils no insight whatever into real *farming*, much less secure the confidence of *farmers*.

It is the pleasure of some writers on experimental farms, to institute a comparison, or even strict analogy, betwixt them and experimental gardens. As the latter have improved the art of gardening, they argue, so would the former improve Agriculture. But the truth is, there can be no analogy betwixt the introduction into common gardens of the results obtained in experimental gardens, and the results of experiments obtained in such small experimental farms as recommended by agricultural writers, introduced into the common field culture of a farm; because, the experiments in an experimental garden having been made by the spade, may be exactly transferred into almost any common garden, and, of course, succeed there satisfactorily; whereas the experiments made by the spade in a small experimental farm, cannot be performed with the spade on a common farm; they must there be executed by the plow, and, of course, in quite different circumstances. The rough culture of the plow, and most probably in different circumstances of soil, manure, and shelter, cannot possibly produce results similar to the culture of the spade, at least no farmer will believe it; and if *they* put no confidence in experiments, of what avail will experimental farms be? Announcements of such results may gratify curiosity, but no benefit would be conferred on the country by experiments confined within the inclosures of an experimental farm No doubt, a few of the most unprejudiced of the farmers will perform any experiment, with every desire for its success, and there is as little doubt that others will follow the example; and some will be willing to test the worth of even a suggestion; but as these are the usual modes by which every new practice recommends itself to the good graces of farmers, no intervention of an experimental farm is therefore required for their promulgation and adoption. It is the duty of the promoters of experimental farms to disseminate a proved experiment quickly over the country, and the most efficient mode of doing so is to secure the confidence of farmers in it. To ensure *their* confidence, it will be necessary to show them that they can do the same things as have been done on the experimental farm *by the usual means of labor* they possess, and they will then show no reluctance to follow the example. Take the risk, in the experimental farm, of proving results, and show the intrinsic value of those results to the farmers, and the experiments, of whatever nature, will be performed on half the farms of the kingdom in the course of the first season.

For this purpose it is necessary to ascertain the size an experimental farm should be, which will admit of experiments being made on it, in a manner similar to the operations of a farm. The leading operation, which determines the smallest size of the fields of an experimental farm, is plowing. The fields should be of that size which will admit of being plowed in ordinary time, and at the same time not larger than just to do justice to the experiments performed in them. I should say that *five* acres imperial is the least extent of ground to do justice to plowing ridges along, across, and diagonally. *Three* acres, to be of such a shape as not to waste time in the plowing, would have too few ridges for a series of experiments, and to increase their number would be to shorten their length, and lose time in plowing. But even five acres are too small to inclose with a fence; ten acres, a good size of field for small farms, being nearer the mark for fencing. Taking the size of an experimental plot at five acres, the inclosure might be made to surround the divisions of a rotation; that is, of a rotation of four years, let twenty acres be inclosed; of five years, twenty-five acres, &c.; but in this arrangement the experiments would only prove really available to small tenants, who frequently cultivate all their crops within one fence, and the subject thus experimented on would not be individually inclosed within a fence, as is the case with crops on larger farms.

The whole quantity of land required for an adequate experimental farm may thus be estimated. New varieties of seeds would require to be increased by all the possible modes of reproduction. Old varieties should undergo impregnation—be subjected to different modes of culture—be preserved pure from self-impregnation—and be grown in different altitudes. Each variety of seed already cultivated, such as wheat, barley, oats, potatoes, turnips, &c., to undergo these various modifications of treatment on five acres of land, would, including the whole, require an immense extent of ground, and yet, if each kind did not undergo all these varieties of treatment, who could then aver that all our seeds had been subjected to satisfactory *field* experiments? Only *one* kind of grain, treated as variedly as might be, on five acres for each modification of treatment, would occupy *seventy* acres; and were only five kinds of seed taken, and only five varieties of each, and the whole cultivated on both low and high ground, the quantity of ground required altogether would be 3,500 acres. The extent of ground thus increases in a geometrical progression, with an increase of variety of plants. Besides, the numerous useful grasses, for the purposes of being cut green, and for making into hay, would require other 1,000 acres. The whole system of pasturing young and old stock on natural and artificial grasses in low grounds and on high altitudes, and in sheltered and exposed situations, would require at least 3,000 acres. Then, experiments with forest-trees, in reference to timber and shelter in different elevations and aspects, would surely require 1,000 acres. Improvements in bog and muir lands should have other 1,000 acres. So that 9,500 acres would be required to put only a given proportion of the objects of cultivation in this country to the test of full experiment.—Such an extent of ground will, no doubt, astonish those who are in the habit of talking about 200 acres as capable of affording sufficient scope for an experimental farm. Those people should be made to understand that the plow must have room to work, and that there is no other way of experimentizing satisfactorily for *field* culture, on an experimental farm, but by affording it a real field to work in. If less ground be given, fewer subjects must be taken; and if any subject is rejected from experiment, then the system of experimentizing will be rendered incomplete. The system of experimentizing should be carried out to the fullest extent of its capability on experimental farms, or it should be left, as it has hitherto been, in the hands of farmers. The farmers of Scotland have worked out for themselves an admirable system of husbandry, and if it is to be improved to a still higher pitch of skill by experimental farms, the means of improvement should be made commensurate with the object, otherwise there will be no satisfaction, and certain failure; for the promoters of experimental farms should keep in mind that the existing husbandry, improved as it is, is neither in a stationary nor in a retrograding, but in a progressive state toward farther improvement. Unless, therefore, the proposed experiments, by which it is intended to push its improvement still farther toward perfection, embrace every individual of the multifarious objects which engage the attention of agriculturists, that one may be neglected which, if cultivated, would have conferred the greatest boon on Agriculture. I come, therefore, to this conclusion in the matter: that *minute* experiments on the progressive developments of plants and animals are absolutely requisite to establish their excellence or worthlessness, and these can be performed on a small space of ground; but to stop short at this stage, and not pursue their culture on a *scale commensurate with the operations of the farm*, is to render the experimental farm of little avail to practical husbandry, and none at all to interest the farmer.

So large an extent of farm would most probably embrace all the varieties of soil. It should, moreover, contain high and low land, arable, bog, and muir land, sheltered and exposed situations, and the whole should lie contiguous, in order to be influenced by the climate of the same locality. It would scarcely be possible to procure such an extent of land under the same landlord, but it might be found in the same locality on different estates. *Such* a farm, rendered highly fertile by draining, manuring, liming, and labor, and plenished, as an experimental farm should be, with *all* the varieties of crop, stock, implements, and woods, would be a magnificent spectacle worthy of a nation's effort to put into a perfect state for a national object. What a wide field of observation would it present to the botanical physiologist; containing a multiplicity of objects made subservient to experiment! What a laboratory of research for the chemist, among every possible variety of earths, manures, plants, and products of vegetation! What a museum of objects for the naturalist, in which to observe the living habits and instincts of animals, some useful to man, and others injurious to the fruits of his labor! What an arena upon which the husbandman to exercise his practical skill, in varying the modes of culture of crops and live-stock! What an object of intense curiosity and untatisfying wonder to the rustic laborer! But, above all, what interest and solicitude should the statesman feel the appliance of such a mighty engine, set in motion, to work out the problem of agricultural skill, prosperity, and power.*

---

9 OF THE KIND OF EDUCATION BEST SUITED TO YOUNG FARMERS.

> "Between the physical sciences and the arts of life there subsists a constant mutual interchange of good offices, and no considerable progress can be made in the one without, of necessity, giving rise to corresponding steps in the other. On the one hand, every art is in some measure, and many entirely, dependent on those very powers and qualities of the material world which it is the object of physical inquiry to investigate and explain."
>
> HERSCHEL.

WITH respect to the education of young farmers, no course of elementary education is better than what is taught at the excellent parochial schools of this country. The sons of farmers and of peasants have in them a favorable opportunity of acquiring the elements of a sound education, and they happily avail themselves of the opportunity; but, besides elementary education, a classical one sufficiently extensive and profound for farmers may there also be obtained. But there are subjects of a different nature, sciences suited to the study of maturer years, which young farmers should make a point of learning—I mean the sciences of Natural Philosophy, Natural History, Mathematics, and Chemistry. These are taught at colleges and academies. No doubt these sciences are included in the curriculum of education provided for the sons of landowners and wealthy farmers; but every class of farmers should be taught them, not with a view of transforming them into philosophers, but of communicating to them the important knowledge of the nature of those phenomena which daily present themselves to their observation. Such information would make them more intelligent farmers, as well as men. The advantages which farmers would derive from studying those sciences will be best understood by pointing out their nature.

It is evident that most farming operations are much affected by external influences. The state of the weather, for instance, regulates every field operation, and local influences modify the climate very materially. Now it should be desired by the farmer to become acquainted with the causes which give rise to those influences, and these can only be known by comprehending the laws of Nature which govern every natural phenomenon. The science which investigates the laws of these phenomena is called *Natural Philosophy*, and it is divided into as many branches as there are classes of phenomena. The various classes of phenomena occur in the earth, air, water, and heavens. The laws which regulate them, being unerring in their operation, admit of absolute demonstration; and the science which affords the demonstration is called *Mathematics*. Again, every object, animate or inanimate, that is patent to the senses, possesses an individual identity, so that no two objects can be confounded together. The science which makes us acquainted with the marks for identifying individuals is

* Paper by me on the subject in the Quart. Jour. of Agri., vol. vii. p. 536.

termed *Natural History*. Farther, every object, animate or inanimate, cognizable by the senses, is a compound body made up of certain elements. *Chemistry* is the science which makes us acquainted with the nature and combinations of those elements. We thus see how generally applicable those sciences are to the phenomena around us, and their utility to the farmer will be the more apparent, the more minutely each of them is investigated. Let us take a cursory view of each subdivision as it affects Agriculture.

Mathematics are either abstract or demonstrative. Abstract mathematics "treat of propositions which are immutable, absolute truth," not liable to be affected by subsequent discoveries, "but remains the unchangeable property of the mind in all its acquirements." Demonstrative mathematics are also strict, but are "interwoven with physical considerations"—that is, subjects that exist independently of the mind's conceptions of them or of the human will; or, in other words still, considerations in accordance with nature. Mathematics thus constitute the essential means of demonstrating the strictness of those laws which govern natural phenomena. Mathematics must, therefore, be first studied before those laws can be understood.—Their study tends to expand the mind—to enlarge its capacity for general principles, and to improve its reasoning powers.

Of the branches into which Natural Philosophy is divided, that which is most useful to farmers is *Mechanics*, which is defined to be "the science of the laws of matter and motion, so far as is necessary to the construction of machines, which, acting under those laws, answer some purpose in the business of life." Without mechanics, as thus defined, farmers may learn to *work* any machine which answers their purpose; but it is only by that science they can possibly understand the *principles* upon which any machine is constructed, nor can any machine be properly constructed in defiance of those principles. Both machinists and farmers ought to be versed in mechanical science, or the one cannot make, and the other guide, any machine as it ought to be; but, as I have had occasion to express my sentiments on this subject already, I shall abstain from dilating farther upon it here. Mathematical demonstration is strictly applicable to mechanics, whether as to the principles on which every machine operates, or the form of which it is constructed. The *principles* of mechanics are treated of separately under the name of *Dynamics*, which is defined to be "the science of force and motion."

Pneumatics is the branch of natural philosophy which is next to mechanics in being the most useful to the farmer to know. It "treats of air, and the laws according to which it is condensed, rarefied, gravitates." The states of the air, giving a variable aspect to the seasons, as they pursue their "appointed course," endue all atmospherical phenomena with extreme interest to the farmer. Observation alone can render variety of phenomena familiar; and their apparent capriciousness, arising most probably from the reciprocal action of various combinations of numerous elements, renders their complicated results at all times difficult of solution; for all fluids are susceptible of considerable mutations, even from causes possessing little force; but the mutations of elastic fluids are probably effected by many inappreciable causes. Nevertheless, we may be assured that no change in the phenomena of the atmosphere, however trivial, takes place but as the unerring result of a definite law, be it chemical or physical.

Closely connected with pneumatics, in so far as the air is concerned, are the kindred natural sciences of *electricity* and *magnetism*. These agencies, though perfectly perceptible to one or more of the senses, and evidently constantly at work in most of the phenomena of the atmosphere, are mysteriously subtle in their operations. It is extremely probable that one or both are the immediate causes of all the changes which the atmosphere is continually undergoing. It is hardly possible that the atmosphere, surrounding the globe like a thin envelop, and regularly carried round with it in its diurnal and annual revolutions, should exhibit so very dissimilar phenomena every year, but from some disturbing cause, such as the subtile influences of electricity, which evidently bear so large a share in all remarkable atmospherical phenomena. Its agency is the most probable cause of the *irregular* currents of the air called winds, the changes of which are well known to all farmers to possess the greatest influence on the weather.

Natural History comprehends several branches of study. *Meteorology* consists of the observation of the apparent phenomena of the atmosphere. The seasons constitute a principal portion of these phenomena. The clouds constitute another, and are classified according to the forms they assume, which are definite, and indicative of certain changes. The winds constitute a third, and afford subject for assiduous observation and much consideration. Attention to the directions of the wind and forms of the clouds will enable farmers to anticipate the kind of weather that will afterward ensue in a given time in their respective localities. The prevalence of the aqueous meteors of rain, snow, hail, and ice, is indicated by the state of the clouds and winds.

Hydrography is the science of the watery part of the terraqueous globe. It makes us acquainted with the origin and nature of springs and marshes, the effects of lakes, marshes, and rivers, on the air and on vegetation in their vicinity; and the effects of sea air on the vegetation of maritime districts.

Geology is the knowledge of the substances which compose the crust of the earth. It explains the nature and origin of soils and subsoils; that is, the manner in which they have most probably been formed, and the rocks from which they have originated; it discovers the relative position, structure, and direction in which the different rocks usually lie. It has as yet done little for Agriculture; but a perfect knowledge of geology might supply useful hints for draining land, and planting trees on soils and over subsoils best suited to their natural habits, a branch of rural economy as yet little understood, and very injudiciously practiced.

Botany and *botanical physiology*, which treat of the appearance and structure of plants, are so obviously useful to the agricultural pupil, that it is unnecessary to dilate on the advantages to be derived from a knowledge of both.

Zoology, which treats of the classification and habits of all animals, from the lowest to the highest organized structure, cannot fail to be a source of constant interest to every farmer who rears stock. There are few wild quadrupeds in this country; but the insect creation itself would employ a lifetime to investigate.

Anatomy, especially *comparative anatomy*, is highly useful to the farmer, inasmuch as it explains the functions of the internal structure of animals upon which he bestows so much care in rearing. Acquainted with the functions of the several parts which constitute the corporeal body, he will be the better able to apportion the food to the peculiar constitution of the animal; and also to anticipate any tendency toward disease, by a previously acquired knowledge of premonitory symptoms. Comparative anatomy is most successfully taught in veterinary schools.

The only other science which bears directly on Agriculture, and with which the pupil farmer should make himself acquainted, is *Chemistry;* that science which is cognizant of all the changes in the constitution of matter, whether effected by heat, by moisture, or other means. There is no substance existing in nature but is susceptible of chemical examination. A science so universally applicable cannot fail to arrest popular attention. Its popular character, however, has raised expectations of its power to assist Agriculture to a much greater degree than the results of its investigations yet warrant. It is very generally believed, not by practical farmers, but chiefly by amateur agriculturists, who profess great regard for the welfare of Agriculture, that the knowledge derived from the analysis of soils, manures and vegetable products, would develop general principles which might lead to the establishment of a system of Agriculture as certain in its effects as the unerring results of science. Agriculture, in that case, would rank among the experimental *sciences*, the application of the principles of which would necessarily result in increased produce. The positive effects of the weather seem to be entirely overlooked by these amateurs. Such sentiments and anticipations are very prevalent in the present day, when every sort of what is termed *scientific* knowledge is sought after with an eagerness as if prompted by the fear of endangered existence. This feverish anxiety for scientific knowledge is very unlike the dispassionate state of mind induced by the patient investigation of true science, and very unfavorable to the right application of the principles of science to any practical art. Most of the leading agricultural societies instituted for the promotion of practical Agriculture, have been of late assailed by the entreaties of euthusiastic amateur agriculturists, to construct their premiums to encourage only that system of Agriculture which takes chemistry for its basis.

These are the physical sciences whose principles seem most applicable to Agriculture; and being so, they should be studied by every farmer who wishes to be considered an enlightened member of his profession. That farmers are quite competent to attain to these sciences, may be gathered from these observations of Sir John Herschel: "There is scarcely any well-informed person who, if he has but the will, has not the power to add something essential to the general stock of knowledge, if he will only observe regularly and methodically some particular class of facts which may most excite his attention, or which his situation may best enable him to study with effect. To instance one subject which *can* only be effectually improved by the united observations of great numbers widely dispersed: Meteorology, one of the most complicated but important branches of science, is at the same time one in which any person who will attend to plain rules, and bestow the necessary degree of attention, may do effectual service." But in drawing our conclusions, great caution is requisite; for, "In forming inductions, it will most commonly happen that we are led to our conclusions by the especial force of some two or three strongly impressive facts, rather than by affording the whole mass of cases a regular consideration; and hence the need of cautious verification. Indeed, so strong is this propensity of the human mind, that there is hardly a more common thing than to find persons ready to assign a cause for every thing they see, and in so doing, to join things the most incongruous, by analogies the most fanciful. This being the case, it is evidently of great importance that these first ready impulses of the mind should be made on the contemplation of the cases most likely to lead to good inductions. The misfortune, however, is, in natural philosophy, that the choice does not rest with us. We must take the instances as Nature presents them. Even if we are furnished with a list of them in tabular order, we must understand and compare them with each other, before we can tell which are the instances thus deservedly entitled to the highest consideration. And, after all, after much labor in vain, and groping in the dark, accident or casual observation will present a case which strikes us at once with a full insight into the subject, before we can even have time to determine to what class its *prerogative* belongs."*

Many farmers, I dare say, will assert it to be far beyond the reach of their means, and others beyond their station, to bestow on *their* sons so learned an education as that implied in the acquirement of the sciences just now enumerated. Such apprehensions are ill-founded; because no farmer that can afford to support his sons at home, without working for their bare subsistence, but possesses the means of giving them a good education, as I shall immediately prove; and no farmer, who confessedly has wealth, should grudge his sons an education that will fit them to adorn the profession they intend to follow.

It cannot be denied that a knowledge of mathematics and natural philosophy greatly elevates the mind. Those farmers who have acquired these sciences, must be sensible of their tendency to do this; and they will therefore naturally wish *their* sons to enjoy what they themselves do. Those who of themselves do not know these sciences, on being informed of their beneficial tendency, will probably feel it to be their duty to educate their sons, and thereby put it in their power to raise themselves in society and at the same time shed a lustre on the profession of which they are members. The same species of reasoning applies to the acquirement of the peculiar accomplishments bestowed on the mind by a knowledge of natural history and chemistry. Neither the time nor expense of acquiring such an education is of that extent or magnitude as to deter any farmer's son from attempting it, who occupies a station above that of a farm steward. Besides these considerations, a good education, as the trite saying has it, is the best legacy a parent can leave his child; and, on this account, it is better for the young farmer himself to bestow on him a superior education, in the first instance, with a part even of the money destined by his father to stock him a farm, than to plenish for him a larger farm, and stint his education. The larger farm would, no doubt, enable the half-educated son to earn a livelihood more easily; but the well-educated one would be more than compensated in the smaller farm, by the possession of that cultivated intelligence which would induce him to apply the resources of his mind to drawing forth the capabilities of the soil, and making himself an infinitely superior member of society. Were industrious farmers as eager to improve their sons' minds by superior education, as they too often are to amass fortunes for them—a boon unprofitably used by uncultivated minds—they would display more wisdom in their choice. No really sensible farmer should hesitate to decide which course to take, when the intellectual improvement of his family is concerned. He should never permit considerations of mere pelf to overcome a sense of right and of duty. Rather than prevent his son having the power to raise himself in his profession, he should scrupulously economize his own expenditure.

I shall now show that the time occupied in the acquisition of those sciences which are expedient for the farmer to learn, is not lost when compared with the advantages which they may bestow. *Part of three years* will accomplish all, but three years are doubtless an immense time

* Discourse on the Study of Natural Philosophy, pp. 133, 182.

for a young man to *lose!* So it would be; but, to place the subject in its proper light, I would put this statement and question for consideration—Whether the young farmer's *time*, who is for years constantly following his father's footsteps over the farm, and only superintending a little in his absence, while the father himself is, all the time, quite capable of conducting the farm, is not as much *lost*, as the phrase has it, as it would be when he is occupied in acquiring a scientific education at a little distance from home? Insomuch as the young man's *time* is of use to the *farm*, the two cases are nearly on a par; but, in as far as both cases affect himself, there is no question that science would benefit him the more—no question that a superior education would afterward enable him to learn the practical part of his profession with his father, with much greater ease to himself. The question is thus narrowed to the consideration of the alternative of the cost of keeping the son at home, following his father as idly as his shadow, or of sending him to college. Even in this pecuniary point of view, the alternative consists merely of the difference of maintenance at home, and that in a town, with the addition of fees. That this difference is not great, I shall now show.

Part of three years, as I have said, would accomplish all amply, and in this way: the first year to be devoted to mathematics, the second to natural philosophy, and the third to natural history and chemistry; and along with these principal subjects, some time in both years should be devoted to geography, English grammar and composition, book-keeping, and a knowledge of cash transactions. The two months' vacation in each year could be spent at home. There are seminaries* at which these subjects may† be studied, at no great distance from every farmer's home. There are, fortunately for the youths of Scotland, universities, colleges, and academies, in many parts of the country. Edinburgh, Glasgow, Aberdeen, and St. Andrews, can boast of well-endowed universities and colleges; while the academies at Dundee, Perth, Ayr, Dollar, and Inverness, have been long famed for good tuition.

---

10. OF THE DIFFERENT KINDS OF FARMING.

"I'll teach you differences."
LEAR.

PERHAPS the young farmer will be astonished to learn that there are many and various systems of farming; yet so in reality is the case, and moreover that they all possess very distinctive characteristics. There are *six* kinds of farming practiced in Scotland alone; and though all are pursued under some circumstances common to all, and each kind is perhaps best adapted to the particular soil and situation in which it is practiced; yet it is highly probable that one of the kinds might be applicable to, and profitably followed, in all places of nearly similar soil and locality. Locality, however, determines the kind of farming fully more than the soil; the soil only entirely determining it when of a very peculiar consistence. The comparative influence of locality over soil in determining this point will be better understood after shortly considering each kind of farming.

[* Or free schools. † Or ought to be. *Ed. Farm. Lib.*]

1. One kind is wholly confined to *pastoral* districts, which are chiefly situated in the Highlands and Western Isles of Scotland—in the Cheviot and Cumberland hills of England—and very generally in Wales. In all these districts, farming is almost restricted to the breeding of cattle and sheep; and, as natural pasture forms the principal food of live-stock in a pastoral country, very little arable culture is there practiced for their behoof. Cattle and sheep are not always both reared on the same farm. Cattle are reared in very large numbers in the Western Isles, and in the *pastoral valleys* among the mountain-ranges of England, Wales, and Scotland.* Sheep are reared in still greater numbers in the *upper parts* of the mountain-ranges of Wales and of the Highlands of Scotland; and on the green round-backed mountains of the south of Scotland and the north of England. The cattle reared in pastoral districts are small sized, chiefly black colored, and horned. Those in the Western Isles, called "West Highlanders," or "Kyloes," are esteemed a beautifully symmetrical and valuable breed of cattle. Those in the valleys of the Highland mountains, called "North Highlanders," are considerably inferior to them in quality, and smaller in size. The black-faced, mountain, or heath, horned sheep, are bred and reared on the upper mountain-ranges, and fattened in the low country. The round-backed green hills of the south are mostly stocked with the white-faced, hornless, Cheviot breed, ; though the best kind of the black-faced breed is also reared in some localities of that district, but seldom both breeds are bred by the same farmer. Wool is a staple product of sheep pastoral farming.

Pastoral farms are chiefly appropriated to the rearing of one kind of sheep, or one kind of cattle; though both classes of stock are bred where valleys and mountain-tops are found on the same farm. The arable culture practiced on them is confined to the raising of provisions for the support of the shepherds and cattle-herds; and perhaps of a few turnips, for the support of the stock during the severity of a snow-storm; but the principal artificial food of the stock in winter is hay, which in some cases is obtained by inclosing and mowing a piece of natural grass on a spot of good land, near the banks of a rivulet, the alluvial soil along the river sides being generally of fine quality. All pastoral farms are large, some containing many thousands of acres—nay miles in extent; but from 1,500

[* In the United States the mountain ranges running from east to west may be considered our "pastoral" or grazing districts.

The farther we go east the more are such lands devoted to sheep husbandry, while in the west and south-west they are given up to the rearing of cattle, to be sold, as lean or stock cattle, to the grazier, who sometimes buys and carries them through the winter on wheat straw, and fattens them on grass against the next autumn. But more generally they are sold in spring, grazed through the summer, and fattened on corn the following winter. Thus prepared for market, they are either killed and packed in the West, or driven thence in spring and summer to the eastern markets. For our pastoral or grazing districts, a comparatively smaller and more thrifty race of cattle, weighing, when at market, from 500 to 700, is most advantageous for all parties, as, with but little exception they have to "shift for themselves" throughout the year, and often get no special feeding.

It is as true now as it was in the time of Sir John Sinclair, that where the surface is barren and the climate rigorous, it is essential that the stock bred and maintained there should be enabled to sustain the severities and vicissitudes of the weather as well as scarcity of food, or any other circumstance in its locality and treatment that might subject a more delicate breed to injury. For the purposes of the cattle breeder in the mountains, it is probable that the hardy middle sized North Devon would be found most eligible; or if it should be deemed expedient to try a foreign cross which we have not tried, obvious reasons suggest that the Polled, or Galloway, and the Scotch Highland races should be had recourse to. *Ed. Farm. Lib.*]

to 3,000 acres is perhaps an ordinary size.* Locality determines this kind of farming.

The *stocking* of a pastoral farm consists of a breeding stock of sheep or cattle, and a yearly proportion of barren stock intended to be fed and sold at a proper age. A large capital is thus required to stock at first, and afterward maintain such a farm; for, although the quality of the land may not be able to support many heads of stock per acre, yet, as the farms are large, the number of heads required to stock a large farm is very considerable. The rent, when consisting of a fixed sum of money, is of no great amount per acre, but sometimes it is fixed at a sum per head of the stock that the farm will maintain.

A *pastoral farmer* should be well acquainted with the rearing and management of cattle or sheep, whichever his farm is best suited for. A know ledge of general field culture is of little use to him, though he should know how to raise turnips and make hay.

2. Another kind of farming is practiced on *carse* land. A carse is a district of country, consisting of deep horizontal depositions of alluvial or diluvial clay, on one or both sides of a considerable river; and may be of great or small extent, but generally comprehends a large tract of country. In almost all respects, a carse is quite the opposite to a pastoral district. Carse land implies a flat, rich, clay soil, capable of raising all sorts of grain to great perfection, and unsuited to the cultivation of pasture grasses, and, of course, to the rearing of live-stock. A pastoral district, on the other hand, is always hilly—the soil generally thin, poor, and various, and commonly of a light texture, much more suited to the growth of natural pasture grasses than of grain, and, of course, to the rearing of live-stock. Soil decides this kind of farming.

Being all arable, a *carse farm* is mostly stocked with animals and implements of labor; and these, with seed-corn for the large proportion of the land cultivated under the plow, require a considerable outlay of capital.—Carse land always maintains a high rent per acre, whether it consists solely of money, or of money and corn valued at the fiars prices. A carse farm, requiring much capital and much labor, is never of large extent—seldom exceeding 200 acres.

A *carse*† *farmer* requires to be well acquainted with the cultivation of grain, and almost nothing else, as he can rear no live-stock; and all he requires of them are a few milch cows, to supply milk to his household and farm-servants, and a few cattle in the straw-yard in winter, to trample down the large quantity of straw into manure—both of which classes of cattle are purchased when wanted.

3. A third sort of farming is that which is practiced *in the neighborhood of large towns*. In the immediate vicinity of London, farms are appropriated to the growth of garden vegetables for Covent-Garden market; and, of course, their method of culture can have nothing in common with either pastoral or carse farms. In the neighborhood of most towns, garden vegetables, with the exception of potatoes, are not so much cultivated as green crops, such as turnips and grass, and dry fodder, such as straw and hay, for the use of cow-feeders and stable-keepers. The practice of this kind of farming is to dispose of all the produce, and receive in return manure for the land. And this constitutes this kind of farming a retail trade like that in town, in which articles are bought and sold in small quantities, mostly

* It is to be regretted that neither the Old nor the New Statistical Account of Scotland gives the least idea of the *size* of the farms in any of the parishes described.

[† What we call a grain farmer. *Ed. Farm. Lib.*]

for ready money.* When there is not a sufficient demand in the town for all the disposable produce, the farmer purchases cattle and sheep to eat the turnips, and trample the straw into manure, in winter. Locality decides this kind of farming.

The chief qualification of an *occupant of this kind of farm* is a thorough acquaintance with the raising of green crops—potatoes, clover, and turnips; and his particular study is the raising of those kinds and varieties that are most prolific, for the sake of having large quantities to dispose of, and which, at the same time, are most suitable to the wants of his customers.

The *capital required for a farm of this kind*, which is all arable, is as large as that for a carse farm. The rent is always high per acre, and the extent of land not large—seldom exceeding 300 acres.

4. A fourth kind of farming is the *dairy* husbandry. It specially directs its attention to the manufacture of butter and cheese, and the sale of milk. Some farms are laid out for the express purpose; but the sale of milk is frequently conjoined with the raising of green crops, in the neighborhood of large towns, whose inhabitants are whence daily supplied with milk, though seldom from pasture, which is mostly appropriated as paddocks for stock sent to the weekly market. But a true dairy-farm requires *old pasture*. The chief business of a *dairy-farm* is the management of cows and of their produce; and whatever arable culture is practiced thereon is made entirely subservient to the maintenance and comfort of the dairy stock.—The milk, where practicable, is sold; where beyond the reach of sale, it is partly churned into butter, which is sold either fresh or salted, and partly made into cheese, either sweet or skimmed. No stock are *reared* on dairy-farms, as on pastoral, except a few quey (heifer) calves, occasionally to replenish the cow stock; nor aged stock *fed* in winter, as on farms in the vicinity of towns. The bull calves are frequently fed for veal, but the principal kind of stock reared are pigs, which are fattened on dairy refuse.—Young horses, however, are sometimes successfully reared on dairy-farms. Horse labor being comparatively little required thereon, mares can carry their young, and work with safety at the same time; while old pasture, spare milk, and whey afford great facilities for nourishing young horses in a superior manner. Locality has decided this kind of farming on the large scale.

The purchase of cows is the principal expense of stocking a *dairy-farm;* and as the purchase of live-stock in any state, especially breeding-stock, is always expensive, and live-stock themselves, especially cows, constantly liable to many casualties, a dairy-farm requires a considerable capital. It is, however, seldom of large extent—seldom exceeding 150 acres. The arable portion of the farm supplying the green crop for winter food and litter, does not incur much outlay, as hay—*that* obtained from old pasture grass—forms the principal food of all the stock in winter. The rent of dairy-farms is high.

A *dairy-farmer* should be well acquainted with the properties and management of milch cows, the manufacture of butter and cheese, the feeding of veal and pork, and the rearing of horses; and he should also possess as much knowledge of arable culture as to enable him to raise those kinds of

[* The facilities afforded by steam for the quick transportation of perishable articles—such as fruit and milk, and the more delicate vegetables—has had the effect of opening market gardens at a comparatively great distance from the large towns. A railroad or a steamboat will bring these articles into market, from a distance of fifty miles, with as little delay, and less injury by transportation, than an ordinary conveyance would bring them ten miles. Obvious as is this fact it is deemed proper to mention it, that it may not be lost sight of in the purchase of farms.

Ed. Farm. Lib.]

green crops, and that species of hay, which are most congenial to cows for the production of milk.

5. A fifth method of farming is that which is practiced in most arable districts, consisting of any kind of soil not strictly carse land. This method consists of a regular system of cultivating grains and sown grasses, with the partial rearing, and partial purchasing, or wholly purchasing, of cattle; and no sheep are reared in this system, they being purchased in autumn, to be fed on turnips in winter, and sold off fat in spring. This system may be said to *combine the professions of the farmer, the cattle-dealer, and the sheep-dealer.**

To become a *farmer* of this *mixed husbandry*, a man must be acquainted with every kind of farming practiced in the country. He actually practices them all. He prosecutes, it is true, each kind in a rather different manner from that practiced in localities where the particular kind is pursued as the only system of farming; because each branch of his farming must be conducted so as to conduce to the welfare of the whole, and, by studying the mutual dependence of parts, he produces a whole in a superior manner.— This multiplicity of objects requires from him more than ordinary attention, and much more than ordinary skill in management. No doubt, the farmers of some of the other modes of farming become very skillful in adapting their practice to the situations in which they are actually placed, but his more varied experience increases versatility of talent and quickness of discernment; and, accordingly, it will be found that the farmers of the *mixed husbandry* prove themselves to be the cleverest and most intelligent agriculturists of the country.

---

11. OF CHOOSING THE KIND OF FARMING.

"Choice, being mutual act of all our souls, makes merit her election."
TROILUS AND CRESSIDA.

THESE are the various kinds of farming pursued in this kingdom; and, if there be any other, its type may, no doubt, be found in the mixed system just described. One of these systems must be adopted by the aspirant pupil for his profession. If he succeed to a family inheritance, the kind of farming he will follow will depend on that pursued by his predecessor, which he will learn accordingly; but if he is free to choose for himself, and not actually restricted by the circumstances of peculiar locality, or soil, or inheritance, then I would advise him to adopt the mixed husbandry, as containing within itself all the varieties of farming which it is requisite for a farmer to know.

If he is at liberty to take advice, I can inform him that the mixed husbandry possesses advantages over every other; and practically thus: in pastoral farming, the stock undergoes minute examination, for certain purposes, only at distantly stated times; and owing to the wide space over which they have to roam for food in pastoral districts, comparatively less attention is bestowed on them by shepherds and cattle-herds. The pastoral farmer has thus no particular object to attract his attention at home between those somewhat long intervals of time; and in the mean while time is apt to hang heavy on his hands. The carse farmer, after the labors of the field are finished in spring, has nothing but a little hay-making

[* Which, in our country, are often combined. *Ed. Farm. Lib.*]

and much bare-fallowing in summer, to occupy his mind until the harvest. Dairy-farming affords little occupation for the farmer in winter. The farmer in the vicinity of large towns has almost nothing to do in summer, from turnip-seed to harvest. Mixed-husbandry, on the other hand, affords abundant and regular employment at all seasons. Cattle and sheep feeding, and marketing grain, pleasantly occupy the short days of winter. Seed-sowing of all kinds affords abundant employment in spring. The rearing of live-stock, sale of wool, and culture of green crops, fill up the time in summer until harvest; and autumn, in all circumstances, brings its own busy avocations at the ingathering of the fruits of the earth. There is, strictly speaking, not one week of real leisure to be found in the mixed system of farming—if the short period be excepted, from assorting lambs in the beginning of August to putting the sickle to the corn—and that period is curtailed or protracted, according as the harvest is early or late.

If the young farmer is desirous of attaining a knowledge of every kind of farm work—of securing the chance of profit every year—and of finding regular employment at all seasons in his profession, he should determine to follow the mixed husbandry. It will not in any year entirely disappoint his hopes. In it, he will never have to bewail the almost total destruction of his stock by the rot, or by the severe storms of winter, as the pastoral farmer sometimes has. Nor can he suffer so serious a loss as the carse farmer, by his crop of grain being affected by the inevitable casualties of blight or drouth, or the great depression of prices for a succession of years. Were his stock greatly destroyed or much deteriorated in value by such casualties, he might have the grain to rely on; and were his grain crops to fail to a serious extent, the stock might insure him a profitable return. It is scarcely within the bounds of probability that a loss would arise in any year from the total destruction of live-stock, wool, *and* grain. One of them may fail, and the prices of all may continue depressed for years; but, on the other hand, reasonable profits have been realized from them *all* in the same year. Thus, there are safeguards against a total loss, and a greater certainty of a profitable return from capital invested in the *mixed*, than in any *other* kind of husbandry at present known.

---

12. OF SELECTING A TUTOR-FARMER FOR TEACHING FARMING.

"These are their tutors, bid them use them well."
TAMING OF THE SHREW.

After resolving to follow farming as a profession, and determining to learn the mixed, as the best system of husbandry, it now only remains for the young farmer to select a farmer who practices it, with whom he would wish to engage as a pupil. The best kind of pupilage is to become a boarder in a farmer's house, where he will not only live comfortably but may learn this superior system of husbandry thoroughly. The choice of locality is so far limited, as it must be in a district in which this particular system is practiced in a superior manner. The qualifications are numerous. The farmer should have the general reputation of being a good farmer; that is, a skillful cultivator of land, a judicious breeder, and an excellent judge of stock. He should possess agreeable manners, and have the power of communicating his thoughts with ease. He should occupy a good farm, consisting, if possible, of a variety of soils, and situate in a tolerably good climate, neither on the top of a high hill, nor on the confines of a large moor or bog, but in the midst of a well cultivated country. These circumstances of soil and locality should be absolute requisites in a farm intended to be made the residence of *pupils*. The top of a hill exposed to every blast that blows, or the vicinage of a bog, overspread with damp vapor, would surround the farm with a climate in which no kind of crop or stock could arrive at a state of perfection; while, on the other hand, a very sheltered spot in a warm situation, would give the pupil no idea of the vexations experienced in a precarious climate. His inexperience in these things will render him unfit to select for himself either a qualified farmer, or a suitable farm; but friends are never wanting to render assistance to young aspirants in such emergencies, and if their opinion is formed on a knowledge of farming, both of the farm and the personal qualifications of the farmer they are recommending, some confidence may be placed in their recommendations.

As a residence of one year must pass over ere the pupil can witness the course of the annual operations of the farm, his engagement at first should be made for a period of *not less* than a year; and at the expiring of that period he will, most probably, find himself inadequate to the task of managing a farm. The entire length of time he would require to spend on a farm, must be determined by the paramount consideration of his having acquired a competent knowledge of his profession.

---

13. OF THE PUPILAGE.

"A man loves the meat in his youth that he cannot endure in his age."
MUCH ADO ABOUT NOTHING.

HAVING settled these preliminaries with the tutor-farmer, the pupil should enter the farm—the first field of his anticipations and toils in farming—with a resolution to acquire as much professional knowledge, in as short a time as the nature of the business which he is about to learn will admit of.

The commencement of his tuition may be made at any time of the year; but since farming operations have a regular beginning and ending every year, it is obvious that the *most* proper time to *begin* to view them is at the *opening of the agricultural year*, that is, *in the beginning of winter*. It may not be quite congenial to the feelings of him who has perhaps been accustomed to pass his winters in a town, to participate for the first time in the labors of a farm on the eve of winter. He would naturally prefer the sunny days of summer. But the beginning of winter being the time at which every important operation is *begun*, it is essential to their being understood throughout, to *see them begun*, and in doing this, minor inconveniences should be willingly submitted to, to acquire an intimate knowledge of a profession for life. And, besides, to endeavor to become acquainted with complicated operations, after the *principal arrangements* for their accomplishment have been *completed*, is purposely to invite wrong impressions of them.

There is really nothing disagreeable to personal comfort in the business of the farm in winter. On the contrary, it is full of interest, inasmuch as the well-being of living animals then comes home to the attention more forcibly than the operations of the soil. The totally different and well-marked individual characters of different animals, engage our sympathies in different degrees; and the more so, perhaps, of all of them, that they appear more domesticated when under confinement than at liberty to roam about in quest of food and seclusion. In the evening, in winter, the hospitality of the social board awaits the pupil at home, or at a friend's house, after the labors of the day are over. Neighbors interchange visits at that social season, when topics of conversation common to all societies are varied by remarks on professional occurrences and management, elicited by the modified practices of the different speakers, from which the pupil may pick up much useful information. Or should society present no charms to him, the quieter companionship of books, or the severer task of study, is at his command. In a short time, however, the many objects peculiar to the season which present themselves in the country in winter, cannot fail to interest him.

The very first thing to which the pupil should direct his attention on entering the farm, is to become well acquainted with its *physical geography*—that is, its position, exposure, extent; its fences, whether of wall or hedge; its shelter, in relation to rising grounds and plantations; its roads whether public or private; its fields, their number, names, sizes, relative positions, and supply of water; the position of the farm-house and steading of farm-stead. Familiar acquaintance with all these particulars will enable him to understand more readily the orders given by the farmer for the work to be performed in any field. It is like possessing a map of the ground on which certain plans of operations are about to be undertaken. A plan of the farm would much facilitate an introduction to this familiar acquaintance. The *tutor*-farmer should be provided with such a plan to give to each of his pupils, but if *he* have it not, the pupil himself can set about constructing one which will answer his purpose well enough.

14. OF DEALING WITH THE DETAILS OF FARMING.

"Oh! is there not some patriot . . .
To teach the lab'ring hands the sweets of toil?
Yes, there are such."

THOMSON.

THE principal object held in view, while making the preceding observations, was the preparation of the mind of the young person, desirous of becoming a farmer, into such a state as to enable him, when he enters a farm as a pupil, to anticipate and overcome what might appear to him great difficulties of practice, which, with an unprepared mind, he could not know existed at all, far less know how to overcome; but, on being informed that he must encounter them at the very outset of his career, he might use the means pointed out to him for meeting and overcoming them. These difficulties have their origin in the pupil seeing the operations of the farm, of whatever nature, performed for the first time, in the most perfect manner, and always with a view to accomplishment at some *future* period. The only mode of overcoming such difficulties, and thereby satisfying his mind, is for the pupil to ascertain by inquiry the *purport* of every operation he sees performing; and though he may feel that he does not quite comprehend that purport, even when informed of it, still the information will *warn* him of its approaching consummation, and he will not, therefore, at any time thereafter be taken by surprise when the event actually arrives. If I show the pupil the importance of making inquiry regarding the purport of every operation he sees performing, I see no better mode of rendering all farming operations intelligible to his mind. In order to urge him to become familiar with the purport of everything he sees going on around him, I have endeavored to point out the numerous *evils* attendant on farmers, landowners, and emigrants neglecting to become thoroughly acquainted with practical husbandry, before attempting to exercise their functions in their new vocations. And, in order that the young person desirous of becoming a farmer may have no excuse for not becoming *well* acquainted with farming, I have shown him where, and the manner how, he can best become acquainted with it; and these are best attained, under present circumstances, by his becoming an inmate for a time in a farm-house with an intelligent farmer. Believing that the foregoing observations, if perused with a willing mind, are competent to give such a bias to his mind as to enable the pupil, when he enters a farm, to appreciate the importance of his profession, and thereby create an ardent desire for its attainment, I shall now proceed to describe the details of every operation as it occurs in its due course on the farm.

The description of these details, which are multifarious and somewhat intricate, will compose by far the most voluminous portion of this work, and will constitute the most valuable and interesting part of it to the pupil. In the descriptions, it is my intention to go very minutely into details, that no circumstance may be omitted in regard to any of the operations, which may have the appearance of presenting a single one to the notice of the pupil in an imperfect form. This resolution may invest the descriptions with a degree of prolixity which may, perhaps, prove tiresome to the general reader; but, on that very account, it should the more readily give rise to a firm determination in the pupil to follow the particulars of every operation into their most minute ramifications; and this because he cannot be too intimately acquainted with the nature of every piece of work, or too much informed of the various modifications which every operation has frequently to undergo, in consequence of change in the weather, or the length of time in which it is permitted by the season to perform it. Descriptions so minute will answer the purpose of *detailed instructions* to the pupil; and, should he follow them with a moderate degree of application through one series of operations, he will obtain such an insight into the nature of field labor as will ever after enable him easily to recognize a similar series when it is begun to be put into execution. Unless, however, he bestow considerable attention on *all* the *details* of the descriptions, he will be apt to let pass what may appear to him an unimportant particular, but which may be the very keystone of the whole operation to which they relate. With a tolerable memory on the part of the pupil, I feel pretty sure that an *attentive perusal of the descriptions* will enable him to identify any piece of work he afterward sees performing in the field. This achievement is as much as any book can be expected to accomplish.

In describing the details of farming, it is necessary to adhere to a determinate method; and the method that appears to me most instructive to the pupil is to follow the usual routine of operations pursued on a farm. It will be requisite, in following that routine implicitly, to describe every operation from the *beginning;* for it must be impressed on the mind of the pupil that farm operations are not conducted at random, but on a tried and approved system, which commences with preparatory labors, and then carries them on with a determinate object in view throughout the seasons, until they terminate at the end of the agricultural year. The preparatory operations commence immediately after harvest, whenever that may happen, and it will be earlier or later in the year according as the season is early or late; and as the harvest is the consummation of the labors of the year, and terminates the autumnal season, so the preparatory operations begin with the winter season. Thus the winter season takes the precedence in the arrangements of farming, and doing so, that should be the best reason for the pupil commencing his career as an agriculturist in winter. In that season he will have the advantage of witnessing every *preparation as it is made* for realizing the future crops—an advantage which he cannot enjoy if he enter on his pupilage at any other season; but it is a great advantage, inasmuch as every piece of work is much better understood, when viewed from its commencement, than when seen for the first time in a state of progression.

Having offered these preliminary remarks respecting the condition of the agricultural pupil when about to commence learning his profession, I shall now proceed to conduct him through the whole details of farming, as they usually occur on a farm devoted to the practice of the mixed, or, in other words, of the most perfect system of husbandry known; while, at the same time, he shall be made acquainted with what constitute differences from it in the corresponding operations of the other modes of farming, and which are imposed by the peculiarities of the localities in which they are practiced. These details I shall narrate in the order in which they are performed, and for that purpose will begin with those of Winter—the season which commences the agricultural year—for the reason assigned in the paragraph immediately preceding this one.

WINTER.

> "All nature feels the renovating force
> Of WINTER, only to the thoughtless eye
> In ruin seen. The frost-concocted glebe
> Draws in abundant vegetable soul,
> And gathers vigor for the coming year."
>
> THOMSON.

THE subjects which court attention in Winter are of the most interesting description to the farmer. Finding little inducement to spend much time in the fields at this torpid season of the year, he directs his attention to the more animated portions of farm-work conducted in the steading, where almost the whole stock of animals are collected, and where the preparation of the grain for market affords pleasant employment for work-people within doors. The progress of live-stock to maturity is always a prominent object of the farmer's solicitude, but especially in winter, when the stock are comfortably housed in the farmstead, plentifully supplied with wholesome food, and so arranged in various classes, according to age and sex, as to be easily inspected at any time.

The labors of the field in winter are confined to a few great operations. These are, plowing the soil in preparation of future crops, and supplying food to the live-stock. The plowing partly consists of turning over the ground which had borne a part of the grain crops; and the method of plowing this *stubble land*—so called because it bears the straw that was left uncut of the previous crop—is determined by the nature of the soil. That portion of the stubble land is first plowed which is intended to be first brought into requisition for a crop in spring, and the rest is plowed in the same succession that the different crops succeed each other in the ensuing seasons. The whole soil thus plowed in the early part of winter in each field (where the farm is subdivided with fences), or in each division (where there are no fences), is then neatly and completely provided with channels, cut with the spade in suitable places, for the purpose of permitting the water that may fall from the heavens to run quickly off into the ditches, and thereby to maintain the soil in as dry a state as is practicable until spring. Toward the latter part of winter, the newest grass land—or *lea*,* as grass land is generally termed—intended to bear a crop in spring, is then plowed; the oldest grass land being earlier plowed, that its toughness may have time to be meliorated by spring by exposure to the atmosphere. ....When the soil is naturally damp underneath, winter is the season selected for removing the damp by draining. It is questioned by some farm-

[* Every agricultural student and reader would do well to notice these peculiar terms employed by English agricultural writers, because it is in that country, above all others, that the spirit of investigation is constantly at work. It is there that the progress of discovery is most steady. and publication most prompt and diffusive, and that, above all, in our own mother tongue. *Ed. Farm. Lib.*]

ers whether the winter is the best season for draining, as the usually rainy and otherwise unsettled state of the weather then renders the carriage of the materials for draining very laborious. On the other hand, it is maintained by other farmers that, as the quantity of water to be drained from the soil determines both the number and size of the drains, these are thus best ascertained in winter; and, as the fields are then most free of crop, they are in the most convenient state to be drained. Truth may perhaps be found to acquiesce in neither of these reasons, but rather in the opinion that draining may be successfully pursued at all seasons.....Where fields are uninclosed, and intended to be fenced with the thorn-hedge, winter is the season for performing the operation of planting it. Hard frost, a fall of snow, or heavy rain, may put a stop to the work for a time, but in all other states of the weather it may proceed in perfect safety.....When meadows for irrigation exist on any farm, winter is the season for beginning the irrigation with water, that the grass may be ready to mow in the early part of the ensuing summer. It is a fact well worth keeping in remembrance, in favor of *winter* irrigation, that irrigation in winter produces wholesome, and in summer unwholesome, herbage for stock. On the other hand, summer, not winter, is the proper season for forming water-meadows.....Almost the entire live-stock of an arable farm is dependent on the hand of man for food in winter. It is this circumstance which, bringing the stock into the immediate presence of their owner, creates a stronger interest in their welfare then than at any other season. The farmer then sees them classed together in the farmstead according to their age and sex, and delights to contemplate the comparative progress of individuals or classes among them toward maturity. He makes it a point to see them provided at all times with a comfortable bed or lair, and a sufficient supply of clean food at appointed hours in their respective apartments. The feeding of stock is so important a branch of farm business in winter that *it* regulates the time for prosecuting several other operations. It determines the quantity of turnips that should be carried from the field for the cattle in a given time, and causes the farmer to consider whether it would not be prudent to take advantage of the first few dry fresh days to store up a quantity of them, to be in reserve for the use of the stock during the storm that may be at the time portending—for storms like other

"Coming events cast their shadows before."

It also determines the quantity of straw that should be provided from the stack-yard, in a given time, for the use of the animals; and upon this, again, depends the supply of grain that can be sent to the market in any given time. For although it is certainly in the farmer's power to thresh as many stacks as he pleases at one time, provided the machinery for the purpose is competent for the task—and he is tempted to do so when prices are high—yet, as new threshed straw forms superior provender for live-stock confined in the farmstead, its supply, both as litter and fodder, is therefore mainly dependent on its use by the stock; and as its consumption as litter is greater in wet than in dry weather, and wet weather prevails in winter, the quantity of straw used in the course of that season must always be very considerable, and so, therefore, must the quantity of grain ready to be sent to market. All the stock in the farmstead in winter, that are not put to work, are placed under the care of the *cattle-man*.....The feeding of that portion of the sheep-stock which are barren, on turnips in the field, is a process practiced in winter. This forms fully a more interesting object of contemplation to the farmer than even the feeding of cattle—the behavior of sheep in any circumstances being always fascinating. Sheep being put on turnips early in winter, a favorable opportunity is thereby afforded the farmer, when clearing the field partially of turnips for the sheep (in a manner that will afterward be fully described to the pupil), to store a quantity of them for the cattle in case of an emergency in the weather, such as rain, snow, or frost. This removal of the surplus turnips that are not used by the sheep confined on the land renders sheep-feeding a process which, in part, also determines the quantity of that root that should be carried from the field in a given time.....The flock of ewes roaming at large over the pastures requires attention in winter, especially in frosty weather, or when snow is on the ground, when they should be supplied with hay, or turnips when the former is not abundant. The *shepherd* is the person who has charge of the sheep flock.....The large quantity of straw used in winter causes, as I have said, a considerable quantity of grain to be sent at that season to market. The preparation of grain for sale constitutes an important branch of winter farm-business, and should be strictly superintended. A considerable portion of the labor of horses and men is occupied in carrying the grain to the market-town, and delivering it to the purchasers—a species of work which jades farm-horses very much in bad weather. ....In hard frost, when the plow is laid to rest, or when the ground is covered with snow, and as soon as,

"—— by frequent hoof and wheel, the roads
A beaten path afford,"

the farm-yard manure is carried from the courts, and deposited in a large heap, in a convenient spot near the gate of the field which is to be manured with it in the ensuing spring or summer. This work is carried on as long as there is manure to carry away, or the weather continues in either of those states.....Of the implements of husbandry, only a few are put in requisition in winter: the plow is in constant use when the weather will permit; the threshing-machine enjoys no sinecure; and the cart finds periodic employment.

The *weather* in winter is of the most precarious description, and, being so, the farmer's skill to anticipate its changes in this season is severely put to the test. Seeing that all operations of the farm are so dependent on the weather, a familiar acquaintance with the local prognostics which indicate a change for the better or worse is incumbent on the farmer. In actual rain, snow, or hard frost, none but in-door occupations can be executed; but, if the farmer have wisely "discerned the face of the sky," he can arrange the order of these in-door operations, so as they may be continued for a length of time, if the storm threaten a protracted endurance, or be left without detriment, should the strife of the elements quickly cease.

The winter is the season for *visiting the market town* regularly, where the surplus produce of the farm is disposed of—articles purchased or bespoke for the use of the farm, when the busy seasons arrive—where intermixture with the world affords the farmer an insight into the actions of mankind—and where selfishness and cupidity may be seen to act as a foil to highten the brilliancy of honest dealing.

Winter is to the farmer the season of *domestic enjoyment*. The fatigues of the long summer-day leave little leisure, and much less inclination, to tax the mind with study; but the long winter evening, after a day of bracing exercise, affords him a favorable opportunity, if he have the inclination at all, of partaking in social conversation, listening to instructive reading, or hearing the delights of music. In short, I know of no class of people more capable of enjoying a winter's evening in a rational manner, than the family of the country gentleman or the farmer.*

Viewing winter in a higher and more serious light—in the repose of nature, as emblematical of the mortality of man—in the exquisite pleasures which man in winter, as a being of sensation, enjoys over the lower creation—and in the eminence in which man, in the temperate regions, stands, with respect to the development of his mental faculties, above his fellow-creatures in the tropics: in these respects, winter must be hailed by the dweller in the country, as the purifier of the mental as well as of the physical atmosphere.

On this subject, I cannot refrain from copying these beautiful reflections by a modern writer, whose great and versatile talents, enabling him to write well on almost any subject, have long been known to me. "Winter," says he, "is the season of Nature's annual repose—the time when the working structures are reduced to the minimum of their extent, and the energies of growth and life to the minimum of their activity, and when the phenomena of nature are fewer, and address themselves less pleasingly to our senses than they do in any other of the three seasons. There is hope in the bud of Spring, pleasure in the bloom of Summer, and enjoyment in the fruit of Autumn; but, if we make our senses our chief resource, there is something both blank and gloomy in the aspect of Winter.

"And if we were of and for this world alone, there is no doubt that this would be the correct view of the winter, as compared with the other seasons; and the partial death of the year would point as a most mournful index to the death and final close of our existence. But we are beings otherwise destined and endowed—the world is to us only what the lodge is to the wayfaring man; and while we enjoy its rest, our thoughts can be directed back to the past part of our journey, and our hopes forward to its end, when we shall reach our proper home, and dwell there securely and forever. This is our sure consolation—the anchor of hope to our minds during all storms, whether they be of physical nature, or of social adversity. . . . .

"We are beings of sensation certainly; many and exquisite are the pleasures which we are fitted for enjoying in this way, and much ought we to be grateful for their capacity of giving pleasure, and our capacity of receiving it; for this refined pleasure of the senses is special and peculiar to us out of all the countless variety of living creatures which tenant the earth around us. They eat, they drink, they sleep, they secure the succession of their race, and they die; but not one of them has a secondary pleasure of sense beyond the accomplishment of these very humble ends.† We

[* Especially if reared in a love of books, and the study of the natural history of all around them. *Ed. Farm. Lib.*]

[† It strikes us as a gloomy and mistaken view to say that in the whole range and variety of creation, man should be the only being endowed with susceptibility to social pleasures. Who has not witnessed with admiration not merely the force of conjugal, parental and filial ties between animals and birds, but the social affections also—the sentiments of friendship and hospitality, of jealousy, revenge, and of triumph! We may mention an instance under our own observation, of friendship and hospitality displayed between two dogs. A lady residing in Baltimore petted a magnificent Newfoundland dog, *Pelham*, while her mother, residing at Annapolis, bestowed her friendship on a small terrier, whose name was *Guess*. A steamboat plied between the two cities, and Pelham often accompanied his mistress on her visits to her mother, and there formed an intimacy with Guess. When the boat was leaving for Baltimore, Guess was sure to accompany his mistress to see her friends off; and on one occasion was left on board and carried to Baltimore, where he was landed among strangers, not knowing where to put his head. Pel-

stand far higher in the mere gratifications of sense; and in the mental ones there is no comparison, as the other creatures have not an atom of the element to bring to the estimate.

"The winter is, therefore, the especial season of man—*our own season*, by way of eminence; and men who have no winter in the year of the region in which they are placed, never of themselves display those traits of mental development which are the true characteristics of rational men, as contrasted with the irrational part of the living creation. It is true there must be the contrast of a summer, in order to give this winter its proper effect, but still, the winter is the intellectual season of the year—the season during which the intellectual and immortal spirit in man enables him most triumphantly to display his superiority over 'the beasts that perish.'"*

---

15. OF THE STEADING OR FARMSTEAD.

"When we see the figure of the house, then must we rate the cost of the erection."
HENRY IV. Part II.

(1.) BEFORE proceeding to the consideration of the state in which the pupil should find the various *fields* at the beginning of winter, it will tend to perspicuity in the furnishing of a farm to let him understand, in the first instance, the *principles* on which a *steading*, or *onstead*, or *farmstead*, or *farm-offices*, or *farmery*, as it has been variously styled, intended for a farm conducted on the mixed husbandry, should be constructed, and also to enumerate its *constituent parts*. This explanation being given, and got quit of at once, the names and uses of the various parts of a farmstead will at once become familiarized to him. And before beginning with the description of anything, I may here express it as my opinion that my descriptions of all the farm operations will be much more lucid and graphic if addressed personally to the pupil.

ham by chance met him in the street, was transported with joy at the sight of him, learned how by accident he had arrived, and soon persuaded him to go home with him, where he knew his mistress would kindly entertain him for her mother's if not for his own sake, until the boat should return. It was an instance of cordial hospitality such as towns' gentlemen are not *always* ready to reciprocate with their friends from the country.

No one, in fact can be at a loss for examples to show that Providence has kindly blessed inferior beings with capacity for other than mere brutal enjoyments. The congregation of various birds is a remarkable indication of the spirit of sociality among the feathered tribe of creation so animals herd together under the same love of company. The strongest fences cannot confine some horses in a field alone. Cattle will not fatten in the finest pastures without society; nor is this propensity confined to animals of the same species. A charming naturalist says he knew a doe, then still alive, that was brought up from a little fawn among dairy cows. With them it went afield, and with them it returned to the cow-yard. The dogs of the house took no notice of the deer, being used to her; but if strange dogs came by, a chase ensued, and while the master would look on and smile to see his favorite securely leading her pursuers over hedge, or gate, or stile, till she returned to the cows, they, with fierce lowings and menacing horns, would drive the assailants quite out of the pasture.

This complete degradation of all other created things, placing such a vast abyss between them and man, seems to detract from the benevolence of a common Father over all. Let us felicitate ourselves on the superiority of our physical structure and reasoning faculties, and the improvements and the power thence derived; but let us also remember in humility if not in shame, that of all animals, not one is more prone than man, to the wanton abuse of his strength.

Ed. Farm Lib.]

* Mudie's Winter, Preface, p. 3—5.

(2.) To present a *description* of a steading in the most specific terms, it will, in the first place, be necessary to assume a size which will afford accommodation for a farm of given extent. To give full scope to the mixed husbandry, I have already stated that a farm of 500 imperial acres is required. I will therefore assume the steading, about to be described, to be suited to a farm of that extent. At the same time you should bear in mind that the *principles* which determine the arrangement of this particular size, are equally applicable to much smaller, as well as much larger steadings; and that the mixed husbandry is frequently practiced on farms of much smaller extent.

(3.) It is a requisite condition to its proper use, that every steading be *conveniently placed* on the farm. To be *most* conveniently placed, in theory, it should stand in the *center* of the farm; for it can be proved in geometry that of any point within the area of a circle, the center is the nearest to every point in its circumference. In practice, however, circumstances greatly modify this theoretical principle upon which the site of all steadings should be fixed. For instance, if an abundant supply of water can be easily obtained for the moving power of the threshing-machine, the steading may be placed, for the sake of thus economizing horse labor, in a more remote and hollow spot than it should be in other circumstances. If wind is preferred, as the moving power, then the steading will be more appropriately placed on rising ground. For the purpose of conveying the manure down hill to most of the fields, some would prefer the *highest* ground near the center of the farm for its site. Others, on the contrary, would prefer the *hollowest* point near the center, because the grain and green crops would then be carried down hill to the steading, and this they consider a superior situation to the other, inasmuch as the grain and green crops are much more bulky and heavy than the manure. In making either of these choices, it seems to be forgotten that loads have to be carried both *to* and *from* the steading; but either position will answer well enough, provided there be no steep ascent or descent to or from the steading. The latter situation, however, is more consonant to experience and reasoning than the other; though level ground affords the easiest transit to wheel-carriages. It is also very desirable that the farm-house should be so situated as to command a view of every field on the farm, in order that the farmer may have an opportunity of observing whether the labor is prosecuted steadily; and if other circumstances permit, especially a plentiful supply of good water, the vicinity of the farm-house should be chosen as the site for the steading; but if a sacrifice of the position on the part of either is necessary, the farm-house must give way to the convenience of the steading.

(4.) As a farm of mixed husbandry comprises every variety of culture, so its steading should be constructed to *afford accommodation for every variety of produce.* The grain and its straw, being important and bulky articles, should be accommodated with room as well after as before they are separated by threshing. Room should also be provided for every kind of food for animals, such as hay and turnips. Of the animals themselves, the horses being constantly in hand at work, and receiving their food daily at regular intervals of time, should have a stable which will not only afford them lodging, but facilities for consuming their food. Similar accommodation is required for cows, the breeding portion of cattle. Young cattle, when small of size and of immature age, are usually reared in inclosed open spaces, called courts, having sheds for shelter and troughs for food and water. Those fattening for sale are either put into smaller courts with troughs called hammels, or fastened to stakes in byres or feeding-houses, like the cows. Young horses are reared either by themselves in courts with sheds and mangers, or get leave to herd with the young cattle. Young pigs usually roam about everywhere, and generally lodge among the litter of the young cattle, while sows with sucking pigs are provided with small inclosures, fitted up with a littered apartment at one end, and troughs for food at another. The smaller implements of husbandry, when not in use, are put into a suitable apartment; while the carts are provided with a shed, into which some of the larger implements which are only occasionally used, are stored by. Wool is put into a cool, clean room. An apartment containing a furnace and boiler to heat water and prepare food when required for any of the animals, should never be wanting in any steading. These are the principal accommodations required in a steading where live-stock are cared for; and when all the apartments are even conveniently arranged, the whole building will be found to cover a considerable space of ground.

(5.) The leading *principle* on which these arrangements is determined is very simple, and it is this: 1. Straw being the bulkiest article on the farm, and in daily use by every kind of live-stock, and having to be carried and distributed in small quantities by bodily labor though a heavy and unwieldy substance, should be centrically placed, in regard to all the stock, and at a short distance from their respective apartments. The position of its receptacle, the *straw-barn*, should thus occupy a central point of the steading; and the several apartments containing the live-stock should be placed equidistant from the straw-barn, to save labor in the carrying of straw to the stock. 2. Again, applying the principle, that so bulky and heavy an article as straw should in all circumstances be moved to short distances, and not at all, if possible, from any other apartment but the straw-barn, the *threshing-machine*, which deprives the straw of its grain, should be so placed as at once to throw the straw into the straw-barn. 3. And, in farther application of the same principle, the *stack-yard* containing the unthreshed straw with its corn, should be placed contiguous to the threshing-machine. 4. Lastly, the passage of straw from the stack-yard to the straw-barn through the threshing-machine being directly progressive, it is not an immaterial consideration in the saving of time to place the stack-yard, threshing-mill, and straw-barn in a right line.

(6.) Different classes of stock require different quantities of straw, to maintain them in the same degree of cleanliness and condition. *Those classes* which require the *most* should therefore be placed *nearest* the *straw-barn*. 1. The younger stock requiring most straw, the courts which they occupy should be placed contiguous to the straw-barn, and this can be most effectually done by placing the straw-barn so as a court may be put on each side of it. 2. The older or fattening cattle requiring the next largest quantity of straw, the hammels which they occupy should be placed next to these courts in nearness to the straw-barn. 3. Horses in the stables, and cows in the byres, requiring the smallest quantity of straw, the stables and byres may be placed next farthest in distance to the hammels from the straw-barn. The relative positions of these apartments are thus determined by the comparative use of the straw. 4. There are two apartments of the steading whose positions are necessarily determined by that of the threshing-machine; the one is the upper-barn, or threshing-barn, which contains the unthreshed corn from the stack-yard, ready for threshing by the mill; and the other the corn-barn, which is below the mill, and receives the corn immediately after its separation from the straw by the mill to be cleaned for market. 5. It is a great convenience to have the granaries in direct communication with the corn-barn, to save the labor of carrying the clean corn to a distance when laid up for future use. To confine the space occupied by the steading on the ground as much as practicable for utility, and at the same time insure the good condition of the grain, and especially this latter advantage, the granaries should always be elevated above the ground, and their floors then form convenient roofs for either cattle or cart-sheds. 6. The elevation which the granaries give to the building should be taken advantage of to shelter the cattle-courts from the north wind in winter; and for the same reason that shelter is cherished for warmth to the cattle, all the cattle-courts should always be open to the sun. The courts being thus open to the south, and the granaries forming a screen from the north, it follows that the granaries should stretch east and west on the north side of the courts; and, as has been shown, that the cattle-courts should be placed one on each side of the straw-barn, it also follows that the straw-barn, to be out of the way of screening the sun from the courts, should stand north and south, or at right angles to the south of the granaries. 7. The fixing of the straw-barn to the southward of the granaries, and of course to that of the thresh-ing-machine, necessarily fixes the position of the stack-yard to the north of both. Its northern position is highly favorable to the preservation of the corn in the stacks. 8. The relative positions of these apartments are very differently arranged from this in many existing steadings; but I may safely assert, that the greater the deviation from the *principle* inculcated in paragraphs (5) and (6) in the construction of steadings, the less de-sirable they become as habitations for live-stock in winter.

(7.) This leading principle of the construction of a steading which is in-tended to afford shelter to live-stock during winter, is as comprehensive as it is simple, for it is applicable to *every size of steading*. Obviously correct as the principle is, it is seldom reduced to practice, possibly because arch-itects, who profess to supply plans of steadings, must be generally unac-quainted with their practical use. There is one consideration upon which architects bestow by far too much attention—the constructing of steadings at the least possible cost; and, to attain this object by the easiest method, they endeavor to *confine* the various apartments in the least possible space of ground, as if a few square yards of the ground of a farm were of great value. No doubt, the necessity of economy is urged upon them by the grudging spirit of the landlord when he has to disburse the cost, and by the poverty of the tenant when that burden is thrown upon him. Now, economy of construction should be a secondary consideration in compari-son with the proper accommodation which should be afforded to live-stock. Suppose that, by inadequate accommodation, cattle thrive by 10s. a-head less in the course of a winter than they would have done in well construct-ed courts and hammels (and the supposition is by no means extravagant), and suppose that the farmer is prevented realizing this sum on three lots of twenty cattle each of different ages, there would be an *annual loss* to him of £30, from want of proper accommodation. Had the capital sum of which the annual loss of £30 is the yearly interest, been expended in constructing the steading in the best manner, the loss would not only have been averted, but the cattle in much better health and condition to slaugh-ter, or to fatten on grass. Economy is an excellent rule to follow in farm-ing, but it should never be put in practice to the violation of approved principles, or the creation of inconveniences to live-stock, whether in the steading or out of it. I regret to observe both errors too prevalent in the construction of steadings. For example: It is undeniable that as cattle occupy the courts only in winter, when the air, even in the best situations, is at a low temperature, and the day short, they should in such circumstances enjoy as much light and heat from the sun as can be obtained. It is quite practicable to afford them both in courts facing fully to the south, where these influences may be both seen and felt even in winter. Instead of that, cattle-courts are very frequently placed within a quadrangle of buildings, the southern range of which, in the first instance, eclipses the winter's sun of even his diminished influence; and the whole of which, besides, converts the chilling air, which rushes over the corners of its roof into the courts, into a whirlwind of starvation, which, if accompanied with rain or sleet, is sure to engender the most insidious diseases in the cattle. Beware, then, of suffering loss by similar fatal consequences to your cattle; and, to prompt you to be always on your guard, impress the above simple principle of the construction of steadings firmly upon your minds. Rest assured that its violation may prove in the end a much greater loss by preventing the cattle thriving, than the paltry sum saved at first in the outlay of the buildings can possibly ever recompense you for that loss.

(8.) Fig. 1, Plate I. gives an *isometrical view* of an existing steading suitable for the mixed husbandry, somewhat though not on the precise principles which I have inculcated just now, but rather on the usual plan of huddling together the various parts of a steading, with a view of saving some of its original cost.* There are many steadings of this construction to be found in the country, but many more in which stalls for feeding cattle are substituted for hammels. The north range *a a* represents the granaries with their windows, *b* the upper barn, *c c* the arches into the sheds for cattle under the granaries. The projecting building *d* in the middle is the straw-barn, which communicates by a door in each side with the court *e* or *f* for the younger cattle. The projecting building *g*, standing parallel with the straw-barn on the right hand side of the court *f*, is the stable for the work-horses; and the other projecting building *h*, also parallel with the straw-barn on the left-hand side of the court *e*, is the cart-shed. The cow-byres *i*, and hammels *k* for feeding cattle, are seen stretching to the right in a line with the north range *a*, but too far off from the straw-barn *d*: *l* are hammels for a bull and queys: *m*, sheds for shepherds' stores: *n*, stack-yard with stacks: *o*, turnip stores: *p*, piggeries: *q*, calves' court: *r*, implement-house: *s*, boiling-house: *t*, horse-pond: *u*, hen-house: *v*, liquid manure tank: *w*, hay-loft: *x*, out-houses: *y*, slaughter-house: and *z*, hammels for young horses. This is a common disposition of the prin-

[* The reader will, probably, find nothing in "THE BOOK OF THE FARM" which, at first view, may seem more obnoxious than this to the appearance of being on a scale of accommodation and expense unsuited to American farmers and American husbandry. And yet, when he comes to examine the observations of the Author in all their details, they will be found to be replete with practical instruction and directions, which may be heeded with profit in the construction and arrangement of all buildings, on whatever scale, designed for the shelter of domestic animals, the care and distribution of their food, and the preservation of farm vehicles and implements—such buildings as in England are termed the "*steading or farmstead.*"

Some there may be, and doubtless are—such as wealthy merchants on retiring from the cares and vicissitudes of commerce—who unite the means with the desire to have their farmstead as complete as the best architectural design and materials can make it. To all such the plans here presented may serve as models, while they, and others with less means at command, may so modify them as to suit all difference of circumstances—avoiding some portions, and yet seeing much in parts of them that is eligible and in accordance with their own views and means.

But, without desiring to prejudice the judgment of the reader, we may ask him at least to admit, in the costliness of the illustrations connected with this part of the Book of the Farm, some evidence that the Publishers desire not to shun any outlay that may be necessary to make the FARMERS' LIBRARY worthy of public patronage, and fitted to fulfill their own promises.

Ed. Farm. Lib.]

cipal parts of a *modern improved* steading; and a slight inspection of the plate will convince you that in the arrangement of its different apartments is exhibited much of the principle which I have been advocating. Many *modifications* of this particular arrangement may be observed in actual practice:—1. such as the removal of the straw-barn *d* into the north range *a*, and the placing of hammels, such as *k*, into the courts *e* and *f*, and the conversion of one of the sheds *c* into cart-sheds. 2. Another modification encloses a large court divided into two, within a range of buildings forming three sides of a quadrangle, and retaining the north range for the granaries, of a higher altitude than the rest. 3. While another comprises two large courts, each surrounded by three sides of a quadrangle, the range in the middle occupied by the threshing-mill and straw-barn being retained at a higher altitude than the rest. 4. Another completes the quadrangle around one court. 5. While another surrounds a large court, divided into two, with a quadrangle. 6. And the last modification surrounds two separate courts, each with a quadrangle, having a common side. These modifications are made to suit either large or small farms; but they all profess to follow the same plan of arrangement. In truth, however, so varied is the construction of steadings, that, I dare say, no two in the country are exactly alike. Modifications in their construction in obedience to influential circumstances may be justifiable, but still they should all have reference to the *principle* insisted on above.

(9.) Fig. 2, Plate II. contains an engraving of a *ground-plan* of the steading represented by the isometrical view in Plate I. It is unnecessary for me to describe in detail all the component parts of this plan, as the names and sizes of the various apartments are all set down. A short inspection will suffice to make you well enough acquainted with the whole arrangement. This plan has been found, by extensive use, to constitute a commodious, convenient, and comfortable steading for the stock and crop of 500 acres, raised by the mixed husbandry; and those properties it possesses in a superior degree to most similar existing steadings of the same extent in this country, and in a much greater degree than any of the modified plans to which I have just alluded.

(10.) The *steading I would desire to see erected* would be exactly in accordance with the principle I have laid down. I do not know one, nor is there probably in existence one exactly on that principle, but I have seen several, particularly in the north of England, which have impressed me with the belief that there *is* a construction, could it be but discovered, which would afford the most excellent accommodation, the greatest convenience, and the utmost degree of comfort to live-stock; and live-stock being the principal inhabitants of steadings, too much care, in every respect, cannot, in my opinion, be bestowed on the construction of their habitations, so as to insure them in the inclement season the greatest degree of comfort. I shall describe both an isometrical view and ground-plan of a steading of imaginary construction, in strict accordance with the above principle—the principle itself having been brought out by the promptings of experience. I shall minutely describe these plans, in the sanguine hope that the obvious advantages which they exhibit will recommend their construction for adoption to all proprietors and tenants who feel desirous of obtaining a plan of a steading for crop and stock, the arrangements of which have been suggested by matured practical experience. The size of *these particular plans* is not suited to any farm, whereon the mixed husbandry is practiced, of *less* extent than 500 acres; because, in order to illustrate their *principle*, it was necessary to fix on some *definite* size, that the *relative* sizes and positions of the different apartments might be definitively set down; but the whole *arrangement* of the apartments is suited to any size of plan, as the size and number of the apartments may be enlarged or diminished according to the extent of the farm.

(11.) Fig. 3, Plate III. represents an *isometrical view* of such a steading, and keeping the principle upon which it is constructed in mind, you will find that this view illustrates it in every respect that has been stated. 1. A A is the principal or north range of building, of two stories in hight, standing east and west. It contains two granaries, A and A, the upper-barn C, which is also the site of the threshing-machine; the corn-barn being immediately below it is of course invisible, the sheds D D are under the granaries; E is the engine-house, and F the steam-engine furnace-stalk, where the power employed to impel the threshing-machine is steam, G the implement-house entering from the west gable, and H the hay-house, under a granary. These several apartments, while occupying the north range, are greatly serviceable in sheltering the young stock in the large courts I and K from the north wind. 2. Immediately adjoining to the south of the corn-barn, upper-barn, and threshing-machine, is the straw-barn L, standing north and south, contiguously placed for the emission of straw from either side into the courts I and K. 3. It is also conveniently situated for supplying straw to the feeding hammels M, to the right or eastward of the large court K, and equally so for supplying it to those at N, to the left or westward of the large court I. 4. It is accessibly enough placed for supplying straw to the work-horse stable O, and the saddle-horse stable P, to the right or eastward in a line of the principal range A. It is equally accessible to the cow-byre Q, and calves'-cribs R, to the left or westward, in a line of the principal range A. S is the stack-yard, from which the stacks are taken into the upper-barn C, by the gang-way T; U is the boiling-house; V the cart-shed, opposite and near the work-house stable O; W is the wool-room, having a window in the gable, and its stair is from the straw-barn L; XX comprise two small hammels for bulls; Y is the servants' cow-house, in the hammels N; Z is the gig house, adjoining to the riding-horse stable P. *a* are four sties for feedi pigs therein; *b* is a small open court, with a shed for containing *young* pigs after they have just been weaned; *c* are two sties for brood-sows while lying-in. *d* are three apartments for the hatching and rearing of fowls. *e* and *f* are turnip-stores for supplying the hammels M; *g* is the turnip-store for supplying the large court K; *h*, that for small hammels X, and the servants' cow-house Y; *i*, that for the large court I; and *p* and *q* are those for the hammels N. *k* is the open court and shed, with water-trough for the calves; *l* the open court, with water-trough for the cows. *m* is the turnip-shed for the cow-house Q, and calves'-cribs R. *n* is the hay-stack built in the stackyard S, near the hay-house H. *o* and *o* are straw-racks for the center of the large courts I and K. *u* is the ventilator on the roof of the boiling-house U; *r* that on the cow-house Q; *s*, that on the calves'-cribs R; *t* and *w*, those on the roof of the work-horse stable O; and *y*, that on the riding-horse stable P. *x* is the liquid manure-well to which drains converge from the various parts of the farmstead. *z* are feeding-troughs, dispersed in the different courts and hammels. *v* is the open court for the servants' cows. And *f' and f'* are potato stores.

(12.) A very little consideration of the arrangement just now detailed, will suffice to show you that it completely illustrates the principle I have been advocating for the construction of farmsteads. Still, looking at the isometrical view, in fig. 3, Plate III., it will be observed that the threshing-machine C—the machinery for letting loose the straw—is situated in the middle of the great range A, ready to receive the unthreshed crop behind from the stack-yard S, and as ready to deliver the straw threshed into the straw-barn L standing before it. The store of straw in L, being placed exactly in the center of the premises, is easily made available to the large courts I and K and the sheds D and D by its four doors, two on each side. The straw can be carried down the road on the right of the straw-barn L, to the hammels M; and along the farther end of the court K, through the gate at H to both the stables O and P. It can with as much facility be carried across the eastern angle of the large court I, through the gate at the bull's hammels X, to the range of hammels at N, and to the servants' cow-house Y, by its door near the turnip-store *h*. It can also be carried right across the same court I, through the gate behind Y to the cow-house Q, and the calves'-cribs R. The hammels X, the pigs in *a b* and *c*, and the fowls in *d*, can easily be supplied with straw. You may observe in the arrangement of these apartments, that the stables O and P, and the cow-house Q, and the calves'-cribs R, are situated *behind* the hammels M and N, and they are there for these reasons: Hammels for feeding cattle requiring much more straw than stables and byres, according to the foregoing theory, should be placed *near* the straw-barn; and hammels, moreover, being only occupied in winter by stock, should derive, during that season, the fullest advantage that can be given them of the light and heat of the sun. The servants' cow-byre Y being placed nearer the straw-barn than the hammels N, may seem to contravene the principle laid down; but the cow-byre, if desired, may be removed to the other end of the hammels, though in the case where young horses and queys in calf are intended to occupy the small hammels N, it may conveniently remain where it is, as they do not require so much straw as cows. If these hammels are to be destined to the accommodation of *feeding* stock, then the byre ought to be removed to the extreme left of the building. This form of steading is amply *commodious*, for it can accommodate *all* the working and breeding stock, together with four generations of young stock in different stages of growth. A more *convenient* arrangement than this for a farmstead, as I conceive, can scarcely be imagined, and all the parts of it are of such a magnitude as not only to afford ample room for every thing accommodated within it, but with proper fittings up, the arrangement is capable of conferring great *comfort* on its inmates. Its commodiousness will be the more apparent after the ground-plan has been considered in detail.

(13.) Fig. 4, Plate IV. is the *ground-plan* of the steading, of which the preceding plate that has just been described is the isometrical view. The *straw-barn* L is seen at once, running north and south. It is purposely made of the hight of the upper barn to contain a large quantity of straw, as it is often convenient in bad weather to thresh out a considerable quantity of corn, when no other work can be proceeded with, or when high market prices induce farmers to reap advantage from them. There is another good reason for giving ample room to the straw-barn. Every sort of straw is not suited to every purpose, one sort being best suited for litter, and another for fodder. This being the case, it is desirable to have always both kinds in the barn, that the fodder-straw may not be wasted in litter, and the litter-straw given as fodder to the injury of the bestial. Besides, the same sort of straw is not alike acceptable as fodder to every class of animals. Thus wheat-straw is a favorite fodder with horses, as well as oat-straw, while the latter only is acceptable to cattle. Barley-straw is only fit for litter. To give access to litter and fodder straw at the same time, it is necessary to have a door from each kind into each court. Thus four doors, two at each side near the ends, are required in a large straw-barn. Slit-like openings should be made in its side-walls, to admit air and promote ventilation through the straw. A sky-light in the roof at the end nearest the threshing-machine, is useful in giving light to those who take away and store up the straw from the threshing-machine when the doors are shut, which they should be whenever the wind happens to blow too strongly through them into the machine against the straw. Instead of dividing straw-barn doors into two vertical leaves, as is usually done, they should be divided horizontally into an upper and a lower leaf so that the lower may always be kept shut against intruders, such as pigs. while the upper admits both light and air into the barn. One of the doors at each end should be furnished with a good stock-lock and key and thumb-latch, and the other two fastened with a wooden hand-bar from the inside. The floor of the straw-barn is seldom or never flagged or causewayed, though it is desirable it should be. If it were not so expensive, the asphaltum pavement would make an excellent floor for a straw-barn. Whatever substance is employed for the purpose, the floor should be made sc firm and dry as to prevent the earth rising and the straw moulding. Mouldy straw at the bottom of a heap superinduces throughout the upper mass a disagreeable odor, and imparts a taste repugnant to every animal. That portion of the floor upon which the straw first alights on sliding down the straw-screen of the threshing-machine, should be strongly boarded to resist the action of the forks when removing the straw. Blocks of hard-wood, such as the stools of hard-wood trees, set on end, causewaywise, and sunk into the earth, form a very durable flooring for this purpose. Stone flagging in this place destroys the prongs of the pitchforks. The straw-barn should communicate with the chaff-house by a shutting door, to enable those who take away the straw to see whether the chaff accumulates too high against the end of the winnowing-machine. The communicacation to the wool-room in this plan is by the straw-barn, by means of the stair c', made either of wood or stone. The straw-barn is represented 72 feet in length, 18 feet in breadth, and 15 feet in hight to the top of the side walls.

(14.) C is the *corn-barn*. Its roof is formed of the floor of the upper barn, and its hight is generally made too low. The higher the roof is the more easily will the corn descend to be cleaned from the threshing-machine down the hopper to the winnowing-machine. Nine feet is the least hight it should be in any instance. The plan gives the size of the corn-barn as 31 feet by 18 feet, but taking off 5 feet for partitioning off the machinery of the threshing-mill, as at *s*, the extent of the workable part of the barn-floor will be 26 feet by 18 feet. In that space I have seen much barn-work done, but it could be made more by diminishing the size of the shed D of the court K. The corn-barn should have in it at least two glazed windows to admit plenty of light in the short days of winter, and they should be guarded outside with iron stanchions. If one window cannot be got to the south, the door when open will answer for the admission of sunshine to keep the apartment comfortably dry for the work-people and the grain. The door is generally divided into upper and lower halves, which, as usually placed, are always in the way when the winnowing-maahine is used at the door. A more convenient method is to have the door in a whole piece, and when opened, to fold back into a recess in the outer wall, over the top of which a plinth might project to throw off the rain. In this case the ribets and lintel must be giblet-checked as deep as the thickness of the door, into which it should close flush, and be fastened with a good lock and key, and provided with a thumb-latch. The object of making the corn-barn door of this form is to avoid the inconvenience of its opening into the barn, where, unless it folds wholly back on a wall, is frequently in the way of work, particularly when winnowing roughs, and taking out sacks of corn on men's backs. As to size, it should not be less in the opening than 7½ feet in hight and 3½ feet in width. A light half-door can be hooked on, when work is going on, to prevent the intrusion of animals, and the wind sweeping along the floor. The floor of the corn-barn is frequently made of clay, or of a composition of ashes and lime; the asphaltic composition would be better than either; but in every instance it should be made of wood—of sound, hard red-wood Drahm battens, plowed and feathered, and fastened down to stout joists with Scotch flooring sprigs driven through the feather-edge. A wooden floor is the only one that can be depended on being constantly dry in a corn-barn; and in a barn for the use of corn, a dry floor is indispensable. It has been suggested to me that a stone pavement, square-jointed, and laid on a bed of lime over 9 inches of broken stones; or an asphaltum pavement, laid on a body of 6 inches of broken stones, covered with a bed of grout on the top of the stones, would make as dry and more durable barn-floor than wood, and which will not rot. I am aware that stone or asphaltum pavement is durable, and not liable to rot; but there are objections to both, in a corn-barn, of a practical nature, and it is certain that the best stone pavement is not proof against the undermining powers of the brown rat; while a wooden floor is durable enough, and certainly will not rot, if kept dry in the manner I shall recommend over the page. The objections to all stony pavements as a barn-floor are, that the scoops for shoveling the corn pass very harshly over them—that the iron nails in the shoes of the work-people wear them down, and raise a dust upon them—and that they are hurtful to the bare hands and lighter implements, when used in taking up the corn from the floor. For true comfort in all these respects in a barn-floor, there is nothing like wood. The walls of this barn should be made smooth with hair-plaster, and the joists and flooring forming its roof cleaned with the plane, as dust adheres much more readily to a rough than to a smooth surface. The stairs to the granaries *s* and *s* should enter from the corn-barn, and a stout plain-deal door with lock and key placed at the bottom of each. And at the side of one of the stairs may be inclosed on the floor of the barn a space, *t*, to contain light corn to be given to the fowls and pigs in summer when this sort of food is scarce about the steading.

(15.) As the method of hanging doors on a giblet-check should be adopted in all cases in steadings where doors on outside walls are likely to meet with obstructions on opening inward, or themselves becoming obstructive to things passing outward, the subject deserves a separate notice. In fig. 5, *a* is a strong door, mounted on crooks and bands, fully open, and thrown back into the recess of the wall *b*; the projecting part of the lintel *c* protecting it effectually from the rain; *d* is the giblet-check in the lintel, and *e* that in the ribets, into which the door shuts flush; *f* is the light movable door used when work is going on in the corn-barn.

Fig. 5.

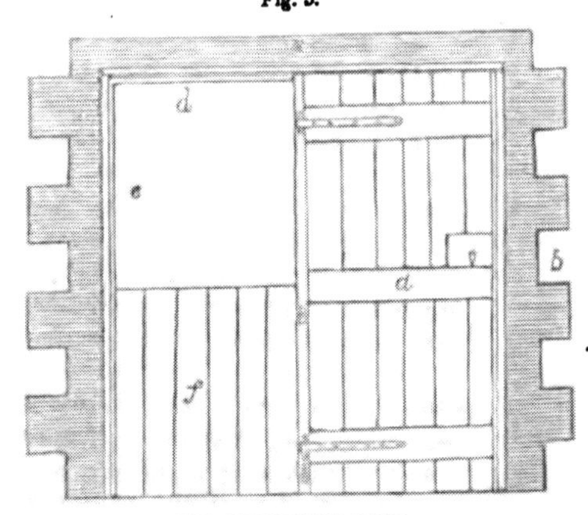

THE CORN-BARN DOOR.

(16.) The wooden floor of the corn-barn is liable to decay unless precautions are used to prevent it, but a much too common cause of its destruction is vermin—such as rats and mice. It is discreditable to farmers to permit this floor to remain in a state of decay for any length of time, when an effectual preventive remedy is within their reach; and the more certainly preventive that remedy is, the more it should be appreciated. I used a most effectual method of preventing the destructive ravages of either vermin or damp, by supporting the floor in the particular manner represented in fig. 6. The earth, in the first instance, is dug out of the barn to the depth of the foundations of the walls, which should be two feet below the door soles; and, in the case of a new steading, this can be done when the foundations of the walls are taken out. The ground is then spread over with

Fig. 6.

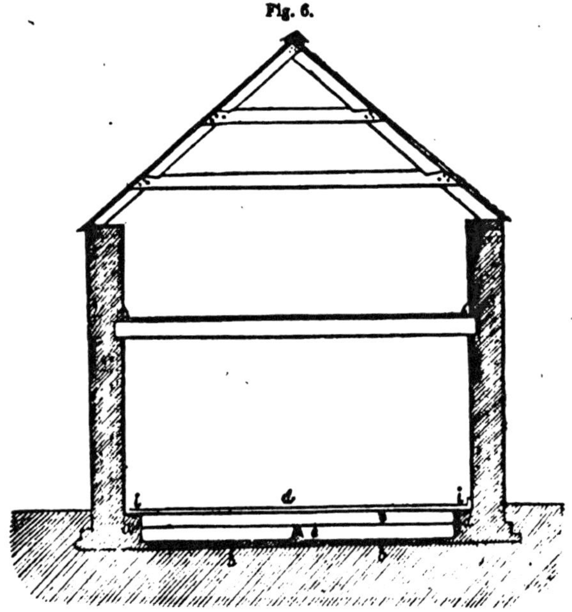

SECTION OF THE CORN BARN FLOOR.

a layer of sand, sufficient to preserve steadiness in the stout rough flags *b b*, which are laid upon it and jointed in strong mortar. Twelve-inch thick sleeper walls *a a*, of stone and lime, are then built on the flags, to serve the purpose of supporting each end of the joists of the floor. The joists *c*, formed of 10 by 2½ inch plank, are then laid down 16 inches apart, and the spaces between them filled up to the top with stone and lime. The building between the joists requires to be done in a peculiar way. It should be done with squared rubble stones, and on no account should the mortar come in contact with the joists, as there is nothing destroys timber, by superinducing the dry rot, more readily than the action of mortar upon it.—For this reason great care should be observed in building in the joists into the walls—in placing the safe-lintels over the doors and windows, the stones being dry-bedded over them—and in beam-filling between the couple-legs. The floor *d* is then properly laid on a level with the door-sole, and finished with a neat skifting board *i i* round the walls of the barn. By this contrivance the vermin cannot possibly reach the floor but from the flags, which are nearly 2 feet under it. A hewn stone pillar *e*, or even two, are placed on the flags under each joist to support and strengthen the floor. This construction of floor admits of abundance of air above and below to preserve it, and affords plenty of room under it for cats and dogs to hunt after the vermin. This figure also gives a section of the building above the corn-barn, including the floor of the upper barn, the outside walls, and the coupling, slating, and ridging of the roof of the middle range of building.

(17.) The *chaff-house*, *r*, stands between the corn and straw barns. It is separated from the former by a wooden partition, and from the latter by a stone-wall. Its hight is the same as that of the corn-barn, the floor of the upper barn forming a roof common to both. It is 18 feet in length and 14 feet in width. It contains the winnowing-machine or fanners of the thresh-ing-machine, from which it receives the chaff. It has a thin door with a thumb-latch into the straw-barn, for a convenient access to adjust any of the gearing of the fanners; as also a boarded window hung on crooks and bands, fastened in the inside with a wooden hand-bar, and looking into the large court K; but its principal door, through which the chaff is emptied, opens outward into the large court I. This door should be giblet-checked, and fastened from the inside with a wooden hand-bar. The space between the head of the fanners and the wall should be so boarded up as not to interfere with the action of the fanner-belts, but merely prevent the chaff being scattered among the machinery, and any access by persons being effected by the machinery into the upper barn.

(18.) D D are two *sheds* for sheltering the cattle occupying the courts I and K from rain and cold, by night or day, when they may choose to take refuge in them. The shed of the court I is 52 feet in length by 18 feet in width, being a little longer than that of the court K, which is 47 feet in length and 18 feet in width, and their hight is 9 feet to the floor of the granaries, which forms their roof. The access to these sheds from the courts is by arched openings of 9 feet in width, and 7½ feet in hight to the top of the arch. There should be a rack fastened against one of the walls of each shed to supply fodder to the cattle under shelter in bad weather, as at *h'*. As when a large number of cattle are confined together, of whatever age, some will endeavor to obtain the mastery over the others, and to prevent accidents in cases of actual collision, it has been recommended to have two openings to each shed, to afford a ready means of egress to the fugitives; and, as a farther safety to the bones and skins of the unhappy victims, the angles of the hewn pillars which support the arches should be chamfered. In my opinion, the precaution of two openings for the reason given is unnecessary, inasmuch as cattle, and especially those which have been brought up together, soon become familiarized to each other; and two openings cause draughts of air through the shed. If holes were made in the faces of the pillars opposite to each other in the openings, so as bars of wood could be put across them, the cattle could at any time be kept confined within the sheds. This might at times be necessary, especially when the courts are clearing out of the manure. The shed of the court K has a door *d'* in the back wall for a passage to the work-people when going from the corn to the upper barn, by the gangway T.

(19.) E is the engine-house for the steam-engine, when one is used. It is 18 feet in length and 8 feet in width, and the granary-floor above forms its roof. It has a window looking into the large court I, and a door into the boiler and furnace-house F, which house is 24 feet in length and 8 feet in width, and has an arched opening at the left or west end. The chimney-stalk is 6 feet square at the base, and rises tapering to a hight of 45 feet. If wind or horses are preferred as the moving power, the windmill-tower or horse-course would be erected on the site of F.

(20.) G is the *implement-house* for keeping together the smaller implements when not in use, when they are apt to be thrown aside and lost.—The intrinsic value of each implement being small, there is too generally less care bestowed on them than on those of more pecuniary value; but in use each of them is really as valuable as the most costly, and even their cost in the aggregate is considerable. The implement-house is 18 feet in length by 14 feet in width, and its roof is formed of the granary-floor.—This house should be provided with a stout plain-deal door with a good lock and key, the care of which should only be entrusted to the farm-steward. It should also have a partly glazed window like that of the cow-house, as sometimes this apartment may be converted into a convenient work-shop for particular purposes. The floor should be flagged, or laid with asphaltum pavement. Besides the implements, this apartment may contain the barrel of tar, a useful ingredient on farms where sheep are reared, and where cart-naves require greasing; the grindstone, a convenient instrument on a farm on many occasions for sharpening edge-tools, such as scythes, axes, hay-knife, dung-spade, &c. A number of wooden pins and iron spikes, driven into the walls, will be found useful for suspending many of the smaller articles upon. The walls should be plastered.

(21.) H is the *hay-house* at the east end of the north range A, and corresponding in situation to the implement-house. It is 18 feet in length, 17 feet in width, and its roof is also formed of the floor of the granary above. Its floor should be flagged with a considerable quantity of sand to keep it dry, or with asphaltum. It should have a giblet-checked door to open outward, with a hand-bar to fasten it by in the inside; it should also have a partly glazed window, with shutters, to afford light when taking out the hay to the horses, and air to keep it sweet. As the hay-house communicates immediately with the work-horse stable O by a door, it can find room for the work-horse corn-chest *y*, which may be there conveniently supplied with corn from the granary above by means of a spout let into the fixed part of the lid. For facilitating the taking out of the corn, the *end* of the chest should be placed against the wall at the side of the door which opens into the stable, and its back part should be boarded up with thin deals to the granary-floor, to prevent the hay coming upon the chest. Its walls should be plastered. This hay-house is conveniently situated for the hay-stack *n* in the stack-yard S.

(22.) The *form* of the *corn-chest*, *y*, is more convenient and takes up less room on the floor, when high and narrow, than when low and broad.—When of a high form, a part of the front should fold down with hinges, to give easier access to the corn as it gets low in the chest. Part of the lid should be made fast, to receive the corn-spout from the granary, and to lighten its movable part, which should be fastened with a hasp and padlock, and the key of which should be constantly in the custody of the farm-steward, or of the man who gives out the corn to the plowmen, where no farm-steward is kept. A fourth part of a peck-measure is always kept in the chest, for measuring out the corn to the horses. You must not imagine that, because the spout supplies corn from the granary when required, it supplies it without measure. The corn appropriated for the horses is previously measured off on the granary-floor, in any convenient quantity, and then shoveled down the spout at times to fill the chest; besides, lines can be marked on the inside of the chest indicative of every quarter of corn which it can contain.

(23.) O is the *stable* for the *work-horses*. Its length, of course, depends on the number of horses employed on the farm; but in *no instance should its width be less* than 18 feet, for comfort to the horses themselves, and convenience to the men who take charge of them. This plan, being intended for a definite size of farm, contains stalls for 12 horses, and a loose box besides—the whole length being 84 feet. Few stables for work-horses are made wider than 16 feet, and hence few are otherwise than hampered for want of room. A glance at the particulars which should be accommodated in the width of a work-horse stable will show you at once the inconvenience of this narrow breadth. The entire length of a work-horse is seldom less than 8 feet; the extreme width of the hay-rack is about 2 feet; the harness, hanging loosely against the wall, occupies about 2 feet; and the gutter occupies 1 foot: so that in a width of 16 feet there are only 3 feet left from the heels of the horses to the harness, on which to pass backward and forward to wheel a barrow and use the shovel and broom. No wonder, when so little space is left to work in, that cleanliness is so much neglected in farm-stables, and that much of the dung and urine are left to be decomposed and dissipated by heat in the shape of ammoniacal gas, to the probable injury of the breathing and eye-sight of the horses, when shut up at night. And, what aggravates the evil, there seldom is a ventilator in the roof; and, what is still worse, the contents of the stable are much contracted by the placing of a hay-loft immediately above the horses' heads.—Whatever may be the condition of a work-horse stable in reference to size and room, its walls should always be plastered with good haired plaster, as forming the most comfortable finishing, and being that most easily kept clean.....Some people imagine that twelve horses are too great a number to be in one stable, and that two stables of six stalls each would be better. Provided the stable is properly ventilated, there can no injury accrue to a larger than to a smaller number of horses in a stable; and, besides, there are practical inconveniences in having two work-horse stables on a farm. The inconveniences are that neither the farmer nor farm-steward can personally superintend the grooming of horses in two stables; that the orders given to the plowmen by the steward must be repeated in both stables; and that either all the plowmen must be collected in one of the stables to receive their orders, or, part of them not hearing the orders given to the rest, there cannot be that common understanding as to the work to be done which should exist among all classes of work-people on a farm.

(24.) Another particular in which most work-horse stables are improperly fitted up, is the narrowness of the *stalls*, 5 feet 3 inches being the largest space allowed for an ordinary sized work-horse. A narrow stall is not only injurious to the horse himself, by keeping him peremptorily confined to one position, in which he has no liberty to bite or scratch himself, should he feel so inclined, but materially obstructs the plowman in the grooming process, and while supplying the horse with food. No work-horse, in my opinion, should have a narrower stall than 6 feet from center to center of the travis, in order that he may stand at ease, or lie down at pleasure with comfort. If "the laborer is worthy of his hire," the work-horse is deserving of a stall that will afford him sound rest.

(25.) It is a disputed point of what form the *hay-racks* in a work-horse stable should be. The prevailing opinion may be learned from the general practice, which is to place them as high as the horses' heads, because, as it is alleged, the horse is thereby obliged to hold up his head, and he cannot then breathe *upon* his food. Many more cogent reasons, as I conceive, may be adduced for placing the racks low down. In the first place, a work-horse does not require to hold his head up at any time, and much less in the stable, where he should enjoy all the rest he can get. 2. A low rack permits the position of his neck and head, in the act of eating, to be more like the way he usually holds them, than when holding them up to a high one. 3. He is not nearly so liable to pull out the hay among his feet from a low as from a high rack. 4. His breath cannot contaminate his food *more* in a low than in a high rack, because the greatest proportion of the breath naturally ascends; though breathing is employed by the horse to a certain degree in choosing his food by the sense of smell. 5. He is less fatigued eating out of a low than from a high rack, every mouth-full having to be pulled out of the latter, from its sloping position, by the side of the mouth turned upward. 6. Mown-grass is much more easily eaten out of a low than a high rack. 7. And lastly, I have heard of peas falling out of their straw, when eaten out of a high rack, into the ears of the horse, and therein setting up a serious degree of inflammation.

(26.) The *front rail* of the *low-rack* should be made of strong hard-wood, in case the horse should at any time playfully put his foot on it, or bite it when groomed. The front of the rack should be sparred for the admission of fresh air among the food, and incline inward at the lower end, to be out of the way of the horses' fore-feet. The bottom should also be sparred, and raised about 6 inches above the floor, for the removal of hay seeds that may have passed through the spars. The *corn-trough* should be placed at the near end of the rack, for the greater convenience of supplying the corn. A spar of wood should be fixed across the rack from the front rail to the back wall, midway between the travis and the corn-trough, to prevent the horse tossing out the fodder with the side of his mouth, which he will sometimes be inclined to do when not hungry. The *ring* through which the stall collar-shank passes, is fastened by a staple to the hard-wood front rail. I have lately seen the manger in some work-horse stables in steadings recently erected made of stone, on the alleged score of being more easily cleaned than wood after the horses have got prepared food. From my own observation in the matter, I do not think wood more difficult of being cleaned than stone at any time, and especially if cleaned in a proper time after being used—daily, for instance. As plowmen are proverbially careless, the stone-manger has perhaps been substituted on the supposition that it will bear much harder usage than wood; or perhaps the landlords, in the several instances in which stone-mangers have been erected, could obtain stone cheaper from their own quarries than good timber from abroad: but either of these reasons are poor excuses for the carelessness of servants on the one hand, or the parsimony of landlords on the other, when the well-being of the farmers' most useful animals is in consideration; for, besides the clumsy appearance of stone in such a situation, and its comfortless feel and aspect, it is injurious to the horses' teeth when they seize it suddenly in grooming, and it is impossible to prevent even some work-horses biting any object when groomed; and I should suppose that stone would also prove hurtful to their lips when gathering their food at the bottom of the manger. I have no doubt that the use of stone-mangers will have a greater effect in grinding down the teeth of farm-horses, than the "tooth of old Time" itself.

(27.) The *hind posts* of *travises* should be of solid wood rounded in front, grooved in the back as far as the travis boards reach, sunk at the lower ends into stone blocks, and fastened at the upper ends to battens stretching across the stable from the ends of the couple legs, where there is no hay-loft, and from the joists of the flooring where there is. The *head-posts* are divided into two parts, which clasp the travis boards between them, and are kept together with screw-bolts and nuts. Their lower ends are also sunk into stone blocks, and their upper fastened to the battens or joists. The *travis boards* are put endways into the groove in the hind-post, and pass between the two divisions of the head-post to the wall before the horses' heads; and are there raised so high as to prevent the contiguous horses troubling each other.

(28.) The *floor* of all stables should be made hard, to resist the action of the horses' feet. That of a work-horse stable is usually causewayed with small round stones, embedded in sand, such as are to be found on the land or on the sea-beach. This is a cheap mode of paving. When these cannot be found, squared blocks of whinstone (trap rock, such as basalt, greenstone, &c.) answer the purpose fully better. Flags make a smoother pavement for the feet than either of these materials, and they undoubtedly make the cleanest floor, as the small stones are very apt to retain the dung and absorb the urine around them, which, on decomposition, cause filth and constant annoyance to horses. To avoid this inconvenience in a great degree, it would be advisable to form the gutter behind the horses' heels of hewn freestone, containing an entire channel, along which the urine would flow easily, and every filth be completely swept away with the broom. The channel should have a fall of at least 1½ inch to the 10 feet of length. The paving on both sides should incline toward this gutter, the rise in the stalls being 3 inches in all. In some stables, such as those of the cavalry and of carriers, the floor of the stalls rise much higher than 3 inches, and on the Continent, particularly in Holland, I have observed the rise to be still more than in any stables in this country. Some veterinary writers say that the position of the feet of the horse imposed by the rise, does not throw any unnecessary strain on the back tendons of the hind-legs.* This may be, but it cannot be denied that in this position the toes are raised above the heels much higher than on level ground. I admit that a rise of three inches is necessary in stalls in which geldings stand, as they throw their water pretty far on the litter; but in the case of mares, even this rise is quite unnecessary. It is indisputable that a horse always prefers to stand on level ground, when he is free to choose the ground for himself in a grass-field, and much more ought he to have level ground to stand on in a stable, which is his place of *rest*.

(29.) Fig. 7 gives a view of the particulars of such a *stall for work-horses* as I have described. *a a* are the strong hind-posts; *b b* the head-posts, both sunk into the blocks *c c c c*, and fastened to the battens *d d*, stretching across the stable from the wall *e* to the opposite wall; *f f* the travis-boards let into the posts *a a* by grooves, and passing between the two divisions of the posts *b b*; the boards are represented high enough to prevent the horses annoying each other; *g g* curb-stones set up between the hind and fore posts *a* and *b*, to receive the side of the travis-boards in grooves, and thereby secure them from decay by keeping them beyond the action of the litter; *h* is the sparred bottom of the hay-rack, the upper rail of which holds the ring *i* for the stall collar-shank; *k* the corn-manger or trough; *l* the bar across the rack, to prevent the horse tossing out the fodder; *m* the pavement within the stall; *n* is the freestone gutter for conveying away the urine to one end of the stable; *o* the pavement of the passage behind the horses' heels; *p* are two parallel spars fastened over and across the battens, when there is no hay-loft, to support trusses of straw or hay, to be given as fodder to the horses in the evenings of winter, to save the risk of fire in going at night to the straw-barn or hay-house with a light.

(30.) The *harness* should all be hung against the wall behind the horses, and none on the posts of the stalls, against which it is too frequently placed to its great injury, in being constantly kept in a damp state by the

* Stewart's Stable Economy, p. 17.

horses' breath and perspiration, and apt to be knocked down among the horses' feet. A good way is to suspend harness on stout hard-wood pins driven into a strong narrow board, fastened to the wall with iron hold-fasts; but perhaps the most substantial way is to build the pins into the wall. The harness belonging to each pair of horses should just cover a space of wall equal to the breadth of the two stalls which they occupy, and when windows and doors intervene, and which, of course, must be left free, its arrangement requires some consideration. This mode of arrangement I have found convenient. A spar of hard-wood nailed firmly across the upper edge of the batten *d*, fig. 7, that supports both the posts

Fig. 7.

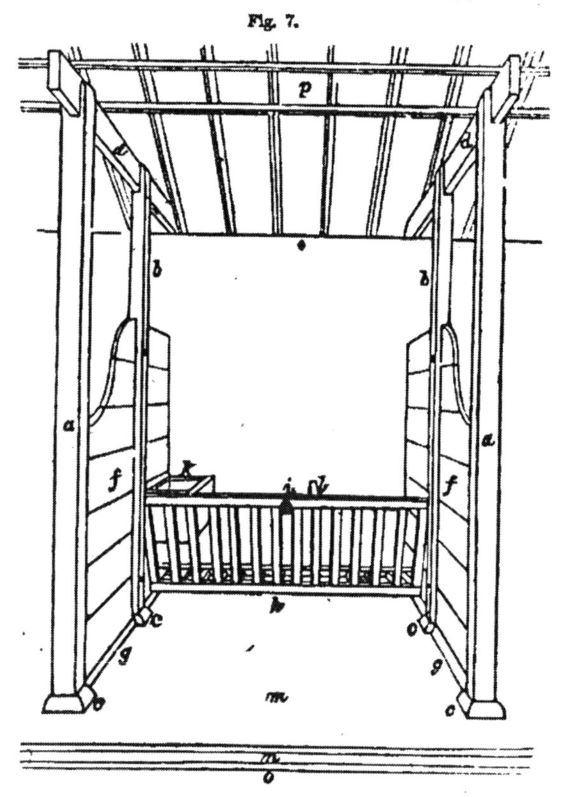

STALL FOR WORK-HORSE STABLE.

of a stall, will suspend a collar on each end, high enough above a person's head, immediately over the passage. One pin is sufficient for each of the cart-saddles; one will support both the bridles, while a fourth will suffice for the plow, and a fifth for the trace-harness. Thus 5 pins or 6 spaces will be required for each pair of stalls, and in a stable of 12 stalls, deducting a space of 13 feet for 2 doors and 2 windows in such a stable, there will still be left, according to this arrangement, a space for the harness of about 18 inches between the pins. Iron hooks driven into the board betwixt the pins will keep the cart-ropes and plow-reins by themselves. The curry-comb, hair-brush, and foot-picker may be conveniently enough hung up on the hind-post betwixt the pair of horses to which they belong, and the mane-comb is usually carried in the plowman's pocket.

(31.) Each horse should be bound to his stall with a *leather stall-collar*, having an iron-chain collar-shank to play through the ring *i* of the hay-rack, fig. 7, with a turned wooden sinker at its end, to weigh it to the ground. Iron-chains make the strongest stall-collar-shanks, though certainly noisy when in use; yet work-horses are not to be trusted with the best hempen cords, which often become affected with dry-rot, and are, at all events, soon apt to wear out in running through the smoothest stall-rings. A simple stall-collar, with a nose-band, and strap over the head, is sufficient to secure most horses; but as some have a trick of slipping the strap over their ears, it is necessary to have either a throat-lash in addition or a belt round the neck. Others are apt, when scratching their neck with the hind-foot, to pass the fetlock joint over the stall-collar-shank, and finding themselves thus entangled, to throw themselves down in the stalls, bound neck and heel, there to remain unreleased until the morning, when the men come to the stable. By this accident, I have seen horses get injured in the head and leg for some time. A short stall-collar-shank is the only preventive against such an accident, and the low rack admits of its being constantly used.

(32.) The *roof* of a *work*-stable should always be open to the slates, and not only that, but have openings in its ridge, protected from the weather by a particular kind of wood-work, called a ventilator. Such a thing as a ventilator is absolutely necessary on the roof of a work-horse stable. It is distressing to the feelings to inhale the air in some farm stables at night, particularly in old steadings economically fitted up. It is not only warm from confinement, moist from the evaporation of perspiration, and stifling from sudorific odors, but cutting to the breath, and pungent to the eyes, from the decomposition of dung and urine by the heat. The windows are seldom opened, and many can scarcely be opened by disuse. The roof in fact is suspended like an extinguisher over the half stifled horses. But the evil is still farther aggravated by a hay-loft, the floor of which is extended over and within a foot or less of the horses' heads. Besides the horses being thus inconvenienced by the hay-loft, the hay in it, through this nightly roasting and fumigation, soon becomes dry and brittle, and contracts a disagreeable odor. The only remedy for these inconveniences in work-horse stables is the establishment of a complete ventilation through them.

(33.) Fig. 8. represents one of these *ventilators*, in which the Venetian blinds *a* are fixed, and answer the double purpose of permitting the escape of heated air and effluvia, and of preventing the entrance of rain or snow. The blinds are covered and protected by the roof *b*, made of slates and lead; *c* is an apron of lead. Such a ventilator would be more ornamental to the steading than fig. 8, and more protective to the blinds, if its roof projected 12 inches over. One ventilator of the size of 6 feet in length, 3 feet in hight in front, and 2 feet above the ridging of the roof,

Fig. 8.

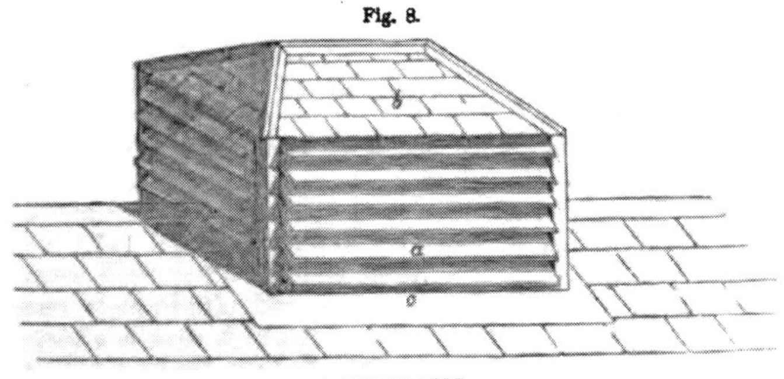

A VENTILATOR.

for every four horses in a work-horse stable, may perhaps suffice to maintain a complete ventilation. But openings in the roof will not of themselves constitute ventilation, unless there be an adequate supply of fresh air from below, to enforce a current; and this supply should be obtained from openings in the walls, including the chinks of doors and windows when shut, whose gross areas should be equal to those of the ventilators. The openings should be formed in such situations, and in such numbers, as to cause no draught of cold air to be directed against the horses. They might conveniently be placed, protected by gratings of iron on the outside to prevent the entrance of vermin, in the wall immediately behind the harness, through which the air would pass, and cross the passage toward the horses; and the air on thus entering the stable should be made to deflect to both sides of each opening, by striking against a plate of iron placed before the opening, at a short distance from the wall. I observe other forms of ventilators in use in steadings, one consisting of large lead pipes projected through the roof, with the ends turned down; and another having a portion of the slating or tiling raised up a little, and there held open. Either of these plans is much better than no ventilation at all, and I dare say either mode may be cheaply constructed; but neither is so effectual for the purposes of ventilation as the one I have figured and described.

(34.) Besides the ordinary stalls, a *loose-box*, *u*, will be found a useful adjunct to a work-horse stable. A space equal to two stalls should be railed off at one end of the stable, as represented at *u* on the plan, fig. 4, Plate IV. It is a convenient place into which to put a work-mare when expected to foal. Some mares indicate so very faint symptoms of foaling that they frequently are known to drop their foals under night in the stable, to the great risk of the foal's life, where requisite attention is not directed to the state of the mare, or where there is no spare apartment to put her into. It is also suitable for a young stallion, when first taken up and preparing for traveling the road; as also for any young draught-horse taken up to be broke for work, until he become accustomed to a stable. It might be, when unfortunately so required, converted into a convenient hospital for a horse, which, when seized with an unknown complaint, might be confined in it, until it is ascertained whether the disease is infectious, and then he should be removed to an out-house. Some people object to having a loose-box in the stable, and would rather have it out of it; but the social disposition of the horse renders one useful there on the occasions just mentioned. It is, besides, an excellent place in which to rest a fatigued horse for a few days.

(35.) Adjoining to this I have placed the *stable* for *riding-horses*, as at P on the plan, fig. 4, Plate IV, not that those stables should always be together, for the riding-horse stable can be placed at any convenient part of the farmstead or near the farm-house. It may be fitted up in the form of three stalls of 6 feet each, or two loose-boxes of 9 feet each, according to inclination, that is, a size of 18 feet square will afford ample room for all the riding-horses a farmer will require. The high rack is always put up in riding-horse stables, to oblige the horse to keep up his head, and maintain a lofty carriage with it. The long manger, stretching from one travis to another, is frequently used where the high rack is approved of. But the neatest mode of fitting up the stall of a riding-horse stable is with a hay-crib in one of the corners, and a corn-box in the other, both being placed at convenient hights from the ground. The stall-posts in riding-horse stables are fastened into the ground in a body of masonry, and not to the roof, as that should be made as lofty as the hight of the balks of the couples will admit, and it should be lathed, and all the walls plastered, for the sake of appearance, warmth, and cleanliness. The corn-chest may be placed either in the recess of the window, where its lid might form a sort of table, or in a corner. One door and a window are quite enough for light and entrance. The door should open outward on giblet-checked rnoets, and provided with a good lock and key, and spring-latch with a handle, so as not to catch the harness. The hight and width of both riding and work-horse stable-doors are usually made too low and too narrow for the easy passage of ordinary sized horses in harness; 7½ feet by 3½ feet are of the least dimensions they should ever be made. A ventilator is as requisite in a riding as a work-horse stable, and, to promote ventilation, the under part of the window should be provided with shutters to open. The neatest floor is of droved flags; though I have seen in stables for riding-horses very beautiful floors of Dutch clinkers.

(36.) The lowest part of a *high rack* is usually placed about the hight of a horse's back, in contact with the wall, and the upper part projecting about 2 feet from it. This position is objectionable, inasmuch as the angle of inclination of the front with the wall is so obtuse as to oblige the horse to turn up the side of his mouth before he can draw a mouthfull of provender out of it, though the front be sparred at such a width as to permit hay and grass to pass easily through. A better plan is to have the front nearly parallel with the wall, and the bottom sparred to admit the falling out of dust and seeds.

(37.) The *long manger*, which is always used with the high rack, is chiefly useful in permitting the corn to be thinly spread out, and making it more difficult to be gathered by the lips of the horse, and on that account considered an advantageous form of manger for horses that are in the habit of bolting their corn. I doubt whether horses really masticate their corn more effectually when it is spread out thin, though no doubt they are obliged to take longer time in gathering and swallowing it, when in that state.

(38.) The *hay crib* fixed up in one corner of the stall, usually the far one, is not large enough to contain fodder for a work-horse, though amply so for a riding-horse. A work-horse will eat a stone of hay of 22 lbs. every day, which, when even much compressed, occupies about a cubic foot of space. To make a quadrifid hay-crib contain this bulk, would require the hay to be hard pressed down, to the great annoyance of the horse, and the danger of much waste by constant pulling out. Plowmen require no encouragement by small racks to press fodder hard into racks. This they usually do, with the intention of giving plenty of it to their horses; but were racks generally made capacious enough, they would have less inducement to follow a practice which never fails to be attended with waste of provender. Such hay-cribs are usually made of iron.*

[* Not so yet in the United States, although iron is being more and more substituted for wood, for various purposes, and would be for many more, if iron-masters were farmers, or, *vice versa*, to make both more familiar with the numerous purposes to which it might be economically applied.

Among those who will read this, there are yet doubtless some who well remember when rope traces and wooden mould-boards and hay-forks were used almost exclusively. Iron might be employed to advantage for a great variety of new uses. It needs for this and other improvements that Americans be brought, as they will be by degrees, to disburden themselves of the party demagogues by whom they are ridden, and learn, instead of being absorbed by party politics, to turn their attention and studies to their own true and peculiar interests.

Few things serve better to distinguish the habits and even the character of the progeny from the parent stock—the Americans from their English ancestors—than the more perfect finish and durability of all *their* mechanical works, machinery and buildings.

(39.) With regard to the *relative advantages* of *stalls* and *loose-boxes* in riding-stables, there is no doubt that, for personal liberty and comfort to the horse, the latter are much to be preferred, as in them he can stand, lie down, and stretch him out in any way he pleases; but they require more litter and a great deal of attention from the groom to keep the skin of the horse clean, and preserve the horse-clothes from being torn—considerations of some importance to a farmer who has little use for a regularly-bred groom to attend constantly on his riding-horse; unless he be a sportsman.*

(40.) The *floor* of the riding-horse stable may be paved either with small stones, and a gutter of freestone to carry off urine, like the work-horse stable, or, what is better, with jointed flags; but the neatest form of floorings is of jointed *droved* flags, grooved across the passage from the door to the stalls, to prevent the slipping of the horses' feet. This plan has also the advantage of being the cleanest as well as the neatest, but it is obviously more suited to the stables of the landlord than the tenant.

(41.) If you use a wheeled vehicle of any kind, the *coach-house* should adjoin the riding-horse stable. Of 18 feet square in size it will contain two light-wheeled carriages, and afford ample room besides for other purposes, such as the cleaning of harness, &c. As the utmost precautions of ventilation and cleanliness cannot prevent deposition of dust in a riding-horse stable, the harness should be placed beyond its reach in the coach-house, where it should be hung upon pins against a boarded wall. To keep it and the carriages dry in winter, there should be a large *fire-place* in the coach-house. The floor should be flagged, and the roof and walls lathed and hair-plastered. A door should open from the riding-horse stable, provided with lock and key, and the large coach-house door should open outward on a giblet-check, and be fastened with bolts and a bar in the inside. Z in plan fig. 4, Plate IV. is the coach-house, with the large fire-place *i* in it. Coach-houses having to be kept dry in winter, to prevent the moulding of the leather-work, are frequently kept so by *stoves*, which, when not in use in summer, become rusted and out of working order; and when again lighted in that state, never fail to smoke and soil every thing with soot.†

There things are made to endure; here they are made to answer the purposes of the day There railroads often cost *one hundred and fifty thousand dollars a mile!* but when they are done, *they are done!* On the other hand, Americans beat the world in ingenuity and in readiness to imitate and improve. Short apprenticeships, slighted and imperfect structures, unseasoned and perishable materials hastily put together, and even the restless and roaming temper of our population, may be regarded as the natural growth of our freer and looser form of government, and we must take the bitter with the sweets. *Ed. Farm. Lib.*]

[* "Horse cloths" or coverings are not generally used or needed on American farms. Horses are in no danger from *cold*. Stables should be always dry, but well ventilated, and care should always be taken, when horses are *heated*, not to leave them *at rest*, in a cold wind or current of air uncovered. No man of any consideration or mercy would do it, or allow it to be done.

Ed. Farm. Lib.]

[† This suggestion of a fire-place in a coach-house may be regarded as another English refinement, and like many things in this book, which must be published (from their inseparable connection with others that are practicable and expedient) is not therefore to be considered as recommended for imitation.

A fire-place in a house connected with the farmstead, to prevent the mould on harness, which in England is the result of the dampness of the climate, implies more capital and more careful servants than we have at command in this country. Approved English or foreign servants, as they are styled abroad, on coming to America either go at once, with their means, and for a few dollars, several hundred miles west, and there buy government lands at $1 25 an acre, or they

(42.) The *cow-house* or *byre*, Q, is placed on the left of the principal range, in a position corresponding with that of the work-horse stable. It is 53 feet in length and 18 feet in width. The stalls of a cow-house, to be easy for the cows to lie down and rise up, should, in my opinion, never be less than 5 feet in width. Four feet is the more common width, but that is evidently too narrow for a large cow, and even 7 feet are considered by some people as a fair-sized stall for two cows; though, in my opinion, every cow should have a stall for herself, for her own comfort when lying or standing, and that she may eat her food in peace.* The width of the byre should be 18 feet; the manger is 2 feet in width, the length of a large cow about 8 feet, the gutter 1 foot, leaving 7 feet behind the gutter for the different vessels used in milking the cows and feeding the calves. The ceiling should be quite open to the slates, and a ventilator, moreover, is a useful apparatus for regulating the temperature and supplying fresh air to a byre. A door, divided into upper and lower halves, should open outward to the court on a giblet-check, for the easy passage of the cows to and from the court, and each half fastened on the inside with a hand-bar. Two windows with glass panes, with the lower parts furnished with shutters to open, will be quite sufficient for light, and, along with the half-door, for air also. The walls should be plastered for comfort and cleanliness.

(43.) The *stalls* are most comfortably made of wood, though some recommend stone, which always feels hard and cold. Their hight should be 3 feet, and in length they should reach no farther than the flank of the cow, or about 6 feet from the wall. When made of wood, a strong hard-wood hind-post is sunk into the ground, and built in masonry. Between this post and the manger should be laid a curb-stone, grooved on the upper edge to let in the deals of the travis endways. The deals are held in their places at the upper ends by a hard-wood rail, grooved on the under

remain in the towns, under much higher wages than the *American Farmer* can afford to give taxed as he is to support enormously expensive military and civil establishments. Who would believe, for example, that in Maryland the farmers and planters, asking so little, and getting so much less, from Government, pay 100 men $4 per day each, and even the postage on all their political and private as well as public correspondence, for the space of *three months every year, to make new laws and patch up old ones!*

Would any cultivated agricultural community, educated as they ought to be, with an understanding of *their own* true interests and just power, submit for one year to be thus humbugged and fleeced? *Ed. Farm. Lib.*]

[* This would all be very well if the American farmer had capital to build, for better accommodation, on any scale, however expensive. But where he is forced, according to a common saying, to "cut his coat according to his cloth," less roomy stalls must answer. In our best dairy establishments, as at Morrisania and others, the partition between the stalls is usually very short, just sufficient to prevent the heads of the cows from coming in contact, leaving the space open between their bodies, the width of the stall being often not more than three feet in the clear, and these seem to answer well.

In some of these best milk establishments, strong *tubs*, which are easily removed to be cleaned or filled, are in use for giving short provender, cut hay or straw, or corn fodder, as the case may be, wet and mixed with bran shorts or meal of some sort, leaving the long provender to go into the manger, which runs from one end of the stable to the other, sometimes resting on the floor. These tubs are filled in the feeding passage, from which also the long provender is supplied to the mangers. Usually this feeding passage is between two rows of stalls in which the cows stand with their heads to the passage. We shall hereafter give exact plans, where it may be deemed necessary, on a scale suited to American farmers; but it is deemed best here not to disturb the copy before us, as every part in the plan has some connection with some other part. From the whole the reader proposing to build may easily select such portions as he may like, and re-combine them to meet and satisfy his own views. *Ed. Farm. Lib.*]

side, into which the ends of the deals are let, and the rail is fixed to the back of the hind-post at one end, and let into the wall at the other and there fastened with iron holdfasts. Stone travises are no doubt more durable, and in the end, perhaps, more economical, where flag-stones are plentiful; but I would in all cases prefer wood, as feeling warmer, being more dry in winter, and less liable to injure the cows coming against them, and within doors will last a long time. The plan of the stalls may be seen at Q and Y in the plan fig. 4, Plate IV.

(44.) The *mangers* of byres are usually placed on a level with the floor, with a curb-stone in front to keep in the food, and paved in the bottom.—This position I conceive to be highly objectionable, inasmuch as, when breaking the turnips, the head of the animal is depressed so low that an undue weight is thrown upon the fore-legs, and an injurious strain induced on the muscles of the lower jaw. A better position is, when the bottom of the manger, made of flag-stones or wood, resting on a building of stone and mortar, is raised about 20 inches from the ground, and a plank set on edge in front to keep in the food. This plank should be secured in its position with iron rods batted into the wall at one end, and the other end passed through the plank to a shoulder, which is pressed hard against the plank on the opposite side by means of a nut and screw. This form of manger may be seen in fig. 18, p. 110. In this position of the manger, the cow will eat with ease any kind of food, whether whole or cut, and all feeding-byres for oxen should also be fitted up with mangers of this construction. Mangers are generally made too narrow for cattle with horns, and the consequence is the rubbing away of the points of the horns against the wall.

(45.) The method of *supplying green food* to cattle in byres may be various, either by putting it into the manger from the inside, or from the outside through holes in the wall made exactly opposite their heads. Either way is equally serviceable to the cattle, but the latter is the more convenient for the cattle-man. Its construction may be easily understood by fig. 9, which represents the door shut in the opening of the wall on the outside. But, convenient as this mode of supplying food is, I prefer giving it by the stall, when that is as wide as 5 feet, because, in cold weather in winter, the draught of air occasioned by the opening of the small doors at the heads of cows may endanger their health. There is another method by having a passage of 3 feet in width betwixt the stalls and the wall, from which both turnips and fodder may be supplied to the cows. In this case the space behind the cows is reduced to 4 feet in width.

Fig. 9.

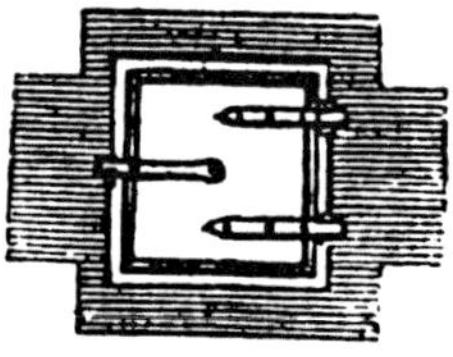

DOOR THROUGH WHICH TO SUPPLY MANGERS WITH TURNIPS.

(46.) The *floor* of byres should be paved with small round stones, excepting the gutter, which, being as broad as an ordinary square-mouthed shovel, should be flagged at the bottom, and formed into the shape of a trough by two curb-stones. A gutter of this form can be quickly cleaned out. A similarly formed gutter, though of smaller dimensions, should run from the main one through the wall to the court, to carry off the urine.—The causewaying of the stalls of a cow-house should go very little farther up than the hind-posts, because in lying down and rising up, cattle first kneel on their fore-knees, which would be injured in the act of being pressed against any hard substance like stones. This inner part of the stall should be of earth, made softer by being covered with litter. The urine gutters may be seen in the plan at Q and Y in fig. 4, Plate IV.

(47.) Fig. 10 represents a section of a *travis and manger of a byre*, where *a* is the wall, *b* the building which supports the manger *c*, having a front of wood, and bottomed with either flags or wood, *d* the hard-wood hind-post, sunk into the ground, and there built in with stones and mortar, *e* the hard-wood top-rail, secured behind the post *d*, and let into and fixed in the wall *a* with iron holdfasts, *f* the stone curb-stone, into which the tra-

Fig. 10.

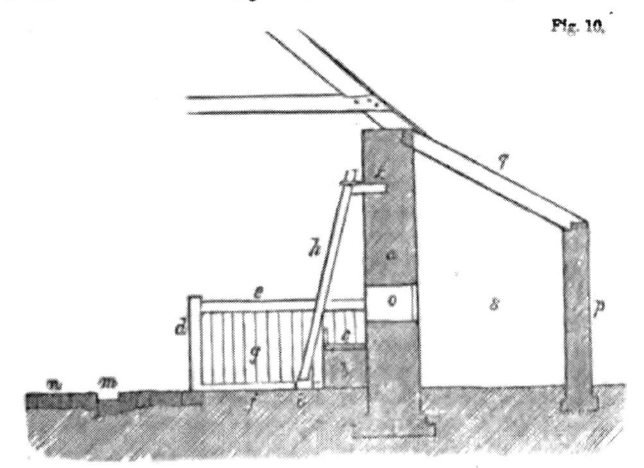

BYRE TRAVIS, MANGER, AND STAKE.

vis-board is let; *g* the boarding of wood, let endways into the curb-stone below, and into the top-rail above, by a groove; *h* is a hard-wood stake, to which the cattle are fastened by binders, the lower end of which is let into a block of stone *i*, and the upper fastened by a strap of iron to a block of wood *k* fixed into the wall *a*; *m* is the gutter for the dung, having a bottom of flag-stones, and sides of curb-stones; *n* the paved floor; *o* the opening through the wall *a* by which the food is supplied into the manger *c* to the cattle, from the shed *s* behind. This shed is 8 feet wide, *p* being the pillars which support its roof *q*, which is just a continuation of the slating of the byre roof, the wall *a* of which is 9 feet, and the pillars *p* 6 feet, in hight. But where no small doors for the food are used, the shed *s*, pillars *p*, and roof *q*, are not required—a small turnip store being sufficient for the purpose, and to which access may be obtained by the back door, seen in Q, at the right hand of the stalls in fig. 4, Plate IV.

Fig. 11.

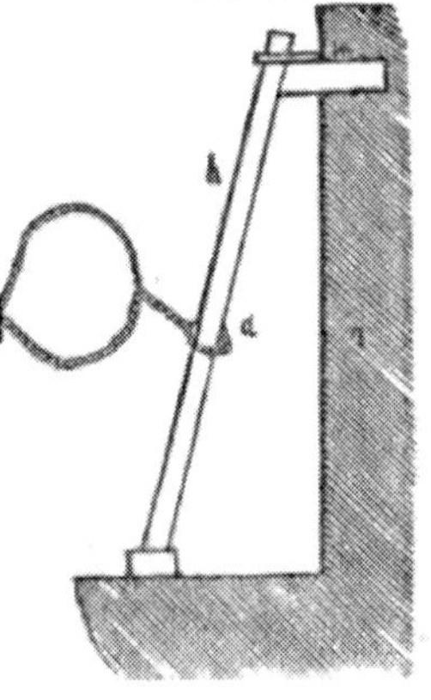

CATTLE SEAL OR BINDER.

(48.) Cattle are bound to the stake in various ways. 1. One way is with an iron chain, commonly called a *binder* or *seal*. This is represented in fig. 11, where *a* is the large ring of the binder which slides up and down the stake *h*, which is here shown in the same position as it is by *h* in the section of the stall in fig. 10. The iron chain being put round the neck of the beast, is fastened together by a broad-tongued hook at *c*, which is put into any link of the chain that forms the gauge of the neck, and cannot come out again until turned on purpose edgeways in reference to the link of which it has a hold. This sort of binder is in general use in the midland and northern counties of Scotland. 2. Another method of binding is with the *baikie*, which is made of a piece of hard wood *e*, fig. 12, standing upright and flat to the neck of the beast; a rope *g*, fastens the lower end of it to the stake, upon which it slides up and down by means of a loop which the rope forms round the stake. This rope passes *under* the neck of the animal, and is never loosened. Another rope *k*, is fastened at the upper end of the piece of wood *e*, and, passing *over* the neck of the animal, and round the stake, is made fast to itself by a knot and eye, and serves the purpose of fastening and loosening the animal. The neck being embraced between the two ropes, moves up and down, carrying the baikie along with it. This method of binding animals to the stake, though quite easy to the animals themselves, has this objectionable property, which the *seal* has not, of preventing the animals turning round their heads to lick their bodies, which they can do with the seal pretty far back, and yet are unable to turn round in the stall. The seal being made of iron, is more durable than the baikie. The top of the stake of the seal is inclined toward the wall *n*, and fixed as represented by *m* in fig. 11; the baikie stake is held perpendicular, and is fixed to a log of wood *m*, fig. 12, stretching parallel to the wall *o*, across the byre, of which log the cross section only is here shown. The seal-stake is placed in an inclined position to allow its top to be fastened to the wall, and in regard to it the animal is comparatively loose; but as the neck is always held close to the baikie-stake, that stake must be placed in a perpendicular position to allow the animal to move its neck up and down to and from the manger.

Fig. 12.

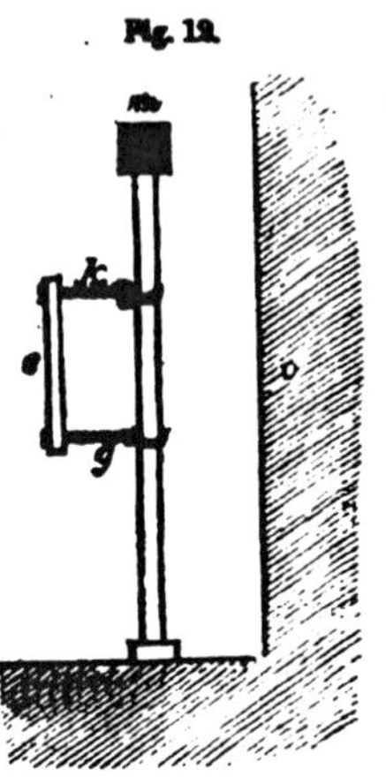

A BAIKIE.

(49.) This construction of the byre with its fittings up, is quite as well suited to *fatten oxen* as to accommodate *milch cows*. Feeding byres are usually constructed much too small for the number of oxen confined in them. When stalls are actually put up, they seldom exceed 4 feet in width; more frequently two oxen are put into a double stall of 7 feet, and not unfrequently travises are dispensed with altogether, and simply a tri angular piece of boarding is placed across the manger against the wall, to divide the food betwixt such pair of oxen. In double stalls, and where no stalls are used, even small-sized oxen, as they increase in size, cannot all lie down together to chew their cud and rest, whereas, the fatter they become, they require more room and more rest; and large oxen are hampered in them from the first. In such confined byres, the gutter, moreover, is too near the heels of the oxen, which prevents them standing back when they desire. Short stalls, to be sure, save the litter being dirtied, by the dung dropping from the cattle directly into the gutter, and this circumstance no doubt saves trouble to the cattle-man; but in such a case the litter is saved by the sacrifice of comfort to the animals. Such considerations of economy are quite legitimate in cowkeepers in town, where both space and litter are valuable, but that they should induce the construction of inconvenient byres in farmsteads indicates either parsimony on the part of the landlord or ignorance on that of the architect; and no farmer who consults the well-being of his animals, and through them his own interest, should ever originate such a plan, or sanction it where he finds it to exist. The truth is, these confined structures are ordered to be erected by landlords unacquainted with Agriculture, to save a little outlay at first. Expenditure to them is a tangible object; but in dealing thus with their tenants, they seem not to be aware they are acting with shortsightedness toward their own interests; for want of proper accommodation in the farmstead certainly has, and should have, a considerable influence on the mind of the farmer, when valuing the rent of the farm he wishes to occupy. Should you have occasion to fit up a byre for the accommodation of milch cows or feeding oxen, bear in mind that a small sum saved at first, may cause you to incur a yearly loss of much greater amount than the saving, by not only preventing your feeding cattle attaining the perfection which a comfortable lodging would certainly promote in them; but in affecting the state of your cows by want of room, the calves they bear in such circumstances are sure to prove weak in constitution.

(50.) Immediately adjoining the cow-house should be placed the *calves' house*. This apartment is represented at R of the plan in fig. 4, Plate IV. fitted up with cribs. It is 35 feet in length, and 18 feet in width, and the roof ascends to the slates. Calves are either suckled by their mothers, or brought up on milk by the hand. When they are suckled, if the byre be roomy enough, that is, 18 feet in width, stalls are erected for them against the wall behind the cows, in which they are usually tied up immediately behind their mothers; or, what is a less restrictive plan, put in numbers together in loose boxes at the ends of the byre, and let loose from both places at stated times to be suckled. When brought up by the hand, they are put into a separate apartment from their mothers, and each confined in a loose-box or crib, where the milk is given them. The superiority of separate cribs over loose boxes for calves is, that calves are prevented sucking one another, after having got their allowance of milk, by the ears, or teats, or scrotum, or navel; by which malpractice, when unchecked, certain diseases may be engendered. The crib is large enough for one calf at 4 feet square and 4 feet in hight, sparred with slips of tile-lath, and having a small wooden wicket to afford access to the calf. The floor of the cribs may be of earth, but the passage between them should be flagged or of asphaltum. Abundance of light should be admitted, either by windows in the walls, or sky-lights in the roof; and fresh air is essential to the health of calves, the supply of which would be best procured by a ventilator, such as is represented in fig. 8, p. 95, already described. There should be a door of communication with the cow-house, and another in two divisions, an upper and a lower, into a court furnished with a shed, as *k* in fig. 4, Plate IV. which the calves may occupy until turned out to pasture. The cribs should be fitted up with a manger to contain cut turnips, and a high rack for hay, the top of which should be as much elevated above the litter as to preclude the possibility of the calves getting their feet over it. The general fault in the construction of calves' houses is the want of both light and air—light being cheerful to creatures in confinement, and air particularly essential to the good health of young animals. When desired, both can be excluded. The walls of the calves' house should be plastered, for the sake of neatness and cleanliness. Some people are of opinion that the calves' house should not only have no door of communication with the cow-house, but should be placed at a distance from it, in order that the cows may be beyond the reach of hearing the calves. Such an objection could only have originated from an imperfect acquaintance with the nature of these animals in the circumstances. A young cow even that is at once prevented smelling and suckling her calf, does not recognize its voice at any distance, and will express no uneasiness about it after the first few minutes after parturition, and after the first portion of milk has been drawn from her by the hand.

(51.) The front of one of these *calves cribs* is represented by fig. 13, in which *a* is the wicket-door which gives access to it, *b b* are the hinges, and *c* is a thumb-catch to keep the door shut. You will observe that this kind of hinge is very simple and economical. It consists of the rails of the wicket being a little elongated toward *b*, where they terminate in a semi-circular form, and the lower face of which is shaped into a pin which fills and rotates in a round hole made in a billet of wood, seen at the lower hinge at *b*, securely screwed to the upright door-post of the crib. Another billet *d* is screwed immediately above the lower rail *b*, to prevent the door being thrown off the hinges by any accident. Cross-tailed iron hinges, of the lightness suited to such doors would soon break by rusting in the dampness usually occasioned by the breath of a number of calves confined within the same apartment.

Fig. 12.

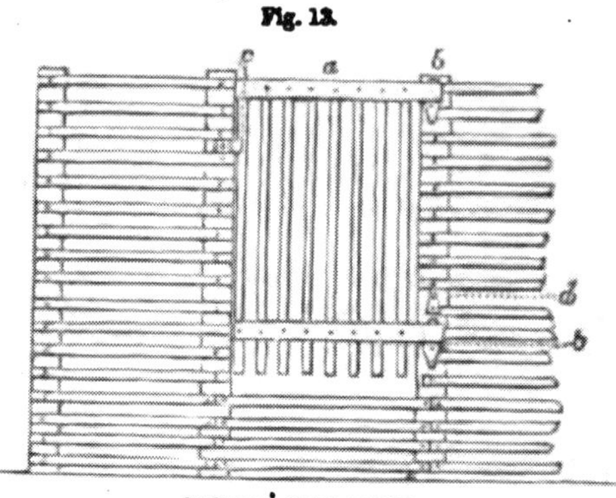

CALVES' CRIB DOOR.

(52.) A pretty large *court* should be attached to the *cow-house*, in which the cows can walk about for a time in the best part of the day in winter, basking in the sun when it shines, rubbing against a post that should be set up for the purpose, drinking a little water provided for them in a trough *w*, and licking themselves and one another. Such a court is, besides, necessary for containing the manure from the byre, and should have a gate by which carts can have access to the manure: *l* is such a court on the plan, fig. 4, Plate IV. being 58 feet in length by 30 feet in width.

(53.) *k* in the plan, fig. 4, Plate IV. is the *court* attached to the *calves' house*, 30 feet in length by 25 feet in width, in which should be erected, for shelter to the calves in cold weather, or at night before they are turned out to pasture, or for the night for a few weeks after they are turned out to pasture, a *shed k*, 30 feet in length by 12 feet in width, fitted up with mangers for turnips, and racks for hay. A trough of water, *w*, is also requisite in this court, as well as a gateway for carts by which to remove the dung.

(54.) On the left of the cow-house is the *boiling-house* U, for cooking food in, and doing everything else that requires the use of warm water.—The boiler and furnace *b'* should be placed so as to afford access to the boiler on two sides, and from the furnace the vent rises to the point of the gable. A fire-place *a'* is useful for many purposes, such as melting tar, boiling a kettle of water, drying wetted sacks, nets, &c. One door opens into the byre, and another, the outer one, is in the gable, through which access to the byre may be obtained, or, if thought better, through the gate and court of the byre. There should be a window with glass, and shutters in the lower division, to open and admit air, and a ventilator *u*, fig. 3, Plate III. on the roof may be advisable here as a means at times to clear the house of steam. The walls of the boiling-house should be plastered. As proximity to water is an essential convenience to a boiling-house, water is quite accessible in the trough of the cows' court *l*, or, what is still better, in a trough connected with it outside, as at *l'*, in fig. 3, Plate III. or *w*, in fig. 4, Plate IV.

(55.) Windows should be of the form for the purpose they are intended to be used. On this account windows for stables, and for other apartments, should be of different forms. 1. Fig. 14 represents a window for a stable. The opening is 4½ feet in hight by 3 feet in width. The frame-work is composed of a dead part *a* of 1 foot in depth, 2 shutters *b b* to open on hinges, and fasten inside with a thumb-catch, and *c* a glazed sash 2 feet in hight, with 3 rows of panes. When panes are made under 8 inches square, there is a considerable saving in the price of glass. The object of this form of a stable window is, that generally a great number of small articles are thrown on the sole of a work-horse stable window, such as short-ends, straps, &c. which are only used occasionally, and intended to be there at hand when wanted. The consequence of this confused mixture of things, which it is not easy for the farmer to prevent, is that, when the shutters are desired to be opened, it is scarcely possible to do it without first clearing the sole of everything; and, rather than find another place for them, the window remains shut. A press in a wall might be suggested for containing these small articles; but in the only wall, namely, the front one of the stable O, in which it would be convenient to make such a press, all its surface is occupied by the harness hanging against it; and besides, no orders, however peremptory, will prevent such articles being, at throng times, thrown upon the window-soles; and where is the harm of their lying there at hand, provided the windows are so constructed as to admit of being opened when desired? When a dead piece of wood, as *a*, is put into such windows, small things may remain on the sole, while the shutters *b b* are opened over them. 2. In other apartments, such as byres, corn-barn, calves' house, boiling-house, implement-house, hay-house, where there is no chance of an accumulation of sundry articles in the window-sole, the shutters of the windows, *if desired*, may descend to the bottom of the frame, as in fig. 15. The size of the window may still be the same, 4½ feet in hight and 3 feet in width. The frame consists of two shutters *a a* 2 feet in hight, with a glazed sash *c* 2½ feet in hight, having 4 rows of panes.—Such a form of window will admit a great deal of light and air.

Fig. 14.

STABLE WINDOW.

Fig. 15.

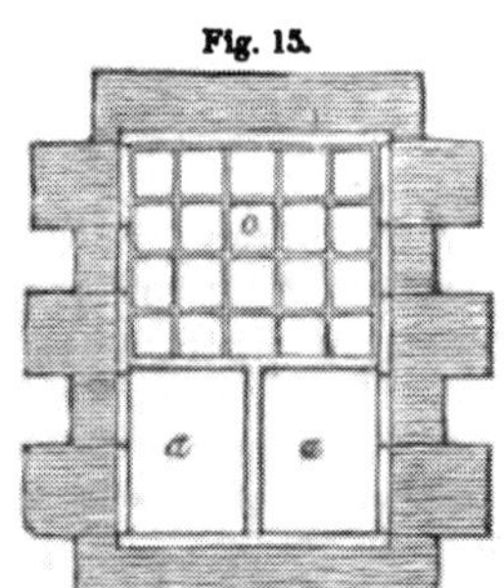

BYRE, &C. WINDOW.

(56.) The *upper barn* B, as seen in fig. 16, occupies the whole space above the corn-barn and chaff-house. It is 32 feet in length and 30 feet in breadth, and its roof ascends to the slates. It has a good wooden floor like the corn-barn, supported on stout joists. It contains the principal machinery of the threshing-machine, and is wholly appropriated to the storing of the unthreshed corn previous to its being threshed by the mill. For the admission of barrows loaded with sheaves from the stack-yard, or of sheaves direct from the cart, this barn should have a door toward the stack-yard of 6 feet in width, in two vertical folds to open outward, on a giblet-check—one of the folds to be fastened in the inside with an iron cat-band, and the other provided with a good lock and key. It is in this barn that the corn is fed into the threshing-mill; and, to afford light to the man who feeds in, and ample light to the barn when the door is shut—which it should be when the wind blows strongly into it—a sky-light should be placed over the head of the man. The large door should not be placed immediately behind the man who feeds in, as is frequently the case in farmsteads, to his great annoyance when the sheaves are bringing in. There should be slits in the walls for the ventilation of air among the corn-sheaves, which may not at all times be in good order when taken into the barn. A hatchway

Fig. 16.

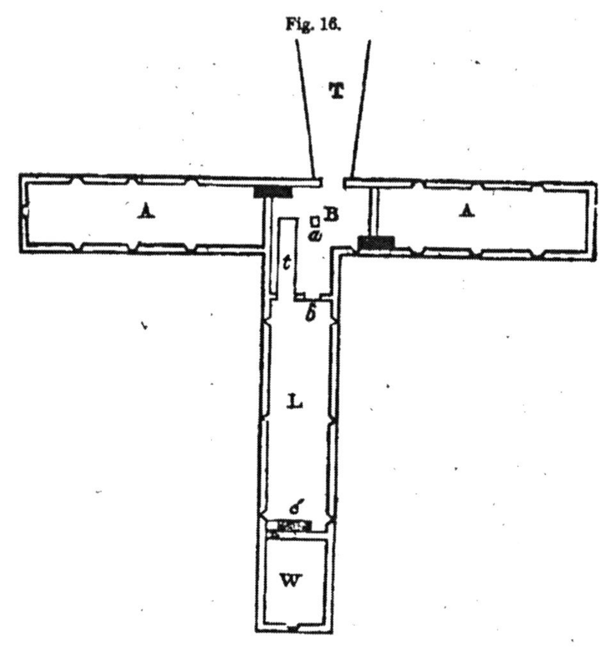

PLAN OF UPPER BARN, GRANARIES, AND WOOL-ROOM.

a, 3 feet square, in the floor, over the corn-barn below, is useful when any corn or refuse has to be again put through the mill. Its hatch should be furnished with strong cross-tailed hinges, and a hasp and staple, with a padlock and key, by which to secure it from below in the corn-barn. An opening *b*, of 4 feet in hight and 3 feet in width, should be made through the wall to the straw-barn, for the purpose of receiving any straw from it that may require to be put through the mill again. This opening should be provided with a door of one leaf, or of two leaves, to fasten with a bar, from the upper barn. The threshing-machine is not built on the floor, but is supported on two very strong beams extending along the length of the barn: *t* is the site of the threshing-machine in the figure.

(57.) Immediately in connection with the upper barn is the *gangway*, T, fig. 4, Plate IV. and fig. 16. It is used as an inclined plane, upon which to wheel the corn-barrows, and form a road for the carriers of sheaves from the stack-yard. This road should at all times be kept hard and smooth with small broken stones, and at the same time sufficiently strong to endure the action of barrow-wheels. Either common asphaltum or wood pavement would answer this purpose well. To prevent the body of the gangway affecting the wall of the corn-barn with dampness, it should be kept apart from that wall by an arch of masonry. Some farmers prefer taking in the corn on carts instead of by a gangway, and the carts in that case are placed alongside the large door, and emptied of their contents by means of a fork. I prefer a gangway for this purpose; because it enables the farmer to dispense with horse-labor in bringing in the stacks if they are near at hand, and they should always be built near the upper-barn for convenience. Barns in which flails alone are used for threshing the corn, are made on the ground, and the barn-door is made as large as to admit a loaded cart to enter and empty its contents on the floor.

(58.) In fig. 16, AA are two granaries over the sheds DD, implement-house G, and hay-house H, in fig. 4, Plate IV. That on the left is 76 feet in length and 18 feet in width, and the other 65 feet in length and 18 feet in width. The side walls of both are 5 feet in hight. Their roofs ascend to the slates. Their wooden floors should be made strong, to support a considerable weight of grain; their walls well plastered with hair plaster; and a neat skifting-board should finish the flooring. Each granary has 6 windows, three in front and three at the back, and there is one in the gable, at the left hand over the door of the implement-house. These windows should be so formed as to admit light and air very freely, and I know of no form of window so capable of affording both, as this in fig. 17, which I have found very serviceable in granaries. The opening is 4½ feet in length and 3 feet in hight. In the frame *a* are a glazed sash 1 foot in hight, composed of two rows of panes, and *b* Venetian shutters, which may be opened more or less at pleasure: *c* shows in section the manner in which these shutters operate. They revolve by their ends, formed of the shape of a round pin, in holes in the side-posts of the frame *d*, and are kept in a parallel position to each other by the bar *c*, which is attached to

Fig. 17.

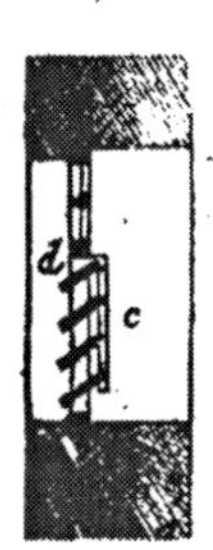

GRANARY WINDOW AND SECTION OF SHUTTERS.

them by an eye of iron, moving stiff on an iron pin passing through both the eye and bar *c*. The granary on the right hand being the smallest, and immediately over the work-horse corn-chest, should be appropriated to the use of horse-corn and other small quantities of grain to be first used. The other granary may contain seed-corn, or grain that is intended to be sold when the prices suit. For repairing or cleaning out the threshing-machine, a large opening in the wall of this granary, exactly opposite the machinery of the mill in the upper barn, will be found convenient. It should be provided with a large movable board, or folding doors, to close on it, and to be fastened from the granary. This opening is not shown in fig. 16.

(59.) At the end of the straw-barn L is the wool-room W, fig. 16, its site being indicated by W on the roof of the isometrical view, fig. 3, Plate III. It just covers the small hammels X, and is therefore 25 feet in length and 18 feet in breadth. It enters from the straw-barn L by means of the stone or wooden trap-stair *c'*. Its floor should be made of good wood, its walls and roof lathed and hair-plastered. Its window should be formed like that of the byres, with a glazed sash above, and opening shutters below. A curtain should be hung across the window to screen the light and air from the wool when desired. The door need not exceed 6 feet in hight, but should be 3½ feet in width, to let a pack of wool pass easily through. As the wool is most conveniently packed in this room, there should be provided in the roof two strong iron hooks, for suspending the corners of the pack-sheet in the act of packing it, and another from which to suspend the beam and scales for weighing the fleeces. Although the wool will usually occupy this room only when the cattle are in the field, yet in case it should be found expedient to keep it over year, or have animals in the small hammels X in summer, and in case their breath should ascend into the wool through any openings of the joinings of the deals of the floor, it will be a safe precaution for preserving the wool in a proper state, to have the roof of the hammels below lathed and plastered. This room could be entered by a door and stone hanging-stair in the gable.

(60.) M and N in the plan, fig. 4, Plate IV. are *hammels* for the feeding of cattle, rearing of young horses, and tending of queys in calf until they are tied up in the cow-house. 1. Hammels consist of a shed, and an open court, communicating by a large opening. The shed part need not be so wide as the rest of the apartments in the farmstead, in so far as the comfort of the animals is concerned; and in making them narrower considerable saving will be effected in the cost of roofing. 2. There is no definite rule for the *size* of hammels; but as their great convenience consists in conferring the power to assort cattle according to their age, temper, size and condition, while at liberty in the fresh air, it is evident that hammels should be much smaller than courts, in which no assortment of animals can be attempted. 3. The courts of hammels, from which the dung is proposed to be taken away by horse and cart, should not be less than 30 feet in length by 18 feet in breadth, and their entrance gates 9 feet in width; and this size of court will accommodate 4 oxen that will each attain the weight of 70 stones imperial. This is the size of the courts of the hammels M. Should it not be thought inconvenient to take the dung out of the courts with barrows, then they need not be made larger than 20 feet in length by 17 feet in width, and this is the size of the courts of the hammels N, which will accommodate 3 oxen of the above size. 4. The sheds to both sizes of courts need not exceed 14 feet in width, and their length will be equal to the width of the courts. Of these dimensions 4 oxen in the larger will have just the same accommodation as 3 oxen of the same size in the smaller hammels. 5. All hammels should have a trough, *z*, for turnips, fitted up against one of the walls of the court. The side-wall is the most convenient part, when a large gate is placed in front, through which the carts are backed to clear away the dung from the courts. In the case of the smaller courts, the turnips may be supplied to the trough over the top of the front wall. 6. To give permanency to hammels, the shed should be roofed as effectually as any of the other buildings, though to save some expense at first, many farmers are in the habit of roofing them with small trees placed close together on the tops of the walls of the sheds, and of building thereon either straw, corn, or beans. This is certainly an excellent place upon which to stack beans or peas; but the finished building is that which should be adopted in all cases. Temporary erections are constantly needing repairs, and in the end actually cost more than work substantially executed at first. 7. The division betwixt the shed and court forms the front wall of the shed, through which an opening forms the door betwixt them. This door, 6 feet in width, should always be placed at one side and not in the middle of the hammel, to retain the greatest degree of warmth to the interior of the shed. The corners of its scutcheon should be rounded off to save the cattle being injured against sharp angles. The divisions betwixt the respective courts should be of stone and lime walls, 1 foot in thickness, and 6 feet in hight. Those with in the sheds should be carried up quite close to the roof. Frequently they are only carried up to the first balk of the couples, over which a draught of air is generated along the inside, from shed to shed, much to the discomfort of the animals; and this inconvenience is always overlooked in hammels which are built with the view of saving a little cost in building up the inside division walls to the roofs. 8. Racks for fodder should be put up within the sheds, either in the three spare corners, or along the inner end. 9. In my opinion there is no way so suitable for feeding oxen, bringing up young horses in winter, or taking care of heifers in calf, as hammels; and of the two sizes described above, I would decidedly prefer the smaller, as permitting the fewer number of animals to be put together. 10. XX are two small hammels at the end of the straw-barn L for accommodating a bull, or stallion, or any single animal that requires a separate apartment for itself. These are each 18 feet in length and 12 feet in width within the sheds, the roofs of which are formed of the floor of the wool-room W; and 29 feet in length and 12 feet in width in the courts. The doors into them should be made to open outward, on giblet-checks. The courts are furnished with turnip-troughs, *z*, and one water-trough, *w*, will serve both courts, as shown in the plan, figure 4, Plate IV. A rack should be fitted up for fodder in the inside of each shed.

(61.) It should be observed that a part of the hammels N is fitted up as a byre Y. This byre is intended to accommodate the servants' cows.— There are 8 stalls—6 for the plowmen's, 1 for the farm-steward's, and 1 for the shepherd's cows—and they are nearly 5 feet in width. The length of the byre is 38 feet, and its width is only 14 feet, which gives a rather small space behind the cows; but, as servants' cows are generally small, and the milk from them immediately carried away, if there is just sufficient room for feeding and milking them, and adequate comfort to the cows themselves, a large space behind them is unnecessary. This byre has a ventilator *r'*. The cows are furnished with an open court *v*, 38 feet in length and 20 feet in width, and a water-trough *w*.

(62.) I and K of the plan, fig. 4, Plate IV. are *two large courts for young cattle*, both in the immediate vicinity of the straw-barn L, and both having a shed D under one of the granaries; I is 84 feet in length and 76 feet in width, and K 84 feet in length and 77 feet in width. Troughs for turnips should be fitted up against one or more of the walls surrounding the courts in the most convenient places, such as at *z* in both courts. Besides racks for fodder, *h'*, against one of the walls within the sheds D D, there should racks be placed in the middle of the courts, that the cattle may stand around and eat out of them without trouble. The square figures *o o* in the middle of the courts I and K indicate the places where the racks should stand, and their form may be seen at *o o* in the isometrical view, fig. 3, Plate III. Around two sides of K is a paved road *e'*, 13 feet in width, for carts going to be loaded with grain at the door of the corn-barn C. Though the cattle have liberty to walk on this pavement, it should be kept clean every day. Such courts are quite common in steadings for the rearing of young cattle in winter, and even for feeding large lots of cattle together, as is practiced by most farmers who do not rear calves; but, for my part, I prefer hammels for all classes and ages of cattle; for, although cattle are restricted in them in regard to space, still the few in each hammel have plenty of room to move about. There is no hardship to the animals in this degree of confinement, while they have the advantage of quietness among themselves in the open air, produced by being assorted according to temper, size, sex, and age. On abolishing large courts altogether out of steadings,

I would substitute in their place hammels of different sizes, and convert the cattle-sheds D and D into cart-sheds and receptacles for the larger class of implements. It is probable that the use of large courts will not soon be dispensed with in farmsteads, and for that reason I have retained them in the plan; but I have no doubt that a period will arrive when farmers, *to insure* to themselves larger profits from cattle, will see the advantage of taking the utmost care of them, from the period of their birth until disposed of in a ripe condition at an early age; and then hammels will be better liked than even courts are at the present day, and farmers will then universally adopt them.

(63.) Fig. 18 represents a *trough for turnips suited both for hammels and courts*, where *a* is the wall against which the trough is built, and *b* uild-

Fig. 18.

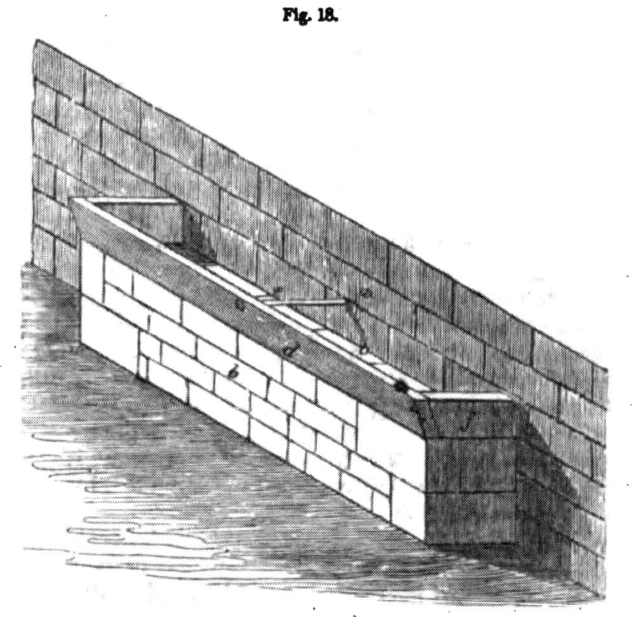

TURNIP TROUGH.

ing of stone and lime 2 feet thick. The lime need not be used for more than 9 inches in the front and sides of the wall, and the remaining 15 inches may be filled up with any hard material; *c* is the flagging placed on the top of the wall, and forming the bottom of the trough. Some board the bottom with wood, where wood is plentiful, and it answers well enough; but, of course, flags, where easily obtained, are more durable, though wood is pleasanter for the cattle in wet and frosty weather in winter. *d* is a plank, 3 inches thick and 9 inches in depth, to keep in the turnips. Oak planking from wrecks, and old spruce trees, however knotty, I have found to make cheap and very durable planking for the edging of turnip troughs. The planks are spliced together at the ends, and held on edge by bars of iron *e* batted with lead into the wall, in the manner already described in (44,) p. 100; and the figure clearly shows this mode of fastening the plank. The masonry represented in the figure is finer than need be for the purpose; and the trough, though here shown short, may extend to any length along the side or sides of a court.

(64.) The *straw-racks for courts* are made of various forms. 1. On farms of light soils, where straw is usually scarce, a rack of the form of fig. 19, having a movable cover, will be found serviceable in preserving the straw from rain, where *a a* is the bottom inclined upward to keep the straw always forward to the front of the rack in reach of the cattle.—Through the apex of the bottom, the shank which supports the movable cover *b* passes, and this cover protects the straw from rain. The shank with its cover is worked up and down, when a supply of straw is given, by a rack and pinion *c*, to which pinion is attached the lying shaft, on which is shipped a handle *d*. A rack of this kind is made of wood, and should be 5 feet square, and 5 feet in hight to the top of the corner posts; and sparred all round the sides, as well as the bottom, to keep in the straw. 2. A more common kind of rack is represented by fig. 20, which is of a square form, and sparred all round the sides to keep in the straw. The cattle draw the straw through the spars as long as its top is too high for them to reach over it; but after the dung accumulates, and the rack thereby becomes low, the cattle get at the straw over the top. This kind is also made of wood, and should be 5 feet square and 4 feet in hight.—3. Fig. 21 represents a rack made of malleable iron, intended to supply the straw to the cattle always over its top, and is therefore not sparred, but rodded, in the sides, to keep in the straw. In use it remains constantly on the ground, and not drawn up as the dung accumulates, as in the case of the other kinds of racks described. This kind is 5½ feet in length, 4½ feet in breadth, and 4½ feet in hight; the upper rails and legs are made of iron 1 inch square, and the other rails ¾ inch. Iron is, of course, the most durable material of which straw-racks for cattle can be made.

Fig. 19.

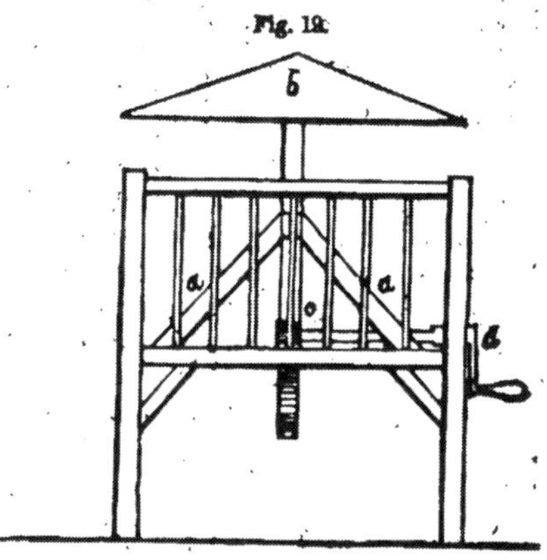

COVERED STRAW-RACK.

Fig. 20.

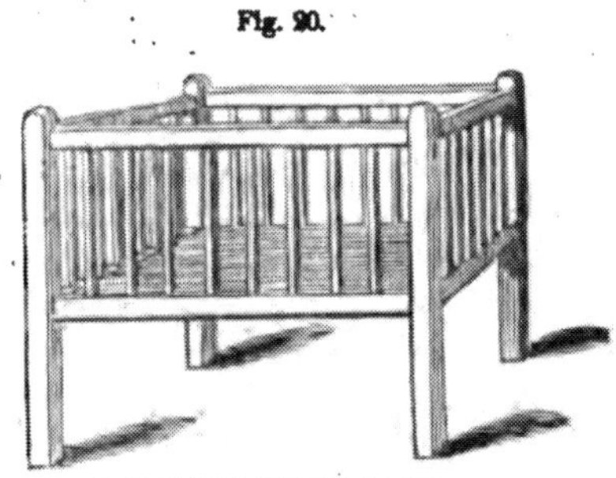

WOODEN STRAW-RACK.

Fig. 21.

IRON STRAW-RACK.

(65.) There are few things which indicate greater care for cattle when housed in the farmstead, than the erection of *places for storing turnips* for their use. Such stores are not only convenient, but the best sort of receptacles for keeping the turnips clean and fresh. They are seen in the isometrical view, fig. 3, Plate III. and in the plan, fig. 4, Plate IV. at *e* and *f* for the use of the hammels M; at *g* for that of the court K; at *h* for the hammels X, and servants' cow-house Y; at *i* for the use of the court I; at *m* for that of the cow-house Q, and calves' cribs R; and at *p* and *q* for the hammels N. The walls of these turnip-stores should be made of stone and lime, 8 feet by 5 feet inside, and 6 feet in hight, with an opening in front, 2 feet and upward from the ground, for putting in and taking out the turnips thereat; or they may be made of wood, where that is plentiful, with stout upright posts in the four corners, and lined with rough deals. They may be covered with the same material, or with straw, to protect the turnips from frost. They should be placed near the apartments they are intended to supply with turnips, and at the same time be of easy access to carts from the roads. These receptacles may, of course, be made of any convenient form.

(66.) The supply of *water* to all the courts where as many turnips as they can eat are not given to the cattle, is a matter of paramount consideration in the fitting up of every farmstead. In the plan, fig. 4, Plate IV. troughs for water are represented at *w*, in the large courts I and K, in those of the cow-houses *l* and *v*, and calves' cribs *k*, as well as in those of the bulls' hammels X. The troughs may be supplied with water either directly from pump-wells, or by pipes from a fountain at a little distance—the former being the most common plan. As a pump cannot conveniently be placed at each trough, there is a plan of supplying any number of troughs from one pump, which I have found to answer well, provided the surface of the ground will allow the troughs being *nearly* on the same level, when placed within reach of the animals. The plan is to connect the bottoms of any two or more troughs, set on the same level, with lead pipes placed under ground; and, on the first trough being supplied direct from the pump, the water will flow to the same level throughout all the other troughs. There is, however, this objection to this particular arrangement, that, when any one of the troughs is emptying by drinking, the water is drawn off from the rest of the troughs, that it may maintain its level throughout the whole; whereas, if the trough which receives the water were placed a few inches *below* the top of the one supplying it, and the lead pipe were made to come from the bottom of the supply trough over the top of the edge of the receiving one, the water would entirely be emptied from the trough out of which the drink was taken, without affecting the quantity in any of the others. 1. To apply these arrangements of water-troughs to the plan fig. 4, Plate IV. Suppose that a pump supplies the trough *w*, in the court I, direct from a well beside it—a lead pipe passing, on the one hand, from the bottom of this trough under ground to the bottom of the trough *w*, in the court K, and, on the other hand, to that of the trough *w* in the calves' court *k*, and thence to that of the trough *w* in the court *l* of the cow-house Q; and, in another direction, to the bottom of the trough *w* in the court *v* of the servants' cow-house Y; and suppose that the troughs in K and *k* and *l* and *v* are placed on the same level as the supply trough in the court I, it is obvious that they will all be supplied with water as long as there is any in the supply trough, and the emptiness of which will indicate that the water from it had been drawn off by the other troughs, and that the time had fully arrived when it was necessary to replenish the trough in the court I direct from the pump. The supply trough, in such an arrangement, should be larger than either of the other troughs. The trough of the bulls' hammels X might be supplied by a spout direct from the pump in the court I. In this way a simple system of watering might be erected from one pump to supply a number of troughs in different courts. It may be proper to illustrate this mode of connecting water-troughs by a figure. Let *b*, fig. 22, be the supply trough at the pump, and *f* the receiving one, and let both be placed on the same level; then let *g* be a lead pipe connecting the bottoms of both troughs, the ends of which are protected by hollow hemispherical drainers, such as *c*. It is here obvious, from the law which regulates the equilibrium of fluids, that the water, as supplied by the pump to *b*, will *always* stand at the same hight in *f*. 2. I shall now illustrate the other method of supplying troughs also by a figure. Let *a*, fig. 22, be the supply trough immediately beside the pump; let *b* be the trough in any other court to be supplied with water from *a*, and for that purpose it should be placed 3 inches below the level of *a*. Let a lead pipe *d* be fastened to the under side of the bottom of *a*, the orifice of which, looking upward, to be protected by the hemispherical drainer *c*. Let the lead pipe *d* be passed under ground as far as the trough *b* is situated from *a*, and emerge out of the ground by the side of and over the top of *b* at *e*. From this construc-

Fig. 22.

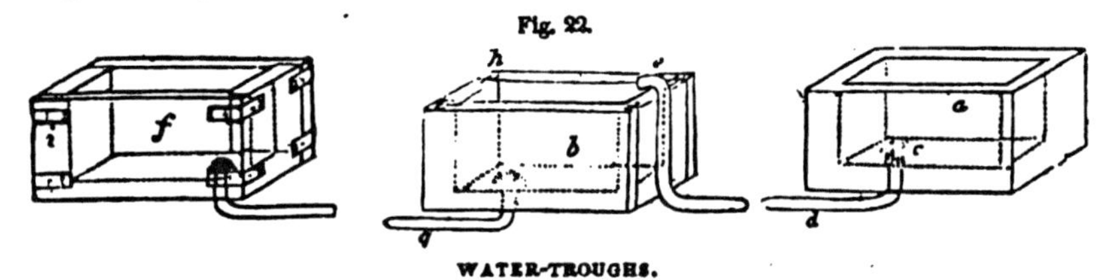

WATER-TROUGHS.

tion it is clear that, when *a* is filling with water from the pump, the moment the water rises to the level of the end of the pipe at *e*, it will commence to flow into *b*, and will continue to do so until *b* is filled, if the pumping be continued. The water in *a*, *below* the level of the end of the pipe at *e*, may be used in *a* without affecting that in *b*, and the water in *b* may be wholly used without affecting that in *a*. 3. Water-troughs may be made of various materials; the form of *a* is that of one hewn out of a solid block of freestone, which makes the closest, most durable, and best trough; that of *b* is of flag stones, the sides of which are sunk into the edges of the bottom in grooves filled with white lead, and there held together with iron clamps *h*. This is a good enough kind of trough, but is apt to leak at the joints. Trough *f* is made of wood dove-tailed at the corners, and held together by clamps of iron *i*. These troughs may be made of any size and proportions. 4. In some steadings, the water-troughs are supplied from a large cistern, somewhat elevated above their level, and the cistern is filled with water from a well either by a common or a force-pump. But, in this arrangement, either a cock, or ball and cock, are requisite at each trough: in the case of a cock, the supply of water must depend on the cock being turned on the trough in due time; and, in that of a ball and cock, the supply depends on the cistern being always supplied with water from the pump. There is great inconvenience and expense in having a ball and cock at each trough. 5. In steadings where there is an abundant supply of water from natural springs, accessible without the means of a pump, lead pipes are made to emit a constant stream of water into each trough, and the surplus is carried away in drains, perhaps to the horse-pond. 6. There is still another mode which may be adopted where the supply of water is plentiful, and where it may flow constantly into a supply-cistern. Let the supply-cistern be 2 feet in length, 1 foot wide, and 18 inches in depth, and let it be provided with a ball and cock, and let a pipe proceed from its bottom to a trough of dimensions fit for the use of cattle, into which let the pipe enter its end or side a little way, say 3 inches, below the mouth of the trough. Let a pipe proceed from this trough, from the *bend* of the pipe, as from the bend of the right-hand pipe *e* at the bottom of the trough *b*, fig. 22, to another trough, into whose end it enters in like manner to the first trough, and so on into as many succeeding troughs, from trough to trough, on the same level, as you require; and the water will rise in each as high as the mouth of the pipe, and, when withdrawn by drinking from any one of them, the ball and cock will replenish it direct from the supply-cistern; but the objection to a ball and cock applies as strongly to this case as to the other methods, although there is economy of pipe attending this method.

(67.) In most farmsteads a *shed for carts* is provided for, *though many farmers are too regardless of the fate of these indispensable machines by permitting them to be exposed to all vicissitudes of weather.* The cart-shed is shown at V in the isometrical view fig. 3, Plate III. and by V in the plan fig. 4, Plate IV. immediately behind the hammels M, facing the work-horse stable O, and looking to the north, away from the shrinking effects of the sun's heat. It is 80 feet in length, 15 feet in width within the pillars, and 8 feet in hight to the slates in front. The roof slopes from the back-slating of the hammels, and is supported at the eave by a beam of wood resting on 7 stone and lime pillars, and a wall at each end. The pillars should be of ashler, 2 feet square, and rounded on the corners, to avert their being chipped off with the iron rims of the wheels by the carelessness of the plowmen, when backing the carts into the shed. For the same purpose, a pawl-stone should be placed on each side of every pillar. This shed is longer than what is actually required where double-horse carts are only used, 6 ports being sufficient for that number, but single-horse carts are now so much in use, that more of these are required, perhaps not fewer than 8. Two single-horse carts can stand in each port, one in front of the other. Any spare room in the shed may be employed in holding a light cart, the roller, the grass-seed machine, the turnip-sowing machine, the bodies of the long carts, and other articles too bulky to be stowed into the implement-house G.

(68.) Though swine are usually allowed to run about the steading at pleasure, yet, to do them justice, they should be accommodated at times with protection and shelter, as well as the rest of the live-stock. *Piggeries* or *pig-sties* are therefore highly useful structures at the farmstead. They are of three kinds: 1. Those for a *brood-sow with a litter of young pigs.* This kind should have two apartments, one for the sow and the litter to sleep in, covered with a roof, and entered by an opening, the other an open court in which the feeding-trough is placed. For a breeding-sty each apartment should not be less than 6 feet square. This kind of sty is represented by *c c* in the plan, fig. 4, Plate IV. and at *c* in the isometrical view, fig. 3, Plate III. 2. Those for *feeding-pigs:* these should also have two apartments, one with litter for sleeping in, covered by a roof and entered by an opening; the other an open court for the troughs for food. A sty of 4 feet square in each apartment, will accommodate 2 feeding-pigs of 20 stones each Of this kind of sty 4 are represented at *a* in the isometrical view, fig. 3,

Fig. 23.

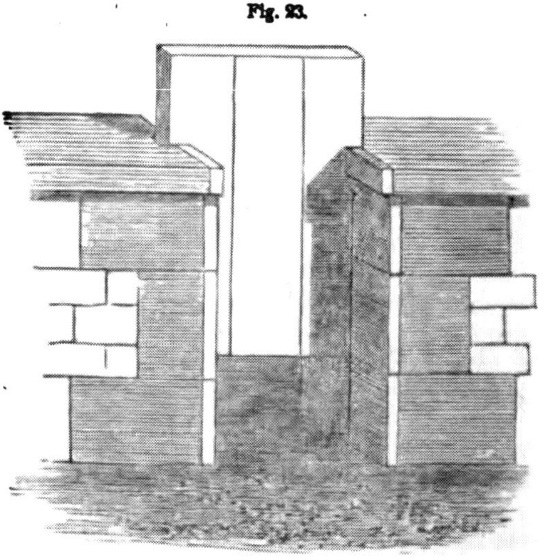

DOOR OF PIG-STY.

Plate III. and in the plan fig. 4, Plate IV. These two sorts of sties may each have a roof of its own, or a number of them may have a large roof over them in common. The former is the usual plan, but the latter is the most convenient for cleaning out, and viewing the internal condition of the sties. 3. The third kind is for the accommodation of weaned young pigs, when it is considered necessary to confine them. These should have a shed at one end of the court, to contain litter for their beds. The court and shed are represented at *b* in the plan, fig. 4, Plate IV. and isometrical view, fig. 3, plate III. They extend 25 feet in length and 21 feet in width. 4. The floors of all these kinds of sties should be laid with stout flags to prevent every attempt of the swine digging into the ground with their snouts. 5. As swine are very strong in the neck, and apt to push up common doors, the best kind of door which I have found for confining them by, is that formed of stout boards, made to slip up and down within a groove in hewn stones forming the entrance in the outside wall. This form of door may be seen in fig. 23, and seems to require no detailed description.

(69.) *Domestic fowls* require accommodation in the steading as well as other stock. 1. They should be provided with *houses for hatching their eggs in*, as also for *roosting* in undisturbed, and both kinds should be constructed in accordance with the nature of the birds, that is, those fowls which roost on high should be kept in a different house from those which rest on the ground. The roosts should be made of horizontal round spars of wood, and spaces of 18 inches cube should be made of wood or stone at a hight of 1 foot or 18 inches from the ground, to contain the straw nests for those which are laying. The hatching-houses should be fitted up with separate compartments containing large nests elevated only 3 or 4 inches above the level of the floor. 2. The foundations should be of large stones and the flooring of strong flags, firmly secured with mortar above a body of small broken stones, and the roofs completely filled in under the slates to prevent the possibility of vermin lodging either above or below ground. 3. Good doors with locks and keys should be put on the houses, windows provided for the admission of light and air, and an opening made in the outer wall of the roosting-house, 4 feet above the ground, to admit the fowls which roost on high by a trap-ladder resting on the ground, as at *d* in fig. 3, Plate III. The roof should be water-tight at all times, and lathed and plastered in the inside, for warmth and cleanliness. The fowls' house may be seen at *d* in the isometrical view, fig. 3, Plate III, and in the plan, fig. 4, Plate IV. 4. It is not absolutely necessary that either the hen-houses or pig-sties should be placed where they are represented in the plan; but as they do not there interrupt the free entrance of the sun into the court K, and therefore do not interfere with the comfort of more important stock, they are there of easy access, themselves quite exposed to the sun, which they should always be, and they square up the front of the farmstead. The hen-house has been recommended by some people to be built near the cow-byre to derive warmth from it; but all the heat that can be obtained from mere juxtaposition to a byre is quite unimportant, and not to be compared to the heat of the sun in a southern aspect.

(70.) S, in both the plan, fig. 4, Plate IV. and isometrical view, fig. 3, Plate III. is part of the *stack-yard*. 1. As most of the stacks must stand on the ground, the stack-yard should receive that form which will allow the rain-water to run off and not injure their bottoms. This is done by ridging up the ground. The minimum breadth of these ridges may be determined in this way. The usual length of the straw of the grain crops can be conveniently packed in stacks of 15 feet diameter; and as 3 feet is little enough space to be left on the ground between the stacks, the ridges should not be made of less width than 18 feet. 2. The stack-yard should be inclosed with a substantial stone and lime wall of 4½ feet in hight. In too many instances the stack-yard is entirely uninclosed and left exposed to the depredation of every animal. 3. It is desirable to place the outside rows of the stacks next the wall on *stools* or *stathels*, which will not only keep them off the wet ground, should they remain a long time in the stack-yard, but in a great measure prevent vermin getting into the stacks. These stathels are usually and most economically made of stone supports and a wooden frame. The frame is of the form of an octagon, under each angle and centre of which is placed a support. The frame-work consists of pieces of plank, *a a*, fig. 24, one of which is 15 feet, and the others 7½ feet in length, 9 inches in depth, and 2½ inches in thickness; and the supports consist of a stone, *b*, sunk to the level of the ground, to form a solid foundation for the upright support, *c*, 18 inches in hight, and 8 inches square, to stand upon, and on the top of this is placed a flat rounded stone

Fig. 24.

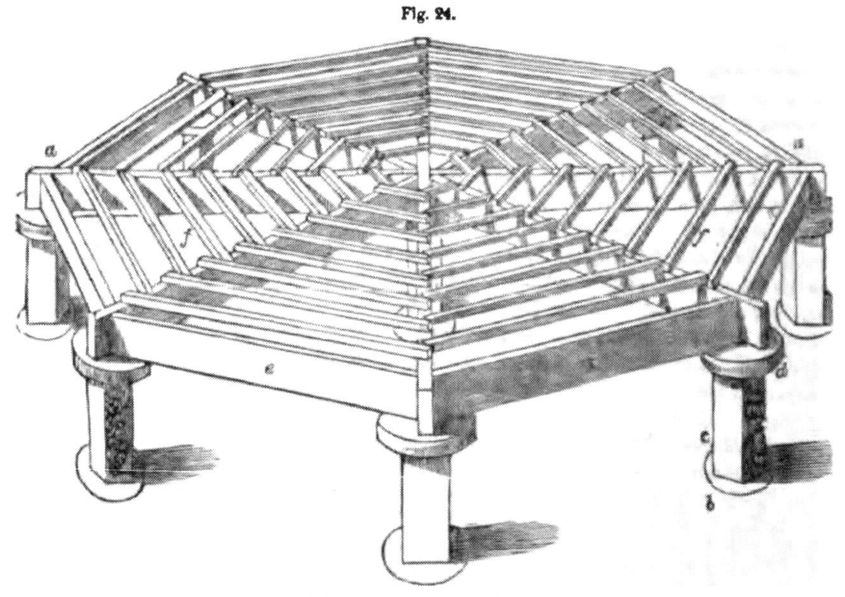

A STATHEL FOR STACKS.

or bonnet, *d*, of at least 2 inches in thickness. The upright stone is bedded in lime, both with the found and bonnet. All the tops of these stone supports, 8 in number around the ninth in the center, must be on the same level. Upon them are placed on edge the scantlings *a*, 9 inches in depth, to each side of which are fastened with strong nails the bearers *e e*, also 9 inches in depth and 2 inches in thickness. In this way each support bears its share of the frame-work. The spaces between the scantlings *a* are filled up with fillets of wood, *f f*, nailed upon them. If the wood of the frame-work were previously preserved by Kyan's process, it would last perhaps twenty years, even if made of any kind of home timber, such as larch or Scotch fir. 4. There should be a wide gateway into the stack-yard, and where the corn is taken on carts to the upper barn to be threshed, the same gateway may answer both purposes, but where there is a gangway to the upper barn, the gate may be placed in the most convenient side of the stack-yard. Where carts are solely used for taking in the corn to the upper barn, the rows of stacks should be built so widely asunder as to permit a loaded cart to pass at least between every two rows of stacks, so that any particular stack may be accessible at pleasure. When a gangway is used, this width of the arrangement of the stacks is not necessary, the usual breadth of 3 feet between the stacks permitting the passage of corn-barrows, or of back-loads of sheaves. Thus, where a gangway is used, the stack-yard is of smaller area to contain the same bulk of grain. 5. Stack-stools, or *stathels*, or *staddles*, as they are variously called, are sometimes made of cast-iron; but these, though neat and efficient, are very expensive and liable to be broken by accidental concussion from carts. Stacks on stathels are represented in fig. 3, Plate III. by figures of stacks, and in fig. 4, Plate IV. by circles. Stathels should also be placed along the stack-yard wall from *m'* to *n'* and from *n'* to *o'* in fig. 3, Plate III.

(71.) A *pigeon-house* is a necessary structure, and may be made to contribute a regular supply of one of the best luxuries raised on a farm. As pigeons are fond of heat at all seasons, there seems no place in the farmstead, especially in winter, better suited for the accommodation of their dwelling than the upper part of the boiling-house. A large pigeon-house is not required, as, with ordinary care, pigeons being very prolific breeders, a sufficient number for the table may be obtained from a few pairs of breeding birds. I have known a pigeon-house not exceeding 6 feet cube, and not very favorably situated either for heat or quietness, yield 150 pairs of pigeons in a season. For a floor, a few stout joists should be laid on the tops of the walls. The flooring should be strong and close, and the sides front and roof, in the inside, lathed and plastered. A small door will suffice for an entrance, to which access may be obtained from the boiling-house by a ladder. The pigeon-holes may be seen in the gable of the boiling-house U, in the isometrical view, fig. 3, Plate III. They may be formed of wood or stone, and should always be kept bright with white paint. The cells in this sort of pigeon-house should be made of wood, and placed all round the walls. I think that 9 inches cube are large enough for the cells. Another site for a pigeon-house may be chosen in the gable of the hay-loft above the riding-horse stable, in fig. 1, Plate I.

(72.) Although potatoes are best kept in winter in pits, yet an apartment to contain those in use for any of the stock, will be found very convenient in every steading. For convenience the *potato-store* should be near the place of their consumption or their preparation into food. In the latter case, proximity to the boiling-house is convenient. 1. Accordingly one potato-store will be found at *f'*, just at the door of the boiling-house U, in the isometrical view, fig. 3, Plate III. and plan fig. 4, Plate IV. It is 30 feet in length and 10 feet in width, and its door being placed in the center, two kinds of potatoes may conveniently be stored in it at the same time, without the chance of admixture. The door should be provided with a good lock and key. 2. Another store of potatoes may be placed in the apartment *f'* next the cart-shed V, 18 feet by 15 feet, to supply them to the feeding beasts in the hammels M, or to the young stock in the courts; but should this apartment not be required for this, it can be used for any other purpose.

(73.) *Rats and mice* are very destructive and dirty vermin in steadings, and particularly to grain in granaries. Many expedients have been tried to destroy them in granaries, such as putting up a smooth triangular board across each corner, near the top of the wall. The vermin come down any part of the walls to the corn at their leisure, but when disturbed run to the corners, up which they easily ascend, but are prevented gaining the top of the wall by the triangular boards, and on falling down either on the corn or the floor, are there easily destroyed. But preventive means in this case are much better than destructive, inasmuch as the granaries are thereby always kept free of them, and consequently always sweet and clean. 1. The great means of prevention is, to deprive vermin of convenient places to breed in above ground, and this may be accomplished in all farmsteads by building up the tops of all the walls, whether of partitions or gables, to the sarking, or the slates, or tiles, as the case may be, and beam-filling the tops of the side walls, between the legs of the couples, with stone and mortar; taking care to keep the mortar from contact with the timber. These places form the favorite breeding-ground of vermin in farmsteads, but which delightful occupation will be put a stop to there, when occupied with substantial stone and mortar. The top of every wall, whether of stables, cow-houses, hammels, and other houses, should be treated in this manner; for, if one place be left them to breed in the young fry will find access to the corn in some way. The tops of the walls of old as well as of new farmsteads can be treated in this manner, either from the inside, or, if necessary, by removing the slates or tiles until the alteration is effected. One precaution only is necessary to be attended to in making beam-fillings, especially in new buildings, which is, to leave a little space *under* every couple face, to allow room for subsidence or the bending of the couples after the slates are put on. Were the couples, when bare, pinned firmly up with stone and lime, the hard points would act as fulcra, over which the long arm of the couple, while subsiding, with the load of slates new put on, would act as a lever, and cause their points to rise, and thereby start the nails from the wall plates, to the imminent risk of pushing out the tops of the walls, and sinking the top of the roof. 2. But besides the tops of the walls, rats and mice breed under ground, and find access into apartments through the floor. To prevent lodgment in those places also, it will be proper to lay the strongest flagging and causewaying upon a bed of mortar spread over a body of 9 inches of small broken stones, around the walls of every apartment on the ground-floor where any food for them may chance to fall, such as in the stables, byres, boiling-house, calves'-house, implement-house, hay-house, pig-sties, and hen-house. The corn-barn has already been provided for against the attacks of vermin; but it will not be so easy to prevent their lodgment in the floors of the straw-barn and hammels, where no causewaying is usually employed. The principal means of prevention in those places are, in the first place, to make the foundation of the walls very deep, not less than two feet, and then fill up the interior space between the walls with a substantial masonry of stone and lime mixed with broken glass; or perhaps a thick body of small broken stones would be sufficient, as rats cannot burrow in them as in earth.

(74.) It is very desirable, in all courts occupied by stock, to prevent the farther discharge of rain-water into them, than what may happen to fall upon them directly from the heavens. 1. For this purpose all the eaves of the roofs which surround such courts should be provided with *rain-water spouts*, to carry off the superfluous water, not only from the roofs, but to convey it away in drains into a ditch at a distance from, and not allow it to overflow the roads around, the farmstead. 2. With a similar object in view, and with the farther object of preserving the foundations of the walls from damp, *drains* should be formed along the bottom of every wall not immediately surrounding any of the courts. These drains should be dug 3 inches below the foundation-stones of the walls, a conduit formed in them of tile and sole, or flat stones, and the space above the conduit to the surface of the ground filled up with broken stones. These broken stones receive the drop from the roofs, and carry away the water; and, should they become hardened above the drains, or grown over with grass, the grass may be easily removed, and the stones loosened by the action of a hand-pick. Rain-water spouts should be placed under the front-eaves of the building A A, and on both sides of the straw-barn L, and along the front-eaves of the stables O and P, of the byre Q, calves' cribs R, and of the hammels M and N. These lines of eaves may easily be traced in the isometrical view, fig. 3, Plate III. The spouts may be made either of wood or cast-iron, the latter being the more durable, and fastened to the wall by iron holdfasts. Lead spouts are, I fear, too expensive for a steading, though they are by far the best. The positions of the rain-water drains around the steading may be traced along the dotted lines, and the courses the water takes in them are marked by arrows, as in the plan, fig. 4, Plate IV.

(75.) But it is as requisite to have the means of conveying away superfluous water *from* the courts, as it is to prevent its discharge *into* them. 1. For this purpose, a *drain* should enter into each of the large courts, and one across the middle of each set of hammels. The ground of every court should be so laid off as to make the lowest part of the court at the place where the drain commences or passes; and such lowest point should be furnished with a strong block of hewn freestone, into which is sunk flush an iron grating, having the bars only an inch asunder, to prevent the passage of straws into the drain. Fig. 25 gives an idea of such a grating, made of malleable iron, to bear rough usage, such as the wheel of a cart passing over it; the bars being placed across, with a curve *downward*, to keep them clear of obstructions for the water to pass through them. A writer, in speaking of such gratings, says, "they should be strong, and have the ribs well bent *upward*, as in that form they are not so liable to be choked up."* This remark is quite true in regard to the form gratings should have in the sewers of towns, for with the ribs bent *downward* in such a place, the accumulated stuff brought upon them by the water would soon prevent the water getting down into the drains; but the case is quite different in courts where the straw covers the gratings from the first, and where being loose over the grating whose ribs are bent downward, it acts *as a drainer*, but were the gratings bent upward, as recommended, the same straw, instead of acting as loose materials in a drain through which the water percolates easily, would press hard against the ribs, and prevent the percolation of water through them. Any one may have perceived that the straw of dunghills presses much harder against a raised stone in the ground below it, than against a hollow. The positions of these gratings are indicated in the plan, fig. 4, Plate IV. by x in the different courts; and in fig. 2, Plate II. they are seen at the origin of all the liquid manure drains, in the form of small dark squares. 2. Drains *from* the courts which convey away *liquid manure* as well as superfluous water, should be of a different construction from those described for the purpose of carrying away rain-water. They should be built with stone and lime walls, 9 inches high and 6 inches asunder, flagged smoothly in the bottom, and covered with

Fig. 25.

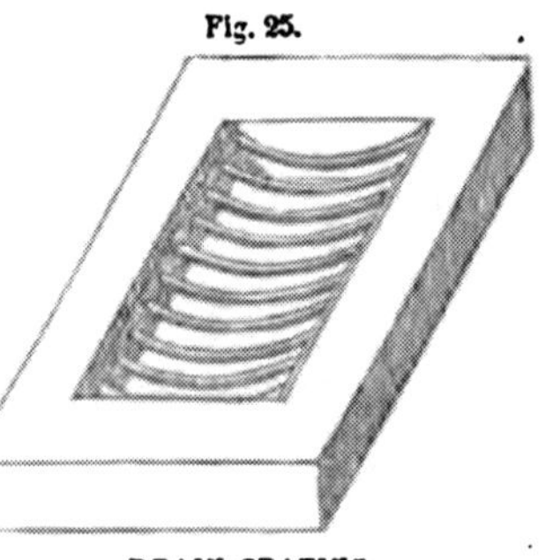

DRAIN GRATING.

* Highland and Agricultural Society's Prize Essays, vol. viii. p. 375.

single stones. Fig. 26 shows the form of this sort of drain, and sufficiently explains its structure. As liquid manure is sluggish in its motion, the drains conveying it require a much greater fall in their course than rain-water drains. They should also run in direct lines, and have as few turnings as possible in their passage to the *reservoir* or *tank*, which should be situate in the lowest part of the ground, not far from the steading, and at some convenient place in which composts may be formed. One advantage of these drains being made straight is, that, should any of them choke up at any time by any obstruction, a large quantity of water might be poured down with effect through them, to clear the obstruction away, as none of them are very long. These drains may be seen in the plan, fig. 4, Plate IV. to run from x in their respective courts in straight lines to the tank k'. It would be possible to have a tank in each set of hammels and courts, to let the liquid manure run directly into them; but the multiplicity of tanks which such an arrangement would occasion, would be attended with much expense at first, and much inconvenience at all times thereafter in being so far removed from the composts. Were the practice adopted of taking the liquid manure to the field at once, and pouring it on the ground, as is done by the Flemish farmers, then a tank in every court would be convenient.

Fig. 26.

LIQUID MANURE DRAIN.

(76.) The *liquid manure tank* should be built of stone or brick and lime. Its form may be either round, rectangular, or irregular; and it may be arched, covered with wood, left open, or placed under a slated or thatched roof—the arch forming the most complete roof, in which case the rectangular form should be chosen. I have found a tank of an area of only 100 square feet, and a depth of 6 feet below the bottoms of the drains, contain a large proportion of the whole liquid manure collected during the winter, from courts and hammels well littered with straw, in a steading for 300 acres, where rain-water spouts were used. The position of the tank may be seen in the plan, fig. 4, Plate IV. at k'. It is rectangular, 34 feet in length and 8 feet in width, and might be roofed with an arch. The tank z, in the isometrical view, fig. 3, Plate III. is made circular, to show the various forms in which tanks may be made. A *cast-iron pump* should be affixed to one end of the tank, the spout of which should be as elevated as to allow the liquid to run into the bung-hole of a large barrel placed on the framing of a cart.

(77.) *Gates* should be placed on every inclosed area about the steading. Those courts which require the service of carts should have gateways of not less width than 9 feet; the others proportionally less. 1. The more common form of gate is that of the five-barred, and which, when made strong enough, is a very convenient form. It is usually hung by a heel-crook and band. I am not fond of gates being made to shut of themselves, particularly at a steading; for, whatever ease of mind that property may give to those whose business it is to look after the inclosure of the courts, it may too often cause neglect of fastening the gate after it is shut; and, unless gates are constantly *fastened* where live-stock are confined, they may nearly as well be left altogether open. The force of the contrivance of gates to shut of themselves has often the effect of knocking them to pieces against the withholding-posts. 2. Sometimes large boarded doors are used as gates in courts and especially in a wall common to two courts. They are, at best, clumsy looking things, and are apt to destroy themselves by their own intrinsic weight. 3. Sometimes the gate is made to move up like the sash of a window, by the action of cords and weights running over pulleys on high posts—the gate being lifted so high as to admit loaded carts under it. This may be an eligible mode of working a gate betwixt two courts in the peculiar position in which the dung accumulating on both sides prevents its ordinary action, but in other respects it is of too complicated and expensive a construction to be frequently adopted. I shall have occasion afterward to speak at large on the proper construction of gates.

(78.) I wish to *suggest some slight modifications* of this plan of a steading, as they may more opportunely suit the views of some farmers than the particular arrangements which have been just described. 1. I have already suggested that, if the large courts I and K are to be dispensed with and hammels adopted in their stead, the hammels M could be produced toward the left as far as the causeway *e′*, on the right hand of the straw-barn L; and so could the hammels N be produced toward the right as far as the south gate of the court I. By this arrangement the cart-shed V, and store-houses *g* and *f′*, would be dispensed with, and the cattle-sheds D D converted into cart-sheds and a potato-store. 2. The piggeries *a*, *b*, and *c*, could then be erected in the middle of the court at K, and the hen-houses in the middle of the court I, respectively, of even larger dimensions than I have given them in the places they occupy. 3. If desired, the work-horse stable O might be separated from the principal range A by a cart-passage, as is the case with the byre-range Q, by which alteration the hay-house and stable would have doors opposite, and the present north door of the hay-house dispensed with. It would be no inconvenience to the plowmen to carry the hay and corn to the horses across the passage. 4. If the stable were disjoined, the right-hand granary may have a window in the east gable, uniform with that in the west over the implement-house G. 5. It may be objected to the boiling-house U being too far removed from the work-horse stable O. As there is as little inherent affinity betwixt a boiling-house and byre as betwixt one and a stable, the boiling-house might be removed nearer to the stable, say to the site of the riding-horse stable P, and the coach-house Z could then be converted into a potato-store, with a common door. 6. The gig-house and riding-horse stable could be built anywhere in a separate range, or in conjunction with the smithy and carpenter's shop, should these latter apartments be desired at the steading.—7. The servants' cow-byre Y could be shifted to the other end of the hammel range N, to allow the hammels to be nearest the straw-barn. 8. Any or all of these modifications may be adopted, and yet the principle on which the steading is constructed would not be at all affected. Let any or all of them be adopted by those who consider them improvements of the plan represented on Plate IV.

(79.) As I have mentioned both a *smithy* and *carpenter's shop* in connection with the steading, it is necessary to say a few words regarding them. It is customary for farmers to agree for the repairs of the iron and wood work of the farm with a smith and carpenter respectively at a fixed sum a-year. When the smithy and carpenter's shops are near the steading, the horses are sent to the smithy, and every sort of work is performed in the mechanic's own premises; but when they are situate at such a distance as to impose considerable labor on horses and men going to and from them, then the farmer erects a smithy at the steading for his own use, fitting it up with a forge, bellows, anvil, and work-bench. Such a smithy, to contain a pair of draught-horses when shoeing, would require to be 24 feet in length and 15 feet in width, with a wide door in the center, 7½ feet high, and a glazed window on each side of it. As the time of a pair of horses is more valuable than that of a man, a smithy is often erected at the steading, while the carpenter's shop is at a distance.

(80.) All the roads around the steading should be properly made of a thick bed, of not less than 9 inches, of small broken whinstone metal, carefully kept dry, with proper outlets for water at the lowest points of the metal bed, and the metal occasionally raked and rolled on the surface until it becomes solid.

(81.) The best way of building such a steading as I have just described is, *not to contract for it in a slump sum;* because, whatever alterations are made during the progress of the work, the contractor may take advantage of the circumstance, and charge whatever he chooses for the extra work executed, without your having a check upon his charges. Nor, for the same reasons, should the mason, carpenter, or slater work be contracted for separately in the slump. The *prices per rood or per yard, and the quantities of each kind of work, should be settled beforehand between the employer and contractor.* The advantage of this arrangement is, that the work is finished according to the views and tastes of the individual for whose use the farmstead has been built—he having had the power of adopting such slight modifications of the plan, during the progress of the work, as experience or reflection may have suggested. The contractor is paid according to the measurement of the work he thus executes. A licensed surveyor, mutually chosen by both parties, then measures the work, and calculates its several parts according to the prices stipulated for betwixt the contractor and his employer, and draws up a report of the value of each kind of work, the total sum of which constitutes the cost of the farmstead. Installments of payment are, of course, made to the contractor at periods previously agreed upon. This plan may give you no cheaper a steading than the usual one of contracting by a slump sum, but cheapness is not the principal object which you should have in view in building a steading.—Your chief object should be the convenience of your work-people, and the comfort of your live-stock. This plan enables you to erect a steading in accordance with your own views in every respect; and you can better judge, in the progress of the work, of the fitness of the plan for the accommodation required, than by any study of the plans on paper—which, upon the whole, may appear well enough adapted to the purposes intended, but may, nevertheless, overlook many essential particulars of accommodation and comfort.

(82.) What I mean by essential particulars of accommodation and comfort in a steading are such as these: In giving a foot or two more length to a stable or byre, by which each animal may have two or three inches more room laterally, more ease would be given to it, and which is a great comfort to working stock: A window, instead of looking to the cold north, may be made with as much ease to look to the warm south: A sky-light in the roof, to afford a sufficient light to a place that would otherwise be dark: An additional drain to remove moisture or effluvia, which, if left undisturbed, may give considerable annoyance: A door opening one way instead of the other, may direct a draught of air to a quarter where it can do no harm: These little conveniences incur no more cost than the incongruities of arrangement which are often found in their stead, and though they may seem to many people as trifles unworthy of notice, confer, nevertheless, much additional comfort on the animals inhabiting the apartments in which they should be made. A door made of a whole piece, or divided into leaves, may make a chamber either gloomy or cheerful; and the leaves of a door formed either vertically or horizontally, when left open, may either give security to an apartment, or leave it at liberty to the intrusion of every passer by. There are numerous such small conveniences to be attended to in the construction of a steading before it can be rendered truly *commodious* and *comfortable*.

(83.) Before the prices of work to be executed can be fixed on between the employer and contractor, *minute specifications* of every species of work should be drawn up by a person competent for the task. A vague specification, couched in general terms, will not answer; for when work comes to be executed under it, too much liberty is given to all parties to interpret the terms according to the interest of each. Hence arise disputes, which may not be easily settled even on reference to the person who drew up the specifications, as he possibly may by that time have either forgotten his own ideas of the matter, or, in adducing his original intentions under the particular circumstances, may possibly give offence to one party, and injure the other; and thus his candor may rather widen than repair the breach. Whatever are the ideas of him who draws up the specification, it is much better to have them all embodied in the specifications, than to have to explain them afterward.

(84.) The *principle* of *measuring* the whole work after it has been executed, is another consideration which it is essential you should bear in mind. It is too much the practice to tolerate a very loose mode of measuring work; such as measuring *voids*, as the openings of doors and windows are termed, that is, on measuring a wall, to include all the openings in the rubble-work, and afterward to measure the lintels and ribets and corners. In like manner, chimney-tops are measured all round as rubble and then the corners are measured also as hewn work. Now the fair plan obviously is to measure every sort of work as it stands by itself; where there *is* rubble let it be measured for rubble, and where there *is* hewn work let it be measured for as such. You will thus pay for what work is actually done for you, and no more; and more you should not pay for, let the price of the work be what it may. This understanding regarding the principle of measurement should be embodied in the specifications.

(85.) To see if the *principle* I have endeavored to enforce in the arrangement of the component parts of a steading for the mixed husbandry be applicable to steadings for other modes of husbandry, you have only to apply it to the construction of steadings usually found in the country.

(86.) In *pastoral* farms, the accommodation for stock in the steading is generally quite inadequate to give shelter, in a severe winter and spring, to the numbers of animals reared on them. For want of adequate accommodation, many of both the younger and older stock suffer loss of condition—a contingency much to be deprecated by the store-farmer, as the occurrence never fails to render the stock liable to be attacked by some fatal disease at a fu-

Fig. 27.

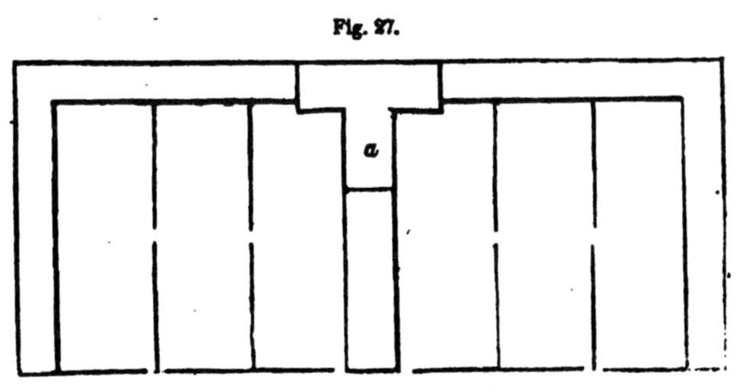

PASTORAL FARM-STEADING.

ture period. In the steadings of such farms, the numerous cattle, or still more numerous sheep, as the stock may happen to be, should have shelter. The cattle should be housed in sheds or hammels in stormy weather, supplied with straw for litter and provender, or, what is still better, supported on hay or turnips. For this purpose their sheds should be quite contiguous to the straw-barn. Sheep should either be put in large courts bedded with straw, and supplied with hay or turnips, or so supplied in a sheltered spot, not far distant from the steading. The particular form of steading suitable to this species of farm seems to be that which embraces three sides of a double rectangle, having the fourth side open to the south, each rectangle enclosing a large court, divided into two or more parts, on each side of the straw-barn, which should form a side common to both rectangles. This form answers to the modification pointed out at 3, in paragraph (8.) p. 82, and it is shown in fig. 27, where *a* is the straw-barn, with but the courts placed on each side of it.

(87.) In the steadings of *carse* farms, comfortable accommodation for stock is made a matter of secondary import. In them it is not unusual to see the cattle-courts facing the north. As there is, however, great abundance of straw on such farms, the stock seem to be warm enough lodged at night. Where so much straw is required to be made into manure, the courts and stables should be placed quite contiguous to the straw-barn. The form of steading most suitable to this kind of farm seems to be that of three sides of a rectangle, embracing a large court, divided into two or three parts, facing the south, and having the upper and corn-barn projecting behind into the straw-yard, as described in modification 2, (8.) p. 82, and shown in fig. 28, where *a* is the straw-barn, near the courts, and contiguous to which should be the byres and stables.

Fig. 28.

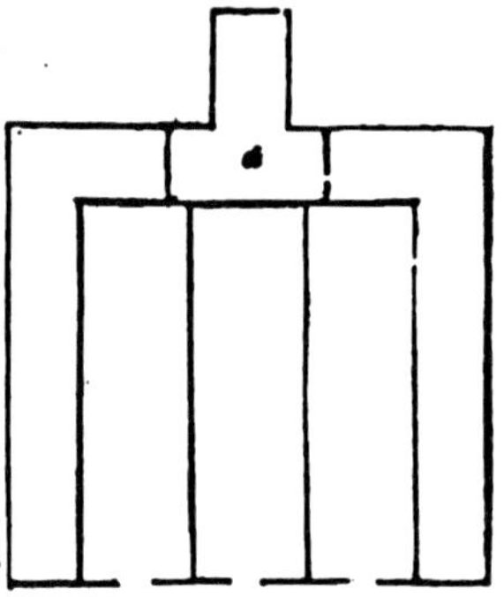

CARSE FARM-STEADING.

(88.) In farms in the *neighborhood of towns*, the cow-houses, feeding-byres, or hammels, being the only means of converting the straw into manure, which is reserved for home use from the sale of the greatest part to the cow-feeders and stablers in towns, should be placed nearest the straw-barn. The very confined state in which cows are usually kept in the byres of such farms, and especially in those near the largest class of towns, makes them very dirty, the effects of which must injure the quality of the dairy produce. In constructing a steading for a farm of this kind, such an inconvenience should be avoided. The most convenient form of steading is that of the three sides of a rectangle, embracing within it a set of feeding-hammels facing the south; the threshing-mill and straw-barn being in the north range, the work-horse stable in one of the wings, and the cow-byre in the other, from both of which the dung may be wheeled into their respective contiguous dunghills, as is described in modification 1, (8.) p. 82, and shown graphically in fig. 29, where *a* is the straw-barn, on both sides of which are the byres and stable, and *c* are hammels inclosed within the rectangle.

Fig. 29.

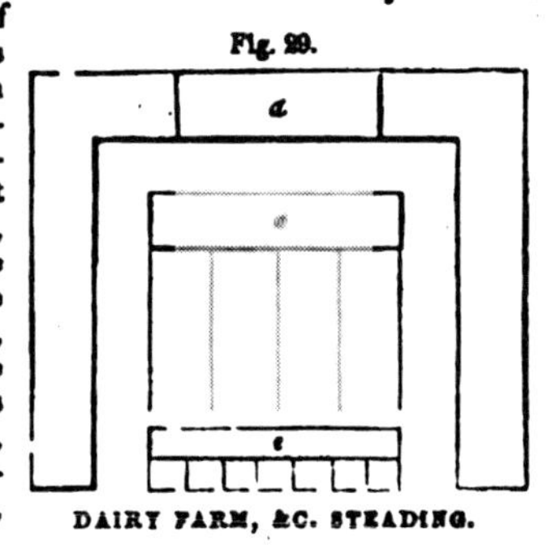

DAIRY FARM, &C. STEADING.

(89.) In *dairy* farms, the cows being the greatest means of making manure, their byres, as well as the hammels for the young horses and young queys, and the sties for the swine, should be those most contiguous to the straw-barn. It should be the particular study of the dairy-farmer to make the byre roomy and comfortable to the cows, the thriving state of that portion of his stock being the source from which his profits are principally derived. The form of steading recommended for farms in the neighborhood of towns seems well adapted to this kind of farming, in which the hammels could be occupied by the young horses and young queys, and beside which the pig-sties could also be placed, such as are shown in fig. 29, where *c* are the hammels, and *e* the hog-sties, but which may be placed elsewhere if desired.

(90.) It may prove of service to inquire whether this principle of constructing steadings for every sort of farm is inculcated by the most recent or authoritative writers on Agriculture. 1. In the collection of designs of farm-buildings, in the Prize Essays* of the Highland and Agricultural Society of Scotland, the absolute necessity for the contiguity of cattle-sheds, hammels, and stables to the straw-barn, is a matter not sufficiently attended to. When hammels are placed in front of the principal buildings, as in No. 1 of the designs, doors are required in the back of the hammels for taking in the straw. These doors not only incur additional cost in the making, but, being placed in the shed, induce the animals to escape through them, and, when open, occasion an uncomfortable draught of air. The openings, too, betwixt the sheds and courts of the hammels being placed in the center, cold easily circulates through the sheds. And the separation of the calves'-house from the cow-byre as in

* Prize Essays of the Highland and Agricultural Society, vol. viii. p. 363.

design No. 2, must be very inconvenient in rearing calves. 2. In "British Husbandry," the principle of constructing a steading is thus laid down: "The position of a threshing-mill should decide that of almost every other office; for it cuts, or ought to cut, the hay into chaff, together with much of the straw; and the house that immediately receives this chaff ought to be so placed as to admit of a convenient delivery to the stalls and stables. Thus the straw-barn, chaff-house, ox-stalls, and horse-stables, with the hay-stacks and the sheep-yard (if there be any), should be dependent on the position of the threshing mill, as they will be attended with waste and expense of labor."* If the chaff-cutting machine is to be employed for preparing much of the straw for the use of the stock, it should be placed in the *straw-barn*, otherwise the straw must be carried to it, which would entail a considerable deal of labor. It is thus the position of neither the chaff-cutting or threshing-machine that should determine the site of the rest of the steading. The threshing-machine cannot conveniently be placed near the centre of a steading, because it would then be necessarily removed to a distance from the stack-yard, and the carriage of the sheaves from which would also entail considerable labor. In the examples of existing steadings given in this recent work from pages 85 to 109, being chiefly the plans of steadings on the properties of the Duke of Sutherland, the position of the straw-barn seems in them to be considered a matter of secondary importance.—In the plans in pages 85, 86, 100, 103, 107, 108, and 109, the straw-barn is surely placed at an inconvenient distance from the apartments occupied by the live-stock, and the carriage of straw from it to them must "be attended with waste and expense of labor." 3. Professor Low inculcates the principle more correctly where he says, "Barns, being the part whence the straw for fodder and litter is carried to the stables, feeding-houses, and sheds, they should be placed so as to afford the readiest access to these different buildings. It is common to place them as near the centre of the range as the general arrangement of the other buildings will allow."† This is quite correct in principle; but, in referring to the figure, it is said that, "In the design of the figure, in which are represented the barns, this principle of arrangement is observed;" yet, on inspecting the figure at p. 624, it will be observed that the feeding-hammels are placed at a greater distance from the straw-barn than even the pig-sties and poultry-yards. It does not appear that the yard behind the sties is intended to be occupied by anything but manure; so, if the hammels had occupied the more eligible site of the hog-sties, they would not have interposed betwixt the sun-light and any stock. It may also be observed that the cow-houses, which require less straw than feeding stock, are placed nearer the straw-barn than the hammels on the right. 4. Mr. Loudon, in treating of the "fundamental principles for the construction of the various parts which compose a farmery," recommends the houses for the various kinds of stock to be constructed according to the size and shape of the animals to be accommodated; and, *assuming* the horse, the ox and the sheep to be of the form of a wedge, he draws these two conclusions: "First, that the most economical mode of lodging the first two of these quadrupeds must be in houses the walls of which form concentric circles, or segments of circles parallel to each other; . . . . and, secondly, that in all open yards where quadrupeds are allowed to run loose, and eat from racks and mangers, ... when the rack or manger is to be in a straight line, the breadth of the broad end of the wedge must be allowed for each animal; and, when it is to be curved, the radius of the curve must be determined by the breadth of the smaller end of the wedge.—From this theory it may also be deduced that there must be one magnitude, as well as one form, more economical than any other, for lodging each of these animals; and that this magnitude must be that circumference of a circle which the narrow ends of the wedges completely fill up, and no more."‡ And figures are given of both curved and straight mangers and racks, to illustrate these principles. Now, independent of the acknowledged inconvenience of accommodating any circular or curved form of apartment or building in a steading, as is universally felt in regard to the usually circular form of the horse-course of a threshing mill, the very data on which this theory is founded are incorrect; for, although it is true that many horses and oxen are of the form of a wedge, yet the higher bred and better stock, to which all improving breeders are desirous of assimilating their own, are not wedge-shaped. The Clydesdale draught-horse, the short-horn ox, and the Leicester sheep, the nearer they attain perfect symmetry of form, the nearer they approach the form of a parallelopipedon, instead of a wedge, in the carcass. This theory is, therefore, not universally applicable.—Indeed, Mr. Loudon afterward says (p. 375) that "these principles for the curvilinear arrangement of stalls, racks, and troughs, we do not lay down as of very great importance, but rather with a view to induce the young architect to inquire into the reasons of things, and to endeavor in everything to take principles into consideration rather than precedents." The object is laudable, but its aim will scarcely be attained by the young architect having his attention directed to questionable data. He is at all times much more disposed to follow his own crude fancies, in the construction of steadings, than to improve on precedents suggested by the farmer's experience.

(91.) It may be interesting to inquire *why* the quadrangular form of steading was so much in vogue some years ago. It was, doubtless, adopted on account of its compactness of form,

* British Husbandry, vol. i. p. 97. † Low's Elements of Practical Agriculture, 2d edition, p. 623.
‡ Loudon's Encyclopædia of Architecture, p. 373.

admitting it to be erected at a considerable saving of expense, at a time—during that of the war—when building-materials of every kind, and wages of every description of artisans, were very high. I do not believe that the value of all the ground on which the largest steading could stand formed any inducement for the adoption of the compact form of the quadrangle, but rather from the wish of the landlord to afford no more than bare accommodation to the tenant's stocking. An economical plan, furnished by an architect, would thus weigh more strongly with him than a mere regard for the comfort of his tenant's live-stock, whose special care he would consider more a tenant's than a landlord's business. It is not so easy to account for the tenant's acquiescence in such a form of steading; for, although it must be owned that, at that period, very imperfect notions were entertained of what were requisite for the comfortable accommodation of animals, yet the tenant's own interest being so palpably involved in the welfare of his stock, might have taught him to desire a more comfortable form of steading. Thus an imperfect state of things originated in the parsimony of landlords, and was promoted by the heedlessness of tenants. The consequences were that cattle were confined in courts inclosed all around with high buildings, eating dirty turnips off the dunghill, and wading or standing mid-leg deep in dung and water; and frequently so crowded together, and stinted of food, that the most timid among them were daily deprived of their due proportion of both food and shelter. Is it any matter of wonder that cattle at that time were unequally and imperfectly fed? In the steadings of the smaller tenants, matters were, if possible, still worse. The state of the cattle in them was pitiable in the extreme, whether in the courts, or while "cabined, cribbed, and confined" in the byres. Though those in the latter were, no doubt, under the constant shelter of a roof, they were not much better off as to cleanliness and food; and much worse off for want of fresh air, and in a state of body constantly covered with perspiration. But these unmerited hardships, which the cattle had to endure every winter, have been either entirely removed or much ameliorated, within these few years, by the adoption of conveniences in the construction of steadings on the part of landlords, and superior management, acquired by experience, on the part of tenants.—Troughs are now erected along the walls of courts, at convenient places, for holding turnips, now given clean to the cattle. Rain-water spouts are now put along the eaves of the houses surrounding the courts. Drains are now formed to carry off the superfluous moisture from the courts. The courts themselves are opened up to the meridian sun, and really made comfortable for cattle. And hammels are now built for cattle in steadings where they were before unknown. Still, notwithstanding the decided improvement which has undoubtedly taken place in the construction of steadings, there are yet many old steadings which have not been amended, and too many modern ones erected in which all the improvements that might have been have not been introduced. Should it be your fate to take a farm on which an old steading of the quadrangular form is standing, or a new one is proposed to be built, in repairing the one, and constructing the other, be sure never to lose sight of *the leading principles* of construction inculcated above, and insist on their being put into practice. A little pertinacity on your part on this point will, most probably, obtain for you all your wishes, and their attainment to the full will vindicate you in offering a higher rent for the farm, without incurring risk of loss.

(92.) It is now time to enter minutely into the *specifications* upon which every kind of work in the construction of a substantial steading should be executed, and those below will be found applicable to every size and plan of steading. As they accord with my own experience and observation in these matters, and both have been considerable, I offer them with the greater degree of confidence for your guidance. They embrace the particulars of mason-work, carpenter-work, slater-work, plumber-work, smith-work, and painter and glazier-work; but they are not drawn up in the formal way that specifications are usually done, the various subjects as they are specified being illustrated by examples and the elucidation of principles.

(93.) Of the specifications of *mason-work*, the first thing to be done is the *digging of the foundations of the walls*. When the site of the steading is not obliged to be chosen on a rock, the depth of the foundations of all the outside walls should never be less than 2 feet. Judging by usual practice, this may be considered an inordinate depth, and as incurring much expense in building an unnecessary quantity of foundation walls, which are immediately after to be buried out of sight; but this depth is necessary on account of the drains which should be made around the outside walls, to keep all the floors dry in winter, and it is scarcely possible to keep them dry with drains of less depth than 27 inches, which afford the water a channel of only 3 inches below the bottom of the foundations. The ground-floor of dwelling-houses may be kept in a dry state by elevating it a considerable hight above the ground; but such an expedient is impracticable in a steading where most of the apartments, being occupied by live-stock, must be kept as near as possible on a level with the ground; and it is not wood-floors alone that must be kept dry, but those of sheds, barns, and byres, whether made of composition, or causeway, or earth. The injurious effects of damp in the floors of stables, byres, and hammels, on the condition of the animals inhabiting them in winter, or of barns on the state of the straw, corn, or hay in them, are too much overlooked. Its malign influences on the health of animals, or in retarding their thriving, not being apparent to the senses at first sight, are apt to be ascribed to constitutional defect in the animals themselves, instead of, perhaps, to the truer cause of the unwholesome state of the apartments which they occupy. The truth is, the floor of every apartment of the steading, whether accommodating living creatures, or containing inanimate things, cannot *be too dry;* and, to render them as much so as is practicable, there seems no way of attaining the end so effectually as to dig the foundations of the walls deep, and to surround them with still deeper drains. This position I shall here endeavor to prove to you satisfactorily. There are many substances upon which walls are usually founded, which, from their nature, would make walls constantly damp, were expedients not used to counteract their natural baleful properties. Amorphous rocks, such as granite, which are impervious to water; whinstone rocks, which, though frequently containing minute fissures, being delinquescent, become very damp in wet weather; clay, and tilly clay even more than the unctuous, retains a great deal of water—all these substances form objectionable ground upon which to found any building. Stratified rocks, such as sandstone, not retaining the water long, form drier substances for a foundation than any of the amorphous rocks or clays. Pure sand is not always dry, and it is apt to form, in some situations, an insecure foundation. Pure gravel is the driest of all foundations, but not the most secure. From the nature of these various substances, excepting the gravel, it would appear that no wall founded on them can assuredly be kept dry at all seasons; and therefore drains are necessary to render and keep them dry at all seasons. Moreover, a foundation made in a bank of even the driest gravel will prove damp, unless the precaution of deep draining betwixt the foundation and the rise of the bank is resorted to. Rather than choose a site for your steading which is overhung by a bank, make a deeper foundation on more level ground, and drain it thoroughly, or even build some hight of waste wall, and fill up a part of the ground that is low around the steading. I have experienced the bad effects of digging a foundation for a steading in a rising ground of tolerably dry materials,and also the good effects of filling up low ground at a part of another steading, and have found the air in the apartments of the latter at all seasons much more agreeable to the feelings than in the former. The bad effects of the former I endeavored to counteract by deep draining, though not so effectually as in the latter case. I am therefore warranted in concluding that dry apartments are much more healthy for animals, and better for other things, than are those which feel cold and damp. A circling, however, of substantial drains around the steading, between it and the bank, will render the apartments to the feelings, in a short time, in a comparatively comfortable state.

(94.) The *outside walls should be founded with stones* 3 feet in length, 2 feet in breadth, and 8 or 9 inches in thickness, so laid, in reference to the line of foundation, as to form a scarcement of 6 inches on each side of the wall above them. The low walls may stand on one course of such foundation, while the higher walls should have two such courses.

(95.) *All the walls*, both external and internal, *should be built of the best rubble-work*, the stones being squared, laid on their natural beds, closely set in good lime mortar, and well headed and packed. Headers should go through the thickness of the walls at not more than 5 feet apart in every third course. The walls should only be built one course in hight on side, before the other side is brought up to the same level, the first of the courses to go through two-thirds of the wall, besides the headers or band-stones.

(96.) The *external walls should be 2 feet in thickness*, and the *internal* division-walls, *as* also the walls composing the fronts and subdivisions of the courts and hammels, 1 *foot*. The low external walls should be raised 9 feet, and the high external walls of the middle range, as well as that of the straw-barn, 15 feet above the ground. All the gables of the external walls, and all the internal division walls, should rise to the pitch of their respective roofs, and be entirely filled up to the sarking or tiles, as the case may be. The front and side walls of the large courts and bulls' hammels, and the subdivision walls of the courts of the hammels, should be raised 6 feet, and the front walls of the hammels, as also those of the cows' and calves' courts and pig-sties, 5 feet above the ground. All the walls which carry roofs should be beam-filled with rubble-work, with the precaution given in (73.) p. 118.

(97.) The *external fronts* of all the outside walls, as well as those of the front walls of the courts and hammels, should be faced with *hammer-dressed rubble in courses*, not exceeding 6 inches in thickness, with the vertical and horizontal joints raised or drawn in hollow. The tops of the front and subdivision walls of the courts and hammels should be finished with a *coping* of hammer-dressed round-headed stones, 12 inches in diameter, firmly set close together in good lime mortar.

(98.) To *test* if rubble masonry is well built, step upon a leveled portion of any course, and, on setting the feet a little asunder, try by a searching motion of the legs and feet whether any of the stones ride upon others. Where the stones ride, they have not been properly bedded in mortar. To ascertain if there are any hollows, pour out a bucketfull of water on the wall, and those places which have not been sufficiently packed or hearted with small stones, will immediately absorb the water.

(99.) The *width of all the doors* should be 3 feet 6 inches, and their hight 7 feet, with the exception of those of the work-horse stable, corn-barn, straw-barn, and saddle-horse stable, which should be 7 feet 6 inches. The width of the arches of the cattle-courts should be 9 feet; that of those of the hammels 6 feet, and that of the ports of the cart-shed 8 feet, and all 7 feet 6 inches in hight. The *width of all the windows* should be 3 feet, and their hight

4 feet, with the exception of those of the granaries, which should be 4 feet in width and 3 feet in hight. The windows should have a bay inside of 6 inches on each side. Slits of 1 foot 3 inches in hight and 3 inches in width in front, with a bay inside like the windows should be left in the walls of the straw and upper barns for the admission of air to the straw and the corn in the straw. All the voids should have substantial discharging arches over the timber-lintels to be able to support the wall above, even although the timber-lintels should fail.

(100.) All the *door-soles* should be laid 3 inches above the ground or causeway, and those of the stables and byres and calves'-house should be beveled in front, that the feet of the animals going out and in may not strike against them.

(101.) The *corners* of the buildings should be of *broached ashler*, neatly squared, 2 feet in length, 12 inches of breadth in the bed, and 12 to 18 inches in hight, having 1 inch chisel draught on both fronts. The *windows and doors should have ashler ribets*—the outbands 2 feet in length, and the inbands at least two-thirds of the thickness of the walls, and both 12 inches of breadth in the beds, and 14 or 15 inches in hight. They should have 1 inch of the front, 5 inches of ingoings, and 4 inches of checks, clean droved. The tails of the outband ribets should be *squared and broached*. The doors of the work-horse and saddle-horse stables, upper and corn barns, hay-house and bulls' hammels, should have droved giblet-checks, to permit them opening outward. The *window-sills* should be droved, projecting 1½ inches, and 6½ or 7 inches in thickness. The *lintels* of both the doors and windows should have 1 inch of the front, 5 inches of ingoings clean droved, and be from 14 to 15 inches in hight.—The *skews* should be broached when such are used, having 1 inch chisel-draught on both margins of the front, and the inner edge with a 4-inch check-plinth, having an inch back-rest under it. The *holes in the byre-wall*, through which the turnips are supplied, should be 20 inches square, with ashler ribets, flush sills and lintels, having broached fronts and droved giblet-checks to receive their shutters. The side *corners of the arched openings* of the cattle-courts and hammels, and those of the ports of the cart-shed, should be regular out and inband, 2 feet in length, 12 inches of breadth in the bed, and 12 inches in hight, and dressed in a manner similar to the other corners, but should be chamfered on the angles. The *arches* should be elliptical, with a rise of 2 feet, with broached soffits on both fronts, an inch-droved margin, and radiated joints. In the plan, fig. 4, Plate IV. the cart-shed ports are not arched, there being no room for such a finishing in the peculiar form of the roof. The *pillars* of the cart-shed, the byre, turnip-shed, and calves'-shed, should be 2 feet square in the waist, of broached ashler, with inch-droved margins, and built of stones 12 inches in hight. Those of the two former should have a droved base course, 12 inches in depth, with 1½ inches washing, chamfered on the angles. The *tops of the walls* of the pig-sties, calves'-shed, hen-house, and potato-store, should have a 6-inch droved plinth, 12 inches in the bed. The *fire-places* in the boiling-house and coach-house should have a pair of droved jambs and a lintel, 3 feet 6 inches of hight in the opening, and a droved hearth-stone 5 feet in length and 3 feet in breadth. The *boiler* should have a hearth-stone 4 feet 6 inches in length, and 2 feet 6 inches in breadth, and it should be built with fire-brick, and have a cope of 4 inches in thickness of droved ashler. The *flues* from both the fire-places and the boiler should be carried up 12 inches clear in the opening, and should have chimney-stalks of broached ashler, 2 feet in hight above the ridges of the respective roofs, 2 feet square, and furnished with a droved check-plinth and block 12 inches in depth. The *gates* of all the cattle and hammel courts should be hung on the droved ashler corners when close to a house, but on droved built pillars when in connection with low court-walls. The riding-horse stable, if laid at all with *flags*, should have them 4 inches thick, of droved and ribbed pavement behind the travis-posts, having a curved water-channel communicating with a drain outside. The travis-posts of the work-horse stable should be provided with droved stone *sockets*, 12 inches in thickness, and 18 inches square, founded on rubble-work, and a droved *curb-stone* should be put betwixt the stone sockets of each pair of head and foot travis-posts, provided with a groove on the upper edge to receive the under edge of the lower travis-board. For the better riddance of the urine from the work-horse stable, there should be a droved curved *water-channel*, 6 inches in breadth, wrought in freestone, all the length of the stable, with a fall at least of 1½ inches to every 10 feet of length. The water-channel in the cow-byres and feeding-houses should be of droved curb-stones, 6 inches thick, 12 inches deep, and laid in the bottom with 3-inch thick of droved pavement, placed 6 inches below the top of the curb-stones. If stone is preferred for *water-troughs*, which it should always be when easily obtained, the troughs should not be of less dimensions than 3½ feet in length, 2 feet in breadth, and 18 inches in depth over all; or they may be made of the same dimensions of pavement-flags put together with iron-batts. Wood may be substituted for stone when that cannot be easily obtained. The liquid-manure *drain* should be 9 inches in hight and 6 inches in width in the clear, with droved flat sills and hammer-dressed covers. A stone 2 feet in length, 18 inches in breadth, and 8 or 9 inches in thickness, with an opening through it, giblet-checked, will contain a *grating* 15 inches in length and 9 inches in breadth, with the bars one inch asunder, at the ends of the liquid-manure drains in the courts. The liquid-manure *tank*, sunk into the ground, will be strong enough with a 9-inch brick or rubble wall of stone and lime-mortar, having the bottom laid with jointed flag-pavement. If the ground is gravelly, a puddling of clay will be requisite behind the walls, and below the pavement of the bottom. The bottom of the *feeding-troughs* in the byres, courts, and hammels, should be of 3-inch thick of flag-pavement, jointed and scabbled on the face, or of wood. All the *window-sills* in the inside should be finished with 3-inch droved or scabbled pavement.

(102.) The *walls in the front of the courts* are intended to be quite *plain;* but, should you prefer ornamental structures, their tops may be finished with a 6-inch droved cope, 15 inches in breadth, with a half-inch washing on both fronts; and with a droved base-course 12 inches in depth, having a washing of 1½ inches. The pillars of the gates to the larger courts may be of droved ashler, in courses of an octagonal form, of 15 inches in thickness, and 2 feet by 2 feet, with 12-inch base, and a 12-inch checked plinth and block, built at least 18 inches higher than the wall. And if you prefer an outside hanging-stair to the upper-barn instead of the gangway, or to the wool-room, the steps should be droved 3 feet 6 inches clear of the wall, with 6 inches of wallhold. And, farther, you may substitute droved crow-steps on the gables for the broached skews, with an inch back-rest under them. These crow-steps, in my opinion, are no ornaments in any case in a steading. They are only suited to a lofty, castellated style of building.

(103.) The floors of the cow-byres, work-horse stable, stalls of the riding-horse stable, passage of the calves'-house, coach-house, boiling-house, implement-house, hay-house, and turnip and potato stores, should be laid in *causeway* with whinstone, or with small land stones, upon a solid stratum of sand, with the precaution of a bed of broken stones under the flagging as formerly recommended in (73.) p. 118. A causeway, 13 feet in breadth, should also be made in the large court K to the corn-barn door, round to the gate at H, for the use of loaded carts from the barn, with a declivity from the wall to the dung area of 2 inches in the 10 feet. Causeways are usually formed in steadings with round hard stones found on the land, or in the channels of rivers, or on the sea-shore, imbedded in sand. In those situations the stones are always hard, being composed of water-worn fragments of the primitive and secondary as well as of trap-rocks; but round boulders of micaceous sandstone, usually found in gravel pits, are unfit for the purpose of causeways, being too soft and slaty. A more perfect form of causeway is made of squared blocks of trap, whether of basalt or greenstone, imbedded in sand, such as is usually to be seen in the streets of towns. The ready cleavage of trap-rocks into convenient square blocks renders them valuable dépôts, where accessible, of materials for causeways and road metal. The floors of the pig-sties and poultry-yards should be laid with strong, thick-jointed stones imbedded in lime mortar, having broken glass in it, upon a bed of 9 inches thick of small broken stones, to withstand not only the digging propensities of the pigs on the surface, but also to prevent vermin gaining access from below through the floor to the poultry. The areas of the cattle-courts, and floors of the sheds, hammels, and cart-shed, will be firm enough with the earth beaten well down.

(104.) There is a plan of making the floors of out-houses, recommended by Mr. Wadell of Berwickshire, which deserves attention. It is this: Let the whole area of the apartment be laid with small broken stones to the depth of 9 inches. Above these let a solid body of masonwork, of stone and lime properly packed, be built to the hight of 12 or 14 inches, according to the thickness of the substance which is to form the upper floor. The lime, which is applied next the walls, should be mixed with broken glass. If a composition is to form the floor, it should be laid on 3 inches in thickness above the masonry; but if asphaltum, 1 inch thick will suffice, the difference in the hight being made up in the masonry.* This plan of Mr. Waddell's seems well adapted for making a solid and secure foundation against vermin, for the causewaying of the several apartments mentioned above; but it is not so well adapted for wood-floors either as a preservative against damp, or preventive against vermin, as the plan described at p. 88, (16.)

(105.) While treating of the subject of causewaying, I may as well mention here the various sorts of flooring and pavement which may be formed of other materials than those already mentioned; and the first is *concrete*, which forms a very good flooring for indoor use. It is formed of a mixture of coal-ashes obtained from furnaces, and from a fourth to a third part or more, according to its strength, of slaked lime, and worked into the form of paste with water. A coating of clay of 2 or 3 inches is first laid on the ground leveled for the purpose, and upon the clay, while in a moist state, the concrete is spread two or three inches in thickness, and beaten down with a rammer or spade until the under part of the concrete is incorporated with the upper part of the clay. The surface of the concrete is then made smooth by beating with the back of a shovel, and when left untouched for a time, that substance assumes a very hard texture. This is a cheap mode of flooring, labor being the principal expense attending it.

(106.) Another sort of pavement is that of *asphaltum*, suitable either for indoor use, or for outdoor purposes, where no cartage is to be employed upon it. It is a composition of bitumen, obtained from coal-tar after the distillation of naphtha, and small clean gravel. When applied, the bitumen and gravel in certain proportions are melted together in a pot over a fire, and when sufficiently liquified and mixed, the composition is poured over the surface of the ground to be paved, which is previously prepared hard and smooth for the purpose,

* Prize Essays of the Highland and Agricultural Society, vol. viii. p. 373.

about an inch or more in thickness, and is spread even and smoothed on the surface with a heated iron roller. When completely dry, the asphaltum becomes a perfect pavement, as hard as stone, and entirely impervious to water. It would form an excellent flooring for the straw-barn, servants'-house, boiling-house, potato-stores, and the passages in the cow-byre and calves'-house. It might also make roofing to out-houses, where there is no chance of the roof being shaken. As made at the Chemical Works at Bonnington near Edinburgh, it costs 5d. per square foot when laid down, which makes it an expensive mode of paving. Whether this asphalte will bear heat, or the trampling of horses' feet, I do not know; but it seems there is a sort of asphaltum pavement in France which will bear 100° of heat of Fahrenheit, and is employed in flooring the cavalry barracks of that country. The substance of which this pavement is made, is called "The Asphaltic Mastic of Seyssel," and for the manufacture and sale of which a company has been formed in Paris to supply pavement for various purposes. The substance is a natural asphalte found at Pyrimont, at the foot of the eastern side of Mount Jura, on the right bank of the river Rhone, one league north of Seyssel. In chemical composition this asphalte contains 90 per cent. of pure carbonate of lime, and 9 or 10 per cent. of bitumen. To form the asphalte into a state fit for use, it is combined with mineral pitch, obtained at the same place, in the proportion of 93 per cent. of the asphalte to 7 per cent. of the mineral pitch. The pitch when analyzed contains of resinous petroliferous matter from 69 to 70 per cent. and of carbon from 30 to 35 per cent. The preparation of this asphalte being tedious, its cost is greater than that mentioned above. For foot-pavements or floors it is about 6½d. and for roofs 8½d. per square foot.*

(107.) Another mode of causewaying is with blocks of wood, commonly called *wood-pavement*. Portions of the streets of London have been laid with this kind of pavement, the blocks having been previously subjected to the process of Kyanizing, and they are found to make a smooth, clean, quiet, and durable causewaying. This would be a desirable method of paving the road round the large court K, Plate IV. the straw-barn, work-horse stable, hay-house, cow-byres, passage in the calves'-house, riding-horse stable, coach-house, and potato-stores. It would be expedient, when used in a stable or byre, that some other substance than sand be put between the blocks, for that is apt to absorb urine too readily. Grout formed of thin lime and clean small gravel, or asphalte poured in between the blocks, might repel moisture. This latter expedient has already been tried, as may be seen at page 14 of Mr. Simm's observations on asphalte. There are various methods of disposing of the blocks of wood so as to make a steady and durable pavement. 1. The earliest plan adopted in London, in 1838, was that of Mr. Stead, a specimen of laying which I had an opportunity of seeing in the Old Bailey, London, in 1839. It consisted of hexagonal blocks of wood set on end upon a sandy substratum. The blocks had the Kyan stamp on their side Since then the substratum upon which the blocks rest has been made of Roman cement and what is called Thames ballast, which I suppose means Thames river sand. The cost of this mode is 9s. the square yard for 6-inch blocks, and 2s. the yard for the concrete. 2. Another plan is that of Mr. Carey, which consists of setting cubical blocks on end, a mere modification of that of Mr. Stead. The cost is for 8-inch blocks 12s. 6d.; 9-inch blocks 13s. 6d.; and 10-inch blocks 14s. 6d. the square yard. 3. Mr. Grimmans's is another mode of wood-paving. It consists of the blocks forming oblique parallelopipedons at an angle of 77°, and they are so cut as to set from right to left and from left to right, presenting a sort of herring-bone work. The blocks are chamfered at the edges to prevent the slipping of horses' feet. With the concrete of Roman cement and Thames ballast, this paving is charged 12s. the square yard. 4. Mr. Rankin's method secures the safety of the horses' feet in slipping, but is too elaborate a mode for general adoption. It consists of a number of small blocks, cut out of the same piece of wood, lying above one another in a complicated fashion. With concrete, its cost is 16s. the square yard. 5. Of all the modes of wood-paving yet invented, that of the Count de Lisle is the best. It consists of placing beside each other oblique cubes of 6 inches, having an inclination of 63° 26′ 5 8-10′, a number derived by calculation from the stereotomy of the cube. "These blocks are cut and drilled by machinery, mathematically alike; and are so placed in the street that they rest upon and support each other from curb to curb, each alternate course having the angle of inclination in opposite directions These courses are connected to each other, side and side, by dowels, which occupy the exact centers of two isosceles [equilateral?] triangles, into which each block is divisible. This arrangement affords the means of connecting every block with four others, and prevents the possibility of one being forced below the level of another. Pressure and percussion are therefore distributed, in effect over large surfaces, and a perfect cohesion established. Nor is this cohesion advantageous only as a means of resistance against superincumbent force. It is of equal value in withstanding any effort to break up the uniformity of surface by undue expansion. The concrete foundation having a slight elliptical curve given to it, and the wood-paving being so laid as to correspond with that curve, for the purposes alike of strength and surface drainage, there is naturally a slight tension on the dowels in an upward direction, which the pressure from above tends to relieve; while the lower ends of the blocks abut so closely together in one direction, and every block is so kept in its position by two

* Simm's Practical Observations on the Asphaltic Mastic or Cement of Seyssel, p. 3.

dowels on each side in the other direction, that the whole mass will take any increased curve consequent upon expansion, without the slightest risk of either partial or general displacement." There is much facility in replacing these blocks, especially since "the doweling of them together at the manufactory in panels of 24 each, 6 in length by 4 in width, the blocks at the sides of which being connected by iron cramps. Thus prepared, the process of covering a street is exceedingly rapid and simple. One end of a panel is cut off at an angle to agree with that of the curb and the curve of the street, and is then abutted against it; each panel containing four courses in alternate angles, another dove-tails precisely with the first, and thus panel after panel is laid until the street is crossed, and the last cut off to abut against the other curb." To prevent slipping, grooves are cut across the street at about 6 inches apart, and others are formed along the street, to prevent rutting, and the joinings of the longitudinal grooves are broken. The substratum upon which this mode of wood-paving is made to rest is a concrete formed of "blue lias lime, a metallic sand, and Thames ballast," which becomes permanently solid and impervious to water after two or three days, by the oxidation of the metallic sand. The cost of this mode is 13s. the square yard for 6-inch blocks and concrete complete, and 6d. a yard every year for keeping it in perfect repair for 10 or 20 years; 12s. for 5-inch blocks; 11s. for 4-inch blocks, and proportionately for repairs.

(108.) Of these various modes of wood-paving, the following are the quantities of each which have been tried in London up to November, 1841, viz:

Of Stead's hexagons	8,710	sq. yds.
De Lisle's oblique cubes	19,838	..
Carey's squares	1,750	..
Of Grimman's oblique parallelopipedons	650	sq. yds.
Rankin's inverted pyramids	492	..
	31,440	..

The Metropolitan Wood-paving Company have adopted De Lisle's system.

(109.) As to the durability of wood-paving, it is reasonable to suppose that "a structure of wood, instead of resisting the pressure or percussion of passing vehicles, like such an incompressible substance as granite, yields to it sufficiently to counteract friction, from its inherent property of elasticity. Hence in Whitehall, where the blocks have been down about two years, they are not reduced in depth ⅛ of an inch on an average; and this reduction, being more the result of compression than of abrasion, is not likely to continue even at that ratio; for the solidity of the blocks is increased even if the volume be thus slightly reduced. Indeed, paradoxical as it may at first appear, the traffic, which is destructive of wood-paving in one way, contributes to its preservation in another; and may thus be explained: The wood-paving is put down in a comparatively dry state, and, if it were always perfectly dry, would be much more susceptible of destruction from accidental or mechanical, as well as from natural causes. But, soon after it is constructed, it becomes perfectly saturated from rain and other causes, and continual pressure forces more and more water into the blocks, until every pore is completely filled. In this state, the water assists in supporting superincumbent weight, while it effectually preserves the wood from decay. For, in fact, of the 6 sides of a block of the given form, only the upper one is exposed to the action of the atmosphere; below the surface the whole mass is as thoroughly saturated as if it were immersed in water; and the surface itself becomes so hardened by pressure and the induration of foreign substances, such as grit and sand, as to be impervious to the action of the sun, especially in a northern climate; and that water is a preservative against decay may be proved in a variety of cases. . . . . . . Dry rot, therefore, can never affect good wood-paving, nor can any other secondary process of vegetation, in consequence of the preservative qualities of water—the shutting out, in short, of atmospheric influence; and it is questionable if, under other circumstances, the incessant vibration to which the blocks are subjected, by traffic, would not have a strong preservative tendency."

(110.) On the comparative cost of laying down and maintaining wood-paving with other sorts, a statement which has been made regarding wood-paving and paving with granite, in the parish of St. Mary le Strand, in London, for the last 7 years, tells in favor of the wood. It is this:

Granite-paving and concrete cost				£0	12	6	the square yard
Repairs for 7 years at 3d. the yard				0	1	9	.. ..
				0	14	3	.. ..
Deduct the value of the stones for streets of lesser traffic				0	3	0	.. ..
Actual cost of 7 years				0	11	3	.. ..
Wood-paving cost	£0	13	0 the sq. yd.				
Repairs for 7 years at 6d. the yard	0	3	6				
	0	16	6				
Deduct value of the wood for paving streets of lesser traffic	0	3	0				
				0	13	6	.. ..
				£0	2	3	.. ..

Giving an apparent advantage of 2s. 3d. the square yard to the granite-paving for the first 7 years; but, were the comparison continued for an indefinite period onward, it would be found that the same blocks of wood would last longer than the same blocks of granite, and hence the wood would be cheaper in the long run.

(111.) On comparing its cost with macadamization, it is found that macadamized roads of much traffic, such as Oxford-street, Piccadilly, cost from 2s. 6d. to 3s. the square yard every year, besides the expense of the original formation: whereas wood-paving can be laid down and kept in repair for a rent-charge of 2s. 3d. the square yard every year—being a saving of from 10 to 30 per cent. per annum.*

(112.) I have dwelt the longer on the subject of wood-paving, because I am persuaded that it would make a much more durable road about steadings, and to the fields of farms, than the materials usually employed for such purposes; and, as to their comparative condition under traffic, there would be a decided superiority on the side of the wood-paving, for farm roads are usually in the most wretched state of repair—every hour of time and every ton of metal expended on them being grudged, as if they were an item with which the farmer had nothing whatever to do. I do not say that wood-paving would be cheap where wood is scarce and carriage long, and of course dear; but in those parts of the country where larch wood is in abundance, and where it realizes low prices, it might be, I conceive, profitably employed in not only making farm roads, but in paving every apartment in the steading.

(113.) Another method still of causewaying is with *Dutch clinkers*—a kind of very hard brick made in Holland, of about the breadth and thickness of a man's hand. They are used in paving roads and streets in that country. They are set lengthways on edge and imbedded in sand, and are laid so as to form a slight arch across the road. Most of the great roads in Holland are paved with this brick, and more beautiful and pleasant roads to travel on cannot be found anywhere, except perhaps in the heat of summer, when they become oppressively hot. I had an opportunity of seeing a part of the road near Haarlem laid with these clinkers, and observed, as a part of the process, that, as a certain piece of the causewaying was finished, bundles of green reeds were laid lengthways across the road over the new laid bricks, to temper the pressure of the wheels of carriages upon the bricks on going along the roads, until the bricks should have subsided firmly into the stratum of sand. As these clinkers are small, they can be laid in a variety of forms, some as a beautiful kind of Mosaic work. The import duty on Dutch clinkers was reduced to 3s. per 1,000 on 1st January, 1834: in 1819 it was 16s. 8d.† The present price of clinkers (1842) in London is 35s. the 1,000.

(114.) Fine smooth durable pavement is made of the beautifully stratified beds of the inferior gray sandstone, a rock nearly allied to graywacke. It is a rock of fine texture, hard and perfectly impervious to water. It occurs in abundance in the south-east part of Forfarshire, and, being chiefly shipped at Arbroath in that county, it has received the appellation of "*Arbroath Pavement*." Hard flags from the counties of Caithness and Orkney also form very durable, though not always smooth, pavement. Some, however, of this, as well as of the Arbroath pavement, requires very little, if any, dressing with tools on the face. The Caithness pavement is cut on the edge with the saw, the Arbroath pavement with common masons' tools. In a paper read to the British Association at their meeting at Glasgow in 1840, Professor Traill described this flag as belonging to the red sandstone series, although its appearance as pavement would lead one to suppose it to belong to an older formation. Pavement is also formed of the stratified portions of the sandstone of the coal-formation. Most of the foot-pavement of the streets of Edinburgh is of this kind. Its face requires to be wrought with tools, and its texture admits water. Arbroath pavement costs from 2d. to 4d. per square foot at the quarry, according to thickness. Both it and Caithness pavement cost 10d. and common stone pavement 6d. per foot in Edinburgh. When jointed and droved, the cost is 9d. per square foot additional.

(115.) In connection with the subject of masonry, I may advert to the *sinking of wells* for a supply of water. 1. In trap and other amorphous rocks, little water may be expected to be found, and the labor of sinking by blasting with gunpowder renders a well sunk in these substances a very expensive undertaking. When there is probability of finding water in stratified rocks under trap, the latter may be penetrated by boring with a jumper, with the view of forming an artesian well; but before such a project is undertaken, it should be ascertained beforehand that stratified rock or diluvium exists below the trap, and that the dip of either is toward the site of the well. Of so much importance is one good well on a farm, that a considerable expense should be incurred rather than want, at any season, so essential a beverage as water to man and beast. When insuperable objects exist against finding water on the spot, perhaps the better plan will be either to go a distance to a higher elevation, where a common well may succeed in finding water, and then convey it to the steading by a wood or iron or lead pipe; or to descend to a lower site and throw the water up to the steading by means of a force-pump. Either of these plans may be less expensive, or more practicable than the boring through a hard rock to a great depth. The well in Bamborough Castle, in Northumberland, was sunk upward of 100 feet through trap to the sandstone below; and at Dundee, a bore was made through trap, 300 feet, to the inferior sandstone below, by means of a steam-engine, to obtain water for a spinning-mill. 2. In gravel and sand, a well may be sunk to a considerable depth before finding water. Being

* Stevens's Wood-Paving in London. † McCulloch's Commercial Dictionary, Art. *Tariff.*

desirous of a supply of water to three adjoining fields of dry turnip land, resting on a deep bed of pure gravel, and which had no watering-pool. I fixed on the most likely spot to contain water, near the foot of a rising ground of diluvial clay, in which to dig a well, and it happened to be a spot common to all the fields. After persevering to the depth of 22 feet without success, at the imminent hazard of overwhelming the men with gravel, as a despairing effort, at night-fall I caused a foot-pick to be thrust down into the bottom of the pit as far as the handle, and on withdrawing the instrument, water was seen to follow it. Next morning three feet more were dug, when the water excavating the gravel around the bottom of the pit rendered farther digging a dangerous operation for the men, so the ring of the well was there begun to be built with stones. The water afterward would rise no higher in the well than the level where it was first found, but the supply, nevertheless, was sufficient for the use of three fields. On finding water in this case, in the midst of very hopeless symptoms, I would recommend perseverance to diggers of wells, and success will most probably reward their efforts. 3. In very unctuous clay, such as is found in carse land, water is difficult to be obtained by digging to ordinary depths; but as such a country is usually situate near a large river, or on the side of a broad estuary, by digging to the depth of the bed of the river, some sand will most probably be found through which the water will find its way to the well; and though brackish in the estuary, it may come into the well sweet enough for all domestic purposes. 4. Wells dug in stratified rocks, such as sandstone, may be supplied with water at a moderate depth, perhaps 6 or 8 feet; but among regular strata there is as much risk of losing water as there is ease in obtaining it. To avoid disappointment, it will be necessary to puddle the seams of the rock on that side of the well in which it dips downward. 5. The substance which most certainly supplies water on being dug into is diluvial clay, a substance which forms the subsoil of the greatest extent of arable land in this kingdom. This clay is of itself impervious to water, but it is always intersected with small veins of sand frequently containing mica, and interspersed with numerous small stones, on removing which, water is found to ooze from their sites, and collect in any pit that is formed in the clay to receive it. The depth to be dug to secure a sufficiency of water may not be great, perhaps not less than 8 feet or more than 16 feet; but when the clay is homogeneous and hard, and there is little appearance of water, digging to upward of 40 feet in depth will be required to find water. I knew a remarkable instance of a well that was dug in such clay in Ireland, in which 40 feet were penetrated before any water was found; but immediately beyond that depth, so large a body of pure water was found in a small vein of sand, that the diggers escaped with difficulty out of the well, leaving their tools behind. A force-pump was obtained to clear the well of water, in order to allow the ring to be built; but it was unable to reduce the bulk of water, so that the ring remains unbuilt to this day; the water always stands within three feet of the top of the well, and the clay is not much affected by it. 6. Suppose, then, that this wall is to be dug in clay containing small stones and veins of sand. Let a circle of 8 feet in diameter be described on the surface of the ground, from whose area let the ground-soil be removed to be used elsewhere. After throwing out a depth of 8 or 9 feet with the spade, let a winch and rope and bucket be set up to draw the stuff out of the well. While the digging is proceeding, let a sufficient quantity of flat stones be laid down near the winch, by which to let them down to build the ring. A depth of 16 feet will most probably suffice, but if no water is found, let the digging proceed to the requisite depth. A ring of 3 feet in diameter will be a large enough bore for the well, the rest of the space to be filled up with dry rubble masonry, and drawn in at the top to 2 feet in diameter. Whenever the building is finished, the water should be removed from the well with buckets, if the quantity is small, and with a pump if it is large, to allow the bottom to be cleared of mud and stones. A thick flat stone, reaching from the side of the ring to beyond the center, should be firmly placed on the ground at the bottom of the well, for the wooden pump to stand upon, or for the lead pipe to rest on. If a wooden pump is used, a large flat stone, having a hole in it to embrace the pump, should be laid on a level with the ground upon the ring of the well; but if a lead pipe is preferred, the flat stone should be entire and cover the ring, and the *clayey* earth thrown over it. The cost of digging a well in clay, 8 feet in diameter and 16 feet deep, and building a ring 3 feet in diameter with dry rubble masonry, is only £5, exclusive of carriage and the cost of the pumps. A wooden-mounted larch pump of from 15 to 20 feet in length costs from £3 to £3 10s. and a lead one £2 16s. with 1s. 2d. per lineal foot for pipe of the depth of the well. The wooden pump will last perhaps twenty years, and the lead one a lifetime, with ordinary care, and the lead at all times is worth something.

(116.) The making of the well naturally suggests the subject of *water*. The different kinds of water receive names from the sources from which they are derived. Thus there is *sea-water*, the water of the ocean; *rain-water*, the water which falls from the atmosphere; *river-water*, the water which flows in the channels of rivers; *spring-water*, the water as it naturally issues from the ground; *well-water*, the water collected in wells; *pond-water*, the water collected in an artificial hollow formed on the surface of the ground; and *marsh-water*, the stagnant water collected in swamps and bogs. All these sorts of water possess different properties, acquired from the circumstances from which each is derived.

(117.) *Pure water* is not found in nature, for all the sorts of water accumulated on or near the surface of the earth, though differing in purity in regard to each other, are none of them pure in the chemical sense of the term; that is, free of the admixture of other matter, such as gases, salts, earth. Pure water is colorless, and insipid to the taste. Its specific gravity is 1,000 ounces per cubic foot. It is made the standard of gravity, 1 being its equivalent mark. It is an inelastic fluid. It consists of hydrogen and oxygen, the combination by weight being 8 of oxygen and 1 of hydrogen—by volume, 1 of oxygen to 2 of hydrogen—and by equivalent or atom, 1 of hydrogen with 1 of oxygen; its chemical symbol being H+O or HO. Pure water is obtained by the distillation of rain or river water, and, to retain it so, it must be kept in closed bottles filled to the stopper, as it has a strong affinity for common air, oxygen, and carbonic acid gas.

(118.) Water from the condensed vapor of fresh water is the purest that can be obtained by natural means. Hence, rain-water collected after rain has fallen for a time, at a hight above the ground, in the country, and at a distance from any dwelling of man, or new-fallen melted snow, is the purest water that can be collected in a natural state; but, nevertheless, it is not pure, inasmuch as it contains oxygen, nitrogen, carbonic acid, and earthy matter, which it has met with in the atmosphere, besides nearly as much common air as it can absorb. Procured from the roofs of buildings, rain-water is always contaminated with many additional impurities, derived from the channels through which it has flowed. It is generally very dark-colored, and, when allowed to stand, deposits a quantity of earthy ingredients. It is not in a proper state for domestic purposes until it has got quit of as much of these impurities as it can by deposition.

(119.) Rain-water for domestic purposes is collected in cisterns. The form of a *rain-water cistern*, represented by fig. 30, I have found an useful one for allowing the undisturbed deposition of impurities, and at the same time the quick flowing off of the purer water, without disturbing the deposition. Let *a b b c* be a cistern of stone or wood, placed at a convenient spot of the steading or farm-house, for the reception of rain-water. I have found that such a cistern, of the capacity of 12 cubic feet, holds a sufficient quantity of rain-water for the domestic purposes of an ordinary family. A cistern of 2 feet square at the base, and 3 feet in hight, will just contain that quantity; but, as the size of an ordinary wash-tub is 2 feet in diameter, the space betwixt *d* and *d* must be made 2 feet 6 inches at least, and the hight of the cistern *b* could be 2 feet; but if more water is required than 12 cubic feet, then the hight should be 3 feet, which gives a capacity to the cistern of 18 cubic feet. Suppose the cistern represented in the figure to contain 18 cubic feet, then the area of *a* will be 2½ feet square, and *b* 3 feet in hight, supported on two upright stones *d d* of the breadth of the cistern and 2 feet high. The cistern may either be made of a block of freestone hewn out to the dimensions, or of flags, of which the sides are let into grooves in the bottom and into each other, and imbedded in white-lead, and fastened together with iron clamps, having a stone movable cover *c*. Or it may be formed of a box of wood, securely fastened at the corners to be water-tight, with a cover of wood, and resting on the stone supports *d d*. Stone being more durable, is, of course, preferable to wood for a cistern that stands out in the open air. A hollow copper cylinder *g* is fastened perpendicularly into the bottom *a*, having its lower end projecting 1 inch

Fig. 30.

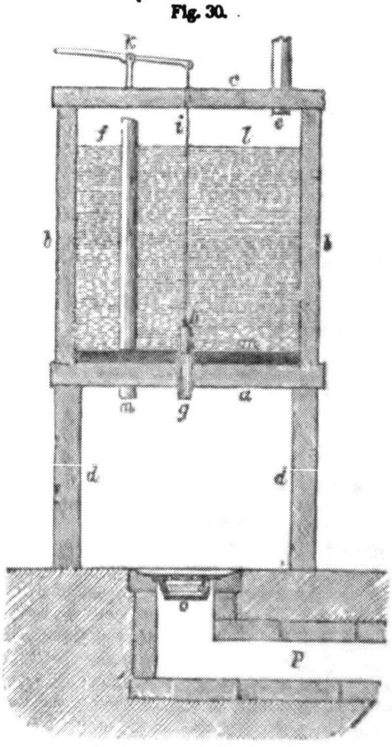

RAIN-WATER CISTERN.

below, and its upper 3 inches above, the respective surfaces of the bottom. The upper end of the copper cylinder is formed to receive a ground truncated cone of copper called a plug or stopper, which is moved up and down with the lever *k*, by means of the stout copper rod *i*. The plug must be made watertight with grease, the rod of which passes through a hole in the cover, to be connected with the lever, whose support or fulcrum is fixed on the cover. These parts are all made of copper, to withstand rusting from the water, with the exception of the lever, which may be of iron, painted. The rain-water is supplied to the cistern by the pipe *e*, which descends from the rain-water conductor, and is let through a hole in the cover. The water is represented standing as high as *l*, but, in case it should rise to overflow, it can pass off by the lead waste-pipe *f*, which is secured and movable at pleasure in a ground-washer *n*, whose upper end is made flush with the upper surface of the bottom *a*. After the water has entered the cistern, it gets leave to settle its sediment, which it may do to the hight of the upper end of *g*. The sediment is represented by *m*, and, when it accumulates to *h*, the cover *c* should be taken off, and the waste-pipe *f* removed, and it can then be cleaned completely out by the washer *n*. The waste water runs away through the air-trap *o*, and along the drain *p*. It is more convenient to have two small than one large cistern, as, while the water is rising in the one, that in the other gets leave to settle. The cost of such a cistern, with droved stones, and to contain 18 cubic feet, with the proper mountings, may be about £5. I think it right to say, in commendation of this form of water-cistern, that in no case have I known the water about the plug to be frozen—in consequence, perhaps, of the non-conducting power of the mud in the bottom of the cistern. The rod *i* has sometimes become fast to the ice on the top of the water, but a little boiling water poured down by the side of the rod through a funnel soon freed it from restraint.

(120.) Rain-water, besides containing gases in solution, becomes impregnated with many saline substances in its passage through the ground; and hence the water of springs and rivers always contains many ingredients. The purest spring-water is that which has passed through gravelly deposits, such as of granite, sandstone, quartz; because the component parts of those stony substances being insoluble, the water cannot take up much of them. In the same way the water of old wells is purer than that of new, because the long continued action of the water has removed or gradually dissolved the soluble matters in the same passages through the ground to the well. "The matters generally contained in spring, well, and river water," says Mr. Reid, "are carbonate of lime, sulphate of lime, muriate of lime, sulphates of potash and soda, muriate of soda, and sometimes a little magnesia. 'In rain-water,' says Dr. Murray, 'the muriates I have found generally to form the chief impregnation, while in spring-water the sulphates and carbonates are predominant, and in the former the alkalies,' potash and soda, 'are in larger quantity, while the earths, particularly lime, are more abundant in the latter.'"* It is in its combination with one or more of these salts that water becomes *hard*, chiefly with the sulphate of lime or gypsum, and the carbonate of lime or limestone. Water is said to be hard when it will not dissolve but decompose soap. *Soft* water, on the other hand, does not decompose, but combines easily with soap and dissolves it. Hard water is not so fit as soft for many culinary purposes, such as making tea and boiling vegetables. It is, therefore, of importance for you to know when water is in a hard or soft state. By placing a few thin slices of white soap in a clean tumbler of the water to be examined, its *hardness* will be indicated by *white flakes* or *curdy particles around the soap*, the effect of decomposition—the acids of the salts in the water combining with the alkali of the soap and leaving the fatty matter. A very small quantity of either of the salts enumerated above will render water hard. Water can dissolve 1-500 part of its weight of gypsum; but, according to Dr. Dalton, 1-1000 part is sufficient to render it hard; and Mr. Cavendish says that 1200 grains of water containing carbonic acid will hold in solution 1 grain of limestone. Limestone is insoluble in pure water; but water containing carbonic acid in solution can dissolve it.

(121.) "To discover whether the hardness be owing to the presence of limestone or gypsum, the following chemical tests," says Mr. Reid, "may be applied. A solution of the nitrate of barytes will produce a white precipitate with water containing either gypsum or limestone; if limestone have been present in the water the precipitate will be dissolved, and the liquid rendered clear on adding a few drops of pure nitric acid; if the presence of gypsum caused the precipitate, this will not be dissolved by the nitric acid. A solution of the sugar of lead may be used in the same way, but the nitrate of barytes is preferred."†

(122.) As to a practical remedy for hard water, boiling will remove the lime. The carbonic acid in excess in the water is converted into the gaseous form, and the carbonate of lime then becoming insoluble, falls to the bottom of the vessel. Hence the incrustation of tea-kettles. If the hardness is caused by gypsum, a little pearlash or soda (carbonate of potash or carbonate of soda) will remove it, and the lime of the water will also be precipitated with the carbonic acid of the pearlash or soda.

(123.) River-water is always softer than spring or well water, because it deposits its earthy ingredients when flowing in contact with common air, which it absorbs in considerable quantity. By analysis, the water of the river Clyde yielded 1-35 of its bulk of gases, of which 19-20 were common air. "All that is necessary," remarks Mr. Reid, "in order to render river-water fit for use is to filter it. This is rather a mechanical than a chemical operation, and is done by causing the water to pour through several layers of sand, which intercepts the muddy particles as the liquid passes through. Filtering stones, made of some porous material, such as sandstone, and hollowed out so as to be capable of containing a considerable quantity of water, have sometimes been employed to purify water. Compressed sponges have also been employed for this purpose. Sand and charcoal form the chief elements in the construction of the filters now so much employed for purifying water, the powdered charcoal acting not only mechanically in detaining any muddy particles but having a chemical effect in sweetening the water (rendering it fresh) if it be at all tainted or even in retarding putrefaction, if it have any tendency that way."‡

* Reid's Chemistry of Nature, p. 195. † Ibid. p. 199. ‡ Ibid. 20.

(124.) Water, as a beverage, would be insipid or even nauseous without the gases and saline matters usually found in it. They give a natural seasoning and a sparkling appearance to it, thereby rendering it agreeable to the taste. Every one knows the mawkish taste of boiled water when drank alone.

(125.) As I am on the subject of water, a few words should here be said on the making of *horse-ponds*. The position of the horse-pond will be seen in figs. 1 and 2, in Plates I. and II. When a small stream passes the steading, it is easy to make a pond serve the purpose of horses drinking and washing in it, and the water in such a pond will always be pure and clean. But it may happen, for the sake of convenience, when there is no stream, that a pond should be dug in clay, in which case the water in it will always be dirty and offensive, unless means are used to bring water by a pipe from a distance. If the subsoil is gravelly, the water will with difficulty be retained on it, on which account the bottom should be puddled with clay. Puddling is a very simple process, and may be performed in this manner: Let a quantity of tenacious clay be beaten smooth with a wooden rammer, mixing with it about one-fourth part of its bulk of slaked lime, which has the effect of deterring worms making holes in it. After the mass has lain for some time souring, let large balls of it be formed and thrown forcibly on the bottom of the pond, made dry for the purpose, and beaten down with the rammer or tramped with men's feet, until a coating 6 or 7 inches in thickness is formed, or more, if there is plenty of clay. Then let a quantity of clean gravel be beaten with the rammer into the upper surface of the clay before it has had time to harden. Should the pond be large, and the weather at the time of making it so dry as to harden the clay before its entire bottom can be covered with it, let the puddling and graveling proceed together by degrees. Above the coating of gravel, let a substantial causeway of stones and sand be formed to resist the action of the horses' feet, and which, if properly protected at the ends, and finished on the open side of the pond, will withstand that action for a long time. I have seen a sort of pond recommended to be made, into which the horses enter at one end, and pass through it by the other. This is a convenient shape of pond, in as far as it admits of the uninterrupted passage of the horses *through* the pond, but it is liable to serious objections. Being contracted laterally, the pair of horses which first descend to drink will occupy the greatest proportion of its whole breadth, and, while in that position, the succeeding pair must drink the muddy water at their heels; and, as the contracted form precludes easy turning in the deepest part of the water, none of the rest of the horses can be permitted to block up the opposite or open end of the pond. A much better form of pond, I conceive, is with an open side, having the opposite side fenced, and the water supplied clean at the upper end, and made to flow immediately away by the lower. At such a pond a number of horses can stand in a row to drink at the same time, and easily pass each other in the act of washing the legs after drinking. As to the depth, no horse-pond should ever exceed the hight of the horses' knees. The water should on no account reach their bellies; for although I am quite aware of plowmen being desirous to wade their horses deep, and of even wishing to see their sides laved with water, to save themselves some trouble in cleaning, that is no reason why you should run the risk of endangering the health of your horses by making the pond deeper than the knee

(126.) With regard to the *kind of stone* which should be employed in the building of a steading, it must be determined by the mineral product of the locality in which it is proposed to erect it. In all localities where stone is accessible, it should be preferred to every other material; but where its carriage is distant, and of course expensive, other materials, such as brick or clay, must be taken. In large flat tracts of country, stone is generally at too great a distance; but in those situations, clay being abundant, brick may be easily made, and it makes an excellent building material for walls, and far superior to the old-fashioned clay walls which were in vogue before brick became so universally used for building. Of stone, any kind may be used that is nearest at hand, though some rocks are much better adapted for building purposes than others. 1. Of the primitive rocks, gray granite forms a beautiful and durable stone, as is exemplified in the buildings in Aberdeenshire, Cornwall, and Newry in Ireland. Gneiss, micaslate, and clayslate, do not answer the purpose well. They give a rough edgy fracture, frequently rise too thin in the bed, especially in the case of clayslate; are not unfrequently curved in the bed, and at the same time difficult to be dressed with the hammer. 2. Of the transition series, graywacke makes a beautiful building-stone, as may be seen in the houses at Melrose. The old red sandstone, though a good building-stone, has a disagreeably sombre aspect, as seen at Arbroath; but the inferior gray sandstone which prevails in the neighborhood of Dundee, is a beautiful and durable building-stone. 3. All the sandstones of the coal formation form excellent materials for building, as is exemplified in Edinburgh and many other places. 4. The limestone, from marble to the mountain carboniferous limestone, make fine building-stone, as at Plymouth; but in case of fire they are apt to be calcined by heat, as exemplified in the cathedral at Armagh before it was repaired. And 5. Even the trap-rocks are employed in building houses where sand stones are scarce. Though the two classes of rock are frequently located together Whinstone is objectionable, inasmuch as it throws out dampness in wet weather, and the walls require to be lathed and plastered on the inside, to render the house even comfortable. Frequently where whinstone is near at hand, and sandstone can be obtained at a little distance, the latter is employed as corners ribets and lintels, though the contrast of color betwixt them is too violent to be pleasant to the eye. If sandstone, therefore, can be procured at a reasonable cost of carriage, you should give it the preference to whinstone, for the sake of comfort to your live-stock in their habitations in wet weather. You may, indeed, choose to incur the expense of lathing and plastering all the insides of the walls of the steading; but a lathed wall in any part of a steading would be apt to be broken by every thing that came against it, and is, on that account, an unsuitable finishing for a steading. 6. The worst sort of building-stone are landfast boulders of the primitive and trap rocks, which, although reduceable by gunpowder, and manageable by cleavage into convenient shaped stones, incur great labor in their preparation for building; and even after the stones are prepared in the best manner they are capable, their beds are frequently very rough, and jointings coarse, and the variety of texture and color exhibited by them, render them at the best unsightly objects in a building. They are equally unsuitable for dry-stone dykes as for buildings, for in the case of dykes, they must be used very nearly in their natural state, as the usual charge for such work will not bear labor being bestowed on the preparation of the material. Still, after all, if no better material for building houses is near at hand than those boulders, they must be taken as the only natural product the country affords. There is a class of boulders, composed chiefly of micaceous sandstone, found in banks of gravel, which answer for dry stone dykes admirably, splitting with ease with a hand-pick into thin layers, and exhibiting a rough surface on the bed, very favorable to their adherence together in the wall. This species of building material is abundant in Forfarshire, where specimens of dry-stone building may be seen of a superior order. In these remarks of the general choice of building-stones by Mr. G. Smith, architect in Edinburgh, there is much truth: "The engineer and architect," says he, "go differently to work in choosing their stones. The former, in making his experiments for his piers and bridges, selects the strongest and hardest as most suited to resist great pressure. The latter, for his architectural decorations, chooses not only the most beautiful as to texture and uniformity of color, but those which may be easily cut into the most delicate mouldings, and which, moreover, will stand the winter's frost and the summer's heat. It may be remarked that the hardest stones are not always those which hold out the best against the effects of the weather."*

(127.) I may here observe, in concluding my observations on the specifications of masonry, that any lime that is used on a farm, for the purpose of steeps for grain or for mortar, gets leave to lie about in the most careless manner, either under a shed, or at some place contiguous to water, where it had been made up into mortar. In either case there is waste of a useful article; and in many parts of the country, where carriage is far distant, it is a high-priced article. The lime that is to be used in a dry state should be kept under cover; and all that is required in a season could be held in a cask or small hogshead to stand in a corner of the cart-shed or potato-store, but not in the straw-barn, where a little damp may cause it to ignite the straw. With regard to mortar, no more should be made at a time than is used, or it should be carefully heaped together in a convenient place, and covered with turf

(128.) In Sweden, mortar is made and kept in a convenient form of cart, represented by fig. 31, a practice which might with propriety be followed in this country. The cart consists of a cube *a b c d* of a side of 3 feet made of 2-inch thick battens. The wheels are formed of the two sides of the cube, on which are fixed circular segments such as *e* and *f*, made of strong battens 3 inches thick, secured by a screw at each end into the side of the

Fig. 31.

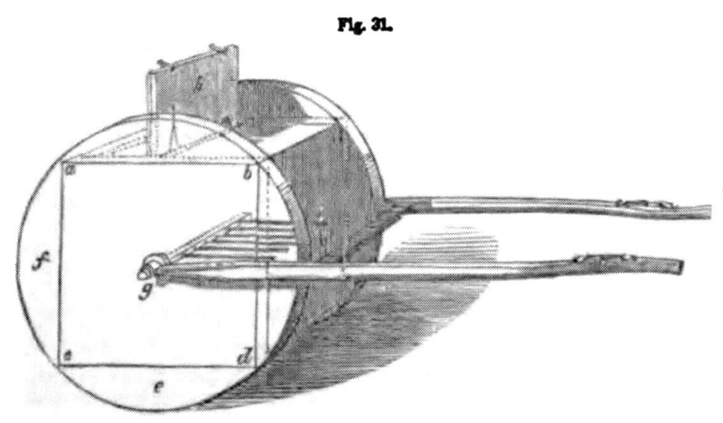

SWEDISH MORTAR-CART.

* Prize Essays of the Highland and Agricultural Society, vol. x. p. 85.

cube, and the circles are shod with iron as common wheels. The axle *g*, inside and outside is closely passed through the cube, so as not to allow any of the mortar to come out. The axle moves in a small iron nave attached to the shafts of the cart. On it are screwed iron bars *i*, which pass through one of the sides of the cube, and fastened to it by screws *k*. The use of these bars is to break the mortar when too tough; and if one set of bars is found insufficient for that purpose, similar ones should be put through the opposite side of the cube. A lid *h* is well secured to the cube by hinges, and kept fast by means of a hasp. When the shafts are drawn, the whole cart revolves with the wheels.

(129.) The lime is put into the cart by the lid, and sprinkled over with a little water, about half a gallon of which to the bushel of lime will be enough the first time. The cart is then driven round a while; and when the driver, who must often look to the mortar, finds that all the water is imbibed, a little more must be poured in, and the cart again driven round. Water is poured in in small quantities until the lime forms coagulated masses or balls, and then it is worked until no dry lime is seen in the mass. The success of making good mortar depends on the skill of the driver, who will soon learn to do it well after a cart-ful or two of driving. Three bushels of lime and sand can be prepared in this way in a short time, but the sand should not be put in till after the lime has been sufficiently wrought with water.*

(130.) Of the specification of *carpenter-work*, the first timber that is used in building consists of *safe-lintels*, which should be 4 inches thick, of such a breadth as to cover the space they are placed over, and they should have a solid bearing at both ends of 12 inches.

(131.) The *scantlings* or *couples* for the roofs vary in size with the breadth of the building. When the building is 18 feet wide, the scantlings should be 8 inches wide at bottom, 7 inches at top, and 2½ inches thick. Those for 15 feet wide buildings should be 7½ inches wide at bottom, and 6½ inches at top. All scantlings should be placed 18 inches apart from center to center, upon *wall-plates* 8 inches wide by 1½ inches thick, firmly secured to *bond-timber* built into the tops of the walls. These dimensions of scantlings are suitable for a roof of blue slates. For a tile-roof the scantlings are placed 2 feet apart from center to center.—For roofing with gray-slates, which are very heavy, the scantlings should be 3 inches thick With tiles and gray-slates the roofs require a higher pitch than with blue slates, and this is given by making the scantlings 1 foot longer.

(132.) The *balks* of an 18 feet wide building should be 7½ inches broad by 2½ inches thick, and, for the 15 feet one, 7 inches by 2½ inches. In both cases the balks should be of the length of one of the scantlings, which will bring its position so low down on the scantlings as to be only a little more than 3 feet above the wall-heads It is generally supposed that one balk is sufficient for the support of the scantlings; but it will be seen in fig. 6, p. 88, that I have represented a vertical section of the principal range of the steading with two balks, because I would always prefer two balks to one, and the only objection to the two is the expense. When two balks are employed, the lower one will be about 2 feet, and the upper one about 5 feet, above the wall-heads.

(133) If a slated roof is adopted, there should be a *ridge-tree* 10 inches broad by 2 inches thick, and the tops of the scantlings should be bound with *collar-pieces*, 5 inches broad and 2 inches thick, half checked into the scantlings. If a tile-roof is preferred, it is sufficient that the tops of the scantlings be checked in with collar-pieces, as just described.

(134.) The whole roof should be covered with *sarking*, ⅝ inch thick, and clean jointed. A tile-roof requires *tile-lath*, 1½ inches square, and 11 inches apart, excepting at the eaves, which should have a boarding from 12 inches to 15 inches broad, and ⅝ inch thick for slates. Tile-lath is also employed with gray-slates.

(135.) The *peands* and *flankers* should be 9 inches broad at bottom, and 7 inches at top, and 3 inches thick, properly backed to receive the sarking or tile-lath of the respective sorts of roofs.

(136.) The *joists* of the flooring in the part of the buildings that is 18 feet wide should be 10 inches deep by 2½ inches in thickness, placed 18 inches asunder from center to center, and having a wall-hold or rest of 12 inches at each end. When the bearings of joists exceed 8 feet, it is a more secure and economical plan to have beams, instead of battens, laid across the building, 13 inches deep, and 6½ inches in width, with a wall-hold of 12 inches at each end. Upon these should rest joists 7 inches deep, and 2½ inches in breadth, and not more than 16 inches apart from center to center, dove-tailed into the beams with a hold of 9 inches at each end. These joints are best cut out of Memel log of first or second quality, the difference of price between the two qualities being 2d. the cubic foot.

(137.) The *floors* of the upper and corn-barn and granaries should be of 1½ inches thick, of red or white wood battens, grooved and tongued, and well seasoned when wrought and laid. The under side of the floor, and the joists which support the floor of the upper-barn, forming the roof of the corn-barn, should be clean dressed. to prevent the adherence of dust.

(138.) In some parts of the country, and especially in East-Lothian, the floor of the corn-barn is made of composition; but, in order to leave a part of the floor clean, upon which to winnow the grain, a space, 12 feet square, is usually left in the middle of the floor This

* Quarterly Journal of Agriculture vol. xi. p. 345.

space is laid with sleeper-joisting, 7 inches deep by 2½ inches thick, and 18 inches apart from center to center, supporting a flooring of deal 2 inches thick, grooved and tongued. As a precaution against vermin, as well as the enjoyment of cleanliness while winnowing and otherwise handling the grain, I would always recommend an entire wooden floor for the corn-barn, to be laid down in the manner described in (16.) and represented in fig. 6, p. 88

(139.) The *windows* of the stables should be of the form of fig. 14, p. 105. Those of the other apartments of the steading, with the exception of the granaries, should be of the form of fig. 15, p. 105; and those of the granary should be of the form of fig. 17, p. 107. The *astragals*, if not made of wood, may be of cast-iron or zinc. Cast-iron astragals cost 1s. and zinc 9½d. the square foot.

(140.) The *exterior doors*, 7½ feet high, should be of 1¼ inch deal, grooved, and tongued, and beaded, having three back-bars, 7 inches broad by 1¼ inches thick; those of the corn-barn, cow-byre, and boiling-house, being in two horizontal leaves, that of the upper-barn in two vertical leaves, and those of the rest of the apartments being entire.

(141.) If desired, small windows of one or two rows of panes may be placed above all the outside doors; in which case, the voids of these doors should be made proportionally high, say 8 feet.

(142.) The *inside doors* should be 7 feet high, of ¾ inch deal, with three back-bars 6 inches broad and 1 inch thick, grooved, and plowed, and beaded. They should have checks 6 inches broad by 2½ inches thick, and keps and facings 4½ inches broad by ¾ inch thick.

(143.) The *travis boarding* of the work-horse stable should be 1¼ inches thick, 9½ feet long, 7 feet 6 inches high at the fore and 4 feet 6 inches high at the heel posts, doweled in the joints with oak pins, and of an ogee form on the top, let into a 2-inch deep groove in the heel-post, and coped with beading. The *heel-posts* should be 6 inches square, beaded, the *fore-posts*, on both sides, 5 inches by 2½ inches, and both fixed at the top to *runtrees*, 6 inches deep by 2 inches broad. The side walls of the end-stalls should be finished in the same manner, and firmly secured to wall-straps and bond-timbers.

(144.) The travis-boarding of the riding-horse stable should be of the same strength as just described; but the heel-posts should be turned 5 feet high above the ground, with moulded caps and balls, and let from 18 inches to 2 feet into the ground, through a stone frame 18 inches square and 12 inches thick, firmly built with stone and mortar. The fore-posts should be 3 inches in diameter on both sides to the hight of the travis-boarding. Heel-posts are also made of cast-iron, which cost 22s. each.

(145.) The *hay-racks* of the work-horse stable should have a hard-wood rail, 3 inches deep by 2½ inches wide, and the spars of fir, 2 inches broad by 1½ inches thick, placed 2½ inches apart. These spars should be put on both front and bottom.

(146.) The hay-racks of the riding-horse stable should be of hard-wood, and placed high up, with rails, 3 inches deep by 2½ inches wide, and turned rollers, 2 inches of diameter, set 2½ inches apart. Cast-iron racks are frequently used in the corner of the stall, and they cost 10s. each.

(147.) The *mangers* of the riding-horse stable should be of rounded battens in front, of full breadth of the stalls, placed at a convenient hight above the floor, and bottomed and lined with 1¼-inch deal.

(148.) In the work-horse stable, *corn-boxes* are placed in the near angle of the hay-racks.

(149.) The *stalls* of the cow or feeding byres should be made of 1¼-inch deal, beaded, grooved, and tongued. They should be 6 feet long, and 4 feet high, with 1-inch beaded coping, let into stakes or heel-posts, 5 inches to 6 inches diameter, and held to the wall at the head with a 2-inch fillet, and iron hold-fast on each side. The heel-posts should either be taken to the hight of the byre-wall, and secured to runtrees, 6 inches deep by 2 inches broad, or fastened into the ground with masonry like those of the riding-horse stable.

(150.) The doors of the *feeding-holes* of the byres should be of 3-inch deal, of two thicknesses, crossed.

(151.) The *stairs* from the corn-barn to the granaries, if of wood, should have 11 inches of tread and 7 inches of hight of steps. A stair or trap of similar dimensions may lead to the wool-room.

(152.) The floors of the granaries, upper and corn-barns, and wool-room, should have an angular *skirting*, 3 inches by 3 inches, around them.

(153.) Should the upper-barn, or granaries, or wool-room, be ascended by outside stone stairs, they should be furnished with plain ¾-inch iron *railing*, carried around the outer edge of the steps and platform, with a hard-wood hand-rail, or be inclosed with ¾-inch deal lining, the whole hight above the steps, and properly framed.

(154.) The interior of the *hen-house* should be fitted up with rough ¾-inch deal shelves and divisions, and roosting-trees 3 inches deep by 2 inches broad.

(155.) The doors of the hen-house should be of 1¼-inch deal, beaded, grooved and tongued.

(156.) *Wooden ventilators* should be placed upon the roof above every alternate pair of horses and cattle, of the form and dimensions of fig. 8, p. 95; or they may consist of ¾-inch deal, 6 inches square, in an opening above every alternate stall, and furnished on the upper part above the roof, with bent tubes of lead, 6 lbs. to the square foot, or with zinc ones of the same dimensions. The zinc ventilators vary in price, according to size, from 4s. to 7s. each

(157.) The ceilings of the stables, boiling-house, granaries, where tile are used for roofing, wool-room, and hen-house, should be *lathed* with Baltic split-lath 3-16 of an inch in thickness. "Laths are sold by the bundle, which is generally called a hundred; but 7 score, or 140, are computed in the 100 for 3-feet laths; 6 score, or 120, in such as are 4 feet; and for those which are denominated 5 feet, the common 100, or 5 score."* Lath is also made of home wood, usually Scots fir, sawn up into ¾-inch plank, and split irregularly with the ax, and, when nailed on, the splits are kept open by means of a wedge. The duty on foreign lath-wood is from £4 5s. to £3 12s. and on that from the colonies from 15s. to 25s. on the bulk of 6 feet wide by 6 feet high, according to the length of the timber.

(158.) The riding-horse stable should have *saddle-brackets* of ⅞-inch deal, firmly supported, and two pins let into rails 6 inches wide and 1½ inches thick, for each horse. The work-horse stable should have two similar rails, with large and small pins for each horse.

(159.) Every court and hammel should be provided with a *gate*, the forms and dimensions of which I will afterward give, when I come to speak of the subject of gates in general, in spring.

(160.) The entrance to the piggeries should be furnished with doors of 1-inch deal, of two thicknesses, crossed, as represented in fig. 23, p. 114.

(161.) All the varieties of *fir timber* imported into the country are employed in the building of steadings, and those kinds are most used in localities which are obtained from the nearest sea-ports. For example, along the east coast of this country Memel logs and Baltic battens are used for all rough purposes, while on the west coast no timber is to be seen in the construction of steadings but what is brought from America. 1. Norway and St. Petersburg battens being cut to proper lengths and breadths, form cheap and very durable timber for all farm purposes. The price is, for red from 3d. to 3¼d., for white from 2¾d. to 3d. the lineal foot. The Norway battens are a shade cheaper. The red or white-wood battens make excellent floors, and plain deal doors for inside use. Such flooring is beautifully dressed by planing machinery at Mr. Burstall's mills at Leith. 2. Memel logs are admirably fitted for joisting, windows, outside doors, and all outside work, it being composed of strong and durable fibre, surrounded with resinous matter. It sells for from 2s. 4d. to 2s. 6d. the cubic foot. The greatest objection to its use for small purposes is its knottiness, on which account the Norway battens make handier small scantlings and cleaner door-work. 3. The American red-pine is excellent timber, being clean, reedy, and resinous. It is seldom or never of so large dimensions as Memel log. It fetches from 2s. to 2s. 2d. the cubic foot. It is fitted for beams, joists, scantlings, windows, and outside doors. 4. American yellow-pine is well suited to all inside work, and especially that which requires the highest finish, such as bound-doors, window-fittings, and mantel-pieces. There is no wood that receives paint so well. The logs are generally of immense sizes, affording great economy of timber in cutting them up. Its price is, for small sizes 1s. 8d. and for large 2s. 3d. the cubic foot. 5. Swedish 11-inch plank is good and useful timber, but its scantlings are not very suitable for farm-buildings. I have seen stout joists for granaries made of it, with a ¾ draught taken off the side for sarking. It forms excellent planking for wheeling upon, and for gangways. It sells, the white-wood for from 5d. to 6d. and the red from 6d. to 7d. the lineal foot.

(162.) In the interior of the country, at a distance from sea-ports, *home* timber is much used in farm buildings. Larch forms good scantlings and joists, and is a durable timber for rough work, and so does well grown Scots fir of good age, and cut down in the proper season; but its durability is not equal to larch, or generally any good foreign timber for rough purposes.†

(163.) All the timber I have referred to is derived from the trees belonging to the natural order of *Coniferæ*, or cone-bearing trees. 1. The Scots fir, *Pinus sylvestris*, is a well known tree in the forests of this country, and few new plantations are made without its aid, as a nurse for hard-wood trees. In favorable situations it grows to a large size, as is evidenced in the Memel log, which is just the produce of the Scots fir from the forests of Lithuania. I have seen Scots fir cut down at Ardovie, in Forfarshire, of as good quality and useful sizes as the best Memel. 2. The Swedish plank is of the spruce, *Abies excelsa*, or *communis*—a tree which, as it is treated in this country, comes to little value, being rough and full of knots. Inspection of a cargo from Sweden, which arrived at Hull in 1808, convinced Mr. Pontey that the white deal, which fetched at that time from £14 to £15 10s. the load of 50 cubic feet, was of common spruce, the planks having been recently sawn, and a small branch left attached to one of them.‡ 3. Whether the Norway pine is the same species as the pine found in some of the forests of the north of Scotland, I do not know. I observe that some writers speak of the Norway batten as of the Norway spruce, called by them *Pinus Abies*

* McCulloch's Dictionary of Commerce, art. *Lath*.

† In vol. ix. p. 165, of the Prize Essays of the Highland and Agricultural Society, you will find a long account of the Larch Plantations of Atholl, drawn up by me from the papers of the late Duke of Atholl; and in vol. xii p. 122, of the same work, is an account of the native pine forests of the north of Scotland, by Mr John Grigor, Forres.

‡ Pontey's Profitable Planter, p. 41, 4th edition, 1814; and at p. 56 he relates an anecdote of a person who, though long accustomed to attend on sawyers, was deceived by some Scots fir, which he considered excellent foreign plank.

It may be that the white-wood battens are derived from that tree; but the red-wood kind has, very probably, the same origin as the red-wood of the north of Scotland, which is from a variety of the *Pinus sylvestris*, or *horizontalis* of Don.* 5. The red pine of Canada is the *Pinus resinosa*. 6. And the yellow pine is the *Pinus variabilis* or *Pinus mitis* of Michaux, which towers in lofty hight far above its compeers. It grows to the gigantic hight of 150 feet, and must require great labor to square it to the sizes found in the British market, large as these sizes unquestionably are. 7. The larch, *Larix europæa*, is a native of the ravines of the Alps of the Tyrol and Switzerland, where it shoots up, as straight as a rush, to a great hight.

(164.) In regard to the composition of wood, and its chemical properties, "It is considered by chemists that dry timber consists, on an average, of 96 parts of fibrous and 4 of soluble matter in 100; but that their proportions vary somewhat with the seasons, the soils, and the plant. All kinds of wood sink in water when placed in a basin of it under the exhausted receiver of an air-pump, showing their specific gravity to be greater than 1,000," and varying from 1.46 (*pine*) to 1.53 (*oak*). . . . "Wood becomes snow-white when exposed to the action of chlorine; digested with sulphuric acid, it is transformed first into gum, and, by ebullition with water, afterward into grape sugar. . . . . Authenreith stated, some years ago, that he found that fine sawdust, mixed with a sufficient quantity of wheat flour, made a cohesive dough with water, which formed an excellent food for pigs—apparently showing that the digestive organs of this animal could operate the same sort of change upon wood as sulphuric acid does. . . . . The composition of wood has been examined by Messrs. Gay-Lussac and Thenard, and Dr. Prout. According to Dr. Prout, the oxygen and hydrogen are in the exact proportions to form pure water; according to the others, the hydrogen is in excess."†

(165.) "When minutely divided fragments of a trunk or branch of a tree," as M. Raspail observes, "have been treated by cold or boiling water, alcohol, ether, diluted acids and alkalies, there remains a spongy substance, of a snow-white color when pure, which none of these reägents have acted on, while they have removed the soluble substances that were associated with it. It is this that has been called *woody-matter*, a substance which possesses all the physical and chemical properties of cotton, of the fibre of flax, or of hemp."

(166.) "On observing this vegetable *caput mortuum* with the microscope, it is perceived to be altogether composed of the cells or vessels which formed the basis or skeleton of the living organs of the vegetable. They are either cells which, by pressing against each other, give rise to a net-work with pentagonal or hexagonal meshes; or cells with square surfaces; or else tubes of greater or less length, more or less flattened or contracted by drying—sometimes free and isolated, at other times agglomerated and connected to each other by a tissue of elongated, flattened, and equilateral cells; or, lastly, tubes of indefinite length, each containing within it another tube formed of a single filament spirally rolled up against its sides, and capable of being unrolled under the eye of the observer, simply by tearing the tube which serves to support it. We find the first in all young organs, in annual and tender stems, in the pith of those vegetables that have a pith, and always in that of the monocotyledons. It is in similar cells that the fecula is contained in the potato. The second is met with in all the trunks and woody branches of trees. The tubes and the spirals (*tracheæ*) are found in all the phanerogamous plants. These are the organs which constitute the fibre of hemp, of flax, &c."

(167.) "Experiment in accordance with the testimony of history, proves that, if excluded from the contact of moist air, woody matter, like most of the other organized substances, may be preserved for a indefinite period." The plants found in coal mines, the wood, linen cloths, bandages, and herbs and seeds found in the coffins of Egyptian mummies, have all their characters undecayed, and yet these tombs are in many cases nearly 3000 years old. "But, if the woody matter be not protected against the action of air and moisture, the case is very different. By degrees its hydrogen and oxygen are disengaged, and the carbon predominates more and more. Thus the particles of the texture are disintegrated gradually, their white color fades, and passes through all the shades till it becomes jet-black; and if this altered woody matter be exposed to heat, it is carbonized without flame, because it does not contain a sufficient quantity of hydrogen. Observe, also, that the cells of woody matter contain different sorts of *substances tending to organize*, and that these are mixed and modified in many different ways." . . . "Woody matter, such as I have defined it, being formed of one atom of carbon and one atom of water, as soon as it is submitted to the action of a somewhat elevated temperature, *without the contact of air*, experiences an internal reäction, which tends to separate the atom of water from the atom of carbon. The water is vaporized, and the carbon remains in the form of a black and granular residue"‡

(168.) Now, if any means could be devised by which the substances in the cells of woody matter could be deprived of their tendency to organize, when in contact with common air, wood might be rendered as permanently durable, and even more so, than the grains of wheat which have been found undecayed in Egyptian mummies. This discovery seems to

* See Quarterly Journal of Agriculture, vol. xi. p. 530. † Ure's Dictionary of the Arts, art. *Wood.*
‡ Raspail's Organic Chemistry, translated by Henderson, pp. 141-164.

have been made by Mr. Kyan. In contemplating the probability of the use of home timber being much extended in the construction of steadings, when the young woods at present growing shall have attained their full growth, it may be proper that the growers of wood, and the farmers on the estates on which wood is grown, be made aware of this mode of preventing timber being affected by the dry-rot. What the true cause of dry-rot is, has never yet been determined, but it frequently shows itself by a species of mildew, which covers the timber, and the action of which apparently causes the wood to decay, and crumble down into powder. The mildew, however, is neither the dry-rot nor its cause, but its effect. It is distinctly seen by the microscope to be a fungus; and as the fungus itself is so minute as to require the aid of the microscope to be distinctly seen, its seeds may be supposed to be so very minute as to be taken up by the spongeoles of trees. But whatever may be the *cause* of dry-rot in timber, there is not a doubt now of the fact, after years of successful experience, that the process discovered by Mr. Kyan of simply steeping timber in a solution of corrosive sublimate, bi-chloride of mercury, preserves timber from dry-rot.

(169.) The principle upon which the chemical action of the corrosive sublimate upon vegetable matter, preserves the timber is easily explained. All plants are composed of cellular tissues, whether in the bark, alburnum, or wood. The tissue consists, as you have seen, of various shaped cells; and although they may not pass uninterruptedly along the whole length of the plant, as M. de Candolle maintains, yet air, water, or a solution of any thing, may be made to pass through the cells in their longitudinal direction. Experiments with the air-pump have proved this beyond dispute. Those cells, and particularly those of the alburnum, contain the sap of the tree, which, in its circulation, reaches the leaves, where its watery particles fly off, and the enlarging matter of the tree, called the albumen, remains. Albumen is the nearest approach in vegetables to animal matter, and is, therefore, when by any natural means deprived of vitality, very liable to decomposition, particularly that which is connected with the alburnum, or sap-wood. Now, corrosive sublimate has long been known to preserve animal matter from decay, being used to preserve anatomical preparations; and even the delicate texture of the brain is preserved by it in a firm state. The analogy between animal and vegetable albumen being established, there seems no reason to doubt the possibility of corrosive sublimate preserving both substances from decay; and, accordingly, the experiments of Mr. Kyan, with it, on albuminous and saccharine solutions, have confirmed the correctness of this conjecture. The prior experiments of Fourcroy, and especially those of Berzelius, in 1813, had established the same conclusions, though neither of these eminent chemists had thought of their practical application to the preservation of timber. Berzelius found that the addition of the *bi-chloride* (corrosive sublimate) to an albuminous solution produced a *proto-chloride* of mercury (calomel), which readily combined with albumen, and produced an insoluble precipitate. This precipitate fills up all the cellular interstices of the wood, and becomes as hard as the fibres."*

(170.) Even after timber has been subjected to this process, it is requisite to give the air free access to it by means of ventilation, and for that purpose, where timber is covered up, which it is not likely to be in a steading, small openings, covered and protected by cast-iron gratings in frames, should be made through the outside walls.

(171.) With regard to the expense of this process, which is a material consideration to those who have large quantities of timber to undergo the treatment, it costs for steeping £1 the load of 50 cubic feet. But persons having tanks for their own use only, and not for the purposes of trade, pay 5s. for each cubic foot of the internal contents of the tank. A tank, fitted up to steep large scantlings and logs, costs about £50, and the process may cost 3d. or less the cubic foot to those who construct a tank for themselves.

(172.) Other means have been devised for preserving timber from decay, such as pyroligneous acid, derived from the smoke of burning wood; naphtha, obtained by distillation of coal-tar; and in 1839 a patent was taken out by Sir William Burnett, of the medical department of the Navy, for steeping wood in a solution of the chloride of zinc;† but experiment has not yet had time to decide whether any of these methods possesses any superiority over the valuable process practiced by Mr. Kyan.

(173.) The pine tribe, of which I have been speaking as of so much use in our farm buildings, is also highly useful in the arts. It is from the *Pinus sylvestris* and the *Abies excelsa* that *tar* is obtained in the largest quantities, for the use of all nations; and it is a substance which is of great utility in a farm, though not requisite in large quantity. The tar of the north of Europe is of a much superior description to that of the United States. It is obtained by a process of distillation, which consists of burning, in a smothering manner, roots and billets of fir-timber, in pits formed in rising ground for the purpose, and covered with turf.

(174.) The quantity of tar imported into this country in 1837, was 11,480 lasts, of 12 barrels per last, each barrel containing 31½ gallons. The duty is 15s. per last, 12s. upon tar from the British possessions, and 2s. 6d. per cwt. upon Barbadoes tar. "Tar produced or manufactured in Europe is not to be imported for home consumption, except in British ships,

* See paper by me on this subject in vol. viii. p. 385 of Quarterly Journal of Agriculture.
† Repertory of Patent Inventions, New Series, vol. xii. p. 346.

or in ships of the country of which it is the produce, or from which it is imported, under penalty of forfeiting the same, and £100 by the master of the ship."*

(175.) Besides tar, most of the pines afford one or other of the turpentines. Common turpentine is extracted by incision from the *Abies excelsa* and the *Pinus sylvestris*.

(176.) Of the *specifications of plumber-work*, the kind of work done after the carpentry the *flanks* and *peands* should be covered with sheet-lead, weighing 6 lbs. to the square foot, 18 inches broad. The ridges should be covered either with droved angular freestone ridge-stones, or with 6-lb. lead, 18 inches broad, supported on 2½ inches in diameter of ridge-rolls of wood. Platforms and gutters should have 7-lb. lead. In cisterns, it should be 8-lb. in the bottom and 6-lb. in the sides. Rain-water spouts of 4½ inches in breadth, and conductors of 2½ or 3 inches diameter, should be of 6-lb. lead.

(177.) The lead of commerce is derived from the ore *galena*, which is a sulphuret, yielding about 87 per cent. of lead and 13 of sulphur. Galena is found in greatest quantity in transition rocks, and of these the blackish transition limestone contains the largest. The ore is more frequent in irregular beds and masses than in veins. The galena lead-mines of Derbyshire, Durham, Cumberland, and Yorkshire, are situate in limestone, while those of the Leadhills, in Scotland, are in graywacke. Great Britain produces the greatest quantity of lead of any country in the world, the annual produce being about 32,000 tons, of which the English mines supply 20,000. The rest of Europe does not supply 50,000 tons. The export of lead has fallen off considerably, and its price has experienced a corresponding depression for some years past, on account of the greatly increased production of the lead-mines of Adra in Granada, in Spain.†

(178.) As *zinc* has been substituted in some cases for lead in the covering of buildings, although sufficient experience has not yet been obtained as to their comparative durability, it may be proper to give here the sizes and prices of covering flanks, peands and ridges with zinc. The flanks are covered with zinc, weighing 16 ounces to the square foot, at a cost of 6½d. the square foot. The peands and ridges are covered with 12-inch sheet zinc, weighing 18 ounces in the square foot, at a cost of 7d. the square foot. The zinc covers for the peands and ridges are so prepared that they clasp by contraction, and thereby hold fast by the wooden ridge-rolls, and this is so easily done that any mechanic may put them on. Where soldering is required in zinc-work, such as the laying on of platforms on roofs, the cost of the sheet of 18 ounces to the square foot is enhanced to 9d. the square foot. Zinc in all jobs costs about half the price of lead.

(179.) Zinc is not very suitable for gutters and platforms, on account of its thinness—the wood below warping in warm weather, and tearing up the sheets of zinc.

(180.) Zinc is an ore which occurs in considerable quantity in England. It is found in two geological localities, in the mountain limestone and in the magnesian limestone. It occurs in veins, and almost always associated with galena or lead-glance. It is of the greatest abundance in the shape of a sulphuret or blende, or *black-jack*, as the miners call it. There is also a siliceous oxide of zinc, and a carbonate, both called calamine. In North America, the red oxide of zinc is found in abundance in the iron mines of New-Jersey. The zinc of commerce is derived, in this country, from the blende and calamine. It is naturally brittle, but a process has been discovered by which it is rendered malleable, and it retains its ductility ever after. It is this assumed ductility which renders the metal useful for domestic purposes. "It is extensively employed for making water-cisterns, baths, spouts, pipes, plates for the zincographer, for voltaic batteries, filings for fire-works, covering roofs, and a variety of architectural purposes, especially in Berlin; because this metal, after it gets covered with a thin film of oxide or carbonate, suffers no farther change from long exposure to the weather. One capital objection to zinc as a roofing material is its combustibility."‡

(181.) The most malleable zinc is derived from Upper Silesia, under the name of *spelter*, which is sent by inland traffic to Hamburgh and Belgium, where it is shipped for this country. The quantity imported in 1831 was 76,413 cwt. and in 1836 it had fallen off to 47,406 cwt. A considerable portion of these quantities was exported to India and China, amounting, in 1831, to 62,684 cwt. The duty is £2 a ton on what is formed into cakes, and 10s. per cwt. on what is not in cakes.‖

(182.) The *slater-work* is then executed. Of its specifications, if blue slates are to be employed, they should be selected of large sizes, well squared, and have an overlap of ⅗, gradually diminishing to the ridge, and well bedded and shouldered with plaster-lime. The slates are fastened to the sarking with malleable iron nails, weighing 15 lbs. to the 1000, after being steeped when heated in linseed oil. These nails cost 3s. 4d. the 1000, 1300 being required for a rood of 36 square yards. Cast-iron nails were used for slating until a very few years ago, and which were also boiled in oil.

(183.) Slating is performed by the rood, and from 1000 to 1200 blue slates should cover a rood. The cost of the slates, in towns, including carriage, and putting them on with nails, is £4 4s. the rood.

* McCulloch's Dictionary of Commerce, art. *Tar*.
† See Ure's Dictionary of the Arts, and McCulloch's Dictionary of Commerce, arts. *Lead*.
‡ Ure's Dictionary of the Arts, art. *Zinc*. ‖ McCulloch's Dictionary of Commerce, art. *Zinc*.

(184.) Blue slate is derived from the primitive rock clay-slate. It occurs in large quantities through the mountainous parts of the kingdom. Good slate should not absorb water, and it should be so compact as to resist the action of the atmosphere. When it imbibes moisture, it becomes covered with moss, and then rapidly decays.

(185.) The principal blue slate quarries in Great Britain are in Wales, Lancashire, Westmoreland, Cumberland, Argyle, and Perthshires. The most extensive quarry is in Caernarvonshire, in Wales, near the town of Bangor, on the Penrhyn estate. It employs 1500 men and boys. The Welsh slate is very large and smooth, and much of it is fit for putting into frames for writing-slates. When used very large, being thin, it is apt to warp on change of temperature. The English slates at Ulverstone, in Lancashire, and in the counties of Westmoreland and Cumberland, are not so large as the Welsh, but equally smooth and good.—The Easdale slates, in Argyleshire, are small, thick, waved on the surface, and contain many cubical crystals of iron-pyrites, but its durability is endless. Being a small and heavy slate, it requires a stout roofing of timber to support it. The Ballihulish slates are rather smoother and lighter than the Easdale, though also small, and containing numerous crystals of iron-pyrites, and is equally durable. The slates in Perthshire are of inferior quality to either of these. "The ardesia of Easdale," says Professor Jameson, "was first quarried about 100 years ago; but was for a long time of little importance, as sandstone flags and tiles were generally used for roofing houses. As the use of slates became more prevalent, the quarries were enlarged, so that 5,000,000 slates are annually shipped from this island. The number of workmen is at present (in 1800) about 300, and they are divided into quarriers and day-laborers. The quarriers are paid annually at a certain rate for every 1000 slates, from 10d. to 15d. I believe, as their work has been attended with more or less difficulty. The day-laborers are employed in opening new quarries, and have from 10d. to 1s. a day."*

(186.) Slates are assorted into sizes at the quarry. The sizes at Bangor vary from 36 inches in length to 5½ inches in breadth. Their weight varies from 82 to 12 cwt. the 1000, and the prices from 140s. to 10s. the 1000 for the smaller, and from 55s. to 35s. the ton for the larger sizes.

(187.) Cisterns, with sides and ends 1 inch thick, 1s. 10d. the cubic foot contents.
" " " 1½ " 2s. 2d. " "

(188.) The *export* of slates from England to foreign ports has increased from 2,741 tons in 1828, to 6,061 tons in 1832. That of framed slates has decreased in number, in the same period, from 37,034 to 15,420.

(189.) The shipping expenses of slates at Bangor are 6d. the ton, and bills of lading 3s. 6d.†

(190.) When the roof is to be covered with *tile*, it should be laid with lath 1½ inches square to a gauge of 10 or 11 inches. There should be 3 or 4 courses of slates along all the eaves. The flanks, peands and ridges should be covered with tile. The whole under joints of the tiling should be pointed with plaster-lime.

(191.) Tiling is executed by the rood of 36 square yards; and, as pan-tiles are obliged to be made of a certain size, namely, 13½ inches long, 9½ inches wide, and ½ inch thick, by 17th Geo. III. c. 42, under a penalty of 10s. for every 1000, a rood will just contain 576 tiles.—Tiles should be smooth on the surface, compact, and ring freely when struck, when they will resist water. When they imbibe moisture by porosity, they soon decay in winter by the effects of rain and frost.

(192.) There were, in 1830, 5,369 brick and tile manufacturers in England and Wales, and 104 in Scotland, and must have greatly increased since.

(193.) The duty on tiles was abolished in 1833, the revenue derived from that source being very trifling. The duty on foreign pan-tiles is £15 the 1000. The export of tiles is inconsiderable, not having exceeded, in 1830, 803,742.‡

(194.) *Gray-slates* require the roof to be lathed in the same manner as tile, but, not being of an uniform size like tile, they are assorted to sizes in the quarry. The larger and heavier slates are put next the eaves, and gradually diminish in size to the ridge. The course at the eaves is laid double, slate above slate. Every slate is hung upon the lath by a wooden pin being passed through a hole at the upper end, and, on being laid on, the slates are made to overlap at least ⅓. Gray-slates should either be bedded and shouldered in plaster-lime, or laid on moss, the latter making the warmer roof.

(195.) Gray-slates are pretty smooth on the surface, and, when so compact in texture as to resist moisture, form a durable though very heavy roof.

(196.) The flanks are made of slate, but the ridge is covered with droved angular ridge-stone of freestone. As this species of roofing is not adapted to pavilion roofs, the peands should be covered with lead, but the safest form of roof with gray-slates is with upright gables.

(197.) The cost of gray-slating depends on the locality where it is wished to be done. At Edinburgh it costs £6 a rood; whereas in Forfarshire, the matrix of the grap-slate, it can be done, exclusive of carriage, for £2 10s. the rood. In Forfarshire the slates cost £4 per

* Jameson's Mineralogy of the Scottish Isles, vol. i. p. 195.
† McCulloch's Dictionary of Commerce, art. *Slates*.
‡ Ibid. art. *Bricks and Tiles*.

per 1,000; 360 are required for a rood; the putting them on, including dressing, boling, pins for the slates, and nails for the laths, costs only 15s.; and with moss for bedding 1s.. and lime for teothing 3s., 22s. the rood. The droved angular freestone ridging-stone, including carriage, costs 6d. a lineal foot, or 10s. the rood.

(198.) Gray-slates are obtained in best quality from gray slaty inferior sandstone belonging to the old red sandstone series. They are derived from the same quarries as the far-famed Arbroath pavement, being, in fact, formed by the action of frost on pavement, set on edge for the purpose. A mild winter is thus unfavorable to the making of slates. From Carmylie to Forfar, in Forfarshire, is the great field for the supply of gray-slates; and as blue-slates can only be obtained there by sea and long land carriage, and there is little clay fit for tiles, they constitute the chief roofing of cottages and small farm-houses in that part of the country, their aspect being cold and unpicturesque, though snug enough.

(199.) Of all sorts of slating, there is none equal to blue-slate for appearance, comfort, and even economy in the long run. When a blue-slate roof is well executed at first, with good materials, it will last a very long time. Tile roofs are constantly requiring repairs, and the employment of gray-slate is a sacrifice of, and a burden upon, timber. Of the blue-slate the Welsh give the cheapest roofing, being larger and much lighter than Scotch or English slates.

(200.) As the *plaster-work* of a steading does not require to be of an ornamental nature, its *specifications* should be simple. The ceilings of the riding-horse stable, boiling-house, wool-room, hen-house, and granaries, when tile-roofing is employed, should be finished with two coats of the best haired plaster, hard rubbed in. The walls of the granaries, corn-barn, work-horse stable, cow-byre, boiling-house, calves'-house, wool-room, gig-house, and hen-house, should be finished with one coat, hard rubbed in. The walls of the riding-horse stable should have three coats, hard rubbed in. Plaster-work is measured by the square yard, and costs for one coat 3d., for two coats from 4d. to 4½d., and for three coats from 5d. to 6d. the square yard.

(201.) It is necessary to say something regarding the *specifications of smith-work*, although there is not much of this kind of work required in a steading. All the outside doors, including those on the feeding-holes at the byre, should be hung with crooks and bands; the crooks should be fastened into the ingoings of the ribets with melted lead. The larger crooks and bands cost 10s. and the smaller 5s. the pair. The inside doors should be hung with T hinges, 18 inches long, and the opening parts of the windows with 9-inch T hinges. The former are 1s. and the latter 9d. a pair. The outside doors should have good 10-inch stock-and-plate locks, which cost 2s. 6d. each, except where there are more than one outside door to the same apartment, in which case all the doors but one can be fastened by bars from the inside The inside doors should have the same sort of locks; the common stock-lock, which cost 1s. 6d. each, not being worthy of commendation. Thumb-latches are convenient for opening and keeping shut doors that do not require to be constantly locked, such as the doors of the corn-barn, granary, boiling-house, cow-byre, and hen-house. These latches cost from 5d. to 7d. each. A wooden bar of hard wood, to open and shut from both sides, is a convenient mode of fastening inside doors. The upper barn-door, of two vertical leaves, requires an iron stay-band to fasten it with. The doors of the riding-horse and work-horse stables should be provided with sunk flush ring-handles and thumb-latches, to be out of the way of catching any part of the harness. The mangers of the riding-horse stable, and the upper rail of the hay-rack of the work-horse stable, should be provided with rings and staples for the stall-collar shanks to pass through. These cost 1d. each.

(202.) Various descriptions of *nails* are used for the different parts of work in a steading. The scantlings of the roofs are fastened together with *double-doubles*, which cost 5s. per 1,000. Deals of floors are fastened down with flooring-nails, 16-lb. weight, and 4s. 6d per 1,000. The bars of the plain-deal doors are put on with 10-lb. nails, which cost 3s. 6d. the 1,000. For finishing, *single-flooring* nails at 2s. 6d., and 2-inch springs at 2s. to 2s. 3d. the 1,000 are used.

(203.) As a security against robbery, iron stancheons, ⅞ inch in diameter, should be fixed on the outside of the low windows of the corn-barn and implement-house. Such stancheons cost 3d. per pound.

(204.) Iron is chiefly found among the members of the coal formation, in bands composed of nodules, which are called compact clay-ironstone, a carbonate of iron. It is abundant in the west of Scotland and in South Wales. Its annual worked production is probably not less than 1,000,000 tons.*

(205.) The windows of all the apartments should be *glazed* with best 2d crown-glass, fastened in with fine putty. Glazing is executed for 2s. the square foot.

(206.) A *skylight* in blue slating is made of a frame fastened to the sarking. In the roofing, tiles are made on purpose to hold a pane of glass. In gray-slating, a hole is made in the slate to suit the size of the pane. A dead skylight of zinc, to answer any kind of roofing, costs 4s.

(207.) There is a duty of 73s. 6d. the cwt. on good window, and 30s. on broad or inferior window-glass, which is returned in drawback on exportation to foreign countries. When

* Ure's Dictionary of the Arts, art. *Iron*.

glass intended for exportation is cut into panes, it must be in panes of less than 8 inches in the side to enable it to claim the drawback.

(208.) Glass of small sizes, though of good quality, such as is fit for glazing hot-houses and forcing-frames, costs only from 8d. to 10d. the square foot; while in ordinary sized panes it costs 1s. 3d., and in still larger sizes it is charged 1s. 6d. the square foot. I am not sure but the sort fit for hot-houses would answer the purpose of glazing the windows of a steading.

(209.) "The researches of Berzelius having removed all doubts concerning the acid character of silica, the general composition of glass presents now no difficulty of conception. This substance consists of one or more salts, which are silicates with bases of potash, soda, lime, oxide of iron, alumina, or oxide of lead; in any of which compounds we can substitute one of these bases for another, provided that one alkaline base be left. Silica, in its turn, may be replaced by the boracic acid, without causing the glass to lose its principal characters."*

(210.) Rain-water spouts, or *runs* as they are technically termed, may be made of wood, cast-iron, lead, or zinc. Wooden ones may be made out of the solid or in slips nailed together. When made out of the solid, with iron hold-fasts, they cost 1s. and when pieced together 6d. the lineal foot. The conductors from both kinds cost 8d. the lineal foot. Wooden spouts should be pitched inside and painted outside. Cast-iron ones are heavy, but they cost no more than 2s. a yard if of 4½ inches diameter, and the conductors, of from 2 to 4 inches diameter, from 8s. to 18s., of 9 feet in length each. Lead makes the best spout, but it is very expensive, being 1s. 6d. a foot. Zinc ones, on the other hand, are very light. Stout 4-inch zinc spouts cost 9½d. the foot, and a 2½ pipe as conductor, 7½d. the foot. The lowest part of this pipe is made strong enough to resist accidents. Every sort of water-spout should be cleaned at least once a year, and the wooden ones would be the better for an annual coat of paint.

(211.) The outsides of all the outside doors and windows, all the gates of the courts and hammels, and the water-troughs in the various courts, if made of wood, should receive *three coats of good paint*. Painting costs 3d. or 4d. the square yard, but three coats can be done for 8d. the square yard. The best standing colors, and they happen to be the cheapest too, are gray, stone, or slate-blue; the last seems to be most commonly preferred. Green is dear and soon fades, and red seems very distasteful in buildings. But the truth is, that white-lead and oil are the principal ingredients in paint, and all the coloring matter has no power to preserve timber from the effects of the weather. A substance called *lithic paint* has recently been found to answer well for country purposes. The lithic, which costs 2½d. per lb. is ground to powder, and mixed, in a certain proportion, with cold coal-tar, and the mixture is applied with a brush. This paint deprives the coal-tar of its noxious smell, and hardens it into a durable paint in a few days.

(212.) White lead of commerce is a *carbonate of lead*, or *ceruse*, as it is called, artificially formed from pure lead. It has long been made with great success at Klagenfurth, in Carinthia, and large quantities are made in England. The compound is 1 equivalent of lead, 1 of oxygen, and 1 of carbonic acid; or by analysis, of lead 77·6, oxygen 6, and carbonic acid 16·4 in 100 parts. White lead, when it enters the human system, occasions dreadful maladies. Its emanations cause that dangerous disease the *colica pictonum*, afterward paralysis, or premature decrepitude and lingering death. All paints are ground into fine powder in a mill, as being a safer plan for the operator, as well as more expeditious, than by the hand.†

(213.) I have said (81.), (82.), (83.), that when the building of a steading is to be measured, the work that has actually been executed should alone be measured, and no *allowances*, as they are called, should on any account be permitted to increase the amount of cost. The correctness of this rule will appear obvious, and its adoption reasonable, after you have learnt the sort of claims for allowances made by tradesmen in various sorts of work.

(214.) In the first place, in regard to *masonry*, double measure is claimed on all circular work. Claims are made for allowance on all levelings for joists, bond-timbers, and wall-heads. The open spaces or voids left in the walls for doors and windows, are claimed to be measured along with rubble-work. Girthing around the external walls of rubble-work is claimed in measurement, the effect of which is, to measure the square pieces of building in each corner twice over. Scontions of all voids are claimed to be measured over and above the rubble-work. The ashlar for the hewn-work is first measured with the rubble, and then it is claimed to be measured by itself. In like manner, chimney-tops are first measured as rubble, and then claimed to be measured again as ashlar. In short, wherever any sort of mason-work differs from the character of the general work under the contract, allowances are claimed.

(215.) In regard to *carpentry*, the claims are equally absurd. For the cuttings connected with the peands and flanks of roofs, 18 inches of extra measurement are claimed. The same extent is claimed for angles in the flooring, and in all such unequal work. In window-making a claim is made for 3 inches more than the hight, and 4 inches more than the width of windows, which is more than the voids; whereas the measurement should be confined

* Ure's Dictionary of the Arts, art. *Glass*. † Ibid. arts. *White-lead*, *Paint*.

to the mere daylight afforded by the windows. In many instances 1¼, and even double measure, is claimed for round work, according to its thickness. Where plain deal is cleaned on both sides, such as the under part of the floor of the upper-barn, which forms the roof of the corn-barn, or shelving, 1¼ measure is claimed.

(216.) In *slating*, claims are made on the making of peands and flanks, from 18 inches to 3 feet in width, and for eaves, from 12 inches to 18 inches in width, more than the actual work done. For all circular work, such as the slating of a round horse-course of a threshing-machine, double measure is claimed.

(217.) In *plaster-work*, double measure is claimed for all circular work. There is an allowance made in plastering which is, however, quite reasonable, and that is, in the case where new work is joined to old, an allowance of one foot is made around the new work, as the old part has to be wetted and prepared for its junction with the new.

(218.) A perusal of these statements naturally suggests the question, how could such claims have originated? If a workman execute the work he agreed to undertake, and gets payment for what work he executes, he is not entitled to ask more. But what proves an aggravation of such demands is, that modes of measurement differ in different counties—that different allowances are made on different kinds of work—and that those allowances differ n different counties. So it appears that those allowances are based on no principle of equity But it may be urged in justification of these allowances that the prices of work, as usually estimated, are too low to remunerate the contractor for his labor, and that allowances are therefore requisite to insure him against loss. To this specious statement it may be replied by asking, why should any *honest* contractor estimate work at such rates that he knows will not remunerate him? A rogue will do so, because he wishes to have possession of a job at all hazards, in order to make up his foreseen loss by exorbitant claims for allowances. If employers will not pay sufficiently for good work, as is alleged against them, and perhaps with truth, let them understand that they shall receive insufficient work as an equivalent for *their* stinted money. But it is very unfair to take advantage of an honorable employer, by capricious and absurd allowances, when he is all the while desirous to pay his workmen well for their labor. So much dependence is sometimes placed on allowances by contractors, that I have heard of a case where a surveyor was obliged to reduce the claims made against a single steading, to the extent of £800! Such a fraudulent system ought to be entirely abolished, and it is quite in the power of those who employ tradespeople to abolish it

(219.) It would be completely abolished were contracts to contain stringent clauses prohibiting all allowances whatsoever; and to consist of detailed measurements, and specified prices for every species of work to be executed. If more work happens to be executed than was expected, its value can easily be ascertained by the settled measurements and prices, and if less, the contractor is still paid for what he has actually executed. Were such a form of contract uniformly adopted, proprietors and farmers could measure the work done as well as any surveyor, whose services might, in that case, be dispensed with; but, what would be still better, the measurements of the surveyor could be checked by the proprietor or the tenant if either chose to take the trouble of doing it. Where any peculiar kind of work is desired to be executed, it could be specified in a separate contract.

(220.) Having thus amply considered all the details which should form a part of all specifications of the different kinds of work required to build a steading, I shall now give the particulars which should be specified in all contracts, and that these may not be imaginary, but have a practical bearing, I shall take the steading as shown in the plan, fig. 4, Plate IV. as the example. In order that the data furnished in the proposed specifications shall be generally applicable, I shall first give the measurements of the various kinds of work proposed to be executed—then the quantity of materials required for constructing the same—and lastly, the prices paid for the different sorts of work in Edinburgh, both including and excluding the cost of carriages, that you may have a criterion by which to judge of the cost of doing the same kind of work in other parts of the country. You may reasonably believe that the prices of labor and materials are higher in Edinburgh than in the country; but, on the other hand, you must consider the superiority of the workmanship obtained in so large a town. These must affect the total amount of the estimate to a certain extent, but to what exact per centage I cannot say. I am told that carpenter-work is very little dearer in Edinburgh than in the country, but that mason-work, smith-work and plaster-work are all considerably higher; but of smith-work, as I have already said, little is required in building a steading.*

[* Now, although the suggestions and reasoning of this chapter apply to a country where the kind and cost of the materials, and yet more the cost of labor, differ very materially from such as are in use or paid in this country, yet the reasoning and the rules laid down are of universal application; and how would it be possible to omit them, without impairing essentially the value of the work in hand?

All who have had much experience in building have found it to be difficult to guard against imposition; and this is especially the case with men who have not been qualified, either by education or experience, to *judge for themselves*. By education we mean instruction at school or

MEASUREMENT OF THE PLAN OF A PROPOSED STEADING IN FIG. 4, PLATE IV.

Mason-work.

6225 Cubic yards of Foundations, and wheeling the earth not farther than 60 yards distance
207 .. Drains with sills and covers.
85 Cubic roods of Rubble-walls, 2 feet thick.
47 .. Division rubble-walls, 12 to 15 inches thick, including dykes.
42 Lineal feet of Chimney-vents.
400 .. Corners of buildings.
80 .. Corners for archways.
59 .. Arched lintels for archways.
1528 .. Ribets, sills, lintels, and steps.
75 .. Arched lintels over doors.
24 .. Ringpens of archways to granary.
80 .. Corners, sills, and lintels of feeding-holes of byres.
60 .. Corners of gateways to courts.
286 .. Corners or hammer-dressed scontions for gates in dykes.
90 .. Coping of chimney-stalks.
110 .. Ashlar pillars for sheds, from 18 to 20 inches square.
294 .. Skews on gables.
1671 .. Semi-circular hammer-dressed coping on dykes.
100 .. Gutters in byres.
94 .. Coping round liquid manure tank.
300 .. Steps of stairs to granaries.
45 .. Brick stalk for steam-engine, 6 feet square at the base.
152 Square roods of Rubble-causeway.
287 Lineal yards of Causewayed gutters around the buildings outside.
2 Pairs of jambs and lintels.
Building in boiler, including boiler and furnace complete.
17 Droved stones, with gratings for liquid manure drains.
8 Water-troughs in courts.
31 Stones for heel-posts of stalls.
31 Stones for curbs of stall-boardings.

Carpenter-work.

540 Square feet of 4-inch thick safe lintels.
2768 Square yards of Roofing, with balks and sarking.
563 .. Joisting and flooring of granaries and corn-barn.
762 Lineal feet of Ridge-battens.
192 .. Dressed beams for pillars of roofs of sheds.
1141 .. Door-checks or fixings, 6½ inches by 2½ inches.
1366 .. Door-keps or stops and facings.
2132 Square feet of 1½-inch deal doors.
1360 .. 1½-inch divisions of stalls.
829 Lineal feet of Heel and fore-posts.
18 .. Manger in riding-horse stable.
18 .. Hay-rack in ditto.
96 .. Hay-racks, low, in work-horse stable.
84 .. Feeding-troughs in byres.
670 .. .. .. courts.
36 .. Racks in cattle-sheds.
432 Square feet of Daylight of windows.
760 .. Sparred divisions of cribs for calves.
669 Lineal feet of Rian-water spouts.
87 .. Conductors from ditto.
10 Small doors of feeding-holes of byre.
14 Corn-boxes for work-horse stable.
2 Square racks for center of courts.
1 Corn-chest for work-horses.
1 .. for riding-horse stable.
7 Luffer-board ventilators for roofs.
8 Sparred gates, from 9 feet to 10 feet wide.
12 .. .. 5 ..
Rails, harness-pins, and saddle-trees.
Stathel-frames for stacks.
Pump with mounting.

Slater-work.

77 Square roods of Blue-slating, gray-slating, or tiling.

elsewhere in the rules of mensuration and the principles of mechanics far enough to know—what every school-boy might easily be taught—enough of architecture to know the *names* of every part and piece of timber employed in building, and the manner of measuring carpenters' and bricklayers' and plasterers' work—a sort of useful practical information which any young man might acquire in a few days of earnest, ardent study.

One important point to be guarded against is the liability to be imposed upon by exorbitant charges for every, the very slightest addition to, or departure from the plan of building agreed upon. If such chapters as these have no other effect, we may hope they will assist in impressing upon the mind of the farmer the obligation he is under as a parent and a friend, to see that his son is so educated as to enable him to form his own correct opinion, to the end that while he should be at all times ready to do full justice, and even to be as liberal as he can afford to be to mechanics, he shall be prepared to detect and resist all attempts at imposition.

Plumber-work.

1084 Square feet of Lead on ridges, flanks, and peands.
669 Lineal feet of Lead rain-runs or spouts.
87 .. Lead-pipes or conductors from runs.

Plaster-work.

1807 Square yards of 1st, 2d, and 3d coat plaster.

Smith-work.

22 Stock-and-plate locks.
28 Pairs of crooks and bands.
9 Pairs of cross-tailed hinges.
35 .. .. .. small.
2 Sets of fastenings for double doors.
3 Locks for small courts.
10 Pairs of crooks and bands for feeding-holes.
10 Sneck-fastenings for ditto.
33 Thumb-latches.
18 Manger-rings.
17 Seals for fastening cows, or feeding cattle.
Stanchions for windows.
Cast-iron runs and conductors, } when used.
.. travis-posts, }
.. hay-racks for riding-horse stable, }
.. window-sashes, }
Boiler and furnace.
Mounting for gates.

QUANTITIES OF MATERIALS AND NUMBERS OF CARRIAGES IN STEADING.

108¼ Cubic roods of 2-feet walls, each rood containing 36 cubic yards of building, requiring 40 cart-loads of rubble-stones, 2 cart-loads of lime, and 4 or 5 cart-loads of sand, besides water
710 Ashlar corners.
1004 Ribets.
100 Sills and lintels, from 4 ft. to 4½ ft. long.
20 .. .. 30 inches long.
31 Steps, from 3½ feet to 4 feet long.
60 .. .. 4½ .. 5 ..
20 Lineal feet of Coping of chimney-stalks.
2 Pairs of chimney-jambs, 3½ ft. by 2 ft. long.
2 Lintels for ditto, from 3½ ft. to 4 ft. long.
110 Ashlar stones for pillars, from 18 inches to 20 inches square.
294 Lineal feet of Skews.
280 .. Curbstones.
100 .. Sills for gutters in byres.
94 .. Coping round liquid manure tank.
17 Stones for gratings to drains.
31 .. heel-posts.
8 .. water-troughs.
31 .. curbstones below boarding in stables and byres.
100 Square roods of Causeways.
77 .. Slating.
136 Loads of Timber.
386 Square feet of Glass for windows.

On ascertaining the quarry-mail, or prime cost of the stones, and the cost of carriage, in the locality in which you intend to build your steading, the cost of each of the above quantities of materials will easily be ascertained.

(221.) The following schedule gives the prices of those materials in Edinburgh, and they are stated both inclusive and exclusive of carriages.

Mason-work.	*Including Carriage.*			*Excluding Carriage.*		
	L.	S.	D.	L.	S.	D.
Digging foundations..............................per cubic yard.	0	0	6	0	0	4
Rubble-foundations, reduced to 2 feet thick..............per rood of 36 cubic yards.	10	0	0	8	0	0
Rubble building, 2 feet thick..............................	8	8	0	7	0	0
.. .. 18 inches thick and under, reduced to 1 foot thick..............................	5	0	0	4	6	0
Rubble drains, with dressed flags, sills and covers, 12 inches square in the opening..............................per lineal yard.	0	3	0	0	2	4
Ditto, 15 inches by 18 inches in the opening.............................. ..	0	5	0	0	4	0
Hammer-dressed coursed work, with raised or hollow joints.........per square foot.	0	0	3			
Where bricks are used for building the walls, the prices are for—						
9 thick brick on edge walls..............................per square yard.	0	1	9	0	1	6
4½ .. .. bed	0	2	0	0	2	6
6	0	5	0	0	4	8
Chimney-vents, plastered ....per lineal foot.	0	0	6			
Droved ashlar, from 7 to 8 inches thick..............................per square foot.	0	1	2	0	1	0
Broached	0	1	0	0	0	10
.. corners, averaging 3 feet girth..............................per lineal foot.	0	2	6			
.. supports for stacks, from 2 feet to 2½ feet in girth.......... ..	0	2	0			
Droved ribets, front and ingoing with broached tails, 2 feet long and 1 foot in the head	0	2	6			

Mason-work (continued).		Including Carriage.			Excluding Carriage.		
		L.	S.	D.	L.	S.	D.
Droved projecting sills, 7 inches thick	per lineal foot.	0	2	0			
Sills and lintels dressed similar to the ribets	..	0	1	6			
Droved cornices for chimney-stalks, 6 to 7 inches thick	..	0	1	6			
Droved block-course for chimney-stalks, 6 inches deep	..	0	0	9			
Droved skews, 2½ to 3 inches thick	per square foot.	0	0	9			
Broached	..	0	[illegible]	8			
Corners for coach-house doors, with droved giblet-checks	per lineal foot.	0	1	8			
Elliptical arched lintels for ditto	..	0	1	9			
Segmental	..	0	1	9			
Broached pillars for cart-sheds, &c.	per square foot.	0	0	10			
Droved jambs and lintels	..	0	0	9			
.. Arbroath pavement and hearths	..	0	0	9	0	0	8
.. freestone pavement	..	0	0	9	0	0	8
Broached	..	0	0	8	0	0	7½
Dressed and jointed flagging	..	0	0	7	0	0	6½
.. hanging steps, ordinary sizes	per lineal foot.	0	2	4	0	1	3
.. common steps, ..	..	0	1	6	0	1	3
.. plats of hanging stairs, single measure	per square yard.	0	1	5½	0	1	4½
.. stone-skirtings, 4½ inches deep	per lineal foot.	0	0	4	0	0	3½
.. ridge-stones, common form	..	0	0	7½	0	0	7
.. socket-stones for travis-posts	each.	0	5	6	0	5	0
.. feeding-troughs	per square foot.	0	1	1	0	1	0
.. stone water-troughs	..	0	1	0	0	0	11
Curb-stones for gutters in byres	per lineal foot.	0	0	6	0	0	5½
Droved curb-stones for stalls	..	0	0	6	0	0	5
Semi-circular coping for dykes, hammer-dressed, from 12 inches to 14 inches diameter	..	0	0	6½	0	0	6
Square dressed whinstone-causeway	per rood of 36 square yards.	7	7	0	7	0	0
Rubble causewaying ..	..	2	14	0	2	7	0
When ornamental masonry is introduced into steadings, these are the prices:							
Droved base-course and belts, 12 inches deep	per lineal foot.	0	1	1			
.. wall-head plinths, 6 inches thick	..	0	0	10			
.. cornices, 9 to 10 inches thick	..	0	2	6			
.. block-course, 12 inches deep	..	0	1	1			
.. checked plinth and block for chimney-stalks, 1 foot deep	..	0	1	2			
Polished hanging steps, ordinary sizes	..	0	2	6	0	2	5
Polished plats of hanging stairs, single measure	per square foot.	0	1	6	0	1	4
Broached copings, with droved edges, for dykes	..	0	0	9	0	0	8
Droved pillars for small gates to hammels, &c.	..	0	0	10	0	0	9
Building in boiler and furnace complete		0	18	0			
Bricks	per 1000.	1	17	0	1	10	0
Rubble stones	per load.	0	3	0	0	0	6

Carpenter-work.

Item		L.	S.	D.	L.	S.	D.
Safe-lintels and rough beams	per cubic foot.	0	3	6	0	3	4
Dressed beams	..	0	4	0	0	3	10
Scantling for roofs, 7 inches by 2½ inches, and 18 inches from center to center	per square yard.	0	2	4	0	2	2
Scantling for roofs, 7½ inches by 2½ inches, and 18 inches from center to center	..	0	1	6	0	1	2
Balks, 6 inches by 2 inches, and 18 inches from center to center	..	0	1	9	0	1	7
.. 5 .. 2	..	0	1	6	0	1	4
Wall-plates for roofing, 7 inches by 1½ inches	per lineal foot.	0	0	4	0	0	3½
Ridge trees, 10 inches by 2 inches	..	0	0	9	0	0	8
Ridge and peand battens, 2½ inches diameter	..	0	0	3	0	0	2½
¾-inch thick Baltic sarking	per square yard.	0	1	10	0	1	9
Tile-lath, 1½ inches square, and 11 inches apart	..	0	0	6	0	0	5½
Bond-timber, 3½ inches by 1½ inches, and 20 inches apart	..	0	0	6	0	0	5½
Baltic split lath, 3-16 inch thick	..	0	0	6	0	0	5½
Plain joisting, 7 in. by 2½ in. and 18 inches from center to center	..	0	2	4	0	2	2
.. .. 8 in. by 2½ in.	..	0	3	6	0	3	4
.. .. 9 in. by 2½ in.	..	0	4	0	0	3	10
.. .. 10 in. by 2½ in.	..	0	4	6	0	4	4
.. .. 12 in. by 2½ in.	..	0	5	0	6	4	10
1½ Batten flooring, grooved and tongued	..	0	3	4	0	3	2
Door-checks, 6 inches by 2½ inches	per lineal foot.	0	0	6	0	0	5
Checked window grounds, 2 inches by 1½ inches	..	0	0	2½			
Finishing grounds, 2 inches by ½ inch	..	0	0	1½			
Windows for barns and byres, of the form in fig. 15	per square foot.	0	1	8	0	1	6
.. stables, of the form in fig. 14	..	0	1	8	0	1	6
.. granaries, of the form in fig. 17	..	0	1	6	0	1	4

		Including Carriage		
1½ Travis-boarding for riding-horse and work-horse stables, doweled	per square foot	0	0	8
1½ travis-boards, grooved and tongued and beaded, for byres	..	0	0	6
1½-inch deal lining, grooved and tongued, for end stalls of riding-horse stable, with fixtures	..	0	0	5
¾-inch deal linings, beaded in walls, over and under the mangers in the riding-horse stable	..	0	0	3
Turned travis-posts, for riding-horse stable	each	0	8	0
Beaded travis-posts, fore-posts, and runtrees, for work-horse stable, reduced to 3 inches square	per lineal foot	0	0	5
Stakes and runtrees of byres, 4 inches to 5 inches in diameter	..	0	0	6

(294)

Carpenter-work (continued).

	Including Carriage	L.	s.	D.
Hard-wood high hay-racks, with turned rollers 2 inches diameter and 2½ inches apart, for riding-horse stable	..	0	3	0
Fir sparred low hay-racks for work-horse stable	..	0	1	0
Mangers for riding-horse stable	..	0	1	6
Corn-boxes for work-horse stable	each	0	3	0
1½-inch deal beaded outside doors, with 3 backbars	per square foot	0	0	7
¾-inch deal beaded inside doors, with three 1-inch backbars	..	0	0	6
Sparred calves'-cribs	..	0	0	5
Facings, keps, skirting, and coping, reduced to 4 inches broad	per lineal foot	0	0	3
Ogee copings for travises	..	0	0	2
1 inch beaded coping for lining	..	0	0	2½
Rain-spouts of wood, out of the solid	..	0	1	6
.. .. .. when pieced	..	0	0	6
Conductors from rain-spouts	..	0	0	8
Small doors for feeding-holes of byres	each	0	1	0
Racks for center of courts	..	0	12	0
Corn-chest for work-horses		0	15	0
Stout 5-barred gates, 9 feet wide, for courts	each	1	10	0
.. 4 .. 5 .. .. hammels	..	0	14	0
Rails, harness-pins, and saddletrees		2	10	0
Luffer-board ventilators, 6 feet long by 4 feet wide, and 2½ feet high in front	per square foot	0	1	6
Octagonal stathel-frames for stacks, 15 feet diameter	each	1	15	0
Pump with mounting, 20 feet long		3	10	0

Slater-work.

Item		L.	s.	D.
Blue-slating	per rood of 36 square yards	4	4	0
Gray	..	2	11	0
Tiling	..	2	10	0
Blue slates	per 1,000	3	10	0
Gray slates	..	4	10	0
Tiles	..	2	17	0

Plumber-work.

Item		L.	s.	D.
6-lb. per square foot lead on peands, flanks and ridges (25s. per cwt.)	per square foot	0	1	3½
5-lb. lead for aprons to ventilators, &c.	..	0	1	1
Mastic for raglets	per lineal foot	0	0	1½
Rain-water pipes of 6-lb. lead	..	0	1	6
6-inch open runs of 6-lb. lead, supported with iron straps or holdfasts, 2 feet apart	..	0	1	6
Lead-pump, with mounting	..	2	10	0
Lead-pipe for ditto	..	0	1	2

Smith-work.

Item		L.	s.	D.
Cast-iron travis heel-posts	each	1	2	0
.. corner hay-racks for riding-horse stable	..	0	10	0
.. pump for liquid-manure tank, with 6-feet pipe		3	0	0
Stock and plate-locks for outside doors, 10 inches long	..	0	2	6
18-inch cross-tailed hinges	per pair	0	1	3
9-inch	..	0	0	9
Thumb-latches	each	0 to 0	0 to 0	5 to 7
Manger rings	..	0	0	1
Seals for binding cattle	..	0	2	6
Cast-iron rain-spouts, 4½ inches diameter	per lineal yard	0	2	0
Pipes from ditto, 2 inches diameter, 9 feet long	each	0	8	0
.. .. 4 inches diameter, 9 feet long	each	0	18	0
Gate-mountings	..	0 to 0	12 to 15	0 to 0
36-inch boiler, with furnace complete, or 14s. per cwt.		2	10	6
30-inch .. .. .. or 14s. per cwt.		2	7	0
24-inch .. .. .. or 14s. per cwt.		2	4	0
Crooks and bands for outside doors	per pair	0	10	0
.. .. feeding-hole doors in byres	..	0	5	0
Stanchions, ¾ inch diameter	per pound	0	0	3
Cast-iron window sashes	per square foot	0	1	0

Plaster-work.

Item		L.	s.	D.
Best 3-coat plaster	per square yard	0 to 0	0 to 0	5 to 6
.. 2	..	0	0	4½
.. 1	..	0	0	3

Glazier-work.

Item		L.	s.	D.
Best second crown-glass in small panes	per square foot	0	0	10
.. .. .. large panes	..	0	1	2

Painter-work.

Item		L.	s.	D.
White lead, colored gray, stone, or slate-blue, 3 coats	per square yard	0	0	8

(222.) There is a simple rule for determining the *pitch* which a roof should have for the various sorts of slating. In blue-slating the rule is, that the roof should be in hight ⅓ of the breadth of the building. Suppose that a building is 18 feet inside in width like the middle range of the steading, the walls are each two feet thick, which gives a breadth of 22 feet over walls. Deduct 6 inches on each wall for an escarpment on its top, upon which the scantlings or couple-legs rest upon the wall-plates, and ⅓ of 21 feet gives 7 feet for the hight of the roof above the walls. Old fashioned houses have a pitch of the square, that is, the hight is equal to half the breadth, which, in the supposed case, would be 10½ feet. In gray slating the pitch is fixed at 1 foot below the square, or the hight would be 9½ feet. In tiling, the pitch may be lower than even in blue slating, and it is determined according to circumstances; and even blue slate roofs are made as low in the pitch as ¼ of the breadth, that is with large Welsh slates. Taking the rise at 7 feet, the scantlings should be 13 feet long each, and the balk, of course, as long. Taking the rise at 9½ feet, the scantlings should be 14 feet long. (131.) and (132.), p. 138.

(223.) A liquid manure tank can be constructed at little cost. An excavation being made in clay, a lining should be built all round. The lining may be either of rubble masonry, of stone and mortar, or of brick and mortar. If the subsoil is not of a retentive nature, a plastering of Roman cement will suffice to render the building retentive. A 9-inch wall, or a brick in length, will make a lining of sufficient strength to contain the liquid. The tank should be covered over in any of the various ways I have mentioned in (76.), and paved in the bottom with flags or bricks secured by cement. A cast-iron pump should be inserted at one end of the tank when it will be ready for use.

(224.) The cost of constructing such a tank, with brick in length and cement, will be somewhat as under, exclusive of drains:

	Feet.	Inches.		£	s.	d.
Inside length of tank	13	6				
.. width ..	6	6				
.. depth ..	6	0	=19½ cubic yards.			
Cutting the bed of the tank, at 3d. per cubic yard				£0	7	6
Building wall, including bricks and mortar				6	8	0
Plastering and cement				0	16	0
Covering with flags				2	15	0
Total				£10	6	6

Such a size of tank is said to be sufficient for a farm of from 150 to 200 acres, A receptacle of a more simple and unexpensive nature might be constructed, which would answer some of the ends of a more complete tank. It might be made under a shed, and composed of walls of clay, and covered with slabs of boarding. The expense of such a receptacle would be somewhere as under, the dimensions being as in the preceding case:

	£	s.	d.
Cutting the clay, at 3d the cubic yard	£0	7	[illegible]
Clay and carting	0	14	[illegible]
Boards, and expense of covering	0	5	[illegible]
Total	£1	6	[illegible]

Such a tank, however, would suffer in frost or drouth. A cask sunk into the ground, [illegible] open channels to it, forms a sufficient tank for a cottager.*

[* There are few instances in which gross neglect of valuable resources within their [illegible] so glaring on the part of American agriculturists as in the general failure to collect an abu[illegible] supply of water for all purposes from the roofs of their houses, and especially where [illegible] any considerable size, and yet more when they are so contiguous that the rain which falls on several of them might be collected into one common reservoir.

Having ascertained, as any one may do by reference to meteorological tables kept anywhere in his County or State, what is the annual quantity of rain-water which falls in that region, it is easy to calculate what measure of water may be gathered into a cistern from roofing of a given surface. Waistell urges the importance of placing spouts round all the buildings of the farm, observing that, besides the value of the supply of water thus obtained, the buildings will be benefited by the walls and foundations being kept drier than when the water from the roof is suffered to fall upon them. He states that the quantity of water that falls annually in his county upon every 100 superficial feet, or 10 feet square, of building, is about 1,400 imperial gallons.

In Mississippi, from dire necessity, such cisterns are in common use; but how many farmers over the country might, in this way, save the immense labor bestowed and the time lost in bringing water by hand from a distance, and in sending their domestic animals to it, while the remedy is so close at hand! By a simple process of filtration, and the use of ice, water so collected makes the purest and best drinking-water.

The tank or cistern may be puddled round with clay, to avoid the expense of Roman cement. But the yet more valuable use of tanks is the one referred to by our author for *collecting the*

16. THE FARM-HOUSE.

"Do you but mark how this becomes the house."
LEAR.

(225.) In alluding to the *farm-house* at all, it is not my intention to give a full plan of one, as I have given of the steading; because its internal arrangements are generally left to the fancy of architects or of its occupiers, and with little regard to their adaptation to a farm. Any specific plan which I would recommend of a farm-house would therefore, I fear, receive little attention from either landlord or tenant. But the part of it which is exclusively devoted to labor has so intimate a connection with the management of the farm that I must give my opinion upon it. The part I mean includes the kitchen and dairy, and their accompanying apartments. Now, it may frequently be seen in the plans furnished by architects, that, to give the farm-house a fashionable and airy appearance, the working portion of it is too often contracted and inconveniently arranged. The principle of its construction should be to make this part of the house thoroughly commodious in itself, and at the same time prevent its giving the least annoyance to the rest from noise or disagreeable effluvia, which cannot at all times be avoided in the labors of the kitchen. Both objects would be accomplished by placing it independent of the main body of the house, and this is best effected by a jamb. Whatever may be the external form given to the house, the relative positions of its two parts may easily be preserved whether in the old-fashioned form of a front tenement and back jamb, or the more modern and beauteous form of the Elizabethan style.

(226.) The ground-plan which I recommend of the kitchen and the other parts of the farm-house in which work is performed, may be seen in fig 32, where *a* is the *kitchen*, 18 feet in length, 16 feet in breadth, and 10 feet in hight, provided with a door to the interior of the house, in the wall nearest to you, another to the kitchen pantry *k* and dairy *m*, and a third to the scullery *d* and porch *p*. It contains two windows, one on each side of *g* on the left, a large kitchen range, oven, and furnace-pot at *b*, a commodious lock-up closet *c*, a wall-press *h*, and a dresser and table *g*. There is a stair at *c* to the servants' and other apartments above, and which also leads to the principal bed-rooms in the upper story of the house. Beyond the kitchen is the *scullery d*, which contains a large furnace-pot *e*, a sink in the window *f*, a wall-press *h*, and a dresser *g*. This apartment is 18 feet

---

liquid manure; and we are assured that in France and Germany, and more especially in Belgium, where manure is saved as we save dollars, a manure-tank is considered as indispensable as any other part of the steading or farm-buildings.

Their size will depend, of course, on the number of animals which the system he pursues may invite the farmer to keep. Instead of making them round or oval, it is contended that the best way is to make them into cubes or squares. A tank, says DE RHAM, for a farm of 200 acres of arable land, should be 15 feet wide, 15 deep, and 45 long—giving 3 cubes of 15 feet of liquid.—The rule in Germany is to have tanks large enough to contain ten times as many hogsheads as there are heads of cattle on the place. But as we shall, in the JOURNAL OF AGRICULTURE, treat more fully of this subject, we need not prolong this note. Of the expense of construction, every one must judge for himself, according to the price of labor and materials; but of this we are sure, that these must be exorbitant, where the profit does not very soon afford ample remuneration.

Ed. Farm. Lib.

in length, 10 feet in breadth, and 10 feet in hight. A door from it, and another from the kitchen, open on a lobby common to both, and which lobby gives access by another door to the principal kitchen entrance-door

Fig. 32.

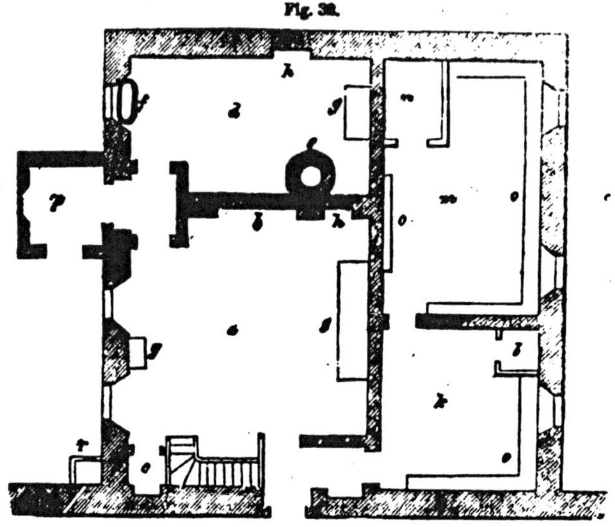

GROUND PLAN OF A KITCHEN, &C. OF A FARM-HOUSE.

through the porch *p*. The porch *p*, 6 feet square, is erected for the purpose of screening both the kitchen and scullery from wind and cold, and it contains the back entrance-door, and is lighted by a window. On the outside, and in front of the porch-door, is *r*, the rain-water cistern, fig. 30, p. 134. On going to the right from the kitchen to the *kitchen-pantry k*, is a wall-press in the passage. The pantry *k* is provided with a door; a window, which should look to the east or north; a larder *l*, and abundance of shelving at *o*; it is 12 feet square, having a roof of 10 feet in hight. Within this pantry is the *milk-house or dairy m*, having two windows also facing to the north or east; a lock-up closet *n*, and shelving *o* around the walls; it is 18½ feet in length, 12 feet in breadth, and 10 feet in hight.

(227.) These are the different apartments, and their relative positions, required for conducting the business of a farm within the house, and in the fitting up of which are many particulars which require attention. The *floor* of the kitchen should be of *flagged pavement polished*, that it may be cleaned with certainty and ease. The outside wall and ceiling should be lathed, and all the walls and ceiling plastered with the best hair-plaster. Iron hooks, both single and double, should be screwed into the joists of the roof, from which may be suspended hams or other articles. The dressers *g* are best made of plain-tree tops and black American birch frames, the chairs of the latter wood, and the stools of common fir. In case of accidents, or negligence in leaving them unfastened at night, it would be well to have the lower sashes of the windows of the kitchen and scullery fast, and the upper ones only to let down for the occasional admission of fresh and the escape of heated air.

(228.) In the *scullery*, the sink *f* should be of polished free-stone, made to fit the window-void, with a proper drain from it, provided with a cesspool. The floor should be of the same material as that of the kitchen, for the sake of cleanliness. The outside wall and ceiling should be lathed, and all the walls and roof plastered. There should be a force-pump in the scullery to fill a cistern with water at the upper part of the house, to contain a constant supply for the sink. A boiler behind the kitchen fire, provided with a small cistern and ball-cock and ball in connection with the upper cistern, for the supply of cold water into the boiler; and a cock from it in the kitchen, and another from it in the scullery, for drawing off warm water when required, could be fitted up at no great cost, and would be found a most serviceable apparatus in a farm-house.

(229.) The large furnace-pot *e* should be built in with fire-brick surrounded with common brick, plastered, and protected with cloth on the outside, rubbed hard into the plaster, and the mouth of the pot protected with a 4-inch pavement polished. To carry off the superfluous steam, a lead-pipe should be fastened into a narrow, immovable portion of the pot-lid, and passed through the wall into the flue. An iron bar should project from the stone-wall about 3 feet or so above the furnace-pot, having a horizontal eye at its end directly over the center of the pot, to be used when making the porridge for the reapers' morning meal in harvest, as shall be described afterward. The dresser *g* should be of the same material as that of the kitchen. There should be iron hooks fastened into the roof for hanging any article thereon. Shelving is also useful in a scullery.

(230.) The outside walls and ceiling of the *kitchen-pantry* should be lathed, and all the walls and ceiling plastered. The flooring should be of the same material as that of the kitchen, or of hard brick. The shelving *o* should be of wood of several tiers, the lowest row being 3½ feet above the floor. The movable portion of the window should be protected with fly zinc-gauze, and so also the side and door of the larder *l*. A few iron hooks in the roof will be found useful for hanging up game or fowls. A set of steps for reaching above an ordinary hight is convenient in a pantry.

(231.) The outside walls and ceiling of the *milk-house* should be lathed, and the walls and roof plastered. The flooring should be of polished pavement, for the sake of coolness. The windows should be protected in the movable part with fly zinc-gauze, which is much better than wire-gauze; and the side and door of the lock-up closet *n* should also be lined with zinc-gauze. The best shelving for a milk-house is marble; and, though this substance may appear extravagant in a farm-house, the price of marble is now so much reduced that it is worth the extra expense, the import of foreign marble being now free. Marble is always cool, and easily cleaned and freed of stains. Scottish marble is hard and unequal of texture. The gray-veined marble from Leghorn is therefore preferable, though the black marble of the county of Galway in Ireland is equally good; but the gray color has a coolness and freshness about it in a dairy, which the black does not possess. Polished pavement is the next best material for coolness, but it is very apt to stain with milk or butter, and the stains are difficult of removal. I speak from experience, and know the labor required to keeping stone-shelving in a milk-house always sweet and clean; and, let me say farther, unless it is so kept, any other material is preferable to it. If marble be rejected on account of expense, I would recommend stout shelving of beech or plane-tree, as being smooth, and hard, and easily kept clean. This shelving should be 2 feet broad, 1½ inches thick, and, to be convenient, should not exceed the hight of 3 feet from the floor.

(232.) It is necessary to make the wall which separates the kitchen and scullery from the milk-house and pantry of brick or stone, to keep the latter apartments more cool, and less likely to be affected by the heat and vapor, which must, of necessity, sometimes escape from the former. It would, no doubt, be *convenient* for the removal of dishes to have a door communicating between the scullery and milk-house; but it is much better to avoid every risk of contamination from a place which must at times be filled with vapors injurious to milk—a substance which is at all times delicately susceptible of injury.

(233.) The windows of all the apartments should be provided with *shutters on the inside;* and it may be a safe precaution against nocturnal intruders to protect those of the milk-house and pantry with iron stancheons on the outside, as they should, occasionally at least, be left open even all night.

(234.) On this side of the kitchen will be observed a stair. It is 4 feet in width, and intended to lead to the story above the kitchen floor, as also to the upper story of the principal part of the house. The story above the kitchen may be subdivided in this way. Let a continuation of the brick or stone wall which separates the kitchen and scullery from the milk-house and kitchen-pantry be carried up, in the form of a partition of lath and plaster, to the roof of the second story, which may be 9 feet in hight, as seen on the right of *g*, fig. 33. The wall of the kitchen flue *b* should, of

Fig. 33.

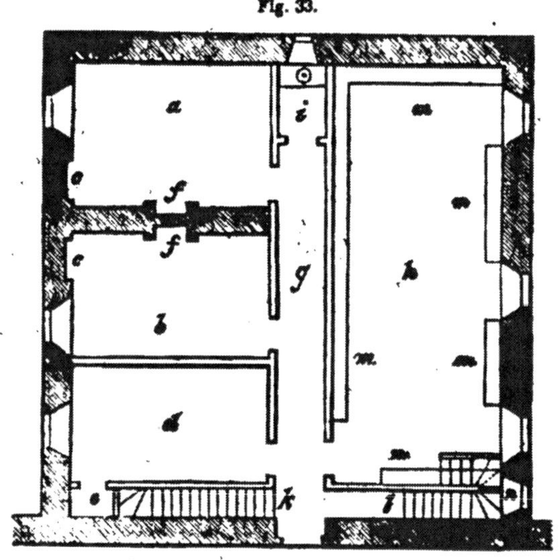

CHEESE-ROOM &C. OF FARM-HOUSE.

course, be carried up to a chimney-stalk above the ridging, containing at least 4 flues from below—one of the kitchen fire, one of the small furnace-pot of the kitchen, one of the oven, and one of the large furnace-pot in the scullery; but there should also be one from the room above the scullery, and one from one of the rooms above the kitchen; and, to render both kitchen and scullery as wholesome by ventilation as possible, there should be a small flue from the ceiling of each to carry off heated air and vapor. The kitchen-stalk would thus contain 6 flues from below and 2 from above.

(235.) The upper story should be partitioned off in the way as seen in fig. 33. Let the apartment *a* above the scullery be fitted up with a fire-place *f* as a *bed-room* for the female servants, having a closet *c* in the outer wall. After taking off a passage *g* of 3½ feet in width along the whole length of this part of the house, this room will be 14 feet long and 10 feet wide. The space above the kitchen may be divided into 2 *bed-rooms*—one *b*, 14 feet in length, by 9 feet in width, and 9 feet in hight, with a fire-place *f* and window, and closet *c*. This might be occupied as a sitting-room and bed-room by the housekeeper, if the services of such a person are required; if not, it might serve as a large store-room, with a fire-place which would be useful for various purposes. The other room *d*, 14 feet in length by 8 feet 3 inches in width, and 9 feet in hight, having a window in it, but no fire-place, might be fitted up as a *bed-room for occasional stranger servants*. This latter apartment has a closet *e* in it, 3 feet in width by 2 feet in depth, directly above the lock-up closet *c* off the kitchen.

(236.) At the end of the passage is a *water-closet i*, lighted by a window in the gable of the jamb. The size of the water-closet is 5 feet 3 inches by 3½ feet. Its cistern is supplied with water from the cistern that supplies the sink in the scullery, and its soil-pipe could descend in an appropriate recess in the wall. The window of the water-closet could give light to the *passage g* by a glass-window above the water-closet door, or the passage could be lighted by a cupola in the roof, or it could be lighted from the cheese-room *h* by two windows in the lath and plaster wall, each of which could have a pane to open into the cheese-room for the purpose of ventilation.

(237.) The entire space above the kitchen-pantry and milk-house may be appropriated to a *cheese-room h*, 29 feet 3 inches in length on the floor, 12 feet in width, and 9 feet in hight, having 3 windows in it. Besides the floor, proper shelving *m* should be put up for the accommodation of the cheese, in its various stages toward maturity; and the lower halves of the windows should be provided with Venetian shutters, outside of the glass, to regulate the air into the room when the windows are opened.

(238.) If there is sufficient room in the roof above these various apartments for a *garret*, access can be obtained to it by a stair at *l*, which would have to return upon itself in ascending the 9 feet, the hight of the story; and both this stair and the one *k* down to the kitchen could be lighted by the window *n*. If there is no garret, then the cheese-room will be 32 feet 3 inches in length, by dispensing with the stair *l*. The window *n* could then be also dispensed with.

(239.) These dimensions of kitchen and other apartments would be suited to the farm-house of a farm of from 500 to 1000 acres, under the mixed husbandry. The milk-house may, perhaps, be large enough for even a dairy-farm of ordinary extent; but should it be too small for that purpose, it might easily be enlarged by increasing either or both the length and breadth of the building.

(240.) In regard to the relative positions which the farm-house and steading should occupy, it has been remarked by a recent writer, that "It is generally advised that the farm-house should be placed directly in front: to which, however, it may be objected that it casts a shade over the southern entrance of the yard if very near, and if too far off, its distance will be found to be inconvenient. Perhaps the best situation is on one side of the farm-yard, with the common parlor and kitchen opening nearly into it: farmers may talk as they like about unhealthy odors arising from the stables and yards, but there never was any one injured by them, and they cannot keep too close an eye upon their servants and stock."* If farmers "cannot keep too close an eye upon their servants and stock," and if the position of their houses will enable them to do so, they should do something more than place them "on one side of the farm-yard:" they must remain constantly in them, and cause "their servants and stock" to be continually in sight in the farm-yard, otherwise their watching will be of no avail; for when the servants come to know that the house has been placed there merely to watch their proceedings, they at least, if not the stock, can and will easily avoid the particular place constantly overlooked by the house. The truth is, and every farmer knows it that it is not the spot occupied by his house, whether here or there, that maintains his authority over his servants; he knows that he himself must be "up and doing" in the fields in the farm-yard, every where—"be stirring with the lark,"

"From morn to noon, from noon to dewy eve"—

ere he can ascertain whether his servants are doing their work well, and his stock thriving

* British Husbandry, vol. i. p. 86.

(301)

well. Inconvenience to himself in going a great distance betwixt his house and the steading, will induce the farmer to place his house *near* rather than at a distance from the steading. He wishes to be within call—to be able to be on the spot in a few seconds, when his presence is required in the farm-yard, the stable, the byre, or the barn; but more than this he does not want, and need not care for. Place your house, therefore, if you have the choice, on some pleasant spot, neither "direct in front" nor much in the rear of the steading. If there be no such spot at hand, make one for your house, place it there, and dwell in it, with the comfortable assurance that your servants will not regard you the less, or your stock thrive the worse, because you happen to live beyond the influence of the "unhealthy odors arising from your stables and yards"—odors, by the way, of the unpleasantness of which I never heard a farmer complain. No one of that class but a sloven would place his house beside a dunghill.

---

17. THE PERSONS WHO LABOR THE FARM.

"*John.* Labor in thy vocation.
Geo. Thou hast hit it; for there's no better sign of a brave mind than a hard hand."
HENRY VI. *Part II.*

(241.) THOSE who labor a farm form the most important part of its material; they are the spirit that conducts its operations. You should, therefore, become early acquainted with those functionaries. They are the farmer himself, the steward or grieve, the plowman, the hedger or laborer,* the shepherd, the cattle-man, the field-worker, and the dairy-maid. These have each duties to perform, which, in their respective spheres, should harmonize and never interfere with each other. Should any occurrence happen to disturb the harmony of labor, it must arise from some misapprehension or ignorance in the interfering party, whose aberrations must be rectified by the presiding power. I shall consider the duties in the order I have mentioned the respective agents.

(242.) And first, those of the *farmer*. It is his province to originate the entire system of management—to determine the period for commencing and pursuing every operation—to issue general orders of management to the steward, when there is one, and if there be none, to give minute instructions to the plowmen for the performance of every separate field operation—to exercise a general superintendence over the field-workers—to observe the general behavior of all—to see if the cattle are cared for—to ascertain the condition of all the crops—to guide the shepherd—to direct the hedger or laborer—to effect the sales of the surplus produce—to conduct the purchases conducive to the progressive improvement of the farm—to disburse the expenses of management—to pay the rent to the landlord—and to fulfill the obligations incumbent on him as a residenter of the parish. All these duties are common to the farmer and the steward engaged to manage a farm. An independent steward and a farmer are thus so far on the same footing; but the farmer occupies a loftier station. He is his own master—makes bargains to suit his own interests—stands on an equal footing with the landlord in the lease—has entire control over the servants, hiring and discharging them at any term he pleases—and possesses power to grant favors to servants and friends. The farmer has not all those duties to perform in any one day, but in the course of their proper fulfillment, daily calls are made on his attention, and so large a portion of his time is occupied by them, that he finds little leisure to go far from home, except in the season when few operations are performed on a

[* Altogether inapplicable to our country. *Ed. Farm. Lib.*]

farm, viz. the end of summer. These are the professional duties of the *farmer*; but he has those of domestic and social life to fulfill, like every other member of society. If a farmer fulfills all his duties as he ought to do, he cannot be said "to eat the bread of idleness."

(243.) The duty of *steward*, or *grieve*,* as he is called in some parts of Scotland, and *bailiff* in England, consists in receiving general instructions from his master the farmer, which he sees executed by the people under his charge. He exercises a direct control over the plowmen and field-workers; and unreasonable disobedience on their part of his commands is reprehended as strongly by the farmer as if the affront had been offered to himself: I say *unreasonable* disobedience, because the farmer is the judge of whether the steward has been reasonable in his demands. It is the duty of the steward to enforce the commands of his master, and to check every deviation from rectitude he may observe in the servants against his interests. Although he should thus protect the interests of his master from the attacks of any servant, yet it is not generally understood that he has control over the shepherd, the hedger, or the cattle-man, who are stewards in one sense, over their respective departments of labor. The farmer reveals to the steward alone the plans of his management; intrusts him with the keys of the corn-barn, granaries and provision-stores; delegates to him the power to act as his representative on the farm in his absence; and takes every opportunity of showing confidence in his integrity, truth and good behavior. When a steward conducts himself with discretion in his master's absence, and exhibits at all times a considerate mind, an active person, and an honest heart, he is justly regarded as a valuable servant.

(244.) Personally, the farm-steward does not always labor with his own hands; verifying, by his judicious superintendence, on a large farm at least, the truth of the adage, that "one head is better than two pair of hands." He should, however, always deliver the daily allowance of corn to the horses. He should, moreover, be the first person out of bed in the morning, and the last in it at night. On most farms he does work: he sows the seed-corn in spring, superintends the field-workers in summer, tends the harvest-field and builds the stacks in autumn, and threshes the corn with the mill, and cleans it with the winnowing-machine, in winter. On some farms he even works a pair of horses, like a common plowman; in which case he cannot personally sow the corn, superintend the workers, build the stacks, or thresh the corn, unless another person takes charge of his horses for the time. This is an objectionable mode of employing a steward; because the nicer operations—such as sowing corn, &c. or the guidance of his horses—must be intrusted to another, and most likely inferior, person. But in by far the greatest number of cases, the steward does not work horses; on the contrary, when a plowman qualifies himself to become a steward, it is chiefly with the view of enjoying immunity from that species of drudgery. In any event, he should be able to keep an account of the work-people's time, and of the quantity of grain threshed, consumed on the farm, and delivered to purchasers.

(245.) Stewards are not required on all sorts of farms. On pastoral farms his species of service would be of no use, as it is on arable land that these are really required. Anywhere, his services are the most valuable where the greatest multiplicity of subjects demand attention. Thus, he is a *more* useful servant on a farm of mixed husbandry, than on one in the neighborhood of a town, or on a carse farm. But even on some farms of

[* Overseer or Manager. *Ed. Farm. Lib.*]

mixed culture, the services of a steward are dispensed with altogether; in which case the farmer himself gives his orders directly to the plowmen, or indirectly through the hedger or cattle-men, as he may choose to appoint to receive his instructions. In such a case the same person is also intrusted to corn the horses; for the plowmen themselves are never intrusted with that business, as they are apt to abuse such a trust by giving too much corn to the horses, to their probable injury. The same person performs other parts of a steward's duty; such as sowing corn, superintending field-workers, and threshing corn; or those duties may be divided betwixt the cattle-man and hedger. On the large farm in Berwickshire on which I learned farming, there was no steward, the cattle-man delivering the master's orders and corning the horses, and the hedger sowing the corn, building the stacks, and threshing the corn. The object of this arrangement was to save the wages of a steward, when the farmer himself was able to undertake the general superintendence. I conducted my own farm for several years without a steward.

(246.) The duties of a *plowman* are clearly defined. The principal duty is to take charge of a pair of horses, and work them at every kind of labor for which horses are employed on a farm. Horse-labor on a farm is various. It is connected with the plow, the cart, sowing-machines, the roller, and the threshing-mill, when horse-power is employed in the threshing of corn; so that the knowledge of a plowman should comprehend a variety of subjects. In the fulfillment of his duties, the plowman has a long day's work to perform; for, besides expending the appointed hours in the fields with the horses, he must groom them before he goes to the field in the morning and after he returns from it in the evening, as well as at mid-day between the two periods of labor. Notwithstanding this constant toil, he must do his work with alacrity and good will; and when, from any cause, his horses are laid idle, he must not only attend upon them as usual, but must himself work at any farm-work he is desired. There is seldom any exaction of labor from the plowman beyond the usual daily hours of work, these occupying at least 12 hours a day for 7 months of the year, that being a sufficient day's work for any man's strength to endure. But occasions do arise which justify the demand of a greater sacrifice of his time, such as seed-time, hay-time, and harvest. For such encroachments upon his time, many opportunities occur of repaying him with indulgence, such as a cessation from labor, especially in bad weather. It is the duty of the plowman to work his horses with discretion and good temper, not only for the sake of the horses, but that he may execute his work in a proper manner. It is also his duty to keep his horses comfortably clean. Plowmen are never placed in situations of trust; and thus, having no responsibility beyond the care of their horses, there is no class of servants more independent. There should no partiality be shown by the master or steward to one plowman more than to another, as it is the best policy to treat all alike who work alike. An invidious and reprehensible practice exists, however, in some parts of the country, of setting them to work in an order of precedency, which is maintained so strictly as to be practiced even on going to and returning from work—one being appointed *foreman*, whose movements must guide those of the rest. Should the foreman prove a slow man, the rest must not go a single bout more than he does; and, if he is active, they may follow as best they can. Thus, while his activity confers no benefit to the farmer beyond its own work, his dullness discourages the activity of the others. This consideration alone should be sufficient ground for farmers to abolish the practice at once, and put the whole of their plowmen on the same footing. I soon saw the evils attending the system, and put an end to it on my own farm. When one plowman displays more skill than the rest, it is sufficient honor for him to be intrusted to execute the most difficult pieces of work; and this sort of preference will give no umbrage to the others, as they are as conscious of his superiority in work as the farmer himself can possibly be. The services of plowmen are required on all sorts of farms, from the carse-farm to the pastoral on which the greatest and the least portion of arable culture are practiced.

(247.) The services of a *shepherd*, properly so called, are only required where a flock of sheep are constantly kept. On carse-farms, and those in the neighborhood of large towns, he is of no use; nor is he required on those farms on which sheep are bought in to be fed off in winter. On pastoral farms, on the other hand, as also those of the mixed husbandry, his services are so indispensable that they could not be conducted without him. His duty is to undertake the entire management of the sheep; and, when he bestows the pains he should on his flock, he has little leisure for any other work. His time is occupied from early dawn, when he should be among his flock before they rise from their lair, and during the whole day, to the evening, when they again lie down for the night. To inspect a large flock at least three times a day, over extensive bounds, implies a walking to fatigue. Besides this daily exercise, he has to attend to the feeding of the young sheep on turnips in winter, the lambing of the ewes in spring, the washing and shearing of the fleece in summer, and the bathing of the flock in autumn. And, over and above these major operations, there are the minor ones of weaning, milking, drafting, and marking, at appointed times; not to omit the unwearied attention to be bestowed, for a time, on the whole flock, to evade the attacks of insects. It will readily be seen, from this summary of duties, that the shepherd has little time to bestow beyond the care of his flock. As no one but a shepherd, thoroughly bred, can attend to sheep, there must be one where a standing sheep-flock is kept, whatever may be the extent of farm. On a *small* farm, his whole time may not be occupied in his profession, when he can make as well as mend nets, prepare stakes for them, and assist the hedger (if there be one) to keep the fences in repair; or he may act as groom, and take charge of the horse and gig, and go errands to the post-town; or he may undertake the duties of steward. On *large* pastoral or mixed husbandry farms, more than one shepherd is required. The establishment then consists of a *head* shepherd, and one or more young men training to be shepherds, who are placed entirely under his control. The office of head shepherd is one of great trust. Sheep being individually valuable, and in most instances consisting of large flocks, a misfortune happening to them, from whatever cause, must incur great loss. On the other hand, the care and skill of the shepherd may secure a good return for the capital invested in sheep. The shepherd acts the part of butcher in slaughtering the animals used on the farm. The only assistance which he depends upon in personally managing his flock is that of his faithful dog, whose sagacity in that respect is little inferior to his own.

(248.) The services of the *cattle-man* are most wanted at the steading in winter, when the cattle are all housed. He has the sole charge of them. It is his duty to clean out the cattle-houses, and supply the cattle with food, fodder, and litter, at appointed hours every day, and to make the food ready for them, should prepared food be given them. The business of tending cattle being matter of routine, the qualifications of a cattle-man are not of a high order. In summer and autumn, when the cows are at grass, it is his duty to bring them into the byre or to the gate of the field, as the case may be, to be milked at their appointed times; and it is also his duty to ascertain that the cattle in the fields are plentifully supplied with food and water. He should see the cows served by the bull in due time, and keep an account of the cows' reckonings of the time of calving. He should assist at the important process of calving. As his time is thus only occasionally employed in summer, he frequently undertakes the superintendence of the field-workers. In harvest, he is usefully employed in assisting to make and carry food to the reapers, and may lend a hand at the taking in of the corn. As cattle occupy the steading in winter on all kinds of farms, the services of the cattle-man appear indispensable; but all his functions may be performed by the shepherd, where only a small flock of sheep are kept. The office of the cattle-man is not one of trust nor of much labor. An elderly person answers the purpose quite well, the labor being neither constant nor heavy, but well-timed and methodical. The cattle-man ought to exercise much patience and good temper toward the objects of his charge, and a person in the decline of life is most likely to possess those qualities.

(249.) *Field-workers* are indispensable servants on every farm devoted to arable culture. They mostly consist of young women in Scotland, but more frequently of men and boys in England; and yet there are many manual operations much better done by women than men. In hand-picking stones and weeds, in filling drains, and in barn-work, they are far more expert, and do them more neatly, than men. The duties of field-workers, as their very name implies, are to perform all the manual operations of the fields, as well as those with the smaller implements, which are not worked by horses. The *manual* operations consist chiefly of cutting and planting the sets of potatoes, gathering weeds, picking stones, collecting the potato crop, and filling drains with stones. The operations with the smaller implements are pulling turnips and preparing them for feeding stock and storing in winter, performing barn-work, carrying seed-corn, spreading manure upon the land, hoeing potatoes and turnips, and weeding and reaping corn-crops. A considerable number of field-workers are required on a farm, and they are generally set to work in a band. They work most steadily under superintendence. The steward, the hedger, or cattle-man, should superintend them when the band is large; but, when small, one of themselves, a staid person, who is capable of taking the lead in work, may superintend them well enough, provided she has a watch to mark the time of work and rest. But field-workers do not always work by themselves; being at times associated with the work of the horses, when they require no particular superintendence. On some farms, it is considered economical to lay the horses idle, and employ the plowmen at their labors rather than engage field-workers. This may be one mode of avoiding a little outlay of money; but there is no true economy in allowing horses "to eat off their own heads," as the phrase has it; and, besides, plowmen *cannot* possibly do light work so well as field-workers. In manufacturing districts field-workers are scarce; but were farmers generally to adopt the plan of employing a few constantly, and hire them for the purpose by the half year, instead of employing a large number at times, young women would be induced to adopt field-labor as a profession, and become very expert in it. It is steadiness of service that makes the field-workers of the south of Scotland so superior to the same class in other parts of the country.

(250.) The duties of the *dairy-maid* are well defined. She is a domestic servant, domiciliated in the farm-house. Her principal duty is, as her name implies, to milk the cows, to manage the milk in all its stages, bring up the calves, and make into butter and cheese the milk that is obtained from the cows after the weaning of the calves. The other domestics gensually assist her in milking the cows and feeding the calves, when there is a large number of both. Should any lambs lose their mothers, the dairy-maid should bring them up with cow's milk until the time of weaning, when they are returned to the flock. At the lambing season, should any of the ewes be scant of milk, the shepherd applies to the dairy-maid to have his bottles replenished with warm new milk for the hungered lambs. The dairy-maid also milks the ewes after the weaning of the lambs, and makes cheese of the ewe-milk. She should attend to the poultry, feed them, set the brooders, gather the eggs daily, take charge of the broods until able to provide for themselves, and see them safely lodged in their respective apartments every evening, and let them abroad every morning. It is generally the dairy-maid, when there is no housekeeper, who gives out the food for the reapers, and takes charge of their articles of bedding. The dairy-maid should be an active, attentive, and intelligent person.

(251.) These are the duties of the respective classes of servants found on farms. You may not require all these classes on your farm, as you have seen that some sorts of farms do not require the services of all. You have seen that a pastoral-farm has no need of a steward, but of a shepherd; a carse-farm no need of a shepherd, but of a steward; a farm in the neighborhood of a town no need of a hedger, but of a cattle-man; and, on a dairy-farm, no need of a shepherd, but of a dairy-maid; but, in the case of a farm of mixed husbandry, there is need of all these classes.

(252.) And now that you have seen how multifarious are the duties of them all, you will begin to perceive how intricate an affair mixed husbandry is, and how well informed a farmer should be of every one of these varieties of labor, before he attempts to manage for himself. To give you a stronger view of this, conceive the quantity and variety of labor that must pass through the hands of these various classes of work-people in the course of a year, and then imagine the clear-headedness of arrangement which a farmer should possess, to make all their various labors coincide in every season, and under every circumstance, so as to produce the most desirable results. It is in its variety that the success of labor is attained: in other words, it is in its subdivision that the facility of labor is acquired, and it is by the intelligence of the laborers that perfection in it is attained. And vain would be the endeavors of any farmer to produce the results he does, were he not ably seconded by the general intelligence and admirable efficiency of his laborers.

---

18. THE WEATHER IN WINTER.

"See, Winter comes to rule the variedyear,
Sullen and sad, with all his rising train;
Vapors, and clouds, and storms. Be these my theme."
THOMSON.

(253.) As the weather, at all seasons, has undeniably a sensible power to expedite or retard the field operations of the farm, it becomes an incumbent duty on you, as pupils of Agriculture, to ascertain the principles which regulate its phenomena, in order to anticipate their changes and avoid their injurious effects. It is, no doubt, difficult to acquire an accurate knowledge of the laws which govern the subtile elements of Nature;

but experience has proved that *accurate observation of atmospherical phenomena* is the chief means which we possess of becoming acquainted with those laws.

(254.) In saying that the weather has power to alter the operations of the farm, I do not mean to assert that it can entirely change any great plan of operations that may have been determined on, for that may be prosecuted even in spite of the weather; but there is no doubt that the weather can oblige the farmer to pursue a different and much less efficient treatment toward the land than he desires, and that the amount and quality of its produce may be very seriously affected by the change of treatment.—For example, the heavy and continued rain in autumn 1839 made the land so very wet that not only the summer-fallow, but the potato-land, could not be seed-furrowed; and the inevitable consequence was that sowing of the wheat was postponed until the spring of 1840, and in many cases the farmers were obliged to sow barley instead of wheat. The immediate effect of this remarkable interference of the weather was restriction of the breadth of land appropriated to autumnal wheat, and the consequent extension of that intended for barley and spring wheat—a change that caused so much work in spring that it had the effect of prolonging the harvest of 1840 beyond the wished-for period, and of otherwise deranging the calculations of farmers.

(255.) Now, when such a change is, and may in any season be, imposed upon the farmer, it becomes a matter of prudence as well as of desire to become so acquainted with usual atmospherical phenomena as to anticipate the nature of the weather that is to come. If he could anticipate particular changes of weather by observing peculiar phenomena, he could arrange his operations accordingly. But is such anticipation in regard to the weather attainable? No doubt of it; for, although it is not as yet to be expected that minute changes of the atmosphere can be anticipated, yet the *kind* of weather which is to follow—whether rainy or frosty, snowy or fresh—may be predicted. We all know the prescience actually attained by people whose occupations oblige them to be much in the open air and to observe the weather. In this way shepherds and sailors, in their respective circumstances, have acquired such a knowledge of atmospherical phenomena as to be able to predict the advent of important changes of the atmosphere; and to show that the sort of knowledge acquired is in accordance with the circumstances observed, it is obvious that, even among these two classes of observers, great difference of acquirements exists on account of diversity of talent for observation. For example: A friend of mine, a commander of one of the ships of the East India Company, became so noted, by observation alone, for anticipating the probable results of atmospherical phenomena in the Indian seas, that his vessel has frequently been seen to ride out the storm, under bare poles, while most of the ships in the same convoy were more or less damaged. As an instance of similar sagacity in a shepherd, I remember in the wet season of 1817, when rain was predicted as inevitable by every one engaged in the afternoon of a very busy day of leading in the corn, the shepherd interpreted the symptoms as indicative of wind and not of rain, and the event completely justified his prediction.

(256.) I conceive that greater accuracy of knowledge in regard to the changes of the weather may be attained on land than at sea, because the effect of weather on the sea itself enters as an uncertain element into the question. It is generally believed, however, that seamen are more proficient than landsmen in foretelling the weather; and, no doubt, when the imminent danger, in which the lives of seamen are jeopardized, is considered, the circumstance may reasonably be supposed to render them peculiarly alive to *certain* atmospherical changes. To men, however, under constant command, as seamen are, it is questionable whether the *ordinary* changes of the atmosphere are matters of much interest. In everything that affects the safety of the ship, and the weather among the rest, every confidence is placed by the crew in the commanding officer, and it is he alone that has to exercise his weather wisdom. On the other hand, every shepherd has to exercise his own skill in regard to the weather, to save himself, perhaps, much unnecessary personal trouble, especially on a hill-farm. Even the young apprentice-shepherd soon learns to look out for himself. The great difference in regard to a knowledge of the weather betwixt the sea-captain and the farmer, though both are the sport of the same elements, consists in this, that the captain has to look out for himself, whereas the farmer has his shepherd to look out for him: the seafaring commander himself knowing the weather, directs his men accordingly; while the farmer does not know it nearly so well as his shepherd, and probably even not so well as his plowmen. See the effects of this difference of acquirement in the circumstances of both. The captain causes the approaching change to be met by prompt and proper appliances; whereas the farmer is too frequently overtaken in his operations from a want of the knowledge probably possessed by his shepherd or plowmen. You thus see the *necessity of farmers acquiring a knowledge of the weather*

(257.) It being admitted that prescience of the state of the weather is essential to the farmer, the question is, how the pupil of Agriculture is to acquire it? No doubt it can best be attained by observation in the field, but as that method implies the institution of a series of observations extending over a long period of years, a great part of the lifetime of the pupil might pass away ere he could acquire a sufficient stock of knowledge by his own experience. This being the case, it is but right and fair that he should know what the experience of others is. This I shall endeavor to communicate, premising that he must observe for himself, after being made acquainted with the *manner of conducting* his own observations.

(258.) The simplest way for me to communicate what has been established in regard to the observation of atmospherical phenomena, is, in the first place, to describe to you the various instruments which have, from time to time, been contrived to indicate those phenomena; and to put these instruments into a right use, you should become well acquainted with their respective modes of action, which are all dependent on strictly scientific principles. All the instruments required are the *barometer, thermometer, weathercock, hygrometer, and rain-gauge.* The principles upon which these instruments operate shall be separately explained; the phenomena of the clouds and winds, upon which the diversity of the states of the atmosphere appear so much to depend, shall be described; and the efficacy of the electric agency, which seems to affect so many of the phenomena observed, shall be noticed. The general principles of atmospherical phenomena being thus considered in this place, I shall have no more occasion to recur to them, but will only have to notice the characteristic phenomena of each season as they occur.

(259.) *Atmospherical phenomena being the great signs by which to judge of the weather*, instruments are used to detect their changes which cannot be detected by the senses. These instruments possess great ingenuity of construction, and they all indicate pretty accurately the effects they are intended to recognize. But though they tell us nothing but the truth, such is the minute diversity of atmospherical phenomena that they do not tell us all the truth. Other means for discovering that must be used; and the most available within our reach is the converting of the phenomena themselves into indicators of atmospherical changes. In this way we may use the transient states of the atmosphere, in regard to clearness and obscurity, dampness or dryness, as they affect our senses of sight and feeling, the shapes and evolutions of the clouds, and the peculiar state of the wind, into means by which to predicate the changes of the weather. But this kind of knowledge can only be acquired by long observation of natural phenomena.

(260.) The most important instrument, perhaps—the most popular, certainly—for indicating the changes of the atmosphere, is the *barometer*, an instrument so universally known and used by farmers, that a particular description of it is here unnecessary. This instrument is formed to be placed either in a fixed position or to be portable. As it is only used in the portable shape to measure the altitude of mountains, the method of using it need not be here described. For a fixed position, the barometer is made either of the figure of an upright column or of a wheel. Whether it is because that the divisions on the large circular disk, pointed out by the long index of the wheel-barometer, are more easily observed than the varieties of the column of mercury in the perpendicular one, is the reason which renders the wheel-barometer more popular among farmers, I know not; but were they to consider that its indications cannot be so delicate as those of the upright form, because of the machinery which the oscillations of the mercury have to put in motion before the long index can indicate any change, the upright form would always be preferred. It is true that the tube of the upright barometer is generally made too small, and is perhaps so made to save mercury and make the instrument cheaper; but a small tube has the disadvantage of increasing the friction of the mercury in its passage up and down the tube. On this account the mercury is apt to be kept above its proper level when falling, and to be depressed below its proper hight when rising. To obviate this inconvenience, a tap of the hand against the case of the instrument is required to bring the mercury to its proper position. The tendency of the mercury to rise may be observed by the convex or raised form of the top of the column; and the hollow or concave form indicates its tendency to fall.

(261.) In observing the state of the barometer, too much regard should not be had to the numerals and words usually written on the graduated scale, placed along the range of the top of the column of mercury; because it is the rising or falling of the mercury alone that is to be taken as indicative of a change of weather, whatever may be its actual hight in the tube. The greatest hight attained by the column is entirely determined by the hight of elevation of the place of observation above the level of the sea. The higher the place is above the sea, the mean hight of the column will be the lower. For example, on comparing two barometers at the same time, at two places of different hights in the same part of the country, and subject to the same climate, one may stand as high as 30 inches, and the other only at 29½ inches. According to the usual markings of barometers, the mercury at the first place would stand at "Fair," whereas, at the other place, it would be at "Changeable." This difference of the mercury is in itself important, but it does not arise from any difference in the state of the air, as indicative of a change of weather, but merely from the difference of elevation of the two places above the level of the sea. The mercury is as near its greatest hight at 29½ inches at the higher place, as it is at 30 inches at the lower place, in reference to their respective positions above the sea; and this being the case, and other circumstances equal, it will be the same weather at both places. This difference of the hight of the mercury is explained in this way. The barometer being the instrument which indicates the weight or pressure of the atmosphere, as its name implies, it is found on trial that the mercury stands highest at the level of the sea, and that it descends as elevation above the sea increases. The depression has been found by experiment to be $\frac{1}{10}$ of an inch for about every 88 feet of elevation, or more correctly as given in this table.*

TABLE SHOWING THE NUMBER OF FEET OF ALTITUDE CORRESPONDING TO DEPRESSIONS OF THE BAROMETER.

Depression.	Altitude in feet.	Depression.	Altitude in feet.
·1	87	·6	527
·2	175	·7	616
·3	262	·8	705
·4	350	·9	795
·5	439	1 inch.	885

(262.) It becomes, then, a matter of some importance for you, in order to place explicit reliance on the changes indicated by your barometer, to ascertain the hight of your farm above the level of the sea. If you know that by other means, namely, by trigonometry, then the allowance in the table will give you its true elevation; but should you not be acquainted with its elevation, which is usually the case with farmers, the mean hight of the barometer can be ascertained by a series of simple observations, made at a given time, over a year or more. For example, "the sum of one year's observations, made at 10 A. M. and 10 P. M. in 1827, was 21615·410 inches, and this number divided by the number of observations, 730, or twice the number of days in that year, gave 29·610 inches as the mean hight or changeable point of the barometer."† Now, taking the mean hight of the barometer at 29·948 inches at the mean level of the sea, where the atmosphere always indicates the greatest density, deduced from nine years' observations at the mean temperature of the air, with a range from 28 inches to 31 inches, it is seen that the instance adduced above of 29·610 inches gives ·338 of an inch less than the mean, which, by the table, indicates an elevation of the place of observation of about 265 feet above the mean level of the sea. It is from the mercury being above or below this point of 29·610 inches in the supposed place of your farm, that you are to conclude what weather may be expected at that place, from the changes of the barometer. From the want of this knowledge, farmers are generally led into the mistake of supposing that the words "Fair," "Change," "Rain," engraved on the scale of the barometer, indicate such weather in all places, when the mercury stands at them. The best way to correct this mistake is to have the words engraved at the hights truly applicable to the particular place of observation. Notwithstanding this source of common error, the barometer is a generally useful instrument, inasmuch as its indications foretell the same results at all seasons, with perhaps only this exception, those of the effects of heat in summer, which cannot of course be noticed in winter.

(263.) The general indications of the barometer are few, and may easily be remembered. A high and stationary mercury indicates steady, good weather. A slow and regular fall indicates rain; and, if during an E wind, the rain will be abundant. A sudden fall indicates a gale of wind, and most probably from the W. Good, steady weather must not be expected in sudden depressions and elevations of the mercury. A fine day

* Quarterly Journal of Agriculture, vol. iii. p. 5. † Ibid, p. 3.

may intervene, but the general state of the weather may be expected to be unsteady. An E. or N. E. wind keeps up the mercury against all other indications of a change. A W. or S. W. wind causes a fall when the wind changes from E. or N. E.; but, should no fall take place, the maintenance of the hight, in the circumstances, is equivalent to a rise, and the reverse of this is equivalent to a fall. The quantity affected by these particular causes may be estimated at $\frac{2}{10}$ of an inch.* The barometer, at sea, is a good indicator of wind, but not of rain. When the barometer is used within doors, the best situation for it is in any room where the temperature is equal, and not exposed to sunshine. The cost of a perpendicular barometer of good workmanship is from £1 11s. 6d. to £2 12s. 6d. according to taste and finish; that of a wheel-barometer from £2 2s. to £5 5s. The barometer was invented by Torricelli, a pupil of Galileo, in 1643.

(264.) Among the variable causes which affect the barometer is the direction of the wind. The maximum of pressure is when the wind is N. E. decreasing in both directions of the azimuth till it reaches the minimum between S. and S. W. This difference amounts to above $\frac{3}{10}$ of an inch at London. The variation occasioned by the wind may be owing to the cold which always accompanies the E. winds in spring, connected as they probably are with the melting of the snow in Norway; but it is not unlikely to be owing, as Mr. Meikle suggests, to its opposition to the direction of the rotation of the earth, causing atmospherical accumulation and pressure, by diminishing the centrifugal force of the aerial particles.†

(265.) The accidental variations of barometric pressure are greatly influenced by latitude. At the equator it may be said to be nothing, hurricanes alone causing any exception. The variability increases toward the poles, owing probably to the irregularity of the winds beyond the tropics. The mean variation at the equator is 2 lines,‡ in France 10 lines, and in Scotland 15 lines, throughout the year—the quantity having its monthly oscillations. These do not appear to follow the parallels of latitude, but, like the isothermal lines, undergo inflections, which are said to have a striking similarity to the isoclinal magnetic lines of Hansteen. If so, it is probably by the medium of temperature that these two are connected.—More lately, M. Kämtz has pointed out the connection of the winds with such changes, and he has illustrated the influence of the prevalent aerial currents which traverse Europe, though not with apparent regularity, yet, at least, in subjection to some general laws.||

(266.) The *sympiesometer* was invented by Mr. Adie, optician in Edinburgh, as a substitute for the common barometer. Its indications are the same, with the advantage of having a longer scale. For the measurement of hights this instrument is very convenient, from its small size admitting of its being carried in the coat-pocket, and not being subject to the same chances of accident as the portable barometer. The hight is given in fathoms on the instrument, requiring only one correction, which is performed by a small table engraved on its case. It is stated to be delicately sensible of changes at sea, particularly of gales. Not being an instrument which has been brought into general use, though Professor Forbes is convinced it might be, I need not allude to it farther here.§

(267.) The next instrument which claims our attention is the *thermometer*. As its name implies, it is a measurer of heat. It is undoubtedly the most perfect of our meteorological instruments, and has been the means

* Quarterly Journal of Agriculture, vol. iii. p. 2.
† Edinburgh New Philosophical Journal, vol. iv. p. 108.
‡ A line = twelfth part of an inch.
|| Forbes's Report on Meteorology, vol. i.
§ See Edinburgh Journal of Science, vol. x. p. 334, for a description of this ingenious instrument; and New Series, vol. iv. pp 91 and 329.

of establishing the most important facts to science; but, being a mere measurer of temperature, it is incapable of indicating changes of the atmosphere so clearly as the barometer, and is therefore a less useful instrument to the farmer. Regarding the ordinary temperature of the atmosphere, the feelings can judge sufficiently well; and, as the condition of most of the productions of the farm indicates pretty well whether the climate of a particular locality can bring any species of crop to perfection, the farmer seems independent of the use of the thermometer. Still, it is of importance for him to know the lowest degree of temperature in winter, as certain kinds of farm produce are injured by the effects of extreme cold, of which the feelings are incapable, from want of habit, of estimating their power of mischief. For this purpose, a thermometer self-registering the lowest degree of cold will be found a useful instrument on a farm. As great heat does no harm, a self-registering thermometer of the greatest heat seems not so useful an instrument as the other two.

(268.) "The thermometer, by which the temperature of our atmosphere was determined," says Mr. John Adie, of Edinburgh, "was invented by Sanctario, in 1590. The instrument, in its first construction, was very imperfect, having no fixed scale, and air being the medium of expansion. It was soon shown, from the discovery of the barometer, that this instrument was acted upon by pressure as well as temperature. To separate these effects, alcohol was employed as the best fluid, from its great expansion by heat, but was afterward found to expand unequally. Reaumur first proposed the use of mercury as the expansive medium for the thermometer. This liquid metal has great advantages over every other medium; it has the power of indicating a great range of temperature, and expands very equally. After its introduction, the melting point of ice was taken as a fixed point, and the divisions of the scale were made to correspond to $\frac{10}{1000}$th parts of the capacity of the bulb. It was left for the ingenious Fahrenheit to fix another standard point, that of boiling water under the mean pressure of the atmosphere, which is given on his scale at 212°; the melting point of ice at 32°. This scale of division has almost universally been adopted in Britain, but not at all generally on the Continent. The zero of this scale, though an arbitrary point adopted by Fahrenheit, from the erroneous idea that the greatest possible cold was produced by a mixture of common salt and snow, has particular advantages for a climate like ours; besides being generally known, the zero is so placed that any cold which occurs very rarely causes the mercury to fall below that point, so that no mistake can take place with regard to noting minus quantities.—The only other divisions of the thermometer between the two fixed points in general use are those of Reaumur and the centesimal: the former divides the space into 80 equal parts; the division of the latter, as indicated by its name, is into 100 parts. In both these scales the zero is placed at the melting point of ice, or 32° Fahrenheit."* The self-registering thermometers were the invention of the late Dr. John Rutherfurd, and his are yet the best. The tube of the one for ascertaining the greatest degree of heat is inclined nearly in a horizontal position and filled with mercury, upon the top of the column of which stands an index, which, on being pushed upward, does not return until made to descend to the top of the mercury by elevating the upper end of the thermometer. This index was first made of metal, which became oxydized in the tube, and uncertain in its motions. Mr. Adie, optician in Edinburgh, improved the instrument, by introducing a fluid above the mercury, in which is floated a glass index, which is free

---

* Quarterly Journal of Agriculture, vol. iii. p. 5

from any action, and is retained in its place by the fluid. "The other thermometer, for registering the lowest degree," says Mr. John Adie, "is filled with alcohol, having an index of black glass immersed in the liquid. This index is always carried down to the lowest point to which the temperature falls; the spirit passes freely upward without changing the place of the index, so that it remains at the lowest point. This instrument, like the other, turns upon a center, to depress the upper end, and allow the index, by its own weight, to come into contact with the surface of the spirit, after the greatest cold has been observed, which is indicated by the upper end of the index, or that farthest from the bulb. In both cases the instruments are to be left nearly horizontal, the bulb end being lowest. This angle is most easily fixed by placing the bulb about ¾ of an inch under the horizontal line."*

(269.) Thermometers of all kinds, when fixed up for observation, should be placed out of the reach of the direct rays of the sun or of any reflected heat. If at a window or against a wall, the thermometer should have a northern aspect, and be kept at a little distance from either; for it is surprising through what a space a sensible portion of heat is conveyed from soil and walls, or even from grass illuminated by the sun. The maxima of temperature, as indicated by thermometers, are thus generally too great; and from the near contact in which thermometers are generally placed with large ill-conducting masses, such as walls, the temperature of the night is kept up, and the minima of temperature are thus also too high. The price of a common thermometer is from 5s. 6d. to 14s. and of Rutherfurd's minimum self-registering thermometer 10s. 6d.

(270.) Many highly interesting results have been obtained by the use of the thermometer, and among the most interesting are those regarding the mean temperature of different localities. In prosecuting this subject, it was found that a diurnal oscillation took place in the temperature as well as the pressure of the atmosphere, and that this again varies with the seasons. Nothing but frequent observations during the day could ascertain the mean temperature of different places; and, in so prosecuting the subject, it was discovered that there were hours of the day, the mean temperature of which, for the whole year, was equal to the mean of the whole 24 hours, which, when established, would render all future observations less difficult. The results exhibit an extraordinary coincidence.

Thus the mean of 1824 gave	13′	past	9	A. M. and	26	past	8	P. M.
.. .. 1825 ..	13′	..	9	.. ..	28	..	8	..
Giving the mean of the 2 years	13′	..	9	.. ..	27′	..	8	..

These results were obtained from a series of observations made at Leith Fort in the years 1824 and 1825 by the Royal Society of Edinburgh.† Some of the other consequences deducible from these observations are, "that the mean hour of the day of minimum temperature for the year is 5 A. M., and that of maximum temperature 40′ past 2 P. M.; that the deviation of any pair of hours of the same name from the mean of the day is less than half a degree of Fahrenheit, and of all pairs of hours, 4 A. M. and P. M. are the most accurate; that the mean annual temperature of any hour never differs more than 3°.2 from the mean of the day for the whole year; that the mean daily range is a minimum at the winter solstice, and a maximum in April; and that the mean daily range in this climate is 6°.065."‡ The mean temperature at Leith Fort for the mean of the two years, at an elevation of 25 feet above the mean level of the

* Quarterly Journal of Agriculture, vol. iii. p. [illegible]

† Edinburgh Philosophical Transactions, vol. x.

‡ Forbes's Report on Meteorology, vol. i.

sea was found to be 48°.36. The mean, taken near Edinburgh, at an altitude of 390 feet above the mean level of the sea, at 10 A. M. and P. M., with a common thermometer, and with the maximum and minimum results of self-registering thermometers, gave these results when reduced to the mean level of the sea: With the self-registering thermometers 48°.415, and with two observations a day with the common thermometer 48°.352, which correspond remarkably with the observations at Leith Fort. These observations were taken at 10 A. M. and 10 P. M., which were found to be the particular hours which gave a near approximation to the mean temperature of the day; but had they been made at the more correct periods of 13′ past 9 A. M. and 27′ past 8 P. M., it is probable that the results with those at Leith Fort would have corresponded exactly.* The mean temperature of any place may be ascertained pretty nearly by observing the mean temperature of deep-seated springs, or that of deep wells. Thus the Crawley Springs, in the Pentland Hills, which supply Edinburgh with abundance of water, situated at an elevation of 564 feet above the level of the sea, give a mean temperature of 46°.3, according to observations made in 1811 by Mr. Jardine, civil engineer, Edinburgh; and the Black Spring, which is 882 feet above the level of the sea, gave a mean temperature of 44°.9, by observations made in the course of 1810–11–15–18–19. A well in the Cowgate of Edinburgh gave a mean temperature of 49°.3, by observations made every month in the year 1794, of which the temperature of the month of June approached nearest to the mean temperature of the year, being 49°.5.†

(271.) The measurement of the humidity of the atmosphere is a subject of greater importance in a scientific than in a practical point; for however excellent the instrument may be for determining the degree of humidity, the atmosphere has assumed the humid state before any indication of the change is noticed on the instrument, and in this respect it is involved in the same predicament as the thermometer, which only tells the existing heat, and both are less useful on a farm than the barometer, which indicates an approaching change. No instrument has yet been contrived by which the quantity of moisture in the air can be ascertained from inspection of a fixed scale, without the use of tables to rectify the observation. The instrument used for ascertaining the moisture of the air is appropriately termed a *hygrometer*. Professor Leslie was the first to construct a useful instrument of this kind. His is of the form of a differential thermometer, having a little sulphuric acid in it; and the cold is produced by evaporation of water from one of the bulbs covered with black silk, which is kept wetted, and the degree of evaporation of the moisture from the bulb indicates the dryness of the air.

(272.) Another method of ascertaining the moisture of the atmosphere, is by the dew-point hygrometer of Professor Daniells; but this instrument is considered rather difficult of management, except in expert hands.

(273.) The best hygrometer is that of Dr. Mason, which consists of two thermometers, fastened upright to a stand having a fountain of water in a glass tube placed betwixt them, and out of which the water is taken up to one of the bulbs by means of black floss silk. When the air is very dry, the difference between the two thermometers will be great; if moist, less in proportion; and when fully saturated, both will be alike. The silk that covers the wet bulb, and thread which conveys the water to it, require renewal about every month, and the fountain is filled when requisite with distilled water, or water that has been boiled and allowed to cool,

* Quarterly Journal of Agriculture, vol. iii. p. 9. † Ibid. p. 10–11.

by immersing it in a basin of the water till the aperture only is just upon the surface, and the water will flow into it. For ordinary purposes of observation, it is only necessary to place the instrument in a retired part of the room away from the fire, and not exposed to weather, open doors, or passages; but for nice experiments the observations should always be made in the open air and in the shade, taking especial care that the instrument be not influenced by the radiation of any heated bodies, or any currents of air. When the hygrometer is placed out of doors in frosty weather, the fountain had better be removed, as the freezing of the water within may cause it to break; in this case a thin coating of ice may soon be formed on the wet bulb, which will last a considerable time wet, and be rewetted when required.

(274.) Very simple hygrometers may be made of various substances, to show whether the air is more or less humid at any given time. One substance is the awn of the Tartarian and wild oats, which, when fixed in a perpendicular position to a card, indicates, by its spiked beard, the degree of humidity. A light hog's bristle split in the middle, and riding by the split upon the stem of the awn, forms a better index than the spike of the awn itself. To adjust this instrument, you have only to wet the awn and observe how far it carries round the index, and mark that as the lowest point of humidity, and then subject the awn to the heat of the fire for the highest point of dryness, which, when marked, will give betwixt the two points an arc of a circle, which may be divided into its degrees. I have used such an instrument for some time. When two or more are compared together, the mean of humidity may be obtained. The awns can be renewed at pleasure. With regard to confiding in the truth of this simple hygrometer, the precaution of Dr. Wells is worth attention. "Hygrometers formed of animal and vegetable substances," he says, "when exposed to a clear sky at night, will become colder than the atmosphere, and hence by attracting dew, or, according to an observation of Saussure, by merely cooling the air contiguous to them, mark a degree of moisture beyond what the atmosphere actually contains. This serves to explain an observation made by M. de Luc, that in serene and calm weather, the humidity of the air, as determined by a hygrometer, increases about and after sunset with a greater rapidity than can be attributed to a diminution of the general heat of the atmosphere."* The principle of this sort of hygrometer may serve to explain a remarkable natural phenomenon. "Hygrometers were made of quills by Chiminello, which renders it probable that birds are enabled to judge of approaching rain or fair weather. For it is easy to conceive that an animal having a thousand hygrometers intimately connected with its body, must be liable to be powerfully affected, with regard to the tone of its organs, by very slight changes in the dryness or humidity of the air, particularly when it is considered that many of the feathers contain a large quantity of blood, which must be alternately propelled into the system, or withdrawn from it, according to their contraction or dilatation by dryness or moisture."† Does Virgil allude to a hygrometric feeling in birds when he says—

> "Wet weather seldom hurts the most unwise,
> So plain the signs, such prophets are the skies:
> The wary crane foresees it first, and sails
> Above the storm, and leaves the lowly vales."‡

(275.) The *Weather-cock* is a very useful instrument to the farmer. It should be erected on a conspicuous part of the steading, which may readily

* Wells on Dew, p. 64. † Edinburgh Encyclopædia, art. *Hygrometry*. ‡ Dryden's Virgil, i. Georgics, 514

be observed from one of the windows of the farm-house. Its position on the steading may be seen in fig. 1, Plate I., and fig. 3, Plate III. Its cardinal points should be marked with the letters N. E. S. W., to show at a glance the true points of the compass. The vane should be fitted up with a ball or box containing oil, which may be renewed when required. There is not a neater or more appropriate form for a vane than an arrow, whose dart is always ready to pierce the wind, and whose butt serves as a governor to direct it to the wind's eye. The whole should be gilt, to prevent the rusting of the iron. Mr. Forster had such a vane erected at his place of residence, which had a small bell suspended from its point which struck upon the arms pointing to the direction of the compass, and announced every change of wind.* Such a contrivance may be considered a conceit, but it has the advantage of letting you know when the wind shifts much about, as when it does there is as little chance of settled weather as in the frequent changes of the barometer. A better contrivance of the bell would be to have a hammer suspended from the dart by a supple spring, and a bell of different tone attached to each of the arms which indicate the point of the compass, and the different toned bells, when struck, would announce the direction in which the wind most prevailed. Besides bells, there is a contrivance for indicating the directions of the wind by an index on a vertical disk, like the dial-plate of a clock, an instance of which may be seen in the western tower of the Register-House in Edinburgh. This would be a very convenient way of fitting up a weather-cock.

(276.) With regard to the origin of the name of *weather-cock*, Beckmann says that vanes were originally cut out in the form of a cock, and placed on the tops of church spires, during the holy ages, as an emblem of clerical vigilance.† The Germans use the same term as we do, *wetterhahn;* and the French have a somewhat analogous term in *coq de clocher*. As the vane turns round with every wind, so, in a moral sense, every man who is "unstable in his ways," is termed a weather-cock.

(277.) In reference to the wind is another instrument called the *anemometer*, or measurer of the wind's intensity. Such an instrument is of little value to the farmer, who is more interested in the direction than the intensity of the wind, as it is that property of it which has most effect in promoting changes of the weather. It must be admitted, however, that the intensity of the wind has a material effect in modifying the climate of any locality, such as that of a farm elevated in the gorge of a mountain pass. Still, even there its direction has more to do in fixing the character of the climate than the intensity; besides, the anemometer indicates no approach of wind, but only measures its force when it blows, and this can be sufficiently well appreciated by the senses. The mean force of the wind for the whole year at 9 A. M. is 0.855, at 3 P. M. 1.107, and at 9 P. M. 0.605.

(278.) The best instrument of this class is Lind's anemometer, which, although considered an imperfect one, is not so imperfect, according to the opinion of Mr. Snow Harris, of Plymouth, who has paid more attention to the movements of the wind than any one else in this country, as is generally supposed. Lind's anemometer "consists of two glass tubes about 9 inches long, having a bore of $\frac{4}{10}$ of an inch. These are connected, at their lower extremities, by another small tube of glass, with a bore of $\frac{1}{10}$ of an inch. To the upper extremity of one of the tubes is fitted a thin metallic one, bent at right angles, so that its mouth may receive horizontally the current of air. A quantity of water is poured in at the mouth, till the

* Forster's researches into Atmospherical Phenomena. † Beckmann's History of Inventions, vol. i.

tubes are nearly half full, and a scale of inches and parts of an inch is placed betwixt the tubes. When the wind blows in at the mouth, the column of water is depressed in one of the tubes, and elevated in the same degree in the other tube; so that the distance between the surface of the fluid in each tube is the length of a column of water, whose weight is equivalent to the force of the wind upon a surface equal to the base of the column of fluid. The little tube which connects the other two is made with a small aperture, to prevent the oscillation of the fluid by irregular blasts of wind. The undulations produced by sudden gusts of wind would be still more completely prevented by making the small tube, which connects the other two large ones, of such a length as to be double between the other two, and be equal to the length of either. The same effect might also be produced by making a thin piece of wood float upon the surface of the fluid in each tube."*

(279.) Another meteorological instrument is the *rain-gauge*. This instrument is of no use to the farmer as an indicator of rain, and, like some of the rest which have been described, only professes to tell the quantity of rain that actually has fallen in a given space, yet even for this purpose it is an imperfect instrument.† "The simplest form of this instrument," says Mr. John Adie, "is a funnel, with a cylindrical mouth, 3 or 4 inches high, and having an area of 100 square inches, made of tinned iron or thin copper. It may be placed in the mouth of a large bottle for receiving the water, and, after each fall, the quantity is measured by a glass jar, divided into inches and parts. A more elegant arrangement of the instrument is formed by placing the funnel at the top of a brass cylindrical tube, having at one side a glass tube, communicating with it at the under part, with a divided scale placed alongside of it. The area of the mouth is to that of the under tubes as 10 : 1; consequently 1 inch deep of rain falling into the mouth will measure 10 inches in the tubes, and 1 inch upon the scale will be equal to a fall of $\frac{1}{10}$ of an inch, which quantities are marked upon the scale, and the water is let off by a stop-cock below. The instrument should be placed in an exposed situation, at a distance from all buildings and trees, and as near the surface of the ground as possible. . . . In cases of snow-storms, the rain-gauge may not give a correct quantity, as a part may be blown out, or a greater quantity have fallen than the mouth will contain. In such cases, the method of knowing the quantity of water is to take any cylindrical vessel, such as a case for containing maps, which will answer the purpose very well; by pressing it perpendicularly into the snow, it will bring out with it a cylinder equal to the depth. This, when melted, will give the quantity of water by measurement. The proportion of snow to water is about 17 : 1, and hail to water 8 : 1. These quantities, however, are not constant, butdepend upon the circumstances under which the snow or hail has fallen, and the time they have been upon the ground."‡ The cost of a rain-gauge, according as it is fitted up, is £1 5s. £2 12s. 6d. and £4 4s.

(280.) These are the principal instruments employed by meteorologists to ascertain atmospherical changes, and seeing their powers and uses, as now described, you can select those which appear to you most desirable to possess. Of them all, only two are *indicators of approaching changes*, the barometer and the weather-cock; and these, of good construction, you will of course have, whichever of the others you may choose to possess.

(281.) Besides these two instruments, there are objects in nature which indicate changes of the weather. Of these the *Clouds* are eminent premon-

* Edinburgh Encyclopædia, art. *Anemometer*. † See Thomson's History of the Royal Society.
‡ Quarterly Journal of Agriculture, vol. iii. p. 13.

itors. It may at first sight be supposed that clouds, exhibiting so great a variety of forms, cannot be subject to any positive law; but such a supposition is erroneous, because no phenomenon in nature can possibly occur, but as the effect of some physical law, although the mode of action of the law may have hitherto eluded the acutest search of philosophical observation. It would be unphilosophical to believe otherwise. We may therefore depend upon it, that every variety of cloud is an effect of a definite cause. If we cannot predict what form of cloud will next ensue, it is because we are unacquainted with the precise process by which they are formed. But observation has enabled meteorologists to classify every variety of form under only three primary figures, and all other forms are only combinations of two or more of these three.*

(282.) 1. The first simple form is the *Cirrus*, a word which literally means a curl, or lock of hair curled. 2. The second is the *Cumulus*, or heap. 3. And the third is the *Stratus*, or bed or layer. Combinations of these three give the four following forms, the names of which at once indicate the simple forms of which they are composed. 1. One is *Cirro-Cumulus*, or combination of the curl and heap. 2. Another is the *Cirro-Stratus*, or combination of the curl and stratus. 3. A third is the *Cumulo-Stratus*, or combination of the heap and the stratus. 4. And, lastly, there is the combination of the *Cumulo-Cirro-Stratus*, or that combination of all the three simple forms, which has received the name of *Nimbus* or rain-cloud. The English names usually given by writers to some of these forms of clouds are very singular, and seemingly not very appropriate. The curl is an appropriate enough name for the cirrus, and so is the rain-cloud for the nimbus; but why the heap should be called the *stacken-cloud*, the stratus the *fall-cloud*, the curled heap the *sonder-cloud*, the curled stratus the *wane-cloud*, and the heaped stratus the *twain-cloud*, is by no means obvious, unless this last form, being composed of *two* clouds, may truly be denominated a *twain*-cloud; but, on the same principle, the cirro-cumulus, and the cirro-stratus, and the cumulo-stratus, may be termed *twain*-clouds. We must, however, take the nomenclature which the original and ingenious contriver of the classification of clouds, Mr. Luke Howard, of London, has given.

(283.) The first form of clouds which demands your attention is the *Cirrus* or curl-cloud. This is the least dense of all clouds. It is composed of streaks of vapor of a whitish color, arranged with a fibrous structure, and occurring at a great hight in the atmosphere. These fibrous streaks assume modified shapes. Sometimes they are like long narrow rods, lying quiescent, or floating gently along the upper region of the atmosphere.—At other times one end of the rod is curled up, and spread out like a feather; and, in this shape, the cloud moves more quickly along than the other, being evidently affected by the wind. Another form is that familiarly known by the "gray mare's tail," or "goat's beard." This is more affected by the wind than even the former. Another form is in thin fibrous sheets, expanded at times to a considerable breadth, like the gleams of the aurora borealis. There are many other forms, such as that of net-work, bunches of feathers, hair, or thread, which may respectively be designated reticulated, plumose, comoid, and filiform cirri.

(284.) In regard to the relative hights at which these different forms of cirri appear, I would say that the fibrous rod assumes the highest position

[* For farther observations on the "Means of Prognosticating the Weather," see last number of Jour. of Ag., page 137. For a valuable work on this subject, as applicable to our own country the reader is referred to Forry on the Climate of the United States. *Ed. Farm. Lib.*]

in the air; the rod with the turned-up end the next highest; the bunch of feathers is approaching the earth; the mare's tail is descending still farther; and the sheet-like form is not much above the denser clouds. Sometimes the fibrous rod may be seen stretching between two denser clouds, and it is then supposed to be acting as a conductor of electricity between them.

(285.) As to their relative periods of duration, the fibrous rod may be seen high in the air for a whole day in fine weather; or it vanishes in a short time, or descends into a denser form. When its end is turned up, its existence is hastening to a close. The plumose form soon melts away; the gray-mare's tail bears only a few hours of pretty strong wind; but the broad sheet may be blown about for some time.

(286.) The sky is generally of a gray-blue when the fibrous rod and hooked rod are seen; and it is of the deepest blue when the plumose watery cirrus appears. It is an observation of Sir Isaac Newton, that the deepest blue happens just at the changes from a dry to a moist atmosphere.

(287.) The cirrus cloud frequently changes into the complete cirro-cumulus, but it sometimes forms a fringed or softened edge to the cirro-stratus; and it also stretches across the heavens into the density of a cirro-stratus. Of all the seasons, the cirrus appears least frequently in winter.

(288.) The *Cumulus* may be likened in shape to a heap of natural meadow hay. It never alters much from that shape, nor is it ever otherwise than massive in its structure; but it varies in size and color according to the temperature and light of the day, becoming larger and whiter as the heat and light increase; hence it generally appears at sunrise, assumes a larger form by noon, often screening the sun from the earth, and then melts away toward night. On this account it has received the designation of the "cloud of day." Its density will not allow it to mount very high in the air; but it is, nevertheless, easily buoyed up for a whole day by the vapor plane above the reach of the earth. When it so rests it is terminated below by a straight line. It is a prevailing cloud in the daytime at all seasons, and is exceedingly beautiful when it presents its silvery tops tinted with sober colors against the bright blue sky. Cumuli sometimes join together and as suddenly separate again, though in every case they retain their peculiar form. They may often be seen floating in the air in calm weather, not far above the horizon; and they may also be seen driving along with the gale at a greater hight, casting their fleeting shadows on the ground. When in motion, their bases are not so straight as when at rest. Cumuli, at times, disperse, mount into the air, and form cirri, or they descend into strati along the horizon; at others a single cumulus may be seen at a distance in the horizon, and then increasing rapidly into the storm-cloud, or else overspreading a large portion of the sky with a dense veil. Does the poet allude to the cumulus, as seen in a summer afternoon, in these breathing words?

> "And now the mists from earth are clouds in heaven,
> Clouds slowly castellating in a calm
> Sublimer than a storm; which brighter breathes
> O'er the whole firmament the breadth of blue,
> Because of that excessive purity
> Of all those hanging snow-white palaces,
> A gentle contrast, but with power divine."*

(289.) The *Stratus* is that bed of vapor which is frequently seen in the valleys in a summer evening, permitting the trees and church spires to stand out in bold relief; or it is that horizontal bank of dark cloud seen to

* Wilson.

rest for a whole night along the horizon. It also forms the thin dry white fogs which come over the land from the sea with an east wind in spring and summer, wetting nothing that it touches. When this dry fog hangs over towns in winter, which it often does for days, it appears of a yellow hue, in consequence, probably, of a mixture with smoke. It constitutes the November fog in London. The stratus is frequently elevated by means of the vapor plane, and then it passes into the cumulus. On its appearing frequently in the evening, and its usual disappearance during the day, it has been termed the "cloud of night." Having a livid gray color when the moon shines upon it, the stratus is probably the origin of those supposed spectral appearances seen at night by superstitious people in days of yore. The light or dry stratus is most prevalent in spring and summer, and the dense or wet kind in autumn and winter.

(290.) "*Cirrus*," remarks Mr. Mudie, "is the characteristic cloud of the upper sky; and no cloud of denser texture forms, or is capable of being sustained there. *Cumulus* is, in like manner, the characteristic cloud of the middle altitude; and although it is sometimes higher and sometimes lower, it never forms at what may be called the very top of the sky, or down at the surface of the ground. *Stratus* is the appropriate cloud of the lower sky, and it is never the first formed one at any considerable elevation; and, indeed, if it appears unconnected with the surface, it is not simple stratus, but a mixed cloud of some kind or other."*

(291.) The forms of the clouds which follow are of mixed character, the first of which that demands our attention is a compound of the cirrus and cumulus, or *cirro-cumulus*, as it is called. The cirrus, in losing the fibrous, assumes the more even-grained texture of the cumulus, which, when subdivided into small spherical fragments, constitute small cumuli of little density, and of white color, arranged in the form of a cirrus or in clusters. They are high in the air, and beautiful objects in the sky. In Germany this form of cloud is called "the little sheep;" which idea has been embodied by a rustic bard of England in these beautiful lines:

> "Far yet above these wafted clouds are seen
> (In a remoter sky, still more serene,)
> Others, detached in ranges through the air,
> Spotless as snow, and countless as they're fair;
> Scattered immensely wide from east to west,
> The beauteous 'semblance of a flock at rest."†

Cirro-cumuli are most frequently to be seen in summer.

(292.) Another form of cloud, compounded of the cirrus and stratus, is called *cirro-stratus*. While cirri descend and assume the form of cirro-cumuli, they may still farther descend and take the shape of cirro-stratus, whose fibres become dense and decidedly horizontal. Its characteristic form is shallowness, longitude, and density. It consists at times of dense longitudinal streaks, and the density is increased when a great breadth of cloud is viewed horizontally along its edge. At other times it is like shoals of small fish, when it is called a "herring sky;" at others, mottled like a mackerel's back, when it is called the "mackerel-back sky." Sometimes it is like veins of wood, and at other times like the ripples of sand left by a retiring tide on a sandy beach. The more mottled it is, the cirro-stratus is higher in the air, and the more dense and stratified, the nearer it is the earth. In the last position, it may be seen cutting off a mountain top, or stretching behind it, or cutting across the tops of large cumuli. Sometimes its striated lines, not very dense, run parallel over the zenith, whose opposite ends apparently converge at opposite points of the hori-

* Mudie's World. † Bloomfield.

zon, and then they form that peculiar phenomenon named the "boat," or "Noah's ark." At times cirro-strati cut across the field of the setting sun, where they appear in well-defined dense striæ, whose upper or lower edges, in reference to their position with the sun, are burnished with the most brilliant hues of gold, crimson, or vermilion. Sometimes the cirro-stratus extends across the heavens in a broad sheet, obscuring more or less the light of the sun or moon, for days together, and in this case a halo or corona is frequently seen to surround these orbs. In a more dense form, it assumes the shapes of some small long-bodied animals, and even like architectural ornaments; and in all its mutations it is more varied than any other form of cloud. The streaked cirro-strati are of frequent occurrence in winter and autumn, whereas the more delicate kinds are most seen in summer.

(293.) A third compound cloud is formed of the cumulus and stratus, called *cumulo-stratus*. This is always a dense cloud. It spreads out its base to the stratus form, and, in its upper part, frequently inosculates with cirri, cirro-cumuli, or cirro-strati. In this form it is to be seen in the plate of the three cows. With all or either of these it forms a large massive series of cumulative clouds which hang on the horizon, displaying great mountain shapes, raising their brilliantly illuminated silvery crests toward the sun, and presenting numerous dusky valleys between them. Or it appears in formidable white masses of variously defined shapes, towering upward from the horizon, ready to meet any other form of cloud, and to conjoin with them in making the dense dark-colored storm-cloud. In either case, nothing can exceed the picturesque grandeur of their towering, dazzling forms, or the sublimity of their masses when surcharged with lightnings, wind, and rain, and hastening with scowling front to meet the gentle breeze, and hurrying it along in its determined course, as if impatient of restraint, and all the while casting a portentous gloom over the earth, until bursting with terrific thunder, scorching with lightning some devoted object more prominent than the rest, deluges the plain with sweeping floods, and devastates the fields in the course of its ungovernable fury. A tempest soon exhausts its force in the temperate regions; but in the tropics it rages at times for weeks, and then woe to the poor mariner who is overtaken by it at sea unprepared. Of the cumulo-stratus the variety called "Bishops' wigs," as represented near the horizon in the plate of the draught-mare, may be seen at all seasons along the horizon, but the other and more imposing form of mountain scenery is only to be seen in perfection in summer, when storms are rife. It also assumes the shapes of larger animals, and of the more gigantic forms of nature and art. Is the cumulo-stratus the sort of cloud described by Shakspeare as presenting these various forms?

> "Sometime, we see a cloud that's dragonish;
> A vapor, sometime, like a bear or lion,
> A tower'd citadel, a pendant rock,
> A forked mountain or blue promontory
> With trees upon't, that nod unto the world,
> And mock our eyes with air:——
> That, which is now a horse, even with a thought,
> The rack dislimns, and makes it indistinct,
> As water is in water."*

(294.) The last compound form of cloud which I have to mention is the *cirro-cumulo-stratus*, called the *nimbus* or rain-cloud. A showery form of the cloud may be seen in the plate of the draught-horse. For my part I cannot see that the mere resolution of a cloud into rain is of sufficient im-

* Anthony and Cleopatra.

portance to constitute the form into a separate and distinct cloud; for rain is not so much a form as a condition of a cloud, in the final state in which it reaches the earth. Any of the three compound forms of clouds just described may form a rain-cloud, without the intervention of any other. Cirro-strati are often seen to drop down in rain, without giving any symptoms of forming the more dense structure of the nimbus; and even light showers fall without any visible appearance of a cloud at all. The nimbus is most frequently seen in summer and autumn.

(295.) There is a kind of cloud, not unlike cumuli, called the *scud*, which is described usually by itself as broken nimbus. It is of dark or light color, according as the sun shines upon it, of varied form, floating or scudding before the wind, and generally in front of a sombre cumulo-stratus stretching as a background across that portion of the sky, often accompanied with a bright streak of sky along the horizon. The ominous scud is the usual harbinger of the rain-cloud, and is therefore commonly called "messengers," "carriers," or "water-wagons."

(296.) On looking at the sky, forms of clouds may be observed which cannot be referred to any of those, simple or compound, which have just been described. On analyzing them, however, it will be found that every cloud is referable to one or more of the forms described. This defectiveness proves two things in regard to clouds. 1. That clouds, always presenting forms which are recognizable, must be the result of fixed laws.— 2. That the sagacity of man has been able to classify those forms of clouds in a simple manner. Without such a key to their forms, clouds doubtless appear, to common observers, masses of inexplicable confusion. Clouds thus being only effects, the causes of their formation and mutations must be looked for in the atmosphere itself; accordingly, it has been found that, when certain kinds appear, certain changes are taking place in the state of the atmosphere; and beyond this it is not necessary for a common observer to know the origin of clouds. It is sufficient for him to be aware of what the approaching change of the atmosphere will be, as indicated by the particular kind of cloud or clouds which he observes; and in this way clouds become guides for knowing the weather. In endeavoring thus to become a judge of the weather, you must become an attentive observer of the clouds. To become so with success in a reasonable time, you must first make yourself well acquainted with the three simple forms, which, although not singly visible at all times, may be recognized in some part of those compound clouds which exhibit themselves almost every day.

(297.) That clouds float at *different altitudes*, and are *more or less dense*, not merely on account of the quantity of vapor which they contain, but partly on account of their distance from vision, may be proved in various ways. 1. On ascending the sides of mountains, travelers frequently pass zones of clouds. Mountains thus form a sort of scale by which to estimate the altitude of clouds. Mr. Crossthwaite made these observations of the altitude and number of clouds in the course of five years:

Altitude of Clouds	Number of Clouds.	Altitude of Clouds.	Number of Clouds.
From 0 to 100 yards	10	From 700 to 800 yards	367
100 to 200 ..	42	800 to 900 ..	410
200 to 300 ..	62	900 to 1000 ..	518
300 to 400 ..	179	1000 to 1050 ..	419
400 to 500 ..	374		3283
500 to 600 ..	486		
600 to 700 ..	416	Above 1050 ..	2098

Hence the number of clouds above 1050 yards were, to the number below, as 2098 : 3283, or 10 : 16 nearly. The nomenclature of Howard not having been known at the time, the forms of the various clouds met with at the different altitudes could not be designated. 2. Another proof of a difference of altitudes in clouds consists in different clouds being seen to move in different directions at the same time. One set may be seen moving in one direction near the earth, while another may be seen through their openings unmoved. Clouds may be seen moving in different directions, at apparently great hights in the air, while those near the ground may be quite still. Or the whole clouds seen may be moving in the same direction with different velocities. It is natural to suppose that the lighter clouds—those containing vapor in the most elastic state—should occupy a higher position in the air than the less elastic. On this account, it is only fleecy clouds that are seen over the tops of the highest Andes. Clouds, in heavy weather, are seldom above ½ mile high, but in clear weather from 2 to 5 miles, and *cirri* from 5 to 7 miles.

(298.) Clouds are often of *enormous size*, 10 miles each way and 2 miles thick, containing 200 cubic miles of vapor; but sometimes are even ten times that size. The size of small clouds may be easily estimated by observing their shadows on the ground in clear breezy weather in summer. These are usually *cumuli* scudding before a westerly wind. The shadows of larger clouds may be seen resting on the sides of mountain ranges, or spread out on the ocean.

(299.) You must become acquainted with the agency of *Electricity* before you can understand the variations of the weather. The subject of atmospherical electricity excited great attention in the middle of the last century by the experiments and discoveries of Franklin. He proved that the electric fluid,* drawn from the atmosphere, exhibits the same properties as that obtained from the electrical machine, and thus established their identity. Since that period, little notice has been taken of its powerful agency in connection with meteorology; but brilliant are the discoveries which have since been made in regard to its powers in the laboratories of Davy, Faraday, and others. They have clearly identified electricity with magnetism and galvanism, and, in establishing this identity, they have extended to an extraordinary degree the field of observation for the meteorologist, though the discovery has rendered meteorology much more difficult to be acquired with exactness. But the science should, on that account, be prosecuted with the greater energy and perseverance.

(300.) It must be obvious to the most indifferent observer of atmospherical phenomena, that the electric agency is exceedingly *active in the atmosphere*, how inert soever may be its state in other parts of the earth. Existing there in the freest state, it exhibits its power in the most sensible manner; and its freedom and frequency suggest the interesting inquiry whence is derived the supply of the vast amount of electricity which seems to exist in the atmosphere?

(301.) Of all investigators of this interesting but difficult inquiry, M. Pouillet has directed his attention to it with the greatest success. He has shown that there are two sources from which this abundant supply is obtained. The first of these is *vegetation*. He has proved, by direct experiment, that the combination of oxygen with the materials of living plants is a constant source of electricity; and the amount thus disengaged may be learned from the fact that a surface of 100 square metres (or rather more than 100 square yards), in full vegetation, disengages, in the course of one day, as much vitrous electricity as would charge a powerful battery.

(302.) That some idea may be formed of the sort of action which takes

* "Electricity, though frequently called a fluid, has but little claim to that designation; in using it, therefore, let it be always understood in a conventional sense, not as expressing any theoretical view of the physical state of electric matter." Dr. Golding Bird

place between the oxygen of the air and the materials of living plants, it is necessary to attend, in the first place, to the change produced on the air by the respiration of plants. Many conflicting opinions still prevail on this subject; but "there is no doubt, however, from the experiments of various philosophers," as Mr. Hugo Reid observes, "that at times the leaves of plants produce the same effect on the atmosphere as the lungs of animals, namely, cause an increase in the quantity of carbonic acid, by giving out carbon in union with the oxygen of the air, which is thus converted into this gas; and it has been also established that at certain times the leaves of plants produce very opposite effects, namely, that they decompose the carbonic acid of the air, retain the carbon and give out the oxygen, thus adding to the quantity of the oxygen in the air. It has not yet been precisely ascertained which of these goes on to the greater extent; but the general opinion at present is, that the gross result o the action of plants on the atmosphere is the depriving of it of carbonic acid, retaining the carbon and giving out the oxygen, thus increasing the quantity of free oxygen in the air."*

(303.) It being thus admitted that both carbonic gas and oxygen are exhaled by plants during certain times of the day, it is important to ascertain, in the next place, whether electricity of the one kind or the other accompanies the disengagement of either gas. Toward this inquiry M. Pouillet instituted experiments with the gold-leaf electroscope, while the seeds of various plants were germinating in the soil, and he found it sensibly affected by the *negative* state of the ground. This result might have been anticipated during the evolution of carbonic gas, for it is known by experiment that carbonic gas, obtained from the combustion of charcoal, is, in its nascent state, electrified *positively*, and, of course, when carbonic gas is evolved from the plant, the ground should be in a state of negative electricity. M. Pouillet presumed, therefore, that when plants evolve oxygen, the ground should be in a positive state of electricity. He was thus led to the important conclusion, that vegetation is an abundant source of electricity.†

(304.) The second source of electricity is *evaporation*. The fact of a chemical change in water by heat inducing the disengagement of electricity, may be proved by simple experiment. It is well known that *mechanical* action will produce electricity sensibly from almost any substance. If any one of the most extensive series of resinous and siliceous substances, and of dry vegetable, animal and mineral produce, is rubbed, electricity will be excited, and the extent of excitation will be shown by the effect on the gold-leaf electroscope. Chemical action, in like manner, produces similar effects. If sulphur is fused and poured into a conical wine-glass, it will become electrical on cooling, and affect the electroscope in a manner similar to the other bodies mechanically excited. Chocolate on congealing after cooling, glacial phosphoric acid on congealing, and calomel when it fixes by sublimation to the upper part of a glass vessel, all give out electricity; so, in like manner, the condensation as well as the evaporation of water, though opposite processes, gives out electricity. Some writers attribute these electrical effects to what they term a change of form or state; but it is obvious that they may, with propriety, be included under chemical action. This view is supported by the fact of the presence of oxygen being necessary to the development of electricity. De la Rive, in bringing zinc and copper in contact through moisture, found that the zinc became oxidized, and electricity was evolved. When he pre-

* Reid's Chemistry of Nature. † Leithead on Electricity.

vented the oxidation, by operating in an atmosphere of nitrogen, no electric excitement followed. When, again, he increased the chemical action by exposing zinc to acid, or by substituting a more oxidable metal, such as potassium, the electric effects were greatly increased. In fact, electrical excitation and chemical action were observed to be strictly proportional to each other. And this result is quite consistent with, and is corroborated by, the necessary agency of oxygen in evolving electricity from vegetation.* But more than all this, "electricity," as Dr. G. Bird intimates, "is not only evolved during chemical decomposition, but during *chemical combination;* a fact first announced by Becquerel. The truth of this statement has been, by many, either altogether denied or limited to the case of the combination of nitric acid with alkalies. But after repeating the experiments of Becquerel, as well as those of Pfaff, Mohr, Dalk, and Jacobi, I am convinced that an electric current, certainly of low tension, is really evolved during the combination of sulphuric, hydrochloric, nitric, phosphoric, and acetic acids, with the fixed alkalies, and even with ammonia."†

(305.) As evaporation is a process continually going on from the surface of the ocean, land, lakes, and rivers, at all degrees of temperature, the result of its action must be very extensive. But *how* the disengagement of electricity is produced, either by the action of oxygen on the structure of living plants, or by the action of heat on water, is unknown, and will perhaps ever remain a secret of Nature. It is easy, however, to conceive how the electricity produced by these and other sources must vary in different climates, seasons and localities, and at different hights in the atmosphere.‡

(306.) It thus appears that the sources of electricity are found to be *evolved in every possible form of action.* It is excited by almost every substance in nature, by friction, which is a mechanical action; it is as readily evolved by chemical action, as you have just learned; as also in the cases of condensation and evaporation of liquids; and it has also been proved to be excited by vital action, as in the case of vegetation; and as the action of oxygen is the same in the animal as in the vegetable function, it is as likely that the respiration of animals produces electricity as that of vegetables. When the sources of this mysterious and subtle agent are thus so numerous and extensive, you need not only not be surprised at its extensive diffusion, but the universality of its presence indicates that its assistance is necessary to the promoting of every operation of Nature. Its identity in all cases is also proved by the fact, that though the means employed for its excitation are various, its mode of action is always the same. In every case of excitation, one body robs the other of a portion of its electricity, the former being *plus* or *positive*, the other *minus* or *negative* in its natural quantity. "The two species, or negative and positive electricity," says Dr. Bird, "exist in nature *combined*, forming a neutral combination (in an analogous manner to the two magnetic fluids) incapable of exerting any obvious physical actions on ponderable matter: by the process of friction, or other mechanical or chemical means, we decompose this neutral combination, the negative and positive elements separate, one adhering to the surface of the excited substance, the other to the rubber; hence in no case of electrical excitation can we obtain one kind of electricity without the other being simultaneously developed. We do not observe any free electricity on the surface of metallic bodies submitted to friction, in consequence of their so readily conducting electricity that the

* Leithead on Electricity.

† Bird's Elements of Natural Philosophy

‡ Forbes's Report on Meteorology, vol. i.

union of the negative and positive fluids takes place as rapidly as they are separated by the friction employed."*

(307.) The *natural state* of every body in regard to its electricity is thus in a *state of quiescence or equilibrium*, but this equilibrium is very easily disturbed, and then a series of actions supervene, which illustrate the peculiar agency of electricity, and continue until the equilibrium is again restored.

(308.) The *force* of the electrical agency seems to be somewhat in the proportion to the energy with which it is roused into action. Dr. Faraday states, that *one grain* of water "will require an electric current to be continued for $3\frac{3}{4}$ minutes of time to effect its decomposition; which current must be strong enough to retain a platina wire $\frac{1}{104}$ of an inch in thickness red-hot in the air during the whole time. "It will not be too much to say, that this necessary quantity of electricity is equal to a very powerful flash of lightning."† When it is considered that, during the fermentation and putrefaction of bodies on the surface of the earth, water is decomposed, and that to effect its decomposition such an amount of electric action as is here related is required to be excited, we can have no difficulty in imagining the great amount of electricity which must be derived from the various sources enumerated being constantly in operation.

(309.) In mentioning the subject of electricity, I will take the opportunity of expressing my opinion that the *electrometer* is a meteorological instrument of much greater utility to you than some of the instruments I have described; because it indicates, with a great degree of delicacy, the existence of free electricity in the air; and as electricity cannot exist in that state without producing some sort of action, it is satisfactory to have notice of its freedom, that its effects, if possible, may be anticipated. The best sort of electrometer is the "*condensing electroscope*:" it consists of a hollow glass sphere on a stand, inclosing through its top a glass tube, to the top of which is affixed a flat brass cap, and from the bottom of which are suspended two slips of gold-leaf. At the edge of the flat brass cap is screwed a circular brass plate, and another circular brass plate, so as to be parallel to the first, is inserted in a support fixed in a piece of wood moving in a groove of the stand which contains the whole apparatus. This is a very delicate instrument, and, to keep it in order, should be kept free of moisture and dust.

(310.) In regard to the *usual state* of the electricity in the atmosphere, it is generally believed that it is positive, and that it increases in quantity as we ascend. In Europe the observations of M. Schübler of Stuttgardt, intimate that the electricity of the precipitating fluids from the atmosphere is more frequently negative than positive, in the proportion of 155 : 100; but that the mean *intensity* of the positive electricity is greater than that of the negative in the ratio of 69 : 43; and that different layers or strata of the atmosphere, placed only at small distances from each other, are frequently found to be in different electric states.‡ It appears, also, from recent observations of M. Schübler, that the electricity of the air, in calm and serene weather, is constantly positive, but subject to two daily fluctuations. It is at its minimum at a little before sunrise; after which it gradually accumulates, till it reaches its first maximum a few hours afterward —at 8 A. M. in May; and then diminishes until it has descended to its second minimum. The second maximum occurs in the evening about two hours after sunset; and then diminishes, at first rapidly, and next in slower progression during the whole of the night, to present again on the follow-

* Bird's Elements of Natural Philosophy. † Faraday's New Researches, 8vo edition
‡ Forbes's Report on Meteorology, vol. I.

ing day the same oscillations. It is probable that the exact time of its increase and decrease is influenced by the seasons. The intensity increases from July to January, and then decreases; it is also much more intense in the winter, though longer in summer, and appears to increase as the cold increases.* These fluctuations may be observed throughout the year more easily in fine than in cloudy weather. "Among the causes modifying the electric state of the atmosphere," observes Dr. Bird, "must be ranked its hygrometric state, as well as probably the nature of the effluvia which may become volatilized in any given locality. Thus, Saussure has observed that its intensity is much more considerable in elevated and isolated places than in narrow and confined situations; it is nearly absent in houses, under lofty trees, in narrow courts and alleys, and in inclosed places. In some places the most intensely electric state of the atmosphere appears to be that in which large clouds or dense fogs are suspended in the air at short distances above the surface of the earth; these appear to act as conductors of the electricity from the upper regions. Cavallo ascertained, from a set of experiments performed at Islington in 1776, that the air always contains free *positive* electricity, except when influenced by heavy clouds near the zenith. This electricity he found to be strongest in fogs and during frosty weather, but weakest in hot weather, and just previous to a shower of rain; and to increase in proportion as the instrument used is raised to a greater elevation. This, indeed, necessarily happens," continues Dr. Bird, "for as the earth's surface is, *cœteris paribus*, always negatively electrified, a continual but gradual combination of its electricity with that of the air is constantly taking place at its surface, so that no free positive electricity can be detected within 4 feet of the surface of the earth."†

(311.) A comparative view of the fluctuations of the barometer and electrometer may tend to show that in their mode of action all the physical agencies may be governed by the same law. The mean results of many observations by various philosophers are as follows:

	1st Maximum.	1st Minimum.	2d Maximum.	2d Minimum.
Density	10 A. M.	4–5 P. M.	10–11 P. M.	4–5 A. M.
Electricity	8–9 A. M.	4 P. M.	9 P. M.	6 A. M.

(312.) These are all the general remarks which are called for at present on the subject of atmospherical electricity. As electrical phenomena exhibit themselves most actively in summer, observations on particular ones will then be more in season than in winter; and the only electrical excitation that is generally witnessed in winter is the *aurora borealis* or northern lights, or "merry dancers," as they are vulgarly called. It mostly occurs in the northern extremity of the northern hemisphere of the globe, where it gives almost constant light during the absence of the sun. So intense is this radiance, that a book may be read by it, and it thus confers a great blessing on the inhabitants of the Arctic Regions, at a time when they are benighted. The aurora borealis seems to consist of two varieties; one a luminous, quiet light in the northern horizon, gleaming most frequently behind a dense stratum of cloud; and the other of vivid coruscations of almost white light, of a sufficient transparency to allow the transmission of the light of the fixed stars. They are sometimes colored yellow, green, red, and of a dusky hue. The coruscations are generally short, and confined to the proximity of the northern horizon; but occasionally they reach the zenith, and even extend to the opposite horizon; their direction being from N. W. to S. E. It seems now undeniable, that the aurora borealis

* Journal of Science and the Arts, No. IV.

† Bird's Elements of Natural Philosophy.

frequently exercises a most marked action on the magnetic needle; thus affording another proof of the identity of the magnetic and electric agencies.

(313.) It is not yet a settled point among philosophers, whether the aurora borealis occurs at the highest part of the atmosphere, or near the earth. Mr. Cavendish considered it probable, that it usually occurs at an elevation of 71 miles above the earth's surface, at which elevation the air must be but $\frac{1}{148567}$ time the density of that at the surface of the earth, a degree of rarefaction far above that afforded by our best constructed air-pumps. Dr. Dalton conceives, from trigonometrical measurements made by him of auroral arches, that their hight is 100 miles above the earth's surface. His most satisfactory measurement was made from that of the 29th March, 1826. As the peculiar appearance of aurora and its coruscations precisely resemble the phenomena which we are enabled to produce artificially by discharges of electricity between two bodies in a receiver through a medium of highly rarefied air, the opinion of Lieut. Morrison, R. N. of Cheltenham, a profound astronomer and meteorologist, is deserving of attention, as regards the position of the aurora at the time of its formation. He states that long, light clouds ranging themselves in the meridian line in the day, at night take a fleecy, aurora-like character. "I believe," he says, "that these clouds are formed by the discharges and currents of electricity, which, when they are *more decided*, produce aurora." Mr. Leithead conjectures that the aurora becomes "visible to the inhabitants of the earth upon their entering our atmosphere."* If these conjectures are at all correct, the aurora *cannot be seen beyond our atmosphere*, and therefore cannot exhibit itself at the hight of 100 miles, as supposed by Dr. Dalton, since the hight of the atmosphere is only acknowledged to be from 40 to 50 miles. This view of the hight of the aurora somewhat corroborates that held by Rev. Dr. Farqharson, Alford, Aberdeenshire, and which has been strongly supported by Professor Jameson.†

(314.) There are other atmospherical phenomena, whose various aspects indicate changes of the weather, and which, although of rarer occurrence than the clouds or electricity, are yet deserving of attention when they appear. These are, Halos around the disks of the sun and moon; Coronæ or *broughs*, covering their faces; Parhelia, or mock suns; Falling Stars; Fire-balls; and the Rainbow. Of these, the halo and corona only appear in winter; the others will be noticed in the course of the respective seasons in which they appear.

(315.) A *halo* is an extensive luminous ring, including a circular area, in the center of which the sun or moon appears. It is formed by the intervention of a cloud between the spectator and the sun or moon. This cloud is generally the denser kind of *cirro-stratus*, the refraction and reflection of the rays of the sun or moon at definite angles through and upon which, is the cause of the luminous phenomenon. The breadth of the ring of a halo is caused by a number of rays being refracted at somewhat different angles, otherwise the breadth of the ring would equal only the breadth of one ray. Mr. Forster has demonstrated mathematically the angle of refraction, which is equal to the angle subtended by the semidiameter of the halo.‡ Halos may be double and triple; and there is one, which Mr. Forster denominates a *discoid* halo, which constitutes the boundary of a large corona, and is generally of less diameter than usual, and often colored with the tints of the rainbow. "A beautiful one appeared at Clapton on the 22d December, 1809, about midnight, during the passage of a

* Leithead on Electricity. † Encyclopædia Britannica, 7th edition, art. *Aurora Borealis*

‡ Forster's Researches into Atmospherical Phenomena.

cirro-stratus cloud before the moon."* Halos are usually pretty correct circles, though they have been observed of a somewhat oval shape; and they are generally also colorless, though they sometimes display faint colors of the rainbow. They are most frequently seen around the moon, and acquire the appellation of *lunar* or *solar* halos, as they happen to accompany the particular luminary.

(316.) The *corona* or *brough* occurs when the sun or moon is seen through a thin *cirro-stratus* cloud, the portion of the cloud more immediately around the sun or moon appearing much lighter than the rest. Coronæ are double, triple, and even quadruple, according to the state of the intervening vapors. They are caused by a similar refractive power in vapor as the halo; and are generally faintly colored at their edges. Their diameter seldom exceeds 10°. A halo frequently encircles the moon, when a small corona is more immediately around it.

(317.) Hitherto I have said nothing of rain, snow, wind, or hail—phenomena which materially affect the operations of the farmer. Strictly speaking, they are not the cause, but only the effects, of other phenomena; and on that account, I have purposely refrained alluding to them, until you should have become somewhat acquainted with the nature of the agencies which produce them. Having heard of these, I shall now proceed to examine particularly the familiar phenomena of *rain*, *snow* and *wind*. Rain and wind being common to all the seasons, it will be necessary to enter at once into a general explanation of both. Snow is peculiar to winter, and will not again require to be alluded to. And hail will form a topic of remark in summer.

(318.) You must be so well acquainted with the phenomenon of *rain*, that no specific definition of it is here required to be given. It should, however, be borne in mind, that the phenomenon has various aspects, and the variety indicates the peculiar state of the atmosphere at the time of its occurrence. Rain falls at times in large drops, at others in small, and sometimes in a thick or thin drizzle; but in all these states, it consists of the descent of water in drops from the atmosphere to the earth. In reflecting on this phenomenon, how is it, (you may ask yourselves) that the air can possibly support *drops of water*, however minute? The air cannot support so dense a substance as *water*; and it is its inability to do so, that causes the water to fall to the ground. The air, however, can support vapor, the aggregation of the particles of which constitutes rain or water. Vapor is formed by the force of the heat of the sun's rays upon the surface of land, sea, lakes and rivers; and from its easy ascent into the atmosphere, it is clear that water is rendered lighter than air by heat, and, of course, vastly lighter than itself. The weight of one cubic inch of distilled water (with the barometer at 30 inches, and the thermometer at 62° Fahrenheit) is 252·458 grains; that of 1 cubic inch of air is 0·3049 of a grain; of course, vapor must be lighter than this last figure. Heat has effected this lightness by rendering vapor highly elastic; and it is not improbable that it is electricity which maintains the elasticity, after the vapor has been carried away beyond the influence of its generating heat, and there keeps it in mixture with the air. The whole subject of evaporation is instructive, and will receive our attention in summer, when it presents itself in the most active condition to our view, and is intimately connected with the phenomenon of dew.

(319.) The *quantity of vapor in the atmosphere is variable.* This Table shows the weight in grains of a cubic foot of vapor, at different temperatures, from 0° to 95° Fahrenheit:

* Forster's Researches into Atmospherical Phenomena.

Temperature.	Weight in grains.	Temperature.	Weight in grains.	Temperature.	Weight in grains.	Temperature.	Weight in grains.
0	0·856	24	1·961	48	4·279	72	8·924
1	0·892	22	2·028	49	4·407	73	9·199
2	0·928	26	2·096	50	4·535	74	9·484
3	0·963	27	2·163	51	4·684	75	9·780
4	0·999	28	2·229	52	4·832	76	10·107
5	1·034	29	2·295	53	5·003	77	10·387
6	1·069	30	2·361	54	5·173	78	10·699
7	1·104	31	2·451	55	5·342	79	11·016
8	1·139	32	2·539	56	5·511	80	11·333
9	1·173	33	2·630	57	5·679	81	11·665
10	1·208	34	2·717	58	5·868	82	12·005
11	1·254	35	2·805	59	6·046	83	12·354
12	1·308	36	2·892	60	6·222	84	12·713
13	1·359	37	2·979	61	6·399	85	13·081
14	1·405	38	3·066	62	6·57[illegible]	86	13·458
15	1·451	39	3·153	63	6·794	87	13·877
16	1·497	40	3·239	64	7·013	88	14·230
17	1·541	41	3·371	65	7·230	89	14·613
18	1·586	42	3·502	66	7·447	90	15·005
19	1·631	43	3·633	67	7·662	91	15·432
20	1·688	44	3·763	68	7·899	92	15·786
21	1·757	45	3·893	69	8·135	93	16·186
22	1·825	46	4·022	70	8·392	94	16·593
23	1·893	47	4·151	71	8·658	95	17·009

Dr. Dalton found that the force of vapor in the torrid zone varies from 0·6 of an inch to 1 inch of mercury. In Britain it seldom amounts to 0·5 of an inch, but is sometimes as great as 0·5 of an inch, in summer; whereas, in winter, it is often as low as 0·1 of an inch of mercury. These facts would enable us to ascertain the absolute quantity of vapor contained in the atmosphere at any given time, provided we were certain that the density and elasticity of vapors follow precisely the same law as that of gases, as is extremely probable to be the case. If so, the vapor will vary from $\frac{1}{60}$ to $\frac{1}{100}$ part of the atmosphere. Dalton supposes that the medium quantity of vapor in the atmosphere may amount to $\frac{1}{70}$ of its bulk.*

(320.) The *theory* propounded by Dr. Hutton, that rain occurs from the mingling together of great beds of air of unequal temperatures differently stored with moisture, is that which was adopted by Dalton, Leslie, and others, and is the current one, having been illustrated and strengthened by the clearer views of the nature of deposition which we now possess.

(321.) On the connection of rain with the *fall of the barometer*, Mr. Meikle has shown that the change of pressure may be a cause as well as an effect; for the expansion of air accompanying diminished pressure, being productive of cold, diminishes the elasticity of the existing vapor, and causes a deposition.†

(322.) M. Arago has traced the progress of *decrease* in the annual amount of the fall of rain *from the equator to the poles*; and these are the results obtained by various observers at the respective places:

Coast of Malabar, in	Lat. 11° 30′	N. the quantity is	135·5 inches.
At Grenada, Antilles	12°	..	126· ..
At Cape François, St. Domingo	19° 46′	..	120· ..
At Calcutta	22° 23′	..	81· ..
At Rome	41° 54′	..	39· ..
In England	53°	..	32· ..
At St. Petersburgh	59° 16′	..	16· ..
At Ulea	65° 30′	..	13·5 ..

On the other hand, the number of rainy days *increases from the equator to the poles*, according to the observations of M. Cotte. Thus.

* Philosophical Magazine, vol. xxiii. p. 353.

† Royal Institution Journa.

From N. lat. 12° to 43°,	there are 78	rainy days.
.. .. 43° to 46°,	.. 103	..
.. .. 46° to 50°,	.. 134	..
.. .. 50° to 60°,	.. 161	..

(323.) There is a great variation in the quantity of rain that falls in the same latitude on *the different sides of the same continent*, and particularly of the same *island*. Thus, to confine the instances to our own island, the mean fall of rain at Edinburgh, on the east coast, is 26 inches; and at Glasgow, on the west coast, in nearly the same latitude, the amount is 40 inches. At North Shields, on the east coast, the amount is 25 inches; while at Coniston in Lancashire, in nearly the same latitude on the west coast, it is as great as 85 inches.*

(324.) A remarkable variation takes place in the fall of rain at *different hights;* the quantity of rain that falls on high ground exceeding that at the level of the sea. This fact may be easily explained by the influence of a hilly country retaining clouds and vapor. At Lancaster, on the coast, the quantity that falls is 39 inches; and at Easthwaite, among the mountains in the same county, the amount is 86 inches. By a comparison of the registers at Geneva and the convent of the Great St. Bernard, it appears that at the former place, by a mean of 32 years, the annual fall of of rain is 30·70 inches; while at the latter, by a mean of 12 years, it is 60·05 inches. Dr. Dalton clearly points out the influence of hot currents of air ascending along the surface of the ground into the colder strata which rest upon a mountainous country. The consequence is, that although neither the hot nor the cold air was accompanied with more moisture than could separately be maintained in an elastic state, yet when the mixture takes place, the arithmetical mean of the quantities of vapor cannot be supported in an elastic state at an arithmetical mean of the temperatures; since the weights of vapor which can exist in a given space increase nearly in a geometrical ratio, while the temperatures follow an arithmetical one.† But the amount of rain at stations abruptly elevated above the surface of the earth, diminishes as we ascend. For example, at Kinfauns Castle, the seat of Lord Gray, on the Tay, in Perthshire, by a mean of 5 years, 22·66 inches of rain fell; while on a hill in the immediate neighborhood, 600 feet higher, no less than 41·49 inches were collected, by a mean of the same period. This is an instance of a high elevation rising pretty rapidly above the castle, but in a natural manner; and it is adduced as a contrast with an artificial elevation of a rain-gauge at the observatory at Paris, when the rain that fell on the town, at a vertical hight of 28 metres (rather more than as many yards), was 50·47 inches, while, according to the observation of M. Arago, it was 56·37 inches in the court below.‡

(325.) The variation in the amount of rain in the *seasons* follows, in a great measure, the same law as that propounded by Dalton in reference to the hights of mountains. The greatest *quantity* of rain falls in autumn, and the least in winter. Thus, according to M. Flaugergues, taking the mean amount as 1—

In winter, there falls	0·1937 inches,	including	December, January and February
In spring, ..	0·2217 ..	..	March, April and May.
In summer, ..	0·2001 ..	..	June, July and August.
In autumn, ..	0·3845 ..	..	September, October and November

It may be useful to give the proportional results of each month. Again, taking the mean amount of the year as 1, the proportional result for

* Table of the quantity of Rain that falls in different parts of Great Britain. By Mr. Joseph Atkinson, Harraby, near Carlisle.
† Manchester Memoirs, New Series vol v.
‡ Forbes's Report on Meteorology, vol. i.

(332)

January, is	0·0716	July	0·0544
February	0·0541	August	0·0679
March	0·0557	September	0·1236
April	0·0802	October	0·1370
May	0·0847	November	0·1250
June	0·0765	December	0·0693

As M. Flaugergues observes, the maximum belongs to October and the minimum to February, and May comes nearest to the mean of 40 years.* Taking these proportional results by the months which constitute the seasons of the agricultural year as I have arranged them, the mean of the seasons will be respectively thus :

Winter,	November	0·1250	Summer,	May	0·0847
	December	0·0693		June	0·0765
	January	0·0716		July	0·0544
Total		0·2659	Total		0·2156
Spring,	February	0·0541	Autumn,	August	0·0679
	March	0·0557		September	0·1236
	April	0·0802		October	0·1370
Total		0·1900	Total		0·3285

This method of division still gives the maximum of rain to autumn, though it transfers the minimum from the winter to the spring ; which, as I think, approaches nearer to the truth in reference to Scotland than the conclusions of M. Flaugergues, which specially apply to France.

(326.) The last table but one gives the proportional amount of rain that fell, in a mean of 40 years, in each month. It may be useful to know the mean *number of rainy days in each of the months.* They are these :

In January	14·4 days,	In July	16·1 days.
February	15·8 ..	August	16·3 ..
March	12·7 ..	September	12·3 ..
April	14·0 ..	October	16·2 ..
May	15·8 ..	November	15·0 ..
June	11·8 ..	December	17·7 ..

These tables show that though the *number* of rainy days is nearly equal in the vernal and autumnal equinoxes, the *quantity* of rain that falls in the autumn is nearly double of that in spring. If this last table is arranged according to the months of the agricultural seasons, the number of rainy days in each season will stand thus :

In Winter,	November	15·0 days.	In Summer,	May	15·8 days.
	December	17·7 ..		June	11·8 ..
	January	14·4 ..		July	16·1 ..
Total		47·1 days.	Total		43·7 days.
In Spring,	February	15·8 days.	In Autumn,	August	16·3 days.
	March	12·7 ..		September	12·3 ..
	April	14·0 ..		October	16·2 ..
Total		42·5 days.	Total		44·8 days.

In all 178·1 days of rain. This arrangement shows that the greatest number of rainy days is in the agricultural winter, and the least number in the spring, which seems to agree with experience.

(327.) Mr. Howard remarks, that, on an average of years, it rains every other day ; and, by a mean of 40 years at Viviers, M. Flaugergues found 98 days of rain throughout the year.†

(328.) With regard to the question, Whether *more rain falls in the night than in the day?* Mr. Howard's statement bears, that of 21·94 inches—a mean of 31 lunar months—rain fell in the day to the amount of 8·67 inches, and in the night to 13·27 inches. Dr. Dalton also says, that more rain falls when the sun is under the horizon than when it is above it.‡

* Encyclopædia Metropolitana, art. *Meteorology.* † Ibid. ‡ Ibid.

(329.) It has not been ascertained whether, on the *whole amount over the globe, rain is increasing or diminishing in quantity.* As M. Arago justly observes, it is very difficult to know how many years of observations are necessary to get a mean value of the fall of rain, the amount being extremely variable. There are, no doubt, several causes which may tend to change the amount of rain in any particular spot, without forming part of any general law, such as the destruction or forming of forests, the inclosure and drainage of land, and the increase of habitations. M. Arago has shown that the fall of rain at Paris has not sensibly altered for 130 years, and that although an *increase* was supposed to have been proved at Milan, by observations for 54 years, yet the extremes of the annual results between 1791 and 1817 were 24·7 and 58·9 inches. The observations of M. Flaugergues, at Viviers, establish an increase there in 40 years. The number of rainy days throughout the year is 98, but dividing the 40 years into decades, the number sensibly increases. Thus—

From 1778 to 1787, there were 830 days.
1788 to 1797, .. 947 ..
1798 to 1807, .. 1062 ..
1808 to 1817, .. 1082 ..

But this result must arise from local circumstances, as at Marseilles there has been a striking *decrease* in 50 years.

(330.) Notwithstanding the *enormous annual fall of rain at the equator*, particular instances of a great depth of rain in a short time have occasionally occurred in Europe, which probably have seldom been equaled in any other part of the globe. At Geneva, on the 25th October, 1822, there fell 30 inches of rain in one day. At Joyeuse, according to M. Arago, on the 9th October, 1827, there fell 31 inches of rain in 22 hours.* With regard to remarkable variations in the quantity of rain in different places, among the Andes it is said to rain perpetually; whereas in Peru, as Ulloa affirms, it never rains, but that for a part of the year the atmosphere is obscured by thick fogs called *garuas*. In Egypt it hardly ever rains at all, and in some parts of Arabia it seldom rains more than two or three times in as many years, but the dews are heavy, and refresh the soil, and supply with moisture the few plants which grow in those sunny regions.

(331.) According to a statement of observations by Mr. Howard, there appears a *relation to exist betwixt the winds and the annual amount of rain.* This is his statement:

Year.	Wind.				Calm days.	Annual rain in inches.
	N. E.	E. S.	S W	W. N.		
1807	61	34	113	114	43	20·14
1808	82	38	108	103	35	23·24
1809	68	50	123	91	33	25·28
1810	81	72	78	83	41	28·07
1811	58	59	119	93	36	24·64
1812	82	66	93	91	34	27·24
1813	76	53	92	124	20	23·56
1814	96	65	91	96	17	26·07
1815	68	36	121	107	33	21·20
1816	64	66	106	102	28	32·37
	74	54	105	100	32	25·18

The remarks which this statement seems to warrant are, that in regard to the E. winds, in the dry year 1807, the class of N.—E. winds is nearly double of the class of E.—S. winds; in 1815, the next driest year, is the same result; and in 1808, the next driest to that, the result is rather more than double. Still farther in regard to E. winds, in the wettest year,

* Forbes's Report on Meteorology, vol I.

(334)

1816, the class of E.—S. winds exceeds that of N. E.; in 1814 they were $\frac{3}{5}$ of the latter; in 1812, $\frac{3}{4}$, and 1810, $\frac{7}{8}$. With regard to the class of W. winds, the class of W. N. winds falls off gradually from 1807 to 1810 inclusive, while the annual amount of rain increases from year to year, and in three of the six remaining years the amount is drier than the average in the dry years, and wetter than the wet ones.

(332.) Mr. Howard says, that 1 year in every 5 in this country may be expected to be extremely dry, and 1 in 10 extremely wet.

(333.) The mean annual amount of rain and dew for England and Wales, according to the estimate of Dr. Dalton, is 36 inches. The mean quantity of rain falling in 147 places, situated between north lat. 11° and 60°, according to Cotte, is 34·7 inches. If the mean fall over the globe be taken at 34 inches, it will, perhaps, not be far from the truth.*

(334.) The *influence of the lunar periods on the amount of rain* deserves attention. Professor Forbes believes that there is some real connection between the lunar phases and the weather. M. Flaugergues, who has observed the weather at Viviers with the greatest assiduity for a quarter of a century, marked the number of rainy days corresponding with the lunar phases, and found them at a maximum at the first quarter, and a minimum at the last.

(335.) It almost always happens that rain *brings down foreign matter from the air*. It is known that the farina of plants has been carried as far as 30 or 40 miles, and the ashes of volcanoes have been carried more than 200 miles. We can conceive that when the magnitude of the particles of dry substances is so reduced as to render them incapable of falling in any given velocity, that their descent may be overcome by a very slight current of the air; but even in still air a sphere of water of only the almost inconceivable size of $\frac{1}{800000}$ part of an inch in diameter falls 1 inch in a second, and yet particles of mist must be much larger than this, otherwise they could not be visible as separate drops; the least drop of water that is discoverable by the naked eye falls with a velocity of 1 foot in the

[* It is said that, on an average, half as much more rain falls in England than on the Continent of Europe. In Ireland, says Doctor KANE, a very able and profound writer on the Industrial Resources of that country, there is probably not more rain than in England, but there is *more damp*. Long since, Arthur Young, he says, noticed the difficulty of drying agricultural produce in Ireland, and to this humidity he attributed the rapid vegetation which clothes that island with natural herbage, even where there is scarcely a trace of soil, and causes it to be likened to an "emerald" set in the ring of the sea. Like causes produce like effects in our country—hence on the hills and small mountains about Lebanon, N. Y. the Messrs. Tilden carry through the driest summers their flock of 1,000 of the best Saxonies in the finest condition; those hills are ever green. The average quantity that falls over the entire surface of Ireland is put down at 36 inches. Thus, if all that falls in the year were collected at one time, it would cover the whole Island to the depth of 3 feet; and as the area of Ireland amounts to 80,208,271 square acres, containing 100,712,631,640 square yards, there are this number of square yards of water precipitated on the island in every year.

For youthful readers we may be allowed in this place to transcribe from the admirable author above mentioned the following brief and simple explanation of the origin and formation of clouds, rain and rivers, and "water power:"

"The land being placed on the surface of our globe at a level superior to that of the ocean, by which its coasts are washed, there is produced continually, by atmospherical conditions, a circulation of the mass of water, which, evaporating from the surface, ascends as vapor to the higher and colder regions of the air, where it is condensed into clouds. These float until the electrical condition which characterizes their peculiar molecular state being dissipated, they fall as rain, as hail, or snow, and the water thus regaining the solid or liquid form, tends continually by its gravity, to a lower level, until it gains the general mass of ocean from whence it had been originally derived. The rain or snow thus falling into the interior and elevated districts of country, forms at first rivulets, then streams, finally rivers; and the force of the descending water is capable of giving motion to machinery: it is the source best known, and most simply applicable, of *water power*."
Ed. Farm. Lib.]

second, when the air is still. Although it is probable that the resistance opposed to the descent of small bodies in air, may be considerably greater than would be expected from calculation, still the wonder is how they are supported for any length of time.* In this difficulty there is much inclination to call in the aid of electricity to account for the phenomenon. Mr. Leithead accounts for it in this way: "When the earth is positive and the atmosphere negative, the electric fluid, in endeavoring to restore its equilibrium, would cause a motion among the particles of the air in a direction from the earth toward the higher region of the atmosphere; for the air being a very imperfect conductor, the particles near the earth's surface can only convey electricity to the more remote particles by such a motion. This would, in effect, partly *diminish* the downward pressure of the air, which is due to its actual density;" and, in doing this, might it not, at the same time, counteract in some degree the gravity of any substance in the air by surrounding it with an electrical atmosphere? "When, on the contrary," continues Mr. Leithead, "the earth is negative and the air positive, this motion of the particles will be reversed; thus increasing the pressure toward the earth, and producing the same effect as if the air had actually *increased in density*;"† and would it not thereby be more capable of supporting any foreign body in it?

(336.) Rain falls at all seasons, but *snow* only in winter, and it is just frozen rain; whenever, therefore, there are symptoms of rain, snow may be expected if the temperature of the air is sufficiently low to freeze vapor. Vapor is supposed to be frozen into snow at the moment it is collapsing into drops to form rain, for we cannot suppose that clouds of snow can float about the atmosphere any more than clouds of rain. Snow is a beautifully crystalized substance when it falls to the ground, and it is probable that it never falls from a great hight, otherwise its fine crystalline configurations could not be preserved.

(337.) The *forms of snow* have been arranged into five orders. 1. The *lamellar*, which is again divided into the *stelliform, regular hexagons, aggregation of hexagons*, and *combination of hexagons* with radii, or spines and projecting angles. 2. Another form is the *lamellar* or *spherical nucleus* with spinous ramifications in different places. 3. Fine *spiculæ* or 6-sided prisms. 4. *Hexagonal pyramids*. 5. *Spiculæ*, having one or both *extremities* affixed to the *center* of a *lamellar* crystal. There are numerous varieties of forms of each class.‡ All the forms of crystals of snow afford most interesting objects for the microscope, and when perfect no objects in nature are more beautifully and delicately formed. The crystals ramify from a center, or unite with one another under the invariable angle of 60°, or its complemental angle of 120°. The lamellated crystals fall in calm weather, and in heavy flakes, and are evidently precipitated from a low elevation. The spiculæ of 6-sided prisms occur in heavy drifts of snow accompanied with wind and intense cold. They are formed at a considerable elevation; and they are so fine as to pass through the minutest chinks in houses, and so hard and firm that they may be poured like sand from one hand into another, with a jingling sound, and without the risk of being melted. In this country they are most frequently accompanied with one of the varieties of the lamellar crystals, which meet their fall at a lower elevation; but in mountainous countries, and especially above the line of perpetual snow, they constitute the greatest bulk of the snow, where they are ready at the surface to be blown about with

* Polehampton's Gallery of Nature and Art, vol. iv.

† Leithead on Electricity, p. 374. This explanation Mr. Leithead also gives to account for the changes in the density of the atmosphere, as indicated by the oscillations of the barometer.

‡ Encyclopædia Metropolitana, art. *Meteorology*.

the least agitation of the air, and lifted up in dense clouds by gusts of wind, and precipitated suddenly on the unwary traveler like a sand-drift of the torrid zone. These spiculæ feel exceedingly sharp when driven by the wind against the face, as I have experienced on the Alps. How powerless is man when overtaken in such a snow-storm, as

> ————"down he sinks
> Beneath the shelter of the shapeless drift,
> Thinking o'er all the bitterness of death!"*

The other forms of snow are more rare.

(338.) All other things being equal, Professor Leslie supposes that a flake of snow, taken at 9 times more expanded than water, descends three times as slow.

(339.) From the moment snow alights on the ground it begins to undergo certain changes, which usually end in a more solid crystalization than it originally possessed. The adhesive property of snow arises from its needly crystaline texture, aided by a degree of attendant moisture which afterward freezes in the mass. Sometimes, when a strong wind sweeps over a surface of snow, portions of it are raised by its power, and, passing on with the breeze under a diminished temperature, become crystalized, and, by attrition, assume globular forms. Mr. Howard describes having seen these snow-balls, as they may be termed, in January, 1814; and Mr. Patrick Shirreff, when at Mungoswells, in East Lothian, observed the like phenomenon in February, 1830.† I observed the same phenomenon in Forfarshire, in the great snow-storm of February, 1823.

(340.) During the descent of snow, the *thermometer sometimes rises*, and the *barometer usually falls.* Snow has the effect of retaining the temperature of the ground at what it was when the snow fell. It is this property which maintains the warmer temperature of the ground, and sustains the life of plants during the severe rigors of winter, in the Arctic Regions, where the snow falls suddenly, after the warmth of summer; and it is the same property which supplies water to rivers in winter, from under the perpetual snows of the alpine mountains. While air, above snow, may be 38° below zero, the ground below will only be at zero.‡ Hence the fine, healthy, green color of young wheat and young grass, after the snow has melted off them in spring.

(341.) In melting, 27 inches of snow give 3 inches of water. Rain and snow-water are the *softest* natural waters for domestic purposes; and are also the purest that can be obtained from natural sources, provided they are procured either before reaching the ground, or from newly fallen snow. Nevertheless, they are impregnated with oxygen, nitrogen, and carbonic acid, especially with a considerable quantity of oxygen; and rain-water and dew contain nearly as much air as they can absorb.‖ Liebig maintains that both rain and snow-water contain ammonia.§

(342.) Snow reflects beautifully blue and pink shades at sunset, as is observed with admiration on the Alps of Switzerland. It also reflects so much light from its surface as to render traveling at night a cheerful occupation; and in some countries, as in Russia and Canada, it forms a delightful highway when frozen.

(343.) *Hoar-frost* is defined to be frozen dew. This is not quite a correct definition; for dew is sometimes frozen, especially in spring, into globules of ice which do not at all resemble hoar-frost—this latter substance being beautifully and as regularly crystalized as snow. The formation of hoar-frost is always attended with a considerable degree of cold.

* Thomson. † Encyclopædia Metropolitana, art. *Meteorology.* ‡ Phillip's Facts.
‖ Reid's Chemistry of Nature. § Liebig's Organic Chemistry.

because it is preceded by a great radiation of heat and vapor from the earth, and the phenomenon is the more perfect the warmer the day and the clearer the night have been. In the country, hoar-frost is of most frequent occurrence in the autumnal months and in winter, in such places as have little snow or continued frost on the average of seasons; and this greatly from great radiation of heat and vapor, at those seasons occasioned by a suspension of vegetable action, which admits of little absorption of moisture for vegetable purposes.*

(344.) Dr. Farquharson, Alford, Aberdeenshire, has paid great attention to the subject of hoar-frost or rime, which frequently injures the crops in the northern portion of our island long before they are ripe. The results of his observations are very instructive. 1. He has observed that the mean temperature of the day and night at which injurious hoar-frosts may occur, may be, relatively to the freezing-point, very high. Thus, on the nights of the 29th and 31st August, 1840, the leaves of potatoes were injured, while the lowest temperatures of those nights, as indicated by a self-registering thermometer, were as high as 41° and 39° respectively.—2. Hoar-frost, at the time of a high daily mean temperature, takes place only during calm. A very slight, steady breeze will quickly melt away frosty rime. 3. The air is always unclouded, or nearly all of it so, at the time of hoar-frost. So incompatible is hoar-frost with a clouded state of the atmosphere, that on many occasions, when a white frosty rime has been formed in the earlier part of the night, on the formation of a close cloud at a later part, it has melted off before the rising of the sun. 4. Hoar-frosts most frequently happen with the mercury in the barometer at a high point and rising, and with the hygrometer at comparative dryness for the temperature and season; but there are striking exceptions to these rules. On the morning of the 15th September, 1840, a very injurious frost occurred, with a low and falling barometric column, and with a damp atmosphere. 5. In general, low and flat lands in the bottom of valleys, and grounds that are in land-locked hollows, suffer most from hoar-frost, while all sloping lands, and open uplands, escape injury. But it is not their relative elevation above the sea, independently of the freedom of their exposure, that is the source of safety to the uplands; for, provided they are inclosed by higher lands, without any wide, open descent from them on some side or other, they suffer more, under other equal circumstances, than similar lands of less altitude. 6. A very slight inclination of the surface of the ground is generally quite protective of the crops on it from injury by hoar-frost, from which flat and hollow places suffer at the time great injury. But a similar slope downward in the bottom of a narrow, descending hollow does not save the crop in the bottom of it, although those on its side-banks higher up may be safe. 7. An impediment of no great hight on the surface of the slope, such as a stone-wall fence, causes damage immediately above it, extending upward proportionally to the hight of the impediment. A still loftier impediment, like a closely-planted and tall wood or belt of trees, across the descent, or at the bottom of sloping land, causes the damage to extend on it much more. 8. Rivers have a bad repute as the cause of hoar-frosts in their neighborhood, but the general opinion regarding their evil influence is altogether erroneous; the protective effect of *running* water, such as waterfalls from mill-sluices, on pieces of potatoes, when others in like low situations are blackened by frost, is an illustration which can be referred to. 9. The severity of the injury by hoar-frost is much influenced by the wetness or dryness of the soil at the

* Mudie's World.

place; and this is exemplified in potatoes growing on haugh-lands, by the sides of rivers. These lands are generally dry, but bars of clay sometimes intersect the dry portions, over which the land is comparatively damp.—Hoar-frost will affect the crop growing upon these bars of clay, while that on the dry soil will escape injury; and the explanation of this is quite easy. The mean temperature of the damp lands is lower than that of the dry, and, on a diminution of the temperature during frost, it sooner gets down to the freezing point, as it has less to diminish before reaching it.—10. Hoar-frost produces peculiar currents in the atmosphere. On flat lands, and in land-locked hollows, there are no currents that are at all sensible to the feelings; but on the sloping lands, during hoar-frosts, there is rarely absent a very sensible and steady, although generally only feeble, current toward the most direct descent of the slope. The current is produced in this way. The cold first takes place on the surface of the ground, and the lower stratum of air becoming cooled, descends to a lower temperature than that of the air immediately above, in contact with it. By its cooling, the lower stratum acquires a greater density, and cannot rest on an inclined plane, but descends to the valley; its place at the summit of the slope being supplied by warmer air from above, which prevents it from getting so low as the freezing temperature. On the flat ground below, the cool air accumulates, and commits injury, while the warmer current down the slope does none; but should the mean temperature of the day and night be already very low before the calm of the evening sets in, the whole air is so cooled down as to prevent any current down the slope. Injury is then effected both on the slope and the low ground; and hence the capricious nature of hoar-frost may be accounted for.*

(345.) *Frost* has been represented to arise from the absence of heat; but it is more, for it also implies an absence of moisture. Sir Richard Phillips defines cold to be "the mere absence of the motion of the atoms called heat, or the abstraction of it by evaporation of atoms, so as to convey away the motion, or by the juxtaposition of bodies susceptible of motion. Cold and heat are mere relations of fixity and motion in the atoms of bodies."† This definition of heat implies that it is a mere property of matter—a point not yet settled by philosophers; but there is no doubt that, by motion, heat is evolved, and cold is generally attended by stillness or cessation of motion.

(346.) Frost generally originates in the upper portions of the atmosphere, it is supposed, by the expansion of the air carrying off the existing heat, and making it susceptible of acquiring more. What the cause of the expansion may be, when no visible change has taken place, in the mean time, in the ordinary action of the solar rays, may not be obvious to a spectator on the ground; but it is known, from the experiments of Lenz, that electricity is as capable of producing cold as heat, to the degree of freezing water rapidly.‡

(347.) The most intense frosts in this country never penetrate more than one foot into the ground, on account of the excessive dryness occasioned in it by the frost itself withdrawing the moisture for it to act upon. Frost cannot penetrate through a thick covering of snow, or below a sheet of ice.

(348.) *Ice* is water in a solid state, superinduced by the agency of frost. Though a solid, it is not a compact substance, but contains large interstices filled with air or other substances that may have been floating on the surface of the water. Ice is an aggregation of crystals subtending with one another the angles of 60° and 120°. It is quickly formed in shallow, but

* Prize Essays of the Highland and Agricultural Society, vol. xiv.

† Phillips's Facts.

‡ Bird's Elements of Natural Philosophy.

takes a long time to form in deep water, and it cannot become very thick in the lower latitudes of the globe, from want of time and intensity of the frost. By 11 years' observations at the Observatory at Paris, there were only 58 days of frost throughout the year, which is too short and too desultory a period to freeze *deep* water in that latitude.

(349.) The freezing of water is effected by frost in this manner. The upper film of water in contact with the air becomes cooled down, and when it reaches 39°.39 it is at its densest state, and of course sinks to the bottom through the less dense body of water below it. The next film of water, which is now uppermost, undergoes the same condensation, and in this way does film after film in contact with the air descend toward the bottom, until the whole body of water becomes equally dense at the temperature of 39°.39. When this vertical circulation of the water stops, the upper film becomes frozen. If there is no wind to agitate the surface of the water, its temperature will descend as low as 28° before it freezes, and on freezing will start up to 32°; but, should there be any wind, then the ice will form at once at 32°, expanding at the same time ⅑ larger than in its former state of water.

(350.) It is worth while to trace the progress of this curious phenomenon—the expansion of ice. In the first place, the water *contracts in bulk* by the frost, until it reaches the temperature of 39°.39, when it is in its state of greatest density, and then sinks. It then resists the freezing power of frost in a calm atmosphere, until it reaches 28°, *without decreasing more in bulk*, and it remains floating on the *warmer* water below it, which continues at 39°.39. When so placed, and at 28°, it freezes, and suddenly starts up to the temperature of 32°, and as suddenly *expands* ⅑ more in bulk than *at its ordinary temperature*, and of course more than that when in its most condensed state at 39°.39. It retains its assumed enlarged state of ice until it is melted.

(351.) So great is the force of water on being suddenly expanded into ice, that, according to the experiments of the Florentine Academy, every cubic inch of it exerts a power of 27,000 lbs. This remarkable power of ice is of use in Agriculture, as I shall illustrate when I come to speak of the effects of frost on plowed land.

(352.) It is obvious that no large body of *fresh* water, such as a deep lake or river, can be reduced in temperature below 39°.39, when water is in its densest state, as what becomes colder only floats upon and covers the denser, which is at the same time warmer, portion; and as ice is of larger bulk, weight for weight, than water, it must float above all, and, in retaining that position, prevent the farther cooling of the mass of water below 39°.39. On the other hand, *sea*-water freezes at once on the surface, and that below the ice must retain the temperature it had when the ice was formed. Frost in the polar regions becomes suddenly intense, and the polar sea becomes as suddenly covered with ice, without regard to the temperature of the water below. The ice of the polar sea, like the snow upon the polar land, thus becomes a protective mantle against the intense cold of the atmosphere, which is sometimes as great as 57° below zero.—In this way, sea animals, as well as land vegetables, in those regions, are protected against the effects of the intensest frosts.

(353.) Ice *evaporates moisture as largely as water*, which property preserves it from being easily melted by any unusual occurrence of a high temperature of the air, because the rapid evaporation occasioned by the small increase of heat superinduces a greater coldness in the body of ice.

(354.) The *great cooling powers of ice* may be witnessed by the simple experiment of mixing 1 lb. of water at 32° with 1 lb. at 172°, the mean temperature of the mixture will be as high as 102°; whereas 1 lb. of ice at 32°, on being put into 1 lb. of water at 172°, will reduce the mixture to the temperature of ice, namely, 32°. This perhaps unexpected result arises from the greater capacity of ice for caloric than water at the temperature of 32°; that is, in other words, more heat is required to break up the crystalization of ice than to heat water.

(355.) It may be worth while to notice that *ponds and lakes are generally frozen with different thicknesses of ice*, owing either to irregularities in the bottom, which constitute different depths of water, or to the existence of deep springs, the water of which, as you have seen, seldom falls below the mean temperature of the place, that is, 40°. Hence the unknown thickness of ice on lakes and ponds, until its strength has been ascertained; and hence also the origin of most of the accidents on ice.

(356.) The phenomenon of *Fog* or *Mist* occurs at all seasons, and it appears always under the peculiar circumstances explained by Sir Humphry Davy. His theory is, that radiation of vapor from land and water sends it up until it meets with a cold stratum of air, which condenses it in the form of mist—which naturally gravitates toward the surface. When the radiation is weak, the mist seems to lie upon the ground; but when more powerful, the stratum of mist may be seen elevated a few feet above the ground. Mist, too, may be seen to continue longer over the water than the land, owing to the slower radiation of vapor from water; and it is generally seen in the hollowest portions of ground, on account of the cold air, as it descends from the surrounding rising ground and mixes with the air in the hollow, diminishing its capacity for moisture.

(357.) Mist also varies in its character according to its electric state; if negatively affected, it deposits its vapor more quickly, forming a heavy sort of dew, and wetting everything like rain; but if positively, it continues to exist as fog, and retains the vapor in the state in which it has not the property of wetting like the other. Thin, hazy fogs occur frequently in winter evenings after clear cold weather, and they often become so permanently electric as to resist for days the action of the sun to disperse them. Thick, heavy fogs occur also in the early part of summer and autumn, and are sometimes very wetting.

(358.) The *fogs in hollows* constitute the true stratus cloud. We see vapor at a distance in the atmosphere, and call it cloud; but when it sinks to the earth, or will not rise, and we are immersed in it, we call it mist or fog. When immersed in a cloud on a mountain, we say we are in a mist; but the same mist will be seen by a spectator, at a distance in the valley, as a beautiful cirro-stratus resting on the mountain.

(359.) The *magnifying power of mist* is a well-known optical illusion.—Its *concealing* and *mistifying effects* may have been observed by every one; and its causing distant sounds to be heard as if near at hand, may also have been noticed by many. The illusive effects of mist are very well described in these lines:

"When all you see through densest fog is seen,
When you can hear the fishers near at hand
Distinctly speak, yet see not where they stand,
Or sometimes them and not their boat discern,
Or half concealed some figure at the stern;
Boys who, on shore, to sea the pebble cast,
Will hear it strike against the viewless mast;
While the stern boatman growls his fierce disdain
At whom he knows not, whom he threats in vain."*

* Crabbe

19. CLIMATE.

"Betwixt th' extremes, two happier climates hold,
The temper that partakes of hot and cold."
DRYDEN.

(360.) This seems a favorable opportunity for saying a few words on climate—a most interesting subject to the farmer, inasmuch as it will enable him to discover the favorable and unfavorable particulars connected with the site of the farm which he may wish to occupy. This is a point, in looking at farms, which I am afraid is entirely overlooked by farmers, much to their disappointment and even loss, as I shall have occasion to observe when we come to be on the outlook for a farm. Meantime let us attend to a few general principles.*

[* It would be vain to attempt to make *notes* on a subject so comprehensive, with a view to adapt the observations of the author in hand to any peculiar circumstances as connected with the climate of the United States, and its connection with the health and agricultural industry of its inhabitants. To do so, it would be necessary to write a book, and that has been done already by the late SAMUEL FORRY, M. D., with a degree of ability and in a spirit that do honor to his memory.

The meteorological phenomena established by observations at our military posts, taken and collected and published under the direction of our accomplished Surgeon General LAWSON, form the basis of Dr. Forry's book on "THE CLIMATE OF THE UNITED STATES," published in 1842.

Well aware that terrestrial temperature, in its effect on the animal and vegetable kingdom, is modified as well by local causes as by the position of the sun, the author of this highly interesting and valuable work has adopted a classification of climates based on physical geography, without reference to latitude.

The military posts from which the facts are supplied for the basis of his deductions, are divided into Northern, Middle, and Southern. The first embracing posts on the coast of New-England, extending as far south as the harbor of New-York—posts on the northern chain of lakes, and posts remote from the northern and inland seas. The Middle embracing the Atlantic coast from Delaware Bay to Savannah, and interior stations. And the Southern, the posts on the Lower Mississippi, and posts in the peninsula of East Florida. The last comprehending a region characterized by the predominance of low temperature—the Southern a high temperature, and the Middle phenomena vibrating to both extremes.

It is to be lamented that a sufficient number of thermometrical observations have not been made through the range of our mountain regions, to determine more exactly the influence of altitude as well as of latitude; but as at such interior and elevated points we have no occasion for military stations, it would not accord with the policy of this Republican Government to make provision for the collection of facts to enlighten and give more activity and profit to mere industrial pursuits. Tabular abstracts presented in the work of Dr. Forry embrace the condensed results of observations made at various posts between 24° 33′ and 46° 39′ of north latitude, and between 67° 4′ and 95° 43′ of longitude west of Greenwich, embracing an extent of 22° 6′ of latitude and 28° 39′ of longitude.

To any one having a just apprehension of what is needed to a rational education of young men intended to be cultivators of the soil, with a knowledge of subjects which it becomes every gentleman to know something about, we need not say how proper and useful it would be to place such books as this in all our country schools. This we can aver with the less hesitation, inasmuch as, in the attempt to extract some passages for the edification of young readers, we find it difficult to make choice of a portion, where all is alike instructive. On the general subject of climate, perhaps the best paper is to be found in the Encyclopædia Britannica, art. Climate. In that article, Professor Leslie estimates that the diminution of temperature of 1° of Fahrenheit's scale corresponds to an extent of 300 feet. But this, says Brande, will hold true only of moderate eleva

(361.) Climate may be divided into *general* and *local*. The former affects alike all places in the same parallel of latitude; it is measured from the equator to the polar circles in spaces, in each of which the longest day is half an hour longer than that nearer the equator; and from the polar circles to the pole, it is measured by the increase of a month. It is obvious that the breadth assumed for those spaces is quite arbitrary, and it is equally clear that each space is subject to a different temperature. In fact, a difference of temperature constitutes the chief distinction in the general climate of places; and it is this great distinction which has given rise to the division of zones on the surface of the globe into the torrid, temperate, and frigid—names indicative of different degrees of temperature.

(362.) The *torrid zone* embraces that space of the globe on both sides of the equator in which the sun passes across the zenith during the year. Being under the perpendicular direction of the sun's rays, this is the hottest portion of the globe. It comprehends 23½° on each side of the equator, or 47° in all.

(363.) The *temperate zones* extend 43° on each side of the torrid, and being acted upon by the sun's rays in an oblique direction, are not so

---

tions. At the altitude of 1 mile, 2 m. 3 m. 4 m. and 5 miles, the increase of elevation corresponding to 1° Fahr. will be respectively 295, 277, 252, 223, and 192. CONFIGURATION, too, has a powerful influence on temperature. The form of the limits of any large mass of land, as determined by its contact with the ocean—that is to say, the greater or less extent of coast it possesses in proportion to its area—exercises, says Brande, and as is well known, a considerable influence on climate. The small amount of variation in the temperature of the ocean tends to equalize the periodic distribution of heat among the different seasons of the year; and the proximity of a great mass of water moderates, by its action on the winds, the heat of summer and the cold of winter. Hence the great contrast between the climate of islands and of coasts, and the climate of the interior of vast continents. Europe presents a remarkable example of this. From Orleans and Paris to London, Dublin, Edinburgh, and even farther north, the mean temperature of the year diminishes very little, notwithstanding the increase of latitude; while in the eastern part of the Continent each degree of latitude, according to Humboldt, produces a variation of 1°.1 Fahr. in the mean temperature.

We must take room here for another extract from Forry:

"Where, indeed, do we not meet the evidences of design? As temperature decreases progressively with the elevation of land, great varieties of vegetation are presented in the same region. While the flowers of spring are unfolding their petals on the plains of Northern France, Winter continues his icy reign upon the Alps and Pyrenees. By this beneficent appointment of Nature, the torrid zone presents many habitable climates. On the great table-plain of Mexico and Gautemala, a tropical is converted into a temperate clime. As the vernal valley of Quito lies in the same latitude as the destructive coasts of French Guiana, so the interior of Africa may possess many localities gifted with the same advantages. In our own country, reference has already been made to the marked contrast between the Atlantic Plain and the parallel mountain ridges; but it is in the geographical features of Columbia, in South America, that we find most strikingly displayed the physical phenomenon of *hight* producing the effect of *latitude*—a change of climate, with all the consequent revolutions of animal and vegetable life, induced by local position. It is on the mountain slopes of from 3,000 to 7,000 feet, beyond the influence of the noxious miasmata, that man dwells in perpetual summer amid the richest vegetable productions of Nature.—In the mountains of Jamaica, at the hight of 4,200 feet, the vegetation of the tropics gives place to that of temperate regions; and here, while thousands are cut off annually along the coast by yellow fever, a complete exemption exists. In these elevated regions, the inhabitants exhibit the ruddy glow of health which tinges the countenance in northern climes, forming a striking contrast to the pallid and sickly aspect of those that dwell below. In ascending a lofty mountain of the torrid zone, the greatest variety in vegetation is displayed. At its foot, under the burning sun, ananas and plantains flourish; the region of limes and oranges succeeds; then follow fields of maize and luxuriant wheat; and, still higher, the series of plants known in the temperate zone. The mountains of temperate regions exhibit, perhaps, less variety, but the change is equally striking. In the ascent of the Alps, having once passed the vine-clad belt, we traverse in succession those of oaks, sweet chesnuts, and beeches, till we gain the region of the more hardy pines and stunted birches. Beyond the elevation of 6,000 feet, no tree appears. Immense tracts are then covered with herbaceous vegetation, the variety in which ultimately dwindles down to mosses and lichens, which struggle up to the barrier of eternal snow. In the United States proper we have at least two summits, the rocky pinnacles of which shoot up to the altitude, perhaps, of 6,500 feet. Of these, Mount Washington, in New-Hampshire, is one. Encircling the base is a

warm in any part of them as the torrid, and the temperature of their several parts, of course, decreases as they are situated farther from the torrid. Besides this, these zones being entirely intercepted by the torrid, the temperature of their northern and southern divisions is hotter and colder as the sun is farther off or nearer to the northern or southern extremes of his declination.

(364.) The *frigid zones* extend from the temperate to the poles. They are intercepted by both the torrid and temperate zones. The sun's rays affect them at a still more oblique angle than the temperate, and, of course, their mean temperature is yet lower than theirs. They are so far removed from the sun, that in winter the sun is never seen in them above the horizon, while in summer he is never under the horizon; and it is this accumulation of the sun's rays in summer that in a degree compensates for the

---

heavy forest—then succeeds a belt of stunted firs—next a growth of low bushes—and still farther up only moss or lichens, or lastly a naked surface, the summits of which are covered, during ten months of the year, with snow. Of the snow-capped peaks of Oregon, we possess no precise knowledge."

As it is, even if it were proper to make more extended extracts, we have room only for the following—recommending the book itself to every one who would desire to possess some acquaintance with a matter that equally affects his own health and the growth of every thing around him to which his labor is applied, and on which he depends for his subsistence; for truly the investigation of the laws of climate embraces almost every branch of natural philosophy—constituting, as it does, according to the broad and true definition of Dr. F., "*the aggregate of all external physical circumstances appertaining to each locality in its relation to organic nature.*"

Here it may be proper, for the information of the useful reader, to state, as it may serve him in his readings in relation to the geography of climate, that writers illustrating the general laws of temperature have drawn around the globe a series of curves or lines of *equal annual temperature*, called *isothermal* lines; lines of *equal summer* called *isothermal* lines, and lines of *equal winter* called *isochemical* curves. It is pleasing, says Dr. F., to contemplate such a division of the earth—each isothermal belt, as well as those of summer and winter temperatures, representing zones, in which we may trace the causes of the similarity or diversity in animal and vegetable productions; and then again he says, to determine the influence of these zones respectively upon the animal economy in health, and the agency exercised in the cessation of disease, have proved investigations still more interesting.

"For full mental and corporeal development, the due succession of the seasons is requisite.—Those countries which have a marked spring, summer, autumn, and winter, are best adapted by this agreeable and favorable vicissitude for developing the most active powers of man. It is, according to Malte-Brun, between the 40th and 60th degrees of north latitude,* that we find the nations most distinguished for knowledge and civilization, and the display of courage by sea and by land. In countries which have no summer, the inhabitants are destitute of taste and genius; while, in the regions unfavored by winter, true valor, loyalty and patriotism are almost unknown. To this all-pervading agency of atmospheric constitution, must be referred, in a considerable degree, the superiority of the warlike nations of Southern Europe over the effeminate inhabitants of Asia; and to the same cause, in connection with others, is to be ascribed the subsequent conquest of the former by the formidable hordes which issued from Northern Europe. And in regard to the political horizon of North America, if we look upon history as philosophy teaching by example, it requires not the gift of divination to foresee the destiny of Mexico and the States south of it, whose inhabitants, enervated by climate, conjointly with other causes, will yield, by that necessity which controls all moral laws, to the energetic arm of the Anglo-Saxon race. The future history of these States would seem to be typified in that of Texas. * * *

"One of the most interesting problems in history is the geographical distribution of the human family; for the oldest records seldom allude to an uninhabited country. From remote ages, it is well known that the inhabitants of every extended locality have been marked by certain physical, moral and intellectual peculiarities, serving, no less than particularity of language, to distinguish them from all other people; but how far this result ought justly to be ascribed to the agency of climate is still an undetermined point. It may, however, with good reason, be assumed that the physical characteristics which distinguish the primitive races of the human family, usually classed under five varieties, exist independent of external causes; while the various families or nations composing each race owe their similarity of physical and moral character, and of language, to the influence of climate, habits of life, and various collateral circumstances. Political institutions and

* This limitation, no doubt well adapted to Europe, is inapplicable to the United States. This is apparent from the fact that the *isothermal* lines, in being traced around the globe, suffer great depression, as will be shown, on the Atlantic region of North America. The 39d and the 46th parallels would consequently form a reasonable boundary.

entire deprivation of them in winter, and has the effect of raising the mean temperature of those zones to a hight in which both human life and vegetation may exist. The frigid zones extend 47°.

(365.) The three zones occupy these relative proportions of space:

The torrid	47°
The temperate	86°
The frigid	47°
	180° from pole to pole.

"The climates of different parts of the earth's surface are unquestionably owing, in great measure, to their position with respect to the sun. At the equator, where the sun is always nearly vertical, any given part of the surface receives a much greater quantity of light and heat than an equal portion near the poles; and it is also still more affected by the sun's vertical rays, because their passage through the atmosphere is shorter than that of the oblique rays. As far as the sun's mean altitude is concerned, it appears from Simpson's calculations, that the heat received at the equator in the whole year is nearly 2½ times as great as at the poles; this proportion being nearly the same as that of the meridian heat of a vertical sun, to the heat derived, at 23½° from the poles, in the middle of the long an-

---

social organization even struggle successfully against climatic agency; for heroes, men of genius and philosophers have arisen both in Egypt, under the tropics, and in Scandinavia, under the polar circle. Climate, however, modifies the whole nature of man. The powerful influence of locality on human organization is apparent at once in surveying the external characters of the different nations of any quarter of the earth. Even in casting one's eye over our National Legislature, the diversity of physiognomy, caused by endemico-epidemic influences, is so obvious that the general countenance of each State's delegation aords a pretty sure criterion to judge of its comparative salubrity. We can at once distinguih the ruddy inhabitant of that mountain chain where health and longevity walk hand in hand, where Jefferson and Madison inspired its cheerful and invigorating breezes, from the blanched resident of our southern lowlands—those fair and inviting plains, whose fragrant zephyrs are laden with poison, the dews of whose summer evenings are replete with the seeds of mortality. As in the smiling, but malarial, plains of Italy—

"In florid beauty groves and fields appear;
Man seems the only growth that dwindles here."

"Nothing is more obvious, as a general law, than that the animal and vegetable kingdoms have been adapted to particular climates—the effects of which, for example, in cold and warm countries, upon the same animal, are so great that the fleece of the same species of sheep in the former is soft and silky, and, in the latter, coarse, resembling hair.

"As regards vegetation, it is in tropical countries, beneath a vertical sun, that it displays its utmost glory and magnificence. It is there, amid eternal summer, that we find groves ever verdant, blooming, and productive. Advancing to the north or south, we soon discover forests, which, denuded of their leaves, assume, during half the year, the appearance of death; and, still approaching the poles, we meet vegetable life under a variety of stunted forms, which are ultimately superseded by a few coarse grasses and lichens.

"In Agriculture, England has been, and to a certain extent still is, our principal school of instruction; but her lessons must be corrected by observing the difference of climate and collateral circumstances. To effect this purpose, a comparative view of the meteorology of the two countries would avail much. But the science of meteorology concerns more particularly the horticulturist; for Agriculture has for its object the fertilization of the soil and the growth and nourishment of indigenous plants, and such as have, by a long course of treatment, become inured to the climate; while horticulture aims not only at a knowledge of the constitution of soils, but aspires to the preserving and propagating of exotic vegetation.

"So closely identified is this science with the every-day occurrences of life, that man is by nature a meteorologist. The shepherd and the mariner, in ages remote, when philosophy had not yet asserted its noble prerogative of releasing the mind from the bondage of superstition, were wont to look with awe upon the face of heaven as an index to prognosticate future results from present appearances, and to read upon it 'times and seasons.'"

Can we doubt but that we have sufficiently exposed the interesting character of this subject, and its bearing on Agriculture, to show that climate is one of the topics which ought to be treated in a Book of the Farm, and one with which every instructor of youth, and every gentleman and political economist, ought to make himself acquainted so far as that may be done, as it may by giving an hour a day, for a few weeks, to the perusal of works which illustrate the researches of those who have devoted attention to the subject? *Ed. Farm. Lib.*]

nual day at the poles. But the difference is rendered still greater by the effect of the atmosphere, which intercepts a greater proportion of the heat at the poles than elsewhere. Bouguer has calculated, upon the supposition of the similarity of the effects of light and heat, that, in lat. 45°, 80 parts of 100 are transmitted at noon in July, and 55 only in December.—It is obvious that, at any individual place, the climate in summer must approach in some degree to the equatorial climate, the sun's altitude being greater, and in winter to the climate of the polar regions."*

(366.) From what has just been observed, it is obvious that the temperature of the air diminishes gradually from the equator to the poles; and it also becomes gradually colder as we ascend in hight above the surface of the ground. Here, then, are two elements by which to judge of the general climate of different latitudes. Moreover, the diminution of heat from the equator to the poles is found to take place in an arithmetical progression—that is, *the annual temperature of all the latitudes are arithmetical means between the mean annual temperature of the equator and the poles.* This law was first discovered by M. Meyer, but by means of an equation which he founded on it, and afterward rendered more simple, Mr. Kirwan calculated the mean annual temperature of every degree of latitude between the equator and the poles. The results were, that the mean temperature at the equator is 84°, that at the poles 31°, and that in N. lat. 54°, 49°.20.

(367.) From Mr. Kirwan's calculations of the mean temperatures of every month, it appears that January is the coldest month in every latitude, and that July is the warmest month in all latitudes above 48°. In lower latitudes, August is the warmest month; while the difference in temperature between the hottest and coldest months increases in proportion to the distance from the equator. Every habitable latitude enjoys a mean heat of 60° for at least two months; and this heat seems necessary for the production of corn. Within 10° of the poles the temperature differs little, and the same is the case within 10° of the equator. The mean temperature of different years differs very little near the equator, but it differs more and more as the latitudes approach the poles.

(368.) As the temperature of the atmosphere constantly diminishes on ascending above the level of the sea, the temperature of congelation must be attained at a certain hight above every latitude; consequently, mountains which rear their heads above that limit must be covered with perpetual snow. The elevation of the freezing region varies according to the latitude of the place, being at all times highest at the equator, and lowest at the poles. In the higher regions of the atmosphere, especially within the tropics, the temperature varies but little throughout the whole year; and hence, in those brilliant climates, the line of perpetual congelation is strongly and distinctly marked. But, in countries remote from the equator, the boundary of frost descends after the heat of summer as the influence of winter prevails—thus varying its position over a belt of some considerable depth.

(369.) But beyond the line of congelation is another which forms the boundary of the ascent of visible vapor, and this point it is obvious must be less liable to change than the point of congelation. At the equator the highest point of vapor is 28,000 feet, at the pole 3,432 feet, and in N. lat. 54° it is 6,647 feet. In tracing this point successively along every latitude, we learn that heat diminishes, as we ascend, in an arithmetical progression. Hence it follows that the heat of the air above the surface of the earth is

* Polehampton's Gallery of Nature and Art, vol. iv.

not owing to the ascent of hot strata of air from the surface, but to the conducting power of the air itself.*

(370.) The question of *local* climate presents a much greater interest to the farmer than that of the general climate of the country which he inhabits. Local climate may be defined to signify that peculiar condition of the atmosphere in regard to heat and moisture which prevails in any given place. The diversified character which it displays has been generally referred to the combined operation of several different causes, which are all reducible, however, to these two: *distance from the equator*, and *hight above the level of the sea.* Latitude and local elevation form, indeed, the great basis of the law of climate; and any other modifications have only a partial and very limited influence.†

(371.) The climate of every individual country may be considered local in reference to that of all other countries in the same degrees of latitude. Islands are thus warmer than continents. The E. coast of all countries is colder than the W., though the latter is moister. Countries lying to the windward of great ranges of mountains or extensive forests are warmer than those to leeward.‡ Small seas are warmer in summer and colder in winter than the standard portion of great oceans, as they are in some degree affected by the condition of the surrounding land. Low countries are warmer than high, and level plains than mountainous regions. Plains present only one species of climate, which differs in its seasonal characters alone, but mountains exhibit every variety, from their latitude to the pole along the meridian of the quadrant. In this way, high mountains, situate in the tropics, present every variety of climate. "If we take each mountain," says Mr. Mudie, "which rises above the line of perpetual snow, as the index to its own meridian, we shall find that each one expresses, by its vegetation, all the varieties of climate between it and the pole; and thus these lofty mountains become means of far more extensive information than places which are situated near the main level of the sea, and more especially than plains, which, when their surfaces are nearly flat, have no story to tell, but the same uniform and monotonous one, for many miles." But although the high tropical mountains are thus indices of climate reaching from the equator to the pole, they are not subject to the seasonal differences which the climates are along the meridian of the quadrant. "Although," continues Mr. Mudie, "the temperature does ascend and descend a little, even upon the mountains immediately under the equator, and although the seasonal change becomes more and more conspicuous as the latitude increases, either northward or southward, yet, within the whole tropical zone, the seasonal difference is so slight that there is no marked summer or winter apparent in the native and characteristic vegetation. . . . From the small change of seasons in this region, they are almost all plants of uniform growth throughout the year, and have no winter for repose; so that, at great elevations, their growth is at all times much slower than that of plants in polar latitudes, during the perpetual sunshine of the summer there. . . . Say that the altitude of the mountain under the equator, upon which the seasonal action is displayed, is a little more than

[‡ Mr. Jefferson considered the difference as equal to 3° of latitude—for example, that the culture of cotton might be carried 3° farther north on the Mississippi than on the Atlantic; and Volney ascribed this to the influence of south-west winds carrying the warm air of the Gulf of Mexico up the valley of the Mississippi; but Forry contests the truth of the theory of these philosophers, by a train of reasoning, for which we regret we have not room. The reader is referred to the whole of section 3, part 1, of his work. *Ed. Farm. Lib.*]

* For tables of the altitudes of the points of congelation and vapor, see Encyclopædia Britannica, 7th edition, art. *Climate.* † Ibid.

3 miles. Then, estimating in round numbers, 1 foot of altitude on the mountain will correspond to about 16,000 feet on the meridian; that is, a single foot of elevation on the mountain is equivalent, in difference of temperature to about 3 miles, or more nearly 3 minutes of a degree in latitude, and therefore 20 feet are equal to a whole degree; and when one once arrives at the mean temperature of London, 400 feet more of elevation will bring one to the climate of Lapland."*

(372.) From these facts and reasonings, it appears that a slight difference of elevation in a mountainous district of this country, which stands upon so high a parallel of latitude, may make a considerable difference of climate, and that, other things remaining the same, that farm which is situated on a high elevation has a much greater chance of being affected by changes of climate than one at a lower level. Yet certainly local circumstances have a material effect in rendering the general position of any farm less desirable, such as vicinity to a lake or marsh, or a leeward position to a hill or large wood in reference to the direction from which the wind generally blows, as these tend to lower the temperature below that of the mean of the country. So, in like manner, any position in a long, narrow valley, or on the side of a large, isolated hill, or in a pass betwixt two mountains separating plains, is more subject to the injurious effects of wind than the mean of the country, as the wind acquires an accelerated motion in such localities. An elevated table-land is subject to a lower temperature and higher winds than a plain of the same extent on a lower level; hence most situations among hills are colder and more windy than on plains. On the other hand, the being on the windward side of a hill or large wood, or on flat ground backed with hills and woods to the N. and E., or being in the midst of a cultivated country, all insure a higher temperature and less injurious winds than the mean of the country. An extensive plain or valley, through which no large river passes, or in which no large lake or wood exists, is subject to very little violent wind. In the former exposed situations, the snow lies long, and the winds are cutting keen; while in the more sheltered positions the snow soon disappears, and the wind is less violent and keen. All these differences in circumstances have a sensible effect on the local climate of every country, and in a small one like Great Britain, varied as it is in its physical geography, and surrounded on all sides by water, they have the effect of dividing the country into as many climates as there are varieties of surface and differences of position in regard to the sea. These local influences, in most seasons, have a greater effect on the time of growth, quantity and quality of the produce of the earth, than the general climate of the country; although no doubt, the latter exercises such a predominating influence in some seasons, by excessive heat or rain, as to overcome all local influences, and stamp an universality of character over the season. "According to Cotte's aphorisms, local heat becomes greater on plains than on hills; it is never so low near the sea as in inland parts; the wind has no effect on it; its maximum and minimum are about 6 weeks after the solstices, it varies more in summer than in winter; it is least a little before sunrise; its maxims in the sun and shade are seldom on the same day; and it decreases more rapidly in autumn than it increases in summer."†

(373.) Besides all these causes, there is another phenomenon which has a material effect on local climate, and that is, the darting of cold pulsations downward from the upper regions of the atmosphere, and of warm pulsations upward from the earth. This is a different phenomenon from ra-

* Mudie's World.

† Polehampton's Gallery of Nature and Art, vol iv

diant heat. These pulsations of temperature are detected by a new instrument called the *æthrioscope;* and although the experiments with it have as yet not been sufficiently numerous to insure implicit confidence in its results, yet the experience of all who have paid attention to the varieties of circumstances which affect climate, can tell them that many causes are evidently at work in the atmosphere, to produce effects which have not yet been recognized by the instruments in common use. "The æthrioscope opens new scenes to our view. It extends its sensations through indefinite space, and reveals the condition of the remotest atmosphere. Constructed with still greater delicacy, it may perhaps scent the distant winds, and detect the actual temperature of any quarter of the heavens. The impressions of cold which arrive from the north will probably be found stronger than those received from the south. But the facts discovered by the æthrioscope are nowise at variance with the theory already advanced on the gradation of heat from the equator to the pole, and from the level of the sea to the highest atmosphere. The internal motion of the air, by the agency of opposite currents, still tempers the disparity of the solar impressions; but this effect is likewise accelerated by the vibrations excited from the unequal distribution of heat, and darted through the atmospheric medium with the celerity of sound. Any surface which sends a hot pulse in one direction, must evidently propel a cold pulse of the same intensity in an opposite direction. The existence of such pulsations, therefore, is in perfect unison with the balanced system of aerial currents. The most recondite principles of harmony are thus disclosed in the constitution of this nether world. In clear weather, the cold pulses then showered entire from the heavens will, even during the progress of the day, prevail over the influence of the reflex light, received on the ground, in places which are screened from the direct action of the sun. Hence at all times the coolness of a northern exposure. Hence, likewise, the freshness which tempers the night in the sultriest climates, under the expanse of an almost azure sky. The coldness of particular situations has very generally been attributed to the influence of piercing winds which blow over elevated tracts of land. This explication, however, is not well founded. It is the altitude of the place itself above the level of the sea, and not that of the general surface of the country, which will mould its temperature. A cold wind, as it descends from the high grounds into the valleys, has its capacity for heat diminished, and consequently becomes apparently warmer. The prevalence of northerly above southerly winds may, however, have some slight influence in depressing the temperature of any climate. In our northern latitudes, a canopy of clouds generally screens the ground from the impressions of cold. But within the Arctic Circle, the surface of the earth is more effectually protected by the perpetual fogs which deform those dreary regions, and yet admit the light of day, while they absorb the frigorific pulses vibrated from the higher atmosphere. Even the ancients had remarked that our clear nights are generally likewise cold. During the absence of the sun, the celestial impressions continue to accumulate; and the ground becomes chilled to the utmost in the morning, at the very moment when that luminary again resumes its powerful sway. But neither cold nor heat has the same effect on a green sward as on a plowed field, the action being nearly dissipated before it reaches the ground among the multiplied surfaces of the blades of grass. The lowest stratum of air, being chilled by contact with the exposed surface, deposits its moisture, which is either absorbed into the earth, or attracted to the projected fibres of the plants, on which it settles in the form of dew or hoarfrost Hence the utility, in this country, of spreading awnings at night, to screen the tender blossoms and the delicate fruits from the influence of a gelid sky; and hence, likewise, the advantage of covering walled trees with netting, of which the meshes not only detain the frigorific pulses, but intercept the minute icicles, that, in their formation, rob the air of its cold. It has often been observed as an incontrovertible fact, that the clearing of the ground and the extension of Agriculture have a material tendency to ameliorate the character of any climate. But whether the sun's rays be spent on the foliage of the trees, or admitted to the surface of the earth, their accumulated effects, in the course of a year, on the incumbent atmosphere, must continue still the same. The direct action of the light would no doubt more powerfully warm the ground during the day, if this superior efficacy were not likewise nearly counterbalanced by exposure to the closer sweep of the winds, and the influence of night must again re-establish the general equilibrium of temperature. The drainage of the surface will evidently improve the salubrity of any climate, by removing the stagnant and putrefying water; but it can have no effect whatever in rendering the air milder, since the ground will be left still sufficiently moist for maintaining a continual evaporation, to the consequent dissipation of heat."*

(374.) The particulars of the *geographical distribution of plants and animals* tend to show the action of general climate on the vegetable and animal functions; the effects of latitude and of elevation above the earth's surface being similar upon both, although most sensibly felt in the vegetable economy. M. Humboldt, the celebrated philosophical traveler, paid great attention to this subject, and, from his own researches, constructed tabular views of the range of animal and vegetable being in both conditions of the globe; but as his observations are more particularly applicable to America, it is not necessary to repeat them here, interesting as they really are."† ‡

(375.) It has been said already, that "the effect of elevation is equivalent to latitude; but it must be recollected that plants will not thrive equally in places with the same mean temperature. Some require a strong ephemeral heat. Hence, in judging of the aptitude of any place for rearing particular plants, we must compare the mean temperature of the summer, as well as of the whole year, before we decide. Thus, we are enabled to explain why the pistacio nut ripens in Pekin, but will not ripen in France, where the isothermal line for the whole year is the same. But though the Chinese winter be more severe than that of France, the summer heat is far greater. Innumerable other instances might be adduced of the same fact. The moisture of a climate has much influence upon its vegetation. Water is the vehicle of the food of plants, and perhaps yields a great proportion of it; so, if moisture be deficient, plants die; but they require water in very different proportions. Those with broad, smooth, soft leaves, that grow rapidly and have many cortical pores, require much water to maintain their vitality; on the other hand, plants with few cortical pores, with oily or resinous juices and small roots, will generally thrive best in dry situations. Exposure to light is necessary for most plants. The green color of plants is only formed in light, as is shown by blanching; and light appears to be the cause of certain movements which are remarked in the flowers of most plants, and in other parts of some

[‡ Doctor Forry has shown that Humboldt's observations on the physical distribution of plants "convey very erroneous impressions, from the circumstance, mostly, that his limits of the Old World are confined to Western Europe, and of the New World to Eastern America." See Forry on the Climate of the United States, page 76. [*Ed. Farm. Lib.*

* Encyclopædia Britannica, 7th edition, art. *Climate*.

† See Edinburgh Philosophical Journal, vols. iii. iv. and v.

delicately organized individuals, which open and close their leaves according to the degree of light. This last property is chiefly seen in tropical plants. Light appears to be necessary to the decomposition of carbonic acid, and the fixation of carbon in their tissues; and it is indispensable to the right performance of the function of reproduction. The influence of soil on vegetables is seen in the preference which many plants have for a calcareous soil; some affect silicious sands, others clay retentive of water; some plants thrive best in the clefts of slaty rocks; some delight to dwell amid granitic rocks; and others on a saliferous soil. Earthy matters enter largely into the composition of some vegetables; and in the epidermis of the gramineæ, silica is invariably found. The presence of animal matters in soils is necessary to many plants, and is generally nutritive to all Iron and copper are found in small quantities in some plants. The *stations* of particular plants have often been determined by these peculiarities of soil; and when a soil and climate are equally suitable for many *social plants*, we find them growing together, until the strongest obtains the mastery, and chokes the others. The common heath appears to have usurped, in Europe, a space once occupied by other genera, if we may judge from what generally happens on exterminating heath; for then other plants very speedily make their appearance, the seeds of which seem to have long preserved their vitality in the earth, and only to have wanted room to spring into visible existence. A continuation of these causes no doubt influences the distribution of particular species." . . . . "On comparing the two Continents, we find in general, in the New World under the equatorial zone, fewer Cyperaceæ and Rubiaceæ, but more Compositæ; under the temperate zone, fewer Labiatæ and Cruciferæ, and more Compositæ, Ericeæ, and Amentaceæ, than in the corresponding zones of the Old World. The families that increase from the equator to the pole, according to the method of fractional indications, are Glumeaceæ, Ericeæ, and Amentaceæ. The families which decrease from the pole to the equator, are Leguminosæ, Rubiaceæ, Euphorbiaceæ, and Malvaceæ. The families that appear to attain their maximum in the temperate zone are Compositæ, Labiatæ, Umbelliferæ, and Cruciferæ."*

(376.) In regard to the geographical distribution of animals, the slightest acquaintance with zoology is sufficient to show that animals do not indiscriminately spread themselves over every part of the habitable globe. "But the natural limitation of species has been, in some measure, affected by human agency. The domesticated animals have been, by man, imported from different parts of Asia into Europe, and finally into America. At the discovery of that continent, it was without the horse, the cow, the sheep, the hog, the dog, and our common poultry, all which are spread over it in innumerable herds, and in some places have relapsed into the wild state, in countries well suited for their subsistence.† The same use-

[† The first neat cattle, a bull and three heifers, were imported into New-England by Edward Winsboro, one of the founders and then Governor of the Plymouth Colony. They were brought over in the ship Charity, in March, 1624. For the first four years the settlers of the old Colony lived without milk. The first notice of horses is 20 years afterward, in 1644. Before the introduction of horses it was, we are told, no uncommon thing for people to *ride on bulls*. For this we have the authority of a forthcoming work on the Lives of the Governors, by JACOB B. MOORE, a victim of political proscription, removed for opinion's sake from a subordinate post in a Department, though possessing qualities to administer it far superior to its head and to most of its members. With his virtues and talents in the army he would, at his age, be enjoying a high position and the certainty of yet higher promotion and higher pay for life. As this work carefully eschews party politics, we do not name any party thus prostituting the powers of the Govern-

* Encyclopædia Britannica, 7th edition, art. *Physical Geography*.

ful animals have been, by Europeans, within the last half century, carried to the larger islands of the Pacific, where they were previously unknown. How many insects may have been propagated by the cargoes of our ships in distant lands, it is easier to conjecture than to estimate; how many have been imported with the cerealea and other gramineæ of Europe into newly discovered regions, it is impossible to say. Human agency has sometimes been the means of propagating in Europe disgusting or destructive species from foreign regions. Thus, the commerce of the Dutch wafted the *Teredo navalis* to the dyke-defended coasts of Holland, to the imminent hazard of that country; the brown rat and the *blatta*, which now infest this country, are believed to be importations from the East Indies; and the white bug, that now lays waste our orchards, is stated to have reached us with American fruit-trees."*

(377.) The definitions of the limits of the zoological divisions on the globe has first been attempted by Mr. Swainson, an eminent English naturalist. "He contends that *birds* of any district afford a fairer criterion of the limits of a geographical distribution than any other class of animals. Quadrupeds he believes to be too much under the dominion of man, and liable to have their geographic limits disturbed by human interference; and the other classes of animals are either too numerous or too few, to afford the means of determining the limits of such divisions; while birds, though seemingly fitted by nature to become wanderers, are surprisingly steady in their localities, and even in the limits of their annual migrations. These migrations are evidently caused by scarcity of food. Thus, our swallows leave us when their insect-food begins to fail, and they naturally pursue that route which is shortest, and affords subsistence by the way. The distance from the shores of the Baltic to Northern Africa is not half so great as between England and America; and during the migration over land, the winged travelers find food and resting-places as they proceed to more genial climates."† ‡

(378.) Before concluding the subject of climate, I may advert to the very generally received opinion among farmers and others who are much exposed in the air, that the weather of Great Britain has changed materially within the memory of the present generation. I am decidedly of this opinion; and I observe that Mr. Knight, the late eminent botanical physiologist, expressed himself on this subject in these words: "My own habits and pursuits, from a very early period of my life to the present time (1829), have led me to expose myself much to the weather in all seasons of the year, and under all circumstances; and no doubt whatever remains on my mind, but that our winters are generally a good deal less severe than formerly, our springs more cold and ungenial, our summers, particularly the

ment, meaning only to refer to and to denounce the anti-republican policy which everywhere gives the most invidious preference to the military over civil virtues; and so will it ever be until the sons of the cultivators of the soil are differently educated from what they have been.

Ed. Farm. Lib.]

[‡ A work of great and curious research has been published lately in France, by Marcel De Serres, with accompanying maps, on the causes of the migrations of divers animals and particularly of *birds and fishes*. We lament the want of time to translate, and of room to append some extracts. It is another of that catalogue of books which should go to make up the library of the country gentleman—by which we mean, once for all, not the man of fine apparel or of ample fortune, for these may belong to the fool, the upstart, or the demagogue. We mean the man of kind and gentle nature, who would not wantonly give pain to a fly, and who is *eager to acquire and willing to impart information*; men whose gracefulness is in the heart and feeling, rather than in exterior pomp or ostentatious display of wealth. Several such "country gentlemen," in our estimation, have we lately seen and "eaten salt" with, in their *working clothes*. *Ed. Farm. Lib.*]

* Encyclopædia Britannica, 7th edition, art. *Physical Geography*. † Ibid.

latter part of them—as warm at least as they formerly were, and our autumns considerably warmer." He adds, that "I think that I can point out some physical causes, and adduce rather strong facts in support of these opinions."

(379.) Of the physical causes of these changes, Mr. Knight conceives that the clearing of the country of trees and brushwood, the extension of arable culture, and the ready means afforded by draining to carry off quickly and effectually the rain as it falls, have rendered the soil drier in May "than it could have been, previously to its having been inclosed and drained and cultivated; and it must consequently absorb and retain much more of the warm summer rain (for but little usually flows off) than it did in an uncultivated state; and as water, in cooling, is known to give out much heat to surrounding bodies, much warmth must be communicated to the ground, and this cannot fail to affect the temperature of the following autumn. The warm autumnal rains, in conjunction with those of summer, must necessarily operate powerfully upon the temperature of the succeeding winter." Hence, a wet summer and autumn are succeeded by a mild winter; and when N. E. winds prevail after these wet seasons, the winter is always cloudy and cold, but without severe frosts; probably, in part, owing to the ground upon the opposite shores of the Continent and of this country being in a similar state. The fact adduced by Mr. Knight in support of this opinion is that of the common laurel withstanding the winter, notwithstanding its being placed in a high and exposed situation, and its wood not being ripened in November.

(380.) "Supposing the ground," continues Mr. Knight, "to contain less water in the commencement of winter, on account of the operations of the drains above mentioned, as it almost always will and generally must do, more of the water afforded by the dissolving snows and the cold rains of winter will be necessarily absorbed by it; and in the end of February, however dry the ground may have been at the winter solstice, it will almost always be found saturated with water derived from those unfavorable circumstances; and as the influence of the sun is as powerful on the last day of February as on the 15th day of October, and as it is almost wholly the high temperature of the ground in the latter period which occasions the different temperature of the air in those opposite seasons, I think it can scarcely be doubted, that if the soil have been rendered more cold by having absorbed a larger portion of water at very near the freezing temperature, the weather of the spring must be, to some extent, injuriously affected." Hence, the springs are now more injurious to blossoms and fruits than they were thirty years ago. Hence, also, the farmers of Herefordshire cannot now depend on a crop of acorns from their extensive groves of oaks.* †

(† On this question of the stability of climates in Europe and America, and the influence of cultivation on temperature, the reader should not rest satisfied until he turns to the array of historical facts and thermometrical data adduced by Doctor Forry in refutation of the theories maintained by the philosophers of the Old and of the New World—among the latter Jefferson and Rush, the latter of whom says: "From the accounts which have been handed down to us by our ancestors, there is reason to believe that the climate of Pennsylvania has undergone a material change. The springs are much colder, and the autumns more temperate, insomuch that cattle are not housed so soon by one month as they formerly were. Rivers freeze later and do not remain so long covered with ice." Doctor F. gives tables of thermometrical observations made at Philadelphia three years successively, at intervals of 25 years, from 1793 to 1824, and during thirty-three years at Salem, Mass., to show a remarkable uniformity of mean temperature. The following table, by Forry, exhibits the duration of winter at the City of New-York:

* Knight's Horticultural Papers.

20. OBSERVING AND RECORDING FACTS.

"Facts are to the mind the same thing as food to the body. On the due digestion of facts depend the strength and wisdom of the one, just as vigor and health depend on the other. The wisest in council, the ablest in debate, and the most agreeable companion in the commerce of human life, is that man who has assimilated to his understanding the greatest number of facts." BURKE.

(381.) THESE words of "the greatest philosophical statesman of our country," as Sir James Mackintosh designated Burke, convey to the mind but an amplification of a sentiment of Bacon, which says that "the man who writes, speaks, or meditates, without being well stocked with *facts* as landmarks to his understanding, is like a mariner who sails along a treacherous coast without a pilot, or one who adventures on the wide ocean without either a rudder or a compass." The expression of the same sentiment by two very eminent men, at periods so far asunder and in so very different conditions of the country, should convince you of the universal application of its truth, and induce you to adopt it as a maxim. You can easily do so, as there is no class of people more favorably situated for the observation of interesting facts than agricultural pupils. Creation, both animate and inanimate, lies before you; you must be almost always out of doors, when carrying on your operations; and the operations themselves are substantial matters of fact, constantly subject to modification by the state of land and the atmosphere. It is useful to observe facts and t

	First ice formed.	First snow fell.	Last ice formed.	Last snow fell.
1831	Oct. 20	Nov. 3	April 10	April 30
1832	Nov. 3	Dec. 12	April 10	Mar. 17
1833	Oct. 31	Dec. 15	Mar. 29	Mar. 1
1834	Oct. 30	Nov. 15	May 15	April 25
1835	Nov. 13	Nov. 27	April 18	April 16
1836	Oct. 26	Nov. 24	April 12	April 13
1837	Oct. 14	Nov. 14	May 1	April 4
1838	Oct. 31	Oct. 31	April 17	April 24
1839	Nov. 20	Nov. 10	Mar. 31	April 17
1840	Oct. 26	Nov. 18	Mar. 26	April 1

The state of the weather as indicated by the course of the winds, and the proportion of fair and cloudy days, based upon three years' observations, are shown in the following table:

Places of Observation.	WINDS.								Prevailing.	WEATHER.			Prevailing.
	N	N W	N E	E	S E	S	S W	W		fair	cl'dy	rain	
	days	days	days	days	days	days	days	days		days	days	days	
FT. MARION	1·55	2·86	9·08	1·03	10·83	1·11	2·64	1·33	S E	19·02	5·19	6·22	Fair
FT. KING...	1·62	2·79	3·46	3·54	4·37	5·63	5·96	3·08	S W	25·75	2·84	1·89	Fair
FT. BROOKE	1·53	3·72	5·58	2·89	4·44	2·75	6·42	3·17	S W	20·33	4·47	5·64	Fair
KEY WEST	3·20	3·13	10·50	5·37	5·37	0·54	1·67	0.38	N E	21·54	3·08	5·92	Fair

We must dismiss the subject, for the want of room, with the following summary observation of a writer who, had he lived, had given earnest of his capacity to make such contributions to the stock of science as would have done yet more honor to himself and his country:

"No accurate thermometrical observations yet made in any part of the world, warrant the conclusion that the temperature of a locality undergoes changes in any ratio of progression; but conversely, as all facts tend to establish the position that climates are stable, we are led to believe that the changes or perturbations of temperature to which a locality is subject, are produced by some regular oscillations, the periods of which are to us unknown. That climates are susceptible of melioration by the extensive changes produced on the surface of the earth by the labors of man, has been pointed out already; but these effects are extremely subordinate, compared with the modification induced by the striking features of physical geography—the ocean, lakes, mountains, the opposite coasts of continents, and their prolongation and enlargement toward the poles.

familiarize yourself with them, as, when accumulated, they form the stores from which experience draws its deductions. Never suppose any fact too trivial to arrest attention, as what may at first seem trivial, becomes, in many instances, far from being so; it being only by the comparison of one circumstance with another, that their relative value can be ascertained; and familiar knowledge alone can enable you to discriminate between those which influence others and those which stand in a state of isolation. In this point of view, observation is always valuable; because at first the pupil must necessarily look upon all facts alike, whatever may ultimately be found to be their intrinsic or comparative importance. The unfoldings of experience alone can show to him which classes are to be regarded by themselves, and which are not only connected with, but form the character of others. Remember, also, that to observe facts correctly is not so easy a matter as may be at first supposed; there is a proper time for the commencement of the investigation of their history, which, if not hit upon, all the deductions will be erroneous; and this is especially the case when you are performing experiments instituted for the purpose of corroborating opinions already adopted; for, in this way, many an acute experimentalist has been proselytized into an erroneous system of belief. But as *pupils* you should have no preconceived notions to gratify, no leanings to any species of prejudice. Look upon facts as they occur, and calmly, cautiously, and dispassionately contrast and compare them. It is only thus that you will be able to discriminate causes from consequences, to know the relative importance of one fact to another, and to make the results of actual observation in the field subservient to your acquiring a practical knowledge of Agriculture.

(382.) The facts to which you should, in the first instance, direct your attention, are the *effects of the weather at the time*, not only on the operations of the fields and on their productions, but also on the condition of the live-stock. You should notice any remarkable occurrence of heat or cold, rain or drouth, unpleasant or agreeable feeling in the air; the effects following any peculiar state of the clouds, or other meteors in the air, as storms, aurora borealis, halos, and the like; the particular operation of rain in retarding or materially altering the labors of the field, and the length of time and quantity of rain that it has taken to produce such an effect; as well as the effects on the health or growth of plants, and the comfort and condition of animals. The effects of cold, or snow, or drouth, upon the same subjects, deserve equal attention.

(383.) You should particularly observe the *time* at which each kind of crop is committed to the ground; how long it is till it afterward appears above it; when it comes into ear; and the period of harvest. Try also to ascertain the quantity of every kind of crop on the ground before it is cut down, and observe whether the event corroborates your judgment. In the same way, try to estimate the weight of cattle by the eye at different periods of their progress toward maturity of condition, and check your trials by measurements. The very handling of beasts for the purpose of measuring them will convey to you much information regarding their progressive state of improvement. When sheep are slaughtered, attend to the weight of the carcass, and endeavor to correct any errors you may have committed in estimating their weights.

(384.) Keep a *register* of each field of the farm; note the quantity of labor it has received, the quantity of manure which has been applied, and the kind of crop sown on it, with the circumstances attending these operations—whether they have been done quickly and in good style, or interruptedly, from the hinderance of the weather or other circumstances; and whether in an objectionable or favorable manner. Ascertain, in each field, the number of ridges required to make an acre, and whether the ridges be of equal length or not. By this you will the more easily ascertain how much dung the field is receiving per acre, the time taken to perform the same quantity of work on ridges of different length, and the comparative value of crop produced on an acre in different parts of the field. The sub division of the field into acres in this manner will also enable you to compare the relative values of the crops produced on varieties of soil, if any, in the same field, under the same circumstances of treatment.

(385.) The easiest and most satisfactory mode of preserving and record ing all these facts is in the *tabular form*, which admits of every fact being put down under its own proper head. This form not only exhibits a full exposition of the whole facts at a glance, but admits of every one being recorded with the least trouble of writing. The advantage of *writing* them down consists not entirely in recording them, but also of impressing them more strongly on the memory.

(386.) The tables should consist of ruled columns, in a book of sufficient size of leaf to contain columns for every subject. There should also be a *plan of the farm*, with every field represented, having its figure, dimensions, and name, the direction of the ridges, and the number of ridges required to make an acre visibly marked upon it. It would be advisable to enter each field into the book, in which could be noted the various sorts of labor it has received, and the produce it has yielded; so that the whole transactions connected with it for the year could be seen at a glance.—There should also be a *plan of the stack-yard* made every year, with each stack represented in it by a circle, the area of which should contain the name of the field upon which the crop in the stack was grown, the quantity of corn yielded by the stack, and in what way the produce was disposed of; and even the cash (if any) which the produce realized, should be marked down. This plan of the stack-yard should be comprehended on a single page of the book.

(387.) To render the whole system of recording facts complete, a summary of them in regard to the weather in each season, together with the produce and value of the crop and stock, should be made up every year to the end of autumn—the end of the agricultural year. In this way, an immense mass of useful facts would be recorded within the narrow compass of a single book. Comparisons could thus be easily made between the results of different seasons, and deductions drawn which could not be ascertained by any other means.

(388.) The *only objection* you can possibly urge against the adoption of this plan is the time required to record the facts. Were the records to be made twice or thrice a day, like the observations of a meteorological register, the objection would be well founded; because I cannot conceive any task more irksome than the noting down of dry and (in themselves) unmeaning details. But the variations and effects of the weather assume a very different importance, when they possess an overruling influence over the progress of the crops. The recording of these and such like facts can only be required at occasional times, of perhaps an interval of days. The only toil connected with the scheme would be the drawing up of the abstract of the year; but when the task, even if irksome, is for your professional benefit, the time devoted to it should be cheerfully bestowed.

21. SOILS AND SUBSOILS.

"I wander o'er the various rural toil,
And know the nature of each different soil."
GAY.

(389.) HAVING expatiated on every subject with which it seemed to me expedient that you should be acquainted, to prepare your mind for the reception of lessons in practice, we shall now proceed together to study farming in right earnest. The first thing, as regards the farm itself, which should engage your attention, is the kinds of *soil* which it contains. To become acquainted with these, so as to be able to identify them anywhere, you should know the external characters of *every* soil usually met with on a farm; because very few farms contain only one kind of soil, and the generality exhibit a considerable variety.

(390.) *Practically*, a knowledge of the external characters of soils is a matter of *no great difficulty;* for, however complex the composition of any soil appears to be, it possesses a character belonging to its kind, which cannot be confounded with any other. The leading characters of ordinary soils are derived from only two earths, *clay* and *sand*, and it is the greater or less admixture of these which stamps the peculiar character of the soil. The properties of either of these earths are even found to exist in what seems a purely calcareous or purely vegetable soil. When either earth is mixed with decomposed vegetable matter, whether supplied naturally or artificially, the soil becomes a *loam*, the distinguishing character of which is derived from the predominating earth. Thus, there are *clay soils* and *sandy soils*, when either earth predominates; and when either is mixed with decomposed vegetable matter, they are then *clay loams* and *sandy loams*. Sandy soils are divided into two varieties, which do not vary in kind, but only in degree. Sand is a powder, consisting of small, round particles of silicious matter; but when these are of the size of a hazel-nut and larger—that is, gravel—they give their distinguishing name to the soil; they then form *gravelly soils* and *gravelly loams*. Besides these, there are soils which have for their basis another kind of earth—*lime*, of which the *chalky oils* of the south of England consist. But these differ in agricultural character in nothing from either the clay or sandy soils, according to the particular formation from which the chalk is derived. If the chalky soil is derived from flinty chalk, then its character is like that of a sandy soil; but if from the under chalk-formation, its character is like that of clay. Writers on Agriculture also enumerate a peat-soil, derived from peat; but peat, as crude peat, is of no use to vegetation, and, when it is decomposed, it assumes the properties of *mould*, and should be considered as such; and mould, which forms the essential ingredient of loams, is decomposed vegetable matter, derived either from Nature or from artificial application. So, for all *practical* purposes, soils are most conveniently divided into clayey and sandy, with their respective loams.

(391.) *Loam*, in the sense now given, does not convey the idea attached to it by many writers; and many people talk of it as if it must necessarily consist of clay. Thus, Johnson, in defining the verb "to loam," gives as a synonym the verb "to clay;" and Bacon somewhere says that "the mellow earth is the best, between the two extremes of clay and sand, if it be not *loamy and binding;*" evidently referring to the binding property of clay. Sir Humphry Davy defines loam as "the impalpable part of the soil, which is usually called *clay or loam.*"* And Mr. Reid defines the same substance in these words: "The term 'loam' is applied to soils which consist of about one-third of finely-divided earthy matter, containing much carbonate of lime. Other soils are peaty, containing about one-half of vegetable matter."† Professor Low gives a more correct, though, in my opinion, not the exact idea of a loam. "The decomposed organic portion of the soil," he truly says, "may be termed *mould;* but he continues to say, and this is what I doubt, that "the fertility of soils is, *cæteris paribus*, indicated by the greater or smaller proportion of mould which enters into their composition. When soils are thus naturally fertile, or are rendered permanently so by art, they are frequently termed *loams.*"‡ You thus see what diversity of opinion exists as to what loam is. Loam, in my opinion, has changed its meaning so far since the days of Johnson, as to consist of any kind of earth that contains a *large admixture of decomposed vegetable matter*—I say a large admixture of vegetable matter, because there is no soil under cultivation, whether composed chiefly of clay or principally of sand, but what contains some decomposed vegetable matter. Unless, therefore, the decomposed vegetable matter of the soil so preponderates as to greatly modify the usual properties of the constituent earths, the soil cannot in truth be called by any other name than a clayey or sandy soil; but when the vegetable matter so prevails as materially to alter the properties of those earths, then a *clay loam* or a *sandy loam* is constituted—a distinction well known to the farmer. But, if it is necessary that clay should have a preponderance in *loam*, then a *sandy* loam must be a contradiction in terms. Again, a soil of purely vegetable origin—such as crude peat or leaf-mould—cannot be called loam; for admixture of an earth of some sort is required to make loam, under every recorded definition of that term. Nor is the fertility of soils dependent on the greater or smaller proportion of mould or decomposed vegetable matter in their composition; for there are soils with apparently very little mould in them, such as sharp gravels, which are highly fertile; and there are moulds, apparently with very little earth in them, such as deaf black mould, which are far removed from fertility. Thus, then, all soils have the properties of clayey or sandy soils, and a considerable quantity of decomposed vegetable matter converts them into loam. Hence it is possible for husbandry to convert an earthy soil into a loam, as is exemplified in the vicinity of large towns.

(392.) A pure *clay*-soil has very distinctive external characters, by which you may easily recognize it. When fully wetted, it feels greasy to the foot, which slips upon it backward, forward, and sideways. It has an unctuous feel in the hand, by which it can be kneaded into a smooth homogeneous mass, and retain any shape given to it. It glistens in the sunshine. It retains water upon its surface, and makes water very muddy when mixed with it or runs over it, and is long of settling to the bottom. It is cold to the touch, and easily soils the hand and any thing else that touches it. It cuts like soft cheese with the spade, and is then in an unfit state to be worked with the plow, or any other implement. When dry, clay-soil cracks into numerous fissures, feels very hard to the foot, and runs into lumps, which are often large, and both large and small are very difficult to be broken, and indeed cannot be pulverized. It soils the hand and clothes with a dry, light-colored, soft dust, which has no lustre. It is heavy in

* Davy's Agricultural Chemistry, 8vo edit. 1839.
† Reid's Chemistry of Nature.
‡ Low's Elements of Practical Agriculture, 2d edit.

weight, and difficult to labor. It absorbs moisture readily, and will adhere to the tongue. When neither wet nor dry, it is very tough, and soon becomes very hard with a little drouth, or very soft with a little rain. On these accounts, it is the most ticklish of all soils to manage; being, even in its best state, difficult to turn over with the plow, and to pulverize with other implements. A large strength of horses is thus required to work a clay-land farm; for its workable state continues only for a short time, and it is the most obdurate of all soils to labor. But it is a powerful soil, its vegetation being luxuriant, and its production great. It generally occurs in deep masses, on a considerable extent of flat surface, exhibiting only a few undulations. It is generally found near a large river, toward its estuary, being supposed to have been a deposition from its waters. Examples of this kind of soil may be seen in Scotland, in the Carses of Gowrie, Stirling, and Falkirk. It may be denominated a naturally rich soil, with little vegetable matter in it, and its color is yellowish-gray.

(393.) When a little *sand and gravel are mixed with clay*, its texture is very materially altered, but its productive powers are not improved. When such a clay is in a wet state, it still slips a little under the foot, but feels harsh rather than greasy. It does not easily ball in the hand. It retains water on its surface for a time, which is soon partially absorbed. It renders water very muddy, and soils everything by adhering to it; and, on that account, never comes clean off the spade, except when much wetted with water. When dry, it feels hard, but is easily pulverized by any of the implements of tillage. It has no lustre. It does not soil the clothes much, and, though somewhat heavy to labor, is not obdurate. When betwixt the states of wet and dry, it is easily labored, and can be reduced to fine tilth or mould. This kind of soil never occurs in deep masses, but is rather shallow; is not naturally favorable to vegetation, nor is it naturally prolific. It occupies by far the larger portion of the surface of Scotland; much of its wheat is grown upon it, and it may be denominated a naturally poor soil, with not much vegetable matter in it. Its color is yellowish-brown.

(394.) *Clay-loam*—that is, either of those clays mixed with a large proportion of naturally decomposed vegetable matter—constitutes a useful and valuable soil. It yields the largest proportion of the fine wheats raised in this country, occupying a larger surface of the country than the carse-clay. It forms a lump by a squeeze of the hand, but soon crumbles down again. It is easily wetted on the surface with rain, and then feels soft and greasy; but the water is soon absorbed, and the surface is again as soon dry. It is easily labored, and may be so at any time after a day or two of dry weather. It becomes finely pulverized, and is capable of assuming a high temperature. It is generally of some depth, forming an excellent soil for wheat, beans, Swedish turnips and red clover. It is of a deep-brown color, often approaching to red.

(395.) All clay-soils are better adapted to fibrous-rooted plants than to bulbs and tubers; but it is that sort of fibrous root which has also a tap-root, such as is found in wheat, the bean, red clover, and the oak. The crops mentioned bearing abundance of straw, the plants require a deep hold of the soil. Clay-soils are generally slow of bringing their crops to maturity, which in wet seasons they never arrive at; but in dry seasons they are always strong, and yield quantity rather than quality.

(396.) A pure *sandy* soil is as easily recognized as one of pure clay. When wet, it feels firm under foot, and then admits of a pretty whole furrow being laid over by the plow. It feels harsh and grating to the touch. When dry it feels soft; and is so yielding, that every object of the least weight sinks in it: it is then apt to blow away with the wind. In an ordinary state, it is well adapted to plants having fusiform roots, such as the carrot and parsnip. It acquires a high temperature in summer. Sandy soil generally occurs in deep masses, near the termination of the estuaries of large rivers, or along the sea shore; and in some countries in the interior of Europe, and over a large proportion of Africa, it covers immense tracts of flat land, and is evidently a deposition from water.

(397.) A *gravelly* soil consists of a large proportion of sand; but the greater part of its bulk is made up of small rounded fragments of rock brought together by the action of water. These small fragments have been derived from all the rock-formations, while the large bowlders, imbedded principally under the surface, have been chiefly supplied by the older formations. Gravelly deposits sometimes occupy a large extent of surface, and are of considerable depth. Such a soil soon becomes warm, but never wet, absorbing the rain as fast as it falls; and after rain, it feels somewhat firm under foot. It can be easily labored in any weather, and is not unpleasant to work, though the numerous small stones, which are seen in countless numbers upon the surface, render the holding of the plow rather unsteady. As an instance of its dry nature, an old farmer of gravelly soil used to joke with his plowmen, and offer them a "roasted hen" to their dinner on the day they got their feet wet at the plow. This soil is admirably adapted to plants having bulbs and tubers; and no kind of soil affords so dry and comfortable a lair to sheep on turnips, and on this account it is distinguished as "*turnip-soil.*"

(398.) *Sandy and gravelly loams*, if not the most valuable, are certainly the most useful of all soils. They become neither too wet nor too dry in ordinary seasons, and are capable of growing every species of crop, in every variety of season, to considerable perfection. On this account, they are esteemed "*kindly soils.*" They never occur in deep masses, nor do they extend over large tracts of land, being chiefly confined to the margins of small rivers, forming haughs or holms, through which the rivers meander from their source among the mountains toward the larger ones, or even to the sea; and, in their progress, are apt at times to become so enlarged with rain, both in summer and winter, as to overflow their banks to a limited extent on either side.

(399.) These are all the kinds of soil usually found on a farm; and of these, the two opposite extremes of the pure clay and the pure sand may most easily be recognized by you. The intermediate shades in the varieties of soil, occasioned by modifications of greater or smaller quantity of decomposed vegetable matter, it would be impossible to describe. Every soil, however, may be ranked under the general heads of clayey and sandy soils; the gravelly and sandy, as you have learned, constituting differences rather in degree than in kind; and as every soil possesses the property of either clay or sand—be the sand derived from silicious or calcareous deposit—it is useless to maintain the nomenclature of chalky and peaty soils, although these distinctive terms may be retained to indicate the origin of the soils thereby implied by them.

(400.) You are now prepared to consider the question, what constitutes *the soil*—properly so called? You will perceive the propriety of such a question, when you consider the different ideas entertained of soil by persons of different denominations. The geologist considers the uppermost alluvial covering of the earth's crust as the soil, and whatever stratum that rests upon, as the subsoil. The botanist considers as the soil that portion of the earth's surface which supports plants. People generally consider the ground they walk upon as the soil; but none of these ideas define the soil in the *agricultural sense*. In that sense, the *soil* consists only of that *portion of the earth which is stirred by the plow*, and the *subsoil* of that which is found *immediately below the plow's course*. In this way the subsoil may consist of the same kind of earth as the soil, or it may be quite different, or it may be of rock. As it is of importance for you to keep *this* distinction of soil and subsoil always in mind, the subject should be illustrated by a figure. Let *a*, fig. 34, be the surface of the

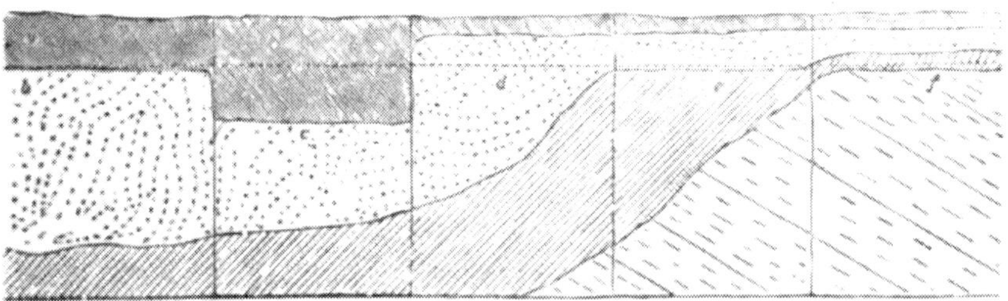

SECTIONS OF SOILS AND SUBSOILS.

ground, the earthy mould derived from the growth and decay of natural plants; *b*, a dotted line, the depth of the plow-furrow. Now, the plow-sole may either just pass through the mould, as at *b*, when the mould will be the soil, and the earth below it the subsoil: Or it may not pass entirely through the mould, as at *c*, when the soil and subsoil will be similar, that is, both of mould: Or it may pass through the earth below the mould, as at *d*, when the soil and subsoil will again be similar, while neither will be mould, but earth: Or it may move along the surface of *e*, when the soil will be of one kind of earth, and the subsoil of another, that is, either an open subsoil of gravel, or a retentive one of clay: Or it may move upon the surface of *f*, when the soil will be earth, or a mixture of clay, sand and mould, and the subsoil rock. These different cases of soil and subsoil are represented in the figure, each in a distinct sectional division.

(401.) The *subsoil*, then, in an *agricultural sense*, is the substance which is found immediately below the line of the course of the plow, be it earth or rock. However uniform in substance, or similar in quality, the subsoil and soil may have been at one time, cultivation, by supplies of vegetable matter, and by presentation of the surface to the action of the air, soon effects a material difference betwixt them, and the difference consists of a change both in texture and color, the soil becoming finer and having a darker tint than the subsoil.

(402.) The *nature* of the subsoil *produces a sensible effect on the condition of the soil above it*. If the soil is clay, it is impervious to water, and if the subsoil is clay also, it is also impervious to water. The immediate effect of this juxtaposition is to render both soil and subsoil habitually wet, until the force of evaporation dries first the one and then the other. A retentive subsoil, in the same manner, renders a sandy or gravelly, that is a porous, soil above it habitually wet. On the other hand, a gravelly subsoil, which is always porous, greatly assists to keep a retentive clay soil dry.—When a porous soil rests upon a porous subsoil, scarcely any degree of humidity can injure either. Rock may be either a retentive or a porous subsoil, according to its structure; its massiveness throughout keeping every soil above it habitually wet; but its stratification, if the lines of stratification dip downward from the soil (as at *f*, fig. 34), will keep even a retentive soil above it in a comparatively dry state.

(403.) These are the *different conditions of soils and subsoils*, considered practically. They have terms expressive of their state, which you should keep in remembrance. A soil is said to be *stiff* or *heavy*, when it is difficult to cut through, and is otherwise laborious to work with the ordinary implements of the farm; and all clay soils are more or less so. On the other hand, it is *light* or *free*, when it is easy to work; and all sandy and gravelly soils, and sandy and gravelly loams, are so. A soil is said to be *wet*, when it is habitually wet; and to be *dry*, when habitually so. All soils, especially clays, on retentive subsoils, are habitually wet; and all soils on porous subsoils, especially gravels and gravelly loams, are habitually dry. Any soil that cannot bring to maturity a fair crop, without an inordinate quantity of manure, is considered *poor;* and any one that does so naturally, or yields a large return with a moderate quantity of manure, is said to be *rich.* Thin, hard clays and ordinary sands are examples of poor soils; and soft clays and deep loams, of rich. A soil is said to be *deep*, when the surface-earth descends a good way below the reach of the plow; and in that case the plow may be made to go deeper than usual, and yet continue in the same soil; and a soil is *thin*, when the plow can easily reach beyond it. Good husbandry can, in time, render a thin soil deep; and bad, shallow plowing may cause a deep soil to assume the character of a thin one. A deep soil conveys the idea of a good one, and a thin, or shallow, or ebb, that of a bad. Carse clays and sandy loams are instances of deep soils, and poor clays and poor gravels those of thin. A soil is said to be a *hungry* one, when it requires frequent applications of a large quantity of manure to bear ordinary crops. Thin, poor gravels are instances of a hungry soil. A soil is said to be *grateful*, when it returns a larger produce than was expected from what was done for it. All loams, whether clayey, gravelly, or sandy—especially the two last—are grateful soils. A soil is said to be *kindly*, when every operation performed upon it can be done without doubt, and in the way and at the time desired. A sandy loam, and even a clay loam, both on porous subsoil, are examples of kindly soils. A soil is said to become *sick*, when the crop that has been made to grow upon it too frequently becomes deteriorated; thus, soils soon become sick of growing red clover and turnips. A *sharp* soil is that which contains such a number of small, gritty stones as to clear up the plow-irons quickly. Such a soil never fails to be an open one, and is admirably adapted for turnips. A fine, gravelly loam is an instance of a sharp soil. Some say that a sharp soil means a *ready* one—that is, quick or prepared to do anything required of it; but I am not of this opinion, because a sandy loam is ready enough for any crop, and it is never called a sharp soil. A *deaf* soil is the contrary of a sharp one; that is, it contains too much inert vegetable matter, in a soft, spongy state, which is apt to be carried forward on the bosom of the plow. A deep, black mould, whether derived from peat or not, is an example of a deaf soil. A *porous* or *open* soil and subsoil, are those which allow water to pass through them freely and quickly, of which a gravelly loam and gravelly subsoil are examples. A *retentive* or *close* soil and subsoil retain water on them; and a clay soil upon a clay subsoil is an instance of both. Some soils are always *hard*, as in the case of thin, retentive clays when dry, let them be ever so well worked; while others are *soft*, as fine, sandy loams, which are very apt to become so on being too often plowed, or too much marled. Some soils are always *fine*, as in the case of deep, easy clay loams; others *coarse* or *harsh*, as in thin, poor clays and gravels. A fine clay is *smooth* when in a wet state, and a thin clayey gravel is *rough* when dry. A soil is said to have a *fine skin* when it can be finished off with a beautifully granulated surface. Good culture will bring a fine skin on many soils, and rich sandy

(410)

and clay loams have naturally a fine skin; but no art can give a fine skin to some soils, such as thin, hard clay and rough gravel.

(404.) The *colors* of soils and subsoils, though various, are limited in their range. *Black* soils are instanced in crude peat and deep vegetable mould; and *white* are common in the chalky districts of England. Some soils are *blue* or *bluish-gray*, from a peculiar sort of fine clay deposited at the bottom of basins of still water. But the most prevailing color is *brown*, from light hair-brown to dark chestnut, the hazel-brown being the most favorite color of the class. The sand and gravel loams are instances of these colors. The browns pass into *reds*, of which there are several varieties, all having a dark hue; such, for instance, are some clay loams. The brown and red soils acquire high degrees of temperature, and they are also styled *warm* in reference to color. There are also *yellow* and *gray* soils, a mixture of which makes a yellowish-gray. They are always *cold*, both in regard to temperature and color; and are the opposite, in these respects, to brown and red soils. Color is indicative of the nature of soils. Thus, all yellow and gray colors belong to clay soils. Gray sand and gray stones are indicative of soils of moory origin. Black soils are deaf and inert; the brown, on the other hand, are sharp and grateful, and many of them kindly; while the reds are always prolific. The color of subsoils is less uniform than that of soils—owing, no doubt, to their exclusion from culture. Some subsoils are very party-colored; and the more they are so, and the brighter the colors they sport, they are the more injurious to the soils above them: they exhibit gray, black, blue, green, bright red, and bright yellow colors. The dull red and the chestnut brown subsoils are good; but the nearer they approach to hazel brown the better. Dull browns, reds, and yellowish grays are permanent colors, and are little altered by cultivation; but the blues, greens, bright reds and yellows become darker and duller by exposure to the air and by admixture with manures.

(405.) These are all the remarks required to be made on soils, in as far as practice is concerned; but a great deal yet remains to be said of them as objects of natural history, and subjects of chemistry, and, above all, as the staple of the farm. Part of the natural history, and part of their chemistry, will appear in the paragraphs immediately below, and part of both will deserve our attention when we treat of the fertility of soils; but the management of soils will occupy our thoughts through every season.

(406.) The external characters of minerals established by Werner, and recognized by mineralogists, have never been used to describe agricultural soils. It would, perhaps, serve no practical purpose to do so; because there are naturally such minute shades in the varieties of soils, and those shades are constantly undergoing changes in the course of good and bad modes of cultivation, that definitions, even when established, would soon become inapplicable. In respect, therefore, to a scientific classification of soils by external characters, there are as yet no data upon which to establish it, and the only alternative left is to adopt such a division as I have endeavored to describe. In adopting that classification I have subdivided it into fewer heads than other writers on the same subject have done. In their subdivisions they include calcareous and peaty soils with the clayey and sandy. Practically, however, calcareous matter cannot be detected in ordinary soils; and, as to chalky soils themselves, their management is so similar to that of light and heavy ordinary soils, according to the formation from which they are derived, that no practical distinction, as I have said need be drawn betwixt them; and in regard to peaty soils, when reduced to earth, which they easily are by cultivation, they partake of the character of mould. The kind of mould which they form you will learn when I come to treat of the fertility of soils.

(407.) In regard to the relation of soils to the subjacent strata, it is held by a recent practical writer on soils that "the surface of the earth partakes of the nature and color of the subsoil or rock on which it rests. The principal mineral in the soil of any district is that of the geological formation under it; hence we find argillaceous soil resting on the various clay formations—calcareous soil over the chalk—and oolitic rocks and silicious soils, over the various sandstones. On the chalk the soil is white; on the red sandstone it is red; and on the sands and clays the surface has nearly the same shade of color as the subsoil."* I do not think that this description of the position of soils is generally correct, because many instances occur to my knowledge of great tracts of soils, including subsoils, having no relation to the "geological formation under them." The fine, strong, deep clay of the Carse of Gowrie rests on the old red sandstone, a rock having nothing in common, either in consistence or color, with the clay above it. The large extent of the gray sands of Barrie, and the great gray gravelly deposits of the valley of the Lunan, in Forfarshire, both rest on the same formation as the carse clay, namely, the old red sandstone; and so of numerous other examples in Scotland. In fact, soils are frequently found of infinitely diversified character, over extensive districts of rock, whose constituents are nearly uniform; and, on the other hand, soils of uniform character occur in districts where the underlying rocks are different as well in their chemical as their geological properties. Thus, an uniform integument of clay rests upon the gray sandstone to the westward of the Carse of Gowrie, in Perthshire, and the same clay covers the Ochil Hills in that county and Fifeshire with an uniform mantle—over hills which are entirely composed of trap. On the other hand, a diversified clay and gravel are found to cover an uniform tract of graywacke in Pertshire. "We have gray sandstone," says Mr. Buist, aptly, when treating of the geology of the north-east portion of Perthshire, "red sandstone, and rock-marl, as it is called, cut by various massy veins of trap or beds of conglomerate and lime; yet I defy any man to form the smallest guess of the rocks below from the soils above them, though the ground is sufficiently uniform to give fair scope for all to manifest the influence possessed by them. There are lands whose agricultural value has been so greatly modified by the presence or withdrawal of a bed of gravel between the arable soil and tilly subsoil, which, when present, affords a universal drain, when absent, leaves the land almost unarable. But if we must show a relation betwixt the sandstone and any of these beds, which of the three," very properly asks Mr. Buist, "are we to select as having affinity with the rock?"†

(408.) In passing from practical to scientific opinion on the origin of soils, we find Mr. De la Beche giving his opinion that "naturally soils are merely decomposed parts of the subjacent rock, mixed with the decomposed portions of vegetable substances which have grown or fallen upon it, and with a proportion of animal substances derived from the droppings of creatures which have fed upon the vegetation, from dead insects and worms which once inhabited the surface, and from the decomposition of animals that have perished on the land and which have not altogether been removed by those quadrupeds, birds and insects that act as natural scavengers."‡ This view of the origin of soils seems to corroborate the opinion of Mr. Morton, quoted above; but if you look more closely into the definitions of the terms used by both writers, you will find there is not that identity of opinion between them which appears at first sight. For "the term rock," says Mr. De la Beche, "is applied by geologists, not only to the hard substances to which this name is *commonly* given, but also to those various *sands, gravels, shales, marls,* or *clays,* which form beds, strata, or masses."‖ Taking this correct geological definition of rock, Mr. De la Beche's view is quite correct in regard to *agricultural* soils, for they certainly are decomposed portions of the rock, geologically speaking—that is, of those "sands, gravels, or clays," upon which they rest; but the impression left on the mind of the reader on perusing Mr. Morton's account of the origin of soils, is that the *rocky* strata, *commonly* so called, because indurated, became decomposed to form soil; and his reference to the various geological formations of England, in explanation of the soils found above them, warrant the correctness of this impression; but it is this very impression which I wish to remove from your minds, because it conveys, in my opinion, an erroneous idea of the origin of soils.

(409.) No doubt the chemical action of the air, and the physical force of rain, frost, and wind, produce visible effects upon the most indurated rocks, but, of course, much greater effects upon incoherent rocks. We know that the action of these agents loads the waters of the Ganges with detritus to the extent of 2½ per cent. of their volume, which is an enormous quantity when we consider that the water discharged by that river into the sea is 500,000 cubic feet per second, although this amount falls far short of Major Rennel's statement of 25 per cent.; yet these agents have not had sufficient power to accumulate, by their own action on indurated rocky strata, all the deposits of clay, gravel, and sand, found accumulated to the depth of many feet. Combined in their action, they could only originate a mere coating of soil over the surface of indurated rock, if the rock were situated within the region of phænogamous vegetation, because then it would be constantly covered with plants. But the plants, in their turn, would protect the rocks against the action of those agencies, and, although they could not entirely prevent, they could at least retard, the accumulation of soil beyond what the decay of vegetation supplied. Even in the tropics, where vegetation displays its greatest luxuriance on the globe, the mould does not increase, though the decay of vegetables every year is enormous. "The quantity of timber and vegetable matter which grows in a tropical forest in the course of a century," says Mr. Lyell, "is enormous, and multitudes of animal skeletons are scattered there during the same period, besides innumerable

* Morton on Soils.
† Prize Essays of the Highland and Agricultural Society, vol. xiii.
‡ De la Beche, How to Observe Geology.
‖ De la Beche's Manual of Geology.

land shells and other organic substances. The aggregate of these materials, therefore, might constitute a mass greater in volume than that which is produced in any coral-reef during the same lapse of years; but, although this process should continue on the land for ever, no mountains of wood or bone would be seen stretching far and wide over the country, or pushing out bold promontories into the sea. The whole solid mass is either devoured by animals, or decomposes, as does a portion of the rock and soil (into their gaseous constituents) on which the animals and plants are supported."* These are the causes of the prevention of the accumulation of soils in the tropics. In colder regions a similar result is thus brought about. "It is well known," continues Mr. Lyell, "that a covering of herbage and shrubs may protect a loose soil from being carried away by rain, or even by the ordinary action of a river, and may prevent hills of loose sand from being blown away by the wind; for the roots bind together the separate particles into a firm mass, and the leaves intercept the rain-water, so that it dries up gradually, instead of flowing off in a mass and with great velocity."†

(410.) Some other agent, therefore, more powerful than the ordinary atmospherical elements, must be brought to bear upon indurated rocks, before a satisfactory solution of the formation of soils can be given. This other agent is water; but the moment that we assent to the agency of water being able by its great abrasive power and great buoyant property, when in motion, to transport the abraded parts of rocks to a distance, and let them fall on coming in contact with some opposing barrier, that moment we must abandon the idea of soils having been universally derived from the indurated rocky strata upon which they are found to rest. I quite agree with Mr. Buist in the conclusions he has drawn in regard to soils, after he had described their relative positions to the rocks upon which they rest in a large and important district of Perthshire, where he says, "that the alluvial matters of these districts, in general, belong to periods much more remote than those ordinarily assigned to them, and came into existence under circumstances prodigiously different from those which presently obtain: that the present causes—that is, the action of our modern rivers, brooks, and torrents, and of the air and water on the surfaces now exposed to them—have had but little share in modifying our alluvial formations, or bringing them into their present form — The doctrine seems to me most distinctly demonstrable, that wherever gravel or clay beds alternate with each other, and wherever bowlder stones prevail remote from the parent rock, or cut off from it by high intervening ridges, that, at the time when the surface of the solid rock became covered with such alluvium, much the greater part of it was hundreds of feet beneath the waves. The supposition of the prevalence of enormous lakes, requiring barriers only less stupendous than our highest secondary mountain-ranges, whose outbursts must have swept every movable thing before them, seems far more untenable than the assumption that the present dry land, at the era of bowlders being transported, was beneath the level of the ocean, from which, by slow elevations, it subsequently emerged. Our newer alluvia, again, which are destitute of erratic bowlders in general, such as our Carse of Gowrie and other clays, must have originated when the sea occasionally invaded the land to such moderate extent that the transportation of rocky masses, from great distances from our mountain-land, had been rendered impossible, by the intervention of elevated ridges, or of secondary mountain-ranges."‡ More than this, is it not probable that, when the stratified rocks were being deposited in water, portions of the matter of which they were about to be formed were carried away by currents, and, by reason of the motion given them, were deposited in eddies in a mechanical state, instead of getting leave to assume the crystaline form of indurated stratified rock? May not all diluvium have thus originated, instead of being abraded from solid strata, although it is possible that some portion may have been derived from the abrasion of rocks? It is also quite conceivable that where indurated rocks, such as chalk, and sandstone, and limestone, were left bare by the subsiding waters, and exposed to atmospherical influences, part of the soil upon them may have been derived at first immediately from them.

(411.) The soil, or incoherent rocks, when complete in all their members, consist of three parts. The oldest or lowest part, not unfrequently termed *diluvium*, but which is an objectionable term, inasmuch as it conveys the idea of its having been formed by the Noachian deluge, which it may not have been, but may have existed at a much older period of the globe. This cannot be called *alluvium*, according to the definition of that deposit given by Mr. Lyell, who considers it to consist of "such transported matter as has been thrown down, whether by rivers, floods, or other causes, upon land not *permanently* submerged beneath the waters of lakes or seas—I say *permanently submerged*, in order to distinguish between *alluviums* and *regular subaqueous deposits*. These regular strata," he continues, "are accumulated in lakes or great submarine receptacles; but the alluvium is in the channels of rivers or currents, where the materials may be regarded as still *in transitu*, or on their way to a place of rest."‖ Diluvium, therefore, should rather be termed subaqueous deposits, and may consist of clay, or gravel, or sand, in deep masses and of large extent. It may, in fact, be transported materials, which, if they had been allowed to remain in their

* Lyell's Principles of Geology, vol. iii. † Ibid.
‡ Prize Essays of the Highland and Agricultural Society of Scotland. vol. xiii.
‖ Lyell's Principles of Geology, vol. iii.

original site, would have formed indurated aluminous and silicious rocks. When such subaqueous deposits are exposed to atmospherical influences, an arable soil is easily formed upon them.

(412.) True *alluvial* deposits may raise themselves by accumulation above their depositing waters, and art can assist the natural process, by the erection of embankments against the encroachments of these waters, and by the casting out of large ditches for carrying them away, as has been done in several places in the rivers and coasts of our country. Atmospherical influences soon raise an arable soil on alluvium.

(413.) The third member of soils is the upper *mould*, which has been directly derived from vegetation, and can only come into existence after either of the other soils has been placed in a situation favorable for the support of plants. Mould, being in contact with air, always exists on the surface, but when either the subaqueous deposit or the alluvium is wanting, the mould then rests upon the one present; or both may be wanting, and then it rests upon the indurated rocky strata.

(414.) When the last case happens, if the rocky stratum is porous, by means of numerous fissures, or is in inclined beds, the arable soil is an earthy mould of good quality for agricultural purposes; such as are the moulds upon sandstones, limestones, and trap, and the upper chalk formation; but if it rest on a massive rock, then the mould is converted into a spongy, wet pabulum for subaquatic plants, forming a marsh, if the site is low, and if high, it is converted into thin peat; and both are worthless soils for Agriculture. When the mould rests immediately upon clay subaqueous deposit, a coarse and rank vegetation exists upon it, and if the water which supports it has no opportunity of passing away, in time a bog is formed by the cumulative growth of the subaquatic mosses.* When mould, on the other hand, is formed on gravelly deposit, the vegetation is short, and dry, and sweet, and particularly well adapted to promote the sound feeding and health of sheep. On such deposits water is never seen to remain after the heaviest fall of rain. When mould rests on alluvial deposit of whatever nature, a rich soil is the consequence, and it will be naturally dry only when the deposit is gravelly or sandy.

(415.) Mr. De La Beche seems to think that farmers do not know the reason why subsoils are favorable or unfavorable to the soil upon them.† I suspect they know more about them than he is aware of. They know quite well that a dry subsoil is more favorable to Agriculture than a retentive one; that gravel forms a drier subsoil than clay; and that the reason why these results should be so is, that clay, or a massive rock, will not let water pass through it so easily as gravel, and I presume no geologist knows more of the matter.

(416.) [We must now observe soils and subsoils in another point of view. A practical outline of the characters of various soils, and the manner in which they may be distinguished one from another, having been already pointed out to you, my intention now is, to consider them scientifically, for the purpose of preparing your minds for following me through the mazy windings of theoretical Agriculture, as developed by the joint application of chemistry, mechanical philosophy, and vegetable physiology. Although, to the contemplative and still more to the speculative student, this branch of the subject will exhibit the greatest charms, still I beg you to bear in mind continually, that it is with *practice* you have to do, and that *theory* must only be used cautiously as an adjunct to well-studied and assiduously-applied *practical knowledge;* and although, by so doing, I fully believe you will not only increase greatly your interest in the whole matter, but will likewise proceed with more rapid strides in the progress of improvement, I feel equally satisfied that an opposite course, viz. the *study of theory antecedent to the application of practice*, will almost invariably be productive of just the opposite effects, viz. the retardation of your real advance in knowledge, and will, moreover, make you run a rea t risk of becoming speculative men, than which nothing can be more inimical to *real* improvement.

(417.) *Soil*, considered scientifically, may be described to be essentially a mixture of an impalpable powder with a greater or smaller quantity of visible particles of all sizes and shapes. Careful examination will prove to us, that although the visible particles have several *indirect* effects, of so great importance that they are absolutely necessary to soil, still the impalpable powder is the only portion which *directly* exerts any influence upon vegetation. This impalpable powder consists of two distinct classes of substances, viz. *inorganic* or *mineral* matters, and *animal* and *vegetable* substances, in all the various stages of decomposition.

(418.) A very simple method may be employed to separate these two classes of particles from each other, viz. the impalpable powder and the visible particles; and, in so doing, we obtain a very useful index to the real value of the soil. Indeed all soils, except stiff clays, can be discriminated in this manner. The greater the proportion of the impalpable matter, the greater, *cæteris paribus*, will be the fertility of the soil. (438.)

(419.) To effect this separation, the following easy experiment may be performed Take a glass tube about 2 feet long, closed at one end; fill it about half full of water, and shake into it a sufficient quantity of the soil to be examined to fill the tube about 2 inches from the bottom; then put in a cork, and having shaken the tube well to mix the earth and water

* For an account of the origin of Bogs, see Alton on Moss.

† De La Beche, How to observe Geology.

thoroughly, set the tube in an upright position, for the soil to settle down. Now, as the larger particles are of course the heavier, they fall first, and form the undermost layer of the deposit, and so on in regular gradation, the impalpable powder being the last to subside, and hence occupying the uppermost portion. Then by examining the relative thickness of the various layers, and calculating their proportions, you can make a very accurate mechanical analysis of the soil.

(420.) The *stones* which we meet with in soil have in general the same composition as the soil itself, and hence, by their gradually crumbling down under the action of air and moisture, they are continually adding new impalpable matter to the soil, and as I shall show you hereafter the large quantity of this impalpable mineral matter which is annually removed by the crops, you will at once perceive that this constant addition must be of great value to the soil. This, therefore, is one important function performed by the stones of soil, viz. their affording a continually renewed supply of impalpable mineral matter.

(421.) When we come to consider the nourishment of plants, we shall find that their food undergoes various preliminary changes in the soil previous to its being made use of by the plants, and the aid of chemistry will prove to us that the effect is produced by the joint action of air and water; it follows, therefore, that soil must be porous. Now, this porosity of the soil is in part produced by the presence of the larger particles of matter, which, being of all varieties of shape, can never fit closely together, but always leave a multitude of pores between them; and in this manner permit of the free circulation of air and water through the soil.

(422.) As the porous nature of soil may, to a certain extent, be taken as an index of its power of retaining moisture, it is advisable to determine its amount. This is effected in the following way: Instead of putting the water first into the tube, as directed above (419), and shaking the soil into it, take a portion of soil dried by a heat of about 200° F. and shake it into the dry tube, and by tapping the closed end frequently on the table, make the soil lie compactly at the bottom; when you have fully effected this, that is, when farther tapping produces no reduction of bulk, measure accurately the column of soil, cork the tube, shake it till the soil becomes again quite loose, and then pour in the water as directed above (419.) After it has fully subsided, tap the tube as before, and re-measure; the increase of bulk is dependent upon the swelling of each particle by the absorption of water, and hence shows the amount of porosity. In very fertile soil I have seen this amount to one-sixth of the whole bulk.

(423.) The functions of the *impalpable* matter are far more complicated, and will require a somewhat detailed description. In this portion of the soil the mineral and organic matter are so completely united that it is quite impossible to separate them from each other; indeed, there are very weighty reasons for believing that they are chemically combined. It is from this portion of the soil that plants obtain all their mineral ingredients, and likewise all their organic portions, in so far as these are obtained by the roots; in fact, plants receive nothing from the soil, except water, which has not been associated with that portion which is at present engaging our attention.

(424.) The particles forming the impalpable matter are in such close apposition that the whole acts in the same way as a sponge, and is hence capable of absorbing liquids and retaining them. It is in this way that soil remains moist so near the surface even after a long continued drouth; and I need not tell you how valuable this property must be to the plants, since by this means they are supplied with moisture during the heat of summer, when otherwise, unless artificially watered, they would very soon wither.

(425.) Another most useful function of this impalpable portion is its power of separating organic matter from water in which it has been dissolved. Thus, for example, if you take the dark brown liquid which flows from a dunghill, and pour it on the surface of some earth in a flower pot, and add a sufficient quantity to soak the whole earth, so that a portion flows out through the bottom of the pot, this latter liquid will be found much lighter in color than before it was poured upon the earth, and this effect will be increased the nearer the soil approaches in its nature to subsoil. Now, as the color was entirely owing to the organic matter dissolved in it, it follows that the loss of color is dependent upon an equivalent loss of organic matter, or, in other words, a portion of the organic matter has entered into *chemical combination* with the impalpable mineral matter, and has thus become insoluble in water. The advantage of this is, that when soluble organic matter is applied to soil, it does not all soak through with the water and escape beyond the reach of the roots of the plants, but is retained by the impalpable portions in a condition not liable to injury from rain, but still capable of becoming food for plants when it is required.

(426.) Hitherto I have pointed out merely the *mechanical* relations of the various constituents of soil, with but little reference to their chemical constitution; this branch of the subject, although by far the most important and interesting, is nevertheless so difficult and complex that I cannot hope for the practical farmer doing much more than making himself familiar with the *names* of the various chemical ingredients, and learning their relative value as respects the fertility of the soil; as to his attempting to prove their existence in his own soil by analysis, I fear that is far too difficult a subject for him to grapple with, unless regularly educated as an analytical chemist.

(427.) Soil, to be useful to the British agriculturist, must contain no less than 12 different chemical substances, viz. silica, alumina, oxide of iron, oxide of manganese, lime, magnesia, potass, soda, phosphoric acid, sulphuric acid, chlorine, and organic matter; each of these substances must engage our attention shortly; and as I by no means purpose to burden your memories by relating all the facts of interest connected with them, I shall confine my observations almost solely to their relative importance to plants, and their amount in soil.

(428.) *Silica.* This is the pure matter of sand, and also constitutes on an average about 60 per cent. of the various clays; so that in soil it generally amounts to from 75 to 95 per cent. In its uncombined state, it has no *direct* influence upon plants, beyond its mechanical action, in supporting the roots, &c.; but, as it possesses the properties of an acid, it unites with various alkaline matters in the soil, and produces compounds which are required in greater or less quantity by every plant. The chief of these are the *silicates of potass and soda*, by which expression is meant the compounds of silica, or, more properly, silicic acid with the alkalies potass and soda.

(429.) *Alumina.* This substance never exists pure in soil. It is the *characteristic* ingredient of clay, although it exists in that compound to the extent of only 30 or 40 per cent. It exerts no *direct chemical* influence on vegetation, and is scarcely ever found in the ashes of plants. Its chief value in soil, therefore, is owing to its effects in rendering soil more retentive of moisture. Its amount varies from ½ per cent. to 13 per cent.

(430.) *Oxide of Iron.* There are two oxides of iron found in soils, namely, the protoxide and peroxide; one of which, the protoxide, is frequently very injurious to vegetation—indeed, so much so, that ½ per cent. of a soluble salt of this oxide is sufficient to render soil almost barren. The peroxide, however, is often found in small quantities in the ashes of plants. The two oxides together constitute from ½ to 10 per cent. of soil. The blue, yellow, red and brown colors of soil are more or less dependent upon the presence of iron.

(431.) *Oxide of Manganese.* This oxide exists in nearly all soils, and is occasionally found in plants. It does not, however, appear to exert any important influence either mechanically or chemically. Its amount varies from a mere trace to about 1½ per cent. It assists in giving the black color to soil.

(432.) These 4 substances constitute by far the greatest bulk of every soil, except the chalky and peaty varieties, but, nevertheless, *chemically speaking*, are of trifling importance to plants; whereas, the remaining 8 are so absolutely essential that no soil can be cultivated with any success unless provided with them, either naturally or artificially. And when you consider that scarcely any of them constitute 1 per cent. of the soil, you will no doubt at first be surprised at their value. The sole cause of their utility lies in the fact that they constitute the *ashes of the plants;* and as no plant can, by possibility, thrive without its inorganic constituents *(its ashes)*, hence no soil can be fertile which does not contain the ingredients of which these are made up. I shall not treat of each separately, but will furnish you with one or two analyses of soil to show their importance, and to impress them more fully on your memory. I regret that I must look to foreign works to furnish these analyses; but the truth is, we have not one single published analysis of British soil by a British chemist which is worth recording. Sir Humphry Davy just analyzed soil to determine the amount of the first 4 substances mentioned, and one or two others, and failed to detect 5 or 6 of the most important ingredients. In fact, the only useful analyses we possess are those performed by Sprengel, and quoted in Dr. Lyon Playfair's second edition of Liebig's Organic Chemistry applied to Agriculture, from which valuable work I quote the following examples.

(433.) Analysis of a *very fertile alluvial soil* from Honigpolder. Corn had been cultivated upon this soil for 70 years without *any manure* having been applied to it, but it was now and then allowed to lie fallow:

	Silica with fine silicious sand	64.800
	Alumina	5.700
	Peroxide of iron	6.100
	.. manganese	0.090
	Lime	5.890
	Magnesia	0.840
	Potass *combined with silica*	0.210
	Soda *combined with silica*	0.393
	Sulphuric acid *combined with lime*	0.210
	Chlorine *in common salt*	0.201
	Phosphoric acid *combined with lime*	0.430
	Carbonic acid *combined with lime*	3.920
Organic matter	Humus	5.600
	Humus *soluble in alkalies*	2.54[illegible]
	Azotized matter	1.58[illegible]
	Water	1.504
		100.000*

* Liebig's Organic Chemistry applied to Agriculture, 2d edit.

(434.) *Alluvial soil* from Ohio, remarkable for its *fertility*—

	Silica with fine silicious sand	79.538
	Alumina	7.306
	Protoxide and peroxide of iron, *with much magnetic iron-sand*	5.824
	Peroxide of manganese	1.320
	Lime	0.619
	Magnesia	1.024
	Potass *combined with silica*	0.200
	Soda	0.024
	Phosphoric acid *combined with lime and iron*	1.776
	Sulphuric acid *combined with lime*	0.122
	Chlorine *in common salt*	0.036
Organic matter,	Humus *soluble in alkalies*	1.950
	Humus *with azotized matter*	0.236
	Resinous matter and wax	0.025
		100.000

(435.) Loamy sand from the environs of Brunswick, *very barren*—

	Silica with coarse silicious sand	95.843
	Alumina	0.600
	Peroxide of iron	1.800
	Peroxide of manganese	a trace.
	Potass and soda	0.005
	Lime *combined with silica*	0.038
	Magnesia *combined with silica*	0.006
	Sulphuric acid	0.002
	Phosphoric acid *combined with iron*	0.198
	Chlorine *in common salt*	0.006
Organic matter,	Humus	0.502
	Humus *soluble in alkalies*	1.000
		100.000

Here the sterility is evidently produced by the small amount of potass, soda, lime, magnesia, and sulphuric acid—all of which are essential for the ashes of most of our usually cultivated crops.

(436.) These analyses will give you some idea of the complex nature of the soil, and the necessity of most *minute* analysis if we wish to ascertain its real value. The reason for such minuteness in analysis becomes obvious when we consider the immense weights with which you have to do in practical Agriculture: for example, every imperial acre of soil, considered as only 8 inches deep, will weigh 1884 tons, so that 0.002 per cent. (the amount of sulphuric acid in the barren soil) amounts to 80.64 lbs. per imperial acre.

(437.) I have purposely avoided saying anything of the *organic* matter of soil, as this is a most complicated subject, and will be far better considered under the head of manures.

(438.) All these substances, except the silica contained in the form of sand, constitute the impalpable matter of soil. It is evident, therefore, that this may differ much in chemical constitution without differing in amount, and yet have the greatest influence upon the fertility of the soil; my design, therefore, of introducing the words "*cæteris paribus*" in paragraph (418) was to induce you to bear in mind that the statement refers *solely* to soil considered *mechanically*. For fear of being misunderstood, therefore, I would paraphrase the sentence thus: Without a certain amount of impalpable matter, soil *cannot possibly be fertile*, yet, while the existence of this material proves the soil to be *mechanically* well suited for cultivation, *chemical* analysis alone can *prove* its absolute value to the farmer.

(439.) *Potass* and *soda* exist in variable quantities in many of the more abundant minerals, and hence it follows that their proportion in soil will vary according to the mineral which produced it. For the sake of reference, I have subjoined the following table, which shows the amount per cent. of alkalies in some of these minerals, and likewise a rough calculation of the whole amount per imperial acre in a soil composed of these, supposing such a soil to be 10 inches deep.

Name of Mineral.	Amount per cent. of Alkali.	Name of Alkali.	Amount per Imperial Acre in a soil 10 inches deep.
Felspar	17.75	Potass	927,360 lbs.
Clinkstone	3.31 to 6.62	Potass and Soda	161,000 to 322,000 lbs.
Clay-slate	2.75 to 3.31	Potass	80,500 to 161,000 lbs.
Basalt	5.75 to 10.	Potass and Soda	37,887 to 56,875 lbs.

(440.) From the above table you see the abundant quantities in which these valuable substances are contained in soil; some, however, of you, who are acquainted with chemistry, will naturally ask the question, How is it that these alkalies have not been long ago washed away by the rain, since they are both so very soluble in water? Now the reason of their not having been dissolved is the following; and it may in justice be taken as an example of those wise provisions of Nature, whereby what is useful is never wasted, and yet is at all times supplied abundantly.

(441.) These alkalies exist in combination with the various other ingredients of the rock in which they occur, and in this way have such a powerful attraction for each other that they are capable of resisting completely the solvent action of water so long as the integrity of the mass is retained. When, however, it is reduced to a perfectly impalpable powder, this attraction is diminished to a considerable extent, and then the alkali is much more easily dissolved. Now this is the case in soil; and, consequently, while the stony portions of soil contain a vast supply of these valuable ingredients in a condition in which water can do them no injury, the impalpable powder is supplied with them in a soluble state, and hence in a condition available to the wants of vegetation.

(442.) In the rocks which we have mentioned, the alkalies are always associated with clay, and it is to this substance that they have the greatest attraction; it follows, therefore, that the more clay a soil contains, the more alkalies will it have, but at the same time it will yield them less easily to water, and through its medium to plants.—H. R. M.]

---

22. PLANTING OF THORN-HEDGES.

"Next, fenc'd with hedges and deep ditches round,
Exclude th' encroaching cattle from thy ground."
DRYDEN'S VIRGIL.

(443.) IMMEDIATELY in connection with the subject of inclosures is the construction of the *fences* by which the fields are inclosed. There are only two kinds of fences usually employed on farms, namely, *thorn-hedges* and *stone-dykes*. As winter is the proper season for planting, or running, as it is termed, thorn-hedges, and summer that for building stone-dykes, I shall here describe the process of planting the hedge, and defer the description of building the dyke until the arrival of the summer season. It may be that the farm on which you have entered as a pupil, or that which you have taken on lease, may not require to be fenced with thorn-hedges. Still it is requisite that you should be made acquainted with the best mode of planting them.*

[* The dryness of our soil and climate, and yet more the want of *persistence* which characterizes American agriculturists, and which is so particularly requisite in the formation of a good hedge, will render the rearing of hedges a work of very limited extent and of doubtful success. Moreover, the liability of all estates to be again and again divided and subdivided, will coöperate with other reasons to the same end. Still, they answer well and are highly ornamental for small inclosures, and for that purpose we are inclined to believe the Maclura, or Osage Orange, will prove valuable as it is beautiful, as any one may see at Mr. Cushing's, near Boston.

As an agricultural topic, we confess we do not regard it as one of general interest, and might have omitted it altogether but for reasons already alleged, in similar cases. The subscriber who reflects that in Stephens's Book of the Farm he is getting a work that would cost him more than $20, will be content to put up with some things that may not have for him immediate value.

On large estates in the South, everything forbids the expectation that hedges will ever be resorted to as division fences; and on small ones in the North, stone supplies a more convenient material. Besides, we anticipate the extension of the soiling system, of which one great benefit will be, that cross fences may be dispensed with, and thus one of the greatest burdens on Agriculture be shaken off.

In the old American Farmer much may be found on the subject of hedges. It was very fully treated by Caleb Kirk, an intelligent practical Quaker farmer of Delaware. But what is now needed to be known by American cultivators who are disposed to make experiments in hedging, is well condensed in the following article from that popular and excellent periodical, the old Albany Cultivator:

HEDGES FOR AMERICA.—A great difference of opinion exists in relation to hedges for this country. There have been some very successful attempts, and there have also been many failures. An examination into the *causes* of this difference of success, in actual experiments, will doubtless be of use, and enable us to judge whether hedges possess advantages over other kinds of fence

(444) The proper time for planting thorn-hedges extends from the fall of the leaf, in autumn, to April, the latter period being late enough. The state of the ground usually chosen for the process is when in lea. I recom-

---

in any case. We lately examined several specimens of successful hedge-making. A part of them were made by John Robinson, of Palmyra, N. Y., a vigorous and enterprising English farmer, whose experiments are of several years' continuance. He has over a hundred rods of hedge in different stages of growth, the management or treatment of which appears to be particularly worthy of attention.

The young thorns are set out in the hedge-row at two years of age, after which they are cut off at the surface of the ground the first year, to cause a thick growth of sprouts; they are again cut off the second year, from four to six inches from the ground, according to their hight and vigor, which causes a second crop of thick sprouts at that hight; the third year they are cut off six or eight inches higher, and so on, rising about at that rate until the hedge is five or six feet high. This mode of treatment, which is well known and often practiced in England, obviates the necessity of *plashing*, if it is successfully performed; the successive crops of thick sprouts thus occasioned, densely interlace each other, and the hedge becomes a thick mass of entangled shoots and branches, which cannot be separated. It is in fact precisely similar to the process of *felting*, but on a larger scale; and when the best specimens thus grown are forcibly shaken at any point, whole rods on either side are shaken with it as in one mass. This felting property thus becomes of more value by far, to the impregnability of the hedge, than the thorns.

One hedge had received three different modes of treatment. A part had been imperfectly cultivated; another portion had been well cultivated for a distance of two feet on each side; and a third stood on ground which was trenched two feet deep before planting. The growth of the second was twice as great as the first, and of the trenched portion still greater. Indeed, one may as well think of raising corn by planting a row in a thick meadow, as to raise a good hedge without keeping the soil constantly mellow about the young trees. A space two feet wide on either side of the hedge is the distance usually kept cultivated.

From six to eight years are needed to make a good, substantial hedge, proof against cattle.

These hedges are set on a bank about eighteen inches above the surface, with a ditch two feet deep serving to carry off surface water on one side. The plants are set six inches apart. If closer, they do not grow so well.

The greatest difficulty which J. Robinson finds, is protecting the young hedge for several years, until it is proof against cattle. For, although it may be placed along the side of a fence, next to crops, or meadow, yet in the course of rotation it is thrown into pasture, and is thus endangered. A longer course of alternating crops would be the remedy in usual cases.

Hedges for *plashing* are not subjected to the successive shortening down which has been just described; but the young stems are suffered to grow until several feet high and an inch or more in diameter, when they are cut partly off near the ground and bent over to an angle of forty-five degrees in the direction of the line of the hedge. A thick growth of branches is not needed before this operation. All the *large* branches should be cut off at the time, but not closely. Young shoots afterward ascend, and growing upright, form cross-bars with the main stems which have been bent over, and interlocking with them produce a sort of lattice-work possessing ultimately great strength. A small portion of the trees are not bent, but remain upright, to stiffen the rest, and slender poles are run along the top, alternating with them, to keep them to their place until the whole is firmly established. These poles being green and of perishable wood, cost little, and rot out when they are no longer needed.

The selection of suitable trees for forming hedges, is of the very first importance. One great reason, without any doubt, why so many have failed in their experiments, is bad selection, or a want of adaptation of certain species to the climate where they were used. The English hawthorn has been found entirely unsuited to most parts of the United States. At Newburgh, according to A. J. Downing, "its foliage becomes quite brown and unsightly after the first of August." He also remarks that it is there extremely liable to the attacks of the borer. Farther south, where the summers are longer and dryer, and consequently more dissimilar to those of England, it is of no value whatever. But in the cooler summers of Western New-York, and where, perhaps, the soil may exert also a favorable influence, it has continued to flourish in *well-managed* hedges for many years. All the hedges of John Robinson, already described, are of this species; a very vigorous hedge, on the grounds of John Baker of Macedon, N. Y., is of the same. We had supposed that moist, rich land would be better suited to this thorn than dry upland; but in the experiments of these intelligent farmers it has been found that good fertile upland is incomparably better.

The sudden failures, however, of this thorn, in some places farther south, should induce its cautious use on a large scale, especially while American species have been found in most parts of the country so much superior. The Washington thorn, *(Cratægus cordata)* is preferred by some, and possesses the advantage of the seeds vegetating freely the first year. But in Pennsylvania and Delaware, where both this and the Newcastle thorn *(C. crus-galli)* have been extensively used for many years, the latter has in all cases been found so decidedly superior in hardiness, vigor and freshness of growth, to the former, as to give it eminently the preference. Indeed, the Newcastle thorn appears to be the only American species extensively tried, which has, *in all cases whatever*, proved to be entirely free from all disease or defect. It is not improbable, however, that the Washington thorn may succeed finely so far north as northern or western New-York, where the English species is itself so much more successful than elsewhere. Its easy growth from seed, besides, renders it worthy of trial. There are other trees, doubtless, of value for this purpose. The Buckthorn has been found perfectly hardy and successful around Boston; and the poisonous character of its bark secures it from attacks of the mice. Its thorns are only

mended lea as the best state for the process, in a paper on thorn-hedges which appeared some years ago;* but experience has since convinced me that this is *not* the best state of the ground for the purpose; because grass grows up from the turf around the young thorn-plants, and cannot be easily removed, but with the removal, at the same time, of a considerable portion of the earth upon which the young plants rest. A much better time, therefore, is after the ground has been thoroughly fallowed during the summer, that is, after it has been perfectly cleared of all weeds; well stirred and commixed with the plow and the harrow, and pulverized, if need be, with the roller; freshened by lengthened exposure to the air; amply manured with good dung, to promote the growth of the young thorn-plants; and sufficiently limed to prevent worms traversing the soil, and, in consequence, moles mining in quest of them. If the field in which the line of hedge is proposed to be planted is not intended to be thoroughly fallowed—that is, by a bare fallow or a crop of potatoes or turnips—the part to be occupied by the hedge should be so treated, in order to render the soil as clean, and fresh, and fertile as possible; and the expense incurred by this treatment of the soil will be repaid by the increased health and strength of the hedge for many years thereafter. There is no doubt that lea-sod affords a firmer bed for the young thorn-plants to rest upon than fallowed ground; but it is of much greater importance to secure the ground from weeds, and health and strength to the young plants, than mere firmness of soil under them, but which peculiar advantage may be attained, too, partly by allowing the fallowed ground to consolidate for a time before commencing the operation, and partly by trampling the soil thoroughly while in the act of planting.

(445.) The ground having been thus prepared, the planting of the hedge may be proceeded with forthwith. If its line of direction is determined by existing fences; that is to say, if one side of a field only requires fencing, then the new fence should be made parallel with the old one that runs N. or S., and it may take any convenient course, if its general direction is E. and W. Should a field, or a number of fields, require laying off anew, the N. and S. fences should run due N. and S., for the purpose of

the pointed ends of the branches, which are hardly sufficient to repel all kinds of intruders. Of its treatment by successive heading down, its *felting* quality, and its capability of plashing, we are not informed, as in nearly if not quite all the specimens we have seen, those operations were omitted.

The expense of a *well-made* hedge, until it is cattle proof, is about fifty cents per rod. Caleb Kirk, of Delaware, who was thorough and successful in his experiments, gave the following as the cost of an excellent hedge thirteen years old:

1,000 quicks, cost from nursery	$5 00	
Planting, man and boy, each two days	2 50	
Dressing, first year, with plow and hoe	1 00	
Expenses first year		$8 50
Dressing for five successive years, plow and hoe		5 00
7th year, trenching with plow, and throwing up ditch, three days	$3 75	
500 stakes (for uprights), cutting, and timber	3 50	
Poles (horizontal), and cutting them	2 00	
One hand three days, at plashing	3 00	
		12 25
8th to 13th year inclusive, one day each year trimming and cleaning		4 50
Expense 13 years, sixty rods		$30 25

It may be questioned whether hedges will ever be extensively used where timber or stones are plenty. But as many places are destitute, or likely to become so, experiments to determine their practicability must become very desirable. The disposition to neglect is so prevalent with most farmers, that *the great care and attention, and constant culture*, so necessary, will not be given, and success cannot take place in such cases. But with skillful management and enterprise they will doubtless be found highly profitable; that if good they will prove a great rural embellishment, we all know; and that those who have fruit gardens to protect from rogues, will find them the greatest security, is equally self-evident.]

* It will be found in the Quarterly Journal of Agriculture, vol. i.

giving the ridges an equal advantage of the sun both forenoon and afternoon. To accomplish this parallelism a geometrical process must be gone through; and to perform that process with accuracy, certain instruments are required.

(446.) In the first place, 3 *poles* at least in number, of at least 8½ feet in length, should be provided. They should be shod and pointed with iron at one end, marked off in feet and half-feet throughout their length, and each painted at the top of a different color, such as white, red, blue, green, or black, so as to form decided contrasts with each other when set in line. Three of such poles are required to determine a straight line, even on level ground; but if the ground is uneven, four or more are requisite. These poles will be found of use, not merely in lining off fences, but they will be required every year on the farm, to set off the breadths of the ridges of fields after being fallowed. 2. An *optical square* for setting off lines at right angles, or a *cross-table*, for the same purpose, should also be provided. The optical square costs 21s. and the cross-table 7s. 6d. 3. You should also have an imperial measuring-chain, of 66 feet in length, which costs 13s., for measuring the breadth or length of the fields, in the process of fencing; or of drills, drains, and any other species of work set by piece to laborers at other times. Iron pins, for marking the number of chains measured, generally go along with the chain.

(447.) Being provided with these instruments, one line of fence is set off parallel to another in this way. Set off, in the first instance, at right angles, a given distance from near one end of the old thorn-fence, if there be one, or of the ditch, and let this distance be 6 feet from the roots of the thorns, so that a space or scarcement of one foot on the edge or lip of the ditch be left, and there plant one of the poles. About 100 yards' distance plant another pole in the same manner, and so on along the length of the fence from which the distances are set off. If there be no fence to set off the distances from, then let a pole be set perpendicularly up in the line the new fence is intended to occupy, and at noon, in a clear day, observe the direction the shadow of the pole takes on level ground, and that is N. and S.; or a pocket-compass can give the direction required, deducting the variation of the needle, which in this country is about 27° W.; but the plan with the pole is the simplest and most handy for work-people. Poles, at about 100 yards' distance, should be set up in the line of the shadow; but you should bear in mind that the first two poles should be set up quickly, otherwise a short lapse of time will make a material difference in the line of direction of the shadow. Twenty minutes make a difference of 5° in the direction of the shadow of the poles, and 5° at the first pole will make a considerable deviation from the true line of N. and S. at the farthest end of the line of the new fence. Adjust the poles with one another to form the straight line, and this line forms the base line of your operations. This line is *c u* in fig. 36, projected by shadow in the manner just described, or set off from the old hedge *a b*. Let *c d* and *e* be 3 poles planted in that line. Let *f* be the cross-table erected in the line betwixt, and adjusted by looking at the poles *c* and *d*. Let *g*, *h*, and *i*, be poles set and adjusted to one another by the cross-table in the line *f k*, which is the breadth of the field, and which distance is measured by the chain to contain a number of ridges of given breadth, as any fractional part of a ridge left at either side of the field afterward proves inconvenient for work. In like manner, let the line *l p* be drawn from the cross-table at *l* by setting the poles *m*, *n*, *o*, *p*. Then set the pole *q* in a line with the poles *k p*, and measure the distance betwixt *q* and *u*, along the line *r s t*, with the chain, which distance, if the two previous operations have been accurately conducted, should be exactly equal to the distance betwixt *f* and *k*, or *l* and *p*; but should it prove greater or less than either, then some error must have been committed, and which can only be rectified by doing the

Fig 36.

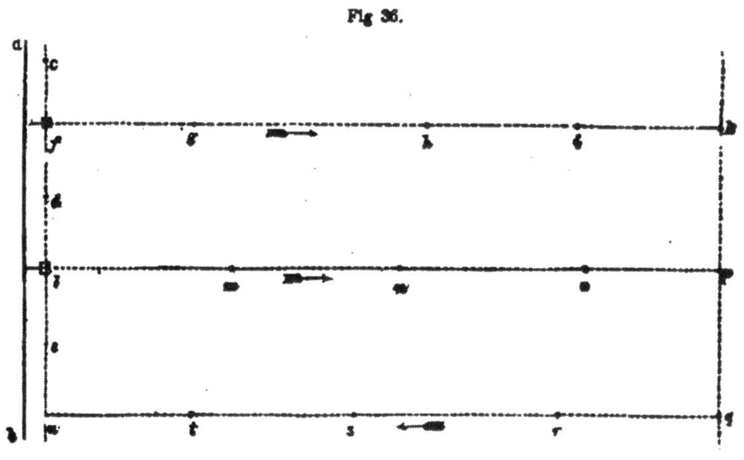

PLAN OF SETTING OFF FENCES PARALLEL TO EACH OTHER.

operation over again. The *arrows* show the directions in which each line should be measured. Great accuracy should be observed in running these lines of fences parallel, for if a similar error is committed at each successive line of fence, the deviation from parallelism may prove very considerable betwixt the first and last lines. Three poles only being employed to set off the lines *f k* and *t p*, the ground may be supposed to be nearly level; but wherever such an inequality of ground is found as to cause you to lose sight of 1 of 3 poles, as many should be employed as to have 3 of them in view at one time. This point should be constantly kept in view in setting the poles.

(448.) A line of fence being thus set off, the next process is to plant it with thorns, and for this purpose certain instruments are required. 1. A strong *garden line* or *cord*, of at least 70 yards in length, having an iron reel at one end, and a strong iron pin at the other. Its use is to show upon the ground the exact line of the fence betwixt the poles. Its cost is, with a common reel and pin, 4s. 2. A few *pointed pins of wood with hooked heads*, to keep the cord in the direction of the line of the hedge, whether that follows a vertical curve or a horizontal one, occasioned by the inequalities of the ground. 3. A *wooden rule*, 6 feet in length, divided into feet and inches, having a piece of similar wood about 2 feet in length, fastened at right angles to one end. Its use is to measure off short distances at right angles. Any country carpenter can make such a rule. 4. No. 5 *spades* are the most useful size for hedging, which cost 4s. 3d. each. 5. A light *hand-pick*, to loosen the subsoil at the bottom of the ditch and to trim its sides, and it costs 5s. 6d. or 6s. 6. An iron *tramp-pick* to loosen the subsoil immediately under the mould, and raise the bowlder stones that may be found in it. In some parts of the country this pick is unknown, but a more efficient implement cannot be employed for the purpose. This pick

Fig. 37.

A TRAMP-PICK.

stands 3 feet 9 inches in hight. The tramp, fig. 37, is movable, and may be placed on either side, to suit the foot of the workman, where it remains firm at about 16 inches from the point, which gradually tapers and inclines a little forward, to assist the leverage of the shank. The shank is ¾ of an inch square under the eye through which the handle passes, and 1½ inches broad at the tramp, where it is the strongest. It costs 6s. 6d. 7. A *ditcher's shovel*, fig. 38. Its use is to shovel the bottom and sides of the ditch, and to beat the face of the hedge-bank. It is 1 foot broad and 1 foot long, tapering to a point, with a shaft 28 inches in length, and its cost is, No. 5, 4s. This is a useful shovel on a farm, cleaning up the bottoms of dunghills in soft ground much better than a spade or square-mouthed shovel; and yet in some parts of the country it is an unknown implement. 8. *Three* men are the most convenient number to work together in running a hedge; and they should, of course, be all well acquainted with spade-work. 9. Should tree-roots be apprehended in the subsoil, a *mattock* for cutting them will be required, and it costs 6s. 6d. 10. A sharp *pruning-knife* to each man, to prepare the plants for planting, which costs 2s. to 3s. each.

Fig. 38.

A DITCHER'S SHOVEL.

Fig. 39.

A THORN-PLANT.

(449.) The *plant* usually employed in this country, in the construction of a hedge, is the common hawthorn. "On account of the stiffness of its branches," says Withering, "the sharpness of its thorns, its roots not spreading wide, and its capability of bearing the severest winters without injury, this plant is universally preferred for making hedges, whether to clip or grow at large."* Thorns ought never to be planted in a hedge till they have been transplanted at least 2 years from the seed-bed, when they will have generally acquired a girth of stem at the root of 1 inch, a length in all of 3 feet, of which the root measures 1 foot, as in fig. 39, which is on a scale of 1¼ inches to 1 foot. The cost of picked plants of that age is 12s. 6d. per 1,000; or, as they are taken out of the lines, 10s. 6d. As thorns are always transplanted too thick in the nursery lines, in order to save room, and draw them up sooner to be tall plants, I would advise their being purchased from the nursery at that age, the year before they are intended to be planted in the fence, and of being laid in lines in ample space in garden mould, or any space of ground having a free, deep, dry soil. By such a process the stems will acquire a cleaner bark and greater strength, and the roots be furnished with a much greater number of minute fibres, which will greatly promote the growth of the young

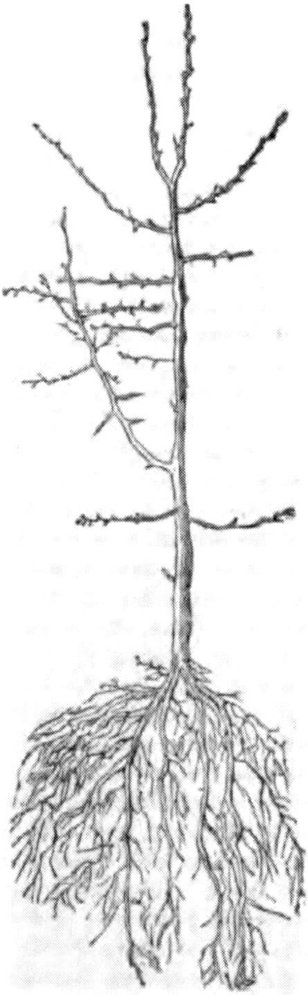

* Withering's Botany, vol. iii.

hedge, and thus amply repay the additional trouble bestowed on the care of the plants. But, whether the plants are so treated before they are planted or not, the bundles, containing 200 plants each, should be immediately loosened out on their arrival from the nursery, and *sheughed in*, that is, spread out upright in trenches in a convenient part of the field, and dry earth well heaped against them, to protect the roots from frost, and to keep them fresh until planted. The plants are taken from the *sheughs* when wanted.

(450.) If the line of fence is to be straight, which should always be the case if natural obstacles do not interfere to prevent it, let the poles be set up in as straight a line as possible from one end of the fence to the other. Should the ground be a plain, this line can be drawn straight with the greatest accuracy; but should elevations, or hollows, or both, intervene, however small, great care is requisite to preserve the straightness of the line, because on such ground a straightness of line, determined by poles, is very apt to advance upon the true line in the hollows, and recede from it in the elevations, especially if the inequalities are abrupt. Surveyors use the theodolite specially to avoid this risk of error, but it may be avoided by using plenty of poles, so that they may not be set far asunder from one another. In case evil disposed persons shift the poles in the night, and thereby alter the line of fence, pins should be driven at intervals, well into the ground, to preserve the marks of the line. Having set plenty of poles, and so as to please the eye, take the reel and cord, and, pushing its pin firmly into the ground at the end of the line of fence where you wish to begin, run the cord out its full length, with the exception of a small piece of twist round the shank of the reel. Be sure to guide the cord exactly along the bottoms of the poles; and should any obstacle to your doing so lie in the way, such as clods, stones, or dried weeds, remove them, and smooth the ground with the spade; and then, with your face toward the cord, draw it backward toward you with considerable force until it has stretched out as far as it can, and then push the shank of the reel firmly into the ground. As the least obstruction on the ground will cause the cord to deviate from the true line, lift up the stretched cord by the middle about 3 feet from the ground, keeping it close to the sides of the poles, and let it drop suddenly to the ground, when, it is probable, it will lie as straight as practicable. Place a rather heavy stone here and there upon the cord to prevent the possibility of its being shifted from its position. With the common spade then cut, or as it is technically termed, *rut* the line of hedge-bed behind the cord, with your face toward the ditch that is to be, taking care to hold the spade with a slope corresponding to that of the sides of the proposed ditch, and not to press upon, or be too far back from, or cut the cord with the spade. Then take the wooden rule, and placing its cross-head along the cord, set off the breadth of the ditch at right angles to the rutted line 4½ feet—first, at both ends of the still stretched cord, and then here and there; and mark off those breadths with wooden pins, which will serve to check any important deviation from the true line at either end of the cord. Now, take up and stretch the cord anew along the other side of the ditch, by the sides of the pins, in the same manner, and with the same precautions as with the hedge-bed, and rut the line with your face toward, and the spade sloping like the side of the ditch. After securing a continuation of the line of the hedge-bed, remove the poles and pins along the length of the cord, and the ditch is thus marked out ready for the formation of the thorn-bed. When about forming the thorn-bed, that end of the line should be chosen for commencing the work which best suits the hand of the workman who is intrusted to make it. The rule for this is, whichever hand grasps the eye of the spade should always be nearest the thorn-bed, and the workman should work backward.

(451.) In forming the *thorn-bed*, raise a large, firm, deep spadeful of earth from the edge of the first rutted line of the hedge, and invert it along that line, with its rutted face toward the ditch. Having placed a few spadefuls in this manner, side by side, beat down their crowns with the back of the spade, paring down their united faces in the slope given to the first rut, and then slope their crowns with an inclination downward and backward from you, forming an inclined bed for the thorn-plant to lie upon as at *b c*, fig. 40. In like manner, place other spadefuls, to the end of the

Fig. 40.

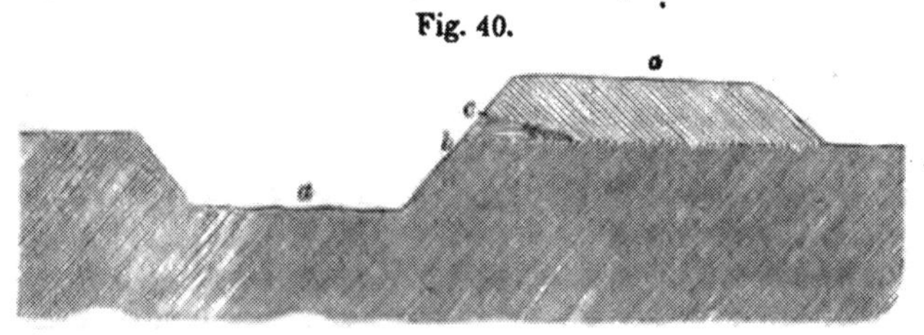

THE THORN-BED.

thorn-bed last made, taking care to join all the spadefuls so as to make one continued bed, and so on to the whole length of the cord of 70 yards.

(452.) While the principal hedger is thus proceeding with the thorn-bed, his two assistants should prepare the thorn-plants for planting. On receiving the thorn-plants from the nursery, the usual practice is to put the bundles of plants into the soil in some convenient corner of the field, until they are wanted for planting. I have recommended the plants being purchased the year before they are to be planted, and transplanted in wide lines in good garden-mould, to enlarge and multiply the root-fibres. And now that the plants are more particularly to be spoken of, I would farther recommend them to be assorted, according to their sizes, as they are taken out of the bundles, and before being transplanted in the lines. The advantage of this plan is this. Plants should be suited to the situation they are to occupy. On examining the bundles, they will be found to contain both stout and weak plants. The stoutest plants cannot derive sufficient nourishment in the poorer class of soils, however well the soils may have been previously treated for their reception; while weak plants will, of course, thrive well in the better soil. From this circumstance, it may be concluded that weak plants are best adapted to all classes of soils. Not so; for however well weak plants may thrive in all soils, stout plants will grow much more rapidly than weak in good soils; and were all the soils good, the most profitable fence would be obtained from the best and picked plants. But as every farm possesses soils of various degrees of fertility, although the class of its soils may be the same; and as plants in a stout and weak state are usually mixed together, the most prudent practice is to put the weaker plants in the best soil, and the stouter plants in the worse kind of soil, thus giving a chance of success to both sorts of plants and soils. Were the plants assorted when placed in transplanted lines, those could be selected which would best suit the soil which was under operation at the time. But should this trouble not be taken at first, still the plants should be assorted when being prepared for planting, according to the nature of the soil, the weaker being taken for the good soil, and the stronger for that of inferior quality. Want of attention to this adaptation of means to ends is one cause of failure in the rearing of thorn-hedges in many parts of the country; and one of those means consists in transplanting the weakest plants in good soil, and allowing them to remain there until they had acquired sufficient strength for being planted out. Although the thorn-plant may truly be said to affect every kind of soil in cultivation, yet the plant, in its different states of growth, will thrive better in one condition or kind of soil than in another; and this discrimination should be exercised by the planter, if he would have a good hedge.

(453.) The prepared thorn-plant is represented by fig. 41; and it is prepared in this way. Grasp the stem of the full plant, immediately above the root, firmly in the hand, and cut it across with a sharp knife, in an inclination toward the top of the plant at *a*; and the cut thus made will be about 6 inches above the root and fibres. Cut away the long parts of the tap-roots *b*, and any other straggling and injured roots, and even injured fibres; but preserve as many of the fibres entire as possible. Burn the tops thus cut off, or bury them deep in the ground; as they will vegetate, and are easily blown about by the wind, and very troublesome to sheep in the wool. Take great care, in frost, to cover up the prepared roots in earth until they are planted, for roots in the least affected by frost will not vegetate. The safest plan, in frosty weather, is to take but a few plants at a time out of the lines. On the other hand, in dry weather in spring, when the hedge is to be planted in dry ground, put the roots of the prepared plants in a puddle of earth and water, in a shady place, for some hours before laying them in the thorn-bed, and their vegetation will thereby be much facilitated.

Fig. 41

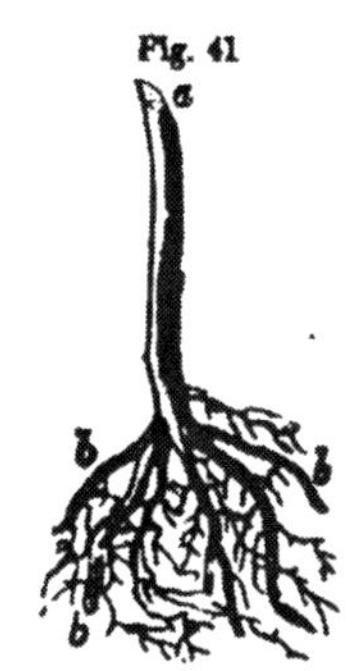

A THORN-PLANT PREPARED FOR PLANTING.

(454.) When both the thorn-bed and plants are prepared, the assistants lay the plants in the bed. This is done by pushing each plant firmly into the mould of the bed, with the cut part of the stem projecting not more than ¼ of an inch beyond the front of the thorn-bed, and with the root-end lying away from the ditch, at distances varying from 6 to 9 inches; the 6 inches being adapted to inferior land, and the 9 inches to good soil. While the two assistants are laying the plants, the hedger takes up all the fine mould nearest the thorn-bed, and, dexterously inverting the shovel-fulls of the mould, places them above the laid plants, and secures them in their places. The two assistants having finished laying the thorns, dig and shovel up with the spade all the black mould in the ditch, throwing it upon the roots and stems of the plants, until a sort of level bank of earth is formed over them. In doing this, one of the assistants lifts the soil across the ditch, moving backward, while the other proceeds forward, face to face, shoveling up all the black mould he can find, whether in a loose or firm state, in the ditch. When the hedger has finished covering the plants with mould, and while the assistants are proceeding to clear all the mould from the ditch, he steps upon the top of the mound which they have thrown up above the plants, and, with his face toward the ditch, firmly compresses, with his feet, the mould above the plants, as far as they extend. By the time the compression is finished, all the mould will have been taken out of the ditch. When the thorns have received this quantity of earth above them, they may be considered in a safe state from the frost; but it is not safe, in frosty weather, to leave them, even for a night, with less earth upon them; for plants may not only be frosted in that short space of time, but the earth may be rendered so hard by frost, as to be unfit for working the next day; and should the frost prove severe and the work be altogether suspended, the plants left at all exposed will inevitably perish. In frosty weather the plants should not be laid on the thorn-bed in the afternoon, but only in the forenoon, as in the afternoon of a short day there probably will not be time to cover the plants with a sufficient quantity of earth. Indeed, in such weather, when the ground continues hard all day, leave the work off altogether—not only because the earth is then in an unfit state for the work, but the frosted earth is apt to chill the tender fibres. On the other hand, if the weather be fresh and not too wet, such as in spring, the plants may be laid in the afternoon in safety. In very wet weather, the work should also be suspended, not only on account of the cloggy state of the ground for good work, but the inability of the men to withstand much rain in winter. The last operation of the ditch and bank will be more uniform and look better, when a considerable length of it is finished at the same time, than when joinings are visible in it at short intervals; but in frosty or in very wet weather, the sooner a piece of it is finished, the better it is for the cleanliness of the laborers and the condition of the work itself. Fig. 39 shows the progress of hedge-planting to the extent described, where *a* is the ditch with the mould all taken out, *b* the thorn-bed sloping inward and downward, *c* the thorn in its bed, with the end of the stem projecting a very little outward, and *d* the mound above in its compressed state.

(455.) The rule observed for the depth of a ditch that stands well, is ½ its breadth, and the width of the bottom ⅙ of the breadth at the top. In the case of hedge-planting, the breadth is 4½ feet; the depth is, of course, 2 feet 3 inches, and the width of bottom 9 inches. The hedge-bank is always broader than the ditch, the soil lying loosely upon it, and in this case is 5 feet; and, of course, the perpendicular hight of the bank is less than the depth of the ditch, being 2 feet. These are, in general, very convenient dimensions for a hedge ditch and bank, where no constant current of water has to be accommodated in the ditch; but should the ditch have to contain a stream of water, though in winter only, it should be made proportionably capacious; for if not so made at first, it will either have to be made so at last, or the force of the water will assuredly make an adequate course for itself, to the danger of destroying the thorn-bed. Ditches that are brought to a point at the bottom are objectionable in shape for many reasons. They do not afford sufficient materials out of them to form a protective bank or mound for the young thorn-plants; they are easily filled up with the mouldering of earth from the sides and tops, and the decay of vegetables; and when any water gets into them, of which there is every chance when there is an overflow of surface-water in the field, they soon get filled up in the bottom with mud. Notwithstanding the commendation of such ditches in works of Agriculture,* they should be avoided when there is the probability of the least quantity of water reaching them; and no ditch in connection with a field can be exempt from the intrusion of water.

(456.) When the work has proceeded to this length, the other implements come into use. If the subsoil of the ditch, however, be a tenacious, ductile clay, the spade alone is best to remove it, as picking is useless in such a substance, especially if somewhat moist; for it will raise no more at a time than the breadth of the face of the pick. But if it consists of hard dry clay interspersed with veins of sand and gravel—which compound forms a very common subsoil in this country—picking is absolutely required; for the spade cannot get through the small stones with effect. In some parts of the country, the handpick is used to loosen such a subsoil, while in others the footpick is employed; and from experience in

* Communications to the Board of Agriculture, vol. ii. Loudon's Encyclopædia of Agriculture.

(427)

both, I would recommend the latter as being by far the more efficient implement for such work, and less laborious to the workman. Let one of the assistants loosen the subsoil with the footpick as deep as he can go for the tramp, with the point of the pick away from him; he then pulls the handle toward him, until he brings it down about half way to the ground, and after that he sits on it, and presses it down with the whole weight of his body, until the subsoil gives way and becomes loose, in which state he leaves it before him, and steps backward. When the picker has thus proceeded a short way, the other assistant lifts up what has been loosened with his spade, and throws it upon the top of the mould above the thorn, taking care to place the subsoil so thrown up continuous with the slope backward, given to the face of the bank. He also throws some to the back of the bank, to cover the whole of the black mould with the subsoil; and endeavors to make the shape of the bank uniform. In doing all this, he works backward with his back to the face of the footpicker, but his back would be to the back of a handpicker, standing upon the subsoil which has been loosened by the footpick. He pares down the side of the ditch nearest his right hand, which, in this case, is the opposite one from the hedge. The hedger follows the last assistant, working toward him face to face, and moving forward, shoveling up all the loose earth left by the assistant's spade, throwing it upon the top and front of the mound, making all equal and smooth, and beating the earth firmly and smooth on the face of the bank. Should the subsoil require no picking at all, the two assistants follow one another, using the spade; and the hedger brings up the rear as before, using the shovel. In this way the hedger throws the earth fully on the face of the bank, even although some should trickle down again into the ditch, rejecting all the larger stones that come in his way, paring down that side of the ditch, giving the proper slope to the bank, and beating the face of the bank with the back of the shovel, and smoothing it downward from its top as far as the black mould is seen on the side of the ditch. The three men thus proceed regularly in their work. Should there be more earth at one place of the ditch than another —which will be the case where there are inequalities in the depth of the ditch—the surplus earth should either be thrown to the back of the bank, rather than its top be made higher at one place than another, or wheeled away to a spot on which a deficiency of earth is apprehended. Besides giving the bank an irregular appearance, it is not desirable to cover the young thorns too heavily with a superincumbent load of earth, so as entirely to exclude the air and moisture from their roots.

(457.) If going along the ditch twice finish the work, the earth in it will have been in a friable state; but with a hard subsoil the work is not so easily done. The handpick is almost always used to raise the last 4 or 5 inches of the bottom of the ditch, and in accomplishing this the same arrangement of the men, and the kind of work performed by each, will have to be gone through; only that, in this case, the assistant uses the hand for the tramp-pick, and works forward. While this last picking and shoveling are proceeding, the hedger again tramps down the top of the bank before throwing up the last portion of earth. The beating with the back of the shovel is absolutely necessary to produce a skin, as it were, on the face of the bank; because the smoothed surface will resist the action of the frost, and thereby prevent the mouldering down of the earth into the ditch. A covering of clay over the bank, and the poorer it is the better for the purpose, is useful in being extremely unfavorable to the vegetation of small seeds. They will readily take root in fine mould, if that formed the external covering, and their eradication afterward would create much trouble and cause much waste of earth. The necessity of beating the clay shows the expediency of projecting the plants but a very short way out of the bank, as that process might wound and injure the points of the stems. Indeed, I would prefer their being nearly buried in the bank, so as the young sprouts had to be relieved from captivity, rather than the points should be injured; but the force of vegetation generally accomplishes their release with ease. While the two assistants are preparing the cord for another stretch, and rutting off both sides of the ditch, the hedger pushes back 2 or 3 inches, less or more, of the crest of the bank with his shovel, in order to make the finished top parallel with the row of thorns, and after he has gently beaten down the front of the top into a rounded form, the process of planting thorns is finished. Fig. 42 gives an idea of a section of the whole work when finished.

Fig. 42.

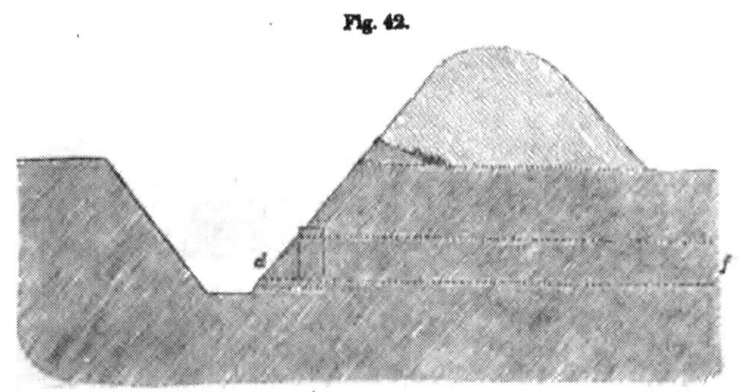

FINISHED HEDGE-BANK.

(458.) Hitherto the work has proceeded quite easily, no obstacles having presented themselves to frustrate or alter the original design of a level fence; but obstacles are sometimes met with, and means should be used to avert or remove them. The obstacles alluded to generally consist of large stones, unequal ground, and stagnant water. 1. Landfast stones are frequently found in clayey subsoils, many of which can be removed with the foot pick, but some are so large and massive as to defy removal but through the assistance of gunpowder. If you should meet with any such enormous masses, and much above ground, it would be better to carry the hedge with a sweep past them, than incur the trouble and expense of removing them with the simplest means. If they lie a short way under the thorn-bed, but have plenty of mould over them, they will do no harm to the hedge above them; but should the earth be scanty over them, it will be proper to make the earth deep enough for thorns above them, if that can be easily done, even although an elevation be thereby caused there, above the general line of hedge. 2. With regard to inequality of surface, when the ground dips in the direction of the hedge, and yet when particular undulations in it are so deep and high as to prevent the flow of water over them in the ditch, the higher parts should be cut the deeper and the hollow parts the less, so as a continuous fall may be obtained for the flow of the water along the bottom of the ditch; but the line of the hedge should be placed on the natural surface of the ground, and thereby partake of its undulations. It is in such cases of compromise that the superabundant earth should be wheeled away from the inordinate depths, to make up for the want of earth in the hollows, and thereby equalize the dimensions of the hedge-bank. Should any hollow be so deep as that the hight on either side will not allow the flow of water, a drain should be made from the hollowest part of the bottom of the ditch down the declination of the adjoining field to some ditch or drain already existing at a lower level. 3. Undulations of the ground cause another inconvenience in hedge-planting, by retaining water in the hollows behind the hedge-bank. Such collections of water, though only of temporary existence, injure much any hedge, but especially a young one. The only effectual way of getting rid of them is fortunately a simple one, which is by constructing a conduit through the hedge-bank from each such hollow to the bottom of the ditch; and as these conduits must be founded upon the subsoil, completely under the black mould, and a little above the bottom of the ditch, they are most conveniently built after the ditch has been entirely dug out; and on this account the thorn-bed cannot be formed across these hollows until after the completion of the ditch and hedge-bank on both sides of them. Some taste and dexterity are required in the hedger to fill up the gaps thus left in the planting of the hedge and finishing them neatly afterward. Fig. 43 will give you an idea how to overcome the in-

Fig. 43.

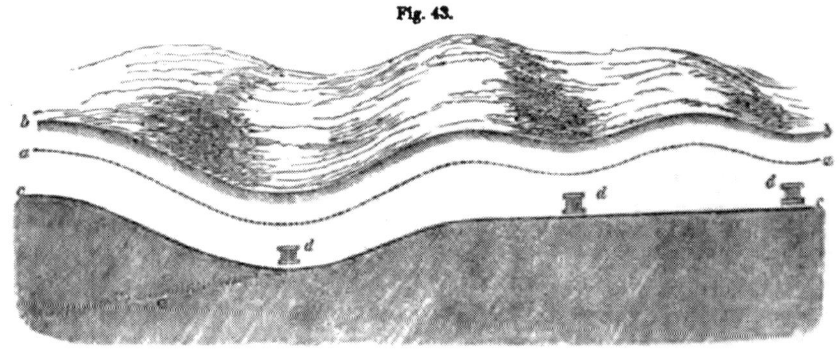

PLAN HOW TO PREVENT WATER LODGING IN HOLLOWS OF FENCES.

convenience created by these hollows, where *a* is the line of hedge upon the natural surface of the undulating ground, *b* the top of the hedge-bank parallel to the hedge, *c* the bottom of the ditch, exposed to view by the entire removal of the ground on this side of the ditch, and which removal also shows the positions of the conduits *d*, which carry the stagnant water away from behind the hedge-bank through below the hedge in the lowest part of the undulations of the ground, and it also shows the position of the drain *e* through the adjacent ground. It will be observed that the bottom of the ditch *c* is not quite parallel with the dotted line of hedge *a*, but so inclined from the right and left, through the hights and hollows of the ground, as to allow the water to flow in a continuous stream toward the lowest part by the drain *e*. Fig. 42 shows by the dotted lines *d* and *f* a vertical section of the position and form of the conduits formed across and below the hedge-bed. The ground behind the hedge-bank is represented in fig. 43 as declining toward the hedge, thereby giving a fall to the surface water in the same direction. To give such water an outlet, a drain should be formed along the head-ridge 2 or 3 yards behind the hedge-bank, so as to be a little out of the way of the roots of the thorns when they push outward, and in connection with all the conduits *d*. This drain should have a conduit at bottom such as drain-tiles afford, and be filled above them with broken stones to about 1 foot from the top.

(459.) In ordinary practice, when two lines of hedges meet, the one terminates against the other, or, crossing each other, form a junction of 4 fields by the corners; and where this latter junction happens, should the land be not of much value, or should the particular situation be much exposed to the weather from an obnoxious quarter, it may be advisable to make a clump of planting of a stellar form. It is necessary, in the first place, to ascertain what quantity of ground can be conveniently spared for the purpose; and that should be determined by the value of the ground, or its exposed situation. If the land is valuable, a smaller piece must suffice; but if shelter only, and not ornament, is the chief requisite, then a larger piece should be appropriated; but whatever may be the object of forming such a clump of planting, it is not worth while to inclose a smaller space of ground than ¼ of an acre, and the largest need not exceed 1 acre in the low country. Supposing the space is determined on, the inclosure of it is gone about in this manner. Ascertain the point where the two lines of hedges would intersect, and fix a pole there, as at *a*, fig. 44; and from it

Fig. 44.

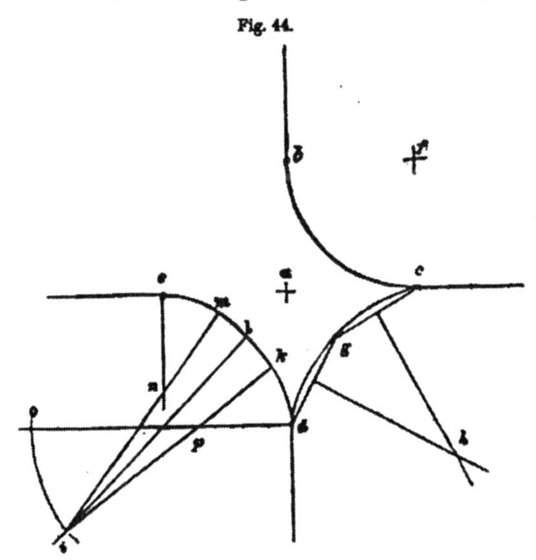

MODES OF DESCRIBING A CURVE IN THE CORNERS OF FIELDS.

measure equal distances with a chain along each line of fence to the points within which is to be included the space of ground allotted for the planting, as from *a* to *b*, *a* to *c*, *a* to *d*, and *a* to *e*. Then there are 3 ways of describing an arc between any two of these outward points. 1. Taking the distance *a b* from *b* as a center, sweep an arc, and from *c* as a center, with the same radius, sweep another arc intersecting that from *b* in *f*; and then from *f* as a center, still with the same radius, sweep the arc *c b*. In like manner an arc of the same radius may be swept betwixt *c* and *d*, *d* and *e*, and *e* and *b*. This rule gives no predetermined arch, but it is one which presents a pleasant curve to the eye. 2. Another plan is to fix the hight of the segment which determines the point, beyond which the hedge shall not approach toward *a*. This is done by at once fixing the point *g*, which gives 3 points, *d*, *g*, and *c*, by which to find the center of the circle *c d*. Join *g d*, which bisect, and from the point of bisection raise a perpendicular; also join *g c*, which bisect, and from the point of bisection raise a perpendicular, and where these two perpendiculars intersect at *h* as a center, sweep the arc *d c*. This rule is founded on the corollary to the 1st problem of the 3d book of Euclid.* A simple rule which practical gardeners employ in drawing one line at right angles to another is this:

* See Duncan's Elements of Plane Geometry.

From the point of bisection, as above, measure 6 feet along the line toward *c* or *g*, from the same point also measure outward 8 feet; from the farther end of the 6 feet measure 10 feet, toward the end of the 8 feet, and where these two lines meet, that is the point in a perpendicular direction from the point of bisection, and a line through which, meeting a perpendicular from the other point of bisection, intersect at the center *h* of the circle *d c*. This rule is directly founded on the celebrated 47th proposition of the 1st book of Euclid. 3. There is still another method of drawing what may be called a compound curve through two given extreme points, and other fixed points between them. The method is this. Let *d* and *e* be the terminations of the straight lines of the fences *d* and *e*, and *l* a point in the intended curve any where beyond the straight line between *d* and *e*, and equidistant from *d* and *e*, but within a quadrant of the two lines of fence; then set off any point *i* also equidistant from *d* and *e*, and join *i l*; from any point on the line *i l*, describe an arc of such radius as shall pass through *l*, but will fall anywhere beyond *d* and *e*. Draw *d o* at right angles to the fence *d*, and make *d o* equal to *i l*, then find a point *p* on the line *d o* equidistant from *o* and *i*. Join *i p*, and produce it to *k*, and from *p* as a center describe the arc *d k*. For, *d o* and *i k* being equal, and *p o*, *p i* being also equal, the remaining *p d* and *p k* must be equal to one another and *i p k* being in a straight line, the circle of which *d k* is an arc, will touch the larger circle, of which *k m* is also an arc, according to Euclid, 3d book, 11 prob. In like manner, the arc *e m* can be described by first drawing *e n*, at right angles to the line of fence *e*, and proceed as before. If the lines of fence run at right angles to each other, the arcs *d k* and *e m* will have equal radii. This is, perhaps, too intricate a mode of drawing such curves for practical purposes, but it is well that your ingenuity be exercised in every possible way, so as you may never be at a loss to apply expedients according to circumstances.

(460.) A very common practice—a much too common one—and recommended by almost every writer on planting hedges, is the leaving a broad scarsement in front of the thorn-bed; and the reason given for adopting the plan is, that it is necessary to supply the young thorns with moisture.

Fig. 45.

EFFECT OF A HEDGE-BANK WITH SCARSEMENT.

It is alleged that the sloping face of the bank conveys away the rain that falls. What although it does? The young thorn does not require to imbibe moisture by the point of its stem, but by its roots, which it can easily do through the mound, as it is loose enough for the admission of rain.—

But, independently of that, it is obvious that a scarsement is so excellent a contrivance for the growth of weeds, that it is impossible to clean a hedge well where there is one. To be sure, earth from the bottom of the ditch may occasionally be thrown upon the scarsement to smother the weeds, but its accumulation there must be limited to the hight of the thorn-bed. Besides, weeds can grow as well upon this earth as upon the scarsement; and, though they may there be mown down at times, the roots of the perennial ones are quite ready to spring up again in favorable weather. The very figure which a thorn-hedge cuts on a scarsement will at once show the impolicy of placing it in such a position. Thus, in the first place, in fig. 45, *a* is the scarsement, on which there is nothing to hinder the weeds *b* to grow in great luxuriance, vying in stature and strength with the young plant *c* itself. How true that there "nothing teems but hateful docks, rough thistles, kicksies, burs, losing both beauty and utility; and our hedges, defective in their natures, grow to wildness."* How is it possible in such a nursery to "deracinate such savagery?" In the next place, such a scarsement holds out a strong temptation to travelers to make it a foot-path, so long as the hedge is young, and when it is situated by the side of a public road. And it invites the poor woman's cow, pasturing on the green road-side, to step upon it and crop the tops of the young hedge along with the grassy weeds; and it makes an excellent run for hares, in the moonlight nights, on passing along which they will not fail to nibble at the young quicks. "Fern is a great enemy to young hedge-plants," says Mr. Marshall; "it is difficult to be drawn by hand without endangering the plants; and, being tough, it is equally difficult to cut it with the hoe; and, if cut, will presently spring up again;" and yet, "in a soil free from stones and other obstructions of the spade," he says, "the planting with an offset (scarsement) is perhaps, upon the whole, the most eligible practice."† Where can fern obtain a better site for growing upon than a scarsement of a young hedge? Such are the inconsistencies into which the acutest writers fall when they relinquish the guidance of common sense.

(461.) Where part of a hedge is desired to be carried across a water-course, an arch or large conduit is often made to span it, and its sides are banked up with sods or earth, and a quantity of mould wheeled upon it, to form the thorn-bed. I have seen such structures, but do not approve of them. If the nature of the ground will at all admit of it, it is far better to plant the thorns on the surface of the natural ground, as near as possible to the water-mark, when the water is flooded. The water-channel, which will probably be dry in summer, when the fields are only used for stock, could be fenced with paling, or, what is a much better fence in such a situation, a stone-wall, if stones can be procured at a reasonable distance, with openings left in it to allow the water to pass through in winter.—These openings could be filled up in summer with a few thorns, to keep in sheep. This latter plan is a much better one than the other, for I have found that hedge-banks on a stone-building do not retain sufficient nourishment in summer to support even young thorn-plants.

(462.) If it is desired to plant a thorn-hedge on the top of a sunk fence, or along the edge of a walk by the side of a shrubbery, or to inclose a shrubbery or a clump of trees in pleasure-ground or lawn, the plants may be assorted and prepared as directed above; but instead of raising a mound, which in such situations would not look well, trench a stripe of ground with the spade, in the intended line of the hedge, at least 3 feet in

* Shakspeare's Henry V. † Marshall on Planting.

breadth, pointing in dung and raking in lime in adequate quantities some time before the time for planting. When that time arrives, stretch the cord in the middle of the stripe, guiding the curves with the wooden pins. First, smoothen the surface of the ground under the cord with a clap of the spade, and then notch deeply with it by the side of the cord, drawing the earth toward you. Into this furrow carefully place the roots and fibres of the thorn-plants, with their cut stems leaning against the cord; and thus, keeping the plants in their places with the left hand, fill up the furrow with earth with a trowel in the right hand. Press the plants firmly against the earth with the outside of the foot placed in a line with the stems, and make the surface level with the spade. After the removal of the cord, press the ground with the row of thorns between your feet, and finish off the work with the rake. In planting ornamental hedges, you should always bear in mind that, for whatever purpose a hedge may be wanted, the thorns should always be planted on the natural surface of the ground; for, if set in traveled earth, unless it is of considerable bulk and depth, they run the risk of either being stunted in growth, or of altogether dying for want of nourishment.

(463.) In setting poles for straight lines, ordinary accuracy of eye will suffice; but in setting them in curves, where geometrical ones cannot be introduced, considerable taste is required by the planner. Such curves can only be formed by setting up large pins, and judge of their beauty by the eye, so that the sweeps may appear naturally to accommodate themselves to the inequalities of the ground, and form, on the whole, a suitable figure for the purpose they are intended to serve. Curves in fields should always be made conformable to the plowing of the adjoining land; for, if such adaptation is not attended to, land may be lost to tillage in the depth or acuteness of the curves. After the large pins are set to show the general form of a long curve, or series of long curves, smaller ones should be employed to fill up the segments between the larger, and the cord then stretched by the side of all the pins, and the beautiful sweep of the curve carefully preserved by the small pins with the hooked heads. If a curved ditch is required, the rutting of the breadth of the ditch, as also the making of the thorn-bed, should follow the cord in its curved position; but great care is required to preserve the two sides of a curved ditch parallel, for if the cross-headed wooden rule is not held at right angles to the line of the hedge, at every point where the breadth of the ditch is measured off—that is, if the cross-head is not held as a tangent to each particular curve—the breadth of the ditch will vary considerably in different places, and, of course, the ditch will there present a twist. There is no error into which laborers are so apt to fall as this: they measure, without thinking of the consequences, at any angle across the ditch; but they should be taught to avoid it, because, if not rectified in time, it will deprive the hedge-bank of essential covering at certain places, on account of the ditch being twisted into broad and narrow portions.

(464.) Where turf is plentiful, it may be employed in this way to fence at once one side of a hedge. Let *a*, fig. 46, be the turf wall 4 feet high, 18 inches broad at the base, and 1 foot at the top, coped with a large turf; *b* the stuff thrown out of the ditch *c*, and inclined upward toward the top of the wall. For keeping in Cheviot or Black-faced sheep, or cattle, a stake and single rail of paling *d*, will be required on the top, but not for Leicester sheep. In Norfolk, a high bank is thrown up, without a wall. from 6 to 7 feet in hight from the bottom of the ditch, and the thorn-plants are set into it as at *b*, fig. 46, among the crude earth taken out of the bottom of the ditch. As might be expected in such a plan, it is no uncommon sight in that county to see the face of the bank, with the quicks in it, washed down by beating rains; and as the roots enlarge and the bank moulders down, the young plants hang their heads downward upon the face of the bank. The reason assigned for the adoption of this objectionable practice is, that there is no wood in that county to form temporary fences until the thorns shall grow, and that being set upon the top of a steep bank, they are out of the reach of cattle at the bottom of the ditch. Even with a wall like *a*, fig. 46, thorns at *b* will never grow so vigorously as when placed at *e*; and in dry weather they are soon stinted

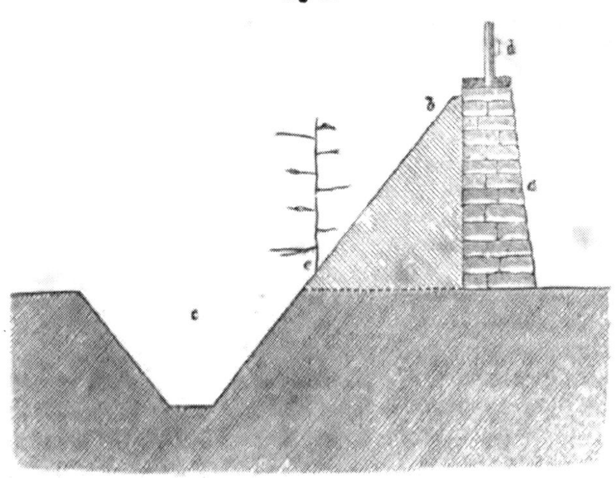

TURF FENCE TO A THORN-HEDGE.

of moisture. Where flat stones are plentiful, a good *sheltering* fence may be formed by inclosing a space of a few feet in breadth between two walls, and on filling it with earth, an upright hedge may be planted in it, where it will thrive very well. Such fences may be seen in Devonshire, where flat stones from the primitive clay-slate formations are obtained in abundance. In connection with the mode of fencing considered in this paragraph is one recommended of building a 2½ feet wall *on the top of the bank behind the hedge* which had been thrown out of the ditch, and to make its coping of turf. There are objections to this plan; in the first place, a turf coping on a stone wall never grows well, and in consequence, turf soon becomes *there* an eyesore. In the next place, a wall founded on earth that has been thrown out of the bottom of the ditch, will not remain even but a very short time, on account of the unequal subsidence of the earth, and the consequent sinking of the stones. A 3 feet stone wall, founded upon the hard ground, on the site of the turf-wall *a* in fig. 46, with a single railed paling raised behind it, until the hedge get up, would make a far better fence both for sheep and cattle. Another mode of planting a thorn-hedge is to build a stone wall as at *a*, fig. 46, in which are left holes, about the position where the letter *a* is situate in the figure, through which the thorns grow which have been planted in the bank of earth *b*. This is also an objectionable mode, inasmuch as the plants, whose roots are ramifying in the bank *b*, have no support from that portion of the stem which has to grow in a horizontal direction through the holes of the wall, and the consequence is, that the leverage of the part of the stem which grows upright in the face of the wall is apt to shake the roots, and should the horizontal portion of the stem rest for support upon the wall within the h e, its weight and motion soon bring down the wall, if it is constructed of dry stones, or shatter it, if built with mortar Thorns have been recommended to be planted at the bottom of a wall, as of *a*, fig. 46, with no bank such as *b* near it, but having the ditch *c* before it as a fence to the hedge, with a paling on its lip. If a *stone* wall is built in such a situation, there seems no use at all of the hedge as a fence, and if a *turf* one, then surely thorns will thrive much better with a bank of earth behind them, such as *b*, than at the bottom of a turf wall.

NOTE—ON SHELTERS.—The employment of artificial shelters, in fields, for plants, and trees, and animals, is carried to a degree of expense, if not of refinement, in England, which is not likely to be extensively imitated in this country. Fig. 35 represents the form of such shelters better than words could well do it.

Not only all along the sea-board of Long Island, but more or less along all our whole sea-coast, fruit and other trees are liable to be blasted and rendered unproductive by the strong blasts which strike them after acquiring a powerful momentum in sweeping over the ocean.

The wall and the wood on the inside and near to it are of the same hight, but, still farther in, the wood rises considerably higher, owing to the peculiar form of the cope of the wall and the shape of the wall itself, being like an isoceles triangle—when the wind strikes its side, it is reflected upward into the air, at the same angle.

Where such shelters can be provided, they are decidedly useful—as all must have perceived the difference, in early spring, between the advance of vegetation on the south and the north side of every inclosure. Even common garden walls in this way afford opportunities of making beds for early plants, of lettuce, cabbage, radishes, and other vegetables. [*Ed. Farm. Lib*

Fig. 35

23. THE PLOW.

"Howsoever any plow be made or fashioned, so it be well tempered, it may the better be suffered." FITZHERBERT.

(465.) THE plow serves the same purpose to the farmer as the spade to the gardener, both being used to *turn over* the soil and the object of doing this is, that this form of operation is the only means known of obtaining such a command over the soil as to render it friable and inclose manure within it, so that the seeds sown into it may grow into a crop of the greatest perfection.*

[* What we may lack, if any, of approbation from the farmer, we shall make up in the approval we challenge from the plow-maker, for the adoption of all that is said by the author in hand in respect of an implement which is almost as indispensable in the manufacture of crops as the stones in the mill for manufacturing the wheat into flour. As of all implements the plow is the most efficient and labor-saving, so on none has the ingenuity of the farmer and the machinist been so much exercised. It would be hazardous to say, that it has been pushed to its *ne plus ultra*, and that nothing now remains but to remove the animal and *hitch on steam* power; but it is not easy to imagine what desideratum remains to be supplied in the construction of the plow. Instead, however, of suppressing anything in the Book of the Farm, we prefer rather to superadd what we find on the subject in a very elaborate essay on the Agriculture of Norfolk, England, to which the Royal Agricultural Society lately awarded a high prize, and paid the compliment of publishing it separately, *in extenso,* with all its illustrations. Among these are the representations of the *prize plow,* which will also be found at the end of this chapter, although we do not perceive any essential difference from or improvement upon the Mid-Lothian plow given by Stephens, the *plate* of which accompanied our last number, and which is *described* in this one.

To go back, as our author does, and bring up the history of its progress to its present excellence of construction, from the rude implement in use by the Romans, will need no justification to the reflecting mind, ready as all such minds will be to draw from it the proud conclusion that the march of improvement has been from as humble beginnings to as high reachings in Agriculture as in other arts. Neither can it fail to inspire the hope that much more may yet be achieved in other departments if not in this. None in fact is yet closed to the career of improvement in the estimation of those who are animated by that spirit for going ahead, without which no melioration would occur in any branch of human industry. The French Vigneron, who, better than any one else understands the culture of the vine, says that after ages of observation, the art of adapting each particular species of vine to the soil most congenial to its culture, *is yet in its infancy!* Let us think so of everything while to improve remains even barely possible!

It may here be mentioned as a curious fact that President Jefferson's explanation and diagrams to illustrate them, on the principles of mechanical philosophy involved in the structure of plows, and especially in their mould boards, have been referred to and quoted by writers of the highest authority on that subject in Europe. It is no less curious, that his son-in-law, Governor Randolph, a man of genius, has the credit of being the inventor of the hill-side plow, with a shifting mould-board. In Mississippi and the South, where lands seem to be peculiarly subject to injury by washing, owing, perhaps, to the suddenness and violence of their showers, hill-side plowing is very extensively practiced, and with great skill and dexterity, by negro plowmen, as any one may see at Mr. Turnbull's, near Bayou Sara and other places. There, on lands very slightly undulating, may be found very perfect and beautiful specimens of this conservative process.

From the "Prize Report" on Norfolk Agriculture.

There is perhaps no implement which has undergone more improvement or more variation than the plow; and a glance at the catalogues which of late years have emanated from the most celebrated implement-makers, will prove of how vast importance it has been considered to obtain such a construction that while lightness and stiffness were insured, vibration in the beam should be avoided, and a perfect action with economy of power, or a diminution of resistance should be secured, for from such a combination true work can alone result. To obtain these

(466.) The *spade* is an implement so simple in construction, that there seems but one way of using it, whatever peculiarity of form it may receive, namely, that of pushing its mouth or blade into the ground with the

great desiderata, appears to have been one of the great objects with Messrs. Ransome, for from no foundry has ever been seen a greater number of these implements, or which combined in a larger extent these leading points; but in every case exhibiting the thought and skill from which they had proceeded. The Reports of the Royal Agricultural Society demonstrate the estimation in which these progressive improvements have been held. It was, however, reserved for these manufacturers to exhibit, at the Southampton meeting of the year 1844, a new plow, which proved itself equally well adapted for light and heavy soils. This implement was chiefly novel in its material, and formation of the handles and beam. These are constructed of the best wrought iron, combining lightness with adequate strength. The beams are made on the "truss principle"—that is, connecting the two sides together in such a manner as to prevent them from giving way to any amount of force, on whichever side it may be applied. The other new point is the mode of fastening the coulter, which facilitates its being placed in any new position with rapidity and ease. The following cut exhibits the figure of the prize implement, either with two or one wheel, and as a swing plow.

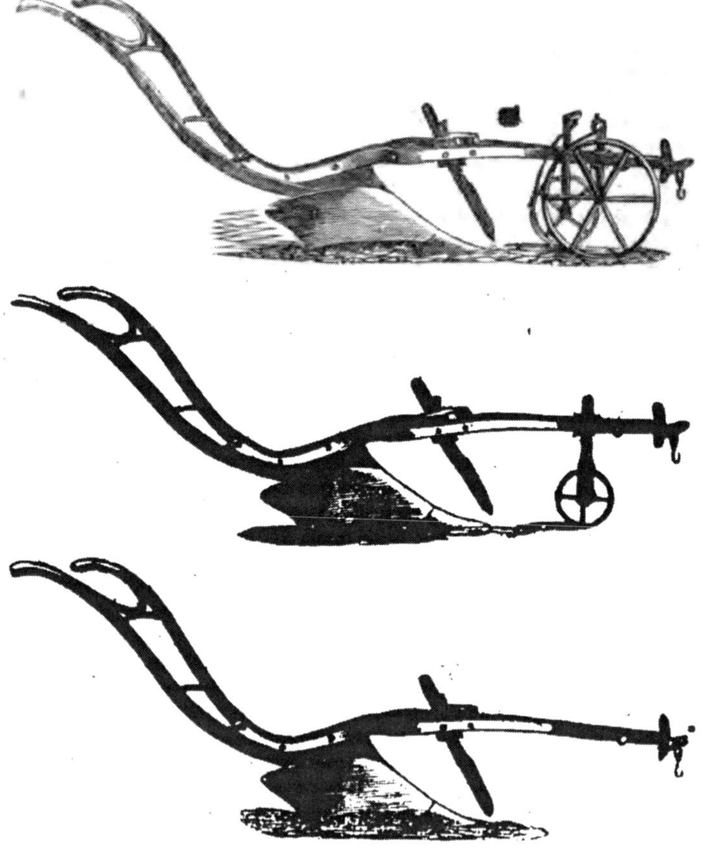

We had seen the plow at work at Mr. Henry Overman's, of Weasenham, in the autumn of 1843, and the opinion of the judges at Southampton has fully borne out the high opinion that a practical farmer gave us of its capability and excellence.

Among the implements of late invention which have deservedly obtained the inventor great credit, and from their utility have come into considerable use—one more particularly than the other—are the subsoil and subturf plows, the invention of Sir Edward Stacey, of Rackheath Hall, Norfolk. Although these implements are so well known, yet any report of Norfolk Agriculture would be incomplete were either a notice or a sketch of one of them omitted. Their effects have been found most beneficial on many soils, where the natural tenacity is increased by a hard substratum. On one farm where the land was subject to suffer from the rains in the autumn, the subsoil plow was passed up the furrows on a turnip fallow previous to the autumn plowing The effect was to free the land from a more than ordinary quantity of moisture which happened

(486)

foot, lifting up as much earth with it as it can carry, and then inverting it so completely as to put the upper part of the earth undermost. This operation, called *digging*, may be done in the most perfect manner; and any attempt at improving it, in so far as its uniformly favorable results are concerned, seems unnecessary. Hitherto it has only been used by the hand, no means having yet been devised to supply greater power than human strength to wield it. It is thus an instrument which is entirely under man's *personal* control.

(467.) The effect attempted to be produced on the soil by the *plow* is an exact imitation of the work of the spade. From the circumstance, however, of the plow being too large and heavy an implement to be wielded by the hand, it is not so entirely under man's control as the spade. To wield it as it should be, he is obliged to call in the aid of horses, which, though not capable of wielding it personally, as man does the spade, can, nevertheless, through the means of appropriate appliances, such as harness, do so pretty effectually. It is thus not so much man himself as the horses which he employs that turn over the ground with the plow, they, in a great measure, becoming his substitutes in performing that operation; and they are so far his superiors, that they can turn over a greater quantity of the soil with the plow in a given time than he can with the spade. Man, however, has this advantage over horses in turning over the soil, that he can do it well with a very simple instrument—the spade; whereas horses require an instrument of more complex structure—the plow—to perform the same sort of work not so well; and the reason is this, that although the spade is really a very simple instrument, the act of digging with it is

to fall that year, and to benefit materially the following crop of turnips. Sir Edward has already described the utility of the former upon the heath-land attached to his estate, as well as the improvement of his park by the subturf plow, in the journals of the Society

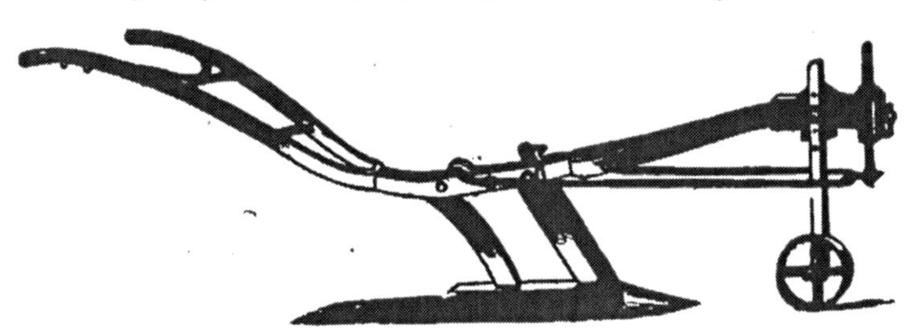

A practice has lately become very prevalent in some parts of the county, to lay the furrows in one direction. This method, if it should be found advantageous, will be greatly aided by Lowcock's new plow. This gentleman is a farmer at Westerland, Devon, and his attention was drawn to its necessity by having found that great injury was sustained in his neighborhood by the currents of air drawn up the furrows when the land was either ridged or thrown into stetches. When the land is laid in one plain surface, it is thought that the seed can be more easily deposited—and that in rainy seasons it will absorb the moisture with greater regularity, and in a dry one would be less injured by drouth. This implement seems to be the combined result of theoretical knowledge and practical experience—Mr. Lowcock farming wet soils.

The mode of adapting it to each furrow is extremely simple. When the plowman has arrived at the end of the furrow, he directs the horses round on the unplowed side of the land, and the draught chain slides on a rod to the other end. While they are moving, he reverses the handles, where a catch drops into a mortice in the beam, and the plow is again ready. When the share and coulter are at work the mould-board flies into its proper direction, in which place the resistance of the newly-cut furrow keeps it. Presuming the conjectures as to the effect of such a system of plowing to be correct, this will become a very valuable implement in Norfolk. Messrs. Ransome obtained a prize for it at Southampton. (See the opposite page.)

The plows in most general use are the Norfolk and the Swing Plow, which have been rendered lighter and steadier than formerly. There are none, perhaps, as a whole, better suited to the soil, although for particular purposes there are some superior, the Rutland having been found from its length of plat to whelm the olland better; while the Norfolk, from its short breast, lays the earth looser and rougher for the operations of the winter.

(487)

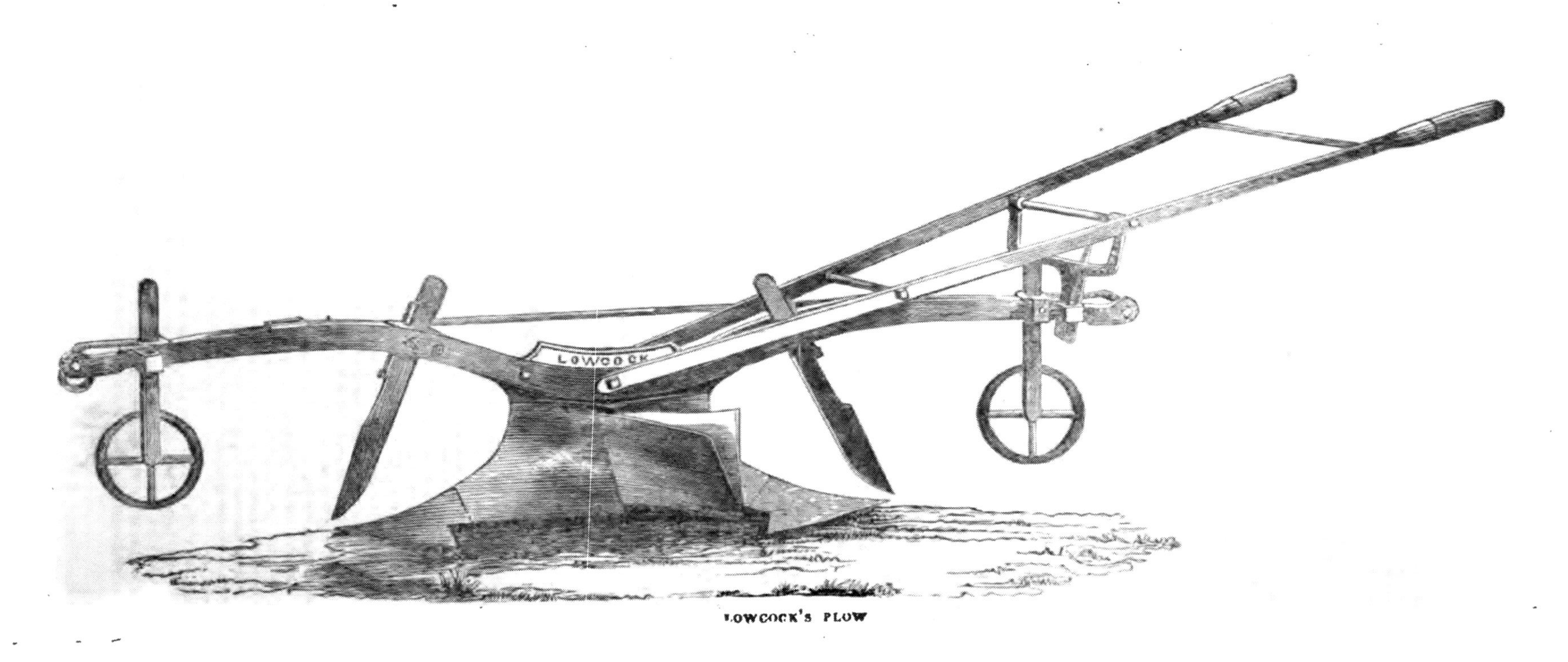

LOWCOCK'S PLOW

not a simple operation, but requires every muscle of the body to be put into action, so that any machine that can imitate work that has called into requisition all the muscles of the body, must have a complex structure. This would be the case even were such a machine always fixed to the same spot, and, for such a purpose, there is little difficulty in practical mechanics in imitating the work of man's hands, by complicated machinery, but it is not so simple a problem in practical mechanics, as it at first sight may appear, to construct a light, strong, durable, convenient instrument, which is easily moved about, and which, at the same time, though complex in its structure, operates by a simple action; and yet the modern plow is an instrument possessing all these properties in an eminent degree.

(468.) The common plow used in Scotland is made either wholly of iron, or partly of wood and partly of iron. Until a few years ago it was universally made both of wood and iron, but now it is generally made entirely of iron. A wooden plow seems a clumsier instrument than an iron one, though it is somewhat lighter. The plow is now made wholly of iron, partly from the circumstance of its withstanding the vicissitudes of weather better than wood; and, however old, iron is always worth something; and partly because good ash timber, of which plows were usually made, is now become so scarce in many parts of the country, that it fetches the large price of 3s. per cubic foot; whereas iron is now becoming more abundant and cheap (204), being no more than £14 per ton for common cast goods, and from £10 to £18 per ton for malleable iron. A wooden plow with iron mountings usually weighs 13 stones imperial, and an iron one for the same work 15 stones. The cost of a wooden one is £3 16s., capable of being serviceable, with repairs, for the currency of a lease of 19 years; that of an iron one £4 4s., which will last a lifetime, or at least many years. Some farmers, however, still prefer the wooden one, alleging that it goes more steadily than the iron. Whatever of prejudice there may be in this predilection for the wooden plow, it must be owned that the iron one executes its work in a satisfactory manner. There is, I believe, no great difference of economy in the use of the two kinds of plows.

(469.) The plow, as it is now made, consists of a number of parts, which are particularly described below at (493), fig. 48, and to which you should immediately refer, in order to become acquainted with them. How well soever these different parts may be put together, if they are not all *tempered*, as it is termed, to one another, that is, if any part has more to do than its own share of the work, the entire implement will go unsteadily. It can be easily ascertained whether a plow goes steadily or not, and the fact is thus practically ascertained; and its *rationale* will be found below.

(470.) On taking hold of the plow by the handles with both hands, while the horses are drawing it through the land, if it have a constant tendency to go deeper into the soil than the depth of the furrow-slice previously determined on, it is then not going steadily. The remedy for this error is twofold, namely, either to press harder upon the stilts with the hands, and, by their power as levers, bring the sock nearer the surface of the ground, and this is called "*steeping;*" or to effect the same thing in another way, is to put the draught-bolt of the bridle a little nearer the ground, and this is called giving the plow "*less earth.*" The pressure upon the handles or stilts should first be tried, as being the most ready remedy at your command; but should it eventually fail of effecting the purpose, or the holding the stilts so be too severe upon your arms, the draught-bolt should be lowered as much as required. But should both these attempts at amendment fail, then there must be some error in another part of the plow. On examining the sock, or share, its point may possibly be found to dip too much below the line of the sole, which will produce in it a tendency to go deeper than it should. This error in the sock can only be rectified at the smithy.

(471.) Again, the plow may have an opposite tendency, that is, a tendency to come out of the ground. This tendency cannot well be counteracted by the opposite method of supporting the stilts upward with the arms, because in this condition of body you cannot walk steadily, having no support for yourself, but rather affording support to the plow. It is for this reason that a very short man can scarcely hold a plow steady enough at any time; and hence such a man does not make a desirable plowman. The draught-bolt should, in the first instance, be placed farther from the ground, and in so doing the plow is said to get "*more earth*." Should this alteration of the point of draught not effect the purpose, the point of the sock will probably be found to rise above the line of the sole, and must therefore be brought down to its proper level and position by the smith (525).

(472.) You may find it difficult to make the plow turn over a furrow slice of the breadth you desire. This tendency is obviated by moving the draught-bolt a little to the right; but in case the tendency arise from some casual circumstance under ground, such as collision against a small stone, or a piece of unusually hard ground, it may be overcome by leaning the plow a little over to the right, until the obstruction is passed. These expedients are said to give the plow "*more land.*"

(473.) The tendency of the plow, however, may be quite the opposite from this—it may incline to take a slice broader than you want; in which case, for permanent work, the draught-bolt should be put a little farther to the left, and for a temporary purpose the plow may be leaned a little over to the left, and which are said to give the plow "*less land.*"

(474.) These are the ordinary instances of uusteadiness in the *going* of plows; and, though they have been narrated singly, two of them may combine to produce the same result, such as the tendency to go deeper or come out with that of a narrower or broader furrow-slice. The remedy should first be tried to correct the most obvious of the errors; but both remedies may be tried at the same time, if you apprehend a compound error.

(475.) Some plowmen habitually make the plow lean a little over to the left, thus giving it in effect less land than it would have, were it made to move upon the flat of the sole; and, to overcome the consequent tendency of the plow to make a narrower furrow-slice than the proper breadth, they move the draught-bolt a little to the right. The plowing with a considerable lean to the *left* is a bad custom, because it makes the lowest side of the furrow-slice, when turned over, thinner than the upper side, which is exposed to view, thereby deluding you into the belief that the land has all been plowed of equal depth; and it causes the horses to bear a lighter draught than those which have turned over as much land in the same time, with a more equal and therefore deeper furrow-slice. Old plowmen, becoming infirm, are very apt to practice this deceptive mode of plowing.—The plow should always move flat upon its sole, and turn over a rectangular furrow-slice; but there are certain exceptions to this rule, depending on the peculiar construction of parts of certain forms of plows, which will be pointed out to you afterward.

(476.) None assume the habit of leaning the plow over to the *right* because it is not so easy to hold it in that position as when it moves upon the sole along the land-side.

(477.) Other plowmen, especially tall men, practice the habit of con stantly leaning hard upon the stilts, or of steeping; and, as this practice has the tendency to lift up the fore point of the plow out of the ground, they are obliged, to keep it in the ground, to put the draught-bolt farther from the ground than it should be. A little leaning of the hands upon the stilts is requisite at all times, in order to retain a firm hold of them, and thereby have a proper guidance of the plow.

(478.) A *good* plowman will use none of these expedients to make his plow go steadily, nor will he fall into any of these reprehensible habits.—He will temper the irons, so as there shall be no tendency in the plow to go too deep or too shallow into the ground, or make too wide or too narrow a furrow-slice, or cause less or more draught to the horses, or less or more trouble to himself, than the nature of the work requires to be performed in the most proper manner. If he have a knowledge of the implement he works with—I mean, a good practical knowledge of it, for a knowledge of its principles is not requisite for his purpose—he will temper all the parts, so as to work the plow with great ease to himself, and, at the same time, have plenty of leisure to guide his horses aright, and execute his work in a creditable manner. I have known such plowmen, and they invariably executed their work in a masterly way; but I never yet saw a plowman execute his work well, who had not acquired the art of tempering the irons of his plow. Until he learns this art, the best made plow will be comparatively worthless in his hands.

(479.) In the attempt to temper the irons, many plowmen adopt a position of the coulter which increases the draught of the plow. When the point of the coulter is put forward in a line with the point of the sock, but a good deal asunder, to the left or land side, in light land that contains small stones, a stone is very apt to be caught between the points of the coulter and sock, and which will throw the plow out of the ground. This catastrophe is of no great consequence when it occurs on plowing land preparatory to another plowing; but it tears the ground on plowing lea, which must be rectified instantly; and, in doing it, there is loss of time in backing the horses to the place where the plow was thrown out. To avoid such an accident *on such land*, the point of the coulter should be put immediately above, and almost close upon, that of the sock; and this is the best temper of those irons, in those circumstances, for lea-plowing. In smooth soils—that is, free of small stones—the relation of the coulter and sock to each other is not of much importance in regard to steadiness; but it is the best practice to cut the soil clean at all times, and the practicabil ity of this should be suited to its nature.

(480.) The *state of the irons* themselves has a material effect on the temper of the plow. If the cutting edge of the coulter, and the point and cutting edge of the sock, are laid with steel, the irons will cut clean, and go long in smooth soil. This is an economical mode of treating plow-irons destined to work in clay-soils. But, in gravelly and all sharp soils, the irons wear down so quickly that farmers prefer irons of cold iron, and have them laid anew every day, rather than incur the expense of laying them with steel, which, perhaps, would not endure work much longer in such soil than iron in its ordinary state. Irons are now seldom if ever steeled; but, whether they are steeled or not, they are always in the best state when sharp, and of the proper lengths.

(481.) An imperfect state of the mould-board is another interruption to a perfect temper of the plow. When new and rough, it accumulates the loose soil upon it, whose pressure against the turning furrow-slice causes the plow to deviate from its right course. On the other hand, when the mould-board is worn away much below, it is apt to leave too much of the crumbled soil in the bottom of the furrows, especially in plowing loose soils. Broken side-plates, or so worn into holes that the earth is easily pressed through them into the bosom of the plow, also cause rough and unequal work; and more or less earth in the bosom affects the balance of the plow, both in its temper and draught. These remarks are made upon the supposition that all plows are equally well made, and may, therefore, be tempered to work in a satisfactory manner; but it is well known that plows sometimes get into the possession of farmers, radically so ill-constructed that the best tempering the irons are capable of receiving will never make them do good work.

(482.) When all the particulars which plowmen have to attend to in executing their work—in having their plow-irons in a proper state of repair, in tempering them according to the kind of plowing to be executed, in guiding their horses, and in plowing the land in a methodical way—when all these particulars are considered, it ceases to surprise that so few plowmen should be first-rate workmen. Good plowmanship requires greater powers of observation than most young plowmen possess, and greater judgment than most will take time to exercise, in order to become familiarized wtth all these particulars, and to use them all to the best advantage. To be so accomplished implies the possession of talent of no mean order. The ship has been aptly compared to the plow, and the phrase, "plowing the deep," is as familiar to us islanders as plowing the land: to be able to put the ship in "proper trim," is the perfection aimed at by every seaman; so, in like manner, to "temper a plow" is the great aim of the good plowman; and to be able to do it with judgment, to guide horses with discretion, and to execute plowing correctly, imply a discrimination akin to sailing a ship.

(483.) [The present age is, perhaps, the most remarkable that time has produced, for the perfection of almost every kind of machine or tool required in the various departments of art and of manufactures. In that most important of all arts—the production of the raw material of human food—something like a corresponding progress has been effected in its machinery and tools, though certainly not to the same degree of perfection as those employed in most of our manufactures, whether they be in animal, vegetable, metallic or mineral productions. Various causes exist to prevent, or at least retard, an equal degree of perfection being arrived at in agricultural machinery, among which may be noticed one pervading circumstance, that affects, more or less, almost every machine or implement employed. This circumstance is, that all the important operations of the farm are performed by *seasons* occupying comparatively short periods of time, and, should the artisan be endeavoring to produce any new or important machine, he can only make trial of it in the proper season. The imperfection of human perception is too well known to leave us in surprise at the first attempt of any improvement turning out more or less a failure. The artisan, therefore, will in all probability find that his project requires amendment; and, before that can be effected, the season is past in which a second trial could be made, and, consequently, must lie over for a year, in the course of which many circumstances may occur to cause its being forgotten or laid aside. Impediments of this kind do not occur to the inventor or improver of manufacturing machinery, where constant daily opportunities are at hand to test the successive steps of his invention. One other general cause, and of another kind, exists, to supersede the necessity, or even the propriety, of employing machinery of such high and delicate finish as we see in the machines of all in-door manufactures. This is the irregularity of the media on which agricultural machinery is employed, and the numerous changes produced on these media—the soils and produce—by vicissitudes of weather and other causes, which not only affect the operation, but also the existence of many of these machines. From this cause, with its train of incidents, it may be inferred that agricultural machinery and tools must, of necessity, be of simple construction, which embrace nothing but the essentials of usefulness · that they have sufficient strength for their intended purpose, and free of any undue weight; that there should be no redundancy nor misapplication of materials; that all materials employed should be of the best quality, and the workmanship plain and sound. These properties, it must be admitted, are of greater importance to agricultural machinery, in general, than the minute delicacy of construction and finish observable in many of those almost intellectual tools employed in some of the other arts and manufactures.

(484.) Although, therefore, agricultural machines in general do not require a high mechanical finish, yet there are among them those which are based on principles implying a knowledge and application of science, as well as mechanical skill, in their construction; and in this class is to be ranked the plow, which, in one word, is the most important of all agricultural machines.

(485.) To the plow, then, our attention is first to be directed, not only as standing at the head of all its fellows in the ranks of the machinery of the farm, but as being the first implement to which the attention of the farmer is called, in the commencement of this the winter season.

(486.) Before entering on the details of the implement as it now appears, it will be interesting to look back for a moment into its history. With the earliest stages of human industry, the tillage of the ground in some shape must be considered as coeval; and in these early attempts, some implement analogous to a plow must have been resorted to. In all ancient figures and descriptions of that implement, its extreme simplicity is to be remarked; and this is but a natural result; but with the progress of human intellect, are to be also observed deviations from the original simplicity, and an increase in the number of its parts, with a corresponding complexity in its structure. The Roman plows, imperfectly as they are described by different Roman authors, is an example of this. And as an example of apparently very remote origin, the *caschrom*, or plow used even at this day, in some portions of the Outer Hebrides and in Skye, forms a very curious and interesting antiquarian relic of the ancient Celtic habits. It is formed, as in fig. 47, of one piece of wood, selected from its

Fig. 47.

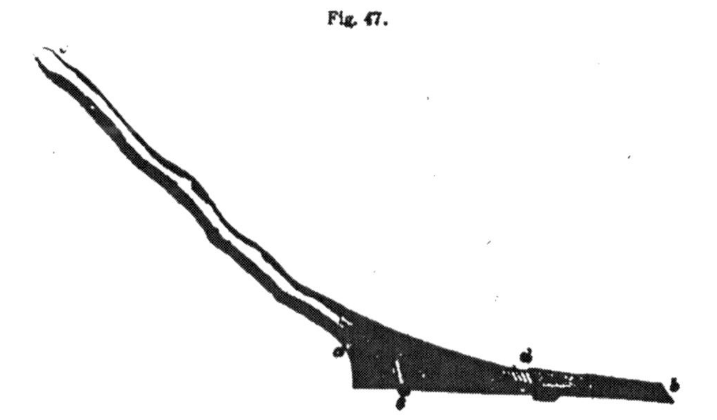

THE CASCHROM

possessing the natural bend at *a*, that admits of the head *a b* assuming a nearly horizontal position, when the handle *c* is laid upon the shoulder of the person who wields the implement. A simple wedge-shaped share, *b d*, is fitted to the fore part of the sole. A wooden peg, *e*, is inserted in the side of the heel at *e*, which completes the implement. On this last member the foot of the operator is applied, to push the instrument into the ground. It is of course worked by the hand alone, and makes simply a rut in the ground. Yet even in this rude implement are to be traced the rudiments of a plow.

(487.) As the cultivation of the soil became more and more an object of industry, corresponding improvements would naturally follow in the implements by which such operations were performed. But in Britain previous to the beginning of the last century, the plow appears to have continued in a very uncouth state. About that period Agriculture seems to have become more an object of improvement. Draining began to be studied, and its effects appreciated. The amelioration of the soil produced by draining would soon call for better modes of dressing such improved soils; hence, still farther improvements in the plow would come into request. In accordance with this, we find the introduction of an improved plow into the northern counties of England, under the name of the Dutch or Rotherham plow. This appears to be the foundation of all the modern improvements, and from the circumstance of engineers and mechanics having been brought from Holland to conduct the draining of the English fens, there is good reason to conclude that the Rotherham plow was originally an importation from Holland, in a similar manner as the barley-mill was, at a later period, borrowed from that country. About the middle of the past century, the Rotherham plow appears to have been partially introduced into Scotland; but until Mr. James Small took up the subject, and, by his judicious improvements gave a decided character to the plow, little or no progress had been made with it.

(488.) Small appears to have been the first who gave to the mould-board and the share a *form* that could be partially imitated by others, whereby, following his instructions, mould-boards might be multiplied, each possessing the due form which he had directed to be given to them. It is to be observed, that when Small first taught the method of construction, mould-boards were really *boards* of wood, and for their defence, were covered with plates of iron. The method of construction being not very clearly defined, and mould-boards being necessarily constructed by many different hands, the improved system, it may be easily conceived, must have been liable to failure in practice. It was, therefore, one of those happy coincidences which now and then occur for the benefit of mankind, that the founding of cast-iron was then beginning to become general. The fortunate circumstance was seized Mould-boards, together with the head or sheath, and the sole and land-side plates, were made of cast-iron; and a model or pattern of these parts having been once formed, any number of duplicates could be obtained, each possessing every quality, in point of form, as perfectly as the original model. The plow, thus in a great measure placed beyond the power of uninformed mechanics to maltreat, came rapidly and deservedly into public esteem, under the name of Small's plow. Though originally produced in Berwickshire, the plow that seems to retain the principal feature of Small's improvements—the mould-board—is now found chiefly in East-Lothian, and, as will appear, differs very sensibly from that now generally used in Berwickshire.

(489.) Other writers, about the same period, published methods for constructing a mould-board on just principles. Among these, the method proposed by Bailey of Chillingham may be mentioned as approaching very near to the true theoretical form. Others less perfect have been proposed, which it is not necessary at present to notice; while several have published general descriptions of their construction of the plow, but have withheld the principles on which their mould-boards are formed.

(490.) While these improvements of the past century were going on, the plow was universally constructed with *wooden framing;* but about the beginning of the present century (the precise year cannot now be well defined), malleable iron began to be employed in their fabrication. The application of this material in the construction of plows came with so much propriety, that it is now, in Scotland, almost universal. It has many advantages; but the most prominent are its great durability under any exposure, and its better adaptation to withstand the shocks to which the implement is frequently liable in the course of working. In a national point of view, it is also deserving of the most extended application, being a produce for which Britain stands unrivaled. This period also, was productive of an innovation on the form of mould-board and share which had been established by Small. The mould-boards hitherto referred to come under the denomination of concave, or more properly straight-lined; when Mr. Wilkie, Uddingstone, near Glasgow, introduced his new form with convex lines, to be afterward more particularly noticed, and which has been adopted in various districts in Scotland, to the exclusion of the concave form.

(491.) At a still later period, a form of plow was brought forward by Mr. Cunningham, Harlaw, near Edinburgh, a practical farmer, in which are combined the properties of Wilkie's, with very slight deviation of form from that of Small's plow—the principal difference being in the form of the share.

(492.) Having, in this short sketch of the progress of the plow, brought it to the point when it has diverged into three varieties, each of which is held in equal estimation in the respective districts in which it is used, it is a remarkable circumstance that each holds its sway in its peculiar locality, to the almost entire exclusion of its compeers. The first two have undergone numerous slight changes, forming sub-varieties, but retaining the respective leading features of the concave and convex mould-boards; and, as they have each spread (especially the first) over a wide extent of country, I purpose to distinguish them by the county in which they are chiefly employed. Thus, the Small's plow shall be denominated the East-Lothian, and Wilkie's the Lanarkshire. plow. The third variety is more limited in its range of application, being almost exclusively confined to Mid-Lothian, and the borders of those counties adjoining to it, throughout which it is known by the name of the Currie plow, but which it is proposed to distinguish here by the name of the Mid-Lothian.

(493.) Before entering upon the detailed description, it will be useful to the agricultural student that a nomenclature be given of the various parts of the plow. Thus, fig. 48, which

Fig. 48.

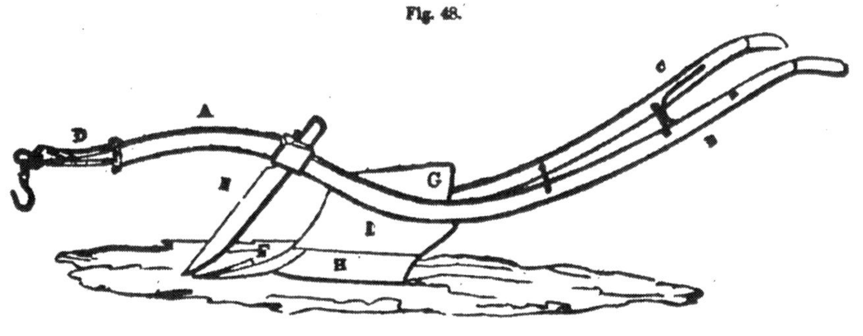

A VIEW OF THE LAND-SIDE OF A PLOW

is a view of a plow in perspective, presents that which plowmen and agricultural mechanics denominate the *land-side*, so called because when in work it is always (except in the case of turn-wrest or right-and-left plows) in contact with the firm or unplowed land. The opposite or right side of the plow, being that which turns over the furrow-slice cut from the firm land, is called the *furrow-side*. That member of the plow to which the animals of draught are yoked, marked A in the figure, is the *beam*. Those parts by which the plowman holds and guides the implement are called the *stilts* or *handles*, B being the *great stilt* or *left handle*, and C the *little stilt* or *right handle;* D is the *muzzle* or *bridle* by which the horses are attached to the beam; E the *coulter* is a cutting instrument that severs the slice from the firm land, and F the *sock* or *share* which cuts the slice below from the subsoil; G is called the *wrest* or *mould-board*. It is probable that the term wrest applied formerly to only a particular portion of the mould-board—the lower portion in the more ancient plow—which was supposed to wrest or turn aside the slice after being cut by the share; thus we find in the Kent turn-wrest plow that the wrest is a simple straight bar of wood. The mould-board, in the improved implement, receives the slice from the share, turns it gradually over, and deposits it continuously at the proper angle. H is the *sole-shoe* on which the plow has its principal support, and on which it moves, and I is the *land-side plate*, only serving to complete the sheathing of the land-side, presenting a uniform smooth surface to the firm land, and preventing the crumbled earth from falling within the body of the plow. These last parts cover the body-frame from view, which will be exhibited among the details.

(494.) Without entering into a description of all the sub-varieties of these plows, it will be sufficient to attend to the type of each variety, and, first, as to their *general qualities and characteristics*.

(495.) *The East-Lothian plow*, figs. 49 and 50, Plate V.—In this plow, the proper lines of the body on the land-side lie all in one plane, which, in working, should be held in the vertical position, or very slightly inclining to the left. The coulter slightly oblique to the land-side plane, the point standing toward the left, the rake of the coulter varies from 55° to 65°. In the mould-board the vertical sectional lines approximate to straight lines, giving the character of apparent concavity, and it is truncated forward. Share pointed, with a feather or cutter standing to the right, having a breadth of at least ¾ the breadth of the furrow, the cutting edge of the feather lying nearly as low as the plane of the sole. The neck of the share is prolonged backward, joining and coinciding with the curve of the mould-board, which curvature is also carried forward on the back of the feather. The character of this plow is to take a furrow of 10 inches in breadth by 7 inches in depth, cut rectangular, leaving the sole of the open furrow level and clean. The resistance to the draught is generally below the average of plows, and this plow is employed for every kind of soil.

(496.) *Lanarkshire plow*, figs. 51 and 52, Plate IX.—In this plow, the proper lines of the land-side lie in different planes; thus when the fore-part of the land-side of the body, taken at the junction of the breast with the beam, is vertical, the hind-part, taken at the heel, overhangs the sole-line ¾ inch, and the beam, at the coulter-box, lies to the *right* of a vertical line from the land-side of the sole about 1 inch, the point of the beam being recurved toward the land-side. In working, the fore-part of the body is held in the vertical line, or slightly inclined to the left. The coulter, by reason of the bend in the beam to the right, and the point being to the left of the land-side, stands very oblique, but nearly coinciding with the land-side, at the hight of 7 inches from the sole. Rake of the coulter from 55° to 65°. The vertical sectional lines of the mould-board are all convex toward the furrow, giving the mould-board the character of convexity, and it is prolonged forward, covering the neck of the share. Share chisel-pointed, with the feather seldom exceeding 5½ inches broad, the cutting edge rising from the point at an angle of 8° till it is 1 inch above the plane of the sole, when it falls into the curve of the mould-board, while the neck passes under the latter. The character of this plow is to take a furrow whose section is a trapezoid, its breadth from 7½ to 9 inches, and greatest depth 6½ inches, the sole of the furrow being not level, and deepest at the land-side. In the finished plowing, the laid-up furrow-slices have the acute angle upward, giving the character which I call *high-crested* to the furrow-slice, especially observable in plowing lea. Resistance to the draught about the average, and it is considered to be well adapted to stiff clay, and to lea-land.

(497.) *Mid-Lothian plow*, figs. 53 and 54, Plate X.—This plow is always worked with a chain bar under the beam. The proper lines of the land-side lie in different planes; thus, when the fore-part of the land-side, taken as in the former case, is vertical, the hind-part, taken at the heel, overhangs the sole line ¾ inch, but the beam is continued straight. In working, the land-side is held vertical, or slightly inclined to the left. The coulter stands rather oblique, and the point about 1½ to 2 inches above the point of the share. Rake of the coulter varying from 56° to 80°. The vertical sectional lines of the mould-board approximate to straight lines, giving the character of concavity, and the mould-board is prolonged forward, covering the neck of the share. The share is chisel-pointed, with feather seldom exceeding 5 inches broad, and, when trimmed for lea-plowing, the cutting-edge rises from the point at an angle of 10° to a hight of 1½ inches above the plane of the sole, when it falls into the curve of the mould-board, while the neck passes under it. The character of this plow is to take a furrow-slice whose transverse section is a trapezoid, having an acute angle with a breadth of 8½ to 9 inches, and usually from 6 to 6½ inches in depth. The sole of the furrow is not level, and is deepest at the land-side. In the finished plowing, the laid-up furrow-slices have the acute angle upward, forming a high crest when plowing lea. Resistance to the draught is about the average, and this plow is considered applicable to every kind of soil, but particularly to plowing lea.

(498.) Before entering upon the *specific details* of the three varieties into which the modern Scotch plows are here divided, it will be necessary to *lay down certain data, on which the details of each variety will be based.* For this purpose, the figures in elevation, figs. 49, 51, and 53, in the plates of the entire plows about to be described, are supposed to stand upon a level plane, the heel and point of the share touching that plane, these being actually the points on which the plow is supported when in motion; this plane shall be called the *base line.* The fore-part of the land-side of the plow's body—standing in the vertical position, as seen in plan figs. 50, 52 and 54, in the plates—is supposed to be placed upon a similar line, touching the land side of the sole-shoe and the point of the share. The base-line is divided into a scale of feet for the convenience of comparison. The zero of the scale is taken at that part of the plow's body, where a vertical transverse section, at right angles to the plane of the land-side, will fall upon a point on the surface of the mould-board, which shall be distant from the land-side plane by a space equal to the greatest breadth of the furrow taken by the respective plows; the hight of this point above the base-line being also equal to the breadth of the slice. Or, the zero is that vertical section of the mould board, which, in its progress under the slice, will just place the latter in the vertical position.—The scale, by this arrangement, counts right and left of the zero. The dotted line marked *surface-line* in figs. 49, 51, 53, in the plates, represents the depth of the furrow taken by the respective plows.

(499.) This zero point has not been fixed on without much consideration; for, having experienced the inconvenience of vague generalities in stating the dimensions of the plow, as given in works on the subject, it has appeared to me desirable that some fixed point should be adopted. and this has been chosen as less liable to change than any other point in the longitude of the plow; all other points of this implement being liable to and may be changed at pleasure, without change of effect. Thus the beam or the handles may be lengthened or shortened, the position of the coulter and the length of the sole may be varied, the mould-board itself may be lengthened or shortened forward without producing any decided change in the working character of the plow, the apparent changes being easily counteracted by a corresponding change in a different direction. The lengthening of the beam, for example, would only require a corresponding change in the hight of its extremity above the base line; an alteration in the length of the share, or in the position of the coulter-box, induces only a corresponding change in the angle which the coulter forms with the base line, which angle, in any case, is liable to change from the wearing of the irons themselves, but which can be rectified as required by shifting the draught-bolt in the bridle. The zero point here proposed can, with tolerable exactness, be determined in any plow with the instruments that every mechanic has in his hands, *squares* and a *foot-rule.*

(500.) It may be well, also, to premise farther, in regard to the contour in elevation of the different plows, that although the hights at the different points throughout the beam and handles are given in detail, as adopted by the best makers, which those unacquainted with the implements may follow with confidence, in the construction of implements of the same character, yet I cannot pass over the circumstance without noticing that, with the exception of one point—the hight of the beam at the draught-bolt—any part of the contour may be altered to the taste of the maker, and even the point of the beam, as already noticed, may be altered, provided the alteration is continued backward or forward in a certain angle. The change in position in the vertical direction of three other points is limited within a certain range—not from principle, however, but for convenience. These points are the *hight of the beam at the coulter-box* and *at the breast-line* of the mould-board; these cannot be brought *lower* than the given dimensions without subjecting the plow to an unnecessary tendency to choke in foul ground, though they may be raised *higher* without injury, provided the corresponding parts—the mould-board and body-frame—are altered in proportion. The third point here alluded to is the *hight of the handles,* which is altogether a point of convenience; but it may be affirmed of this that it is better to be *low* than *high,* since being low places the plow more under the command of the plowman. The different points, as given in plan, being more matters of principle, with exception of the position of the handles, cannot be deviated from without compromising the character of the plow.

(501.) With these preliminary remarks I proceed to the general description of the three varieties, taking first,

(502.) THE EAST-LOTHIAN PLOW.—Fig. 49, Plate V. represents an elevation of this plow, on the furrow side, drawn to a scale of 1 inch to 1 foot, and fig. 50 a horizontal plan of the same. It is found with various shades of difference, but not to the extent or of such a marked character as to require separate description from what follows. The beam and handles or stilts are almost invariably made of malleable iron, the body-frame being of cast-iron, the latter varying slightly with different makers. In its construction, the beam and left handle are usually finished in one continued bar ABC, possessing the varied curvature exhibited in fig. 49, as viewed in elevation. When viewed in plan, as in fig. 50, the *axis* or central line of the beam and *left handle* are in a straight line—though in this arrangement there are some slight deviations among the different makers—the point of the beam being in some cases turned more or less to the right or furrow side, and this is found to vary from ½ inch to 2 inches from the plane of the land-side.

(503.) The *right handle,* DE, is formed in a separate bar, and is attached to the body-frame at its fore end by a bolt, as will be shown in detail, and farther connected to the left handle by the bolts FFF, and the stays GG.

(504.) The *coulter* I is fixed in its box K by means of iron wedges, holding it in the proper position. Its office being that of a cutting instrument, it is constructed with a sharp edge, and is set at an angle of from 55° to 65° with the base-line.

(505.) The *mould-board* L, which is fixed upon the body-frame, and to the right handle, is a curved plate of cast-iron, adapted for turning over the furrow-slice. Its fore-edge or breast MN coincides with the land-side of the plow's body; its lower edge T behind stands from 9½ to 10 inches distant from the plane of the land-side, while its upper edge P spreads out to a distance of 19 inches from B, the land-side plane. In this plow the mould-board is truncated in the fore part, and is met by the gorge or neck of the share, the junction being at the line N.

(506.) The *share or sock* NR is fitted upon a prolongation of the sole-bar of the body-frame, termed the head, and falls into the curves of the mould-board, of which its surface forms a continuation.

(507.) The *bridle C*, or *muzzle*, as sometimes named, is that part to which the draught is applied, and is attached to the point of the beam by two bolts, the one S being permanent, upon which the bridle turns vertically. The other bolt U is movable, for the purpose of varying the *earthing* of the plow; the *landing* being varied by shifting the draught-bolt and shackle V to right or left. The right and left handles are furnished at A and D with wooden helves fitted into the sockets of the handles.

(508.) The *general dimensions* of the plow may be stated thus, as measured on the base-line: From the zero-point O to the extremity of the heel T, the distance is 4 inches, and from O forward to the point of the share R, the distance is 32 inches—giving, as the entire length of sole, 3 feet. Again, from O backward to the extremity of the handles A′ is 6 feet 2 inches, and forward to the draught-bolt V′ 4 feet 7 inches, making the entire length of the plow on the base-line 10 feet 9 inches; but, following the sinuosities of the beam and handle, the entire length from A to C is about 11 feet 3 inches.

(509.) In reference to the *body* of the plow, the center of the coulter-box K is 14½ inches, and the top of the breast-curve M 9 inches before the zero-point, both as measured on the base-line; but, following the rise of the beam, the distance from M to the middle of the coulter-box will be 7 inches.

(510.) The *hights* at the different points above the base-line are marked on the figure in elevation, along the upper edge of the beam and handle; but the chief points in hight are repeated here, the whole of them being measured from the base-line to the upper edge of the beam and handles at the respective points. At the left handle A the hight is 3 feet, at the right handle D 2 feet 9 inches; and a like difference in hight of the two is preserved till the right handle approaches the body at the middle stretcher F; thence the difference increases till it reaches the body. The hight at the point of the beam is 18 inches, and the center of the draught-bolt, at a medium, 17 inches. The lower edge of the mould-board behind, of this plow, at T is usually set about ½ inch above the base-line, and at the junction with the share about the same hight.

(511.) The dimensions in *breadth*, from the land-side line, embrace the obliquity that is given to the direction of the beam and handles, compared with the land-side plane of the body taken at the sole. The amount of obliquity, as exhibited by the dotted line AC, fig. 50, which coincides with the land-side plane of the body, is, that the axis of the beam at the extremity C stands 1¼ inches to the right, and at the opposite end the left handle A stands abont 2 inches to the left of the line. These points may, however, be varied slightly from the dimensions here given. In the first—the point of the beam—it is found in the practice of different makers to range from 1 to 2 inches. In the opinion of some writers and practical men, it is held that the beam should be parallel with the land-side plane of the body. With all deference to such opinions, I apprehend that the direction of the line of draught, in a vertical plane, cannot coincide with the plane of the land-side; for the point of resistance in the plow's body cannot fall in that plane, but will pass through some point to the right of it, and which, from the nature of the subject, cannot be very precisely defined. Both reason and experience, however, point this out to the plow-maker, and especially to the observant plowman. Hence, also, may be remarked, from the instructions laid down by Small* for the formation of the beam—which, in his time, were made of wood—that the land-side of the beam should lie in the plane of the land-side of the body; and, as he directed the beam to be 2½ inches in breadth at that point, its axis must have been 1¼ inches to the right of the land-side plane; and, in all cases, it must be admitted that the resultant of the effect will lie in the axis of the point, provided the draught-bolt is placed in that line. But, for very sufficient practical reasons, the draught-bolt has a range from right to left, by which the effects of variation of soil and other causes can be rectified at pleasure.

(512.) A similar difference of opinion has prevailed in regard to the position of the handles, in reference to the land-side plane. In the plow now under review, the left handle deviates only 2 inches from the line; whereas, as we shall see, another variety has the handlo 7 inches to the left of the line; and this deviation has been advocated on the principle of allowing the plowman to walk right in the middle between the handles, his right and left arms being equally extended.† Now I would again submit whether the man who walks with his arms equally extended, and his body equally distant from either handle, or he who is compelled to have one handle always near his body, whereby he can, on any emergency, bring his body instantaneously in contact with the hand, or that which it grasps—which of these men will have the greatest command over the instrument he guides? Little consideration, I imagine, will be necessary to satisfy the inquirer that the latter will have the advantage.

(513.) The dimensions of the parts of the *frame-work* of the plow are: The beam, at its junction with the mould-board at M, is from 2¼ to 2¾ inches in depth, by 1 inch in breadth—the same strength being preserved onward to the coulter-box K. From the last point a diminution in breadth and depth begins, which is carried on to the extremity C, where the beam has a depth of 1¾ inches, and a breadth of ½ to ¾ inch.

(514.) The *coulter-box* is formed by piercing an oblong mortise through the bar, which has been previously forged with a protuberance at this place, on each side and on the upper edge; the mortise is 2½ by ¾ inches, and the depth 3½ inches.

(515.) From the junction with the mould-board at M backward, the beam decreases gradually till, at the hind palm of the body at B, it is 2 inches in depth, and ¾ inch in breadth, where it merges in the left handle A. This last member retains a nearly uniform size throughout of 2 inches by ¾ inch. The right handle D is somewhat lighter, being usually 1½ inches by ¾ inch, and both terminate in welded sockets, which receive wooden helves, of 6 or 8 inches in length. The stretchers FFF, which support and retain the handles at their due distance apart, are in length suited to their positions in the handles, and their thickness is about ¾ inch diameter, tapering toward the ends, where they terminate in a collar and tail-bolt, with screwed nut. The ap-

* Small's Treatise on Plows.

† Wilkie in Farmer's Magazine, vol. xii.

per stretcher has also a semi-circular stay riveted to its middle, the tails of the stays G G terminating like the stretcher with screwed tails and nuts.

(516.) Having given the general dimensions and outline-description of this plow, there remains to be described the *details of the body-frame and its sheathing*, all the figures of which are on a scale of 1½ inches to 1 foot.

(517.) *The body-frame.*—The different views of the body-frame are exhibited in the annexed cuts, fig. 55 and 56, wherein the same letters refer to the corresponding parts in the different

Fig. 55.

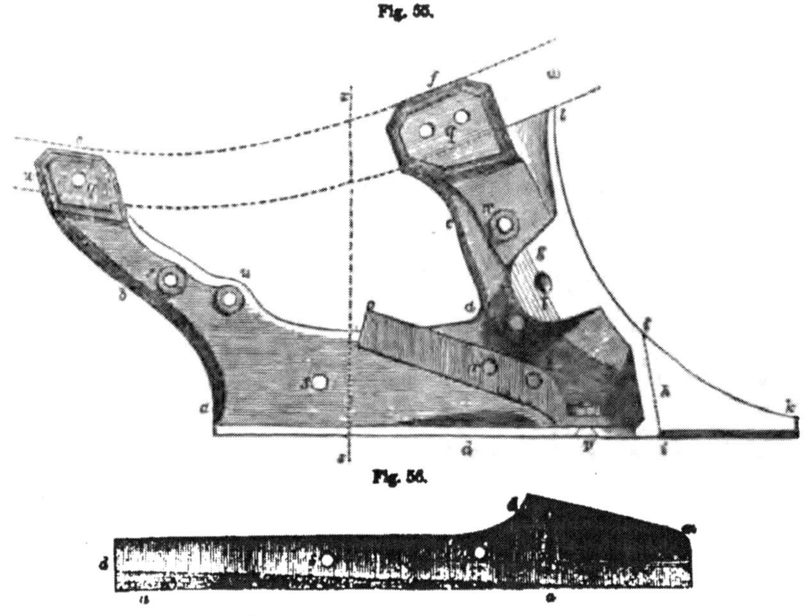

Fig. 56.

THE DETAILS OF THE BODY-FRAME.

figures. Fig. 55 is an elevation of the furrow-side; fig. 56 a plan of the sole-bar of the frame inverted; and a vertical section, on the line *x x*, is given in fig. 57. In all the figures, then, *a a* is the sole-bar, with two arms, *b* and *c*, extending upward, and having at the lower edge a flange *d*, running along the right-hand-side. Each of the arms *b c* terminates in a palm, *e f*, by which it is bolted to the beam. The arm *c* is furnished, beside, with an oblique palm or ear, *g*, upon which the fore edge of the mould-board rests, and to which it is bolted. The sole-bar *a*, with its flange, terminates forward in the head *k*, which is here made to form the commencement of the twist of the mould-board, and upon which the share is fitted, reaching to the dotted line *i i*, fig. 55. The fore edge *k, i, l* of the frame is worked into the curve answering to the oblique section of the fore edge or breast of the mould-board, and serves as a support to the latter throughout their junction. The curvature given to the arm *b* is unimportant to the action of the plow, but the general oblique direction here given to it is well adapted to withstand the thrust constantly exerted in that direction when the plow is at work. In fig. 56, the sloping edge *d m* represents the enlargement of the sole-bar, on which the share is fitted, and where the lower part of the fore edge of the mould-board rests. The depressed portion *m n* is that which is embraced by the flange of the share. In the frame, *o* is the lower extremity of the right handle, broken off at *o*, to show the manner in which it is joined to the sole-flange of the frame by the bolt *p*. The bolt-holes *q q* are those by which the beam is secured to the palms of the frame; *r r* are those by which the land-side plate is attached; and *s s* those of the sole-shoe, *t* being that which secures the mould-board to the ear, and *u* that which receives the lower stretcher of the handles. (See fig. 50, Plate V. at F and O.) The letter *v* marks the second bolt-hole of the mould-board, while its third fixture is effected upon the right handle by the intervention of a bracket, or of a bolt and socket, as seen at O, fig. 50, Plate V. The dotted lines *w w* mark the position of the beam when attached to the body, the beam being received into the seats formed on the land-side of the palms *e f*, as seen more distinctly at *w*, in fig. 57

Fig. 57.

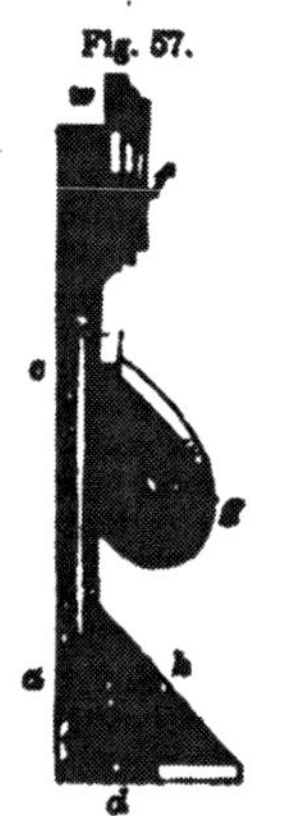

A VERTICAL SECTION OF THE FRAME.

(518.) The body-frame being an important member of the implement, regard is paid to having it as light as may be consistent with a due degree of strength; hence, in the different parts, breadth has been given them in the direction of the strain while the thickness is studiously attenuated in such places as can be reduced with safety. The least breadth of the sole-bar *a* is 3⅜ inches, of the arm *c* 4½ inches, and of *b* 2½ inches. The breadth of the sole-flange is 2 inches, the greatest thickness in any of the parts is ⅝ inch, and the total weight of the frame is 30 lbs.

(519.) *The share.*—Figs. 58–62 are illustrations of the share and its configuration; fig. 58 is a plan, 59 a geometrical elevation of the furrow-side, and 60 a direct end view looking forward, of which *a* is the *boss* adapted to the curvature of the mould-board, *b* the land-side flange which embraces the head on the land-side, *c* the sole-flange, embracing, in like manner, the head below; and these three parts form the neck or socket of the share, fitting closely upon the head, and being, in effect, part of the mould-board. The part *d e f*, fig. 58, forms the share proper, consist-

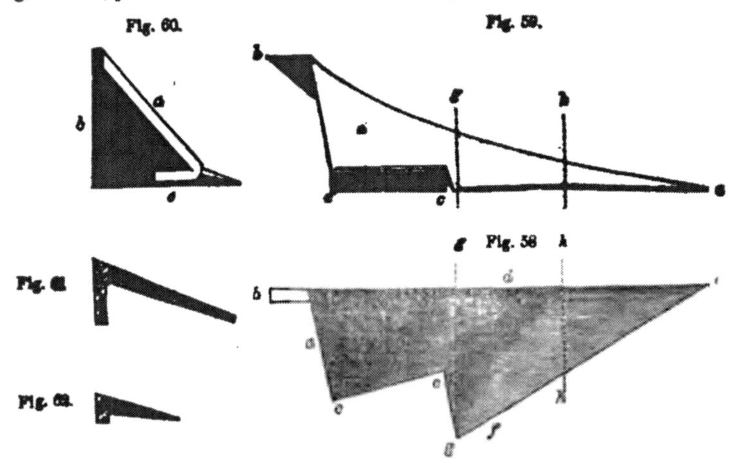

THE DETAILS OF THE SHARE.

ing of *d c e* the *shield*, terminating in the point *e*. and of the part *c g e* the *feather* or *cutter* running off at the point *e*. The extreme *breadth* of the share in this plow, measuring from the land-side to the point *g* of the feather, varies from 6 to 6½ inches; and its *length* in the sole, including the neck, is about 16 inches, the feather being 11 inches. The other figures 61 and 62 are transverse sections of the share on the lines *g g* and *h h* in the respective figures, exhibiting the structure and relation of the shield and the feather, as well as the position of the cutting edge of the feather in relation to the base-line of the plow represented by the line A′V′, fig. 49, Plate V., where, as will be observed, the cutting edge, through its entire length, lies within less than ¼ of an inch of the base-line.

(520.) The share is always formed from a plate forged for the express purpose at the iron-mills, and known in the trade by the term *sock-plate*. Fig. 63 represents the form in which these plates

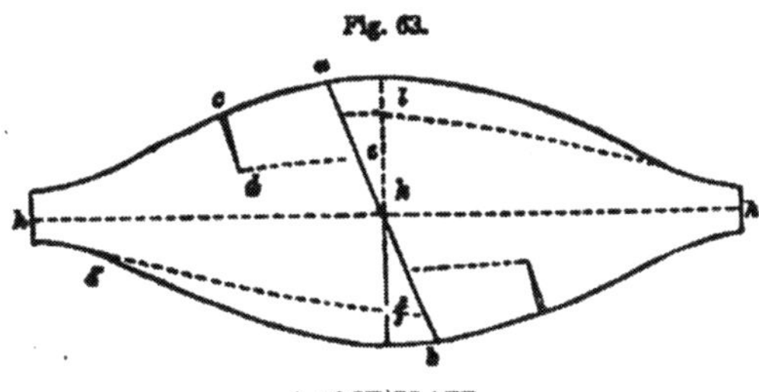

A SOCK-PLATE.

are manufactured, the thickness being from ½ to ¾ inches; they are afterward cut in two through the line *a b*, each half being capable of forming a share. To do this, an incision *c d* is made on the short side to a depth of 2 inches, the part *a c d e* is afterward folded down to form the sole-flange, and the part *b f g* is, in like manner, folded down to form the land-side flange. The point *h* is strengthened, when requisite, to receive the proper form of the shield and point, the latter being tipped with steel. The edge *h c* is extended to the requisite breadth to form the feather. In order to cut a sock-plate at the proper angle, so as to secure a minimum expenditure of labor and material, let a central line *h h* be drawn upon the plate, and bisect this line in the point *k*, the line upon which the plate should be cut will form angles of 70° and 110° nearly, with the line *h h*; or, mechanically, draw *k l*, equal to 5½ inches, at right angles to *h h*, and *l a* parallel to *h h*, mark off 2 inches from *l* to *a*, and, through the points *a k*, draw the line *a b*, which is the proper direction in which the plate should be cut.

(521.) *The sole-shoe.*—The figures 64–66 are illustrative of the *sole-shoe*. Fig. 64 is a plan of the shoe, *a a* being the sole-flange, and *b b* the land-side flange. Fig. 65 is an elevation of the same, and fig. 66 a cross section, showing the filling up of the internal angle opposed to where the greatest wear takes place. The thickness of the sole-flange at the heel *a* is ¾ inch, diminishing forward to ½ inch at 3 inches from the point, and thence it is thinned off to prevent obstruction in its progress through the soil. The breadth of the sole is 2½ inches, and its extreme length 20½ inches. The side-flange is ½ inch thick along the edge by which it is attached to the sole, diminishing upward to ¼ inch at the top edge, the hight being 4¾ inches at the heel and 6 inches at the fore end; weight about 14 lbs. The upper land-side plate is 18 inches in length on the lower edge, being 1½ inch longer than the corresponding edge of the sole-plate, the purpose

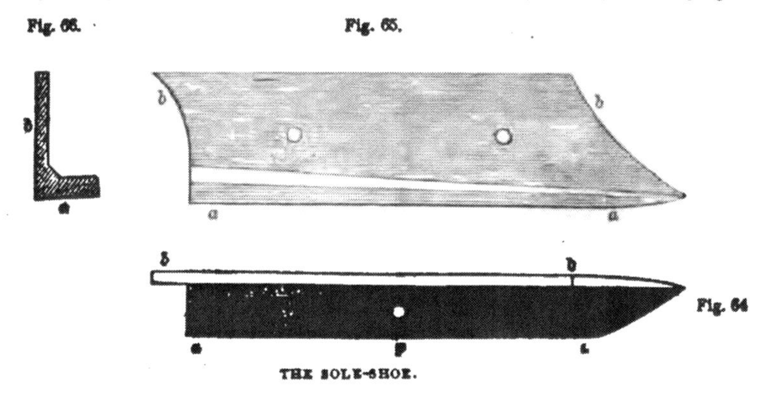

THE SOLE-SHOE.

of which will be seen in the figure of the land-side, fig. 71; the length on the upper edge is 21½ inches. The breadth and the contour of the upper edge must be adapted to the form that may have been given to the beam. The thickness at the lower edge must agree with that of the sole-plate, and be diminished to ¼ inch at the upper edge; weight 9 lbs.

(522.) *The coulter.*—Fig. 67 is an edge and 68 a side view of the coulter of this plow, in which the same letters of reference are applied. The neck *a b* by which it is affixed in the coulter-box, is about 10 inches long, though it may, with all propriety, be extended to *c*; the neck is usually about 2 inches in breadth and ¾ inch in thickness. The blade *b c d* varies in length according to the variety of the plow to which it belongs, from 18 to 22 inches. The breadth of the blade is usually about 3 inches in the upper part, but is curved off behind and terminating in a point at *d*. The thickness of the back at the shoulder *b* is ¾ inch, and tapers gently downward to where the curvature of the back begins; thence it diminishes toward the point to ¼ inch or less. It is formed quite flat on the land-side, and on the furrow-side is beveled off toward the cutting edge, where it is about ⅛ inch in thickness throughout the length of the edge.

(523.) *The bridle.*—Fig. 69 is a plan, and fig. 70 a corresponding elevation of the bridle, and the manner of its attachment to the beam, where *a* is a part of the beam, *b* the cross-head, and *c c* the tails of the bridle, with their arc-heads *d* embracing the beam on the two sides; *e* is the joint-bolt on which the bridle turns for adjustment to *earthing*; *f* is the temper-pin or bolt, and by insertion of it into any one of the holes in the arc-heads, and passing through the beam, which is here perforated for the purpose, the bridle is held in any required position. The draught-shackle *g* is held in its place upon the cross-head *b* by the draught-bolt *h* passing through both parts, and the cross-head being perforated with five or more holes, the bolt and shackle can be shifted from right to left, or from left to right, for the proper adjustment of the *landing* of the plow. To the shackle is appended the swivel-hook *i*, to which is attached the main draught-bar, or swingle-tree of the yoke.

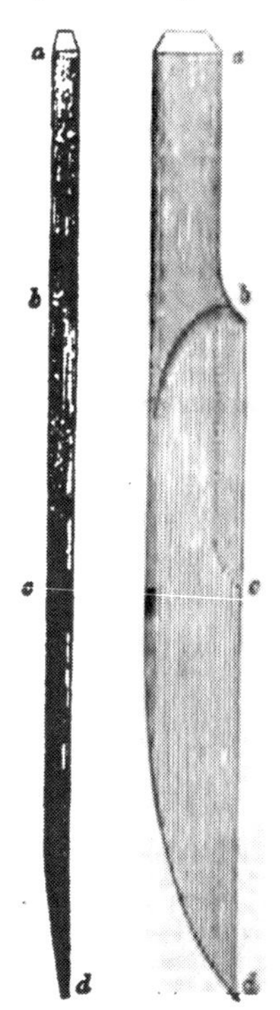

THE COULTER.

(524.) *The land-side.*—Figs. 71 and 72 are illustrations of the land-side—fig. 71 being an elevation of the body of this plow, represented in the working position, but with the extremities cut off. The point of the share and the heel rest upon the base-line at *a* and *b*, and the lines of the sole lying between these points form the very obtuse angle which obtains in the sole of this plow; *a c* is the share, and *d b* the sole-shoe; *e* is the land-side plate, and *f g* a part of the beam. The lines *a d* and *d b*, together with the base-line, form the very low triangle *a d b*, whose altitude at *d* does not exceed ⅜ inch, or by extending the sole-line *b d* to *h* the depression *h a* of the point of the share below this extended line will be ½ inch nearly. Fig. 72 represents a horizontal section of the body, as if cut off at the level of the upper edge of the sole-shoe. Here *a c* is the share, *b d* the sole-flange of the body-frame, the bolt-hole at *b* being that by which the palm of the right handle is fixed to the flange; *e* and *f* the two arms of the frame, as cut across in the section *g i*, the land-side of the sole-shoe coinciding with the land-side plane, the continuation of this line, *g i* to *h*, exhibits the inclination of the share to the land-side, which in this plow may be taken at ½ inch.

(525.) The inclination downward given to the share is intended, and experience confirms the intention, to give steadiness of motion to the implement, by giving it a lengthened base on which to stand. It is evident that if a base the converse of this were given to it—convex instead of concave—so that it should rest on the point *d*, when in motion, the smallest obstruction occurring at the point of the share would give it a tendency to swerve from the horizontal line of progression, and to lose either depth of furrow or be thrown out, thus rendering the management of the plow

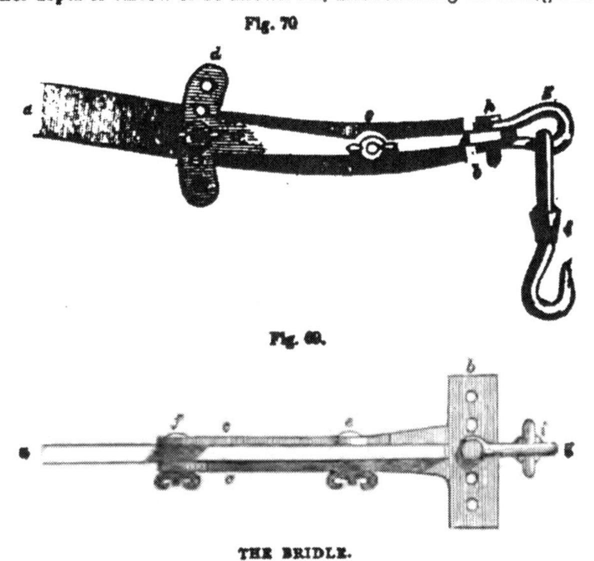

Fig. 70

Fig. 69.

THE BRIDLE.

very difficult and uncertain. Even a perfectly straight base is found not to give the requisite certainty of action, without a greater amount of exertion, as well as closer attention on the part of the plowman. A like reason prevails for this inclination of the share landward, as does for its

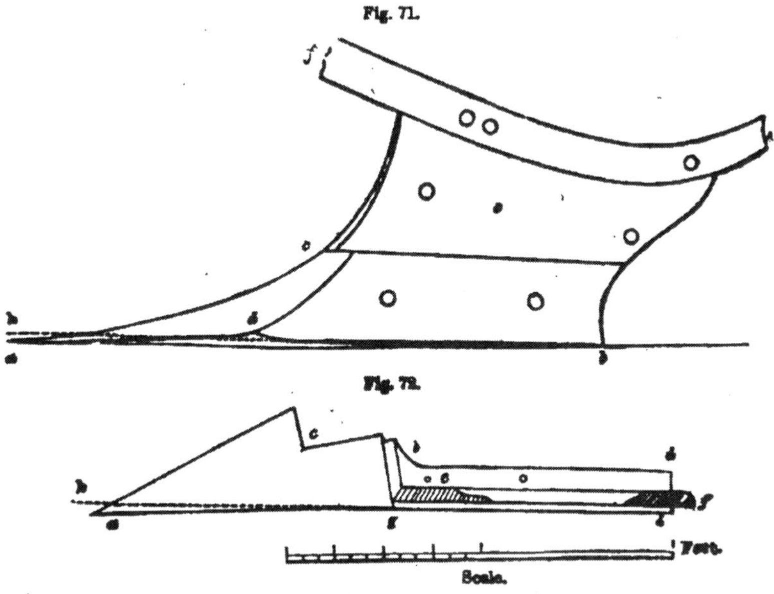

Fig. 71.

Fig. 72.

THE DETAILS OF THE LAND-SIDE.

earthward inclination; and, for the steady motion of the plow, the latter is even more necessary than the former; but there is another reason for this landward inclination, which is, that as the plow is seldom held with its land-side truly vertical, but inclining a little landward, and it being desirable to cut the furrow-slice as near as possible rectangular, the coulter has always a slight tendency landward at the point; hence it becomes necessary to give the share a like bias. By this arrangement of the parts, the incision made by the coulter will be nearly vertical. While it is admitted that these inclinations of the share afford certain advantages in the action of the plow, it must not be concealed that the practice is liable to abuse. It has been stated that, if a different arrangement were followed, a greater degree of exertion and of attention on the part of the plowman would be called forth; thus, if the sole and land-side of the body were perfectly straight, the plow would present the least possible resistance, but, as it would thus be so delicately adjusted, the smallest extraneous obstacle would tend to throw it out, unless a constant, unceasing watch is kept on its movements by the plowman. To obviate this, he gets the share set with a strong tendency to *earth* (for it is this tendency that has most effect), greater than is requisite; and, to prevent the plow taking a too deep furrow, he counteracts this by adjusting the draught-bolt to an opposite tendency; the implement will thus be kept in equilibrium, but it is obtained at an additional expenditure of horse-power. Under any such circumstances, the plow is drawn at a disadvantage to the horses, as will be afterward shown, by reason of an obliquity of the line of draught to the direction of motion, and this disadvantage is augmented by every undue tendency given to the parts by which the obliquity of their action is increased; or, if not so increased, the prevention of the increase will induce a deterioration in the work performed. This point I shall be able also to establish when I come to speak of the action of the plow generally. In the mean time, it may be affirmed that all undue inclination given to the share, but especially in its *earthing*, will either produce an unnecessary resistance to the draught, or it will deteriorate the quality of the plowing. It is, therefore, the interest of the farmer to guard against, and to prevent as much as possible, every attempt at giving any undue bias to this important member of the plow.

(526.) THE LANARKSHIRE PLOW.—The Lanarkshire plow, as constructed by Mr. Wilkie, Uddingstone, is represented in Plate IX.; fig. 51 being an elevation, and fig. 52 a plan. Like the former, it now occurs with various shades of difference, but the leading points remain unchanged; like it, also, its frame-work is invariably made of malleable iron, but, in the construction of this, the application of malleable iron is carried a step farther, as will appear in the details.

(527.) The *beam* and *left handle* are usually finished in one continuous bar, ABC, possessing a still more varied curvature than in the former plow, inasmuch as it is curved horizontally as well as vertically. When viewed in plan, and compared with the land-side plane as applied to the sole-shoe, and the fore-part of the body standing vertical, it is found that the beam, where it meets the breast-curve, coincides with the land-side plane, but at the coulter-box it deviates to the right to the extent of 1½ inches, if measured to the axis of the beam. Instead of continuing to deviate in this direction, the beam returns toward the land-side plane, till at C it is 1 inch to the right.—This formation of the fore-part of the beam gives a position that apparently makes the draught bear from a point within the body of the plow, that may be imagined to approximate to the center of resistance of the body. This is, however, more apparent than real, for the beam in this case acts simply as a bar bent at an angle, and perfectly rigid, on which, suppose a power and resistance applied at its extremities, the resultant of the strain will not follow the axis of the bar through its angular direction, but in the direction of the shortest line between the two points where the power and the resistance are applied. In addition to this horizontal curvature of the beam, it will be observed that the box of the coulter is formed by an increase of thickness on the right side only, while there is even a slight depression on the left side. This double deviation to the right gives an inclination to the plane of the coulter much greater than in any other variety of plow, being about 8° from the vertical. Though this peculiarity in the form of the beam is one of the most decided characteristics of this plow, as we now find it, it does not appear to have been an original element in Wilkie's plow, for the late Mr. Wilkie says, "the beam, which is 6½ feet long, is wrought quite straight on the land-side;"* and, from his data in the same paper, his coulter must have made an angle with the vertical plane of 6°, whereas, by the more modern construction, the angle is 8°. Continuing the comparison with the land-side plane, it will be seen that the left handle, at its junction with the tail of the beam, overhangs the land-side plane to the left by about ½ inch, there being that extent of twist on the surface of the land-side, within the limits of the body, and the same handle continues to recede from that plane till at the helve A it stands 7 inches to the left. This is also a *point* in construction of this plow, though it does not bear upon the principle of its actual working. As before observed regarding the position of the plowman in relation to the handles (512), this point is one that may be liable to be questioned, but, not being an essential point, its determination is of minor importance.

(528.) The *right handle* DE is formed in one bar, and attached to the body-frame, as will appear in detail; and it is connected to the left handle by the stretcher bolts FFF, and the stays GG.

(529.) The *coulter* I is fixed in its box K; the rake or angle at which the coulter stands in this plow, as before stated (266), is from 55° to 65°. The land-side face of the coulter is usually set to form an angle with the land-side plane of the plow, horizontally, of about 4°.

(530.) The *mould-board* L, fixed upon the body-frame and the right handle, is a curved plate of cast-iron, adapted to the turning of the furrow-slice. Its fore-edge or breast MN coincides with the land-side of the body; its lower edge O behind stands from 7½ to 8 inches distant from the land-side, while its upper edge P spreads out to 18 inches from B, the land-side. In this plow, the mould-board is prolonged forward, covering the neck of the share, meeting the shield at the root of the feather Q of the share. At this point NQ, the horizontal breadth of the mould-board is 3 inches: its hight from the base-line, at the same point, is from 2¼ inches to 2½ inches, according as the inclination of the share varies; the length along the lower edge from O to N is 20 inches, and from P to M 23 inches; the extreme length in a straight line from P to N is 33 inches; and the perpendicular hight from the plane of the base-line to P is about 11 inches. Slight deviations from these dimensions of the mould-board are to be found in the numerous sub-varieties of this plow.

(531.) The *share* QR is fitted upon a malleable-iron head, to be afterward described; the neck passing under the mould-board at NQ, and the shield falling into the curve of the mould-board, terminates forward in the chisel-point R.

(532.) The *bridle* C is formed in this plow by the end of the beam being converted into a fork

* Farmer's Magazine, vol. xii.

or sheers, to which is attached the bridle proper S, by means of the draught-bolt U; the sheers forming an adjustment vertically, while the bridle yields it horizontally, by shifting the draught-shackle at S.

(533.) The right and left handles are each furnished at A and D with wooden helves fitted into the sockets of the handles.

(534.) The *general dimensions* of this plow are: From the zero-point O to the extremity of the heel T, 4 inches, and from O forward to the point R of the share is 29 inches—giving, as the entire length of sole, 2 feet 9 inches. Again, from O backward to the extremity of the handles, the distance is 5 feet 6 inches, and forward to the draught-bolt U 4 feet 4 inches, making the extreme length on the base-line 9 feet 10 inches; but following the sinuosities of the beam and handles, the entire length from A′ to U′ is about 10 feet 6 inches. In reference to the body of the plow, the center of the coulter-box is 15 inches, and the point M of the breast-curve 6½ inches before the zero-point O, both as measured on the base-line; but, following the rise of the beam, the distance from M to the middle of the coulter-box will be 10½ inches.

(535.) The *hights* of the different points, as measured from the base line to the upper-line of the beam and handle, are marked on fig. 51; a few only of these may be repeated here. At the helve of the left handle, the hight is 3 feet 2 inches; at the same point in the right, it is 3 feet; at the middle stretcher, the difference in hight is only 1½ inches, but it again increases downward till the right handle meets the sole-bar, to which it is bolted. The hight at the point of the beam is 18 inches. and at the center of the draught-bolt U at a medium 17 inches. The lower edge of the mould-board behind is usually set at ½ inch above the plane of the base-line, and at its junction with the share is from 1½ to 1¾ inches.

(536.) The dimensions of the *frame-work* of this plow are in general as follows: The beam, at its junction with the mould-board at M, is from 2½ to 3 inches in depth, by from 1 to 1¼ inches in breadth, the same strength being preserved onward to the coulter-box K; and thence, forward to the root of the sheers, a gradual diminution goes on to about 2 inches by ¾ inch. The coulter box is formed, as before described, by an oblique mortise being pierced through the beam; which, for this purpose, has been previously forged with a protuberance at this place, to the right side only and upward, giving it a depth of 3 inches. The opening of the coulter-box is about 2½ by ⅞ inches. From the junction with the mould-board, the beam begins to diminish also backward till it merges in the left handle, and here it measures only 2 inches in depth by ¾ inch in breadth. The left handle, where it joins the tail of the beam, has a depth of 2½ inches; and here, also, it forks off into the hind branch of the body; and it diminishes in depth backward to 1¾ inches at the commencement of the helve-socket. The right handle, as in the former case, is somewhat lighter, and is connected with the left by means of stretchers, as already described (528); and both terminate in sockets for receiving the wooden helves.

(537.) *The body-frame.*—This and the succeeding figures of the details of this plow are on a scale of 1½ inches to 1 foot. In the frame of the Lanarkshire plow, as usually constructed, those parts which in the East-Lothian plow I have called the body-frame, are here formed in malleable iron. The two bars or branches of the body are welded to, and form prolongations from, the beam and left handle. Fig. 73 is an elevation of this body-frame; *a a* is a portion of the beam; *b b* a

Fig. 73.

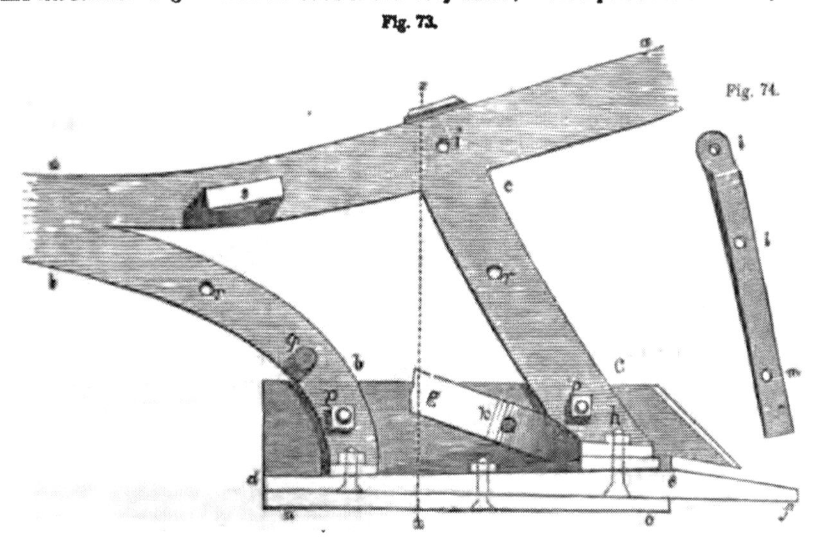

THE DETAILS OF THE BODY-FRAME.

prolongation of the left handle after it merges in the beam, forming the hind-bar of the body-frame; *c c* is the fore-bar falling from the beam; each of those bars is kneed to the right hand at the bottom, forming a palm by which they are bolted to the sole-bar *d e*. This last terminates forward in the *head e f*, upon which the share is fitted. The hind-bar is forged to a breadth of 2 inches, its thickness being ¾ inch. The fore-bar is about 3½ inches broad, and ¾ inch thick; each being respectively thinner than the beam, at the point where they spring from it, by the thickness of the land-side plate. The sole-bar *d e* is made also of malleable iron, and is 15 inches in length in the part from *d* to *e*, with a breadth of 2 inches swelled at *e*, and depth of 1 inch at *e*. The length from *e* to *f* is 8 inches, and in the depth the bar is tapered off from *e* toward *f*, where the depth is ¾ inch. From *e* it tapers backward to ½ inch at *d*. A portion of the right handle is exhibited as broken off at *g*; the lower extremity being twisted to a right angle, so as to lie flat on the sole-bar to which it is bolted, along with the palm of the fore-bar at *h*.

(538.) To determine the position of the points in this body-frame, let the zero-point O, as already fixed, be marked on the beam at 15 inches behind the center of the coulter-box K. and the whole beam curved agreeably to the dimensions given fig. 52, Plate IX.; then, the hight from the bottom of the sole-bar to the top edge of the beam at the zero-point will be 14¾ inches, as before stated, less the thickness of the sole-shoe at that point, or equal to 14 inches. The fore-part of the sole-bar at *e* will have its position determined when a straight-edge applied to its lower side from *d* to *e*, and extending as far as the point of the beam, will place the upper edge of the beam, where it spreads into the sheers, as at C. fig. 51, Plate IX., 18 inches above the *straight-edge*, or line of the sole-bar. The heel *d* of the sole-bar will be 4 inches behind the zero, and its point *f* 19 inches before it, the sole-bar being in all 23 inches. The fore-edge of the fore-bar will be 5 inches before the zero at top where it springs from the beam, and 13 inches at bottom, where it joins the sole-bar. The curvature of the fore-bar is only necessary to prevent its lying in the way of the mould-board, and a radius of 18 inches will effect this.

(539.) The provision for *fixing* the mould-board of this plow consists in a gland, fig. 74, fixed on the body-frame and right handle with bolts at *i k*, supporting the fore-part of the mould-board by means of bolts at *l m*. The remaining fixtures is effected by a bracket H, attached also to the right handle, as seen in fig. 52, Plate IX. The shoe, as seen in position, fig. 73, is marked *n o*, and is secured to the body-frame by the bolts *p p*. The lower stretcher, by which the right handle is connected to the left, is marked *q*, and *r r* mark the bolts for fixing the land-side plate. Fig. 74, already alluded to, is a front view of the gland on which the fore-part of the mould-board is supported, and this is seen also in profile in fig. 75, which is a transverse section of the body-frame on the line *x x*. In this figure *a* is the beam, *c* the fore-bar with its kneed palm at *h*, under which is the sole-bar *e*; *g* is the broken off part of the right handle, terminating in the palm lying over that of the fore-bar; and these three parts are secured by one bolt at *h*. The sole-shoe is seen at *o p*, with its land-side flange, which is fixed by the bolt *p*.

Fig. 75.

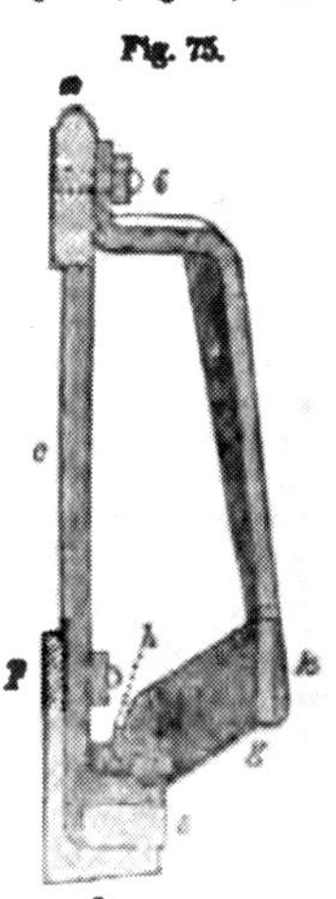

A SECTION OF THE BODY-FRAME.

(540.) *The share.*—The figures from 76 to 82 are illustrative of the shares of this plow, as adapted to both fallow and lea-plowing, where fig 76 is a plan, fig. 77 a geometrical elevation of the furrow-side of the share; and fig. 80 a direct end-view looking forward, in all which *a b* is the neck or socket by which it is attached to the head; *c* is the shield, extending over the body and the feather, but, for distinction, I shall call the portion *e c f* in fig. 76 the body, and *b g h* the feather, *i* being the point of the share, which in this plow is always chisel-shaped. Fig. 78 is an elevation of the furrow-side of the lea-share, and fig. 79 a direct end view of the same. These views have the same letters of reference; and exhibit the rise of the cutting edge of the feather above the plane of the base-line, which, when it reaches the maximum hight, stands 1¼ inches above that plane, which gives an angle equal to 8° or more with the plane of the sole in a transverse direction. The extreme breadth of this share at *e y* is 5½ to 6

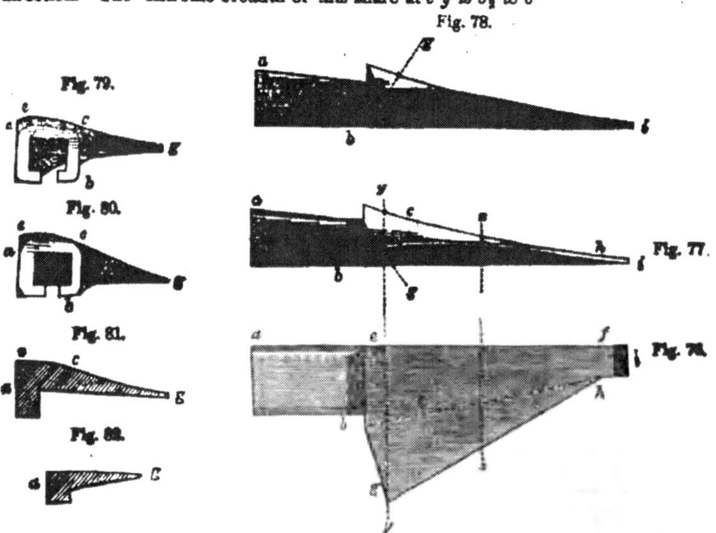

THE DETAILS OF THE SHARE.

inches; the length from the point to the head of the shield *i b*, 10 inches, and again from the point to the extremity of the neck *i a*, is 16 inches. A share thus formed will necessarily cut the furrow lower at the land-side than at the extreme edge of the furrow lower at the land-side, than at the extreme edge of the feather; for, since the share must cut the slice all along its cutting edge at the same instant, that part of the slice which is cut by the chisel-point will be the lowest possible, and every succeeding point backward will be higher and higher till it reach the apex of the curved feather 1¼ inches above the true plane of the sole. Figs. 79 and 80 exhibit the opening of the neck *a b*, which fits upon the head, and *e c g* the outline of the posterior end of the shield and feather of the two shares. Figs. 81 and 82 are transverse sections of figs. 76 and 77 on the lines *y y*, *x x* respectively.

(541.) *The sole-shoe.*—Fig. 83 is a plan of the sole-shoe, where *a b* is the sole-flange with its single bolt-hole, and *c d* the land-side flange. Fig. 84 is an elevation of the same, as viewed on

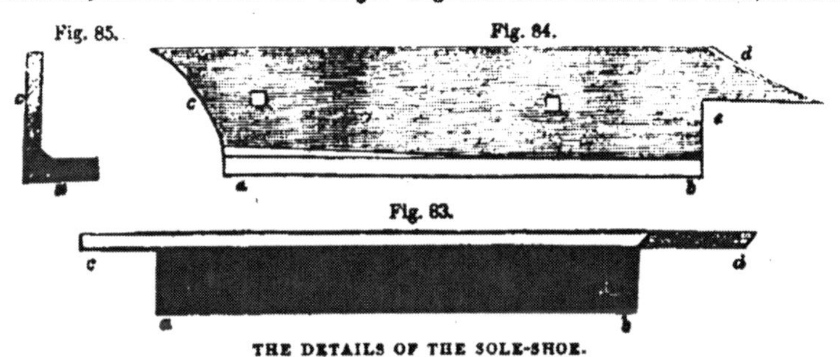

THE DETAILS OF THE SOLE-SHOE.

the furrow-side, wherein *a b* is the sole-flange seen edgewise, and *c d* the side-flange, exhibiting the notch *e*, 2½ inches long and 2 inches deep, adapted to receive the neck of the share, while the slope *d* is adapted to the breast-curve of the mould-board. Fig. 85 is a transverse section of the shoe; *a* the sole, and *c* the land-side, exhibiting also the filling, in the internal angle, opposite to where the greatest wear takes place on the exterior. The land-side flange is 5 inches in hight, and along the line of junction with the sole it is ½ inch thick, lessening upward to ⅜ inch at the upper edge; the sole-flange is ¾ inch in depth at the heel, diminishing forward to ½ inch at the fore-end, and retaining a uniform breadth of 2¼ inches. The length of the sole-flange is 17 inches, and of the land-side flange to the extreme point 20 inches. The upper *land-side plate* in this plow is 15½ inches in length on the lower edge; its upper edge, as exhibited in fig. 90, corresponds in its outline to the beam, joining flush with the left handle. The thickness at the lower edge agrees with that of the upper edge of the sole-shoe, and is diminished at the upper edge to ⅛ inch.

(542.) *The bridle.*—Figs. 86 and 87 are two views of the bridle, the first a plan, the second a

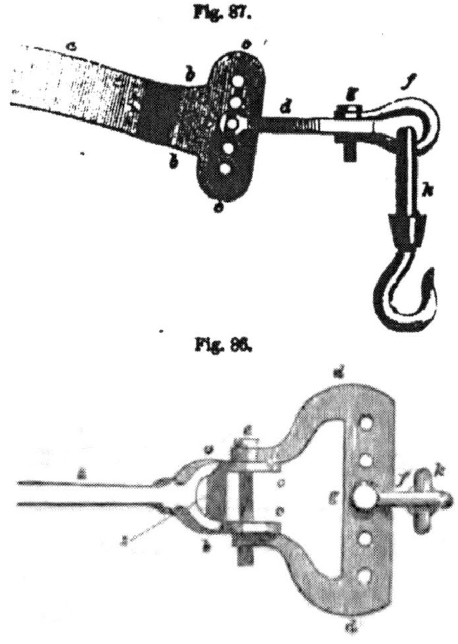

THE DETAILS OF THE BRIDLE.

side view, with the same letters to each. *a* is a portion of the beam, the extremity of which is forked into the sheers *b b*, 2 inches wide, each cheek of the sheers being also spread out into cross-heads *c c*, 5½ inches long, each furnished with four or more perforations; they are also prevented from collapsing by the insertion between them of a stretcher *s*. The bridle *d d* is adapted to the cross-heads of the sheers, and jointed on the draught-bolt *e*. The web *d d* of the bridle, 9 inches in length, is also provided with perforations, and furnished with the shackle *f*, which is attached to it by the bolt *g*. This arrangement affords the usual facility of changing the draught. By shifting the bridle on the cross-heads of the beam, in the vertical direction, the *earthing* of the plow is adjusted, and by the same operation on the shackle of the bridle horizontally, the *landing* is adjusted. The draught swivel-hook *k* is attached to the shackle, as before described, to which are appended the draught-bars afterward described.

(543.) This plow is always provided with a very useful appendage, an *iron hammer*, fig. 88. The head and handle are forged in one piece of malleable iron, the latter part being formed into a nut-key. With this simple but useful tool, the plowman has always at hand the means by which he can, without loss of time, alter and adjust the position of his plow-irons—the coulter and share—and perform other little operations, which circumstances or accident may require—for the performance of which most plowmen are under the necessity of taking advantage of the first *stone* they can find, merely from the want of this simple instrument. The hammer is slung in a staple fixed in the side of a beam in any convenient position, as at *s* in fig. 73. This little appendage is confidently recommended to all plowmen, as an essential part of the furniture of the plow.

Fig. 88.

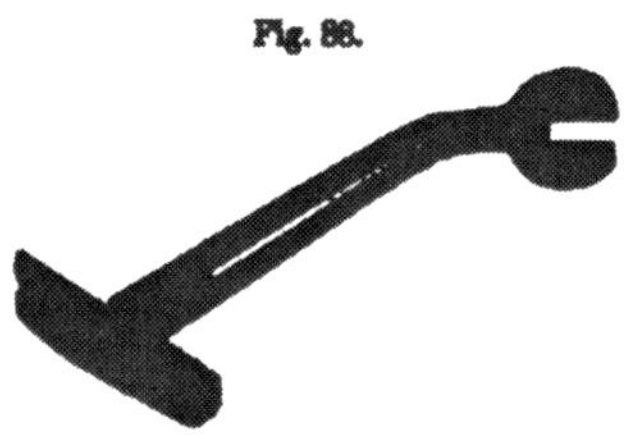

THE IRON HAMMER NUT-KEY.

(544.) *The plow-staff.*—Fig. 89 represents the plow-staff, another and a necessary article of the movable furniture of the plow. It is in form of a small shovel, having a socket, into which a helve of 5 feet in length is inserted, and in some parts of the country this is furnished with an oblique cross-head. Its position in the plow is to lie between the handles, and its use to enable the plowman to remove all extraneous matter, as earth, stubble, roots, weeds, &c., that may accumulate upon the mould-board or the coulter. It is common to all plows.

Fig. 89.

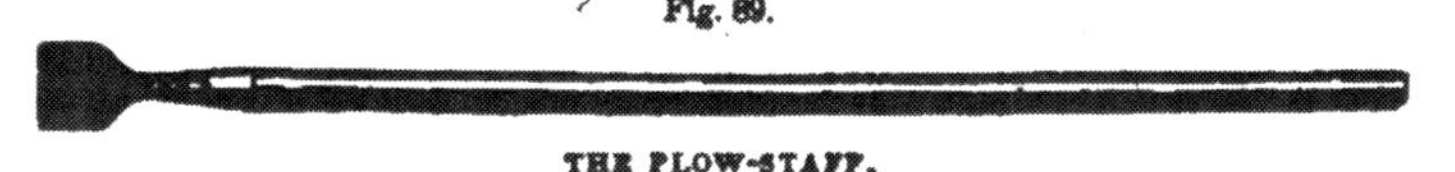

THE PLOW-STAFF.

(545.) *The Land-Side.*—Figs. 90 and 91 are illustrations of the *land-side* of the body of this plow; fig. 90 being an elevation with the extremities cut off, the point of the share, as before, rests upon the base-line at *a* and *b*, and the lines of the sole lying between these points form the obtuse angle in the sole-lines; *a c* is the share, *n* its neck, and *d b* the sole-shoe; *e* is the land-side plate, which is adapted to fill up the entire space between the side-flange of the sole-shoe and the beam; the fore part being adjusted to finish with the edge of the mould-board, while the posteri-

Fig. 90.

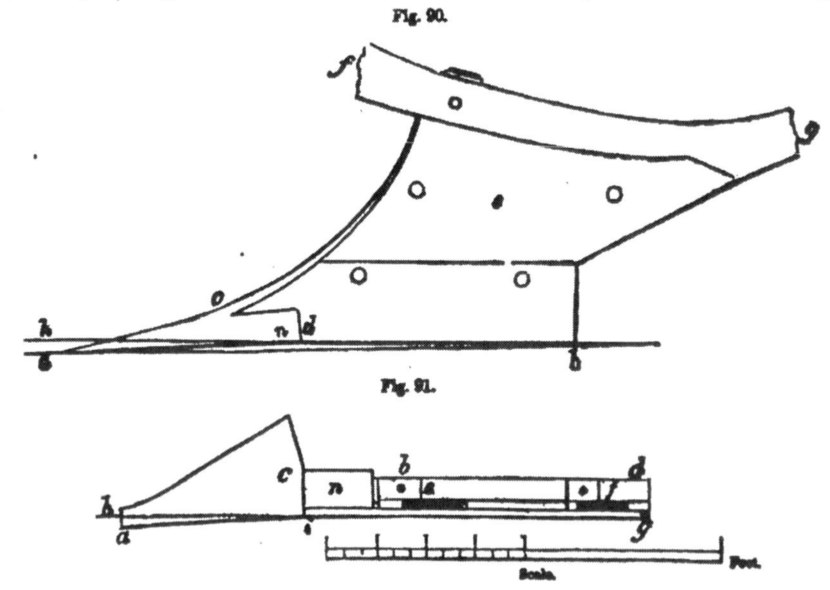

THE DETAILS OF THE LAND-SIDE.

or part may be worked off to the taste of the maker. The lines *a d* and *d b*, together with the base line, form a very low triangle, *a d b*; the altitude at *d* being not more than ¾ of an inch and by extending the side *b d* to *h*, the depression *h a* of the point of the share below the line *b d* thus extended, will be from ½ to ¾ of an inch. Fig. 91 represents a horizontal section of the body, as if cut off at the upper edge of the sole-shoe. Here *a c* is the share, *n* its neck; the line *g h* being a continuation of the land-side plane, indicates the inclination landward of the point of the share, which, in this plow, is usually from ½ to ¾ inch; *b d* is the sole-bar, the bolt-hole at *b* being that by which the right handle is fixed to the bar; *e* and *f* the two arms or bars of the body-frame, as cut across in the section; and *g i* is the land-side flange of the sole-shoe. The line *g i*, continued to *h*, exhibits the inclination of the point of the share to landward of the land-side plane. The same reasoning applies to the inclinations of the share from the sole and land side planes, as has been offered in the case of the East-Lothian plow.

(546.) The Mid-Lothian or Currie Plow.—The Mid-Lothian or Currie plow is delineated in Plate X., where fig. 53 is an elevation of the furrow-side, and fig. 54 a horizontal plan of the entire plow. This variety of the plow, probably from its more recent introduction, has undergone fewer changes than the two former. In one of its essential parts—the mould-board, little or no difference is to be found in all the range of this variety. In the share, greater changes are observable, and also in the coulter, as shall be noticed in due course. In the majority of these plows, a cast-iron body-frame is employed, and in all the mould-board is prolonged forward over the neck of the share; and the draught is applied, through the medium of a chain-bar, placed under the beam. In respect of the mould-board of this plow, it is, in point of curvature, nearly the same as the East-Lothian, though in its prolongation forward, it bears a resemblance to the Lanarkshire, but without possessing that characteristics of that mould-board as will be afterward shown. The share, in so far as it is immediately connected with the mould-board, closely resembles the Lanarkshire, and the external parts of it take also after that plow. The Mid-Lothian plow, therefore, may very appositely be termed a hybrid.

(547.) In the construction of the *frame-work* of this plow, the beam and left handle are usually finished in one continued bar, ABC, possessing the varied curvature exhibited in fig. 53, as viewed in elevation. When viewed in plan, as in fig. 54, the axis of the beam lies in one straight line, though in this there are slight shades of variation, with different makers; and the left handle, from its junction with the tail of the beam, gradually deviates from the line of the beam's axis, till, at the extremity A, it stands 3 inches to the left of the line of that axis. With reference to the plane of the land-side, also, when the fore part of the body is vertical, the point of the beam is inclined to the right of the plane about 1½ inches, and the hind part of the body on the land-side overhangs the edge of the sole ½ inch, there being that extent of twist upon the surface of the land-side, within the limits of the body. Some makers of this plow—and they are those of the greatest eminence—adopt the practice also of throwing the coulter-box to the right hand, in the beam, making the beam plain on the land-side, as in the Lanarkshire plow. This, however, is not universal, many still preferring to have the coulter in the axis of the beam In the first case, the land-side of the coulter stands at an angle of about 7°, and the latter about 5° with the vertical line.

(548.) The *right handle* DE is formed in a separate bar, and attached to the body-frame at its fore end by a bolt, as will be shown in detail; and it is farther connected to the left handle by the stretcher-bolts FFF, and the stays GG.

(549.) The *coulter* I is fixed in its box K by means of iron wedges, which set and retain it in its proper position. The rake or angle that the cutting edge of the coulter in this plow makes with the base-line, takes a greater range than any of the other two, being from 45° to 80°. The land-side face, taken horizontally, is usually set to form an angle of 2° landward, with the land-side plane.

(550.) The *mould-board* L is fixed upon the body-frame, as before described, and is adapted, as in the former cases, to the turning over the furrow-slice. Its fore edge or breast MN coincides with the land-side of the body; its lower edge O, behind, stands from 8½ to 9 inches distant from the land-side; while its upper edge P spreads out to 19½ inches from the land-side. It is, as already observed, prolonged forward, covering the neck of the share, and meeting the shield at the root of the feather Q. At the point NQ, the horizontal breadth of the mould-board is 3 inches, its hight from the base-line, at the same point N, ranges from 2¾ inches to 3¼ inches, according to the degree of inclination that is given to the share; but the real hight from the plane of the sole-shoe is 2½ inches. The length of the mould-board along the lower edge, from O to N, is 23 inches; from P to M, along the upper edge, 26 inches; and the extreme length, from P to N, is 35½ inches. The perpendicular hight, from the plane of the base-line to the upper edge at P, is about 12½ inches, though trifling deviations from these dimensions may be found among the makers of this plow.

(551.) The *share* QR is fitted upon the head, which in general is of cast-iron, as afterward described, the neck passing under the mould-board at NQ; and the shield, falling into the curve of the mould-board, terminates forward in the chisel-point R.

(552.) The *bridle* C of this plow is formed by a pair of straps S, appended to the point of the beam; and from the lower parts of these, the chain-bar H passes to the beam, whereon it is fixed, a few inches before the coulter-box K. The bridle proper, U, is attached by the same bolt that connects the chain to the straps. Shifting the straps S up or down upon the beam, affords the requisite adjustment vertically, and the bridle U gives the horizontal adjustment.

(553.) The right and left handles are each furnished, at A and D, with *wooden helves*, fitted into the sockets of the handles. In this plow, also, there is usually applied a *brace-rod* V, fixed at the fore end to the tail of the beam, and behind to the right handle by a bolt and nut, for the purpose of supporting the right handle.

(554.) The *general dimensions* of this plow are—From the zero point O to the extremity of the heel T, the distance is 5 inches; and from O forward to the point R of the share, is 29 inches; making an entire length of 34 inches on the sole. Again, from O backward to the ex tremity of the handles A′, the distance is 6 feet 2 inches; and forward to the draught-bolt U′, 4 feet 3 inches; making the extreme length on the base-line 10 feet 5 inches; but measuring along the sinuosities of the beam and handle, the entire length from A to U is 11 feet 6 inches.

(555.) In reference to the *body* of the plow, the center of the coulter-box is 16 inches, and the point M of the breast curve 8 inches before the zero-point; both as measured on the base-line; but, in following the rise of the beam, the distance from M to the middle of the coulter-box is 11 inches.

(556.) The *hights* of the different points, from the base-line to the upper edge of the beam and handle, are marked on fig. 53; the chief points only being expressed here. At the helve of the left handle, the hight is 3 feet, the right being 2 inches lower; the difference in hight continuing nearly uniform throughout their length. The hight of the point of the beam at C is 23 inches, and to the center of the draught-bolt at a medium of 16½ inches. The lower edge of the mould-board, behind, is usually set at ½ inch above the plane of the sole; while, at its junction with the share at N, the hight above the base line runs from 1¼ to 1½ inches.

(557.) For the dimensions of all the individual parts of the frame-work of this plow, it is unnecessary to repeat them here, as they correspond so nearly with those already stated in treating of the first two varieties. In this respect, therefore, reference is now made to those before described in paragraphs (513) and (536).

(558.) *The body-frame.*—The Mid-Lothian, like the East-Lothian plow, is usually constructed with a cast-iron body-frame, differing, however, in some respects, from the latter. Fig. 92 is an elevation of the furrow-side of the body-frame. It consists of a plate or web *a b c d* of about ¼ inch thick, upon which is planted the sole-bar *b e f*, the beam-flange *a h*, and also the ribs *b i* and *k l*; these last are for the purpose of strengthening the web. Fig. 93 is a direct view

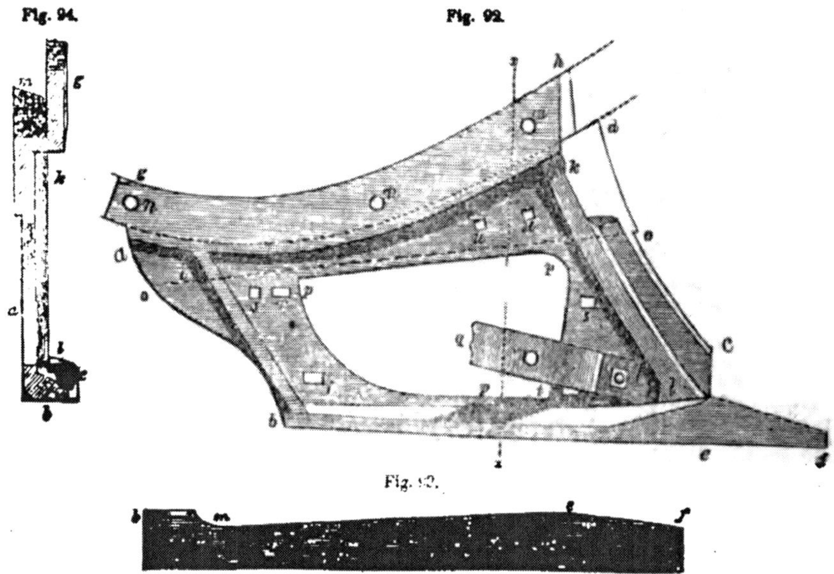

THE DETAILS OF THE BODY-FRAME.

of the under surface of the sole-bar. Its breadth at *b* and *e* is 2½ inches, but from *e* toward *m* it is diminished to 2 inches, where the thickness is ½ inch; but at *e*, where the principal strain falls, through the action upon the share, the depth is increased to 2 inches, from which it tapers forward to *f*, where it measures 1¾ inches in breadth and ¾ inch in depth. From *e* it diminishes also backward; and from *l* to *b* a filling piece is inserted in the pattern, in the angle, as seen at *l*, fig. 94, to increase the strength. A filling piece is also inserted at *k*, fig. 92, to support that point where the strain from the beam falls upon the body, as well as to give a bearing to the breast of the mould-board. Fig. 94 is a transverse section of the body-frame on the line *x x*, looking forward; *a* is the web, *b e* the sole-bar, *k l* one of the ribs in fig. 92, *g* the beam-flange, and *m* the seat into which the beam is received when applied to the frame, and bolted, as at *n n n*. In the best examples of this body-frame, a part of the land-side plating is cast along with the frame; the lower edge of this portion is represented by the dotted line *o o*, fig. 92; and the frame, as here described, is always cast in one piece, but having the perforation *p p p* always formed in it. A broken off portion of the right handle is marked *q*, and is formed at the fore part into a palm, by which it is bolted to the web. The bolt-hole *r* is the place of insertion of the lower stretcher, which connects the right handle to the body-frame; *s s* are the bolts of the land-side plate; *t t* those for the land-side flange of the shoe; *u u* are the bolts for fixing a kneed bracket, on which the upper fore part of the mould-board rests, and is bolted, the lower fixture being at *v*; and a third is obtained through a bracket, bolted upon the right handle, as seen at Y, fig. 54, Plate X. The length of the beam-flange in this frame is from 19 to 19½ inches, and the hight and outline of that part are obtained from the hights marked in fig. 53, Plate X, deducting 1 inch for the thickness of the sole-shoe at the heel, and ½ inch at the point.

(559.) *The sole-shoe.*—Fig. 95 is a plan of the sole-shoe; *a b* the sole-flange 17 inches in length 3 inches in breadth, and 1 inch in depth at *a* the heel, but diminished to ½ inch at *b*; *c d* is the land side flange, ½ inch in thickness at bottom, and ¾ inch at the upper edge, the hight being 4 inches. Fig. 96 is the furrow-side of the shoe, with the same letters of reference; *e* is the notch at the fore

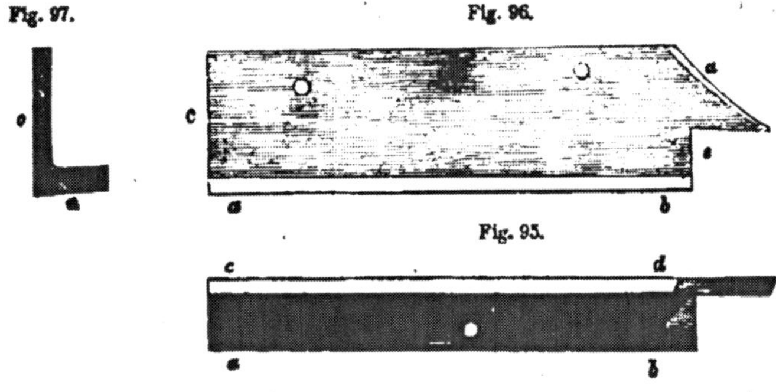

THE DETAILS OF THE SOLE-SHOE.

part, for the passage of the neck of the share; it is 4½ inches in length and 2½ inches in hight, *a* being the curve adapted to the breast of the mould-board. Fig. 97 is a transverse section of of the shoe, *a* the sole, and *c* the side-flange.

(560.) *The share.*—The share of this plow, in principle and construction, is the same as that of the Lanarkshire; but in the present case, the head being of cast-iron, the neck is necessarily some what larger. Fig 98 is a plan, in which *a b* is the neck, *a e d* the land-side, and *e g e* the shield; *b f g*

Fig. 98.

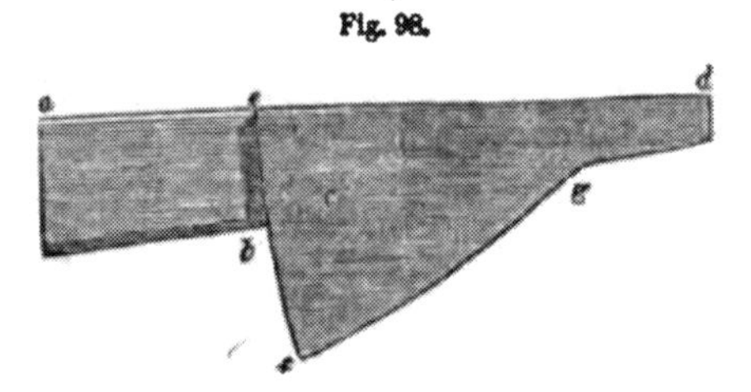

THE SHARE.

is the feather, and *g d* the point of the share, which, in this plow, is usually chisel-pointed, and longer between the termination of the feather and the point, than in the share of the Lanarkshire plow. In farther illustration of this share, reference is made to that of the Lanarkshire plow, where fig. 78 is a direct view of the furrow-side of the share, exhibiting the rise of the cutting edge of the feather above the base-line, which, in the plows considered the most perfect for plowing lea, amounts to a rise of 1¼ to 1½ inches. The extreme breadth of this share over the feather ranges from 4½ to 5½ inches, the length from the point to the head of the shield, at a maximum, is 11 inches, and including the neck, 17 inches; under the same condition the length from the extreme point to the commencement of the feather at *g*, is about 3½ inches. Fig. 79 is an end-view of the share looking forward, in which also the same letters are applied; *a b* is the opening of the neck to receive the head, and *e c g* shows the outline of the posterior extremity of the shield and feather. This, like the Lanarkshire plow, is held as peculiarly adapted to the plowing of lea-land; and as the share just described is that which is adapted for that purpose—for the chief and almost sole difference between the adaptation of these plows for lea and stubble land lies in the configuration of the share—it is necessary to advert to the stubble land or fallow-share. In this the chief, indeed the only, difference lies in the formation of the feather, which for stubble land is made broader, and the cutting edge, instead of rising from the point at an angle of 8°, is formed so as to approach to the plane of the sole, or not exceeding an angle of 4°.

(561.) *The land-side.*—Figs. 99 and 100 are illustrations of the land-side of the body of this plow—the extremities, as in the previous cases, being cut off. Fig. 99 is an elevation, *a b* is the base-line, *a c* the share, *n* its neck, and *d b* the sole-shoe; *e e* are the land-side plates—the upper one, as before stated, being cast as a part of the body, and *f g* is a part of the beam. In the extreme cases of this plow, the altitude of the low triangle *a n b* is ¾ inch; and, when the line of the sole *b n* is extended to *h*, the depression of the point of the share below that line is found to be about 1½ inches. Fig. 100 represents a horizontal section of the body-frame, as if cut off at the upper edge of the sole-shoe; here *a c* is the share, *n* its neck, and *b d* the sole-flange; *e* and *f* are the two bars of the body-frame, and *g i* the land-side of the sole. By continuing the line of the land-side to *h*, the inclination of the share landward is found frequently to be 1 inch.

(562.) *The bridle.*—As has been already noticed, this plow differs from the others in its bridle being connected with a chain-bar, passing under and attached to the beam near the coulter-box; and, for the purpose of receiving this equipage, the point of the beam is elevated to the hight of 22 inches above the base-line. The chain is usually a single rod of iron, with a link and shackle behind, by which it is connected to the beam, by means of a bolt passing through the shackle and

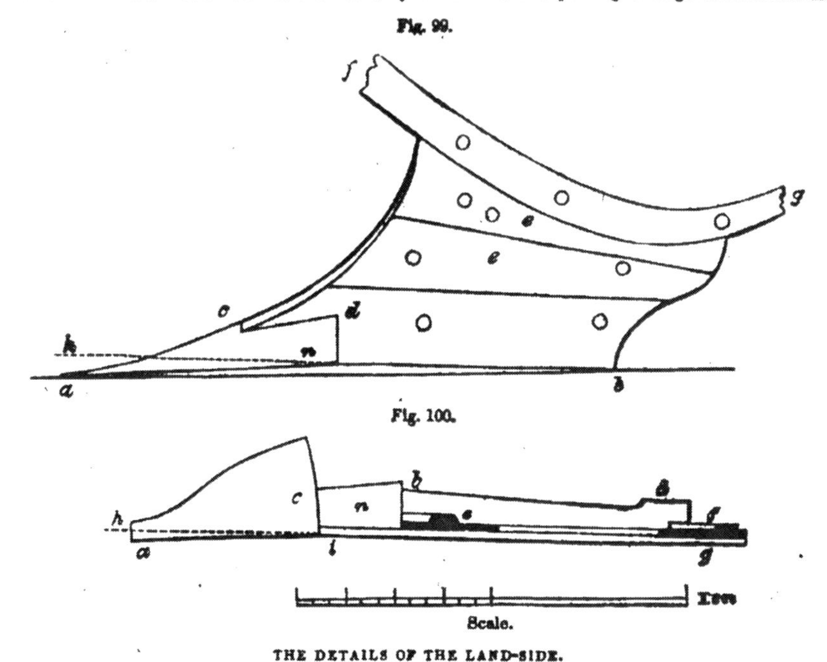

THE DETAILS OF THE LAND-SIDE.

the beam at a point about 3 inches before the coulter-box. The bridle, of which fig. 101 is an elevation and fig. 102 a plan, consists of a pair of iron straps *a b*, 10 inches in length, and $1\frac{1}{4}$ inches

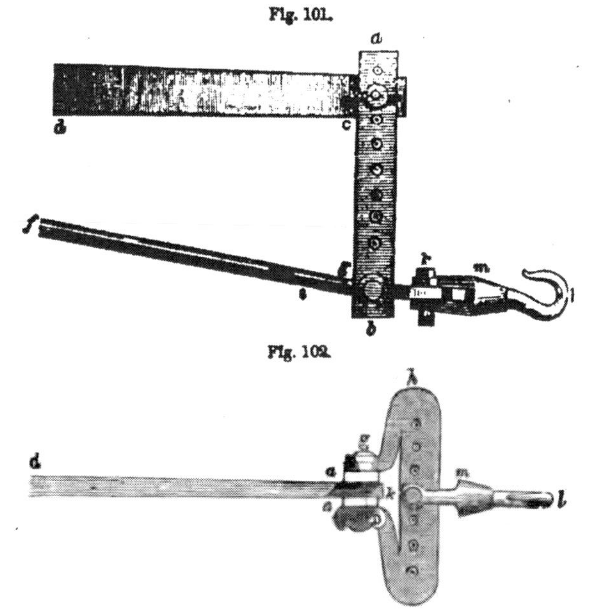

THE DETAILS OF THE BRIDLE.

by $\frac{1}{4}$ inch, each having a number of perforations by which they can be appended to the point of the beam *c d*, by means of a bolt passing through them and the beam; a strap *a b* being on each side of it. The fore end of the chain-bar *f e* is, in like manner, received between the lower ends of the straps at *b*, and secured by the draught-bolt *g*. On the same bolt is appended the bridle proper *h i*, the bolt passing through the whole of the parts. The bridle is formed with a web *h i* in front, 9 inches in length, and 1½ inches in breadth, having also a number of perforations for receiving the shackle-bolt *k*. In this equipage, the draught-swivel hook *l* and the shackle *m* are combined in one, which completes the arrangement. This combination of bridle-mounting gives the same facility as before for shifting the direction of the draught—vertically, by raising or lowering the straps *a b* on the point of the beam, and horizontally, by shifting the shackle-bolt and shackle *k m* right and left.

(563.) OF THE ACTION OF THE PLOW.—The *coulter*, the *share* and the *mould-board* being the principal active parts of the plow, and those which supply the chief characteristics to the implement, it may be useful to the farmer, as well as to the agricultural mechanic, to enter into a more minute descriptive detail of the nature and properties of these members, before entering upon the duties which each in its turn has to perform in the action of cutting and turning over the furrow-slice.

(564.) *The coulter.*—The coulter, in its construction, as well as in the duties it has to perform, is the simplest member of the plow. It is a simple bar, in form as represented by figs. 67 and 68; varying in length, according to the variety of the plow to which it belongs, from 18 to 22 inches. Simple though the form and duties of the coulter may be, there is no member of the plow whereof such a variety of opinions exist as to its position. I have shown that, in practice, the rake or angle which its cutting edge makes with the base-line ranges from 45° to 80°, that of its land-side face from 4° to 8° with the vertical, and that the same face, in the horizontal direction, varies from 0° to 4° with the land-side. The objects of these variations will be duly pointed out, as mere matters of taste and convention among plowmen. Two points alone, in regard to position, should be considered as standard and invariable. These are, 1st, that *the land-side face of the coulter shall be always parallel, in the horizontal direction, to the plane of the land-side of the plow's body*; and, 2d, *that at the hight of 7 inches, or of 6 inches, according to the depth of furrow to which the plow is adapted, the land-side face of the coulter shall be ½ inch to landward, or to the left, of the plane of the land-side of the body.* One other point in position is subject to a great diversity of opinion—that is, the position in which the extreme point of the coulter should stand in relation to the point of the share. In respect to *landing*, or that cause which requires the point of the coulter to be placed to landward of the share, the range of opinion is within moderate bounds, being from 0 to ¾ inch; but, in the vertical direction, the range varies from ½ inch to 2 inches, and in the longitudinal direction a like difference of opinion exists. Thus Small recommends that the point of the coulter should be 2 or 3 inches in *advance* of the point of the share, and ½ or 1 inch *above* the plane of the sole (base-line), while it should be ½ inch or 1 inch to *landward* of the land-side plane.* The first of these propositions, as will be afterward shown, is very much at fault; and the almost universal practice, also, of keeping the two points nearly equal in advance, condemns the practice, and points out equality as the rule. In regard to the position of the point landward, it is liable to considerable variation, partly from the inclination that may be given to the share, and likewise from the degree of obliquity between the coulter and the land-side. This last, indeed, combined with the rule laid down, from the position of the coulter in relation to the land-side, at the hight of 6 or 7 inches, is the true source from which the landward relation of the points of the coulter and share can be ascertained; hence, therefore, in whatever variety of the plow, *the coulter should have its position in regard to land determined first; and the point of the share should take its position from the coulter.* The distance to which the point of the share stands to the right of the coulter should in no case exceed ½ inch, but it were better to confine it to ¼ inch. In the vertical position, the advancing of the point of the coulter to, or retiring from, the share, violates no principle in the relation of the parts; but, to place the coulter at an undue distance above the share leaves that portion of the slice uncut that falls between the two points; which must produce an undue resistance, from the part being forcibly pressed asunder, by a process like clipping, through the inclined action of the share upward. The nature of the soil, whether stony or gravelly, or a loam, will, however, always have an effect on this point of the trimming of the plow; and, as no principle is affected, there is no impropriety in giving a latitude in this direction; though I conceive that a distance of 1 inch between the points of the share and coulter ought to be the maximum, except in cases where the nature of the soil may demand a deviation from that distance.

(565.) The office which the coulter has to perform in the action of the plow is simple and uniform, being merely to make an incision through the soil, in the direction of the furrow-slice that is to be raised. It is a remarkable fact that, in doing this, it neither increases nor decreases the resistance of the plow in any appreciable degree. Its sole use, therefore, is to cut a smooth edge in the slice which is to be raised, and an unbroken face for the land-side of the plow to move against in its continued progress.

(566.) In the early works on the principles of the plow, some misconceptions appear to have been formed of the influence of the coulter, under the supposition that the coulter extending 3 inches in front of the share acted beneficially; and that giving the coulter a great rake, or a low angle with the base, made it cut the soil advantageously, and with less resistance. From a series of experiments, I have satisfied myself that the first of these suppositions is erroneous, and that the projection of the coulter before the share increases the resistance in a very sensible degree. With regard to the second, the resistance seems not to be affected by the angle at which the edge of the coulter stands; and the analogy of a common cutting instrument† does not hold in the case of the coulter of the plow. With a razor or a knife in the hand, we make them pass through any object by *drawing* their cutting edge over the surface to be cut, in the manner as with a saw, which greatly increases the effect without any increase of force; and this holds in all proper cutting instruments; but let the edge of the instrument be placed simply at an angle with the direc-

* Small's Treatise on Plows.

† Ibid.

tion in which the stroke or cut is to be made, and, in making the cut, let this oblique position be retained, so that the cutting edge shall proceed parallel to its original position, without any tendency to *drawing* the edge across the direction of the cut; no saving of force is obtained. This process must be familiar to every one who uses a knife for any purpose whatever. In slicing a loaf, the operator is at once sensible that, by moving the knife gently backward or forward, he is required to exert less force, while he at the same time makes a smoother cut, than he would do by forcing the knife through the loaf, with its edge either at right angles or obliquely to the direction in which the knife proceeds. The coulter of the plow acts in this last position; its cutting edge stands obliquety to the direction of motion, but has no means of drawing or *sliding*, to cross the *forward* motion; it therefore cuts by sheer force of pressure.* Where elastic substances occur, an instrument cutting in this manner has some advantages. In the case of fibrous roots, for example, crossing the path of the coulter—the latter, by passing under them, sets their elasticity in action, by which they allow the edge to slide under them to a small extent, and thus produces the *sawing* effect. In the non-elastic earths, of which soils are chiefly composed, nothing of this kind, it is apprehended, can occur; hence the angle of the coulter, as it affects the force requisite to move the plow, is of little importance.

(567.) I have said that the projection of the coulter in front of the share increases the resistance, and I am borne out in this assertion from the result of experiments not a little inexplicable. On a subject which has of late attracted considerable attention, I was desirous of obtaining information, from experiments alone, on the actual implement; and, to attain this the more fully, I determined on analyzing the resistance as far as possible. With this view, a plow was prepared whose coulter descended 7 inches below the line of the sole, and fitted to stand at any required angle.—This plow, *with its sole upon the surface of two-years' old lea*, and the *coulter* alone in the *soil*, the bridle having been adjusted to make it swim without any undue tendency; the force required to draw this experimental instrument, as indicated by the dynamometer, was 26 imperial stones, or 3½ cwt., and no sensible difference was observed in a range of angles varying from 45° to 70°. This coulter having been removed, the plow was drawn along the surface of the field, when the dynamometer indicated 8 stones, the usual draught of a plow on the surface. Another well-trimmed plow was at work in the same ridge, taking a furrow 10 by 7 inches, and its draught was also 26 stones. On removing the coulter from this plow, and making it take a furrow of the same dimensions, the draught was still the same—namely, 26 stones; the furrow thus taken produced, of course, a slice of very rough plowmanship, and though it exhibited, by a negative, the essential use of the coulter—the clean cutting of the slice from the solid ground—the whole question of the operation and working effects of the coulter are thus placed in a very anomalous position. The question naturally arises, what becomes of the force required to draw the coulter alone through the ground, when, as it appears, the same amount of force is capable of drawing the entire plow, with or without a coulter? A definite and satisfactory answer, it is feared, cannot at present be given to the question, and, until experiments have been repeated and varied in their mode of application, any explanation that can be given is mere conjecture.

(568.) Since we have seen that the *same force is required to draw the plow without a coulter as with it*, and as it has been observed that the work performed without the coulter is very rough, by reason of the slice being in a great measure *torn* from the solid ground, the breast of the plow being but indifferently adapted for cutting off the slice—it is more than probable that the tearing asunder of the slice from the solid ground requires a certain amount of force above what would be required were the slice previously severed by the vertical incision of the coulter. And though we find that the force requisite to make this incision, when taken alone, is equal to the whole draught, yet there appears no improbability in the supposition that the *minus* quantity in the one may just equal the *plus* in the other. Be this as it may, the discovery of the anomaly presents at least a curious point for investigation, and one that may very probably, through a train of careful experiments, point out the medium through which a minimum of draught is to be obtained.

(569.) Regarding the effect of change on the angle of the edge of the coulter, though it does not directly affect the draught of the plow, it is capable of producing practical effects that are of importance. In plowing stubble land, or land that is very foul with weeds, the coulter should be trimmed to a long rake—that is, set at a low angle, say from 45° to 55°; this will give it a tendency to free itself of the roots and weeds that will collect upon it, by their sliding upward on the edge of the coulter; and, in general, will be ultimately thrown off without exertion on the part of the plowman. The accumulation of masses of such refuse on the coulter greatly increases the labor of the horses. The amount of this increased labor I have frequently ascertained by the dynamometer, and have found it to increase the draught of the plows from 26 stones, their ordinary draught when clear, up to 36 stones; and, immediately on the removal of the obstruction, the draught has fallen to an average force of 26 stones. It is unnecessary to add that the prevention of such waste of muscular exertion ought to be the care of the farmer, as far as the construction of his machines will admit of.

(570.) To apply a plow, with its coulter set in the position above described, to lea-land with a rough surface, would produce a kind of plowmanship not approved of; every furrow would be bristling with the withered stems of the unconsumed grasses; for, to plow such land with a coulter set in this way, would cause its partially matted surface to present a ragged edge, from the coulter acting upon the elastic fibres and roots of the grasses, pressing them upward before they could be cut through. The ragged edge of the slice thus produced gives, when turned over, that untidy appearance which is often observable in lea plowing. To obviate this, the coulter should be set at a higher angle, by which it will cut the mat, without tearing it up with a bearded edge. *Crack* plowmen, when they are about to exhibit a specimen of fine plowing, are so guarded

* An ingenious application of the drawing action here illustrated is to be found in the subterranean cutters of Mr. Parkes's steam-plow for plowing moss-land.

against this defect that they sometimes get their coulter kneed forward under the beam so far as to bring the edge nearly perpendicular. The same cause induces the makers of the Lanarkshire plow to set the coulter with its land-side face, not coincident with the land-side plane horizontally but at an angle with it of 4°, thus placing the *right* hand face of the coulter nearly parallel to the land-side plane, and thereby removing the tendency of the ordinary oblique position of the right hand face to produce a rough-bearded edge on the rising slice. The dynamical effect of such a position will be afterward treated of.

(571.) *The share.*—The structure and position of the different shares having been already pointed out (519,) (540,) (560,). and also their relations to the coulter, there remains to make some general remarks on the action of the share, and on the effects resulting from the varieties of that member of the plow.

(572.) We have seen that the coulter performs but a comparatively small portion of the operation required in the turning a furrow-slice. The share, however, takes a more important and much more extensive part in the process; on the functions of the share, in short, depends much of the character of the plow. Its duty is very much akin to that of a spade, if pushed horizontally into the soil with a view to lift a sod of earth; but, as its action is continuous, its form must be modified to suit a continuous action; hence, instead of the broad cutting edge of the spade, which, in the generality of soils, would be liable to be thrown out of its course by obstacles such as stones, the share may be conceived as a spade wherein one of its angles has been cut off obliquely, leaving only a narrow point remaining, adapted to make the first impression on the slice. A narrow point being liable to meet obstruction only in the ratio of its breadth to the breadth of the entire share, the chances of its encountering stones are extremely few; and though the oblique edge, now called the feather, has a like number of chances to come in contact with stones, yet, from its form, taking them always obliquely, and the direct resistance which the body of the plow meets with on the land-side preventing any swerving to the left, such stones as come in contact with the sloping edge of the feather are easily pushed aside toward the open furrow on the right. The share thus acts by the insertion of its point under the slice intended to be raised, and this is followed up by the feather, which continues the operation begun by the point, by separating the slice horizontally from the subsoil or the sole of the furrow; and, simultaneous with this, the coulter separates the slice vertically from the still solid ground. Probably the most natural impression that would occur, at the first thought of this operation, will be that the feather of the share should be of a breadth capable of producing the immediate and entire separation of the slice from the sole; but experience teaches us that such would not fulfill all the requisite conditions of good plowing.—The slice must not only be separated; it must be gradually turned upon its edge, and ultimately still farther turned over until that which was the upper surface becomes the lower, lying at an angle of about 45°. It is found that, *if the slice were cut entirely off from the sole*, the plow would frequently *fail in turning it over* to the position just referred to; it might, in place of this, be only moved a space to the right and fall back, or, at most, it would be liable to remain standing upon its edge; in either case the work would be very imperfect, and it has therefore been found necessary to leave a portion, usually from ¼ to ⅓, of the slice uncut by the feather of the share. This portion of the slice is left to be *torn* asunder from the sole as it rises upon the mould-board, by which means the slice retains longer its hold of the subsoil—turning by that hold, as upon a hinge, till brought to the vertical position, after which it is easily brought into its ultimate place. The breadth of the share is thus, of necessity, limited to ¾ the breadth of the slice at a maximum, though its minimum, as will appear, may not exceed ½.

(573.) The disposition of the *feather* comes next under notice. The feather having to perform the operation of cutting that part of the slice below that lies between the point of the share and the extremity of the feather, it is formed with a thin edge suited to cutting the soil; but the *position of that cutting edge* forms a principal feature in distinguishing the varieties of the plow, as before described. This distinguishing character is of two kinds: 1st, that which has the cutting edge lying parallel, or nearly so, to the plane of the sole, as in the East-Lothian plow; and, 2d, that which has this cutting edge elevated as it retires from the point of the share, rising at an angle with the base-line, which is found to vary from 4° to 8° as in the Lanarkshire and the Mid-Lothian plows; and all the sub-varieties of these plows have their shares coming under one or other of these two divisions.

(574.) The share, in either of the forms above described, passes *under the* furrow slice, making a partial separation of it from the sole of the furrow, rising as the share progresses; the rise, however, being confined entirely to the *land-side edge* of the slice—the furrow edge, as has been shown, remaining still in connection with the solid ground; and the shield and back of the share being a continuation of the mould-board, the latter, in its progress forward, receives the slice from the share and passes it onward, or, more properly speaking, the plow passes under it.

(575.) One important consideration remains to be noticed regarding the practical effects of the two forms of feather. In the first, which has the cutting edge nearly parallel with the plane of the sole, the furrow-slice being cut below at one level over the whole breadth of the share and feather, the slice, when exposed in section, will be perfectly rectangular or very slightly rhomboidal, and the sole of the furrow will be perfectly level across. Such a share, then, will lift a slice of any given breadth and depth, which shall contain a maximum quantity of soil, and this problem can only be performed by a share so constructed.

(576.) In the second case, where the feather rises above the plane of the sole at the angles already named, the feather is found sometimes to attain a hight of 1 inch and 1½ inches above that plane. In all such cases, the feather is also narrow; and, supposing that the part of the slice left uncut by it may be torn asunder, in a continuation of the cut so made, the slice will have a depth at its furrow edge less by about 1½ inches or more than at its land-side edge, as cut by the point of the share. A transverse section of this slice, therefore, fig. 103, would exhibit not a rectangular parallelogram as before, but a trapezoid, whose sides *a b* and *c d* might be each 9 inches, and its sides *b d* and *a c* 6 inches and 4½ inches respectively. A slice of this form would, therefore, be deficient in the quantity of soil lifted, by a quantity contained in the triangle *d c e*, or about 1-7 part of the entire slice; and this deficiency is left by the share in the bottom of the furrow as part of the solid subsoil. The absolute quantity of soil thus left unlifted by shares of this construction will be found to vary with the elevation that is given to the feather; but, wherever this form of share

Fig. 103.

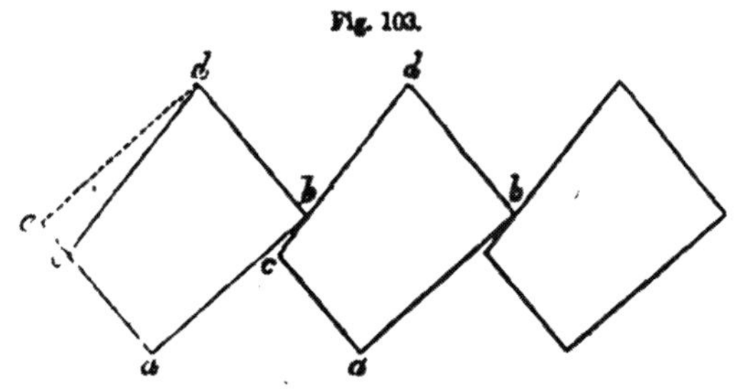

THE CRESTED FURROW-SLICE.

is adopted, results similar to that here described will invariably follow, though they may differ in degree; but the quantity left in the bottom of the furrow will seldom fall short of 10 per cent. of the whole slice. An indirect mode of removing this defect is resorted to in practice, which will be noticed under the head of mould-board.

(577.) The *rule* which I would recommend to be followed in order to secure the maximum of useful effect in the share, as founded on practice and observation, as well as combining the theory of the share and mould-board, is, that the length from the tail of the feather to the point of the share should be from 10 to 11 inches; that the hight of the shield—the surface of the share—on the land-side, opposite to the tail of the feather, be 2¼ inches above the line of the sole-shoe; that the point of the share be ½ inch below the line of the sole-shoe, and not exceeding ¼ inch to landward of the land-side plane, this last point being more properly determinable from the coulter; and, lastly, that no part of the edge of the feather should be more than ¾ inch above the plane of the sole-shoe, that plane being always understood to be at right angles to the land side plane.

(578.) *The mould-board and its action.*—Since the time that Small achieved his great improvement in the formation of the mould-board, that member has been generally held as the leading point in the plow. This, in one sense, is no doubt true; for if there is a spark of science required in the construction of the plow, it is certainly the mould-board that most requires it. Yet, for all this, I have seen a plow making work little, apparently, inferior to the first rate mould-board plows, that had nothing to enable it to turn over the slice but a straight bar of wood.* In this case, however, the work was but apparently well done, there being nothing to consolidate the slices upon each other as they fell over by their own weight. The real state of the case seems to be this, that the share impresses the furrow-slice with its form and character, and the duty of the mould-board is to transmit and deposit that slice in the best possible manner, and with the least possible injury to the character previously stamped upon it by the share. If this view is correct—and there appears no reason why it should be questioned—the mould-board is only a medium through which the slice is conveyed from the share to its destined position. To do this, however, in the most perfect manner, the mould-board has to perform several highly important functions—1st, The transmission of the slice; 2d, Depositing it in the proper position; and, 3d, Performing both these operations with the least possible resistance.

(579.) The *raising* and *transmitting* the slice have frequently been described as if consisting of three or more distinct movements. With all deference to former writers, I conceive it may be viewed as having only two movements, namely, cutting the slice by the share and coulter, and transmitting it to its appointed position through the medium of the mould-board. The first has been already discussed: I now proceed to the second.

(580.) The object of every mould-board is to transmit the slice in the best manner, and with the least possible expenditure of force; but, as might have been expected, we find considerable difference of opinion on both these points, arising from the variations in the form of the mould-board. In a general way, the *transmission of the slice* may be explained in the following manner:

(581.) In fig. 104, *a a* represents a vertical section of part of an unbroken ridge of land, and the parallelogram *a b c d* also a transverse section of an indefinitely short portion of a slice which is proposed to be raised: the breadth *a b* being 10 inches, and depth *a d* 7 inches; the line *d c* will be the bottom of the slice, or the line on which it is separated from the sole by the action of the share. The points of the share and coulter enter at *d*; and, in progressing forward, the slice will be gradually raised at *d*, the point *c* remaining at rest, while the parallelogram revolves upon it as a center. When the share has penetrated to the extent of the feather, the point *d* of the slice will have been raised 2½ to 3 inches. By the continued progress of the plow, the parallelogram representing the slice will be found in the position *c e s f*, and again at *c g h i*. At the fourth stage, when the zero-point of the mould-board has reached the supposed line of section, the slice will have attained the vertical position *c k l m*. During these stages of this uniform process, the slice has been turning on the point *c* as on a pivot, which has retained its original position, while the point *d*, in its successive transitions, has described the quadrant *d e g k*. By the continued progress of the plow, the revolution of the slice will be continued, but it will be observed that, at this stage, it

* A variety of the Kent turn-wrest plow which I have seen in the possession of Mr. Hamilton, of Oarchuie, is an example of this.

changes the center of revolution from *c* to *m*; when the point *k* will have described the arc *k s*, the slice has then reached the position *m n o p*; and ultimately, when the posterior extremity of

Fig. 104.

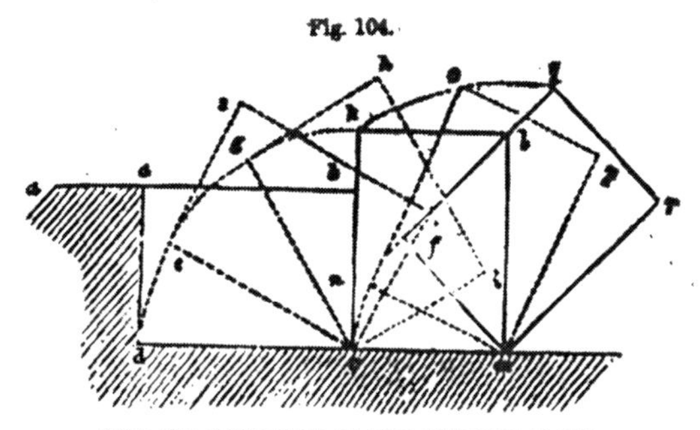

THE TRANSMISSION OF THE FURROW-SLICE.

the mould-board has reached the line of section, the slice will have attained its final position *m f q r*, lying at an angle of 45°, and resting on the previously turned-up slice.

(582.) The *process of turning over the slice*, therefore, approximates to a *uniform motion*, provided the parts of the plow destined to perform the operation are properly constructed. The uniformity, however, is not directly as the rectilineal progress of the plow, but must be deduced from a different function to be afterward explained. And, though the process here described refers only to an indefinitely short slice, it is only necessary to conceive a continuity of such short slices, going to form an entire furrow-slice, extending to the whole length of the field; and the lengthened furrow-slice, being possessed of sufficient tenacity to admit of the requisite and temporary extension which it undergoes, while the plow is passing under and turning it over, is again compressed into its original length, when laid in its ultimate position. The furrow-slice, therefore, under this process, may not inaptly be compared to the motion of a wave in the ocean, keeping in view that the wave of the slice is carried forward in a horizontal direction, whereas the ocean wave is vertical. But in both cases, though the wave travels onward, there is no translation of parts in the direction in which it seems to travel. In the case of the furrow-slice movement, it appears as in the annexed perspective view, fig. 105, where *a b* is the edge of the land as cut by

Fig. 105.

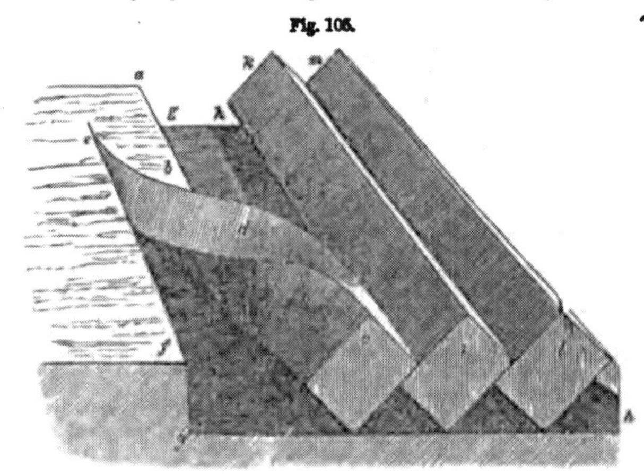

A VIEW OF THE MOVEMENT OF THE FURROW-SLICE.

the preceding furrow; *c d* the slice in the act of turning over, but from which the plow has been removed; *e f*, the edge of the land from which the slice *c d* is being cut; *g h*, the sole of the furrows, and *i k*, *l m*, slices previously laid up. A consideration of this figure will also show that the extension of the slice takes place along the land-side edge *e d* only, from *e* to where the backward flexure is given to it when rising on the mould-board, and where it is again compressed into its original length, by the back parts of the mould-board, in being laid down.

(583.) *Of the furrow-slice.*—To accomplish efficient plowing, the furrow-slice should always be of such dimensions and laid in such position that the two exposed faces in a series of slices shall be of equal breadth; any departure from this rule is a positive fault, whether the object be a seed-furrow or intended for amelioration by exposure to the atmosphere. Furrow-slices laid up agreeably to this rule will not only present the maximum of surface to the atmosphere but they will also contain the maximum of cubical contents, both of which propositions may be illustrated thus by fig. 106. Let *a b* represent the breadth of a 10-inch furrow-slice, and describe the semi-circle *a c b* upon it as a diameter. From the well-known property of the circle, that the angle in a semi-circle is a right angle,* every triangle formed upon the diameter, as a base, will be right angled, and the only isosceles triangle that can be formed within it will be that which has the greatest altitude. The triangle $a\,c\,b$ possesses these properties, for produce $c\,b$ to d, making $c\,d$ equal to $a\,b$—the breadth of the slice, which must always be equal to the distance between the apices of two contiguous furrows. Complete the parallelogram $a\,c\,d\,e$, which will represent the transverse section of a rectangular slice, whose breadth is 10 inches, and whose two exposed faces $a\,c$ and $c\,b$ lie at angles of 45°, and their breadth, as well as the area of the triangle $a\,b\,c$, will be a maximum. In order to prove this, let a section of another slice be formed, whose exposed side $a\,f$ shall be *greater* than the corresponding side $a\,c$ of the former, and let this be taken at 8 inches. From f, through the point b, draw $f\,g$, then will $a\,f\,b$ be a right angle as before; $f\,g$ being also made equal to 10 inches, complete the parallelogram $a\,f\,g\,h$, which will represent the transverse section of a rectangular slice 10 inches by 8 inches, occupying the same horizontal breadth as before, and whose exposed faces will be $a\,f$ and $f\,b$. Draw the line $i\,c\,k$ parallel to $a\,b$, and passing through the apex c of the triangle $a\,c\,b$; and the line $i'\,k'$ also parallel to $a\,b$, passing through the apex f of the triangle $a\,f\,b$. Here the triangles $a\,c\,b$ and $a\,f\,b$ stand on equal bases $a\,b$; but the first lies between the parallels $a\,b$ and $i\,c\,k$, and the second between those of $a\,b$ and $i'\,k'$; the altitude $f\,f'$, therefore, of the triangle $a\,f\,b$, is less than the altitude $c\,c'$ of the triangle $a\,c\,b$. And triangles on equal bases being proportional to their altitudes, it follows that the triangle $a\,f\,b$ is *less* than the triangle $a\,c\,b$, both in area and periphery. Suppose, again, a slice whose side $a\,l$ is *less* than the corresponding side $a\,c$, and let it be 6 inches; from l, through the point b, as before, draw $l\,m$, and construct the parallelogram $a\,l\,m\,n$, we shall have a transverse section of a third slice of 10 by 6 inches, whose exposed faces $a\,l$, $l\,b$, occupy the same horizontal breadth as before. Here the triangle $a\,l\,b$ lies between the parallels $a\,b$ and $i'\,k'$, consequently equal to $a\,f\,b$ and less than $a\,c\,b$.

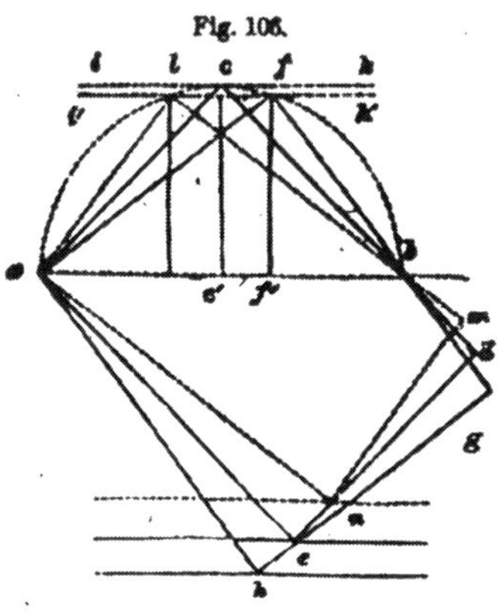

PROPORTIONAL AREAS OF THE FURROW SLICE IN DIFFERENT POSITIONS.

(584.) This simple geometrical demonstration, as applicable to the slice, may be corroborated by the usual formula of the triangle. Thus, the altitude of the triangle $a\,c\,b$ is $= \frac{a\,b}{2} = 5$ inches $= c'\,c$, and the side $a\,c$ or $c\,b$ is $= \sqrt{a\,c'^2 + c\,c'^2}$; of $a\,c'$ and $c\,c'$ being each equal to 5 inches, $a\,c$ or $c\,b$ will $= \sqrt{5^2 + 5^2} = 7{\cdot}071$ inches, which is the depth due to a slice of 10 inches in *breadth*, and the sum of the two exposed faces will be $7{\cdot}071 \times 2 = 14{\cdot}142$ inches.

(585.) In the triangle $a\,f\,b$, $a\,b = 10$ inches and $a\,f = 8$ inches, then $a\,b^2 - a\,f^2 = f\,b^2$ and the $\sqrt{f\,b^2} = 6$ inches. The three sides, therefore, of this triangle are 10, 8 and 6 inches, and the altitude $f'\,f$ is easily found by the principles of similar triangles. Thus, in the similar triangles $a\,f\,f'$, $f\,b\,f'$, $a\,b : a\,f :: f\,b : f\,f'$. The perpendicular $f\,f'$ is therefore $= 4{\cdot}8$ inches; hence the exposed surfaces are as 14·141 : 14, and the altitudes as 5 : 4·8. Since it turns out that $a\,l$ is equal to $f\,b$, and $a\,b$ common to both, it follows that $l\,b$ is equal to $a\,f$, and the periphery and altitude also equal, and *less* in all respects than the triangle $a\,c\,b$. And so of any other position or dimensions.

(586.) The slice which presents a rectangular section is not the only form which is practiced in modern plowing. Of late years, and since the introduction of the improvements by Wilkie on the plow, a system of plowing has been revived† in which the great object seems to be that of raising a slice that shall present a *high shoulder*, as it has been called, or which I have ventured to denominate the *crested furrow*, formerly alluded to. The general impressions that prevail as to the advantages of this mode of plowing are, that the crested furrow affords a *greater surface* to the action of the atmosphere, and a greater quantity of cover to the seed in the case of a seed furrow in lea. As there appears to me some degree of fallacy in the reasonings on this point among practical men, and as it does not appear to have been hitherto sufficiently investigated, I shall venture a few remarks in the hope of leading others to a more full consideration of the points involved in the subject.

(587.) The *crested slice*, instead of the rectangular section of the one already described, presents a rhomboidal but more frequently a trapezoidal section; indeed the latter may be held as inseparable from the practice; but in comparing them I shall first take the exposed surface. In fig. 107, then, let $a\,b\,c\,d$ represent a transverse section of a rectangular slice of 10 by 7 inches, $a\,e$ the base of the triangle, whose sides $a\,b$, $b\,e$ represent the two exposed surfaces of the slice when set up with the sides at angles of 45° to the horizon, its angle at b being 90°—its altitude $b\,f$ will be as before $\frac{a\,e}{2} = 5$ inches. Again, let $g\,h$ be the base of the triangle whose sides $g\,b$, $b\,h$, represent the exposed surfaces of a crested slice—whose base $g\,h$, equal to $g\,d$, may be taken at 9 inches, that being the breadth at which such plows take their furrow. Supposing, also, that the cresting is such as to give an altitude $f\,b$ of 5 inches, as in that of the rectangular slice, we shall have the sides $g\,b$, $b\,h$, from the usual formula, $g\,f^2 + f\,b^2 = g\,b^2$, and $g\,f$ being $4\frac{1}{2}$ inches $f\,b = 5$

* Euclid, 31, iii.
† Blith's "English Improver Improved," p. 266, edit. 1652.

inches, then the $\sqrt{4 \cdot 5^2 + 5^2} = 6 \cdot 72$ inches $= g\,b$ or $b\,h$, being rather more than the best practical authorities for cresting plows give to the depth of a slice; the dimensions recommended being from 8½ to 9 inches broad, and from 6 to 6½ inches in depth. It will therefore always fall short in perpendicular hight of the rectangular slice of 10 by 7 inches. But allowing the hight to be the same, we have two triangles, *a b e* and *g b h* of equal hight but of unequal bases; their areas will therefore be unequal and proportional to their bases.

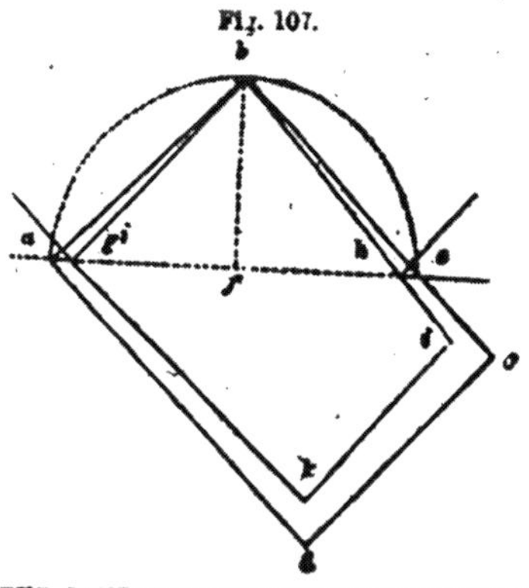

THE COMPARISON OF THE RECTANGULAR AND CRESTED SLICES.

(588.) In bringing these two systems, however, into practice over any extent of surface, suppose a ridge of a field, the *number* of furrows of each required to turn over such ridge, will be exactly in proportion to the length of the base of the triangle, or as 9 to 10. Hence, though the individual crested slices or triangles have an area *less* than that of the rectangular slice in the proportion of 9 to 10; yet the aggregate area of all the triangles over any given breadth of surface, wherever the number of slices of the one exceeds that of the other in the proportion of 10 to 9, will be the same, but no more. The imaginary advantage, therefore, of a greater cover to the seed with a crested furrow falls to the ground, provided the comparison is made with a plow that takes a furrow of 10 inches wide by 7 inches deep, such as the East-Lothian plow.

(589.) It is to be admitted that, were cresting plows that cut their slices 9 inches wide, to take them 7 inches deep, and still preserve the rhomboidal or trapezoidal section, they might, in that case, produce an increase of cover to the seed, as compared with a rectangular slice of 9 by 7 inches. Let us refer again to the last figure, fig. 107, and suppose $g\,b = 7$ inches, $g\,f$ being, as before, 4½ inches, then $g\,b^2 - g\,f^2 = b\,f^2$, or $b\,f$ will be equal to 5·36 inches, while, by the same method the rectangular slice of 9 by 7 inches would give $b\,f$ equal to only 4·39 inches, the crested slice in this case giving a difference of hight of ·97 inch, and ½ of this, or ·48 inch, of greater cover of seed. But this is not a practicable case, inasmuch as the cresting plow cannot be worked in a furrow of 9 by 7 inches, and lay it at an angle that would give equal exposure to both sides of the slice, whether it possess a rectangular or rhomboidal section, the true depth being 6·36 inches nearly, for a slice whose breadth is 9 inches; and the hight *b f* of its triangle would be, if rectangular, only 4·5 inches. Compared with itself, therefore, the plow that takes a 9-inch furrow rectangular yields ½ inch less cover to the seed than when it raises the crested slice; but, even with the advantage of the crest, it is not better than the plow that takes a 10-inch furrow; while, as will appear, the former labors under other disadvantages arising from that peculiarity of structure for which it is valued.

(590.) In order to exhibit the difference of effect of the rectangular and the trapezoidal slices, as lifted and laid on each other by the plow, and as they affect the real intentions of tillage, I shall consider them in separate detail. Fig. 108 is an example of the rectangular slice of 10 by 7 inches, *a b c d* may be taken as a transverse section of the body of the plow, the line *a c* being the terminal outline of the mould-board, *a f* a section of the slice which is just being laid up, and *g h* a slice previously deposited. In the triangle *i g k* the base *i k* is 10 inches, being always equal to the breadth of the slice, the angle at *g* a right angle, and the sides *i g*, *g k* each equal to 7·071 inches, the perpendicular hight *g l* being 5 inches, as before demonstrated. Fig. 109 is a similar representation, of a cresting-plow, with its effects on the slice and the subsoil; *k n o p* is a section of the plow, *k m* a section of a slice in the act of being deposited on the preceding slice *c l*. Here the slices are trapezoidal, as they are always cut by this species of plow; and from this configuration of the slice, the broader sides are not parallel, nor do the conterminous sides of the adjacent slices lie parallel to each other in the transverse direction. The side *b c* lying at an angle of 48° with the base *a b*, while the side *b m* makes the opposite angle at *b* only 41°, the angle at *c* being 84°, and the triangle *a b c* isosceles. The base *a b* of the triangle *a b c* is now supposed to be 8½ inches—the breadth recommended for a seed-furrow—and the side *a c* 6½ inches, the opposite side *l h* being 4½ or 5 inches. The base *a b*, when bisected in *d*, gives $a\,d = 4 \cdot 25$ inches, and since $a\,c^2 - a\,d^2 = c\,d^2$, $c\,d$ will be 4·918 inches, which is less than given by the former demonstration of the crested slice; but I have observed cases still more extreme, where, still referring to the same figure, *a b* was only 7½ inches, but the angle at *c* became so acute as 75°, yet with these dimensions *c d* is still under 5 inches; hence, in all practical cases, with a furrow less than 9 inches in breadth, the result will be a reduction in the quantity of cover for seed.

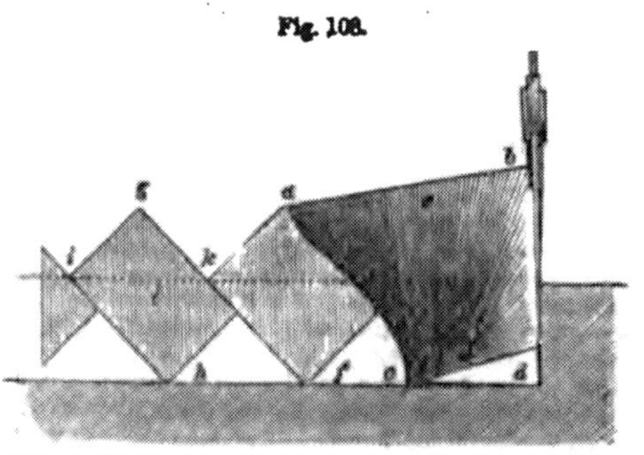

THE EFFECTS OF A RECTANGULAR FURROW-SLICE

(591.) One other point remains to be noticed in reference to the two forms of slice. We have seen that the rectangular slice necessarily implies that the bottom of the furrow shall be cut upon a level in its transverse section, fig. 108; while the slice that is cut by the cresting-plow leaves the bottom of the furrow with a sloping rise from the land-side toward the furrow-side at every slice, and this rise may range from 1 to 1½ inch or more. Returning to fig. 109, the serrated line *f h o* exhibits a transverse section of the surface of the subsoil, from which the soil has been turned up by the cresting plow. The triangular spaces *e f g, g h i* represent the quantity of soil left by such plows at the lifting of each slice. These quantities, which, as before observed, may amount to 1·7 of what the slice ought to be, are thus robbed from it, and left adhering to the subsoil, except in so far as they may be rubbed down by the abrading action of the lower edge of the mould-board, as at *f* and *h*, and the portions of soil so rubbed off are thrust into the spaces under the edge of the slices as they are successively laid up. This last process may be readily observed at any time when the plow is working in tough land or in lea. With a cresting-plow the spaces *f h o* will be seen more or less filled up with crumbled soil, while with the rectangular plow, the corresponding spaces will be left nearly void. I cannot take upon me to say whether or not the filling in of these voids is beneficial to the land in a greater degree than if the 1·7 here left below had been turned up with the slice; but this I can say, that it is more frequently left adhering to the subsoil than it is to be found stuffed under the edge of the slice. Under any view, the system of the crested furrow-plowing is not unworthy the consideration of the farmer.

Fig. 109.

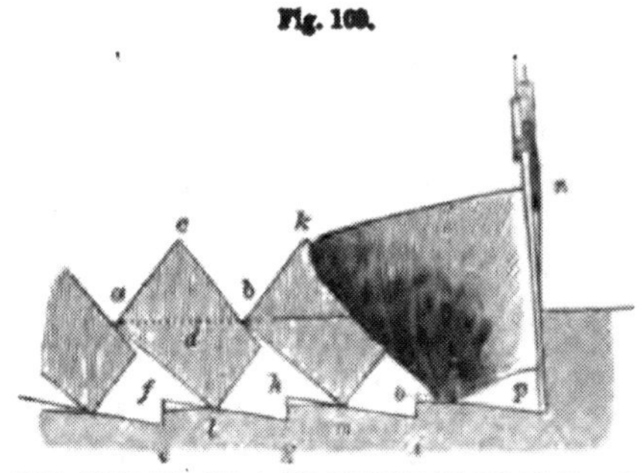

THE EFFECTS OF A TRAPEZOIDAL OR CRESTED FURROW-SLICE.

(592.) In considering the question, there are two points deserving attention. 1st, The immediate effects upon the labor of men and horses. It may be asserted generally, that all plows adapted to form a crested furrow are heavier in draught than those that produce the rectangular furrow. This seems a natural inference from the manner in which they work; the tendency that they all have to *under*-cut by the coulter; the narrow feather of the share leaving more resistance to the body in raising and turning the slice; and not least, the small ridge left adhering to the bottom of the furrow, if rubbed down and stuffed under the slice, is performed by an unnecessary waste of power, seeing that the mould-board is not adapted for removing such adhering obstructions. 2d, The loss of time and labor arising from the breadth of furrow, compared with those plows that take a 10-inch furrow. Thus, in plowing an imperial acre with a 10-inch furrow—leaving out of view the taking up of closings, turnings, &c.—the distance walked over by the man and horses will amount to 9·9 miles nearly; with a 9-inch furrow the distance will be 11 miles; with 8½-inch furrow, it will be 11¼ miles or thereby; and with a 7½-inch furrow 13¼ miles nearly.

(593.) It may, therefore, be of importance for the agriculturist to weigh these considerations, and endeavor to ascertain whether it is more for his interest that his plowing should be essentially well done, and with the least expenditure of power and time, or that it should be done more to please the eye, with a high surface finish, though this may perhaps be gained at a greater expenditure of power and time; while the essentials may in some degree be imperfectly performed.

(594.) On this part of the subject, I cannot refrain a passing remark on the very laudable exertions that have been made all over the country in producing that emulation among our plowmen which has been so successful in producing excellence in their vocation among that useful class of agricultural laborers, as to give them a preëminence over all others of their class in any country. I mean the institution of plowing-matches. While I offer my humble though ardent wishes for a continuance of the means which have raised the character of the Scottish plowmen, I cannot prevent doubts rising in my mind, that, however good and beneficial these competitions are calculated to be, if the exertions of the class are properly directed; yet the best exertions of both the promoters and the actors may be frustrated by allowing a false taste to be engendered among these operatives. That such a false taste has taken root I have no doubt; and the results of it are appearing in the spread of opinions favorable to that kind of plowing which to me appears not much deserving of encouragement—the high-crested system. I have observed, at various plowing-matches, that the palm was awarded to that kind of plowmanship which exhibited the highest surface-finish, without reference at all to the ground-work of it; and I have compared by actual weight, all crumbs included, the quantities of soil lifted by plows that gained prizes with others which did not, because their work was not so well dressed on the surface; and I have found that the one to whom the prize was awarded had not lifted so much soil by 1-10 as some of those that were rejected. I am far from intending, by these remarks, to throw discredit on plowing-matches; on the contrary, I would see them meet with tenfold encouragement, and would also wish to see many more than is usually met with, of the good and the great of the land, assembled at such meetings, to encourage and stimulate by their presence the exertions of the competitors in such interesting exhibitions. With this short digression I leave this subject for the present, with the intention of resuming it in another division of the general subject.

(595.) THE PRINCIPLES AND FORMATION OF THE MOULD-BOARD.—Of the various individuals who have written upon the plow and the formation of the mould-board, Bailey of Chillingham and Small of Berwickshire are perhaps the only two who have communicated their views in a practical shape, and even in their descriptions there is somewhat of ambiguity and uncertainty, but such may be inseparable from the subject. Many other nameless artisans have varied the mould-board until almost every county has something peculiarly its own, and each district claims for its favorite all the advantages due to perfection.

(596.) I have been at great pains to analyze a considerable number of these varieties; and as the subject is not unimportant in a work of this kind, I have selected a few of those best known, and of highest character, as objects of comparison.

(597.) The method adopted to obtain a mechanical analysis of these mould-boards has been simple, but perfectly correct; and as the principle may be applied to the attainment of counterparts of other objects, perhaps more important than a mould-board, it may be deserving of a place here. As a matter of justice, also, to the fabricators of the different mould-boards here exhibited, I am desirous to show the principle on which these transcripts of their works have been thus brought forward in a new form, and in contrast with each other.

(598.) The instrument employed for this purpose is a *double parallelogram* or *parallel ruler*, as represented in fig. 110, which is a perspective of the apparatus. The bars *a b*, *c d*, *e f*, are

Fig. 110.

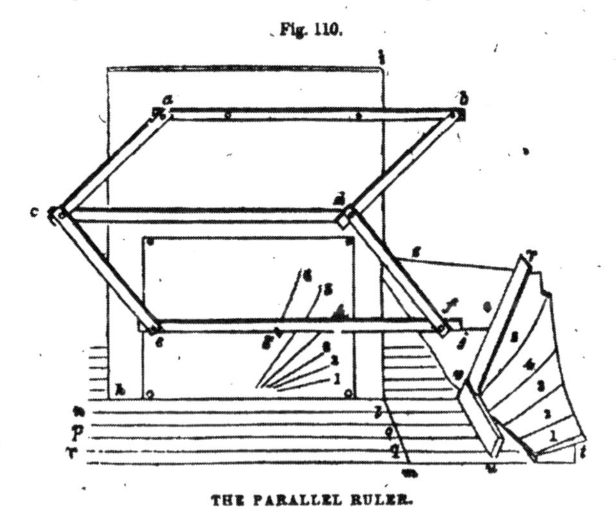

THE PARALLEL RULER.

slips of hard-wood about 3 feet long, or they may be of any convenient length, and 1½ inches broad by ⅜ inch thick; each of which is perforated at *a*, *b*, *c*, *d*, *e*, *f*, the perforations being exactly equidistant in all the bars. Four similar bars, of about half the length, are perforated also at one uniform distance, and the seven bars thus prepared are jointed together upon brass studs, and secured so as to move freely at every joint, but without shake on the studs. The form, when constructed, is that of the two parallelograms *a b c d* and *c d e f*. In the end of the bar *e f* a stout wire pointer *j* is fixed of about 6 inches in length, lying in the plane of the instrument, and parallel to the edge of the bar. In a continuation of this parallel upon the bar *e f*, a socket capable of retaining a pencil or tracer is fixed anywhere at *g*. The instrument so formed is fixed upon a flat board *h i*, of about 3 feet square, by means of two screw nails passing through the bar *a b*, in a position parallel to the lower edge of the board; thus leaving all the other bars at liberty to move upon their joints; which completes the instrument. From the well-known properties of the parallelogram, as applied to the pentograph and the eidograph, it is unnecessary to demonstrate, that whatever line or figure may be traced with the pointer *j*, will be faithfully repeated by the tracing pencil *g*, upon any substance placed before it, and of the same dimensions as the original.

(599.) Another board or table or a level platform, is now to be selected, and a line *l m*, which may be called the fundamental, or leading line, drawn upon it. This line, to an extent of 3 feet or more, is divided into any number of equal parts, but in this case the divisions were 3 inches each; through these points of division are drawn the straight lines *l n*, *o p*, *q r*, &c., indefinitely on each side of the line *l m*, and at right angles to it. The board carrying the instrument is provided with a foot behind, that keeps the face of the board always perpendicular to the platform on which it stands. The plow with the mould-board about to be analyzed is now set upon the table or platform upon which the leading line and the divisions have been laid down; the landside of the plow being set parallel to the leading line, and at any convenient distance from it, suited to the instrument; presenting the mould in the position *s t*, and so placed in reference to the lines of division that the zero line shall coincide with one of them, provided the extremities do not overreach the divisions either way, the landside of the plow being at the same time perpendicular. The instrument is now brought toward one extremity of the mould-board, and placed upon that parallel of the divisions that come nearest to the extremity, as No. 1 in the figure, the edge *l i* of the instrument coinciding with the leading line *l m*. A sheet of paper having been now fixed upon the board *h i* of the instrument, and a tracing-pencil inserted in the socket *g*, the operation of tracing commences. The tracing point is passed in the vertical direction over the surface of the mould-board, tracing along a line No. 1; the pencil at the same time tracing a corresponding line No. 1, on the paper, which will be an exact outline of the face of the mould-board at that division, supposing the mould-board to be cut by a transverse section in that line. The instrument and board are now to be moved one division upon the leading line *l m*, the coincidence of the edge *l i* of the board with that line being still preserved. The tracing point is again made to pass vertically over the face of the mould-board, when the pencil *g* will trace on the paper a second line No. 2. This process, repeated at each successive division, 3, 4, 5, 6, &c., the corresponding lines, 3, 4, 5, 6, &c., on the paper will be traced out, exhibiting a series of perfect sectional lines of the mould-board, each line being that which would arise from an imaginary vertical plane cutting the body of the plow at right angles to its land-side at every 3 inches of its length. To prevent any inaccuracy that might arise from a misapplication of the tracing point to the oblique surface of the mould-board, a straight-edged ruler, in form of a carpenter's square *u v x*, is applied to the mould-board. The stock *u v* of the square being placed on the platform, and parallel to the line *l m*, which brings the edge *v x* always into the vertical plane, and the tracing rod must be kept in contact with this edge, while it traverses the face of the mould-board at each successive section.

(600.) This mode of analysis, it is to be observed, has not been adopted from its having any relation to the principles on which the different mould-boards have been constructed, but because it presents an unerring method of comparing a series of sectional lines of any one mould-board with those of any other; hence it affords a correct system of comparison. But it is not merely a comparative view that is afforded by it, for in the sequel it will be seen that a ground-work is thus afforded from which the mechanic may at any time or place construct a *fac-simile* of any mould-board, the analysis of which has been made after this manner.

(601.) The results of the analysis of a few of the mould-boards from the plows of highest character, as taken by this method, are given in the following figures. Plate XI., fig. 111, is a geometrical elevation in a plane parallel to the land-side of the mould-board of the East-Lothian plow *l d*, being its base-line. The perpendicular lines of division, commencing from the line *o o* the zero, and extending right and left, are the lines of section. Those to the right or fore-end of the mould-board, marked *a a*, *b b*, &c., and those to the left 1 1, 2 2, &c. The curved line *x y z* represents the path described on the face of the mould-board by the lower land-side edge of the furrow-slice, as the mould-board passes under it; this line I shall call the *line of transit*. Fig. 112 is a front view in elevation of the mould-board of the same plow, and corresponding to fig. 111; *k m* is the base-line of the plow; *m g* is the land-side plane in a vertical position, *m* is also the place of the point of the share, and *h i* the line of junction between the neck of the share and the mould-board; the remaining lines beyond *h i* exhibit the outline of all the sections taken by the instrument in reference to the lines in fig. 111. Thus, *o o g m* is the section of the entire body of the plow in the plane of the zero, *o y o* being the outline of the mould-board at this section, and *y* the zero-point; *a a g m* the first section forward from the zero, *b b g m* the second, and so on. In like manner, 1 1 *g m* is the first section backward from the zero, 2 2 *g m* the second, and so on; each section so lettered and numbered having relation to the divisions carrying the corresponding letters and numerals in fig. 111. The entire series of lines 1 1, 2 2, &c. and *a a*, *b b*, &c. thus form a series of profiles of the mould-board, supposing it to be cut vertically by planes at right angles to the land-side of the plow. In fig. 112, also, the dotted line *m x y z* represents the path of the slice or line of transit, as in fig. 111, and *z k* represents a transverse section of the *slice* as finally deposited by the mould-board. Figs. 113 and 114 exhibit, in the same manner, the mould-board of the Currie or Mid-Lothian plow; the divisional and sectional lines being all laid off in the same manner from the zero as in the example just described of the East-Lothian plow, and the zero-point *y* in the line *o y o*, which is 9 inches from the plane of the land-side. Fig. 114 bears also the same relation to fig. 113, and as the letters and numerals in these have the same relation and value as in figs. 111 and 112—the East-Lothian—the description given of that applies not only to the Mid-Lothian, but to the five succeeding figures, viz:

Figs. 115 and 116 represent the Berwickshire plow, being that which has been so successfully adopted by the Marquis of Tweeddale.

Figs. 117 and 118 are of the Lanarkshire plow.

Figs. 119 and 120, Plate XII., are of the Saline or Western Fifeshire plow.

Figs. 121 and 122 are of the FF plow of Messrs. Ransome, of Ipswich.

(602.) With reference to the characters of these different mould-boards, it may be remarked: Of the *East-Lothian* mould-board, fig. 112, Plate XI., that those portions of the sectional lines lying between the lower edge and the line of transit are essentially straight, the two lines beyond the zero backward excepted, these being slightly concave toward the lower edge; and, although the lines before the zero and above the line of transit are concave, that part of the surface has no effect upon the furrow-slice. It is, likewise, to be observed that the parallelogram *k y*, which represents a section of the slice when brought to the vertical position, has its upper angle *y* only touching the zero line, and no other part of the side of the parallelogram in contact with the zero line of section *o y o*; hence the mould-board, by its pressure being exerted chiefly against the upper edge of the slice, will always have a tendency to abrade the crest of its rectangular slice in its progress over the mould-board.

(603.) *In the Mid-Lothian* mould-board, figs. 113 and 114, the lines are also approximating to straight, except in the lower portions of those before the zero, where they produce a convexity of surface, but this part of the mould-board can have little influence. The chief difference, then, lies in those parts of the sectional lines which lie above the path of the slice, and they also have no effect whatever in the formation or the conveyance of the slice; neither can the circumstance of elongation forward in this mould-board have any influence, for the same lines are to be found on the *neck of the share* of the East-Lothian as are here exhibited in the prolongation of the mould-board. We have, therefore, two plows in which the essential lines of the mould-board are the same, but which produce work of an opposite character. It must be kept in view, however, that in the Mid-Lothian the zero-point *y* is only 9 inches from the land-side, while in the East-Lothian it is 10 inches; but the length behind the zero-line being nearly alike in both, and the width at the tail also the same, the difference in distance of the zero-point from the land-side produces a difference in the effect of the pressure of the mould-board on the edge of the slice. This will be perceived from the relation in which the section *k y*, representing the slice, stands to the zero-line *o y o* of the mould-board; for in this case the angle at *y*, formed by the side of the parallelogram and the zero-line, is not more than ¼ of that in the former case. This mould-board, therefore, will convey the slice in whatever form it may be cut, with less risk of injury to the crest than can be expected from the former. But, as these discrepancies cannot produce the marked difference that exists in the appearance of the work performed by these two plows, it is not in the mould-board we are to look for the cause, but in the conformation of the share and the position of the coulter, while the mould-board, from the circumstance last pointed out, is better adapted to convey the slice unaltered.

(604.) *The Berwickshire* mould-board, figs. 115 and 116, which is also truncated forward, has the sectional lines, lying before the zero, nearly straight; but, as they approach the zero, they become gradually and decidedly concave, which increases toward the extremity. This concavity, it will be observed, exists only to a certain extent below the line of transit, and, as the sectional lines approach the line of transit, the curvature is reversed, and the surface becomes convex.—This is a form well adapted to deliver a slice free of injury to the edge or crest, for, from the convexity immediately below the line of transit, the mould-board will never press upon nor abrade the edge of the slice, the pressure being exerted always within the extreme edge, as will be seen from the section *k z* of the slice, as applied at the extremity of the mould-board; though, when in the vertical position, as in *k y*, the section of the slice touches the zero-line at its upper edge, at an angle nearly equal to that in the East-Lothian, showing that it is liable to abrasion until it has passed that line. But this plow, in practice, sets up a furrow of the rectangular species, with its angle or crest better preserved than in many others of this class, while, at the same time, it takes out the sole with the characteristic levelness which belongs to the class.

(605.) The *Lanarkshire* mould-board, figs. 117 and 118, has all its lines convex, the terminal edge excepted, which is nearly straight below, but preserves the convexity as it approaches the line of transit. Even above the line of transit the convexity is continued, and, though not affecting the slice, it gives in appearance a still more decided character of convexity, and, by thus making the upper edge of the mould-board retire, gives a long rake to the breast of the plow. It will be readily conceived that this mould-board, from the convexity of all its sectional lines, is essentially formed for turning up a crested furrow, more especially when the form of its share and the position of its coulter are considered. These last, being formed for cutting the slice with a very acute angle, will deliver it to the mould-board; and, from the form of the latter, the slice will pass over it uninjured; for the pressure upon the mould-board will be always greatest upon those parts of the surface of the slice lying within the edge, preventing thereby the abrasion of that tender part. These circumstances are clearly seen from the relation of the section of the slice *k y* as applied to the zero-line *o y o*, the point of contact lying considerably within the angle at *y* of the slice; and the same relation holds throughout the entire transit, up to the delivery of the slice in the ultimate position *k z*.

(606.) The *Western Fifeshire* mould-board, figs. 119 and 120, Plate XII., it will be readily perceived, belongs to the Lanarkshire class; but in this the convexity is carried so far to an extreme as to round away the lower parts of the mould-board, till, at the lower edge behind, the width is only 6 inches. The terminal line, also, is prominently convex throughout. It differs also from its original type in having those parts of the sectional lines lying above the line of transit tending to recurvature. This, by carrying forward the upper part of the breast, gives the appearance of greater length to the mould-board; but, besides this, the part lying behind the zero is actually longer than in any of the preceding plows, as will appear from the sectional divisions, figs. 111, 113, 115, and 117, Plate XI. As may be anticipated, this variety of the Lanarkshire plow is famed for the acuteness of the furrow which it forms, though, in this respect, it does not excel its prototype. From the way in which the section *k y* of the slice, when in the vertical position, is applied to the zero-line, where the point of contact is seen to lie at ¼ of its breadth within the edge of the slice, it will at once appear how well this mould-board is adapted to transmit an unbroken crest. In every position of its transit up to its ultimate position, the slice will be equally secure from injury in respect to the crest; and, were the crested furrow a true criterion of good plowing, the plow that bears this mould-board, with share and coulter adapted thereto, would be the most perfect; but there are various and important arguments against it.

(607.) In *Ransome's Bedfordshire, or* FF mould-board, figs. 121 and 122, the sectional lines are of a mixed character; those in the fore part being convex, gradually diminishing in convexity to the zero, behind which they become straight lines, tending to concave—the terminal line being slightly so—but becoming convex at the upper edge. It differs from all the Scotch mould-boards in having the terminal edge lengthened out below, instead of the usual shortening, and in having the breast cut away nearly parallel to the line of transit. The plows mounted with this mould-board are generally worked with cast-iron shares, having a wide-spread feather formed for cutting a level furrow-sole. The furrow usually taken with it is shallow, and, when set up, looks flat in the crest; but the work, so far as it goes, is what may be termed, in plow language, *true;* that is to say, the slice is rectangular and cut from a level sole. Though the sectional lines before the zero possess a form that would save the slice from abrasion, yet, at the zero-line and behind it, they have the opposite character in the extreme, and we accordingly find that this mould-board lays a very flat crested furrow, while the share and coulter are perfectly adapted to cut it rectangular.

(608.) With all the foregoing mould-boards, it will be observed that the section of the furrow slice, in its ultimate position, seems to encroach upon the tail of the mould-board; and this is to be understood as arising from the circumstance of the slice being represented as incompressible, and unabraded below or above. In practice, the slice is pressed downward on the angle at *k*, and pressed home upon the preceding slice, so as to bring the face of the slice simply in contact with the terminal line of the mould-board instead of the apparent mutual interpenetration exhibited in the figures.

(609.) From the examples here given of the forms of mould-boards, and the effects which they produce when combined with any particular form of share and position of coulter, it will be easy to draw a conclusion as to the kind of work that will be performed by any plow that comes under our observation, and that without any previous knowledge of its merits; keeping in mind that the *ultimate form of the furrow* will always depend on the form of the share and position of the coulter, that the passage of the slice over the mould-board will have but a very partial effect on the form of the slice, and that this effect will be greater or less according to the form of surface. Thus, a slight convexity of surface, immediately below the line of transit, will with greater certainty secure the transit of the slice without injury to its edge, than may be expected from a surface which has a concavity crossing the line of transit, though it may be obtained, as in the Mid-Lothian and East-Lothian plows, with a straight-lined mould-board; but it will be more certainly obtained if the share is narrow, as in the Mid-Lothian; though this last expedient will induce disadvantages in point of draught, and risk of losing the effect, by any undue placement of the coulter. These disadvantages may arise from the coulter not being sufficiently set to landward, thereby admitting the breast of the plow to scrape upon the land, and send a small portion of earth along the mould-board, accompanying the edge of the slice, which may have the effect of abrading it so much as to injure the *appearance* of the work, though not in fact affecting its efficiency.

(610.) Having thus endeavored to establish some data by which the agricultural mechanist, whether amateur or operative, may be assisted in determining from observation, what practical effects may be expected to result from any form of mould-board and share, I proceed to mention some rules by which he may form a mould-board on what I conceive to be the true principle, but upon which may be engrafted such deviations as taste or other circumstances may require.

(611.) Those writers who contributed to the improvement of the plow in the early stages of its modern history, labored at a time when mould-boards of wood only were employed. Hence, their instructions related to the formation of that material alone, into mould-boards. Later writers have followed in nearly the same course, and have given rules for forming a mould-board, out of a block of wood, of sufficient dimensions to contain all the extremities of the proposed fabric. The change now pervading this branch of mechanics, wherein the introduction of cast-iron has become universal, precludes the necessity of falling back upon any of the old rules; what the agricultural mechanic is now required to furnish being not a mould-board, but, in the language of the foundry, a *pattern*, from which castings are to be obtained perfect facsimiles of the original pattern, and which may be repeated *ad libitum*; from this last circumstance, it follows that the making of a pattern will be a comparatively rare occurrence, and one which he will seldom be called upon to perform. It nevertheless appears desirable that a knowledge of the construction of such a fabric should be communicated in a manner that may enable an ordinarily skilled mechanic to construct a pattern when required, with accuracy and certainty of effect.

(612.) It has been shown that very considerable discrepancies exist in the form given to mould-boards, and there is no doubt that peculiarities of soil may demand variations in form; but the propriety of such wide deviations may be called in question, and the actually required deviation brought within very narrow limits. It appears, indeed, that one form may be brought to answer all required purposes, if aided by a properly adjusted share and coulter.

(613.) From a careful study of the foregoing analytical diagrams, and from comparison of numerous implements and their practical effects, together with a consideration of the dynamical principles on which the plow operates, I have been led to adopt a theoretical form of mould-board, which seems to fulfill all the conditions required in the investigation, and which is capable, by very simple modifications, of adaptation to the circumstances of the medium on which it works. In the outset, it is assumed that the soil is homogeneous, and that it possesses such a degree of tenacity and elasticity as to yield to the passing form of the plow, and to resume, when laid in the due position, that form which was first impressed upon the slice, by the action of the share and coulter; the second consideration being the cutting of a slice from the solid land. In a theoretical view, this must be an operation through its whole depth and breadth; hence the share is conceived to be a cutting edge which shall have a horizontal breadth equal to the breadth of the slice that is to be raised, and that the face or land-side of the coulter shall stand at right angles to this. Another consideration is, that the slice now supposed to be cut has to be raised on one side, and turned over through an angle of 135°, the turning over being performed on the lower right-hand edge, as on a hinge, through the first 90°, the remaining 45° being performed on what was at first the upper right-hand edge. (Fig. 104.) The slice, in going through this evolution, has to undergo a twisting action, and be again returned to its original form of a right prism. To accomplish this last process, it is evident that a *wedge, twisted* on its upper surface, must be the agent; and to find the form and dimensions of this wedge, is solving the problem that gives the surface of the mould-board required.

(614.) We have seen, fig. 104, that the slice, in passing through the first 90°, describes the quadrant with its lower edge, and in doing so, we can conceive a continued slice to form the solid of revolution *a b c d e*, fig. 123, which is a quarter of a cylinder, as shown here in isometrical perspective; the radius *a b* or *a c* being equal to the breadth of the slice. We have next to consider the angle of elevation of the twisted wedge; and in doing this, we must not only consider the least resistance, but also the most convenient length of the wedge. In taking a *low* angle, which would present, of course, proportionally little resistance, it would, at the same time, yield a length of mould-board that would be highly inconvenient, seeing that the generating point, in any section of the slice, must ultimately reach the same hight, whether by a low or a higher angle. From experience, we find that, from the point of the share to that point in the plow's body where the slice arrives at the perpendicular position, and which I have named the zero, that 30 inches form a convenient length. The length *c d* of the solid is therefore made equal to 30 inches or more, and this being divided into 10 equal parts, the parallels 1 1, 2 2, 3 3, &c., are to be drawn upon the cylindrical surface, and between the points *b*, *d*, a curve has to be described that shall be the line of transit of the slice. After investigating the application of various curves to this purpose, I have found that a circular arc is the only one that can be adopted. It presents the least attainable resistance in the first stages of the ascent, where the force required to raise the slice is the greatest, and in the last stages, where the force of raising has vanished, leaving only what is necessary to turn the slice over, there the resistance is at the greatest; and, above all, the circle being of equal flexure throughout, it is in every way best adapted to the objects here required. To determine the radius of curvature of this arc, we must evolve the cylindrical surface *c b d e*, and from it construct the diagram, fig. 124. Draw *e b* equal to *c d* of fig. 123; *e d* equal to the length of the arc *c b* or *d e*, and at right angles to *e b*; divide *e b* into 10 equal parts, and from the points of division draw the ordinates 1 *f*, 2 *g*, 3 *h*, &c., parallel to *e d*; from *b* set off 10 inches for the length of the share along the line *b e*, which will fall 1 inch beyond the division 7, and at this distance draw the dotted line parallel to 7 *m*; upon this set off a distance 7 *m* of 2¼ inches; and through the three points, *d*, *m*, *b*, describe an arc of a circle, whose radius will be found equal to the circumference of the cylinder of which *a b c*, fig. 123, is a quadrant. The circular arc thus found is now to be transferred to the cylindrical surface *c b d e*. The transfer may be performed by drawing the arc on paper, and the paper then laid over the cylindrical surface in such a manner that the points *b*, *m*, *d*, shall be brought to coincide with the points *b*, *m*, *d* of the cylindrical surface; when the remaining points *f*, *g*, *h*, *i*, or any number more, may be marked on the cylindrical quadrant by pricking through the paper with a pointed instrument at short intervals along the arc; or, the length of the ordinates 1 *f*, 2 *g*, 3 *h* of fig. 124, may be transferred to the corresponding parallels of fig. 123, when the lengths of the ordinates will cut the parallels in the points *f*, *g*, *h*, &c. In either case, the curve can now be traced through the points *b*, *p*, *n*, *m*, &c., on the cylindrical surface. Through the points *b*, *p*, *n*, *m*, &c., draw the dotted lines *f f'*, *g g'*, *h h'*, &c., parallel to *c d* or *b e*, and from the center *a* draw the radii *a f'*, *a g'*, *a h'*, &c.; the unequal divisions of the arc *c b* will thus show the proportional angles of ascent of the slice along the line of transit now found, *b*, *p*, *n*, &c., for each division of the length; while the degree of flexure in the curve or line of transit remains uniform by the same, from any one point, to any other equidistant points.

(615.) To convert the prism thus prepared and lined off into that of the twisted wedge, we have only to cut away that portion of it contained within the boundaries *a*, *b*, *c d*, *x*, preserving the terminal edges *a b*, *a x*, and *d x*; and the prism will thus be resolved into a form represented by a portion *a b d x e* of fig. 125, also in isometrical perspective. Of this figure, *a b d x* is the true theoretical surface of the mould-board, from the edge *a b* of the share to the zero-line *d x*; *a b e x* is the sole; the curve *b p n m l*, &c., is the line of transit of the slice; and the triangles 1' *f* 1, 2' *g* 2, 3' *h* 3, 4' *i* 4, &c., are the vertical planes supposed to cut the solid thus reduced in the divisions 1, 2, 3, 4, &c., to the hight of the line of transit, as in the analytical sections of the mould-boards.

(616.) The surface now completed can only raise the slice to the perpendicular position; and to complete the operation, we have to carry the twisted wedge back till it shall place the slice at the angle of 45°. To do this we have to extend the original prism, or suppose it to have been at first sufficiently elongated toward *d d'*, fig. 125, and to superimpose upon its flat side the portion *d d' u x*, or *a d u* of fig. 126. The part *d d' u x* is now to be worked off into a part of a new cylindrical surface, whose radius is *y d* or *y u*, fig. 126, and upon this surface the line *d u*, fig. 125, is to be drawn a tangent to the curve *b d* at *d*. A continuation of the divisions of 3 inches is to be made upon the line *d d'*, and the parallels *a' q'*, *b' r'*, and *u d'*, continued on the cylindrical surfaces. Whatever portion of the superimposed piece *a d' u* may be found to fall within the small arc *a t*, fig. 126, is to be cut away, forming a small portion of an interior cylinder concentric to the point *y*, which being done, the remaining portions of the superimposed piece are to be cut away to the dotted lines *d x*, *a y b z*, *u u'*, of fig. 125, or, what is the same thing, to the lines *d a*, *a' a*, *b' t*, and *u t*, of fig. 126, forming tangents to the curve *a t*, and which will complete the surface of the twisted wedge through its entire length, and to the hight of the line of transit, producing what I conceive to be the *true theoretical surface* of the mould-board.

(617.) Fig. 126 exhibits distinctly, in the quadrant *o b d*, the inequality of the angles of ascent for the slice, where the radii *a p'*, *a n'*, *a m'*, &c., represent the ascents to the corresponding divisions of length in the transit of the slice through the curve *b d u*, which represents the periphery of the cylindrical surfaces at the line of transit. The parts of the figure lying above that line represent those that must be superimposed above the quadrantal portion of the cylinder, to complete the upper regions of the mould-board; these parts acting merely as a preventive against the overfall of soil into the waste of the plow, are of less importance as to form, than those just described, but are quite necessary in the practice of plowing. The parallelogram *y d* exhibits the relation in which the furrow-slice stands to this form of mould-board, when the slice has been raised to the perpendicular, and *y u* in its ultimate position.

(618.) Although I hold this to be a true theoretical form, it is not in this state fit to be employed as a practical mould-board; but the steps to render it so are very simple. The broad shovel-mouth *a b*, fig. 125, would meet with obstructions too numerous to admit for a moment of its adoption in practice; but we have only to remove the right-hand portion of the edge *a b*, in the direction *b q*, making the breadth, *q m*, 6½ or 7 inches broad; that portion also contained 7' *r* 3 is to be cut away, leaving *m r* about 4 inches broad; *b q r m* will then represent the share; the mould-board being thus of the prolonged form in the fore-part. And though this form has no peculiar advantage over the truncated, in respect to working, it is better adapted to admit of the body being constructed of malleable iron, a practice which, though more expensive, is certainly the most preferable, by reason of its greater durability, and being less liable to fracture through the effect of shocks, when stones or other obstructions are encountered.

(619.) Besides the removal of these parts of the theoretical mould-board, other slight modifications are admissible. When the parts have been cut away as described, the edge *b q* of the share will be found too thick for a cutting edge. If brought to a proper thickness, by removing the parts *below*, making the edge to coincide with the curved surface; the share so prepared would have the character that belongs to the cresting plows. The lower edge of the mould-board from *r* to 2 would be also rather high, and would present unnecessary resistance to the lower side of the slice; both parts therefore, require to be reduced. The surface of the feather *b q* is to be sloped down till it become straight between the points *b* and *q*, *q* not being more than ½ inch above the plane of the sole, as at the dotted lines *n z* in fig. 126. The lower edge of the mould-board is also to be rounded off, as shown by the dotted lines along the lower edge from *h* to *o* in fig. 128. To prevent the abrasion of the edge of the slice in passing over the mould-board, it will also be expedient to make the lines from *d* to *u*, in fig. 126, fall in, from below the line of transit upward as shown by the dotted lines at *d′*, *a′*, *b′*, *n*.

(620.) Other modifications may, if required by peculiar taste or otherwise, be given to this form of mould-board. If, after the points *b p n m*, &c. have been determined upon the cylindrical surface, fig. 123, or 125, and, in cutting away the parts above *a b* in the latter figure, instead of reducing the surface to the straight lines 9′ *p*, 8′ *n*, 7′ *m*, &c. we leave the surface slightly convex upon all these lines, a surface will be produced as represented by the dotted sectional lines of fig. 126, or 128, and by becoming slightly either recurved above the line of transit, as in fig. 128, or with continued convexity, as in fig. 118, Plate XI., the surface so produced would deliver the slice without risk of injury to the edge; which, though not of vital importance, is always an object in the estimation of the plowman who performs his work with taste. The same modification would also, in the opinion of many agricultural machine-makers, render the mould-board more efficacious in the working of stiff clay soils.

(621.) Fig. 127, Plate XII., represents an elevation of the new mould-board, as now constructed by me, and fig. 128 the analytical sections of the same, taken in the same manner as described for those preceding, and having the same letters of reference. In the present case, the sectional lines are all straight to the hight of the line of transit; above that line and before the zero they are slightly concave, though, as has been shown, this is not imperative; but, behind the zero, they are convex from a little below the line of transit, as shown by the dotted portions of the lines. The parallelogram *k y*, being a section of the slice when in the vertical position, will be seen to coincide exactly with the zero-line, as it will do through the whole passage of the slice. The letters and numerals in these two figures have the same reference as in the other figures of the mould-board.

(622.) Judging from the trials that have been made of this mould-board, and from the *uniform brightening of its surface* after a few hours' work, it promises to possess a very uniform resistance over its whole surface, which is a principal object to be aimed at in the formation of this member of the plow.

(623.) *The Mould-board Pattern.*—The instructions just given refer solely to the formation of the theoretical surface of the mould-board, including that of the share; but, in the construction of a *pattern* from which mould-boards are to be cast, the process is somewhat different, though based on the principles above laid down.

(624.) In proceeding with this, therefore, the quadrant of the cylinder, upon which the whole problem is grounded, may or may not be prepared. If it is to be employed, then the first process is exactly as before described in reference to the quadrant fig. 123, Plate XII., which must be formed and lined as there described; but the same process may be pursued from lines alone, without the intervention of the solid, and in the following manner: Having described the quadrant of a circle, as *a b c*, fig. 123, of 16 inches radius, construct the diagram fig. 124, as before directed, the entire length *e b* being 30 inches, divided into equal parts of 3 inches each. The arc *b d* is then to be drawn through the points *b p n m*, which points, instead of being a transfer, as before described, from the quadrant, may here be drawn at once with a beam-compass touching the three leading points *b m d*, as before, which will intersect all the divisions, converting them into ordinates 1 *f*, 2 *g*, 3 *h*, &c. to the curve *b d*. The lengths of these ordinates, from the base-line *e b*, are now to be carefully transferred to the quadrant of the circle *b d* of fig. 126, and set off in the circumference thereof; thus the point *b* in fig. 126 corresponds to the termination *b* of the base-line in fig. 124. The first ordinate *g p* is to be set off on the quadrant from *b* to *p*, the second ordinate 8 *n* is set off from *b* to *n*, the third 7 *m* from *b* to *m*, and so on through the entire quadrant of the circle. The radii *a b*, *a p*, *a n*, &c. being now drawn, will furnish the successive angles of elevation, with the sole-plane, for each division of the length throughout the quadrant.

(625.) In applying these to the mould-board, it is to be observed that the first three radii belong to the share, if it is a prolonged mould-board, or the first five if it is truncated. The quadrant, fig. 126, with its radii, being thus completely drawn out at full size upon a board, produce the line *b a* to *y*, and on *y* as a center, with a radius of 7 inches, describe the arc *a t*, and concentric to it the arc *d u*. At an angle of 45° draw *t u* a tangent to the arc *a t*, and the point of intersection of this tangent with the arc will fix the extreme point *u* of the mould-board at the hight of the line of transit; which point will be 19 inches from the land-side plane *b g*, and 12 inches above the plane of the sole, or base-line *y b*. From *d*, lay off divisions of equal parts on the arc *d u*, each equal to 4¼ inches—the diagonal of a square of 3 inches—which completes the lines for the fabrication of the pattern.

Fig 129.

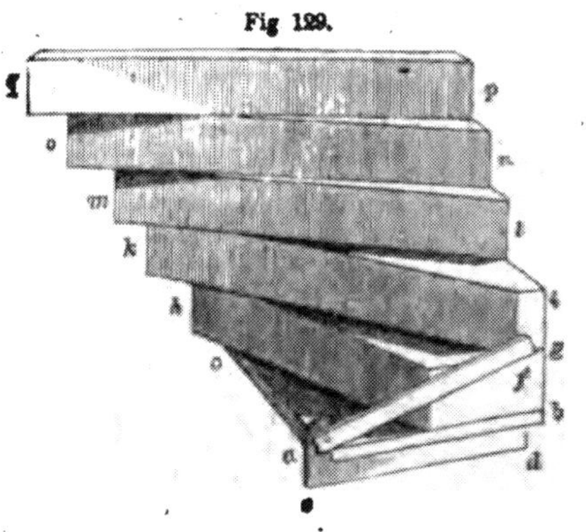

THE BUILDING OF THE BLOCK FOR THE MOULD-BOARD PATTERN.

(626.) The next step in the operation is that of *building a block* out of which the pattern is to be shaped. Provide a deal-board of 3½ feet or thereby in length, with a breadth of 10 inches; have it dressed of uniform thickness, and at least one edge and end straight and right angled, as seen at *a b c*, in the annexed fig. 129, and *a b*, fig. 126, Plate XII., forming a basement to the block, *a* being the right angle, and the continuation of the board being hid from view under the superimposed block. Let the edge *a c* of the board be marked off in equal divisions of 3 inches, agreeing exactly with those of the diagram, fig. 124, marking the divisions with letters or numerals corresponding to the radii of the quadrant, fig. 126, the end *a b* of the board corresponding to the radius *m* of the quadrant, and to the ordinate 7 *m* of the diagram. Provide also a suit-stock or bevel of the form represented by *d e f*, the stock *d e* being a straight bar with a head piece at *e*, fixed at right angles to the stock, and into this the blade *a f* is to be jointed, in such a manner that when the blade and stock are set parallel to each other, they shall just receive the thickness of the basement-board betwixt them, the length of the blade being equal to the breadth of the slice. Five or more pieces of well-seasoned, clean, 3-inch Memel or yellow-pine deal are now to be prepared, each about 30 inches in length, and from 6 to 4 inches in breadth. Set the bevel to the angle *b a m*, fig. 126, and, applying it at the end of the board, as in fig. 129, it will point out the position in which the first block *g h* must be placed on the board in order that it may fill the lines of the pattern. The farther end of the block, being set in like manner to fall within the lines, it is to be firmly attached to the board with screw-nails. The second block *k i* is to be joined to the first by the ordinary method of gluing, being set in the same manner as the first to fill the lines of the pattern at both ends, and this requires its being set obliquely to the first. The third block *l m* is set in like manner, and so on with *n o* and *p q*. The setting of the different blocks will be much facilitated by having the ends *g i l n p* cut off to the plane of the land-side—that is, to coincide vertically with the land-side edge of the board, and by keeping in view that the terminal line *c q* lies at an angle 45°.

(627.) The block being thus prepared, the process of *working it off* is plain and easily performed in this way. Having set the bevel at the angle *b a m*, fig. 126, which answers to the end *a b* of the block, the bevel is applied as in the figure, and the surplus wood is cut away to a short distance within the end *a b* of the board, until the blade of the bevel lies evenly upon the surface, and the kneed head-piece touching the edge of the board. Set the bevel now at the angle *b a l*, and, applying it at the first division on the edge of the board, cut away the surplus wood with a gouge or other tool, in a line parallel to the end of the board, or at right angles to its edge until the edge of the blade *a f* lie evenly on the surface, and the head of the stock touch the edge of the board as before. Repeating this operation at each successive division with the bevel, setting it to the corresponding angle up to the vertical or zero-line, and we have a series of leading lines or draughts, each occupying its true position in the surface of the mould-board to the hight of the line of transit. By continuing these lines, each in the direction already given it, until they terminate in the breast, or in the upper edge of the pattern, we have a corresponding series of points now determined, in the breast and upper edge; and by removing the surplus wood still remaining in the spaces between the lines, and reducing the surface to coincide with them, we have the finished surface from the neck of the share up to the zero.

(628.) To *complete the after portion of the pattern*, we have to form a temporary bevel with a curved blade, adapted to the small arc *a t*, fig. 126, which blade is prolonged in a tangent *t u* at the angle of 45°. With the guidance of this bevel, its stock being still applied to the board, as in fig. 129, cut away all the wood that occurs to interrupt it behind the zero, until it applies every where behind that line without obstruction. At the third division beyond the zero, the pattern may be cut off in a right vertical, though this is not imperative, as the mould-board may be made considerably longer, and even a little shorter, without at all affecting its operation. At whatever distance in length its terminal edge may be fixed, that portion of the line of transit which lies between the zero and the terminus must leave the original curve *h m d*, fig. 123, at a tangent, and it will reach the terminus as such, or it will gradually fall into a reëntering curve, according as the terminus is fixed nearer to or farther from the zero-line; the terminus of the line of transit being always 19 inches distant from the land-side plane. That portion of the surface which now remains unfinished between the arcs *a t* and *d u*, fig. 126, is to be worked off in tangents, applied vertically to the arc *a t*, and terminating in that part of the line of transit that lies between *d* and *u*. Such portions of the interior cylindrical surface as may have been formed under the application of the temporary bevel to the arc *a t*, are now to be also cut away by a line passing through the junction of the tangents *t d′*, *t b′*, *t u*, with the cylindrical arc *a t*, forming a curved termination in the lower part behind—as seen in fig. 127—which completes the surface as proposed.

(629.) The *modifications* formerly pointed out, paragraphs (618.) (619,) and (620,) may now be made upon the lower and the upper parts of the pattern. The breast-curve and the form of the upper edge will now have assumed their proper curvature; and there only remains to have the whole pattern reduced to its due thicknesses. This, in the fore part, is usually about ½ inch, increasing backward below to about 1 inch, and the whole becoming gradually thinner toward the top edge, where it may be 3-16 inch. The perpendicular hight behind is usually about 12 inches, and at the fore part 14 inches.

(630.) Of the Draught of Plows.—From the complicated structure of the plow, and the oblique direction in which circumstances oblige us to apply the draught to the implement, some misconceptions have arisen as to the true nature and direction in which the draught may be applied. The great improver of the plow has fallen into this error, and has, in some measure, been followed by others.* He asserts "that were a rope attached to the point of the share, and the plow drawn forward on a level with the bottom of the furrow, it would infallibly sink at the point." Were this really the case, it would prove that the center of resistance of the plow in the furrow must be somewhere below the level of the sole, which is impossible. As the center of gravity of any body, suspended from a point at, or anywhere near, the surface of that body, will always be found in a continuation of the suspending line, supposing it to be a flexible cord, so, in like manner, the center of resistance of the plow will be always found in the direction of the line of draught. Now if, with a horizontal line of draught from the point of the share, it were found that the point of the share had a tendency to sink deeper into the soil, it would be a clear proof that the plow was accommodating itself to the general law, and that the center of resistance is below the line of the sole. The fallacy of this conclusion is so palpable that it would be an act of supererogation to refute it by demonstration, more especially as it never can be of any utility in a practical point of view. I have thought it necessary, however, to advert to it, as it appears to have

* Small's Treatise on Plows.

aided in throwing a mystery over the mode of applying the *line or angle of draught*, which in itself is a sufficiently simple problem.

(631.) The reasoning hitherto adopted on this branch of the theory of the plow seems to be grounded on the two following data: the *hight*, on an average, of a horse's shoulder, or that point in his collar where the yoke is applied; and the *length* of the draught-chains that will give him ample freedom to walk. It falls out fortunately, too, that the angle of elevation thus produced crosses the plane of the collar as it lies on the shoulders of the horse when in draught, nearly at right angles. It is my purpose, however, in this section to show that (keeping out of view some practical difficulties) the plow may be drawn at any angle, from the horizontal up to a little short of 90°, and that it would require less and less force to draw it as the direction of the line of draught approached the horizontal line. It would, in all cases, be required that the point of the beam, or rather the draught-bolt, should be exactly in the straight line from the center of resistance to the point where the motive force would be applied. If this force could be applied in the horizontal direction, we should have the plow drawn by the minimum of force. This position, however, is impracticable, as the line of draught would, in such a case, pass through the solid land of the furrow about to be raised; but it is within the limits of practicability to draw the plow at an angle of 12°, and, as will be demonstrated, the motive force required at this angle would be 1 stone or 14 lbs. less than is required by drawing at the angle of 20°, which may be held as the average in the ordinary practice of plowing. A plow drawn at this low angle, namely 12°, would have its beam (if of the ordinary length) so low that the draught-bolt would be only 10 inches above the base-line; and this is not an impracticable hight, though the traces might be required inconveniently long. On the same principle, the angle of draught might be elevated to 60° or 70°, provided a motive power could be applied at such high angles. In this, as before, the beam and draught-bolt would have to fall into the line of draught as emanating from the center of resistance. The whole plow, also, under this supposition, would require an almost indefinite increase of weight; and the motive force required to draw the plow at an angle of 60° would be nearly twice that required in the horizontal direction, or 1 16-18 times that of the present practice, exclusive of what might arise from increased weight. We may therefore conclude that to draw the plow at any angle higher than the present practice is impracticable, and, though rendered practicable, would still be highly inexpedient, by reason of the disadvantage of increased force being thus rendered necessary; unless we can suppose that the application of steam or other inanimate power might require it. Neither would it be very expedient to adopt a lower angle, since it involves a greater length of trace-chains, which, at best, would be rather cumbrous; and it would produce a saving of force of only one stone on the draught of a pair of horses. Yet it is worthy of being borne in mind that, in all cases, there is some saving of labor to the horses, whenever they are, by any means, allowed to draw by a chain of increased length, provided the draught-bolt of the plow is brought into the line of draught, and the draught-chains are not of such undue weight as to produce a sensible curvature; in other words, to insure the change of angle at the horse's shoulder, due to the increased length of the draught-chain.

(632.) In illustration of these changes in the direction of the draught, fig. 130, Plate XIII. will render the subject more intelligible. Let *a* represent the body of a plow, *b* the point of the beam, and *c* the center of resistance of the plow, which may be assumed at a hight of 2 inches above the plane of the sole *d e*, though it is liable to change within short limits. The average length of the draught-chains being 10 feet, including draught-bars, hooks, and all that intervenes between the draught-bolt of the plow and the horse's shoulders; let that distance be set off in the direction *b f*, and the average hight of the horse's shoulders where the chains are attached, being 4 feet 2 inches, let the point *f* be fixed at the hight above the base-line *d e*. Draw the line *f c*, which is the direction of the line of draught acting upon the center of resistance *c*; and if the plow is in proper temper it will coincide also with the draught-bolt of the beam; *e c f* being the angle of draught, and equal to 20°. It will be easily perceived, that, with the same horses and the same length of yoke, the angle *e c f* is invariable; and if the plow has a tendency to dip at the point of the share under this arrangement, it indicates that the draught-bolt *b* is *too high* in the bridle. Shifting the bolt one or more holes downward will bring the plow to *swim* evenly upon its sole. On the other hand, if the plow has a tendency to rise at the point of the share, the indication from this is, that the draught-bolt *b* is *too low*, and the rectification must be made by raising it one or more holes in the bridle. Suppose, again, that a pair of taller horses were yoked in the plow, the draught-chains, depth of furrow, and soil—and, by consequence, the point of resistance *c*—remaining the same, we should then have the point *f* raised suppose to *f′*; by drawing the line *f′ c*, we have *e c f′* as the angle of draft, which will now be 22°; and in this new arrangement, the *draught-bolt* is found to be *below* the line of draught *f′ c*; and if the draught-chains were applied at *b*, in the direction *f′ b*, the plow would have a tendency to rise at the point of the share, by the action of that law of forces which obliges the line of draught to coincide with the line which passes through the center of resistance; hence the draught-bolt *b* would be found to rise to *b′*, which would raise the point of the share out of its proper direction. To rectify this, then, the draught-bolt must be raised in the bridle by a space equal to *b b′*, causing it to coincide with the true line of draught, which would again bring the plow to swim evenly on its sole.

(633.) Regarding the relative forces required to overcome the resistance of the plow, when drawn at different angles of draught, we have first to consider the nature of the form of those parts through which the motive force is brought to bear upon the plow. It has been shown that the tendency of the motive force acts in a direct line from the shoulder of the animal of draught to the center of resistance; and referring again to fig. 130, Plate XIII., were it not for considerations of convenience, a straight bar or beam lying in the direction *c b*, and attached firmly to the plow's body anywhere between *c* and *g*, would answer all the purposes of draught, perhaps, better than the present beam. But the draught not being the end in view, but merely the means by which that end is accomplished, the former is made to subserve the latter; and as the beam, if placed in the direction *c b*, would obstruct the proper working of the plow, we are constrained to resort to another indirect action to arrive at the desired effect. This indirect action is accomplished through the medium of a system of rigid angular frame work, consisting of the beam and the body of the plow, or those parts of them comprehended between the points *b*, *h*, *c*, the beam being so connected to the body *a h*, as to form a rigid mass. The effect of the motive force applied to this rigid system of parts at the point *b*, and in the direction *b f*, produces the same result as if *c b* were firmly connected by a bar in the position of the line *c b*, or as if that bar alone were employed, as in the case before supposed, and to the exclusion of the beam *b h*.

(634.) Having thus endeavored to illustrate the causes of the oblique action of the plow, showing that the obliquity is a concomitant following the considerations of convenience and fitness in working the implement, I proceed to show the relative measure of the effects of the oblique action. It is well known that the force of draught required to impel the plow, when exerted in the direction *b f*, may be taken at an average of 24 stones, or 336 lbs. Analyzing this force by means of the parallelogram of forces, if we make the line *b f* to represent 336 lbs., the motive force; and complete the parallelogram *b i f k*, we have the force *b f* held in equilibrium by the two forces *i b* and *k b*; the first acting in the *horizontal* direction to draw the plow forward, the second acting *vertically*, to prevent the point of the beam from sinking, which it would do were a horizontal force only applied to the point of the beam. The relation of these forces *i b* and *k b* to the oblique force will be as the length of the lines *i b* and *k b* to the line *b f*, or the line *i b* will represent 322 lbs., while the oblique force is 336 lbs., and the force *k b* 95 lbs. This last force is represented as *lifting* the beam vertically by suspension, but the same result would follow if the beam were supported by a *wheel* under the point *b*; the wheel would then bear up the beam with the same force as that by which it was supposed to be suspended, 95 lbs. But to carry out the supposition, let the draught now found be applied horizontally from the point *c*. As the plow would then have no tendency either to dip or rise, the force *k b* vanishes, leaving only the direct horizontal force *i b*; hence, were it possible to apply the draught in a horizontal direction from the point of resistance, the resistance of the plow would be 322 lbs. instead of 336 lbs.

(635.) But to return to the previous position of the draught, wherein, still supposing it to be in the horizontal direction, and thereby requiring that the point of the beam have a *support* to prevent its sinking too low. This support may be supposed either a *foot*, as seen in many both ancient and modern plows, or in the shape of a wheel or wheels, so much employed in many of the English plows. We see at once, under this consideration, the office that a wheel performs in the action of a plow. It has been shown, that whether the plow be drawn in the ordinary direction of draught *b f*, in which one oblique propelling force only is exerted, or with two antagonist forces, *b i*, in the horizontal direction, and the upholding force, *b k*, in the vertical, we find that, in the latter, the difference in favor of the motive force is only 1-24 of the usual resistance; but the upholding force is equal to 2-7, while none of these variations has produced any change in the absolute resistance of the plow. The impelling force is theoretically less in the latter case; but since the wheel has to carry a load of 95 lbs., we have to consider the effect of this load upon a small wheel, arising from friction and the resistance it will encounter by sinking less or more into the subsoil. I have ascertained, from experiment, that the difference of force required to draw a wheel of 12 inches diameter, loaded as above described, and again when unloaded, over a tolerably firm soil, is equal to 22 lbs., a quantity exceeding 1½ times the amount of saving that would accrue by adopting this supposed horizontal draught with a wheel. Having thus found the amount of draught at two extremities of a scale, the one, being the oblique draught, in common use at an angle of 20°, the other deduced from this, through the medium of the established principles of oblique forces, and the latter producing a saving of only 1-24 of the motive force, while it is encumbered with an additional resistance arising from the support or wheel; it necessarily follows that, at all intermediate angles of draught, or at any angle whatever, where the principle of the parallelogram of forces finds place—and it will find place in all cases where wheels *yielding any support* are applied to the plow under the beam—there must necessarily be *an increase in the amount of resistance to the motive force.*

(636.) This being a question of some importance, the diagram, fig. 131, will render it more

Fig. 131.

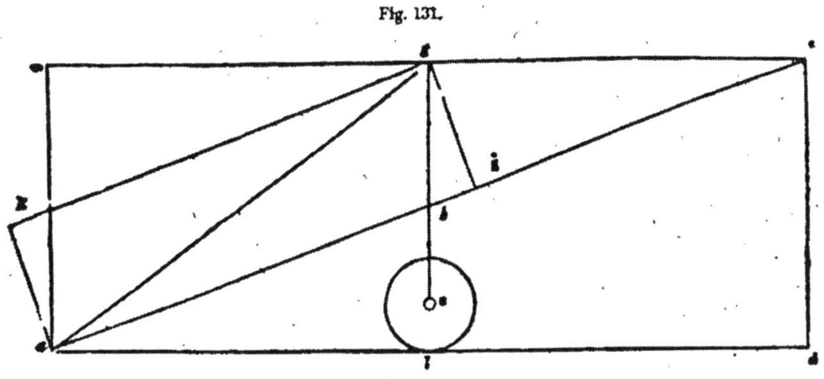

THE DRAUGHT OF WHEEL-PLOWS.

evident. Let *a* be the point of resistance of a plow's body, *b* the point of the beam, *c* the position of the horse's shoulder, and *a d* the horizontal line; then will *c a d* be the angle of draught equal to 20°. Let the circle *e* represent a wheel placed under the beam, which is supported by a stem or sheers, here represented by the line *e b*. In this position the point of the beam, which is also the point of draught, lies in the line of draught; the wheel, therefore, bears no load, but is simply in place, and has no effect on the draught; the motive force, therefore, continues to be 336 lbs. Suppose now the point of the beam to be raised to *g*, so that the line of draught *g c* may be horizontal; and since the line of draught lies now out of the original line *a b c*, and has assumed that of *a g c*—*g* being now supported on the produced stem *c g* of the wheel—draw *g i* perpendicular to *a c*, and complete the parallelogram *a i g k*; the side *a i* will still represent the original motive force of 336 lbs., but, by the change of direction of the line of draught, the required force will now be represented by the diagonal *a g* of the parallelogram, equal to about 351 lbs.; and *g c* is a continuation of this force in a horizontal direction. The draught is therefore increased by 15 lbs. Complete also the parallelogram *a l g m*, and as the diagonal *a g*—the line of draught last found—is equal to 351 lbs., the side *l g* of the parallelogram will represent the vertical pressure of the beam upon the wheel *e*, equal to about 200 lbs., which, from experiments (635), may be valued at 40 lbs. of additional resistance, making the whole resistance to the motive force 391 lbs., and being a total *increase arising from the introduction of a wheel in this position of* 55 lbs. Having here derived a maximum—no doubt an extreme case—and the usual angle of 20° as the minimum, we can predicate that, at any angle intermediate to *l a b* and *l a g*, the resistance can never be reduced to the minimum of 336 lbs. Hence it follows, as a corollary, that *wheels placed under the beam* can never lessen the resistance of the plow; but, on the contrary, must, in all cases, increase the resistance to the motive force more or less, according to the degree of pressure that is brought upon the wheel, and this will be proportional to the sine of the angle in the resultant *a g* of the line of draught.

(637.) The application of a *wheel in the heel* of a plow, does not come under the same mode of reasoning as that under the beam, the former becoming a part of the body, from which all the natural resistance flows; but in viewing it as a part of that body only, we can arrive at certain conclusions which are quite compatible with careful experiments.

(638.) The breadth of the whole rubbing surface in the body of a plow, when turning a furrow, is on an average about 17½ inches, and supposing that surface to be pressed nearly equal in all parts, we shall have the *sole-shoe*, which is about 2½ inches broad, occupying 1-7 part of the surface; and taking the entire average resistance of the plow's body, as before, at 336 lbs., we have 1-7 of this, equal to 48 lbs., as the greatest amount of resistance produced by the sole of the plow. But this is under the supposition that the resistance arises from a uniform degree of friction spread over the whole rubbing surface of the body: while we have seen, on the contrary, that the coulter, when acting alone, presents a resistance equal to the entire plow. It is only reasonable, therefore, in absence of farther experiments, to conclude that the fore parts of the body—the *coulter* and *share*—*yield a large proportion of the resistance when turning the furrow-slice*; but, since we cannot appreciate this with any degree of exactness, let the sole have its full share of the resistance before stated, namely, 48 lbs. If a wheel is applied at or near the heel of the plow, it can only bear up the hind part of the sole, and prevent its ordinary friction, which, at the very utmost, cannot be more than ½ of the entire friction due to the entire sole. A wheel, therefore, placed here, and acting under every favoring circumstance, even to the supposed extinction of its own friction, could not reduce the resistance by more than 24 lbs., being the half of that due to the entire sole, or it is 1-14 of the entire resistance. But we cannot imagine a wheel so placed, to continue any length of time, without becoming clogged in all directions, thereby greatly increasing its own friction; and when it is considered that the necessarily small portion, of any wheel that can be so applied, will sink into the subsoil, to an extent that will still bring the sole of the plow into contact with the sole of the furrow. It will thus be found that the amount of reduction of the general resistance will be very much abridged, certainly not less than one-half, which reduces the whole saving of draught to a quantity not exceeding 12 lbs., and even this will be always doubtful, from the difficulty of keeping such wheels in good working condition. This view of a wheel placed at the heel has been confirmed by actual experiments, carefully conducted, wherein Palmer's patent plow with a wheel in the heel (as patented many years ago), but in this case it was applied on the best principles, gave *indications of increased resistance from the use of the wheel, as compared with the same plow when the wheel was removed*; the difference having been 1¼ stone in favor of no wheel. I hesitate not, therefore, to say that in no case can wheels be of service toward reducing the resistance of the plow, whether they be placed before or behind,* or in both positions, and the chances are numerous that they shall act injuriously. That the use of wheels may, under certain circumstances, bring the implement within the management of less skillful hands than is required for the swing plow, must be admitted; but, at the same time, there may be a question whether, even with that advantage, the practice is commendable. I should be wanting in candor if, for myself, I answered otherwise than in the negative.—J. S.]

* The wheel under the beam, for general use, is thought by many not commendable; but the shoe wheel in the land-side is an improvement, as it diminishes or reduces the resistance materially.

WILKINSON.

24. THE VARIOUS MODES OF PLOWING RIDGES.

. . . . "Your plowshare, ...
Drawn by one pair, obedient to the voice,
And double rein, held by the plowman's hand,
Moves right along, or winds as he directs."
GRAHAM.

(639.) Your knowledge of soils will become more accurate after you have seen them plowed; for as long as a crop, or the remains of one, covers soils, their external characters cannot be fully exposed to view.

(640.) On observing a plow at work, you might imagine that the laying over of a furrow-slice is a very simple process; but it is really not so simple as it appears. You have already seen, in the construction of the plow, that the furrow-slice is laid over by a machine of very complicated structure, though simple in its mechanical action on the soil; and you may learn, by a single trial, that the plow is not in reality so very easily guided as it appears to be in the hands of an expert plowman. You might also imagine that, as the plow can do nothing else but lay over a furrow-slice, the forms of plowing do not admit of much variety; but a short course of observation will convince you that *there are many modes of plowing land.*

(641.) The several modes of plowing land have received characteristic appellations, and these are—gathering-up; crown-and-furrow plowing; casting or yoking or coupling ridges; casting ridges with gore furrows; cleaving down ridges; cleaving down ridges with or without gore furrows; plowing two-out-and-two-in; plowing in breaks; cross-furrowing; angle plowing, ribbing, and drilling; and the preparative operation to all plowing is termed feering or striking the ridges.

(642.) These various modes of plowing are contrived to suit the nature of the soil and the season of the year. Heavy land requires more cautious plowing than light, because of its being more easily injured by rain; and greater caution is required to plow all sorts of land in winter than in summer. The precautions here spoken of allude to the facilities given to surface water to flow away. The different seasons, no doubt, demand their respective kinds of plowing; but some of the modes are common to all seasons and soils. Attention to all the methods will alone enable you to understand which kind is most suitable to particular circumstances of soil, and particular states of season. To give you an idea of all the modes, from the simplest to the most complicated, let the ground be supposed to be even in reference to the state of its surface.

(643.) The supposed flat ground, after being subjected to the plow, is left in the form of *ridges* or of *drills*, each ridge occupying land of equal area, determined by similar lengths and breadths. The ridges are usually made N. and S., that the crop may enjoy the light and heat of the solar rays in an equal degree throughout the day; but they should, nevertheless, traverse the slope of the ground, whatever its aspect may be; and this is done that the surface water may flow easily away.

(644.) Ridges are made of the different breadths of 10, 12, 15, 16, and 18 feet, in different parts of the country. These various breadths are occasioned partly by the nature of the soil, and partly by local custom. With regard to the soil, heavy land is formed into narrow ridges, to allow the rain to flow quickly into the open furrows. Hence, in many parts of England, the ridges are only 10 and 12 feet in width, and in some localities they are in ridglets of 5 or 6 feet. In Scotland, even on the strongest land, the ridges are seldom less than 15 feet; in some localities they are from 16 to 36 feet, and in light soils a not unusual width is 18 feet. In Berwickshire and Roxburghshire, the ridges have for a long period been 15 feet on all classes of soils, being considered the most convenient width for the ordinary manual and implemental operations. In other parts of the country, 16 and 18 feet are more common. More than half a century ago, ridges were made very broad—that is, from 24 to 36 feet—high on the top or crown, and crooked like the letter S, from the mistaken notion that the crook always presented some part of the ridge in a right position to the sun—a form which, although it did, would remove other parts as far away from the sun's influence. In the Carse of Gowrie, such broad, crooked ridges still exist; but the usual practice throughout the country is to have ridges of moderate breadth, straight, and looking to noon-day. In many parts of Ireland the land is not put into ridges at all, being done up with the spade into narrow stripes called *lazy-beds*, separated by deep, narrow trenches. Where the plow is used, however, ridges are always formed, though narrow, but usually of 12 feet. For the sake of uniformity of description, let it be understood, when I speak of a ridge, that an area of 15 feet of width is meant.

(645.) The first process in ridging up land from the flat surface is called *feering* or *striking* the ridges. This is done by planting 3 or more of such poles, graduated into feet and half-feet, as were recommended for setting off the lines of fence (446), and which are used both for directing the plow employed to feer in straight lines, and for measuring off the breadth of the ridges into which the land is to be made up, from one side of the field to the other.

(646.) Land is *feered* for ridging in this way: Let *a b*, fig. 132, represent the S. and E. fences of a field, of which let *x* be the *head-ridge* or *head-land*, of the same width as that of the ridges, namely 15 feet. To mark off its width distinctly, let the plow pass in the direction of *r e*, with the furrow-slice lying toward *x*. Do the same along the other head-land, at the opposite end of the field. Then take a pole and measure off the width of a quarter of a ridge, viz. 3 feet 9 inches, from the ditch lip *a* to *c*, and plant a pole at *c*. With another pole set off the same distance from the ditch *a* to *d*, and plant it there. Then measure the same distance from the ditch at *e* to *f*, and at *f* look if *d* has been placed in the line of *f c*; if not, shift the poles a little until they are all in a line. Make a mark on the ground with the foot, or set up the plow-staff, at *f*. Then plant a pole at *g* in the line of *f d c*. Before starting to feer, the plowman measures off 1¼ ridges—namely, 18 feet 9 inches—from *f* to *k*, and plants a pole at *k*. He then starts with the plow from *f* to *d*, where he stops with the pole standing between the horses' heads, or else pushed over by the tying of the horses. He then, with it, measures off, at right angles to *f c*, a line equal to the breadth of 1¼ ridges, 18 feet 9 inches, toward *t* until he comes to the line of *k l*, where he plants the pole. In like manner he proceeds from *d* to *g*, where he again stops, and measures off 1¼ ridges, 18 feet 9 inches breadth, from *g* toward *v* at a point in the line of *k l*, and plants the pole there. He then proceeds toward the other head-ridge to the last pole *c* from *g*, and measures off 1¼ ridges, 18 feet 9 inches, from *c* to *l*, and plants the pole at *l*. From *l* he looks toward *k* to see if the intermediate poles are in the line *l k*; if not, he shifts them to their proper points as he returns to the head-ridge *x* along the furrow he had made in the line *f c*.

On coming down *c f* he obviates any deviation from the straight line that the plow may have made. In the line of *f c* the furrow-slices of the feering have been omitted, to show you the setting of the poles. It is of much importance to the correct feering of the whole field to have those first two

Fig. 132.

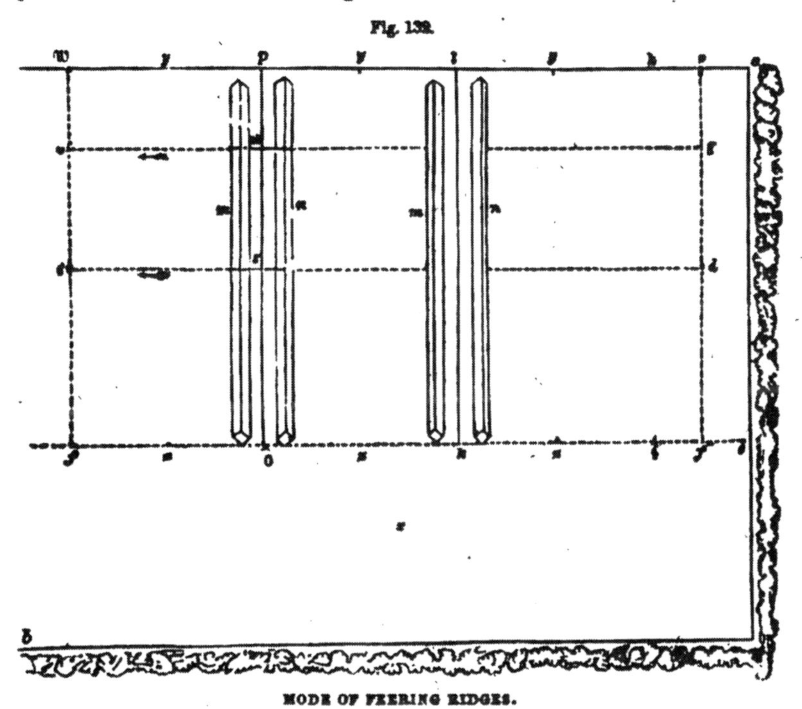

MODE OF FEERING RIDGES.

feerings, *f c* and *k l*, drawn correctly; and, to attain this end, it is proper to employ two persons in the doing of it—namely, the plowman and the farm-steward, or farmer himself. It is obvious that an error committed at the first feerings will be transmitted throughout the whole field. A very steady plowman and a very steady pair of horses, both accustomed to feer, should only be intrusted with the feering of land. Horses accustomed to feer will walk up of their own accord to the pole standing before them. In like manner the plowman proceeds to feer the line *k l*, and so also the line *o p*; but in all the feerings after the first, from *f* to *k*, the poles, of course, are set off to the exact breadth of the ridge determined on—in this case 15 feet, such as from *s* to *t*, *u* to *v*, *p* to *w*, in the direction of the arrows. And the reason for setting off *c l* at so much a greater distance than *l p* or *p w* is, that the ½ ridge *a h* may be plowed up first and without delay, and that the rest of the ridges may be plowed by half-ridges. The half-ridge *a h* is, however, plowed in a different manner from the rest; it is plowed by going round the feering *f c* until the open furrow comes to *a e* on the one side and to *h i* on the other. Then *h i* constitutes the feering, along with *k l*, for plowing the 2 half-ridges *z i* and *z k*, which, when done, the open furrow is left in the line *z y*, corresponding to the open furrow left in the line *e a*, and between which is embraced and finished the full ridge of 15 feet *e z*. The half-ridges *z k* and *z o* are plowed at the same time by another pair of horses, and the open furrow *z y* is left between them, and the full ridge *z k z* is then completed. In like manner the half-ridges *z o* and *z r* are afterward plowed by the same horses, and the open furrow *z y* is left between them, and the full ridge *z o z* is then completed. And so on with every other feering in the field. Had the feering been set off the breadth of a half-ridge—that is, 7½ feet—in the line of *i h*, from *a* to *h* and from *e* to *i*, this half-ridge could only have been plowed by all the furrow-slices being turned over toward *h i*, and the plow returning back empty, thus losing half its time.

(647.) As a means of securing perfect accuracy in measuring off the breadths of ridges at right angles to the feerings, lines at right angles to *f c* should be set off across the field, from the cross-table and poles set at *d* and *g*, in the direction of *d t* and *g v*, and a furrow made by the plow in each of these lines, before the breadths of the feerings are measured along them. Most people do not take the trouble of doing this, and a very careful plowman renders it a precaution of not absolute necessity, but every proficient farmer will always do it, even at the sacrifice of a little time and some trouble, as a means of securing accuracy of work.

(648.) As the plow completes each feering, the furrow-slices appear laid over as at *m* and *n*. While one plowman proceeds in this manner to feer each ridge across the field, the other plowmen commence the plowing of the land into ridges; and, to afford a number of plowmen space for beginning their work at the same time, the feering-plowman should be set to his work at least half a day in advance of the rest, or more if the number of plows is great or the ridges to be feered long. In commencing to plow the ridges, each plowman takes two feerings, and begins by laying the furrow-slices of the feerings together, such as *m* and *n*, to form the crowns of the future ridges. In this way one plowman lays together the furrow-slices of *f c* and *k l*, while another is doing the same with those of *o p* and *r w*. I have already described how the ½ ridge *a h* is plowed, and stated that the rest of the ridges are plowed in ½ ridges. The advantage of plowing by ½ ridges is, that the open furrows are thereby left exactly equi-distant from the crowns; whereas, were the ridges plowed by going round and round the crown of each ridge, one ridge might be made by one plowman a little broader or narrower than the one on each side of it—that is, broader or narrower than the determinate breadth of 15 feet.

(649.) A ridge, *a a*, fig. 133, consists of a crown *b*, two flanks *c*, two furrow-brows *d*, and two open furrows *a a*. An open furrow is finished at the bottom by two mould or hint-end furrows. (Fig. 134.)

(650.) After laying the feering furrow-slices to make the crowns of the ridges, such as at *f c*, *k l*, *o p*, and *r w*, fig. 132, the plan to plow up ridges from the flat ground is to turn the horses toward you on the head-ridges, until all the furrow-slices between each feering are laid over until you reach the lines *y z*, which then become the open furrows. This method

Fig. 133.

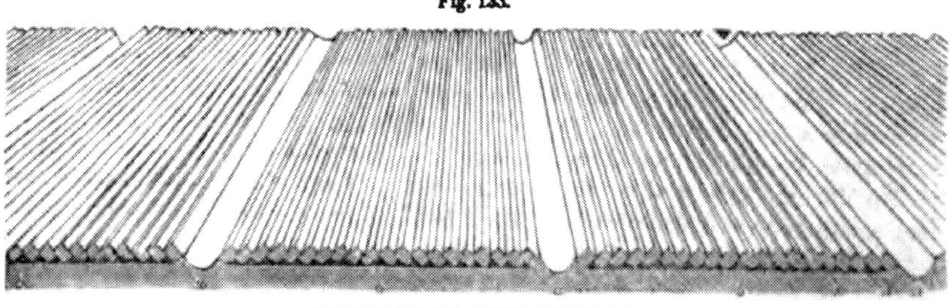

GATHERING UP FROM THE FLAT.

of plowing is called *gathering up*, or gathering up from the flat, the disposition of whose furrows is shown in fig. 133, where *a a a* embrace two whole ridges, on the right sides of which all the furrows lie one way, from *a* to *b*, reading from the right to the left; and on the left sides of which all the furrow-slices lie in the opposite direction, from *a* to *b*, reading from the left to the right; and both sets of furrow-slices meet in the crowns *b b b*. The open furrows *a a a* are finished off with the mould or hint-end furrows, the method of making which is described in the next figure.

(651.) The *mould* or *hint-end* furrow is made in this way: When the last 2 furrow-slices of the ridges *a a*, fig. 134, are laid over, the bottom of the open furrow is as wide as represented by the dotted line *c*, extending

Fig. 134.

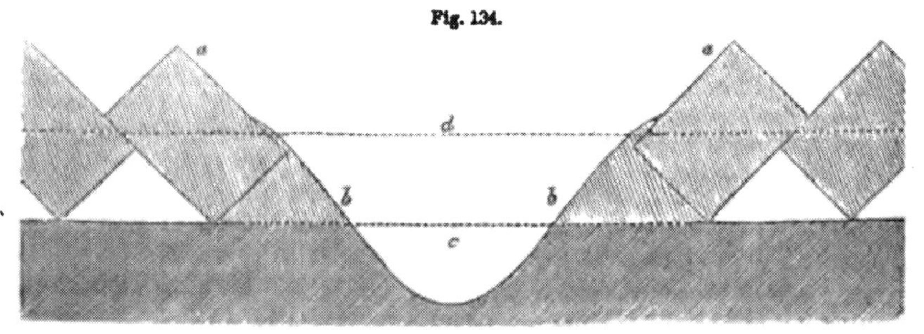

AN OPEN FURROW WITH MOULD OR HINT-END FURROW-SLICES.

from *a* to *a*. The plow goes along this wide space, and first lays over a triangular furrow-slice *b* on one side, and another of the same, *b*, on the other side, up against and covering the lower ends of the last furrow-slices *a a*, and by which operation the ground is hollowed out in the shape represented at *c* by the sole of the plow. The dotted line *d* shows the level of the ground in its former state, before it was begun to be ridged up, and the furrow-slices *a a* show the elevation attained by the land above its former level by plowing.

(652.) A ridge that has been plowed the reverse to gathering up from the flat is said to be *split*, which is the short phrase for crown-and-furrow plowing.

(653.) This kind of plowing of *crown-and-furrow* can easily be performed on land that has been gathered up from the flat. In this case, no feering is required to be purposely made, the open furrows answering that purpose. Thus, in fig. 133, let the furrow-brows *d* be laid over to meet together in the open furrow *a*, and it will be found that they will just meet, since they were formerly separated in the same spot; and so let each successive furrow-slice be reversed from the position it was laid when gathered up from the flat, and as represented in the figure, then *a* will become the crowns of the ridges, and *b* the open-furrows. In this mode, as well as in gathering up, the ridges are plowed by two half-ridges, and in both cases the plowed surface of the ridges is preserved in a flat state; there *should be* no perceptible curvature of the ground, the open furrow only forming a hollow below the level of the plowed surface. When no surface-water is likely to remain on the land, which is the case with light soils, both these are simple modes of plowing land; and they form an excellent foundation upon which to make drills upon stronger soils for turnips.—They are both much practiced in plowing land for barley after turnips.

(654.) But when two plowings are intended to be given to land for barley after turnips, and when it is found inconvenient to cross-furrow the land—which will be the case when sheep on turnips occupy a field of great length in proportion to its breadth, or when the soil or season is too wet to run the risk of letting the land lie any time in a cross-furrow—then the land should be feered so as to allow it to be gathered up from the flat, that the crown-and-furrow plowing may afterward complete the ridges. On looking again at fig. 133, where the ridges are represented complete, it is obvious that, were they plowed from that state into crown-and-furrow by making the open furrows *a a a* the future crowns, a half-ridge would be left at each side of the field—a mode of finishing off a field which no considerate farmer adopts, as it displays great carelessness and want of forethought in forming his plans. The land should, therefore, be so feered at first as to leave a half-ridge next the ditch when gathered up from the flat, and which the subsequent crown-and-furrow plowing will convert into a whole one. Thus, the first feering should be made at *e a*, fig. 132, and every other should be made at the distance of the width of a ridge, namely 15 feet, from the last one, as at *y z*, *y z*, *y z*. On plowing each feering, the open furrows will then be left at *i h*, *k l*, *o p*, and *r w*. These open furrows will form the feerings for, and the crowns of, the future ridges—which, when plowed, the half ridge from *i* to *e* will have to be plowed by itself; thereby, no doubt, incurring some loss of time in laying all the furrow-slices toward the crown *h i*, and returning with the empty plow; but that loss must be endured to get the ridges finished with a perfect form

(655.) I may mention here, that one stretch of the plow with a furrow is called a *landing*, and going and returning with a furrow each way is termed a *bout*.

(656.) Another mode of plowing land from the flat surface is *casting* or *yoking* or *coupling* the ridges. The feering for this mode is done in a different way from either of the two foregoing. The first feering is opened out in the line of *e a*, fig. 132, close to the ditch, and every other is measured off of the width of two ridges from the last—that is, 30 feet asunder—as at *y z*, betwixt *k l* and *o p*, and at half a ridge beyond *r w*. Casting is begun by laying the furrow-slice of the feerings together, and then laying the first furrow-slice toward *e a*, on going up, and toward *y z*, betwixt *l* and *p*, on coming down the bout; and so on, furrow after furrow, turning the horses on the head-ridges always toward you, until the open furrow is left at *y z*, betwixt *k l* and *i h*. The effect of casting is to lay the entire furrow-slices of every ridge in one direction, and in opposite directions on adjoining ridges. The proper disposition of the furrow-slices you will see in perspective in fig. 135, which exhibits three entire ridges, two of them

Fig. 135.

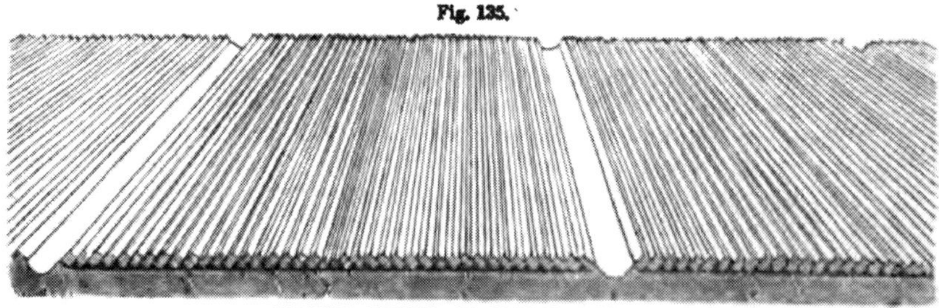

CASTING, YOKING, OR COUPLING RIDGES.

cast or yoked together; that is, the furrow-slices of *a b* meet those of *c b* in *b*, which forms the crown of the double ridge, and those of *c d* lie in the opposite direction from *c b*, and are ready to meet those of the adjoining ridge beyond *d* at *d*, and they leave the open furrow between them at *c*; and so on, an open furrow between every two ridges. Ridges lying thus yoked can easily be recast, by reversing the furrow-slices of *b c* and *c d*, thereby converting the open furrow *c* into a crown of the double ridge, and making the crown *b* an open furrow. Cast ridges keep the land in a level state, and can most conveniently be adopted on dry soils. They form a good foundation for drilling upon, or they make a good seed-furrow on dry land. Lea on light land, and the seed-furrow for barley on the same sort of soil, are always plowed in this fashion. This is an economical mode of plowing land in regard to time, as it requires but few feerings; the furrow-slices are equal, and on even ground; and the horses are always turned inward, that is, toward you. Casting is best performed upon the flat surface, as then the uniform state of both ridges can be best preserved; and should the land be desired to be plowed again, it can be cast the reverse way, and the correct form of the ridges still preserved. In this method of casting, no open furrow is more bare of earth than another.

(657.) Casting ridges is as suitable plowing for strong as light land, provided the ridges are separated by a *gore-furrow*. A gore-furrow is a space made to prevent the *meeting* of two ridges, and as a substitute for an open furrow between them. Its effect is, in so far as the furrow-slices are concerned, like crown-and-furrow plowing, but the difference consists in this, that it turns over a whole ridge, instead of a half-ridge in each feering. It can only be formed where there is a feering or an open furrow. The method of making a gore-furrow is shown in fig. 136. Suppose

Fig. 136.

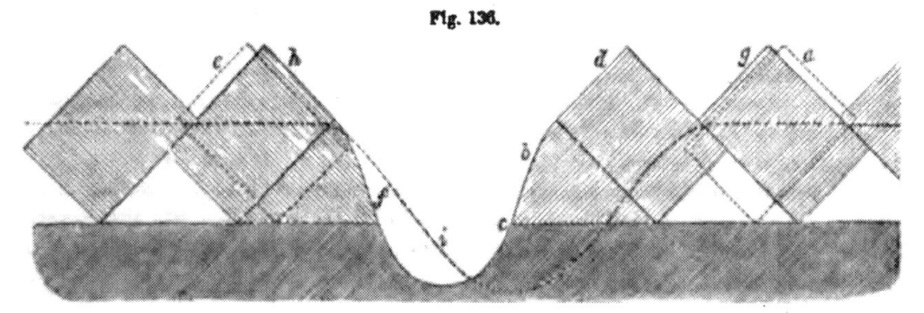

A GORE-FURROW.

that it is proposed to make one in a feering such as is shown by *k l* and *o p* in fig. 132. Let the dotted furrow-slices *a* and *e*, and the dotted line *i* represent an open furrow such as in fig. 136, of which *c* is a point in the middle. Make the plow pass between the center of the furrow-sole *c* and the left-hand dotted furrow-slice *e*, and throw up to the right the triangular-shaped mould-furrow-slice *b*. Then turn the horses sharp round toward you on the head-ridge, and lay the dotted furrow-slice *a* upon *b*, which will then become the furrow-slice *d*, as seen in the fig. at *d*. Again turning the horses sharp round on the head-ridge, take the plow lightly through part of the dotted furrow-slice *e*, and convert it into the triangular shaped mould-furrow-slice *f*, the upper end of *e* being left untouched; but a portion of *f* will trickle down toward *i*. Turn the horses from you this time on the head-ridge, and bring down the plow behind *d*, and lay against it the ordinary furrow-slice *g*. Turning the horses again from you on the near head-ridge, lay the ordinary furrow-slice *h*, by destroying the remainder of the dotted furrow-slice *e* with some more earth, upon the triangular-shaped furrow *f*; which, when done, turn the horses *from* you again on the farther head-ridge for the last time, and come down the open furrow *i*, rubbing the soil up with the mould-board of the plow from *i* against *f*, and clearing out of the furrow any loose soil that may have fallen into it, and the gore-furrow is completed. The dotted line *i* shows the surface of the former state of the land. A gore-furrow is most perfectly formed and retained in clay soil, for one in tender soil is apt to moulder down by the action of the air into the open furrow, which frustrates the purpose of making it a channel for running water; but, indeed, on light soils, gore-furrows are of little use, and, of course, seldom formed.

(658.) When land is cast with a gore-furrow upon gathered ground, it is quite correct to say that the open furrow is more bare of earth than the gore-furrow, as Professor Low intimates, but it is not so correct to say, that "this is an imperfection unavoidable in casting a ridge."* Such a remark is only applicable to cast ridges after they have been gathered up from the flat, and much more so to ridges that have been twice gathered up; but the imperfection does not belong to casting in its most legitimate form, that is, upon the flat ground. Land, in my opinion, should never be cast upon gathered ridges, to remain in a permanent form, but only for a temporary purpose; as in the process of fallowing, for the sake of stirring the soil and overcoming weeds. For, observe the necessary effect of casting a gathered ridge. Suppose the two gathered ridges between *a a a*, fig. 133, were desired to be cast together toward the middle open furrow *a*; the effect would be to reverse the position of the furrows from *a* to *b*, on either side of *a*. They would remain as flat as formerly; but what would be the effect on the furrows on the other halves of the ridges from *b* to *d*? They would be gathered twice, so that the double ridge would have two high furrow-brows by two gatherings, and two low flanks by one gathering. It would, in fact, be unequally plowed, and the open furrow on each side of it would, of course, be bared of earth, having been twice gathered. No doubt, such a distortion might be partially obviated by making the furrow-slices between *a* and *b* on each side of the middle open furrow *a* deeper and larger than those between *b* and *d*, and thus endeavor to preserve a uniform shape to the double ridge; but this would be done by the sacrifice of sterling plowing, and it is much better to confine casting within its own sphere, than practice it in circumstances unfavorable to the land.

(659.) The open furrow in casting does not necessarily bare the earth more than a gore-furrow. It is broader, certainly, from the circumstance of its furrow-slices being laid from each other; but its furrow-sole is not actually plowed deeper than the gore-furrow. In treating of casting, immediately after showing how ridges may be gathered up once and twice, it appears to me that Professor Low seems to intimate, at page 152, that land so gathered up may be cast, and preserve its form; but on this I would observe that casting is almost impracticable after twice gathering; at least it is unadvisable, because, in that case, the effect would be to cleave down the side *b a*, fig. 133, of the ridge on each side of *a*; that is, to throw them down again to the level of the ground; while it would gather up the other two sides *b d thrice*, thereby either making the two sides of each ridge of unequal hights, or, to preserve their level, making the furrow-slices on the same ridges of unequal sizes—practices both undeserving of commendation under any circumstances. Another author, in speaking of casting ridges together, and showing how it may be performed by plowing the furrow-slices of two adjoining ridges in opposite directions, gives the caution that "the inter-furrow, which lies between the two ridges, unavoidably leaves a shoulder or hollow place, of more or less width, according to the expertness of the plowman, in the center of the crown, which defect can only be completely relieved by replowing;"† and informs

* Low's Elements of Practical Agriculture.
† British Husbandry, vol. ii.

us that the defect may be partly prevented by using two plows of different mould-boards. I do not see why plowing two furrow-slices into the open furrow in casting should be more difficult or less sterling than in any other mode of plowing. A good plowman will leave in the crown of the ridge, in either case, neither a shoulder nor hollow place, which are certainly not synonymous terms, as they seem to be represented here, but the opposite.

(660.) Nearly allied to casting is a species of plowing called *two-out-and-two-in*, which can be executed on the flat ground, and requires a particular mode of feering. The first feering should be measured off of the breadth of two ridges, or 30 feet, from the ditch *a e*, fig. 132; and every subsequent feering should be measured at 4 ridges breadth, or 60 feet from the last. The land is plowed in this way. Let *a b*, fig. 137, be the

Fig. 137.

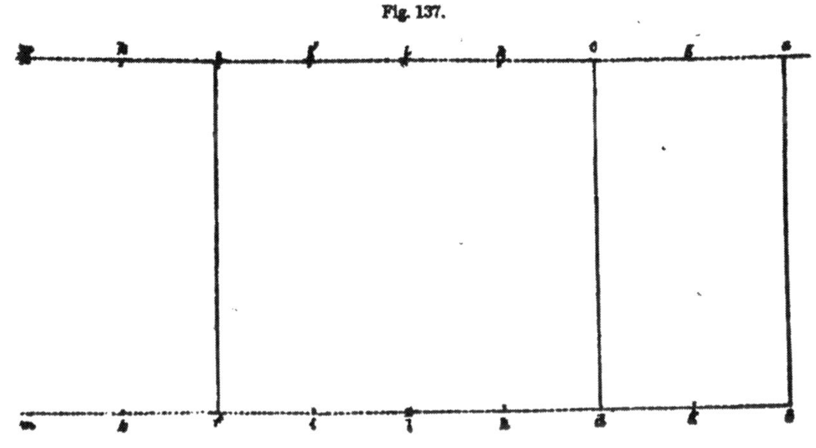

FEERING FOR PLOWING TWO-OUT-AND-TWO-IN.

side of the field, and let *c d* be the first feering of 30 feet from *a b*; and also, let *e f* be the next feering of 60 feet. After returning the feering furrow-slices, begin plowing round the feering *c d*, always keeping it on the right hand, and turning the horses from you, that is, *outward*, on both the head-ridges, until about the breadth of a ridge is plowed on each side of *c d*. These two ridges may be supposed to be represented by *c g* on the one side, and *c h* on the other of *c d*. While this is doing, the two ridges *e i* and *e k* are plowed, in like manner, toward *e f*. At this juncture, open furrows occur at *h h* and *i i*, embracing between them the breadth of 2 ridges, or 30 feet, from *h* to *i*. Then let the plowman who has plowed round *c d*, plow *h* and *i*, always laying the furrow-slices first to *h* and then to *i*, and turning his horses toward him, or *inward*, on both head-ridges, until the ground is all plowed to *l l*, which becomes the permanent open furrow. The next open furrow will be at *m m*, 60 feet or 4 ridges breadth from *l l*. But as yet only 3 ridges have been plowed betwixt *l* and *a*, the fourth ridge *g a* being plowed along with the head-ridges *m a* and *b m*, after all the ridges of the field have been plowed, laying its furrow-slices toward *g g*, and making the open furrow at *a b*. The effect of this mode of plowing is to lay all the furrow-slices in one direction from *a* to *c*, that is, across the 2 ridges *a g* and *g c*, and to lay those from *l* to *c* in the opposite direction, also across 2 ridges *l h* and *h c*, and both double ridges meeting in *c d*, which becomes the crown of the 4 ridges *l h*, *h c*, *g c*, and *a g*. In like manner all the furrow-slices over the ridges *l i* and *i e* on the one hand, and all those over the ridges *m k* and *k e* on the other hand, meet in their crown at *e f*. In plowing by this mode, every plowman takes in a feering of 4 ridges, which he completes before he goes to another. The reason, I suppose, that this mode of plowing has received the appellation of two-out-and-two-in is, that 2 ridges are plowed *toward* the feering and the other 2 *from* the open furrow.

(661.) The appearance of the ground on being plowed two-out-and-two-in is seen in fig. 138, where the space from *a* to *e* is 60 feet, comprehend-

Fig. 138.

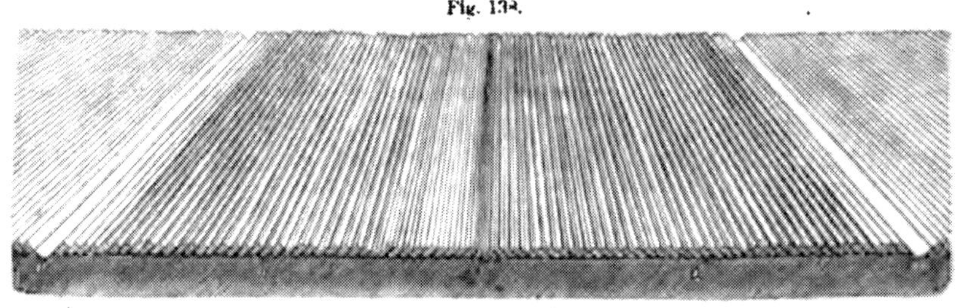

PLOWING TWO-OUT-AND-TWO-IN.

ing 4 ridges, between the open furrows *a* and *e*, 2 of which ridges, *a b* and *b c*, have their furrow-slices lying one way, toward the right, and the other 2, *e d* and *d c*, with theirs lying toward the left in the opposite way, both meeting at *c*. which is the crown of the whole break or division of 4 ridges.

(662.) This method of plowing places the land in large flat spaces, and as it dispenses with many open furrows, it is on this account only suitable for light soils, in which it may be practiced for seed-furrowing. It forms an excellent foundation for drilling upon for turnips, or even for potatoes upon gravelly soils.

(663.) The gore-furrow, described in fig. 136, might be judiciously applied to the plowing of land two-out-and-two-in on the stronger classes of soils; but its introduction changes the character of the ridges altogether, inasmuch as the crown *c* fig. 138, where the furrow-slices *meet*, is not only converted into an open furrow, but the actual crown is transferred from *c* to *b*, and *d*, where the furrow-slices do *not meet* from opposite directions, but lie across the crowns of their respective double ridges in the same direction. Exactly in a similar manner, when the gore-furrow is introduced into cast ridges, as in fig. 135, the crowns *b* and *d* are converted into open furrows, and transferred to *e*, where the furrow-slices lie across the crowns in the same direction on their respective ridges, instead of meeting there.

(664.) A nearly allied plowing to the last is that of *plowing in breaks* or *divisions*. It consists of making feerings at indefinite distances, and plowing large divisions of land without open furrows. Some farmers plow divisions of 8 ridges or 40 yards; but such a distance incurs considerable loss of time to travel from furrow to furrow at the landings. Instead, therefore, of distances of a given number of ridges being chosen, as is the case in the last mode of plowing, two-out-and-two-in, 30 yards are substituted, and this particular breadth answers another purpose of deviating from the sites of the ordinary ridges, which deviation has the advantage of loosening any hard land that may have been left untouched by the plow in any of the sorts of plowings that have yet been presented to your notice. Land is plowed in breaks only for temporary purposes, such as giving it a tender surface for seed-furrowing, or drilling up immediately. You might easily estimate how much time would be lost in plowing land in breaks, were the feerings made at a greater distance than 30 yards, by looking at fig. 137, where the feerings *c d* and *e f* being supposed to be

60 yards asunder, the plows would have to go round *c d* and *e f* until they reached *h* and *i* respectively, in doing which they would have to travel in a progressively increasing distance until its extreme point from *h* to *i* reached 30 yards *for every furrow-slice laid over.* Thus is imposed on men and horses a great deal of traveling for the little work actually done.

(665.) Another mode of plowing, which I shall now describe, is *twice-gathering-up.* Its effect may be seen by looking at fig. 139, where it will

Fig. 139.

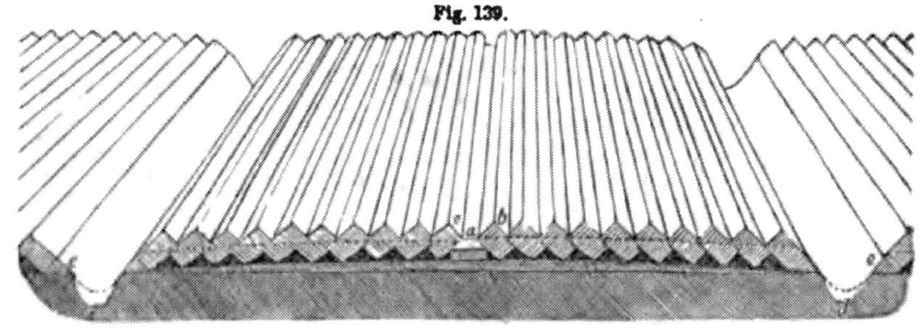

TWICE-GATHERING-UP.

be observed that the furrow-slices rest above the level line of the ground. It may be practiced both on lea and red-land. On red-land that has already been gathered up from the flat, it is begun by making feerings in the crowns of the ridges, as at *b*, fig. 133. The furrow-slices of the feerings are laid together, and the ridges plowed by ½ ridges, in the manner of gathering up from the flat. The ½ ridge left by the feerings at the sides of the field must be plowed by themselves, even at the risk of losing time, because it would not do to feer the first ridge so as to plow the ½ ridge as directed to be done in the first-gathering-up, in fig. 132, around the feering of the ¼ ridge *f c*, because the furrows betwixt *f* and *i*, if plowed in the contrary direction to what they were before, would again flatten the ground, whereas the furrow-slices from *e* to *f* and from *z* to *i*, being plowed in the same direction as formerly, the ground would thereby be raised above the level of *i f*, and disfigure the plowing of the whole ridge *z e*. Gathering up from the flat preserves the flatness of the ground; and the second gathering up would also preserve the land in a flat state, though more elevated, were there depth enough of soil, and the furrow-slices preserved of their proper form, as we have seen in (653), but a roundness is usually given to the ridge in all cases of gathering up furrow-slices toward the crowns, both by the harrowing down of the precipitous furrow-brows and the unequal size of the furrow-slices, from want of soil at the furrow-brows and open furrows. In gathering up lea the second time, no feering is required. The plow goes down a little to the left of the crown of the ridge, and lays over upon the crown a thin and narrow furrow-slice upon its back, as *a*, fig. 139, to serve as a cushion upon which to rest the adjoining furrow-slices. The horses are then turned sharp round from you, and the furrow-slice *b* is laid over so as to rest at the proper angle of 45° upon *a*. Turning the horses again sharp round from you, the furrow-slice *c* is also laid over at the same angle to rest on *a*, but neither *c* nor *b* should approach each other so nearly as to cover *a*, but leave a space of about 3 or 4 inches between them. The object of leaving this small space is to form a receptacle for the seed-corn, which, were *c* and *b* made to meet a sharp angle, would slide down, and leave the best part of the ridge bare of seed. The crown of the rest of the ridges is treated in the same manner, where in fact is constituted its feering, and the ridges are plowed in ½ ridges to the open furrows *d*, which are finished with mould-furrows, but the plowing of these is attended with some difficulty, in order to prevent their gradually mouldering down into the bottom of the open furrows. Twice-gathering-up is only practiced in strong land, and its object is to lift the mould above the cold and wet subsoil. On dry land, no such expedient is required. In fig. 139 the dotted line *e* is meant to represent the former configuration of the ground, and now it may be seen that the open furrow at *d* is deeper than it was with once gathering-up.

(666.) The mode of plowing exactly opposite to twice-gathering-up is that of *cleaving* or *throwing down* land. The open furrows of twice-gathered-up land constitute deep feerings, which are filled up with furrow-slices obtained from the mould-furrows and furrow-brows of each adjoining ridge; and, in order to fill them fully up, the plow should take as deep a hold of the furrow-brow as it can obtain. The furrow-slices are plowed exactly the reverse of those of the twice-gathered-up ridges, and they are also plowed in ½ ridges. The effect of cleaving down is to bring the ground again to the level above which it had been raised by means of the twice-gathering-up. The open furrows are left at the crowns, as at *a*, fig. 139, the mould-furrows of which are seldom stirred, as cleaving down is usually practiced to prepare the land for cross-plowing in the spring. But when heavy land is cleaved down in winter, it is always so with gore-furrows; and these, with open-furrows, afford convenient channels, at every half-ridge, for the water to flow off to the ditches. Since twice-gathering-up is only practiced on strong land, and cleaving down only succeeds

Fig. 140.

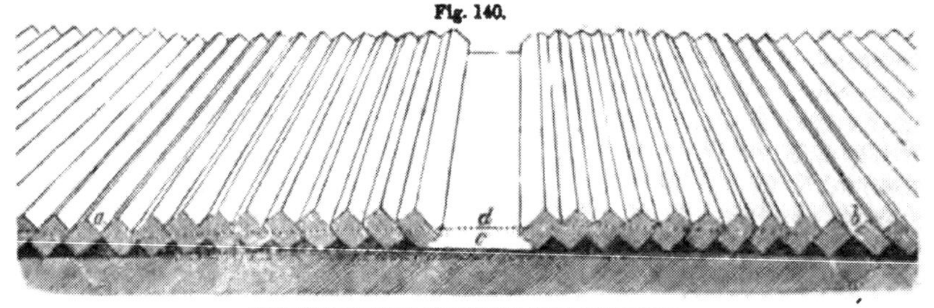

CLEAVING DOWN WITHOUT GORE-FURROWS.

twice-gathering-up, it follows that cleaving down is only practiced on heavy land. The effect produced by cleaving down ground may be seen in fig. 140, which represents it without gore-furrows *b* and mould-furrows *c*; but, in fig. 141, the gore-furrows are shown at *a*, and the open and

Fig. 141.

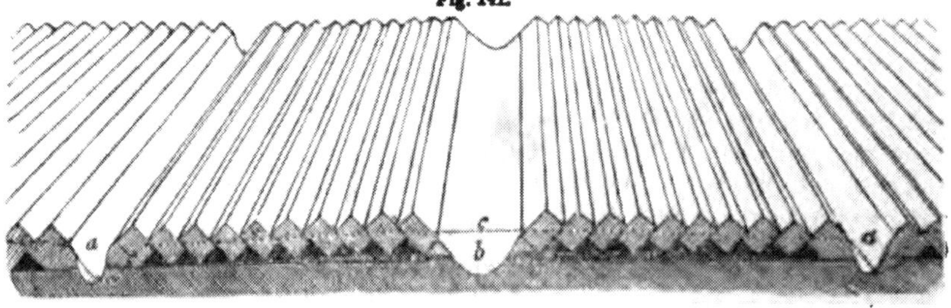

CLEAVING DOWN WITH GORE-FURROWS.

mould-furrows at *b*. The dotted line *d*, in fig. 140, represents the surface of the former state of the ground, as does the dotted line *c*, in fig. 141.—

Below *a* and *b*, fig. 140, are shown the former open furrows by the dotted line, as also does the dotted line below *a*, in fig. 141. In both figures, the ground upon which the furrow-slices rest is made somewhat rounded, to show the effect of twice-gathering it up. In the strict sense, a ridge can only be cleaved after it has been twice plowed. It is, as I think, scarcely correct to say that a ridge is cleaved after one gathering from the flat, for it is then plowed crown-and-furrow. With a strong furrow, a ridge that had been twice-gathered-up can be made flat by one cleaving.

(667.) What is called *cross-plowing*, or the *cross-furrow*, derives its name from plowing right across the furrow-slices as they lie in ridges, in whatever form those ridges may have been formerly plowed. Its object is to cut the existing furrow-slices into small pieces, so that the land may be more easily pulverized and prepared for the future crops. It is usually executed in the spring, and should never be attempted in winter, unless the weather continue so long dry and fresh as to allow the land to be again immediately plowed into ridges in any of the safe forms of plowing I have described. Rain, or snow melting on land lying in the cross-furrow, is at once absorbed and retained in it, and in a short time renders it sour. But, even if cross-furrowing were executed in a proper manner in winter, and the land thereafter safely put into ridges, the land would become so consolidated during winter that it would have to be again cross-furrowed in the spring before it could be rendered friable. The object of cross-furrowing being to pulverize land, it is practiced on every species of soil, and exactly in the same manner. It is plowed in divisions, the feerings being made at 30 yards asunder, and these are plowed in the same manner as two-out-and-two-in, by first going round the feering, turning the horses constantly from you, until about ½ of the division is plowed, and then turning them toward you, still laying the furrow-slices over toward the feerings, until your arrival at the open furrow. In cross-plowing, however, the open furrow is never left open, but is again closed by 2 or 3 of the last furrow-slices being returned, and all marks of it obliterated by the plow shoveling the loose soil into the furrows with the mould-board, which is purposely laid over on its side, and retained in that position by a firm hold of the large stilt. The obliteration of the open furrows is necessary to fill up the hollows that would otherwise be left by them across the ridges after they were formed.

(668.) Another mode of plowing, having a similar object to cross-furrowing—namely, of dividing furrow-slices into pieces—is what is called *angle-plowing* or *angle-furrowing;* and it is so named because the feerings in which it is plowed are made in a diagonal or angular direction across the field. This mode is also plowed in divisions of 30 yards each, and in exactly the same manner as cross-plowing, and with the same precautions as to the season in which it is executed, and the closing of the open furrows. It is never practiced but *after* cross-plowing, and not always then, but on strong land, or unless the cross-plowing has failed to produce its desired effect of comminuting and stirring up the soil. It is chiefly practiced in bare fallowing, and is therefore mostly confined to strong land.

(669.) I have mentioned a mode of plowing called *ribbing* (641). In its best form it is usually performed in spring with the small plow, when it will more appropriately fall to be described than in this place; but there is a species of plowing practiced in some parts of the country in autumn or winter which bears the name of ribbing. I notice it because it is practiced in some parts, not with the view of recommending but of reprobating it. It is, I believe, called *raftering* in England; and is practiced on stubble land, and consists of laying a furrow-slice on its back upon as much of the firm soil as it can cover, as seen in fig. 142, where *a* are the furrow-slices laid over upon the firm soil *b*, and *c* the plow-tracks. The figure represents it more compact, clean, and regular, than it is usually

Fig. 142.

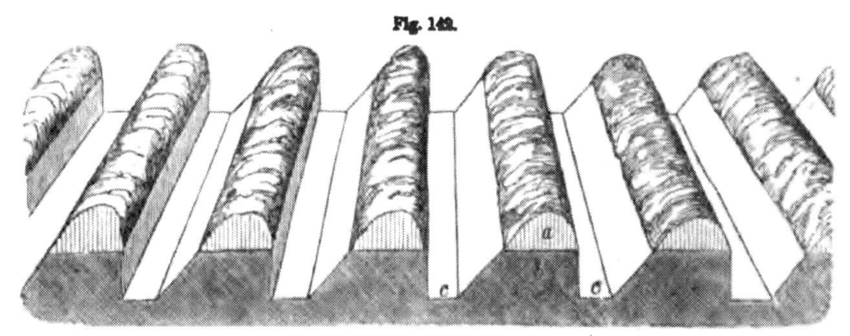

RIB-PLOWING STUBBLE LAND.

found in practice. It is sometimes plowed so as the furrow-slice shall lap and hang over the piece of firm soil upon which it rests, and the plow-tracks are often very crooked. The land lies in this state all winter, and is dry enough; but the greatest proportion of the soil remaining unplowed is none the better for this treatment. It has the advantage of being done in a short time, and without care, as it is generally done in a diagonal direction across the ridges, and without any sort of feering. It is chiefly practiced on land in a very foul state, with the view of destroying the weeds, from its being believed that the under surface of the furrow-slice, where the roots of the weeds are most abundant, is thus more exposed to the action of frost than in any other position; and this opinion is no doubt correct; but, if the plowed portion of the ground is in this manner more exposed to the air, it is evident that the unplowed part cannot be exposed at all, and, as the largest proportion of the land is left unplowed, any advantage attending such a mode of plowing must be greatly counterbalanced by its disadvantages. It is practiced in all sorts of soils. Its practice in Scotland is confined to the north of the Frith of Forth, and even there it is now abandoned on the large farms, though it may still be seen in the fields of the smaller tenants. When a field is so plowed, it has somewhat the appearance of having been drilled.

(670.) The *drilling* of land being confined to summer, I shall defer any remarks on that mode of plowing until its proper season arrives.

---

25. DRAINING.

> "In grounds, by art laid dry, the aqueous bane,
> That marred the wholesome herbs, is turned to use;
> And drains, while drawing noxious moisture off,
> Serve also to diffuse a due supply."
>
> GRAHAM.

(671.) It is barely possible that the farm on which you learn your profession, or the one you may occupy on your own account, may not require draining. Nevertheless, you should be made well acquainted with this essential and indispensable practice in husbandry. But the probability is that, on whatever farm you may pass your life, some part of it at least, if not the whole, will require draining of one sort or another.*

(672.) Draining may be *defined* the art of rendering land not only so free of moisture as that no superfluous water shall remain in it, but that no water shall remain in it so long as to injure or even retard the healthy growth of plants required for the use of man and beast.

(673.) On considering this definition, you may reasonably inquire why water in the soil should injure the growth of useful plants, since botanical physiologists tell us that the greatest bulk of the food of plants consists of, or at least is conveyed to them by water. In what way injury should arise, is certainly not very obvious; but observation has proved that *stag*

[* We have too frequently and too distinctly dwelt upon the importance of divesting land, whether intended for the plow or for grass, of all superfluous moisture, to make it necessary to say more on that point. In England, where the prodigious rental of land, and the value of its products as food for man and beast, instigate the farmer to search for and put in practice every possible mode of augmenting his crops, it is to the condition of the soil with respect to any redundant moisture that he gives his first attention, as we are assured, when examining with a view to rent or purchase.

There the conviction of the importance and the profits of draining, not merely as the basis of permanent improvement, but as a source of immediate remuneration for outlay of capital and labor, is universal; and one of the greatest obstacles at this time to the yet farther progress of agricultural improvement and prosperity in England is, on all hands, admitted to consist in the difficulty of adjusting between landlord and tenant the proportion to be borne by each respectively of the expenses of draining the land, which all admit to be indispensable to the realisation of the profits to which the industry and skill of the tenant entitle him. The landlord refuses to grant a long lease, being eager to get the higher rent which a higher state of improvement and productiveness may enable him to exact, while the tenant justly demands a longer lease to warrant him in the expenditures, especially the draining, by which, and his own skillful management, that higher improvement and greater productiveness are to be achieved.

In our country, the same obstacle to plans of improvement, which require time to carry them out and to reap their profits, does not exist, because here almost every man owns the fee simple of the land he cultivates; but here again there are two yet, perhaps, more formidable obstacles to every plan of well-arranged, permanent improvement, involving heavy outlay in the beginning. These are—want of capital, and the facility of procuring cheap land in the Far West. To these, also, may be added a want of enterprise and a want of knowledge. Our young men are not educated in the principles of their profession, as they are for all *other* professions. Hence, for example, in this very process of draining, some knowledge of geology and of the plain rules and laws of engineering and hydraulics is needed, to carry it forward upon any extended scale. But how many of our young men, except at the military schools (and these at the farmer's expense), are taught even the simplest elements of these sciences? Scarcely enough geology do they know to distinguish gravel from sand, and either from clay, and not enough of mechanics, hydraulics or engineering, to know how to make or to use a common water-level, to ascertain by what line water may be made to flow. If, then, the reader become impatient at seeing so much on one subject, and especially on perceiving that the execution of the plans laid down by the author would be expensive beyond the means of American farmers, there are yet some considerations to reconcile every reasonable mind. First—He should consider that this is but one chapter of, literally, a great BOOK OF THE FARM, so voluminous and comprehensive as necessarily to embrace many things not adapted to all farmers and all localities; and, secondly, that as this work is designed to be studied as well as read, and to serve as well for the young man in his school as for the practical man in the field, it is but fit that the *principles* as well as the *modus operandi* of all agricultural processes should be carefully explained; and we may add, after attentive perusal, that in respect of *draining*, both theory and practice seem here to be so fully described and carefully illustrated as to leave little or nothing to be added in the way of notes.

A long and very valuable essay appears in the last number of the English Farmer's Magazine, on draining as practiced in Aberdeenshire; but we see in it nothing that is not embraced in, if not actually taken from, the work in hand, except some results of particular experiments, and to illustrate the profits of the operation. Well aware are we that they are only our wealthy farmers, such

nant water, whether on the surface or under the ground, does injure the growth of all the useful plants. It perhaps altogether prevents or checks perspiration and introsusception; or it may neutralize the chemical decomposition of substances which largely supply the food of plants. Be it as it may, experience assures us that draining will prevent all these bad effects. You may conceive it quite possible for an obvious excess of water to injure useful plants; because you may have observed that excess of water is usually indicated by the presence, in number and luxuriance, of subaquatic plants, such as rushes (*Juncus acutiflorus* and *J. effusus*), &c., which only flourish where water is too abundant for other plants; and you may even conceive that damp, dark-looking spots in the soil may con-

as have generally acquired fortunes by other pursuits or by inheritance, who can afford such outlays—such outlays, however, it may be added, as will pay well when they are made under judicious circumstances as to locality and markets. How many thousands of acres are there, for instance, along the river and creek shores of the Chesapeake Bay, in Maryland and Virginia, enjoying the advantages of a *practical* contiguity to the best Atlantic markets, which now produce nothing but coarse grass and malaria, which might, if their owners had the energy, the skill and the capital to drain them, produce crops equal to such as have been produced on the Nansemond (Charles county, Md.) estate of B. O. Taylor, Esq.—where, during the past season, under the management of Mr. James K. Nash, a crop of 2,000 barrels of corn has been made and housed; and where, on *two acres*, 125 bushels of good merchantable corn were raised, and that after a fair crop of tobacco the preceding year, and without any unusual application of manure—being 62½ bushels; which, at the present price, 75 cents, is $46 87½ an acre.

Need we ask whether such crops, or say 60 bushels to the acre—at 70 cents a bushel, $42—would not well remunerate a heavy outlay for draining? And do not such crops and such management better deserve the honors bestowed by Agricultural Societies and Institutes, than mere isolated cases of extraordinary crops, produced at enormous expense, and by the fitful efforts to which the most indolent may sometimes be stimulated, in the hope of winning a V. or an X.?—But again the question recurs, how is the capital to be had by those who are without money, and whose lands might be thus drained?—for the tendency of all our legislation that in any way interferes with or acts upon the use and value of labor and capital, owing to the want of spirit and knowledge on the part of the agriculturists, has a tendency to drive capital as well as enterprise into our large cities, and into channels for the benefit of particular and *well-banded classes!*

Suppose a field of say ten acres (and how many there are who own such) to be drained so that it would produce even forty bushels of corn to the acre—these, at 60 cents, would give $24, the interest on $400 at 6 per cent.; leaving the land worth intrinsically more than $100 an acre, and two crops would more than pay the cost. Yet there are thousands on thousands of acres in Maryland and Virginia and North Carolina that might be thus drained and rendered thus productive at an expense of $50 an acre!

But unfortunately if, perchance, a hard-working farmer, toiling through the year, and conducting his household and all other affairs with the severest economy, contrives to have a small surplus beyond his outgoings at the end of the year—even he, instead of investing it in draining his land—in erecting more and better buildings—in providing himself with all necessary implements, and otherwise improving his little estate in appearance and fruitfulness—even he, as such a farmer confessed to us last summer in the State of New-York, sends or goes with it to the *great, all-absorbing Commercial Emporium, to lend it out at 7 per cent. on bond and mortgage.* Hence it is that while no one knows how much of the State is mortgaged, and while every manufacturing village wears the aspect of youth and healthfulness, and larger cities grow into enormous magnitude—the *country*, at least along the great thoroughfares, as far as one can see, sinks lower and lower in its average products, and puts on the appearance of premature decay and consumption. All this would be rectified if the education of the rising generation were such as it should be, because then we should rear up a race of men who would understand, as far as the Government is concerned, what is due to their numbers and the products of their industry; and Agriculture would preside where she ought, at the head instead of at the tail of all Cabinet councils for the general welfare.

Looking carefully through this chapter of the Book of the Farm, on *draining*—the mode of conducting it—the principles by which it is to be guided—the implements employed, and the out-

tain as much water as to injure plants, though in a less degree; but you cannot at once imagine why land *apparently* dry should require draining. Land, however, though it does not contain such a superabundance of water as to obstruct arable culture, may nevertheless, by its inherent wetness, prevent or retard the luxuriant growth of useful plants, as much as decidedly wet land. The truth is, that deficiency of crops on apparently dry land is frequently attributed to unskillful husbandry, when it really arises from the baleful influence of *concealed* stagnant water; and the want of skill is shown, not so much in the management of the arable culture of the land as in neglecting to remove the true cause of the deficiency of the crop, namely, the concealed stagnant water. Indeed, my opinion is—and its conviction has been forced upon me by dint of long

---

lay, under ordinary circumstances, as compared with the results in the account of profit and loss—we see nothing that can be added, except some suggestions as to draining, in a mode and with materials more within the reach of ordinary American farmers, and which appears not to be touched upon in this book. At this there should be the less surprise when we consider that *wood*, or young timber—the material to which we refer—is so scarce and dear in England as almost to place it out of question.

In the Eastern States, and along the high lands stretching far away south, stone is at hand in abundance; and in New-York, and so through the States east of the Hudson, the practice of draining is on the increase, and we believe with universal satisfaction as to the advantages; but in a large portion of intervale and alluvial country south of the Hudson, with which we are better acquainted, there is not stone enough in some districts to make the foundation of a chimney. The situation and circumstances are numerous, however, where farmers are invited by every consideration to drain, with *wood* (if nothing better), which many of them have at command. We believe we have somewhere before adverted to the instance of a Mr. SUMMERS, near Nottingham, Maryland, who, without example or instruction, has reclaimed and brought under the plow spots of land which had before been worthless, and which now yield heavy crops; and which, moreover, give to his fields an air of good management, proclaiming that their proprietor has a *mind to think*, as well as a hand to execute—a consideration which ought to weigh with every farmer of proper spirit, and which it behooves every community to commend more fervently than they now commend only, or first and far above all others, those who are most successful in fields of human destruction.

For those who may be incline[illegible] and there ought to be thousands—to drain the boggy spots on their farms by this method, which [illegible]s within the reach of almost every one, the following observations may prove useful; and the reader will indulge us in this short addition to a long note, when we announce that we shall make no other upon this branch of the Book of the Farm.

"The wood used for this purpose consists of the thinnings of plantations, *i. e.* the small trees commonly converted into paling. Larch is preferable, on account of its greater durability; but Scotch fir, being the cheapest and most abundant kind in this quarter, is generally used. The drains to be filled with wood are usually thirty-two inches in depth, eighteen inches wide at the top, and about six inches at the bottom. It is essential to the efficiency and durability of wooden drains, that the sides be formed with a proper and regular slope from top to bottom. The small trees—or "spars," as they are designated—are prepared for being put into the drain in the following manner: A portion of the butt or thick end of each is sawn off for placing transversely in the drain, about six inches above the bottom; the breadth of the drain at this part may be assumed at nine inches, in which case the length of the cross-bars will require to be about fifteen inches, so as to have three inches resting on each side. They are generally about four inches in diameter, and are placed in the drains at intervals of four feet apart; they are forced firmly into their proper position by a few blows of a heavy mallet, the workman taking care that they are all in the same plane or level. Any earth loosened from the sides in striking down the bars is, of course, thrown out as the work is proceeded with. After the butt-ends of the trees (which are divested of their branches in the wood) are severed, and placed transversely in the drains in the manner just described, the remainder of them are laid longitudinally above the bars, three being commonly placed side by side, and covered with the branches and twigs, or with turf, heath, &c., previous to putting in the earth cast out in opening the drains. It is obvious that this method of draining can be adopted with advantage only in situations where timber is convenient and cheap, and when the subsoil is sufficiently cohesive to afford a proper support to the transverse bars of wood: hence it is inadmissible in the case of boggy lands. The putting in of the wood is accomplished in a very expeditious manner: two persons saw off the butts, and another places them in their proper position in the drain, after which the longitudinal spars are laid on as closely as possible, with the top and butt-ends alternately in the same direction, so as to make them fit the bet-

and extensive observation of the state of the agricultural soil over a large portion of the country—that this is the *true cause of most of the bad farming to be seen*, and that *not one farm* is to be found throughout the kingdom that would *not be much the better for draining*. Entertaining this opinion, you will not be surprised at my urging upon you to practice draining, and of lingering at some length on the subject, that I may exhibit to you the various modes of doing it, according to the peculiar circumstances in which your farm may be placed.

(674.) To the experienced eye, there is little difficulty in ascertaining the particular parts of fields which are more affected than others by superfluous water. They may be detected under whatever kind of crop the field may bear at the time; for the peculiar state of the crop in those

ter. There is thus formed beneath the wood a channel for the passage of water, of about six inches in width and the same in depth.

"The cost of this mode of draining obviously depends much on the price of the wood employed. In most parts of this county the spars used for the purpose are obtainable at from 1s. to 1s. 6d. per dozen; and it requires four dozen, averaging twenty feet in length, to do a hundred yards of drain. Drains thus constructed have been known to last for a very long period; on one farm the writer has been assured that drains formed of wood in the manner just described have been in perfect operation for more than thirty years."

In a recent conversation with Mr. HALL, a clear-headed, practical farmer, near Lebanon, N. Y., he remarked that the practice of draining was extending with himself and neighbors, and, where well executed, was attended with double the amount of produce previously obtained. We are indebted to a valued correspondent (one of his young and promising pupils) for the following note on the system of draining pursued by Mr. WILKINSON, of Dutchess county, N. Y.—one of those farmers whom we are pleased to name as among those who bring their minds to reason and reflect on the principles involved in all their agricultural proceedings.

From a Correspondent.

DUTCHESS AGRICULTURAL INSTITUTE, Aug. 29, 1846.

Dear Sir: Having compared the system of under-draining practiced by Mr. Wilkinson, on the Institute Farm, with that recommended in "Stephens's Book of the Farm," and also with that of many farmers of this State—and being thoroughly convinced that his system is much better adapted to the wants of the farmers of this country generally, on account of the convenience of construction, cheapness, utility, &c—I have been induced to give you this article for publication in your Journal. The course he pursues is as follows: In the first place, in laying out the ditches, he is careful to avoid too great a descent, that the force of the water running in the ditch may not wash the sides or bottom of it, and thus carry the sediment to some more level portion, where it may collect, and obstruct the water, and render the work valueless.—After the ditch is staked out, he plows with a common plow the width required and two furrows in depth, shoveling out the loose earth each plowing. Then a pair of strong oxen are attached to a one-handled subsoil plow, tandem, in single yokes, by which the ditch is plowed to the depth required, which is varied according to circumstances; but the usual depth is from 2½ to 3 feet, about 20 inches wide at the top, and 1 foot at the bottom. He then commences the filling operation, by drawing small stones from where they have been previously heaped for the occasion, or the scattering small stones from the adjacent lots. The stones are drawn by oxen on a low, four-wheeled vehicle, which he calls a truck, or lumber-car, which is about 20 inches high—being a convenient hight for a man to stand beside it on the bank of the ditch, and take the stones from the car and pass them to a man in the ditch, who places them by first paving the bottom with those of a medium size, by setting them on the small end as closely as is convenient, selecting the flat ones, or flags, and placing them next to the bank, to prevent it from caving. The largest are next used and thrown in promiscuously, taking great care not to throw them against the sides of the ditch. The stone-work is then completed by leveling with the smallest ones that can be procured, not filling it nearer than 1 foot from the surface, that the ground may be plowed and subsoiled as well over the ditches as elsewhere. Before filling in the earth, the stones are covered with pine shavings very lightly, which he obtains from the sash-factory for the cartage.

The advantage in the use of shavings over straw, which is generally used, is that the straw entices the ground-moles into the ditch; they burrow in the stones, and dig through to the surface, which forms a passage for surface-water directly among the stones, and will carry with it quantities of fine mould; thus, by frequent rains, the whole work is rendered valueless.

The filling in of the earth is done with a side-hill plow, the team traveling up and down on the bank of earth, plowing a furrow into the ditch—which, Mr. W. says, facilitates the work 300 per cent. over the old process of shoveling it in.

Where the stones are convenient for filling, and taking the average of soil for digging, he thinks his ditches of the above dimensions, completed, cost about 35 cents per rod of 16½ feet

parts, when compared with the others, assists in determining the point. There is a want of vigor in the plants; their color is not of a healthy hue; their parts do not become sufficiently developed; the plants are evidently retarded in their progress to maturity; and the soil upon which they grow feels inelastic, or saddened under the tread of the foot. There is no mistaking these symptoms when once observed. They are exhibited more obviously by the grain and green crops, than by the sown grasses. In *old* pasture, the coarse, hard, uninviting appearance of the herbage is quite a sufficient indication of the moistened state of the soil.

(675.) But there appearances of moistened land, which you may easily observe without any previous tuition; and these are most apparent in soil after it has been plowed, and more apparent still in spring, in the month of March, when the winds become dry and keen. Then you may observe, in a dry day, large patches or stripes, or belts of black or dark-brown colored soil, in the face or near the top of an acclivity, while the rest of the field *seems* quite dry, of a light brown color; or only small spots may be observable here and there; or the flat and hollow parts of the field may be nearly covered with dark-colored soil. You cannot mistake these broad hints of the lurking water below; but, in a few weeks, they may all have disappeared, or be reduced very much in extent, if the weather continues dry, or have become more extended in rainy weather. In the case of their disappearance in dry weather, you may conclude that any wetness of the soil which passes off as the summer advances, can do no harm to cultivated plants, and that the land, in such cases, *does not require to be drained.* Such a conclusion would be very erroneous: because it is on account of the water *remaining in the soil all winter* that the crops receive injury in summer. The amount of *wetness* which you saw pass away first in spring and then in summer, would have done no injury to the crops, for it would be all absorbed, and probably more, in the wants of vegetation; but the wetness remaining in and occupying the pores of the soil and of the subsoil all winter, render the soil so cold, that most of the summer's heat is required to evaporate the superfluous moisture out of it, and, in this very process of drying by evaporation, the heat is dissipated that should be employed in nourishing the crops all summer. No doubt, when the soil and subsoil are put into such a state as that the water that falls upon the soil from the heavens during the winter, on being conveyed quickly away in drains, does take away some of the heat from the soil, but it cannot render it cold or sour. In such circumstances, the natural heat of the weather in spring and summer would have nothing to do but to push forward the growth of the crops to early maturity, to fill them more fully, and make them of finer quality. You thus see how concealed water injures the soil in which it is retained, and you may easily conceive how it may injure the drier soil around it, by its imbibing the water in contact, by capillary attraction. You thus also see the kind of ground that draining effects in soil so situated. Did the symptoms of wet in spring remain as obvious to your senses throughout the summer, you would have no doubt of the land requiring draining; but you may now admit that you may be deceived by land showing even favorable symptoms of drouth. For all that you yet know to the contrary, water may be lurking under what you imagine to be dry soil. Yes, and it does lurk to a very great extent in this country, and will continue to lurk in humid localities and impervious subsoils, until a vent is given to its egress.

(676.) The phenomenon of the dark spots on fields can be satisfactorily explained. Where the surface of the land is at all permeable to water, and where it rests on beds of different depths, of various lengths and breadths, and of different consistence, the water supplied from rain or snow is interrupted in its progress by the retentive beds, and becomes accumulated in them in larger or smaller quantities, according to their form and capacity; and, at length, the superfluous portion is poured from the surcharged strata, and bursts over retentive beds through the surface-soil in the form of land-springs, at a somewhat lower level. Such springs are either concentrated in one place or diffused over a large extent of surface, according as their outlet happens to be extensive or confined, and deep draining is generally required to remove these; for which purpose, deep drains are cut through alternate beds of retentive and permeable matter, and penetrate into the very seats of the springs. It may happen, however, that the surface is as retentive as the subsoil, in which case the water, not penetrating farther than the surface-soil, has a free enough passage between the impervious subsoil and the loose soil; this state of soil requires mere surface-draining. Where the upper soil is pervious, and the subsoil uniformly and extensively retentive, water accumulates on the subsoil, to the injury of plants growing on the surface-soil; and to remove water from such a situation, not deep but numerous drains are required to give sufficient opportunities for it to pass away, and such drains are usually formed in the furrows. Where the soil and subsoil are both porous, the water passes quickly through them, and no draining is required to assist it in flowing away, as the entire subsoil constitutes a universal drain. In this state of soil, water is only held in it by capillary attraction, and what is not so supported sinks down through the porous subsoil by its own gravity. Capillary attraction is quite capable of supporting and bringing as much water through a permeable soil and subsoil, from rain above and sources of water from below, as is useful to vegetation, excepting, perhaps, under the extraordinary occurrence of excessive drouth; and of all the sources from which the soil derives its supplies of water, that from springs is the coldest, most injurious to useful plants, and most permanent in its effects; and hence it is that the abstraction of water from the soil by draining does not necessarily interfere with it as a supporter of plants, as a meliorator of the soil, as a menstruum for the food, as a regulator of temperature to plants.

(677.) These states of water in the soil and subsoil indicate that a knowledge of geology might confer a more perfect understanding of the principles of draining; and, fortunately, practice in this department of rural economy has always been consistent with the facts of geology. But a geological drainer is a character who has not yet made his appearance in the world; because no practical drainer or scientific geologist has yet explored that department of geology which is most useful to Agriculture, in such a manner as to assist the art of draining. Most of our arable soils are contained within the newest rock formations, the intricate relations of which present almost insurmountable obstacles to such a knowledge of them as to be useful in draining. The intricacy of their relations render the operations of draining uncertain; and this uncertainty, I fear, must continue to exist, until the relations of the alluvial rocks are discovered to be as unvarying as those of the more indurated. Perhaps a certainty in the matter is unattainable, because the members of the alluvial formation may not present a strictly relative position to one another. Until the fact, therefore, is ascertained one way or the other, draining must be conducted, in a great measure, by trial or experiment; and in all undertakings on trial, error must be expected to ensue, and unnecessary expense incurred. An unfortunate circumstance, arising from this uncertainty, is the comparative uselessness of the experience acquired in previous operations to guide the drainer himself and others, to the means of securing more certain results in their future efforts at draining. No drainer can affirm that the number and depth, and even the direction, of the drains which he chooses to adopt, are the best suited for drying the field he wishes to drain; nor can he maintain that exactly similar arrangements will produce exactly similar effects in the adjoining or in any other field, at a greater or shorter distance. Every experienced drainer will coincide with the justness of these remarks, and deplore the uncertain nature of his operations; but, nevertheless, the satisfactory consolation is, that as long as he finds draining, even as it is pursued, do good, so long he will continue to practice it. Were geologists to make themselves acquainted with the practical details of draining, and then study that branch of geology which would be of greatest service to draining, it is reasonable to hope that they would confer lasting obligations on the drainer, not only by directing him to a well-grounded certainty in his object, but by showing him how to execute his art with greater simplicity. Were they also to direct particular attention to the relation that subsists, if any, between the surface of the earth's crust and the strata immediately subjacent, their investigations might supply valuable materials for a correct nomenclature and classification of soils.

(678.) You thus perceive that a bare recital of the various modes of draining is not alone sufficient to make you an accomplished drainer; for you should know the principles as well as the practice of the art. The principles can enly be acquired by a knowledge of geology, in as far as it has investigated the structure of the alluvial rocks, which are within your reach everywhere, and entirely within your power on your own farm to investigate. This knowledge, even as it is yet known, is requisite; for any difficulty in draining is found not so much in constructing a drain—most field-laborers can do that—as in knowing *where* to construct it; and a correct knowledge of whether the wetness in the land arises from natural springs or from stagnant water under the surface of the soil, can alone direct you to open the kind of drain required. So generally is the practical part of the operation diffused, that every manager of land conceives he knows the whole subject of draining so correctly, that he will commence his operations with the utmost confidence of success; and this confidence has caused much money to be expended in draining, that has in great part been ill directed; not but that its expenditure has done good, but that it has not done nearly all the good that the means employed might have effected. Much money has thus been expended in many places in making a few scattered deep drains, where a greater number of smaller ones would have answered the purpose much better. A degree of success, however, has attended every attempt at draining, and it is this circumstance, more than any other, that has beguiled many into a belief that they are accomplished drainers; for no one, unfit to direct the operation in a proper manner, would have attempted it at all, unless he had actually experienced injury from wet land; or have attempted it again, unless his attempts had partially, at least, removed the injury; though the results have not been very successful. Were the efforts of ignorance in draining confined to the squandering of money, they might be compensated for by superior management in the other operations of the farm; but, unfortunately, the sinking of valuable capital in injudicious draining cripples the means of the farmer, and at the same time prevents his reaping all the advantages derivable from draining itself. Were draining an operation that could be executed at little cost and trouble, it would be of less importance to urge its prosecution in the most effectual way; but as it is an expensive operation, when conducted in the most economical manner, much consideration should be given to the matter in all its bearings, before attempting to break up ground for draining to any great extent. An examination of the earth's crust, upon which you are to operate, is absolutely necessary to direct your plans aright. Contemplate well, in the first place, the facts which such an examination unfolds to your view, and endeavor by their nature to acquire wisdom to expend your money with prudence as well as skill. Examinations of the soil and subsoil will tell you what kinds require deep draining, and what kinds may be treated with equal success under a different arrangement. Inattention to such distinctions as these has hitherto caused the inordinate application of one general principle, which, as applicable to a particular system, must receive the assent of every drainer who feels the importance of the art, but which, nevertheless, is inapplicable to every case—I mean the system of deep draining.

(679.) You may have observed, from what has been said, that there is more than one species of draining; there is one which draws off large bodies of water, collected from the discharge of springs in isolated portions of ground; and this is called *deep* or *under-draining*, because it intercepts the passage of water at a considerable depth under the surface of the ground; and there is another kind which absorbs, by means of numerous channels, the superabundant water spread over extensive pieces of ground under the surface, and has been called *surface-draining*. This latter kind of draining subdivides itself into two varieties, the one consisting of small open channels formed on the surface of the ground in various directions for the ready use of water flowing upon the land, and this is literally *surface-draining*. The other is effected by means of small drains constructed at small depths in the ground, at short distances from one another, and into which the water as it falls upon the surface finds its way by its own gravity through the loose soil, and by which it is discharged into a convenient receptacle. But for those two species of surface-drains, the water that falls from above would remain stagnant upon the retentive subsoil at the bottom of the plow-furrow. The former kind of surface-draining is called *gaw*-cutting, so named from its resemblance to "a mark or crack left in the soil by a stroke or pressure;"* the latter kind derives its name either from the locality which it occupies, or the arrangement of its lines. From its local position, it has been called *furrow-draining* when it occupies the open furrows of the ridges of a field, though it is not necessary that such drains should always occupy the furrows. It has also been called *frequent-draining*, from the circumstance of the water finding frequent opportunities of escape; but this name, though the original one, is objectionable, inasmuch as the word may imply that the field requires draining frequently, which it certainly will not. From the arrangement of its lines, it has also been denominated *parallel-draining*, on account of the usual parallel position of the drains to one another; and yet it is not absolutely necessary to success that they shall be parallel to one another. As by this kind of draining the land is thoroughly or effectually drained, it has been most appropriately called *thorough-draining;* and this term, as a nomenclature, has the advantage of not committing the drainer to the adoption of any particular form or position of drain, but only to that form or position which renders land thoroughly dry. There are various other modes of draining, such as *wedge-draining*, *plug-draining*, *mole-draining*, each of which will receive consideration in due course.

* See Jamieson's Scottish Dictionary. *Gaw.*

(680.) The most superficial mode of draining is that effected by *open ditches* and *gaw-cuts*, into which the surface-water flows, and is carried off to a distance to some river or lake. This mode of draining does not profess to interfere with any water that exists under the surface of the ground, farther than what percolates through the plowed furrow-slices, and makes its way into the open furrows of the ridges. For the purpose of facilitating the descent of water into the open furrows, the ridges are kept in a bold, rounded form; and that the open furrows may be suitable channels for water, they are carefully water-furrowed, that is, cleared out with the plow after the land has been otherwise finished off with a crop. The *gaw-cuts*, small channels cut with the spade, are carefully made through every natural hollow of the ground, however slight each one may be, and the water-furrows cleared into them at the points of intersection. The gaw-cuts are continued along the lowest head-ridge furrow, and cut across the hollowest parts of the head-ridge into the adjacent open ditch. The recipient ditch forms an important component part of this system of draining, by conveying away the collected waters of the field of which it forms the boundary, and for that purpose is made as much as 4 or 5 feet in depth, with a proportional width. It is immediately connected with a larger open ditch, which discharges the accumulated waters from a number of recipient ditches into the river or lake, or other receptacle which is taken advantage of for the purpose. The large ditch is from 6 to 10 feet in depth, with a proportional width, and, when conveying a full body of water in winter, appears like a small canal. It is evident, from this description, that this is a system of pure *surface*-drainage, and is only applicable to soils that retain water for a long time on the surface, that is, on very tenacious clays; and, accordingly, it is extensively practiced in such districts as the Carse of Gowrie, where it has been so for a very long period. The large ditches there are called *pows*, which literally mean *mires*. The plowmen of the Carse are accustomed to the spade, and are yearly employed, in the proper season, in scouring out the smaller ditches, the larger ones being only scoured occasionally. Whenever a heavy fall of rain occurs in winter or spring, they are employed in clearing out the gaws, and directing the water as fast as possible off the land along the furrows. This operation is a necessary precaution in wet weather upon strong clay land, but it constitutes a very imperfect system of *draining*, and sacrifices a large extent of good surface-soil. It would be better, I think, if the Carse farmers were generally to try the effect of covered drains, which would absorb and carry away surplus water equally well as open ones, and save much time in scouring ditches, besides putting the soil into a fitter state to be worked at any season than it can be done under the present system.

(681.) The *drains* which our *forefathers made in loamy soils, resting on a retentive bottom*, were placed upon the subsoil immediately under the upper soil, where that was deeper than the plow-furrow; but as the arable portion of the soil, when it is quite of a different nature from the subsoil, is never very thick, the drains were necessarily placed at a small depth; and, the cut being so, experience would soon teach drainers the impropriety of placing the materials which are used to fill a drain within reach of the plow, which consisted of very few stones, often not exceeding three, and those not of large size—one being placed on each side of the cut, and another above them, forming a sort of conduit. These conduits being not far from the surface, of small area, and not very numerous in any one place, a small addition of water would carry, and the moles would force a little earth into them, sufficient to obstruct the flow of water in them; and, of course, any drain in that state would produce the very mischief it was intended to remedy. Such paltry drains have evidently been formed on the notion that a simple conduit, placed between a porous soil and retentive subsoil, is sufficient to render the soil permanently dry—a notion the fallacy of which the drainers of the present day are well aware. I have met with several such drains in the course of my draining operations, and they were completely choked up; but on being opened, by the cutting of the new drains, clear water flowed out of them for a considerable time. They were all beyond the reach of the plow, in the manner in which the land had been plowed from time immemorial; but the plowing had consisted of a mere skimming of 4 inches of the soil, and on this account the black mould immediately under the plow-track had been compressed by the sole of the plow into a thin, slaty crust, under which the fine black virgin mould remained untouched, while the plowed surface had become an effete powder by constant cropping.

(682.) Compared with this trifling method, the system of *under* or *deep draining*, being the deepest method of any, is super-excellent. It is technically called Elkington's method, because it was first proposed and practiced by Mr. Joseph Elkington, Princethorp, a farmer in Warwickshire so long ago as 1764. It is related that he discovered the mode of draining, which has since borne his name, by accident. His fields being very wet, and rotting many of his sheep, he dug a trench 4 or 5 feet deep, with the view of discovering the cause of the wetness. While he was deliberating what was to be done, a servant passed by chance with an iron crowbar for fixing sheep-hurdles with in the ground. Having a suspicion that the drain was not deep enough, and desirous to know the nature of the materials under it, he forced the bar 4 feet below the bottom of the trench, and, on pulling it out, to his astonishment, a great quantity of water welled up through the hole it made, and ran along the drain. He was led to infer from this that large bodies of water are pent up in the bowels of the earth, and are constantly injuring the surface soil, but which may be let off by tapping with an auger or rod. This discovery produced a great sensation at the time, and, in fact, introduced a complete revolution in this country in the art of draining. It served to establish draining on correct principles. It was as much more effective a method than the old system, in changing the quality of the soil, as blood-letting from a vein by the lancet affects the constitution in a greater degree than the topical application of leeches. But this method soon underwent modifications in practice.—Casting a drain and tapping with an auger to catch the spring or bed of water, as in the principle of Artesian wells, was the original plan; but when it was found that water did not in every case follow the auger, as it would not when disseminated through a mass of earth, and not subjected to altitudinal pressure, a modification of the plan was inevitable. It was then attempted to run deep trenches through the lowest part of the damp soil to the highest point where the supply of water was supposed to be, or where it made its appearance, and lead it away as it collected by percolation through the soil and subsoil. In order to embrace the whole damp soil of any locality in the drainage, lateral branches were projected on both sides of the main branch, as far as the apparent dampness extended; and, not to omit the smallest extent of the damp soil, tributary branches were sent off to short distances from the lateral ones. The different branches were made of different sizes, according to the quantity of water which each was supposed to have to convey away. This plan of drains, when projected on the surface, looks like the trunk and branches of a tree in winter deprived of its leaves, and it might therefore be called the ramified or dendritic form. This is the plan that has been very extensively pursued since

Elkington's time until about 1824, since which another system has obtained the preference. Many thousands of acres of land had been drained by that method up to that time, and there is no doubt that the country has derived much benefit from the system. I may mention the fact, as an incentive to important discoveries, and as an instance of disinterestedness, that Elkington willingly communicated all his practice to the late Mr. John Johnstone, the eminent drainer, at the request of the Board of Agriculture, through whose influence the British Parliament voted him a reward of £1,000.*

(683.) It will much facilitate your conception of this system of draining, if we consider, in the first place, the *source from which the water that mars the cultivated soil is derived*. When water is evaporated by heat from the sea and land, and conveyed in vapor into an elevated part of the atmosphere, and there retained in an invisible form by the agency of electricity, it remains in that state until a change takes place in the electric equilibrium, when the vapor becomes visible in the form of clouds, which, then becoming independent bodies, become at the same time subject to the laws of physical attraction. Being attracted by the mountains, which are the highest features of the terrestrial portion of the globe, they come into contact with them, give out part of their caloric to them, and, ultimately dissolving, descend upon them in the shape of fog, or rain, or snow. Hence, as you have already seen (324), rain falls much more plentifully upon the mountains than the plains. The rain, as it falls upon the mountains, is absorbed at once by the soil which covers them, and, when it cannot contain any more, the surplus water flows away, and forms streams and rivers.—The portion of water retained by the mountain soil undergoes a very different fate. It is conveyed by its own gravity chiefly, and partly by their capillary attraction, among the mineral strata of which the mountain mass is composed, and continues to seek its way through them until, reaching a point beyond which it meets with no resistance, it comes forth to the day in the shape of a strong spring or springs, or, diffused over the whole surface of the mineral mass, it spreads over a large extent of ground. These different destinations of the same water are occasioned by a difference in the nature and positions of the geological formations of the mountain mass. For example, if the whole rising ground, from its base to the summit, is spread over with a *saddle-shaped covering of tenacious clay*, the water will slide down its face, under the vegetable coating of the surface, as far as the clay descends, which may be to the plain below. This vegetable covering will be permanent grass, if the elevation of the ground is not great; or it will be heath and mosses, if the elevation exceed such a hight as that the mean annual temperature of the air around it does not exceed 40° Fahrenheit; or it may be mould capable of supporting cultivated crops. Thus, in fig. 143, *a* is the clay over the hill; if rain fall on *a*, it will descend on

Fig. 143.

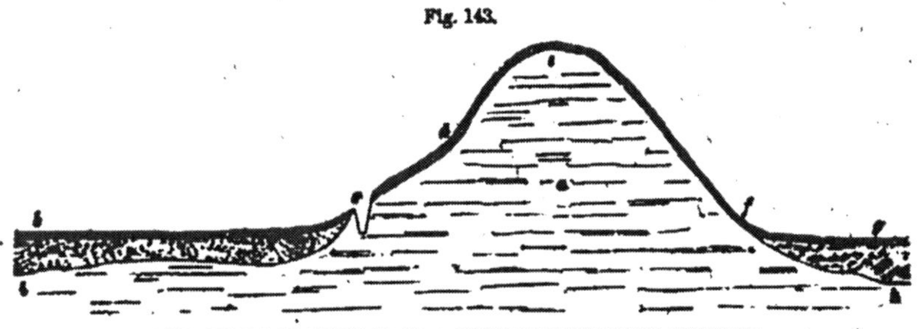

THE ORIGIN OF SPRINGS ON A UNIFORM TENACIOUS SURFACE.

* See Sinclair's Code of Agriculture, notes.

the one side to *d* and on the other to *f*, and, if the temperature of this region is under 40°, *d e f* will form the region of heath and mosses. The water will still pass from *d* to *c*, which is the region of permanent pasture, and it will continue to flow to the plains, to *i* under *c b*, and to *h* under *f g*. The vegetable mould may be traced from *b* by *c*, *d*, *e* and *f*, to *g*. Should the subsoil between the mould *b* and the clay *i* be also retentive, then the water will appear at the surface at *c*, and affect all the space from *c* to *b*; but, should the subsoil from *h* to *g* be porous, then the water will continue to flow from *f* upon the clay *h*, and not affect the surface-mould *f g*.

(684.) Should the mountain, however, consist of *concentric layers of different rocks arranged mantle-shaped around it*, then the water will descend between the lines of junction of the rocks; and should the masses or beds of rock be of different extents, and thickness, and consistence, which is probable, then the water will either appear at the surface of the ground as a spring, from the subjacent rock of a close texture, or it will descend yet lower, and be absorbed by the subjacent rock of a porous texture. In this manner, the harder rocks cause the springs to appear at a high elevation, while the porous ones convey the water to a lower level, until it meets with a resisting substance to cause it to come to the day. In any case the farmer cannot do anything until the water indicates its presence on the surface of the ground, either at a high or low elevation; and then he should take measures accordingly to remove it.

(685.) To illustrate the cases now alluded to, suppose fig. 144 to represent a hill composed of different rocks of different consistence. Suppose

Fig. 144.

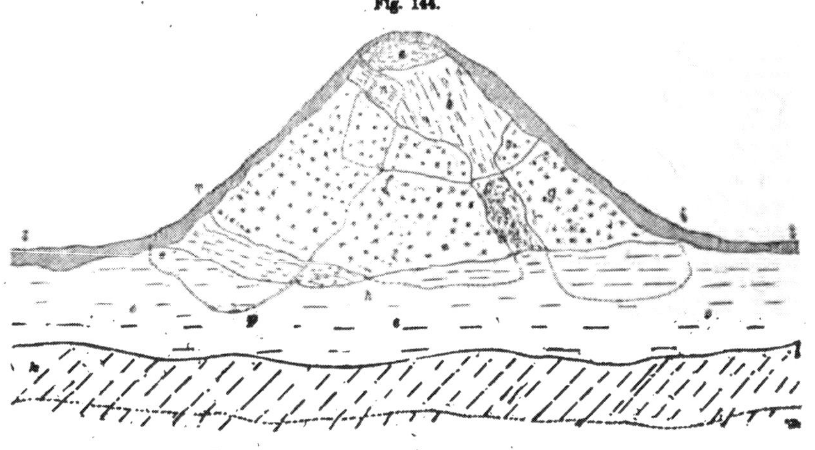

THE ORIGIN OF SPRINGS ON A VARIED SURFACE.

the nucleus rock *a* to be of close texture, when the rain falls upon the summit of the hill, which is supposed not to be covered with impervious clay as in the case above, but with vegetable mould, the rain will not be absorbed by *a*, but will pass down by gravity between *a* and *b*, another kind of rock of close texture. When the rain falls in greater quantity than will pass *between* these rocks, it will overflow the upper edge of *b* and pass over its surface down to *c*, but as *c* is a continuation of the nucleus impervious rock *a*, a large spring will flow down the side of the hill from *c* and render the ground quite wet to *d*, where meeting another large stratum of impervious rock, it will burst out to-day a large spring at *d*, which will be powerful in proportion to the quantity of rain that falls on the mountain. On flowing down *b*, part of the water will be intercepted by the rocks *j* and *g*, both of which being porous, will absorb and retain it until surcharged. The surplus water meeting with the impervious rock *e*, will be partly thrust out to-day along the black line *d h* on the one hand, and *d i* on the other, when the whole line *h i* will present a long dark line of wet oozing out of the soil, with the spring *d* in the center, and which darkness and dampness will extend down the inclined ground as far as the upper line *k l* of that porous stratum of rock. Part of the water absorbed by the porous rocks *f* and *g* will be conveyed under the impervious rock *e*, and come out at their lowest extremities, following the curved dotted lines *h d* and *d i*, and continue to flow on until it reaches the lowest extremity of *e* in the dotted line *k l*, where it will be absorbed by the porous rock *m*.

(686.) By such an arrangement of rocky strata on the side of a mountain range, will be exhibited specimens of both wetness and dryness of soils. The summit *a* will be wet, and so will the surface of *b*, but the surfaces of *f* and *g* will be dry. Again, the surface of *e* will also be wet, but less so than that of *b*, because part of the water is conveyed by *f* and *g* under *e* to the dry stratum *k l*, which being probably thicker, and, at all events, of greater extent, will be drier than either *f* or *g*. On another side of the side of the hill another result will take effect. The rain falling on the summit *a* will descend between *a* and *n*, as far as the lowest extremity of *n* along the dotted line *o p*, which being under the impervious rock *e*, the water will continue to flow out of sight until it descends to *k l*, where it will be absorbed by the porous rock *m*, and thus never appear at all either as a spring or a line of dampness. But should the quantity of rain at any time be greater than what will pass between *a* and *n*, it will overflow *n* and be absorbed in its descent by the porous rock *f*, which, after becoming surcharged, will let loose the superfluous water in the line *h r*, upon the continuation of the rock *n*, part of which will come to-day along the line *h o* of the impervious rock *e*, and part conveyed down by *o p* to the porous rock *k l*, where it will be absorbed. Thus, on this side of the hill, as long as little rain falls, none but its summit will be wet, and all the rest will be dry, though the surfaces of *f* and *k* will always be drier than those of *n* or *e*; but after heavy rains dampness will show itself along the line *h r*, will extend itself even to the line of *k l*, should the rain continue to fall some time.

(687.) The line *s* by the summit *a* to *t* is the mould line pervious to moisture, and which is here represented as is frequently exhibited in nature, namely, a thickness of soil on the southern side of the hill as from *a* to *i*, and a thickness of soil on the northern basis, as from *r* to *s*; but a thinness of soil on the southern face, as from *a* to *r*. It is not pretended that this figure is a truly geological portrait of any mountain. Perhaps no such arrangement of strata actually exists in any single hill, but such overlying and disconnected but conterminous strata do occur over extended districts of hilly country which produce springs much in the way just described. Similar courses of water occur in less elevated districts, though it remains more hidden under the deeper alluvial rocks.

(688.) Now let us apply Elkington's method of draining to these two cases of wetness, and which are of ordinary occurrence. The hill in fig. 143 being supposed to be covered saddle-shaped with an impervious stratum of clay, no water can descend *into* it, but will flow *over* it: *a* is the clay stratum · *b* also an impervious stratum, but not so much so as *a*, containing veins of sand and nodules of stones, and forming a very common subsoil of this country. It is clear that the whole extent of ground from *e* to *b* will be wet on the surface, and the wetness will not exhibit itself in bands, but be diffused in a uniform manner over the whole surface; but as *b*, in this case, is not so tenacious as *a*, the side of the hill from *e* to *c* will always be wetter than the flat ground from *c* to *b*, because some of the water will be absorbed and kept out of sight in the looser clay *b*. The only method of intercepting the large body of water in its descent down *d* is to cut the deep drain at *c*, not only sufficiently large to contain all the water that may be supplied from above *c*, but so deep as to catch any oozing of water from *a* toward *b*. What the depth of this drain should be it is not easy to determine without farther investigation, and to enable that investigation to be made, a large drain should be cut on the flat ground in the line from *b* to *c*, which will also answer the purpose of leading away the water that will be collected by the transverse drain *c*. Suppose the subsoil from *b* to *i* is 4 feet thick, then this leading drain should be made ½ foot deeper, namely 4½ feet, in order that its sole may be placed in impervious matter; and in this case the drain *c*, of the depth of 6 feet, may suffice to keep the flat ground dry. But if from *b* to *i* is 8 or 10 feet in depth, then it would be advisable to make the leading drain from *b* to *c* at least 6 feet deep, in order to drain a large extent of ground on each side of it, and the drain *c* may still do at its former depth, namely 6 feet. Should the bottom of the leading drain get softer and wetter as the cutting descends, its depth should either be carried down to the solid clay at *i*, or perhaps it would be well to try auger holes in the bottom, with the view of ascertaining whether the subjacent water might not rise to and flow along it. The expedient of boring will be absolutely necessary if the depth from *b* to *i* decreases as the distance from the hill increases, for there would be no other way of letting off the water from the basin of the clay from *i* to *c*. Should the flat ground be of considerable extent, or should the face of the plain undulate considerably from right to left, a leading drain will be required in every hollow; and each of them should be made deeper or shallower according as the subsoil is of a drawing texture or otherwise, bearing in mind that the *sole* of the drain should, if possible, rest upon an impervious substance, otherwise the water will escape through the pervious matter, and do mischief at a lower level. The subsoil between *g* and *h* being supposed to be gravel or other porous substance, it is clear that no drain is required at *f* to protect the soil between *f* and *g*, as the porous subsoil will absorb all the water as it descends from *e* to *f*.

(689.) As to the wet surface of the hill itself *c d e f*, it being composed of impervious clay, must be dried on the principle of surface-draining; that is, if the ground is in permanent pasture for the support of sheep, a number of transverse open sheep drains should be made across the face of the hill, and the water from them conveyed in open ditches into the great drain *c*; or if the ground is under the plow, small covered drains will answer the purpose best; and the contents of these can be emptied into the large drain *c*, and conveyed down the large leading drain to *b*. Thus, in fig. 145, *a b* is the main drain along the flat ground into which the large drains *c b* and *d b* flow. It may be observed here that when one large drain enters another, the line of junction should not be at right angles, but with an acute angle in the line of the flow of water, as at *b*. The open surface-drains in permanent pasture exhibit the form as represented in this figure, where the leaders *e f* and *g h* are cut with a greater or less slope down the hill according to the steepness of the acclivity, and the feeders across its face nearly in parallel rows, into their respective leaders. In this way the water is entirely intercepted in its descent down the hill. I may mention that where small drains enter larger, they should not only enter with an inclination, as remarked above, but where they come from opposite sides, as in this case, they should enter at alternate distances, as seen in the case of the three drains above *f*, and not as shown

Fig. 145.

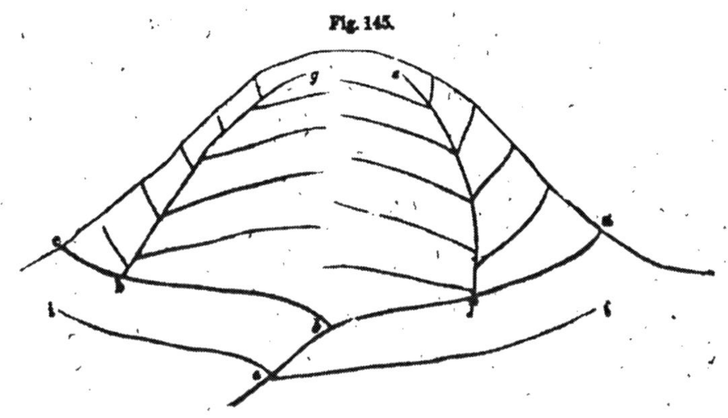

A PLAN OF SHEEP DRAINS ON A HILL OF IMPERVIOUS SUBSOIL.

in the fourth and fifth drains. The large drain *c b d* may either be left open or covered. Should it form the line of separation between arable ground and permanent pasture, it may be left open, and serve to form a fence to the hill-pasture; but should the entire rising ground be under the plow, this, as also the main drain *a b*, and all the small drains, should be covered.

(690.) There are *various ways of making small drains in grass.* One plan is to turn a furrow-slice down the hill with the plow, and make the furrow afterward smooth and regular with the spade. When the grass is smooth and the soil pretty deep, this is an economical mode of making such drains, which have received the appellation of *sheep drains.* But where the grass is rough and strong, and swampy places numerous, the plow is apt to choke with long grass accumulnting between the coulter and beam, and makes very rough work, and the horses are apt to overstrain themselves in the swampy ground. The lines of the drains should all be previously marked off with poles before the plow is used.

(691.) A better though more expensive plan, is to form them altogether with the spade. Let *a*, fig. 146, be a cut thrown out by the spade,

Fig. 146.

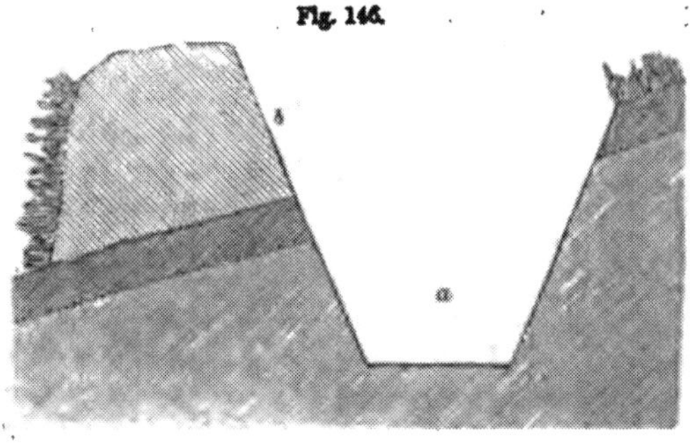

AN OPEN SHEEP DRAIN ON GRASS.

inches wide at bottom, 16 inches of a slope in the high side, and 10 on the low, with a width of 20 inches at top on the slope of the surface of the ground. A large turf *b* is removed by the spade, is laid on its grassy side downward, on the lowest lip of the cut, and the rest of the earth is placed at its back to hold it up in a firm position, the shovelings being thrown over the top to finish the bank in a neat manner. Such a drain catches all the water that descends in the space between it and the drain above, and leads it away to the sub-main drain, such as *e f* or *g h*, which is of similar construction, but of larger dimensions, running up and down the hill, and the lower end of which finds an entrance into the large main drain at the margin of the arable land.

(692.) Another sort of sheep-drain is formed as represented in fig. 147. A cut is made 6 inches wide at bottom, 16 inches deep, and 18 inches wide at top. The upper turf *a* is taken out whole across the cut, as deep as the spade can wield it. Two men will take out such a turf better than one. It is laid on iss grassy face upon the higher side of the drain, and the earth pared away from the other side with the spade, leaving the turf of a trapezoidal shape. While one man is doing this, the other is casting out with a narrow spade the bottom of the cut *b*. The earth and shoveling should be spread abroad over the grass; and the large turf *a* then replaced in its natural position, and tramped down, thereby leaving an open space *b* below it for the water to pass along. This is not so permanent a form of sheep drain as the last, nor is it at all suitable in pasture where cattle graze, as they would inevitably trample down the turf to the bottom of the drain. It is also a temptation for moles to run along; and when any obstruction is occasioned by them or any other burrowing animal, the part obstructed cannot be seen until the water overflows the lower side of the drain, when the turfs have again to be taken up, and the obstruction removed. It forms, however, a neat drain, and possesses the advantage of retaining the surface whole where sheep alone are grazed. Figs. 146 and 147 are drawn on a scale of $\frac{1}{8}$ inch to 2 inches.

Fig. 147.

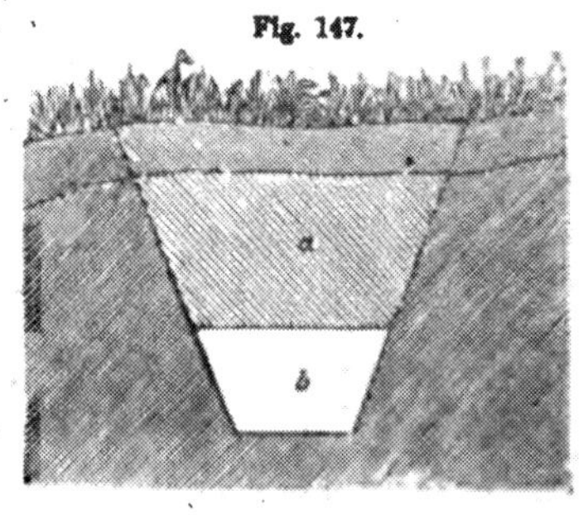

A COVERED SHEEP DRAIN IN GRASS.

(693.) Having described the various modes of pure *surface*-draining, and traced the origin of springs, immediately connected with which arises the necessity for the sheep drains just described, I shall now proceed to you the *deep*, or Elkington's method of draining, which is peculiarly well adapted for draining *isolated hollow spots of ground*. These are usually formed by water standing in winter on an impervious clay subsoil, the water being either entirely derived from rain—in which case the pool becomes dry in summer—but most frequently partly from rain, and partly from springs in the subjacent strata fed from a higher source. Such pools are drained either by boring holes through the impervious clay into a porous stratum below—should such a stratum exist—or by a deep drain, having an efflux at a lower level.

(694.) I shall give you an account of the successful draining of such a pool. It was covered with water in winter to the extent of about 2 acres, in the center of a field of 25 acres; and though no water was visible in summer, its site was always swampy. It obtained the name of the "Duck-mire," wild ducks being in the habit of frequenting it every season. On taking a level from its water, it was found that a drain of 10 feet in depth would be required to carry it away in a 2½-feet-deep drain through the pool. The outlet was on the top of a clay-bank about 150 yards distant from the nearest margin of the pool, rising perpendicularly 40 or 50 feet above the bed of a small river, and was the nearest point for a fall from the pool. The operations were performed in summer, when the pool was comparatively dry. The deep cut of 10 feet was first executed; and to render it ever after secure, a conduit of 9 inches in width by 12 inches in hight was built and covered with land-stones obtained from the field by trench-plowing, above which about 2 feet of stones were placed and covered with turf before the earth was returned into the deep cut. The continuation of the main drain was carried right through the center of the pool, where it could only be formed 30 inches deep, in order to preserve the requisite fall. Another drain, of 3 feet in depth, encircled the area of the pool a little above the water-mark, and was let by each end into the main drain. Both these drains were made 9 inches wide at bottom, to contain a coupled duct of 4 inches in width; and filled with small round stones, admirably adapted for the purpose, none exceeding the size of a goose's egg, gathered from the surface of the field. The stones were blinded with withered wrack, and the earth returned above them, first with the spade and then with the plow. The pool was at once determined to be drained in this manner, because the high bank of clay above the river—and which formed the entire subsoil of the field—forbade any attempt at boring to a porous stratum below.

(695.) But a difficulty occurred in passing the drain through the center of the pool, which was not foreseen. A complete quicksand was met with, the bottom of which, resting on the clay, was much below that of the deep cut. To have effectually drained the quicksand, the cut should have been made at least 13 feet in depth; so that about 2 feet deep of quicksand were obliged to be left beneath the drain; and how to construct a lasting drain upon such a foundation, was a puzzling thing, as the wet sand thrown out by the spade was followed by larger quantities sliding down from each side of the drain, and filling up the emptied space.

(696.) It is here worthy of remark, that I committed a mistake in not ascertaining the existence of the quicksand, and of its depth, before beginning to cast the deep main drain; for I had only thought of making such an outlet as would enable me to make a drain of such a depth through the center of the pool as would drain it; whereas I ought to have ascertained in the first place, the nature of the strata under the pool, which would have made me acquainted with the depth of the quicksand; and the drain and outlet should then have been made of the depth to deprive the quicksand entirely of its water. However, as matters were, I was obliged to do the best to form a drain in the quicksand, and this was found to be a rather troublesome operation. Thick tough turfs were provided, to lay upon the sand in the bottom of the drain, and upon these were laid flat stones, to form a foundation on which to build a conduit of stones, having an opening of 6 inches in width and 6 or 7 inches in hight. The back of the conduit, when building, was completely packed with turf, to prevent the sand finding its way into it from the sides of the drain; and the packing was continued behind the few small rubble stones that were placed over the cover of the conduit. A thick covering of turf was then laid over the stones, so that the whole stones of the drain were completely encased in turf, before the earth was returned upon them. The filling up was entirely executed with the spade, in case the trampling of the horses should have displaced any of the stones; but these extraordinary precautions were only used as far as the quicksand was found to be annoying. After all the drains were finished, a large quantity of water flowed out of the main drain during the succeeding autumn and winter; but by spring the land was quite dry, the blue unctuous clay forming the bottom of the pool became friable, and on the soil and subsoil being intermixed by deep plowing, the new and fresh soil, with proper management, ever after bore fine luxuriant crops.

(697.) A 12-acre field of good, deep land on the farm of Frenchlaw, in Berwickshire, was rendered swampy by springs and oozings of water from the surrounding rising ground being retained upon the clay subsoil. A 4-feet drain was formed all round the base of the rising ground, immediately above the line of wet, and several drains of 3 feet in depth were run through the flat part of the field. The outlet was obliged to be cut through a part of a field on the adjoining farm to the depth of 13 feet, conduited and covered over. The swampiness of the ground was completely removed, and the crops ever after were excellent.

(698.) Another application of Elkington's system may be successfully made in draining the *springs or oozings of water around gravelly eminences standing isolated in single fields, or across more than one field, upon a bed of clay or other impervious matter.* A circumvallation of drain around the base of the eminence, begun in the porous and carried into the impervious substance, having a depth of perhaps from 5 to 7 feet, and connected with a main drain along the lowest quarter of the field, will most effectually dry all the part of it that was made wet by the springs or oozings.

(699.) *Bogs* and *marshes* have been drained with great effect by Elkington's method. These almost always rest on basin-shaped hollows in clay; and, when this is of considerable depth, the only way of draining them is by bringing up a deep cut from the lowest ground, and passing it through the dam-like barrier of clay. But it not unfrequently happens that gravel or sand is found at no great depth below the clay on which bogs rest; in which case, the most ready and economical plan is to bore a hole or holes in the first instance through the clay, with an auger 5 inches in diameter; and, after the water has almost subsided, to finish the work by sinking wells through the clay, and filling them up with small stones to within 2 feet of the top.

(700.) I have never seen an instance of the draining of bog by *boring* or by *wells;* but the late Mr. George Stephens, land-drainer, instances two or three cases of bogs being successfully drained in Sweden by means of bore-holes and wells in connection with drains; and the late Mr. Johnstone adduces many as successful instances in this country.*

(701.) I have seen extensive and successful effects of drying bogs in Ireland by ordinary drains, especially Carrick Bog, at Castle Rattan, belonging to Mr. Featherstonehaugh, in the county of Meath. The plan consists of dividing the bog into divisions of 60 yards in breadth, by open ditches of 4 feet in depth and 4 feet wide at top, allowance being made for the sliding in of the sides and subsidence by drying, and which movements have the effect of considerably diminishing the size of the drains; and these ditches are connected by parallel drains at right angles 3 feet 3 inches in depth, and 18 inches in width. Fig. 148 is a plan of these drains, where *a* are the large ditches and *b* the small drains. The ditch *a* at the bottom is that which takes away all the water to some large ditch, river, or lake. The fall in the ditches and drains is produced by the natural upheaving of the moss above the level of the circumjacent ground, and, of course, this peculiarity causes all the drainage of the bog to flow toward the land.

(702.) The small drains *b*, fig. 148, are made in this manner: A garden line is stretched at right angles across the division from the large open drain *a* to *a*, 60 yards. The upper rough turf is rutted in a perpendicular direction along the line with a short edging-iron. The line is then shifted

* See Stephens's Practical Irrigator and Drainer, edition of 1834; and Johnstone's Systematic Treatise on Draining, 4to edition of 1834—both excellent works on the subject of deep draining.

(608)

18 inches, the width of the top of the drain, and another rut is made by the edging-iron. While one man is employed at this, another cuts out a

Fig. 148.

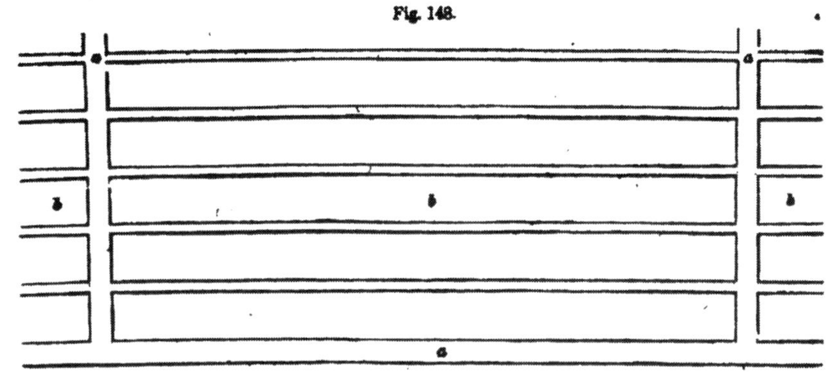

A PLAN OF DRAINS FOR BOGS AS PRACTICED IN IRELAND.

thick turf across the drain with the broad-mouthed shovel, fig. 149; and, if any inequalities or ruggedness are observed in the wet turf, he makes them smooth and square with a stroke or two with the back of the shovel. The drain is thus left for two months to allow the water to run off, the moss to subside, and the turf to dry and harden.

(703.) At the end of that time the long edging-iron, fig. 150, is employed to cut down the sides of the drain in a perpendicular direction 2 feet 3 inches (see fig. 153), and the flat shovel is also again employed to cut the moss into square turfs, which in this case are not thrown out with the shovel—as on account of their wet state they cannot remain on its clear, wet face, when used so far below the hand—but are seized by another man with the small graip, fig. 151, and thrown on the surface to dry. The

Fig. 149. Fig. 150. Fig. 151.

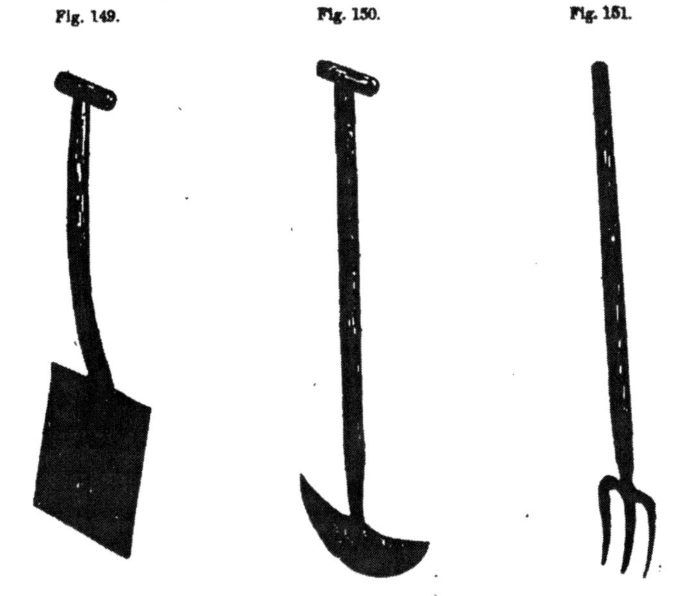

THE BROAD-MOUTHED SHOVEL. THE LONG EDGING-IRON. THE SMALL GRAIP.

work is again left for two months more, to allow time for the water to drain off, and the turfs to dry and harden.

(704.) In these four months the moss subsides about 1 foot. After the two spits of the shovel, the longest edging-iron is again employed to cut down the last spit, which is done by leaving a shoulder *e e*, 5 inches broad, on each side of the drain, fig. 153. The scoop, fig. 152, is then employed

Fig. 152.

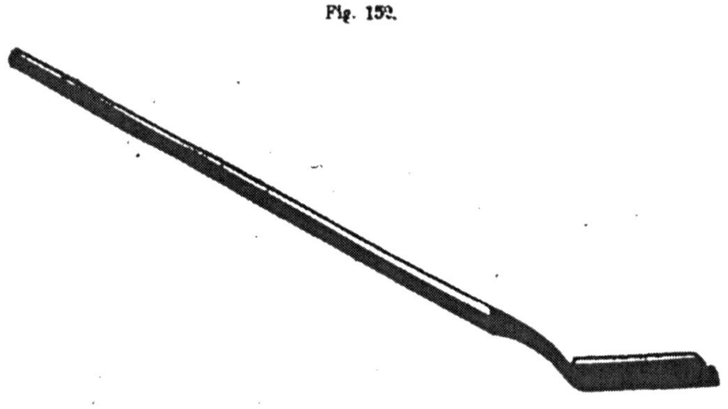

THE BOG DRAIN SCOOP.

to cut under the last narrow spit, which is removed from its position by the small graip. The scoop pares, dresses and finishes the narrow bottom of the drain, with a few strokes with its back, making the duct *d* 1 foot deep.

(705.) The filling of the drain is performed at this time, and it is done in this manner. The large turf *b*, fig. 153, which was first taken out, and is now dry, is lifted by the hand and placed, grass side undermost, upon the shoulders *e e* of the drain, and tramped firmly down with the feet. The second large turf *a*, which is not so dry or light as the first, is lifted by the graip and put into the middle of the drain, and the long, narrow stripes of turf *c c* separated by the scoop from the bottom, along with other broken pieces, are also placed by the graip along both sides and top of the drain, and all the sods just fill up the subsided drain.

Fig. 153.

THE SHOULDERED BOG DRAIN.

(706.) Fig. 153 represents the drain thus finished, which is well suited for the drying of bog, and in its construction possesses the advantage of having all the materials for filling it upon the spot. It is a well-known property of dried moss that it resists the action of water with impunity, and the mode just described of making drains affords ample time for the drying and hardening of the turfs cast out of the drains; but it is not requisite for the efficiency of the turfs that they be dried, as they answer the same purpose quite well in a wet state; but the time allowed for the subsidence of the moss itself is a great advantage to the drain, such materials being never again disturbed in a subsided bog drain. A bog drain requires no other materials, such as wood or tiles, to fill it, there being no material so appropriate or more durable than the moss itself, the slightest subsidence in the drain destroying the continuity of the soles and tiles, whether of wood or clay, while those made of the latter substance will gravitate in the moss by their own weight. To expedite the subsidence of the moss, the cutting of the drains is most successfully practiced in summer, when the drouth not only dries the turfs, but gets quit of a large proportion of the water by evaporation. The scale of this figure is $\frac{1}{8}$ to 2 inches, or $\frac{3}{4}$ inch to 1 foot.

(707.) These are all the cases, as I conceive, in which Elkington's method of draining can be applied, and even in them all it will not be attended with certain success, certainly not with equal success. I have frequently made lines of drains across the spouty sloping faces of fields, to the depth of even 6 feet, and never less than 4, without drying more than the breadth which they covered. In these cases I considered the cost of making them just thrown away; while in other cases 4-feet drains have completely removed the spouts, though the subsoil was apparently identically the same in them all. It is possible that the small veins of sand which were intersected by the cutting might, in the unsuccessful cases, dip away from the drains, and the water in them had perhaps ceased to flow in the same direction after their bisection, and in the successful cases the sand veins may have dipped with even a more favorable inclination for discharging their contents into the drains. Whatever difference of distribution there might have been in the component parts of the strata, in these opposite cases, there is little doubt that it was not sufficiently great to be indicated on the surface of the ground; and it is questionable that even the most minute investigation of apparently similar veins of sand, in similar strata, would acquaint us with their real positions, as it is not impossible that the most trifling difference in the relative positions of such veins may produce very different effects upon the course of the water in the subsoil.

(708.) It is now necessary to describe to you the mode to be adopted in forming drains on Elkington's method. Before determining on the direction in which the lines of drains should run in the field proposed to be drained, it has been recommended to sink pits here and there, of such dimensions as to allow a man to work in them easily, and to a depth which will secure the exposure of the subjacent strata and the greatest flow of water, the depth varying perhaps from 5 to 7 feet. A previous examination of the underground is certainly requisite, and pits will certainly acquaint you with the arrangement of the substrata; and, had I pitted the bottom of the pool, the drainage of which I have described above (696), the depth of the quicksand would have been easily ascertained, and the main drain made commensurate with the circumstances of the case. But I agree with the late Mr. Wilson, of Cumledge, Berwickshire, that the driving of lines of drains from the bottom to the top of the field is the most satisfactory method of obtaining an enlarged view of the disposition of the subjacent strata, and, of course, of the depth to which the drains should be sunk.* Such lines of drains will not be useless; on the contrary they will form the outlets of the system of drains connected with each of them, and for that purpose they should be made in the lowest parts of the field.

(709.) Having thus ascertained the nature of the underground, the lines of the drains which run *across* from the mains should be marked off. This can be done by drawing a furrow-slice along each line; but a neater plan, and one which will not spend the time of horses at all, is to set them off by means of short stakes driven into the ground, or, if the field is in grass, by small holes made in the ground with three or four notches of the spade, and the turf turned over on its grassy face beside each hole.

(710.) It is very desirable that the stones for filling the drains should be laid down along the lines in the order the drains are to be opened up, not

* Prize Essays of the Highland and Agricultural Society, vol. vii.

only on account of having them at hand when the filling process is to commence, and of thereby, perhaps, saving the labor of throwing out the earth that may have fallen down from the sides of the drain, when waiting for the stones, and of procuring additional stones for filling up the spaces thus enlarged, but of saving the horses much trouble in backing and forwarding the cart on the ground when it is necessarily much confined at the side of an opened drain. The stones should be laid down on the upper or higher side of the ground, if there be one, that the earth from the drain may be thrown upon the lower side.

(711.) Suppose that it has been determined to make the drains 6 feet deep. For this depth a width at top of 30 inches, and one at bottom of 18 inches, will be quite sufficient for the purpose of drainage, and for room to men to work in easily. This particular, in regard to the dimensions of the contents of a drain, should always be kept in view when cutting one; as even a small unnecessary addition either to the depth or width, and especially to both, of a deep drain, makes a considerable difference in the quantity of matter to be thrown out, and, of course, in the quantity of stones required for again filling up the excavated space. A simple calculation will at once show you the great difference there is in the contents of a drain a little wider and narrower than another; and the difference is much greater than you would imagine at first sight. A drain of the above dimensions—namely, of 6 feet deep, 2½ feet wide at top, and 1½ feet at bottom—gives an area by a vertical section of 22½ square feet, and, in a rood of 6 yards in length, a capacity of 405 cubic feet; whereas a drain of 6 feet deep, 3 feet wide at top, and 2 feet wide at bottom, as recommended in a particular instance by the late Mr. Stephens,* would give a vertical section of 36 square feet, and a capacity of 644 cubic feet, creating more than 50 per cent. of additional work. And you should bear in mind that, provided the parts of a drain are substantially executed, its *width*, beyond that which will secure efficacy, cannot render it more efficacious. The rule for the width of a drain is very well determined by the ease with which men are able to work at the bottom; and, indeed, men working by the piece, when their work is measured longitudinally, will always prefer narrow to wide drains.

(712.) The cutting of drains should always be contracted for at so much per rood of 6 yards. The size of drain which I have just recommended (711) may be cut for from 1s. 6d. to 2s. per rood, according to the hardness or other difficulties of removing the subsoil. Where clay is very hard and dry, or very spongy, tough and wet, or where many boulders interfere, the larger sum is not too much; but where the subsoil can be loosened with ordinary picking, and is mixed with small sand veins and stones, the smaller sum will suffice. In such a contract, it should always be understood that the first portion of the earth in the refilling shall be returned into the drain by the contractor, and that he shall provide himself with all the tools necessary for the work.

(713.) The first operation in breaking ground is to stretch the garden line for setting off the width of the top of the drain, 30 inches—the drain being begun at the lowest part of the ground—and each division thus lined off consists of about 4 roods, or 24 yards. Three men are the most efficient number for carrying on the most expeditious cutting of drains.—While the principal workman is rutting off the second side of the top of the drain with the common spade, the other two begin to dig and shovel out the mould-earth, face to face, throwing it upon the lower and opposite

* Prize Essays of the Highland and Agricultural Society, vol. vii.

side from the stones. The first spit of the spade most likely removing all the mould, the first man commences the picking of the subsoil with the foot-pick, fig. 37; or, if the mould is too deep to be removed by one spit, and requires no picking, the first man digs and shovels out the remainder of it by himself with the spade. The mould is thus all removed from the lined-off break or division of the drain. When the picking commences, one man uses the foot-pick, working backward; another follows him with his back with a spade, and digs out the picked earth; while the contractor comes forward with the shovel, fig. 38, with his face to the last man, and takes up all the loose earth, and trims the sides of the drain. In this way the first spit of the subsoil is removed. Should the drain prove very wet, and danger be apprehended of the sides falling in, the whole division should be taken out to the bottom without stopping, in order to have the stones laid in it as quickly as possible. Should the earth have a tendency to fall in before the bottom is reached, short, thick planks should be provided, and placed against the loose parts of both sides of the drain, in a perpendicular or horizontal position, according to the form of the loose earth, and there kept firm by short stakes acting as wedges between the planks on both sides of the drain, as represented in fig. 154, where *a a* are

Fig. 154.

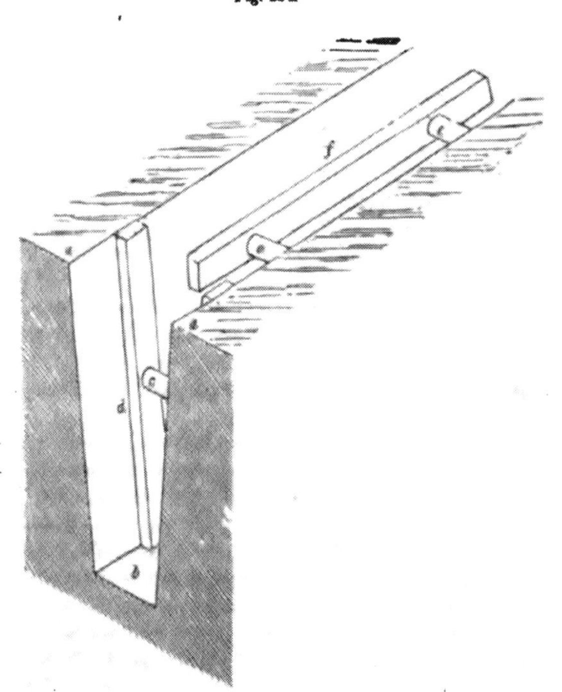

THE POSITION OF PLANKS AND WEDGES TO PREVENT THE SIDES OF DRAINS FALLING IN.

the sides of the drain, *d* planks placed perpendicularly against them, and kept in their places by the short stake or wedge *c*, and where *f* are planks placed horizontally and kept secure by the wedges *e e*.

(714.) But if the earth in the drain be moderately dry and firm, another division of 4 roods may be lined off at top, and the subsoil removed as low as the depth of the former division. Before proceeding, however, to line off a third division, the first division should be cleared out for the builder of the conduit. The object of this plan is to give room to separate the diggers of the earth from the builders of the stones, so as there may be no interference with one another's work, and also to give advantage of the half-thrown-out earth of the second division as a stage upon which to receive the larger stones, such as the covers of the conduit, to their being easily handed to the builder, as he proceeds in the laying of the conduit in the first division. On throwing out the earth to the bottom of the first break, special care should be taken to clear out the bottom square to the sides, to make its surface even, and to preserve the fall previously determined on.

(715.) When a division of the drain has thus been completely cleared out, you yourself, or the farm-steward, should ascertain that the dimensions and fall have been preserved correct as contracted for, before any of the stones are placed in the bottom. I have seen it recommended that the person appointed to build the conduit should ascertain if these particulars have been attended to; but it is always an invidious task for one class of workmen to check the workmanship of another, and on this account such a duty should always be performed by the farmer himself, or by any other authorized person.

(716.) Instead of measuring the dimensions of the drain with a tape-line or foot-rule, which are both inconvenient for the purpose, a *rod* of the form of fig. 155 will be found most convenient, most certain, and most quickly applied. The rod, divided into feet and inches, is put down to ascertain the depth of the drain, and then turned partially round while resting on its end on the bottom of the drain, until the ends of its arm touch the earth on both sides. If the arms cannot come round square to the sides of the drain, the drain is narrower than intended; and if they cannot touch both sides, it is wider than necessary. When the drain is made narrower than intended, you may take it off the contractor's hands, for the men having been able to work in it with ease to themselves, shows that the width is sufficient; but if the drain is wider than necessary, you should object to it to prevent similar enlargements in other places, for although the contract may have been formed by the longitudinal measurement, and not by the cubical contents, the larger space involves you in greater expense to fill up with stones.

Fig. 155.

THE DRAIN-GAUGE.

(717.) All deep drains should be furnished with *built conduits*, that the water may have a free passage in all circumstances, and thereby escape being choked up, and save the consequent expense of relifting and relaying its materials. The relifting of a drain that has *blown*, that is, if one in which the water is forced to the surface of the ground, in consequence of a deposition of mud among the stones preventing its flow under ground, is a dirty and disagreeable business for workpeople, and an expensive one for their employer, as it costs at least 9d., and the filling in again of the earth 1d. more per rood of 6 yards;* besides, additional stones are required to fill the enlarged space occasioned by the unavoidable removal of wet earth along with the stones.

(718.) The *building of the conduit should be contracted for in a separate item from the cutting of the drains*. If both are undertaken by the same party, there is risk of the two sorts of work being so carried on together, to suit the convenience of the contractor and his men, as to deceive even the inspector; whereas, if each sort is inspected and passed before another is allowed to be begun, then both may be executed in a satisfactory manner. The building of the conduit will cost from 1d. to 2d. per rood, ac-

* Stephens's Practical Irrigator and Drainer.

cording to the adaptation of the stones for the purpose.* Flat handy stones can be built firmly and quickly, whereas round shaped ones will require dressing with the hammer to bring them into proper shape, and much pinning to give them stability. The stones are furnished to the builder, and a laborer is also usually provided for him, to supply the stones as he requires them. But circumstances may occur in which it will be more convenient for you to contract with the builder to quarry the stones, supply himself with a laborer and build the conduit, and you to undertake only the carriage of the stones. A dry-stone builder of dykes is a better hand at building conduits for drains than a common mason, as he does not depend upon mortar for giving steadiness to his work.

(719.) Should the ground be firm, and the drain made in summer, and the length of any particular drain not very great, the conduit is most uniformly built when begun at the top and finished at the bottom of the line of drain; but in ground liable to fall down in the sides, or in winter, when the weather cannot be depended upon for two days together, or when the drain extends to many roods in length, the safest plan is to build the conduit immediately after the earth is taken out to the bottom.

(720.) A very convenient article in the building of conduits in a deep drain is a plank of 5 inches in breadth, and of from 6 to 9 feet in length, to put down in the middle of the bottom of the drain, to afford a dry and firm footing to the builder, and to answer the purpose, at the same time, of a gauge of the breadth of the conduit, a space of ½ inch on each side of the plank giving a breadth of 6 inches to the conduit. This plank can be easily removed by two short rope-ends, one attached near each end to an iron staple.

(721.) Suppose the plank set down at the mouth of the drain in the middle of the cut, the dyker begins by leaving a conduit at the mouth of 6 inches wide, having 6 inches of breadth of building on each side of it, and 6 inches high, and using the plank as his foot-board. When the building of these dimensions is finished to the length of the plank, this is carried or pushed by the ropes another length upon the drain, and so on, length after length, until the whole space of drain, when cleared out to the bottom, is built upon. The stones are handed down from the surface to the dyker by the laborer, who, in this case, may be a female field-worker, until the building is finished. The plank is then removed out of the way, the dyker clears the bottom of the conduit of all loose earth, stones, and other matter, with a hand-draw-hoe 5 inches wide in the face. Immediately after this, he lays the flat covers, which extend at least 3 inches on each side over the conduit, they being from 2 to 3 inches in thickness; and they lie ready for him on the half cast out division of the drain, from which they are handed to him as he works backward. The open space left between the meetings of the covers, which will not probably have square ends, should be covered with flat stones, and the space from the ends of the covers and flat stones to each side of the drain should be filled up and neatly packed with small stones. In this way the dyker proceeds to finish the conduit in every division of drain. To keep the finished conduit clear of all impediments, the dyker makes a firm wisp of wheat or oat straw large enough to fill the bore of the conduit; and which, while permitting the water to pass through, deprives it of all earthy impurities.

(722.) Before the conduit is entirely finished, the drainers throw out the earth of the adjoining division of the drain to the bottom, and the conduit is then built upon it in the same manner as the one just described. Should the laborer have any spare time from supplying the builder with materials,

* Prize Essays of the Highland and Agricultural Society, vol vii.

he throws in stones promiscuously upon the covers, until they reach a hight of 2 feet above the bottom of the drain, where they are leveled to a plain surface. They have been recommended to reach the hight of 4 feet, and when the drain is filled with rubble stones entirely, this hight is desirable, to give the water plenty of room to find its way into it; but with a conduit such as in fig. 156, more than 2 feet seems an unnecessary supply of stones, unless in places where water is more than usually abundant. It has also been recommended to break this upper covering of stones as small as road-metal;* but in deep draining, such as this, there seems no good reason for the adoption of such a practice, while it enhances the cost very considerably. Ordinary land stones or quarry rubbish are quite suitable for the purpose, and should any of the stones be unusually large, they can be broken smaller with a sledge-hammer. This the dyker might be employed in occasionally, as he will break stones much more easily than a laborer, and the work might be included in the contract with him. Should the stones be brought as they are required, the process of filling would be greatly expedited were they emptied at once out of the cart into the drain. This could be done by backing the cart to the edge of the drain, and letting the shafts or movable body of the cart rise so gently as to pour out the stones by degrees. To save the edge of the drain, and break the fall of the stones, a strong, broad board should be laid along the side of the drain, with its edge projecting so far as to cause the stones to fall down into the middle of it. A short log of wood placed in front of the board will prevent the wheels of the cart coming farther back than itself. I am aware that this mode of filling drains has been objected to by a competent authority in these matters, the late Mr. Stephens, as being a dangerous practice for the safety of the drain, especially as stones carry much earth along with them.† But in the case of deep and conduited drains I am sure no danger can arise from its adoption; because I have pursued the plan myself to a large extent with perfect impunity, and can vouch for its expedition and economy, and also for its safeness. To prevent the stones doing injury to thin covers, they should not be allowed to fall direct upon *them*, but *upon the end of the stones previously thrown in*, from which position numbers will roll down of themselves upon the covers without force, and the remainder can be leveled down with the hand before the next cart-load is emptied. There is a very considerable saving in the expense of filling drains in this way, provided it be done in the cautious manner just described, compared with the usual plan of laying down the stones when the drain is ready to receive them, and then throwing them singly in by the hand. Were it convenient to lay the stones down before the drain was begun to be cut, the plan would be inapplicable. As to the stones having earth among them, as much care can be taken to avoid that when they are each thrown or shoveled into the cart, as when put into the drain.

(723.) The leveled surface of the stones should be covered with some *dry material* before the earth is put over them. The best substance for the purpose is undoubtedly turf, but it is expensive to prepare and carry from a distance; but, should the field be in grass when it is drained, the turf over the drain could be laid aside at hand by the drainers, and used for covering the stones. Other materials answer well enough, such as withered wrack, dried leaves, coarse grass, broken moss, tanners' refuse bark, or straw; but I much dislike to see good straw wasted for such a purpose, when manure is usually too scanty upon a farm. The object of placing anything upon the stones is to prevent the loose earth finding its

* Prize Essays of the Highland and Agricultural Society, vol. vii.
† Quarterly Journal of Agriculture, vol. iii., note.

way among them; and, although it is not to be supposed that any of the substances recommended will continue long undecomposed, they, however, preserve their consistence until the earth above them becomes so consolidated as to retain its firmness ever afterward. You will learn, in the course of this article, how stones themselves are prepared to answer the purpose of a covering to those below them.

(724.) After the drain has been sufficiently filled with stones, the *earth* which was taken out of it should be *returned* as quickly as possible, in case rain fall and wash the earth down its sides among the stones. The filling in of the first earth of a deep drain is usually included in the contract made with the drainer, and done with the spade, because no horse can assist in that operation until the earth has been put in to such a hight as to enable him to walk upon it nearly on a level with the ground. The men may either put in all the earth with the spade, or they may put in so much as to allow the plow to do the remainder, but in both cases a little is left elevated immediately over the drain, to subside to the usual level of the ground. There will be much less earth left over the filling than you would imagine from the quantity thrown out at first, and the space occupied by the stones; and it soon consolidates in a drain, especially in rainy weather.

(725.) The section of such a drain as I have been describing is seen in fig. 156, where *a* is the opening of the conduit 6 inches square, built with

Fig. 156.

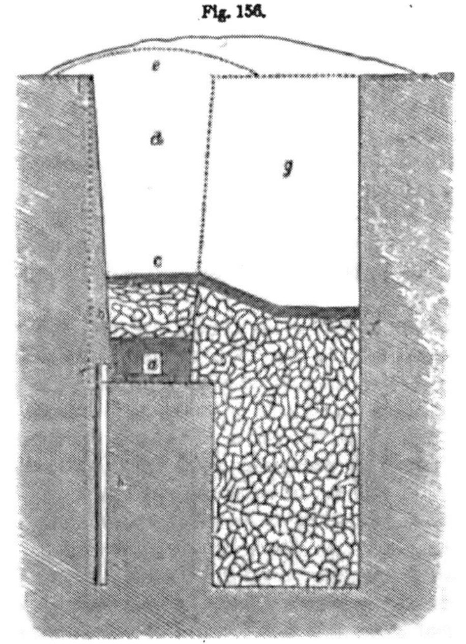

THE DEEP CONDUITED DRAIN, WITH WELL AND AUGER BORE.

dry masonry, and covered with a flat stone at least 2 inches thick; and above it is a stratum of loose round stones *b*, 16 or 18 inches in thickness. The covering above the stones is *c*, and the earth returned into the drain is *d*, with the portion *e* raised a few inches above the ordinary level of the ground. The mouths of such conduits, when forming outlets, should be protected against the inroads of vermin by close iron gratings.

(726.) Should water be supposed or known to exist in quantity below the reach of even a 6-feet drain, means should be used to render the drain available for its abstraction, and these means are, sinking wells and boring holes into the substrata. A *well* is made as represented by a part of fig. 156, where a pit *g* of the requisite depth is cast out on the lower side of the drain *a*, if the ground is not level. A circular or square opening, of 3 feet in diameter, or 3 feet in the side, will suffice for a man to work down several feet by the side of the open drain *d*; and, when the stratum which supplies the water is reached, the well should be filled with small stones to about the hight of those in the drain, as at *f*, and the whole area of the drain and well should be covered with dry substances from *f* to *c*, and the earth is filled in again above all, as at *g*. In making such wells, a small scarcement of solid ground, on a level with the bottom of the building of the conduit *a*, should be preserved, so that the building may have a firm foundation to stand upon, and run no risk of being shaken by the opera tions connected with making the well. I fear this precaution is less attended to in the making of drain wells than it deserves. Such a well should be sunk *wherever* water has been ascertained to be in *quantity* at a lower depth than the drain.

(727.) Or the *auger* may be used instead of the well for the same purpose, by boring through a retentive stratum into a porous, whereby confined water may be brought up into the bottom of the drain, by altitudinal pressure, and escape; or free water may pass down through the bore and be absorbed by the porous stratum below. In the first case, the retreat of the water has to be discovered in making the passage for it to pass away; in the second, it is got rid of by a simple bore. In boring for water at the bottom of a drain, the bore should be made at one side rather than in the middle of the bottom, because any sediment in the water might enter the bore at the latter place and choke it, when the water happened to come up with a small force. In preparation of the bore, let a cut *i k*, fig. 156, be made down the side of the drain, and, inserting the auger at *k*, let the bore be made down through the solid ground, in the direction of *b h*, as far as necessary—the orifice of the bore being made at a little higher level than the bottom of the drain, and an opening left in the building there, to permit the water from the bore to flow easily into and join the water of the drain.

(728.) As *boring-irons* may be as useful to you for finding water for fields, or for draining a bog, or for ascertaining the depth and contents of a moss, as for ordinary draining, it is proper to give a description of them. The *auger*, *a*, fig. 157, is from 2½ to 3½ inches in diameter, and about 16 inches in length in the shell, the sides of which are brought pretty close together; and it is used for excavating the earth through which it passes, and bringing it up. When more indurated substances than earth are met with, such as hardened gravel or thin, soft rock, a *punch b* is used instead, to penetrate into and make an opening for the auger. When rock intervenes, then the chisel or jumper *c* must be used to cut through it; and its face should be of greater breadth than the diameter of the auger used.—There are *rods* of iron *d*, each 3 feet long, 1 inch square iron, unless at the joints, where they are 1½ inch and round, with a male screw at one end, and a female at the other, for screwing into either of the instruments, or into one another, to allow them to descend as far as requisite. The short iron *key e* is used for screwing and unscrewing the rods and instruments when required. A *cross-handle* of wood *f*, having a piece of rod attached to it, with a screw to fasten it to the top of the uppermost rod, is used for the purpose of wrenching round the rods and auger, when the latter only is used, or for lifting up and letting fall the rods and jumper or punch, when they are used. The long iron key *g* is used to support the rods and instruments as they are let down and taken up, while the rods are screwed on or off with the short key *e*. Three men are as many as can conveniently work at the operation of boring drains.

Fig. 157.

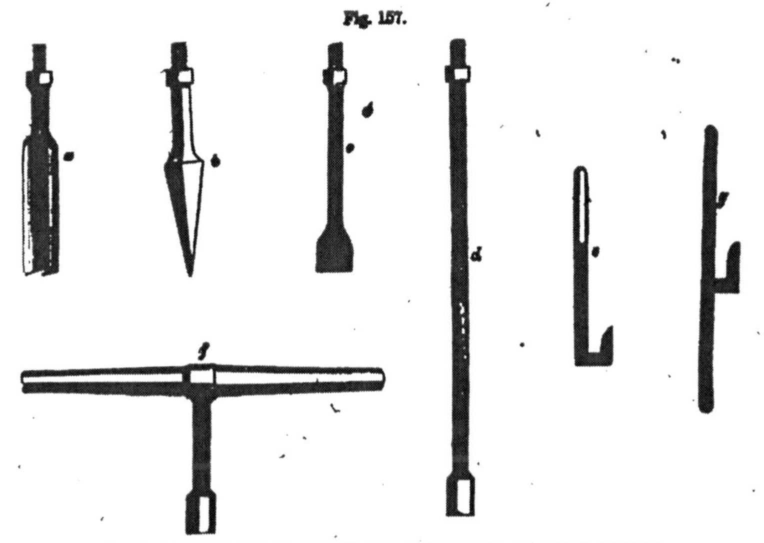

THE INSTRUMENTS FOR BORING THE SUBSTRATA OF DEEP DRAINS

(729.) As I have never witnessed the use of the auger in draining, I will give a description of the manner of using it from a competent authority. "Two men," says Mr. Johnstone, "stand above, one on each side of the drain, who turn the auger round by means of the wooden handle; and, when the auger is full of earth, they draw it out, and the man in the bottom of the drain clears out the earth, assists in pulling it out, and directing it into the hole. The workmen should be cautious, in boring, not to go deeper at a time, without drawing, than the exact depth that will fill the shell of the auger; otherwise the earth through which it is boring, after the shell is full, makes it more difficult to pull out. For this purpose, the exact length of the auger should be regularly marked on the rods from the bottom upward. Two flat boards, with a hole cut into the side of one of them, and laid alongside of one another over the drain, in time of boring, are very useful for directing the rods in going down perpendicularly, for keeping them steady in boring, and for the men standing on when performing the operation."*

(730.) The *principles* of Elkington's mode of draining seem to depend on these three alleged facts. 1. That water from springs is the principal cause of the wetness of land, which, if not removed, nothing effectual in draining can be accomplished. 2. That the bearings of springs to one another must be ascertained before it can be determined where the lines of drains should be opened; and by the bearings of springs is meant that line which would pass through the seats of *true* springs in any given locality. Springs are characterized as true which continue to flow and retain their places at all seasons; and temporary springs consist of bursts of water, occasioned either by heavy rains causing it to appear to the day sooner or at a *higher* level than permanent springs, or by true springs leaking water, and causing it to appear to day at a *lower* level than themselves; and, if such springs are weak, their leakage may be mistaken for themselves. It is evident that, if drains are formed through these *bursts* of water, no effect-

* Johnstone on Elkington's Mode of Draining.

ual draining takes place, and which can only be accomplished by the drain passing through the line of true springs. 3. That tapping the spring with the auger is a necessary expedient, when the drain cannot be cut deep enough to intercept it.* From these three averments it would appear that the seats of true springs are neither at the top nor in the base of a rising ground, but that temporary springs may be at both; and, of course, the more extensive the hight, the more numerous will be the springs, whether true or temporary. In the case of true springs in the side of rising ground, a system of branched drains will be required to remove them; but, in the case of their being situated near the base, their leakage will originate bogs or swampy grounds; and hence Elkington's mode of draining is only adapted to these peculiarities. It has been very extensively and, I must add, successfully practiced in Scotland, for the removal of both these sources of annoyance to land. The system would have ample scope in Ireland, where bog land still exists to an incredible extent; and in England also, where the regularity of alluvial deposits in many of the western and southern counties might give employment to the auger, to great advantage, in removing the largest proportion of the water which is doing injury; but in Scotland the system of tapping is inapplicable in irregular superficial deposits, though it might be tried in the few bogs which rest on regular strata.

(731.) In so far as the soil of Scotland is affected with water, there is no doubt that it is *not now most injured by springs*. What injury it suffered in that way has long been removed, by the extensive application of Elkington's mode of draining; and as in the pursuance of that system experience soon indicated that injury was sustained by the land from other water than that issuing from springs, a modification was introduced into the system, which, not being in accordance with its principles, can excite no surprise that it failed in many instances; and the misguided failures had the effect of bringing disrepute upon otherwise an excellent and efficient mode of draining. The modification I allude to, which brought obloquy upon Elkington's system, was the cutting of *deep* drains in every direction, irrespective of the arrangement of the subjacent strata, and the filling them nearly full of stones of any size and in any order. Much expense was in the first instance incurred by this practice, and when its effects were not commensurate with the outlay, disappointment was the result, and blame was imputed to the system, instead of to the mode of practicing it.

(732.) The chief injury now sustained by the soil of Scotland arises from the *stagnation of rain-water* upon an impervious subsoil. Most of the soil of that country consists of loam, of different consistence, resting on clayey subsoil sufficiently tenacious to retain water, the arable part of which is of unequal depth; where it is shallowest, it is itself injured by the stagnant water immediately below it; and where it is deepest, the plants upon it are injured by chilly exhalations.

(733.) The injury done by stagnant water to arable soil may be estimated by these effects. While hidden water remains, manure, whether putrescent or caustic, imparts no fertility to the soil; the plow, the harrow, and even the roller, cannot pulverize it into fine mould; new grass from it contains little nutriment for live-stock; when old, the finer sorts disappear, and are succeeded by coarse sub-aquatic plants. The stock never receive a hearty meal of grass, hay, or straw, from land in that state; they are always hungry and dissatisfied, and of course in low con

* Johnstone on Elkington's Mode of Draining.

dition. Trees acquire a hard bark and stiffened branches, and became a prey to parasitic plants. The roads in the neighborhood are constantly soft, and apt to become rutted; while ditches and furrows are either plashy, or like a wet sponge, ready to absorb water. The air always feels damp and chilly, and, from early autumn to late in spring, the hoar-frost meets the face like a damp cloth. In winter the slightest frost encrusts every furrow and plant with ice, not strong enough to bear one's weight, but just weak enough to give way at every step, while snow lies long lurking behind the sun in corners and crevices; and in summer musketoes, green-flies, midges, gnats, and gadflies, torment the cattle, and the plowman and his horses, from morning to night; while, in autumn, the sheep get scalded heads, and are eaten up by maggots, during the hot blinks of sunshine. These are no exaggerated statements, but such as I have witnessed in every similar situation; and they may be observed in every county in Scotland, in hill, valley, and plain.*

(734.) The only plan of draining fitted to remove the wetness which produces this state of things, is the one which allows stagnant water to flow easily away through moderately deep and numerous drains; for deep drains cannot take away stagnant water from impervious subsoil at the distances they are usually made. This constitutes the second mode of surface-draining alluded to in a former paragraph, (679), and which has now generally obtained the appellation of *thorough-draining;* and the treatment of which must now receive your attention.

(735.) What should be the exact *size* of these shallow and numerous drains, is not easily determined. It would be one step toward the settlement of this point were the *minimum size* determined, which I shall endeavor to do. A drain is not a mere ditch for conveying away water; were it only this, its size would be easily determined by calculation, or experiment, of the quantity of water it would have to convey in a given time. But the principal function of a drain is to *draw* water toward it from every direction; and its secondary purpose is to convey it away when collected; though both properties are required to be present, to the drain performing its entire functions. These being its functions, it is obvious that the greater the area its sides can present to the matter out of which it draws water, it should prove the more efficacious, and it is also obvious that this efficiency is not so much dependent upon the breadth as upon the depth of the drain, so that, other things being equal, the deeper a drain is, it should prove the more efficient. Now, what are the circumstances that necessarily regulate the depth of drains? In the first place, the culture of the ground affects it; for were land never plowed, but in perpetual pasture, no more earth than would support the pasture grasses would be required over a drain, and this need not, perhaps, exceed 3 inches in depth. The plow, however, requires more room; for the ordinary depth of a furrow-slice is seldom less than 7 inches, and, in cross-furrowing, 8 inches are reached, and 2 inches more than that, or 10 inches in all, may suffice for ordinary plowing; but in some instances, land is plowed with 4 horses instead of 2, in which case the furrow will reach 12 inches in depth, so that 14 inches of depth will be required to place the materials of the drain beyond the danger of an extraordinary furrow. But farther still, subsoil and trench plowing are sometimes practiced, and these penetrate to 16 inches below the surface, so that 18 inches of earth at least, you thus see, will require to be left on the top of a drain, to place its ma-

* See a paper by me on this subject in vol. vi. of the Quarterly Journal of Agriculture.

terials beyond the dangers arising from plowing. This depth having been thus determined by reference to practice, it should not be regarded as a source from which a supply of moisture is afforded to the drain by its drawing power, the water only passing through it by absorption; for it is certain that plowed land will absorb moisture, whether there be any drain below it or n·t. The *drawing* portion of the drain must, therefore, lie entirely below 18 inches from the surface. Now it will be requisite to make the drain below this as deep as will afford a sufficient area for drawing powers of the lowest degree among subsoils. And what data do we possess to determine this critical point? In the first place, it is evident that a subsoil of porous materials will exhaust all its water in a shorter time than one of an opposite nature. Judging from observation, I should say, that 1 inch thick of porous materials will discharge as much water in a given time, as 6 inches of a tilly, or any number of inches of a truly tenacious subsoil. What conclusions, then, ought we to draw from these data? Certainly these, that no depth, beyond the upper 18 inches, farther than what is required for the materials of the drain, will draw water from a truly tenacious subsoil, and that it is therefore unnecessary to go any deeper in such a subsoil; that it is also unnecessary to go any deeper in a subsoil of porous materials, because a small depth in it will draw freely; and that it is only requisite to go deeper in the intermediate kinds of subsoil. Still you have to inquire what should be the specific depths in each of these cases? In the case of really tenacious subsoil, the size of the duct for the water depends on the quantity to pass through it, but, giving the largest allowance of 6 inches with a sole beneath and covering above, 1 foot seems ample depth for these materials to occupy, so that a drain of 2½ feet seems sufficient for the circumstances attending such a subsoil, that is, its minimum depth, which, in such a case, may also be held to be a maximum. In the case of a porous subsoil, it is absolutely necessary for the preservation of its loose materials in their proper position, to have a lining of artificial materials as far as these extend; and as such a lining can hardly be constructed of sufficient strength of less depth than 1 foot, it follows that 2½ feet is the minimum depth also in such a subsoil; but there is this difference betwixt this subsoil and the tenacious one, that the porous may be made as deep as you please, provided you apply sufficient materials for the support of the loose materials. With regard to tilly subsoils, since 1 foot is requisite for the safety of the filling materials, it does not seem an overstretch of liberality to give 6 inches more for extension of the drawing surface, so that the minimum depth in this case seems to be 3 feet, and as much more as the peculiar state of the subsoil in regard to tenacity and porosity will warrant you to go. There is another way of arriving at the same conclusions, and it is this.

(736.) It must be admitted by all drainers, that the part of the drain which is intended to draw water under the earth should be occupied with such loose materials as will easily permit the water to pass through them. It is therefore consonant with reason to give a large area to the sides of a drain in a subsoil that draws water rather slowly, and by consequence a smaller area to one in materials that draw freely, while a drain in pure clay will act chiefly as a channel to convey away the water that is permitted to percolate into it through the superincumbent materials. Keeping these important distinctions in subsoils in view, you shall soon learn what use may be made of them in the construction of efficient drains. I guard myself by saying *efficient* drains; for drains can be ill made, although planned on the most correct principles; and to guard myself against farther misconception, you should bear in mind, that the depths which I have just specified are the *minimum* depths which are considered suitable for the respective circumstances of the drains.

(737.) Viewing drains as mere *channels for the conveyance of water*, it is obvious that the quicker they promote its emission, without injuring themselves or the land, they act the more characteristically; and it is also evident that an open duct will give a freer passage to water than a mass of loose stones, however large or small they may be used. These obvious points being conceded, it follows as a corollary that a drain will act the better of being provided with a duct, along with porous material. Viewing drains as drawers or *gravitators* of water, it is also clear that the more porous the materials, and the greater the quantity used, they will allow the water an easier passage through them. So, it is also requisite on this account to have a duct for the water to pass quickly away. I wish you to pay particular attention to this mode of reasoning in support of the use of ducts, as I conceive that very erroneous opinions prevail among farmers regarding their utility; but I believe such opinions are prompted more on account of the cost incurred by the use of ducts, than from any valid objection that can be urged against their efficacy.

(738.) There are various substances which may be employed as ducts: 1st, dry stones, built as you have seen at *a*, in fig. 156; 2d, a coupling of flat stones set up against each other as a triangle, or in a more rude way two round stones, set one on each side of the drain, with a flat one, or a large round one, to cover them; 3d, tiles made for the purpose. One or all of these forms of ducts answer the purpose well, and should be selected according to the facility of obtaining the materials of which they are composed.

(739.) I must now direct your serious attention to another consideration in the construction of drains. It is a well-known fact that, over whatever kind of substance water flows, it has the power of abrading it; for, besides earthy matter, it will in time wear down by friction the hardest rock. This it is enabled to do, not only by its own physical properties, but by the assistance afforded it by the foreign matters which it almost always holds, both in solution and suspension, so that both physically and chemically it has the power to produce destructive effects. It seems, however, to be a very prevalent opinion among farmers, that hard clay can for any length of time withstand the action of water in a drain. They judge of the hardness of the clay from the state it is in when laid bare to the sight on the drain being opened, imagining that it will remain in the same state, but seeming to forget that water can both soften and scrub against substances. Were clay, indeed, always to retain the hardness it at first exhibits, it would require no protection from the abrading action of water; but, when it is known that it cannot possibly remain so, the safest practice is to afford it protection by a covering, which may be fashioned to suit the purpose, such as a flat stone or tile, both of which obtain the name of *drain-soles*.— If water can affect even the hardest clay, it will, of course, have a much greater effect upon softer earth. The effects usually produced by water on clay subsoils are, that the lower stratum of stones and the tiles become imbedded in it to a considerable depth, as has been found to be the case when drains that have *blown* have been reöpened, and as in the first sets of tile-drains made in Ayrshire. In somewhat softer subsoils, the sandy particles are carried along with the water, and deposited in heaps in the curves and joinings of drains; and, where the subsoil happens to be more sandy than clayey, the foundation which supports the building or tile gives way, and the matter thus displaced forms obstructions at parts which render the drain above them almost useless. Water also carries sand down the sides of the drain, and, where there is no duct, deposits it among the lowest stratum of stones. You thus see that various risks of derangement occur in a drain, where there are no soles to protect its bottom. On this account, I am a strenuous advocate for drain-soles in all cases; and, even where they may really prove of little use, I would rather use too many than too few precautions in draining, because, even in the most favorable circumstances, we cannot tell what change may take place beyond our view of the interior of a drain, which we are never again permitted, and which we have no desire, to see.

(740.) Porous materials, which are the next things he requires for filling drains, are few at the command of the farmer *on his farm*, consisting only, 1st, Of small stones gathered from the surface of the land by the hand; 2d, Small stones so prepared in a quarry by the use of the hammer; and, 3d, Gravel, obtained either from the bed of a river, the sea-beach, or a gravelly knoll.

(741.) Before beginning to break ground for thorough draining, it should be considered what quantity of water the drains will have to convey; and, as the water in the soil is entirely derived from the rain that falls and is absorbed by the soil, its quantity depends upon the climate of the locality in which the drains are desired to be made. Such an investigation is unnecessary in commencing Elkington's mode, as the springs show at once the quantity of water to be conveyed away. In pursuance of the investigation, it is well known that more rain falls on the W. than the E. coast of this country in the ratio 5 : 3; so that, under the same circumstances of soil, nearly double the number or capacity of drains will be required to keep the soil in the same state of dryness in the western as in the eastern coast. With a view to ascertain the quantity, it has been, in the first instance, "ascertained that the water which flows from a drain is considerably less at any one time than what formerly ran on the surface;" and this is an expected result, for evaporation and vegetation together must dissipate much of the water that falls on the ground before it sinks into the soil.

(742.) In order, however, to obtain accurate data on this subject, Mr. James Carmichael, Raploch Farm, Stirlingshire, one of the midland counties, and therefore experiencing about the average fall of rain in Scotland, ascertained that, in a "length of 200 yards, and the distance from drain to drain 18 feet, the *square feet of surface* receiving rain-water for each drain amounts to 10,800; this, at 2 inches of rain in 24 hours, will give 1,800 cubic feet of rain-water, and taking the sectional area of the smallest tile of 2½ by 3 inches at 7.5 inches, and the water moving in this aperture at the rate of 1 mile per hour, the number of cubic feet discharged by the drain in 24 hours will be 6,600, or nearly four times as much as is necessary to carry off so great a fall of rain as 2 inches in 24 hours;" and this besides what would be carried off by evaporation and absorbed by vegetation. Mr. Stirling, of Glenbervie, also in Stirlingshire, has given similar testimony of his experience in regard to the capability of drains to let off water. "I have only three sets of drains," he says, "in which I know the exact fall in the mains near the mouths and the area drained. The land is mostly stiff clay, having in some places a fall of 1 in 6, and for 50 yards from the mouths of the mains only 1 in 140; is drained at 15 feet; the main-tiles are 2¾ by 3½ inches, and the rain which falls on 5 superficial roods is discharged at each mouth. I find the tiles nearly ⅔ full after very heavy rain; therefore that size of tile would, with the same declivity, pass the rain which falls on nearly 2 acres; and, if the fall in the side drains were less, the water would never stand so high in the mains."*

* Prize Essays of the Highland and Agricultural Society, vol. xii.

(743.) It should be borne in mind that these calculations are founded on data obtained from strong clay soil, from which, it may reasonably be supposed, much of the rain that fell had run off, and, consequently, that by a porous soil much more rain will be absorbed; but, although this is doubtless the case, it is obvious that a small orifice will be quite sufficient to carry off much more water than can possibly fall from the heavens in these latitudes in any given time; and that in ordinary rain the drains will be little more than wetted. Still the *drainage should be made to carry off the greatest quantity that falls, although it should occur only once in a lifetime.*

(744.) Having thus calculated the probably greatest quantity of rain that may fall in the locality of your farm, the next step is to *drain each field in succession.* It may seem too indiscriminate an instruction to recommend the draining of every field; for it is possible that some of the fields in your farm may be so dry as not to require entire drainage, but it is scarcely possible but that every field will require draining to a certain degree in some part of it. Be that as it may—in pursuing a system of drainage, every field should be thoroughly examined in regard to its state of wetness throughout the year, for that land is in a bad state which is soaking in winter, though it should be burnt up in summer; but the truth is, burning land requires draining as well as soaked land, because drains will supply moisture to burning land in summer, while it will render soaked land dry in winter. Should your farm be pretty level, it matters not at what side you commence operations; but should it have a decided inclination one way, the lowest portion should first be drained; and, if it inclines in more than one direction, then each plane of inclination should have a system of drains for itself. It deserves consideration, however, in choosing the fields for draining, that as drains are more conveniently made at one member of the rotation of crops than another, it may happen that the field ready in this respect for drainage is not the one situate at the lowest part of the farm; in which case care must be taken to give the water from the drained field such an outlet as will not make the ground below it wetter, and this may be effected either by clearing a ditch along the side of the lower field, or by forming a new ditch, or by leading the water to a ditch, drain or rivulet at some distance.

(745.) The field having thus been fixed upon, the first consideration is the position of those drains that receive the water from the drains that are immediately supplied from the soil; and these are called *main drains.*—In every case they should be provided with a duct, and the ducts may be formed either of stone or of tile—of stone when that material is abundant on the farm, or can be obtained at a short distance—of tile where stone cannot be easily procured; but, if tiles cannot be found at hand, they should be procured from a distance rather than not be obtained at all where stones are scarce.

(746.) *Ducts of stone* may be formed in various ways—the strongest of which are built with masonry and covered with strong flat stones, as in fig. 156.

(747.) Two flat stones, placed against each other at the bottom of the drain, with another covering, both, as at *a*, fig. 158, form an equilateral duct of 6 inches in the side, resting on its apex. It should be held down in its position by small stones *b*, gathered from the land, or broken for the purpose, to a hight of 18 inches; then covered with turf or other dry substance *c*, and the earth *d* returned above them. Where stones are found in sufficient quantity for such a drain, it is highly probable that the subsoil will consist of clay, intermixed with small stones and veins of sand, which, requiring a large area of drawing surface, will fix the depth of the drain at 3 feet. In making this form of duct, the drain will require to be 18 inches wide at top, to allow the drainer room to work while standing on the narrow triangular space at the bottom. Placing the apex of the triangle undermost gives the water power to sweep away any sediment along the narrow bottom; but it possesses the disadvantage of permitting the water to descend by its own gravity, between the joining of the stones, to the subsoil, which runs the risk of being softened into a pulp, or of its sandy portion being carried away; and it is possible for a stone to get jammed in the narrow gutter and form a damming.

(748.) Another form of duct, which I prefer to this, and which is also constructed of stone, may be seen beside it in fig. 159, where *a* is the duct

Fig. 158.

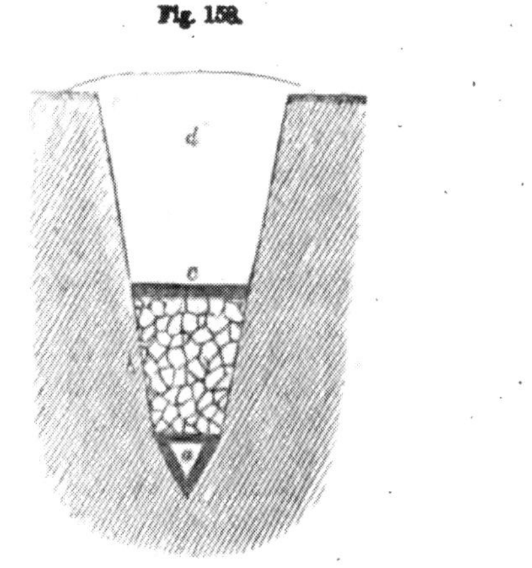

THE TRIANGULAR STONE DUCT.

Fig. 159.

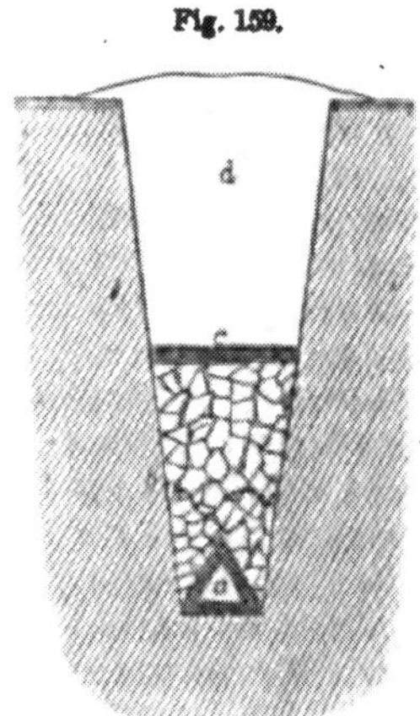

THE COUPLED STONE DUCT.

consisting of a sole lying on the ground, supporting 2 stones meeting at the top, forming an equilateral triangle of 6 inches a side. This form encourages a deposition of sediment to a greater degree than the former, but it prevents, to any dangerous extent, the descent of the water under the sole. Having a flat bottom, the drain can easily be cast out with a width at top of only 15 inches to a depth of 3 feet. The slanting stones of the duct are held in their position by selected stones being placed on each side, which act as wedges between them and the earth; and the whole structure is retained in its place by 18 inches of small stones above them *b*, covered with turf *c*, and the earth *d* returned above them.

(749.) A more perfect duct than either of these is made by a tile and sole. In all main drains, formed of whatever materials, capable of conveying a considerable body of water, a sole is absolutely requisite to protect the ground from being washed away by the water, and a more effectual protection cannot be given to it than by tile and sole. A *main-tile*, which the tiles in main drains are called, of 4 inches wide and 5 inches high, will contain a large body of water; but should 1 such tile be considered insufficient for the purpose, 2 may be placed side by side, as represented by *a* and *b* in fig. 160. Should a still larger space be required, 1 or 2 soles may be placed above these tiles, and other tiles set on them, as *a* and *b* are. Or should a still deeper and heavier body of water be required to pass through a main drain, 1 or 2 tiles can be inverted on the ground on their circular top, as *a*, fig. 161, bearing each a sole *c* upon its open side, and this again surmounted by another tile *b* in its proper position. In such an arrangement, there is some difficulty in making the undermost tile *a* steady on its top; for which purpose, the earth is taken out of a rounded form, and the tile carefully laid and wedged round with stones or earth; but there is greater difficulty in making the uppermost tile *b*

Fig. 160.

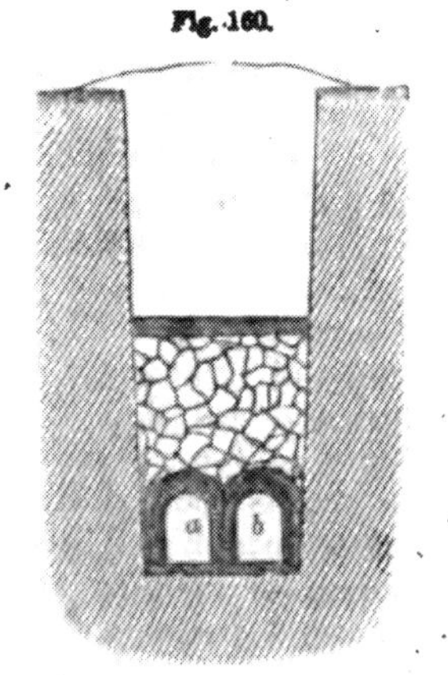

THE DOUBLE TILED MAIN DRAIN.

Fig. 161.

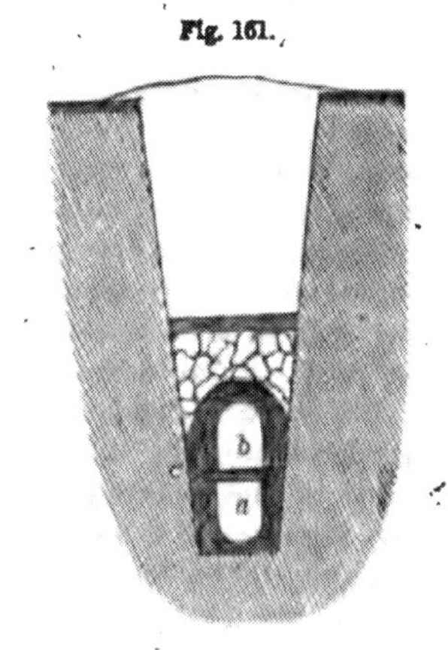

THE INVERTED DOUBLE TILED MAIN DRAIN

stand in that position without a sole, as is recommended by some writers on draining, because the least displacement of either tile will cause the upper one to slip off the edge of the under, and fall into it. In the narrowest of these cases of main drains with tiles, the drains can be easily cut at 15 inches wide at top to the depth of 3 feet. Small stones should be put above the tiles, if at all procurable, to the hight of 18 inches above the bottom; if not procurable, gravel will answer the same purpose; and, if both are beyond reach, they should be enveloped with thin, tough turf, as shown afterward.

(750.) Having thus determined on the construction of main drains, according to circumstances in which the water is to be conveyed away, the next thing is to *fix the place they should occupy in the field.* As they are intended to carry away accumulations of water beyond what they can themselves draw, they should *occupy the lowest parts of a field*, whether along the bottom of a declivity, the end, or the middle of a field. If the field is so flat as to have very little fall, the water may be drawn toward the main drains by making them deeper than the other drains, and as deep as the fall of the outlet will allow. If the field have a uniform declivity one way, one main drain at the bottom will answer every purpose; but, should it have an undulating surface, every hollow of any extent, and every deep hollow of however limited extent, should be furnished with a main drain. No main drain should be put nearer than 5 yards to any tree or hedge, that may possibly push its roots toward it; but although the ditch of a hedge, whose roots lie in the opposite direction, merely receive the surface-water from the field at the lowest end, it should not be converted into a main drain, that should be cut out of the solid ground, and not be nearer than 3 yards to the ditch lip; and the old ditch should be occupied by a small drain, and filled up with earth from the head-ridge.

(751.) As main-drains thus occupy the lowest parts of fields, the *fall* in them cannot be so great as in other parts of the field, though it should be kept quite sufficient for drainage. In the case of a level field, the fall may entirely depend on cutting them deeper at the lowest end than at other places; but, when the fall is small, the duct should be larger than when it is considerable, because the same body of water will require a longer time to flow away. Should the fall vary in the course of the drain, the least rapid parts should be provided with the largest sized tiles; and, in any case, I would recommend an increase of fall on the last few yards toward the outlet, to expedite the egress of the water, and promote an accelerated speed along the whole length of the drain; but, where the fall is rapid enough throughout, there is the less necessity for an increase of acceleration at the termination. It is surprising what a small descent is required for the flow of water in a well-constructed duct. "People frequently complain," says Mr. Smith, "that they cannot find a sufficient fall, or *level*, as they sometimes term it, to carry off the water from their drains. There are few situations where a sufficient fall cannot be found, if due pains are exercised. It has been found in practice that a water-course 30 feet wide and 6 feet deep, giving a transverse sectional area of 180 square feet, will discharge 300 cubic yards of water per minute, and will flow at the rate of 1 mile per hour with a fall of no more than 6 *inches per mile*."* On the principle of the acceleration of water from drains, main drains, where practicable, should be 6 inches deeper than those which fall into them; and the greater depth has the additional advantage of keeping the drains clear of sand, mud, or other substances which might lodge, and not only impede but dam back the water in the drains. Should it so happen, from the nature of the ground, that the fall in the main drains is too rapid for the safety of the materials which construct them, it is easy to cut such a length of the proper fall as the extent of the ground will admit—cutting length after length, and joining every two lengths by an inclined plane.—The inclined planes could be furnished with ducts like the rest of the drain, or, what is better, in order to break the force of the water, like steps of stairs, of brick or stone masonry, built dry. Fig. 162 will illustrate this method at once, where *a b* represents the line of the lowest fall that can be obtained for a main drain in a field; but which, you will observe, is very considerable, and much more so than a main drain should have which has to convey, at any time, a considerable quantity of water. To lessen the fall, let the drain be cut in the form represented by the devious line *c h*, which consists of, first, a level part at the highest end *c d*; then of an inclined plane, *d e*; again of a level part, *e f*; again of an inclined plane, *f g*; and, lastly, of a less level part *g h*, to allow the water to flow rapidly away at the outlet; and this part may be parallel with the inclination of the ground.

(752.) The inclined parts may be filled with materials in different ways. One way is with tiles, as seen from *k* to *l*, where it is obvious that, as drain-tiles are formed square at the ends, those in the inclined plane *k l* cannot conjoin with those on the level above and below, and must, therefore, be broken so as to fit the others at *k* and *l*. In constructing tiles in this way, it is absolutely necessary that the inclined plane be protected with soles, firmly secured from sliding down, at the lowest end at *l*, by having there a strong stone abutting against the lowermost sole; or a better plan would be to line the inclined plane with troughs of hewn stone, which will last for ever.

(753.) Instead of tiles, or hewn troughs, stones may form a conduit upon the inclined plane; and ducts of this material, in such a situation, and built dry with selected stones, would certainly be preferable to tiles, even although they could be obtained of the peculiar form required.

(754.) Or the inclined plane could be conduited with brick, as represented from *r* to *s*. The bricks could be built dry as well as stones, and

* Smith's Remarks on Thorough Draining.

could form either a smooth, inclined sole like tile-soles, or a series of steps, as represented in the figure, where they are set two a side lengthways on bed to form the bottom, as at *o*; set on end upon these for the sides of the

Fig. 162.

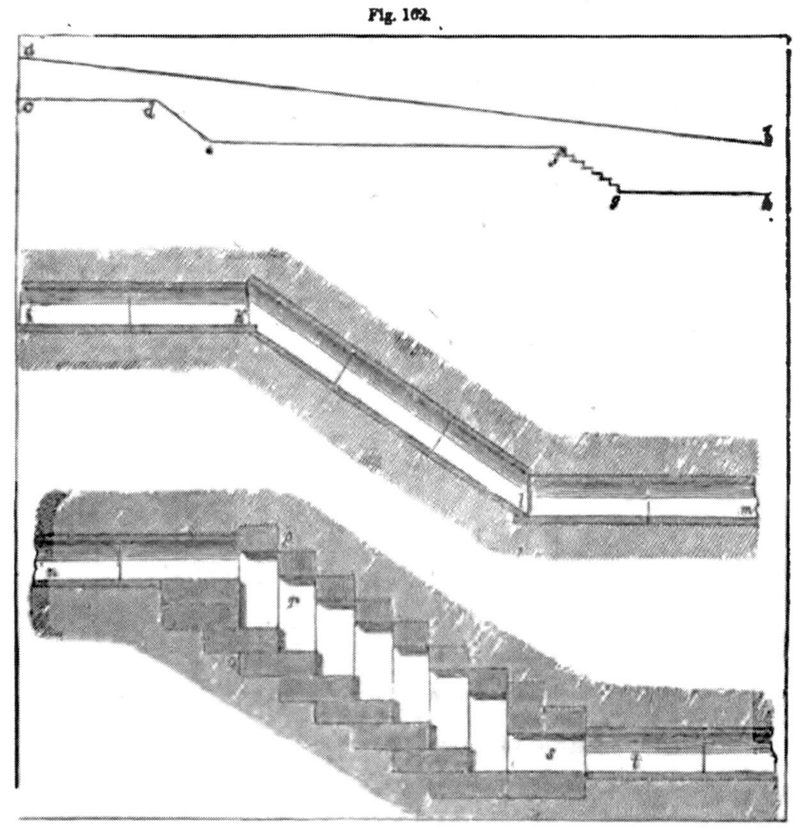

THE DIFFERENT FORMS OF CONDUITS IN THE INCLINED PLAINS OF DRAINS.

conduit, as at *r*; and set lengthways across the conduit upon the upright ones, for the cover, as at *p*. Tiles on the level above connect themselves easily with the bricks, as from *n*; as also on the level below, as at *t*.—Should a considerable run of water be expected at times, the step form is preferable to the smooth, in order to break the fall and impede the velocity of the water, especially toward the lower extremity of the drain, where it may acquire too much momentum without a preventive check of this kind. Although much water is expected to flow through the drain, it would not be prudent to build the steps with lime-mortar, as it is too easily removed, and would not prevent the water finding its way to the foundation; but, in every case, it is proper to build the duct on the inclined planes, with selected materials skillfully put together.

(755.) After having fixed the position of the main drains, and determined their levels and depths as here described, the next thing is the laying off of the *small drains*, which are so placed, or should be so constructed, as to have an easy descent toward the main drains into which they individually discharge their waters. They are usually cut in parallel lines down the declination of the ground; not that all the drains of the same field should be parallel to one another, but only those in the same plane, whatever number of different planes the field may consist of. In a field of one plane, there can be no difficulty in setting off the small drains, as they should all be parallel and all terminate in the same main drain, whether the field is nearly level or has a descent. Thus, in fig. 163, *a* are the

Fig. 163.

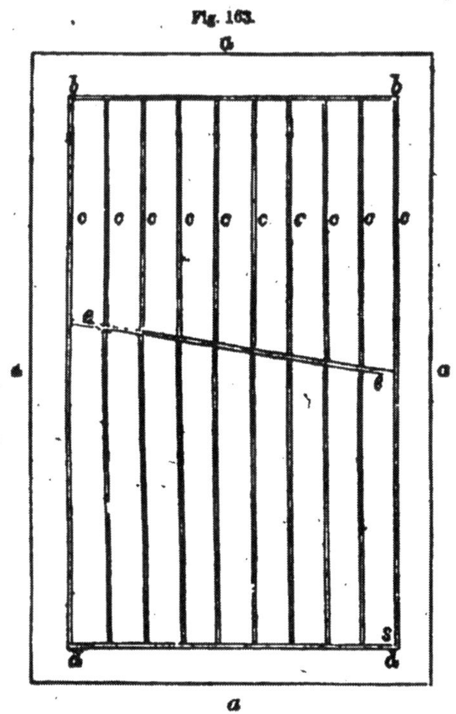

PARALLEL DRAINS IN THE SAME PLANE OF INCLINATION OF THE GROUND.

fences of the field; *d d* is the main drain, whether the field is a level or inclined toward *d*; and *s* is its outlet. In this case, all the drains *c* run parallel to one another, from the one end *b b*, which may be the upper, to the other end *d d*, which may be the lower end; and which convey al the water by the outlet *s*.

Fig. 164.

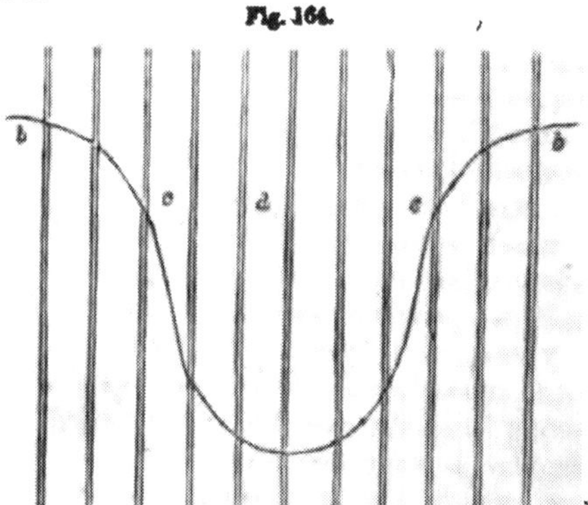

DRAINS IMPROPERLY MADE PARALLEL IRRESPECTIVE OF THE SLOPE OF THE GROUND.

(756.) But when the field has an undulating surface, though the same principle of parallelism is maintained, a different arrangement is followed in regard to it. I have already intimated in a former paragraph (750) that where an undulating surface occurs in a field, a main drain is carried up the hollowest part of it, and the small drains are brought in parallels down the inclination to it This very favorable arrangement for the speedy riddance of water, after it has reached the drain, is not frequently enough attended to. Thus, the common practice is to run the small drains *b c d e b*, in fig. 164, parallel to one another, throughout the whole field, although its undulating surface, as supposed to be represented by the curved line *b c e b*, would cause the so arranged parallelism of the drains at *c* and *e* to run along the sides of the rising ground, where, if any vein of sand occur, it may escape being cut by the drain running parallel along its line either above or below it, instead of being divided across its dip; and even were the sand-vein severed along its length, it would be apt to slip down from the higher side, and render the drain along it inoperative.

(757.) Such drains should be cut, as in fig. 165, up and down the inclined surface *b b*, toward the main drain, which would occupy the line along the

Fig. 165.

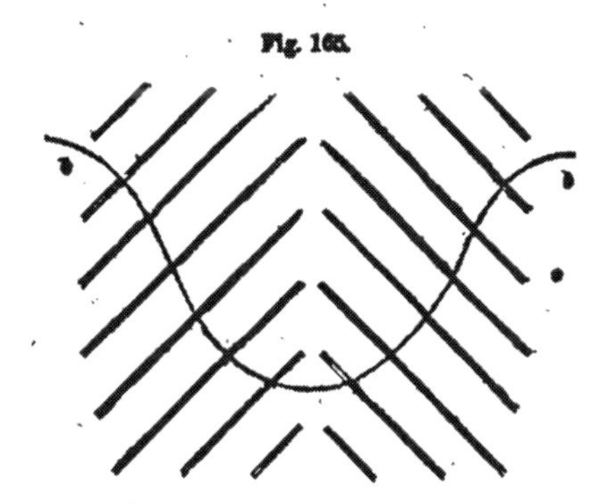

PARALLEL DRAINS IN ACCORDANCE WITH THE SLOPE OF THE GROUND.

points of junction of the drains *b b*. This specific plan is just as easily executed as the other more indiscriminate one of making the direction of every drain of every field alike.

(758.) The next step is to fix the *depth* of drain most suitable for draining the particular field; and this can only be done by having a thorough knowledge of the nature of its subsoil. I have already given reasons for fixing the minimum depth of drains in the different kinds of subsoil, in paragraph (735); but, as the reasoning given there only establishes the principle, it is not sufficient to determine the most proper depth for every peculiarity of circumstances; for this must be determined by the nature of the subsoil which guides the whole affair. If the field present an uniform surface, but inclining, let at least 2 exploratory drains be cut from the bottom to the top of the field, if its extent does not exceed 10 acres, and as many more as it is proportionally larger; and if the subsoil of both is found at once tilly, that is, drawing a little water, let the cut be made 3 feet deep without hesitation. On proceeding up the rising ground, the depth may be increased to 4 feet, to ascertain if that depth will not draw a *great deal more water* than the other. Should the subsoil prove of porous materials, 2½ feet—the minimum—may suffice; though, on going up the rising ground, it may be increased to 3 feet, to see the effect; but should it, on the other hand, prove a pure tenacious clay, 2 feet will suffice at first, increasing the depth in the rising ground to 2½ and even 3 feet; for it may turn out that the stratum under the tenacious clay is porous. Where the surface is in small undulations, the drain should be cut right through both the flat and rising parts. In very flat ground, any considerable variation of depth is impracticable, and only allowable to preserve the fall. From such experimental drains data should be obtained to fix the proper dimensions of the other drains.

(759.) If you find the substratum pretty much alike in all the experimental drains, you may reasonably conclude that the subsoil of the whole field is nearly alike, and that all the drains should be of the same depth; but, should the subsoil prove of different natures in different parts, then the drain should be made of the depth best suited to the nature of the subsoil. A correct judgment, however, of the true nature of the subsoil, cannot be formed immediately on opening a cut; time must be given to the water in the adjoining ridges to find its way to the drain, which, when it has reached, will satisfactorily show the place which supplies the most water; and, if one set of men open all the cuts, by the time the last one has been finished, the first will probably have exhibited its powers of drawing; for it is a fact that drains do not exhibit their powers until some hours after they have been opened. When you are satisfied that the drains have drawn in dry weather as much water as they can, you will be able to see whether or not the shallowest parts have drawn as much as the deepest; and you should then determine on cutting the remainder to the depth which has operated most effectually. If rainy weather ensue during the experiment, still you can observe the comparative effects of the drains, and abide by the results. Never mind though parts of the sides of the cuts fall down during dry or wet weather; they need not be regretted, as they afford excellent indications of the nature of the subsoil, the true structure of which being left by the fall in a much better state for examination than where cut by the spade; and you may then observe whether most water is coming out of the highest or lowest part of the subsoil. It is essential for the durability of drains to bear in mind that they should always stand, if practicable, upon impervious matter, to prevent the escape of the water from the drain by any other channel than the duct.

(760.) You should be made aware that this is not the usual method adopted by farmers for ascertaining the depth to which drains should be cut. The common practice is, knowing that the field stands on tilly bottom, the drains are made of a predetermined depth, and the contract with the laborers is made on that understanding, be the guess true or false, as it may happen. Now, the considerate plan which I have recommended incurs no additional expense, as all the experimental drains will serve their purpose afterward as well as the others; and, even although they should cost more than the same extent of other drains, the satisfaction afforded to the mind of having ascertained the true state of the subsoil more than compensates for any trifling addition of expense which may have been incurred; and be it remembered that any extra expense consists of only scouring out the earth (if any) that may have fallen down, and of supplying more materials to fill up the chasms thereby occasioned. But the ascertainment of the most proper depths for drains in any sort of subsoil is a much more important matter than many farmers seem, to judge from their practice, to be aware of; for by neglecting to descend only ½ a foot, nay, perhaps 3 inches more, many of the benefits of draining may be unattained. I quite agree with the late Mr. Stephens on this subject, when he states that "land may be filled full of small drains, so that the surface shall appear to be dry; but the land thus attempted to be drained will never produce a crop, either in quality or quantity, equal to land that has been *perfectly* drained,"* where a different kind of draining should have been resorted to.

(761.) A very important particular in the art of thorough draining now claims your attention, which is the determining the *distance* that should be left between the drains. It is evident that this point can only be satisfactorily determined after the depths of the drains have been fixed upon, as drains in a porous substratum, which draws water from a long distance, need not, of course, be placed so close together as where the substratum yields water in small quantities, and as drains may be of different depths in the same field, according to the draining powers of the substratum, so they should be placed at different distances in the same field. It is the common practice to fix on the open furrows, between the ridges, for the sites of drains, because the hollow of the open furrow saves a little cutting, though such saving is a trifling consideration compared to the advantage of executing the drains in the best manner. For my part, I can see no greater claim for a drain in the furrow than in any other part of the ridge, especially as most of the water should be received from the subsoil rather than the surface, except in pure clay-soils; and it is, of course, as easy to make them in any other part of the ridge as in the open furrow. These observations of Mr. Smith on this subject court remark. "When the ridges of the field," says he, "have been formerly much raised, it suits very well to run a drain up every furrow, which saves some depth of cutting. The feering being thereafter made over the drains, the hollow is filled up, and the general surface ultimately becomes level." This is all very well for the purpose of leveling the ground, but mark what follows! "When the field is again ridged," he continues, "the drains may be kept in the crowns or middle of the ridges; but, if it be intended to work the field so as to alternate the crowns and furrows, then the ridges should be of a breadth equal to *double* the distance from drain to drain; and by setting off the furrows in the middle, betwixt two drains, the crowns will be in the same position; so that when the furrows take the places of the crowns, they will still be in the middle betwixt two drains, which will prevent the risk of surface-water getting access to the drain from the water-furrows by any direct opening."† No doubt, it is easy to transpose furrows into crowns, and *vice versa;* but how would the transposition be effected in these circumstances, since the drains were made in each former furrow, and it is proposed to make the crowns of the ridges between the drains, the transposition of the crowns could only be effected by adopting the unfarmerlike plan of leaving, in a finished field, a half of the breadth of the ridge adopted at each side; and, rather than practice such slovenliness, would it not be better to cut the drains in the middle of the ridges, and preserve each ridge unbroken?

(762.) With regard to distances between drains, in a partially impervious subsoil, 15 feet are as great a distance as a 3-feet drain can be expected to draw; and, in some cases, I have no doubt that a 4-feet one will be required. In more porous matter, a 3-feet drain will probably draw 20 feet, with as great if not greater effect; and in the case of a mouldy, deep soil, resting on an impervious subsoil—which is not an uncommon combination of soils in the turnip-districts of this country—a drain passing through the mould, and resting perhaps 3 or 4 inches in the impervious clay—which may altogether make it 4 feet deep—will draw, I have no doubt, a distance of 30 feet. More than 30 feet distant, I would feel exceeding reluctance to recommend drains being made, unless the circumstances were

* Quarterly Journal of Agriculture, vol. iii. † Smith's Remarks on Thorough Draining, 4th edition.

remarkably singular, when, of course, a special thing must be done for a special case, such as an entirely porous subsoil containing somewhat indurated portions, when a drain through each of these, at whatever distance, will suffice to keep the whole dry.

(763.) The distance at which ordinary drains in tilly subsoils will not draw is not left to conjecture, but has partially been determined by experiment. Conceiving that a drain in every furrow, in a tilly subsoil, is attended with more expense than any anticipated increase of produce from the soil would warrant, a farmer in East Lothian put a drain in every *fourth* furrow; and that they might, as he conceived, have a chance of drawing at that distance, he caused them to be cut 4 feet deep. A figure will best illustrate the results, where the black lines *a*, fig. 166, are the

Fig. 166.

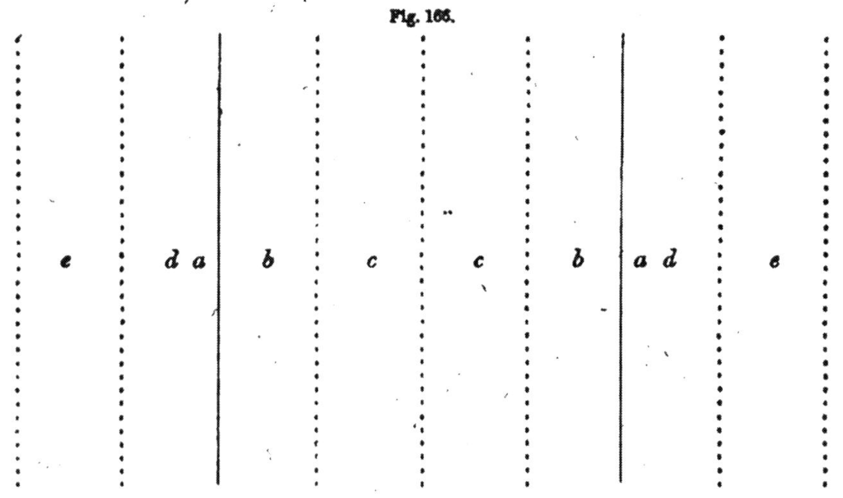

THE EFFECTS OF TOO GREAT A DISTANCE BETWIXT DRAINS.

drains between every fourth furrow, and the dotted lines represent the intermediate undrained furrows; and where it is evident, at the first glance, that the drains *a* have to dry 2 ridges on each side *b c* and *d e*, of which we should expect that the 2 ridges *b* and *d*, being nearest to *a*, should be more dried, in the same time, than the 2 farthest ridges *c* and *e*, and the result agrees with expectation; but still, had the subsoil been of an entirely porous nature, both ridges might have been sufficiently dried by *a*. Trusting to similar contingency, it is not an unusual expectation, entertained by many farmers, that a drain will sufficiently dry 2 ridges on each side, or at least 1 ridge on each side, without ascertaining the exact nature of the subsoil. But mark the results of this particular experiment, which was conducted with the usual expectations. The 2 ridges *b* and *d*, nearest to *a*, actually produced 9 bushels of corn more per acre than the 2 more distant ridges *c* and *e*. This is a great difference of produce from adjoining grounds under the same treatment, and yet it does not show the entire advantage that may be obtained by drained over undrained land, because it is possible that the drain *a* also partially drained the distant ridges *c* and *e*; and this being possible, together with the circumstance that none of the ridges had a drain on each side, it cannot be maintained that either the absolute or the comparative drying power of these 4-feet drains was exactly ascertained by this experiment.* It may be conceived, however, that

* Quarterly Journal of Agriculture, vol. viii.

if the drains had been put into every other, instead of every fourth, furrow, that the produce of all the ridges would have been alike, inasmuch as every ridge would then have been placed in the same relative position to a drain; and the conjecture seems so reasonable that most farmers, from what I observe of their practice, act upon this plan as from a settled opinion. But such a conjecture, not having been founded upon experience, cannot have the force of an opinion, especially when opposed to the great probability that, *cæteris paribus*, land must be more effectually drained by a drain in every furrow than at greater distances; as it is not supposable that the open furrows of *b* and *d* can be so thoroughly drained as the furrow *a*, which contains the drain itself; because it is obvious that the one side of a ridge should be *less* effectually drained than the other, which, if of a *retentive subsoil*, may not be affected at all. All, then, that has been *demonstrated* by this experiment is this, and the proof I consider is important, that a drain—a deep one though it be—will draw water more effectually across one than across two ridges; and it should be useful to you as a guide against imitating the practice of those who seem to believe that a drain cannot have too much to do.

(764.) While taking this view of the subject, I cannot agree in the advice which Mr. Smith gives when he says, "In cases where time or capital are wanting to complete the drainage at once, each alternate drain may be executed in the first instance, and the remainder can be done in the next time the field is to be broken up."* I would much rather use the words of Mr. Stirling, of Glenbervie, where he says that "I think it a *great error* to make the half the number of drains required at first, with the intention of putting one between each at a future period. Let what is drained be done as thoroughly as the farmer's exchequer will allow; the farm will be gone over in as short a time, and much more profitably."—The reason which Mr. Stirling gives for holding this opinion is a true and practical one—namely, because "a tid (or proper condition of the ground for harrowing) cannot be taken advantage of on the drained furrow until the other is dry, and the benefit of an extended period for performing the various operations of the farm is thus lost."† Every farmer who has studied the influence which soil possesses over crops will be ready to allow that wet soil does much more injury to the dry soil in its neighborhood than dry soil does good to the wet. I would, under every circumstance of season and soil, prefer having the half of my farm thoroughly drained, than the whole of it only half drained.

(765.) At whatever distances drains are placed, they should run *nearly at right angles to the main drains.* Excepting in confined hollows, having steep ascents on both sides, the drains should run parallel with the ridges, and always parallel with themselves, in the drainage of the same plane of the field. Drains should be carried through the whole length of the field, irrespective of the wet or dry appearances of parts of it; because uniform and complete dryness is the object aimed at by draining, and portions of land that seem dry at one time may be injuriously wet at others, and these may seem dry on the surface when the subsoil may be in a state of injurious wetness.

(766.) Regarding the *direction which the drains should run in reference to the inclination of the ground*, so as to dry the land most effectually, much diversity of opinion at one time existed; but I believe most farmers are now of the opinion that it should follow the inclination of the ground.—The late Mr. Stephens maintained, and as I think erroneously, that as it

* Smith's Remarks on Thorough Draining, 4th edition.
† Prize Essays of the Highland and Agricultural Society, vol. xii.

is evident that water within the earth, or on the surface, seeks a level where the fall through the porous subsoil is greatest; *therefore* a drain made across the slope or declivity of a field, or any piece of land, will undoubtedly intercept more water than when it is carried straight up the bank or rising ground; and this principle, he says, holds good in every case, whether "the drain be made to receive surface or subterraneous water." I confess I cannot arrive at the conclusion from the premises. He reiterates the same opinion more generally, and apparently more practically, in these words: "Drains winding across the slope or declivity of a field, whatever their number or depth may be, their effect upon tenacious or impervious substrata will be much greater than if they were made straight up and down the slope; and when the soil is mixed with thin strata of fine sand, which is the case in nine times out of ten, the effect will be increased in proportion; and, accordingly, a much less number will answer the purpose, the expense will be greatly lessened, and the land and occupier much more benefited in every respect."* Mr. Smith opposes this opinion, and, what is remarkable, uses the same illustration to refute it, in regard to the property of water and the structure of the substratum, as Mr. Stephens did in support of his views. "Drains," says he, "drawn across a steep, cut the strata or layers of subsoil transversely; and, as the stratification generally lies in sheets at an angle to the surface (see fig. 169), the water passing in or between the strata, immediately below the bottom of one drain, nearly comes to the surface before reaching the next lower drain. But, as water seeks the lowest level in all directions, if the strata be cut longitudinally by a drain directed down the steeps, the bottom of which cuts each stratum to the same distance from the surface, the water will flow into the drain at the intersecting point of each sheet or layer, on a level with the bottom of the drain, leaving one uniform depth of dry soil."† Without taking any other element at present into the argument than the single proposition in hydraulics that water seeks the lowest level in all directions, adduced by Mr. Stephens himself, I shall prove the accu-

Fig. 167.

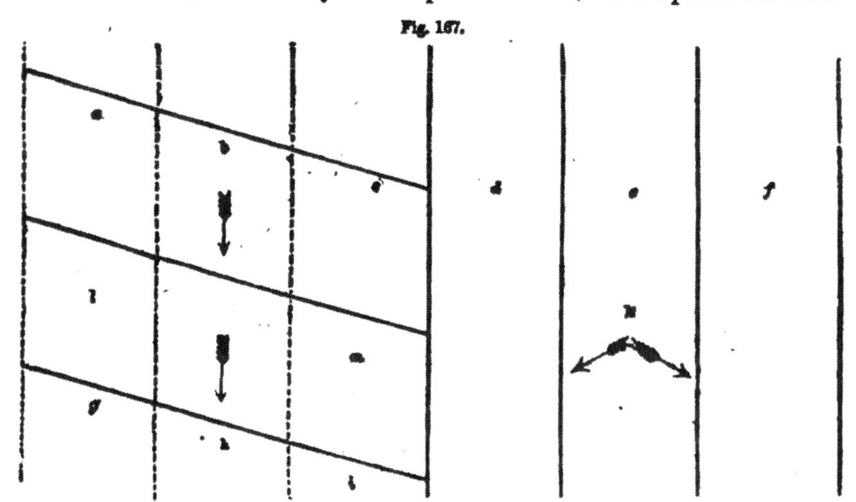

THE COMPARATIVE EFFICACY OF DRAINS ACROSS AND ALONG RIDGES ON A DECLIVITY.

racy of Mr. Smith's conclusions by simply referring to fig. 167, which represents a part of a field all having the same, and that a steep acclivity, and

* Stephens's Practical Irrigator and Drainer.

† Smith's Remarks on Thorough Draining.

which is laid off in the ridges *a b c d e f*, up and down the slope; but the 3 ridges *a b c* have drains across them, and the other 3 ridges have drains parallel with them, the oblique drains being made at the same distance from each other as the up and down ones, whatever that distance may be. Now, when rain falls on and is absorbed by the ridges *a b c d e f*, it will naturally make its way to the lowest level, that is, to the bottom of the drains; and, as the ground has the same declivity, the water will descend according to the circumstances which are presented to it by the positions of the respective systems of drains. On the ridges *d e f*, having the drains parallel to them, and up and down the inclination of the ground, the water will take a diagonal direction toward the bottom of the drains, as indicated by the deflected arrows at *k;* and as ground has seldom only one plane of declination, such as straight up and down, but more commonly two, another in the direction either from *a* to *f* or from *f* to *a*, it follows that the one side, that is, the lower side of a ridge thus situated, will be sooner drained than the other; but both sides will be soon drained, as may be seen in fig. 168, where *a b* are vertical sections of small drains, each 30

Fig. 168.

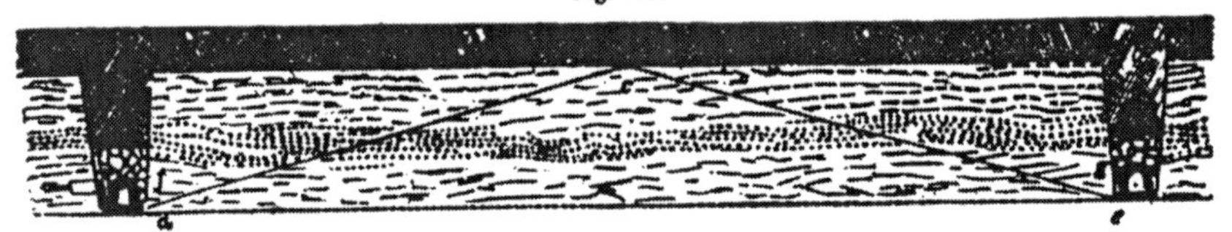

THE DESCENT OF WATER FROM A RIDGE INTO A DRAIN ON EACH SIDE.

inches deep; *c* 1 foot of mould, in which the rain is absorbed as fast as it falls upon the ridge, 15 feet broad, betwixt *a* and *b*. On being absorbed, the rain, seeking the lowest level, will be hastened toward the drains *a* and *b* in the direction of *c d* and *c e*—that is, by a fall of 30 inches in about 8 feet, which is a rapid fall of rather more than 1 in 3, and which rapid fall, as is well known, will clear water quickly, and in the clearance of which the drains have only each to draw a distance of half a ridge, or 7½ feet. Whereas on the ridges *a b c*, which have oblique drains, *a, l, g*, fig. 167, the water will have to run in the direction of the arrows *b* and *h*, in doing which it will have to traverse the entire breadth of the ground betwixt *a* and *l* or *l* and *g*, that is, 15 feet, just double the distance the other drains have. But take the superficial view of the case, and suppose that *d, e, f*, and *a, l, g*, are not drains, but open furrows, it is clear that, when rain falls, the water will flow toward *d, e*, or *f*, as indicated by the arrows at *k*—that is, 7½ feet toward each furrow; whereas the water that falls on *a c, l m*, or *g i*, will have to run across the entire breadth of the ridge from *a* to *l*, or from *l* to *g*, that is, 15 feet, just double the distance of the other, before it can reach the open furrows. Or rather take the more profound case, and trace the progress of the water through the substrata. Mr. Thomson, Hangingside, Linlithgowshire, drained 150 acres of land having an inclination varying from 1 in 10 to 1 in 30. Portions of 3 fields had drains put into them in 1828, 1829 and 1830, in the oblique direction, and, finding them unsuccessful, he put them in the direction of the slope, like the rest of the fields. "In order," says he, "to ascertain the cause of these failures, a cut was made in the field first referred to, entering at a given point, and carrying forward a level to a considerable depth, when it was clearly seen that the substrata, instead of taking in any degree the inclination of the surface, lay horizontally, as represented in fig. 169. It is therefore obvious," he continues, "that in making drains *across* a sloping surface, unless they are put in at the precise point where the substrata crop out (and these are exceedingly irregular in point of thickness), they may in a great measure prove nugatory; because, although one drain is near

Fig. 169.

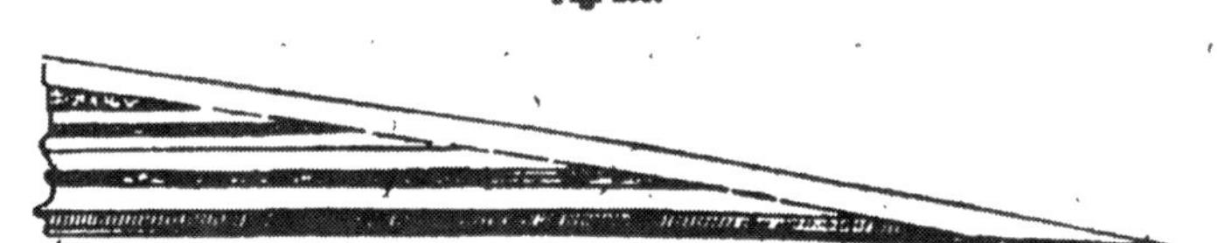

THE USUAL POSITION OF SUBSTRATA IN REFERENCE TO THE SURFACE SOIL.

another, from the rise of the ground, none of them may reach the point sought; whereas, in carrying a drain right up the direction of a slope, it is impossible to miss the extremity of every substratum passed through."* And although a drain in the oblique direction should cut through a vein of sand as at *f*, fig. 168, and thereby carry off the water it contains, yet it cannot be denied that the drains *a* and *b* will also cut through the said vein—which, when they do, what is there to prevent the water in the vein running toward the drains *a* and *b* on each side of the ridge toward *f* and *g*? These observations of Mr. Thomson corroborate Mr. Smith's views, in which I entirely concur.

(767.) In all cases of thorough draining there should a small drain connect the tops of the drains at the upper end of the field. The object of this drain is, in the first instance, to dry the upper head-ridge, and also to protect the upper ends of the ridges from any oozings of water that might come from the fence ditch, or from any rising ground beyond that end of the field. If the fence ditch conveys no current of water, and the hedge-roots lie away from the fiald, and there are no hedge-row trees near at hand, this drain may be made in the ditch itself, and the ends of the furrow-drains brought across the head-ridge to it; but, should water or trees be connected with the ditch, the drain should be made on the head-ridge not nearer than 3 yards from the ditch lip; and it should be of the same depth, though not deeper than the other drains.

(768.) When drains have a course along very long ridges, it is recommended to run a *sub-main* drain in an oblique direction from side to side, or rather only across all the long ridges of the field, as represented by *e e* in fig. 163. The length of any drain, it is maintained, should not exceed 200 yards, without a sub-main drain to assist in carrying off the water; and the reasons assigned by Mr. Carmichael for requiring the assistance of such a drain are, "because, if the fall is considerable, the bottom may be endangered by the velocity and volume of water collected during continued rain; or if the declivity be very limited, and the aperture small, the drain is in danger of bursting from an impeded discharge;" but a complete answer to these apprehensions is found in the very next sentence, namely, "the rule is to apportion the area of all drains to their length, declivity, and distance from each other."† It is quite true what Mr. Smith says on the subject, that "some people are still prone to the practice of throwing in a cross-drain, or to branches going off at right angles, which are of no farther avail in drying the land, while they increase the length of drain without a proportionate increase of the area drained."‡ Should the want, however, of proper sized tiles, in any particular part of a field, where the quantity of water is greater than over the ordinary surface of the farm,

* Prize Essays of the Highland and Agricultural Society, vol. xiii. † Ibid. vol. xii.
‡ Smith's Remarks on Thorough Draining

induce you to incur the expense of a sub-main drain rather than run the risk of injuring the land by the dreaded insufficiency of the drains below, it should be directed across the field as shown by *e* in fig. 163, where, if cut of the same depth as the other small drains, those below should be disjoined from it by a narrow strip of ground in the line of *e* to *e*; but a much better plan is to make the sub-main 6 inches deeper than the rest of the drains, where it can be so deepened, and it will intercept the water coming from the ground above, while the drains will pass continuously over it. In such a case, when the sub-main *e* falls into the small drain *b d* at the side of the field at *e*, that part of the latter below *e* to *s* should be converted into a sub-main, which should be larger than small drains, though sub-mains need not be so capacious as main drains; but, in truth, in such an arrangement as this, sub-mains become mains, inasmuch as they convey as great a quantity of water.

(769.) The *experimental cuts* having been made, you become acquainted with the nature of the subsoil, and determine upon the depth of the drains; then cutting should be proceeded with forthwith, and this part of the work is best and most satisfactorily done by contracting with an experienced spadesman, at so much per rood of 6 yards.* The rates of cutting are generally well understood in the country. Let me impress upon you, in the matter of making a contract, the great satisfaction you will feel in engaging stout, active and *skillful* men; for although you may find men able to work a hard day's work, if they are nevertheless unskillful and inexperienced, you will experience many difficulties. Such men willingly take on work at low rates; but you will find it conducive to your interest rather to give such rates as will enable skillful workmen to earn good wages, than save a little money by employing rough, bungling hands; for there is no comparison between the advantages derivable from good and bad work.

(770.) The cutting of the drains is commenced by that of the main drain which terminates at the outlet, and the operation is commenced at the outlet, or lowest part of the field. The commencement of the operation is done in the same way as pointed out in the drains of the Elkington method—namely, by stretching the garden-line, and rutting off the breadth at top with the common spade by the principal man of the party. A second man then removes the top-mould with the spade; and if the subsoil is of strong clay, or tiles alone are to be used in filling the drains, he lays the mould on one side of the drain, and the subsoil on the other. In other kinds of soils and subsoils, and where stones are to be used in conjunction with tiles, the separation of the soils is not necessary. The reasons for this distinction in the use of the soils will be given a little farther on. The principal man, or contractor, follows, and shovels off all the mould, working with his face to the first man. A third man—for the gang or set of drainers should consist of 3, for expeditious and clean work—loosens the top of the subsoil with the tramp-pick, fig. 37, and proceeds backward with the picking, while the other men are removing the mould along the break or division measured off by the line, perhaps 60 or 70 yards. The second man then removes the loosened subsoil with the spade in fig. 170, which is narrower than the common spade, being 6 inches wide at the point, digging with his back to the face of the picker—that is, working backward; and the leading man follows with a narrow-pointed shovel, fig. 38, called

* It would be extremely convenient and highly satisfactory were the lineal measure of the *rood*, in which all country work is estimated, fixed of the same length throughout the kingdom, as the great diversities existing in this measure are truly perplexing. I cannot see the utility of a general law on weights and measures, if such anomalies as this, and many others, are allowed to exist.

the ditcher's or hedger's shovel, with which he trims the sides of the drain, and shovels out the loose part of the subsoil left by the digger.

(771.) Should the drain be very wet, owing to a great fall of rain, or the cut draw much water from the porosity of the subsoil, to secure a proper consistence to the drain, it is better to leave off the digging at this stage of the work, and proceed to set off another length of line at the top; and, indeed, in such circumstances, it would be expedient to remove the top of the whole length of the particular drain *in hand*, to allow the water time to run off, and the sides of the drain to harden, as perseverance in digging to the bottom, in the circumstances, would be attended with risk of the sides falling in to a considerable extent. This precaution in digging drains is the more necessary to be adopted in digging narrow shallow drains than deep ones, as planks cannot be used in them to support the falling sides, as in fig. 154, because the men could not find room in small drains to work *below* the wedges which keep up the planks. Should the ground be firm, or no inconvenient quantity of water be present in the drain, the digging, of course, may properly be proceeded with to the bottom at once.

Fig. 170.

THE NARROW DRAIN SPADE.

(772.) To effect this, the picking is renewed at the lower part of the drain, and another spit of earth thrown out with a still narrower though of the same form of spade as in the last figure, being only 4 inches wide at the point. The leading man trims down the sides of the drain with this spade, and pulls out the remaining loose earth toward him with the scoop, such as in fig. 171; or throws it out with such a

Fig. 171.

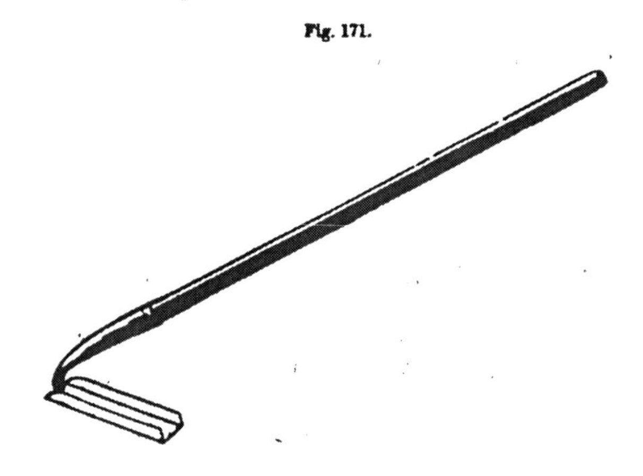

THE EARTH DRAIN SCOOP.

scoop as in fig. 152; and thus finishes the bottom and sides in a neat, even, clean, square, and workmanlike style.

(773.) What with the experimental cuts, and these first two spits of digging below the mould, you will be easily able to determine the drawing property of the subsoil, and, consequently, the depth the drain should go. If the subsoil prove tilly, but still drawing a little water below the mould downward, the drain should certainly be 3 feet deep, and 15 inches wide at top; if of intermixed and minute veins of sand, and otherwise of good drawing materials, then 30 inches of depth will suffice, and 12 inches of width at top; if of quite impervious clay, 2 feet deep and 10 inches of width at top will be found sufficient. It is right to cut the drain a little deeper where there is any sudden rise of the surface, and a little shallower where there are any sudden hollows, than to follow the undulations of the ground where these are trifling. As to the distances betwixt the drains in the first case of a tilly but drawing bottom, 15 feet asunder is, in my opinion, quite wide enough. In the second case of a drawing subsoil, drains at 30 feet asunder will effect as much as in the former case. And, as to pure clays, as 15 feet is too wide a distance, I would prefer 12 feet; but, to suit the ridges, there should be a drain in every open furrow, whatever distance asunder these may be.

(774.) In *filling* drains, it is a common practice with farmers to put in the materials as the drain proceeds in the digging—which, I conceive, is an objectionable proceeding. I think the whole length of the particular drain in hand should be entirely cleared out to the specified dimensions before the filling commence; because it is necessary, in the first place, that the state of the work be inspected, in accordance with the specification, before taking it off the contractor's hands; and inspection implies measurement of the contents in depth and breadth, and the fall of the bottom—whether it be regular throughout, where the slope of the ground is regular, or sufficient, where the general fall of the ground is small; or whether the fall is preserved in all the places where the ground is irregular. These are not trifling considerations, but essential; so much so, indeed, that the very efficacy of a drain as a conductor of water entirely depends upon them.

(775.) The *fall of the ground* can at any time be ascertained by the workmen by a simple contrivance. As the bottom of the drain is cleared out, a damming of 4 to 6 inches high will intercept and collect the water seeking its way along the bottom, and by this it can be seen whether the level line of the water cuts the bottom of the drain as far up as it should do according to the specified fall; and a succession of such dammings will preserve the fall all the way up the drain. When the weather is very dry, and a sufficiency of water wanting in the drain to adopt this mode of testing the fall, a few buckets of water thrown in will detect it, and of course it is only on comparatively level ground that such expedients as these are at all required.

(776.) Another reason for filling drains in this shallow mode of draining, where they are necessarily numerous, is from the upper to the lower end —and not from the lower to the upper, as is too commonly the practice— that the bottom of the drain should be cleared out most effectually with the scoop before the materials are put in, and this is best and most easily done down the natural declivity of the ground; and besides, in doing this it is at once seen whether the fall has been preserved, by the following of the water down the declivity. In deep draining the case is otherwise, because in that case the drains being few in number, and each possessing importance, the falls should be previously determined by leveling, and the amount of each leveling marked, by which means they can be preserved as the filling proceeds; and, besides, there would be risk of a deep drain, which may be of considerable length, and take a long time to throw out, falling in, to allow it to remain open for a length of time.

(777.) Of the materials for filling drains, I shall first notice *stones*, not only because they have hitherto been the most common material, but have been for the longest time employed for the purpose. Drain stones are usually derived from two sources: 1. From the surface of the land; and when they are small and round, not exceeding the size of a goose's egg, no other material is equal to them in durability for the purposes of a drain and, 2. From the quarry, where they must be broken with hammers, like road-metal, to the smallness of from 2½ to 4 inches in diameter. It is a pernicious, and, indeed, an obviously absurd practice, to mix promiscuously stones of different sizes in a drain, as such can never assort together, and nothing can be more absurd than to throw in a stone which nearly fills up the bottom of a drain, and is sure to make a dam across it to intercept water. All large land-stones should be broken into small pieces, and any large, angular piece should not be put near the bottom, which should be kept as open as possible. Stones broken in the quarry are always angular, and in so far they are of an objectionable shape; because on fitting together, face to face, they can become a more compact body than round stones possibly can. No doubt, no ordinary pressure upon a body of earth 18 inches deep could squeeze small, broken stones together so as entirely to compress the spaces between them; but gravity, continually acting on loose bodies, will in time consolidate small stones more and more; and heavy labor on the surface, and subsidence of water through the earth, assist by their action to produce a similar result; and we all know that macadamization makes a much more compact road than the old fashioned large, round stones.

(778.) Stones should never be broken at the side of the drain. I quite agree with Mr. Stirling when he says that "I prefer breaking stones in a *bin*. It is more easy to check the size, and it is done cheaper, as otherwise each heap has to be begun on the sward, and many of the stones are forced into the ground, which adds to the difficulty of lifting them. There will be a saving in carting the stones large, but it will be fully balanced by this disadvantage. I would deprecate of all practice that of breaking the stones in the field, and filling by the chain. This may be contracted for

Fig. 172.

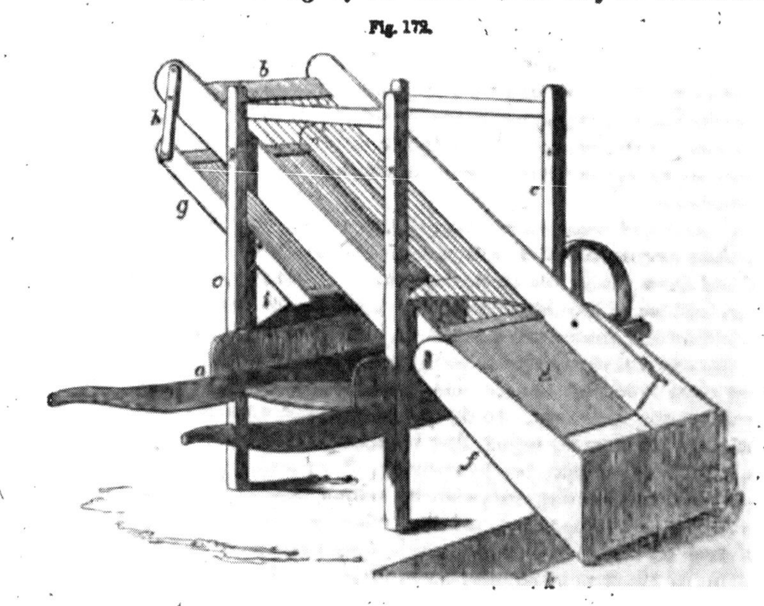

THE DRAIN STONE HARP OR SCREEN.

at a low rate, but it is easy to guess how the contractor makes wages."*
But although I would greatly prefer small, round stones to angular ones

* Prize Essays of the Highland and Agricultural Society, vol. xii.

for drains, yet as the places that afford small, round stones naturally are very limited in number, and draining, if confined to such localities, would be as limited, it is far better to take any sort of quarried stones than leave land undrained, and there is no doubt that almost every sort of stones forms an efficient and durable drain if employed in a proper manner.

(779.) As I am acquainted with no drainer who has bestowed so much pains in the breaking, preparing, and putting in stones into drains, as Mr. Roberton, I shall describe his method of managing quarried stones; and first in regard to the implements used by him for that purpose. 1. There is a portable *screen* or *harp* for riddling and depositing the stones, as seen in fig. 172, which consists of "a wheelbarrow *a*, over and across which is suspended a screen *b*, having the bars more or less apart, according to the description of materials intended to be used. The upper end is hung upon two posts *c c* about 3 feet above the barrow; the lower end rests upon the opposite side of the barrow. To this lower end is affixed a spout *d*, attached about 10 inches from the lower extremity of which is a board *e*, by means of two arms *f*. Another screen *g*, about one-half the length, and having the bars about half an inch apart, is hung parallel, about 10 inches below the larger one. The upper end of *g* is fixed by means of two small iron bars *h* to the upper end of the larger screen; the lower end rests upon a board *i* sloping outward upon the side of the barrow opposite to that on which the spout *d* is situate." 2. A movable trough, or, as it is commonly called, a *tail-board*, *a*, fig. 173, is attached to the hind part of a

Fig. 173.

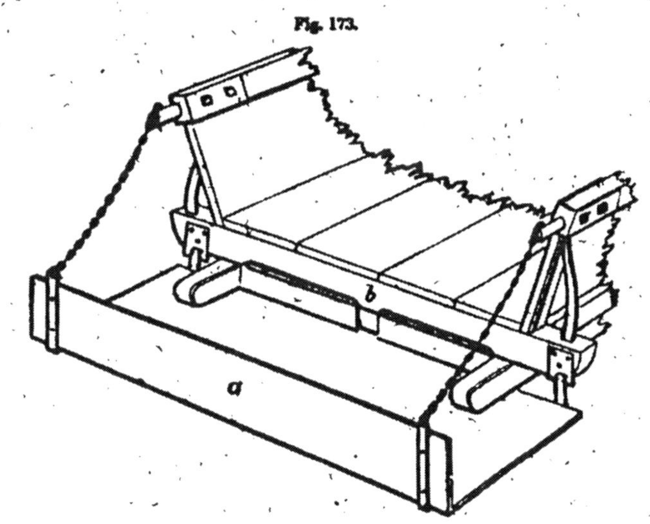

THE TAIL-BOARD TROUGH FOR RECEIVING THE DRAIN STONES IN THEIR FALL.

cart, for the purpose of receiving any stones that may drop while the workmen are shoveling them out of the cart. A portion of the hind part of a cart *b* shows the manner in which it is affixed to it. 3. Fig. 174 is a small

Fig. 174.

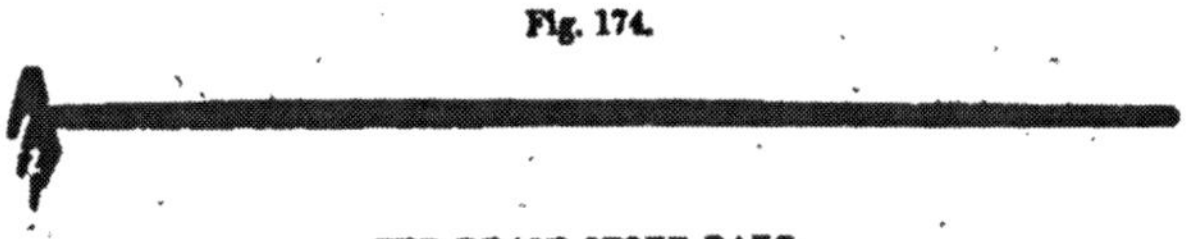

THE DRAIN STONE RAKE.

iron *rake*, used by the workman in charge of the screen, "for the purpose of making the surface of the larger stones of a uniform hight before being covered with the smaller." 4. Fig. 175 is called a "*beater*, which is a square piece of wood the width of the drain, used for beating the smaller stones into the interstices of the larger ones, and thus leveling the surface of the drain."*

Fig. 175.

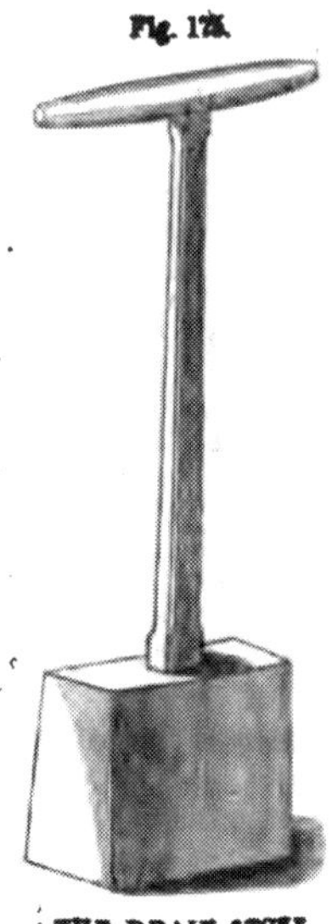

THE DRAIN STONE BEATER.

(780.) The *stones are put in* in this manner: The earth is all put on one side of the drain. The barrow-screen is placed on the other, so as the board *e*, fig. 172, attached to the lower end of the spout *d*, shall reach the opposite side of the drain *k*. The cart, with a load of broken stones from the bin, is brought to the same side of the drain as the barrow, and a little in advance, and there the tail-board *a*, fig. 173, is attached to the hinder part of it. The carter then shovels the stones out of the cart, and empties them over the top of the screen. In doing this, some care is requisite; for, if the stones are thrown over the screen with force, they will not alight sooner than half-way down the screen, and thus its screening efficacy will be impaired. The proper method is to rest the shovel on the top of the screen, which part should be shod with plate-iron, and merely turn it over, by which a separation of the stones is at once effected—the larger ones, rolling down, strike against the board *e*, fig. 172, and drop into the middle of the drain, without disturbing the earth on either side. The smaller ones, at the same time, pass through the upper screen *b*, and, being separated from the rubbish by falling on the lower screen *g*, roll down into the barrow *a*, while the rubbish descends to the ground on the side of the barow farthest from the drain.

(781.) The best form of shovel for putting the stones over the top of the screen is what is called a frying-pan or lime shovel, represented by fig. 176,

Fig. 176.

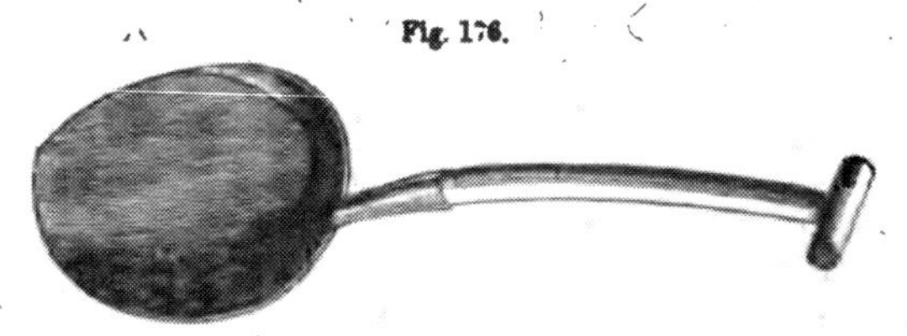

THE FRYING-PAN OR LIME-SHOVEL.

the raised back of which keeps the stones in a collected form until they are turned over the screen, and its point secures an easy access along the bottom of the cart under the stones. Such shovels are much in use for spreading lime and shoveling up the bottoms of dunghills in the border counties of Scotland, and they cost 3s. 10d. each of medium size, ready for use.

(782.) One man takes charge of the filling of the drain. His duties are to move the barrow forward along its side, as the larger stones are filled to the required hight; to level them with the rake, fig. 174; to shovel the smaller stones from the barrow, spread them regularly over the top of the larger, and beat them down with the beater, fig. 175, so as to form a close and level surface through which no earth may pass. When the stones are broken in the quarry so as to pass through a ring 4 inches in diameter, a quarter of them is so small, or should be made so small, as to pass through

* Prize Essays of the Highland and Agricultural Society, vol. xiv.

the wires of the upper screen *b*, fig. 172, which are 1¾ inches apart; and they then will be found sufficient to give the top of the drain a covering of 2 or 3 inches deep, which, being beaten closely down, requires neither straw, turf, or anything else to cover them.

(783.) With regard to *covering with vegetable substances*, Mr. Roberton says, with much probable truth, that "the only possible use of a covering of straw or turf is to prevent any of the earth, when thrown back into the drain, getting down among the stones; but it is evident that such a covering will soon decay, and then it becomes really injurious, because, being lighter (and finer) than the soil, it will, when decomposed, be easily carried down by any water that may fall directly upon the drain; and, if the surface of the stones has been broken so small as to prevent the drain sustaining any injury in this way, then the covering itself must be altogether superfluous. But farther, it will be found that the effect of this practice, in many cases, is still more injurious. When drains are filled in the usual way, whether with land or quarried stones, a man, or sometimes a woman, is appointed to level the surface and put on the straw or turf; and the person appointed to this duty knows that his master expects him to do a certain number of roods per day, and, finding the stones difficult to break, he too frequently contents himself with merely leveling the surface, and, by means of the covering, the fault is effectually concealed. By the method, however, of separating the small stones from the large, the whole expense of this sort of breaking is saved, and a covering is given to the drain on which time will produce no change."* I have often grudged fine straw being wasted in covering drains, when less valuable materials might have been collected for the purpose, such as dry leaves, dry quickens, tanner's refuse bark when near towns, coarse bog hay, broken moss, &c. I never would suffer a particle of good straw to be wasted in covering drains.

(784.) A drain completed in this manner with stones may be seen in fig. 177. The dimensions given by Mr. Roberton are 33 inches deep, 7 inches wide at bottom, and 9 inches wide at the hight of the stones, which is 15 inches; and within these dimensions 15 cubic feet of stones will fill a rood of 6 yards of drain. Mr. Stirling has 30 inches deep in the furrows, 5 inches wide at bottom, and 8 inches wide at 15 inches from the bottom; the contents of a rood of 6 yards being rather more than 12 cubic feet. The figure here represents a drain 36 inches deep, 9 inches wide at bottom, 12 inches at the top of the stones, and the stones 18 inches deep. These dimensions give cubical contents of 23½ feet per rood of 6 yards; that is, about half as many stones more than the drains of Mr. Roberton, and of course so much more expensive. I own I am partial to the breadth of the common spade as a gauge for the width of the bottom of a drain that is to be filled with *stones*, because it gives plenty of room to them to form a durable stony filter, which 7 inches can scarcely do so well, especially when they are broken to 4 inches in diameter. I am quite persuaded, nevertheless, that the permanency of a drain does not depend so much on the quantity as upon the manner in which the stones are put into it; and I am as well persuaded that it is no matter what description of materials are used

Fig. 177.

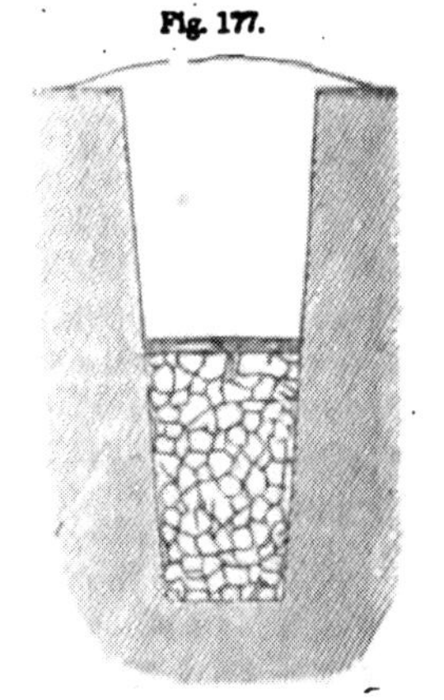

THE SMALL DRAIN FILLED WITH SMALL BROKEN STONES

* Prize Essays of the Highland and Agricultural Society, vol. xiv.

provided there is always left an open and large enough space at the bottom to contain the greatest quantity of water that the drain can possibly have to receive, and provided also that the opening shall be protected from any earth or mud getting in to intercept the flow of water. Yet I agree with Mr. Stirling that our experience is not sufficient to prove what is the *smallest* size that a drain might be to be *permanent*. In which uncertainty it should be of sufficient breadth to prevent moles pushing across it; and this consideration regarding moles acquires greater importance the more the land is drained; for, the deeper we confine water *under* the ground, the deeper will the worms be obliged to go in search of it, and of course the nearer the bottom of the drains will the moles be disposed to burrow in search of their food. Mr. Stirling proposes only to make the bottom of the drain 5 inches, but then he directs the stones to be broken to pass through a ring of 2¼ inches diameter. Such diversity of opinion on the same subject shows you either that experience has not as yet proved what capacity of drain is the best, or that it is immaterial to the draining of land of what breadth drains are made. The principle I maintain in the making of drains is, that, being *permanent* works, they ought to be made in the most substantial manner. It has not yet been ascertained by experiment what dimensions, in given circumstances, afford *sufficient* permanency, and until that point has been settled it is wisdom rather to exceed than to curtail the dimensions; and, although in the mean time the wisdom may be "dear bought," the question of cost is a secondary one to efficiency and permanency.

(785.) With regard to the *quantity of stones used in such drains, and the time required for putting them in*, Mr. Roberton's experience is, in drains of the above dimensions, namely, 33 inches deep, 7 inches wide at bottom, 15 inches filled with stones, and 9 inches wide at the top of the stones—the cubical contents being 15 feet per rood of 6 yards—supposing that a set of carts, driven by boys or women, are able to keep a man employed in unloading them, and another man taking charge of the screen-barrow, 60 to 70 roods can be filled in a summer day of 10 hours; but, as the lineal length depends on the dimensions of the drain, the work, reduced to cubical contents, gives 3½ cubic yards per hour. These data were derived from whole pieces of work, such as in 1840 Mr. Roberton contracted for, for the execution of 4,000 roods, the filling to commence on the 1st July and to be completed on the 12th August. There were 2 sets of carts and 2 screens employed, and the contractors had some stones ready and part of the drains half executed by the 1st July. When the filling commenced 66 roods were finished every day—that is, as it happened, a stretch of drain of exactly 400 yards; but, as the weather was very unfavorable for the work, only 3,300 roods, instead of 4,000, were executed, in which about 2,000 cubic yards of stones were buried. In 1839, of drains of 28 inches deep, 10 to 12 inches of stones in depth, and about 10 cubic feet contents per rood, 2,100 roods, or from 90 to 110 roods per day, were filled, with 1 set of carts and 1 screen, from 1st July to 5th August.

(786.) In Mr. Stirling's case of the drains mentioned above, namely, 30 inches deep in the furrow, 5 inches wide at bottom, and 8 inches at the top of the stones—which were 15 inches deep, and their cubical contents 12.3 feet per rood of 6 yards, the stones being supposed to be carted 1 mile—2 men filled 60 carts of broken stones each day, allowing for loss of time in backing into the bin of stones; a man emptied a cart-load into the drain in 15 minutes, and was ready to return with the cart in 2 minutes more, the horse being supposed to walk at the rate of 3 miles per hour. In this way, a chain of 22 yards, or 3.66 roods, required 3 carts of stones.

(787.) So much for stone, and now for the cost of *tile-draining*. The *dimensions of tile-drains* depend entirely on the mode they are to be constructed. If no soles are to be employed, they may be the narrower; and if nothing else but tile and sole are to be put into them before the earth is returned, they may be the shallower. If the same rule be followed in regard to them as with stone-drains—that is, if 18 inches of earth should be retained over the hard materials, to give liberty to deep plowing—then 18 inches, added to the tile and covering, is the *least depth* that a tile-drain should have; and its *least breadth* is determined by the breadth of sole that is used.

(788.) As the dimensions of these drains depend on the use of *soles*, the necessity for their adoption should be settled at once. It seems to be the uniform opinion of all writers on tile-draining, that, "in hard-bottomed land, the sole-tile is unnecessary; but why unnecessary, as I have before observed, no one has proved to *my* satisfaction. Water being the substance to whose use drains are appropriated, I may mention, in regard to the quantity that may sometimes be found in drains, that Mr. Stirling has found that, after a very heavy fall of rain, tiles of $2\frac{3}{4}$ by $3\frac{1}{2}$ inches are filled with water nearly $\frac{3}{4}$ full;* and yet writers on draining wish to persuade you that such a body of water will not at all affect a clay subsoil or endanger the stability of tiles. I advise you to believe no such assertions, but take for granted that all drains having an earthy bottom of whatever nature, intended to be occupied by tiles, should have soles, or something equivalent—such as slates—under the tiles, to protect the earth from the destructive effects of water. Mr. George Bell, Woodhouselees, Dumfries-shire, has used Welsh slates instead of tile-soles, and found them equally efficacious and much cheaper.† Gray slate and pavement quarries, such as abound in Forfarshire, would supply an abundance of excellent materials for the soles of drains.

(789.) The *breadth of the sole, then, determines the width of the bottom of the drain*; and, should the breadth vary in different parts of the country, the width must in practice be made to suit the sole, but it is probable that soles will be made to suit the proper breadth of drains, when that has been determined by experience. But as that point has not yet been determined by experience, and soles are made of sizes most convenient for their manufacture, the drains must continue to be made of the dimensions suited to the materials by which they are to be filled, until a better order of things arrive. I perceive that the breadth of soles made in the neighborhood of Kilmarnock, at the tile-kilns belonging to the Duke of Portland, in Ayrshire, as well as those made by Mr. Boyle, tile-maker in Ayr, is 7 inches; and this breadth is made to answer tiles varying from 4 to 3 inches in width, inside measure. For a 4-inch tile, a narrower width than 7 inches would not answer; as the tile is $\frac{3}{4}$ of an inch thick, only $\frac{3}{4}$ of an inch is left beyond each side of the tile when placed on the sole, which is as little space as it can stand on securely. For the smaller sized tile of 3 inches, the width is ample; but still it is no disadvantage to a tile to have plenty of room on a sole, as its position can easily be fixed by wedging in stones on each side against the walls of the drain, when stones are used above the tiles; or it leaves sufficient room for a lapping of turf over, and wedging of earth on each side of, the top of the tile. In the case of a 5-inch-wide drain at bottom, the smallest size of tile, $2\frac{3}{4}$ inches wide inside, must be used, as only $\frac{5}{8}$ of an inch would be left on each side of that width of

* Prize Essays of the Highland and Agricultural Society, vol. xii. † Ibid. vol. xiii.

tile. I am aware that to press the tile into the drain, made tight to fit it, without a tile-sole, is a very common practice among drainers; but the practice of pressing hard against the sides of the drain is, in my opinion, objectionable, inasmuch as it is not the hard tile, but the free side of the drain, that draws the water from the land; and to press a hard substance like a tile against the earth in a *shallow* cut, is very like an attempt to curtail the extent of drawing surface. The inducement to use such expedients would be greatly removed were soles made to suit each description of tile; and, what would be still better, were the sizes of tiles more limited in their range, and more uniformly alike; for, as at present made, a great diversity of sizes exists throughout the country, in the area of vertical section as well as in length, so that the prices quoted afford no true criterion of their intrinsic worth.

(790.) Soles are usually made flat, but Mr. Boyle makes them *curved;* not because they are better suited for the purpose, but merely because they are more easily dried in the sheds; but a curved sole is objectionable, as it is more difficult to form a smooth bed for it to lie upon, and it is more apt to break when it happens not to be firmly laid upon its bed than a flat sole.

(791.) As to *tiles*, their perfect form is thus well described by Mr. Boyle: "All tiles should be a *fourth* higher than wide; the top rather quickly turned, and the sides nearly perpendicular. Tiles which are made to spread out at the lower edge, and flat on the top, are weak, and bad for conveying water. Some people prefer tiles with flanges instead of soles; but, if placed even in a drain with a considerably hard bottom, the mouldering of the subsoil by the currents of air and water causes them to sink and get deranged."* Tiles should be smooth on the surface, heavy, firm, and ring like cast-iron when struck with the knuckle. They should be so strong when set as to allow a man not only to stand, but to leap upon them without breaking. The introduction of machinery into the manufacture of drain-tiles, by compressing the clay, and working it thoroughly in a pug-mill to prepare it for being compressed, has greatly tended to increase the strength of tiles. I have seen drain-tiles so rough, spongy, crooked, and thin, as to be shivered to pieces by a night's frost when laid down beside the drain. The use of machinery has caused a great deal more clay to be put into them, and their greater substance has been the cause of improvement in the construction of kilns, in which they are now burned to a uniform texture, as well as some avoidance of breakage in the manufacture; by all which, of course, their cost is lessened. An under-burnt as well as an over-burnt tile is bad, the former being spongy and absorbing water, and ultimately falling down; and the latter is so brittle as to break when accidentally struck against any object.

(792.) The *length* of drain-tiles varies in different parts of the country. Mr. Boyle's are 13 inches; the Duke of Portland's, in Ayrshire, and Mr. Beart's, Godmanchester, Hertfordshire, 12 inches; and those from the Marquis of Tweeddale's machine, 14 inches, when burnt. If the price is the same per 1,000, of course the 14-inch tile is cheaper than the 12-inch, but otherwise the 12-inch is the handiest article in the manufacture, as being less apt to waste in handling, and twist when in the kiln; and their number is much more easily calculated in any given length of drain. The following table shows the numbers of tiles required for an imperial acre of the different lengths made, and placed at the stated distances:

* Prize Essays of the Highland and Agricultural Society, vol. xii.

(696)

	12 in.	13 in.	14 in.	15 in.	
Drains at 12 feet apart require	3630	3351	3111	2904	per acre.
.. 15	2904	2681	2489	2323	..
.. 18	2420	2234	2074	1936	..
.. 21	2074	1914	1777	1659	..
.. 24	1815	1675	1556	1452	..
.. 27	1613	1480	1383	1291	..
.. 30	1452	1340	1245	1162	..
.. 33	1320	1218	1131	1056	..
.. 36	1210	1117	1037	968	..

The numbers of each length of tile required at intermediate distances can easily be calculated from these data.

(793.) I give here a representation of a well-formed drain-tile, and how tiles should be set on soles, as in fig. 178, where *a* and *b* are two 12-inch

Fig. 178.

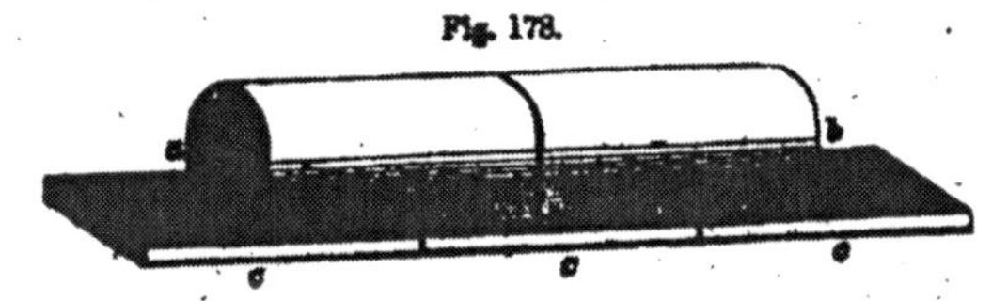

THE DRAIN-TILES PROPERLY SET UPON TILE-SOLES.

tiles, of the correct shape described in paragraph (791) by Mr. Boyle.—They are represented as set upon the sole-tiles *c*; and, to insure a continuation of the same relation between tile and sole, the former should stand upon part of two of the latter, making the joinings of the tiles intermediate with those of the soles, the latter being also 12 inches in length. The drain-tiles used for draining the estate of Netherby, in Cumberland, belonging to Sir James Graham, Bart., and represented in vol. vii. p. 392 of the Prize Essays of the Highland and Agricultural Society, are more pointed in the arch than these, but, on that account, are not so strong in the shoulder to bear a weight upon them. It is the practice of some tile-drainers to put a ½ sole under every joining of 2 tiles, leaving the intermediate space of the bottom without any sole, imagining that this will insure sufficient steadiness to tiles on what they call hard clay, while only half the number of soles are used; but I hope I have said enough on the *hardness* of clay when in contact with water, for you to avoid so precarious a practice.

(794.) There is a mode of joining tiles in drains that meet one another that deserves attention. The usual practice is to break a piece off the corner of 1 or 2 main-drain tiles, where the tiles of the common drains should be connected with them. In breaking off corners, there is risk of breaking the entire tile; and, no doubt, many are broken when subjected to this treatment. Another plan is to set 2 main-drain tiles so far asunder as the inside width of a common-drain tile, and the opening on the other side of the tiles, if not occupied in the same manner by the tiles of another drain, is filled up with pieces of broken tiles or stones, or any other hard substance. It is possible that the broken piece of tile, so placed, may be farther broken or dislodged by the returning of the earth and the action of moles, which may push in earth at that part, and render all above it useless. This is, perhaps, a better plan than running the risk of breaking a number of tiles, and, after all, failing in making the opening suitable for the reception of the adjoining drain-tiles. Both plans, however, are highly objectionable, and should never be resorted to where tiles, formed for the purpose of receiving others in their sides, can be procured. Mr. Boyle, of Ayr, makes main-drain tiles with openings on purpose to receive the shouldered end of the furrow-tiles,* and to answer a similar purpose in particular situations, where such tiles cannot be conveniently joined, he makes ½ and ¼ lengths of main and furrow tiles, which may be so arranged in regard to one another's position as to conjoin the openings of both at the same place. Fig. 179 represents the mode of joining a common drain with a main-drain tile, having an opening in its side. The common tile *b* is not inserted entirely into the main-drain tile *a*, but only placed against it, with a small shoulder, that the openings of both tiles may be always in conjunction.

Fig. 179.

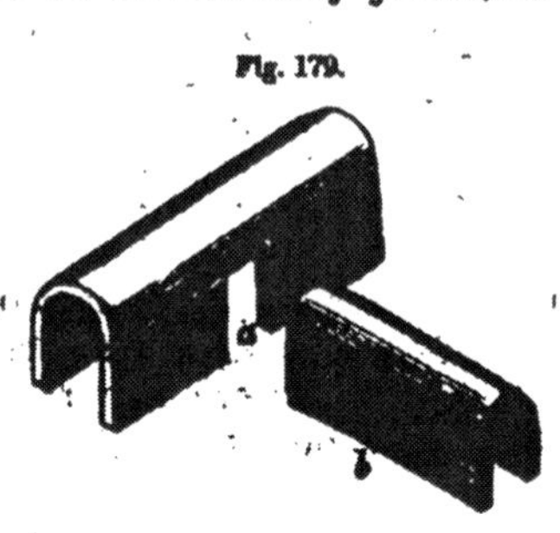

THE JUNCTION OF A COMMON TILE WITH A MAIN-DRAIN ONE.

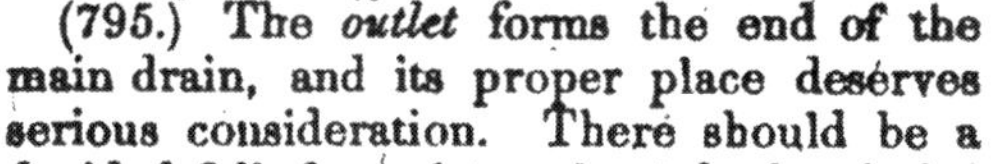

(795.) The *outlet* forms the end of the main drain, and its proper place deserves serious consideration. There should be a decided fall from the outlet, whether it is affected by natural or artificial means. If it be very small—and I have already stated (751) that a small fall is all that is requisite—that is, 1 foot in 150 feet, or 3 feet in the mile, as indicated by the spirit level—the open ditch into which the main drain issues should be scoured deep enough for the purpose, even for a considerable distance; and when this expedient is adopted, it will be requisite to see every year that the outlet is kept open, and the ditch scoured as often as necessary for the purpose.

Fig. 180.

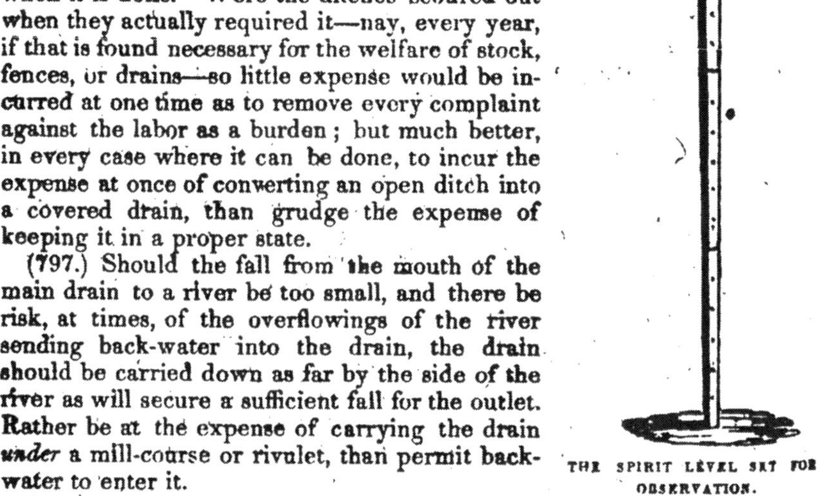

THE SPIRIT LEVEL SET FOR OBSERVATION.

(796.) It is a frequent charge of neglect against farmers, that they allow open ditches almost to fill up before they are again scoured out; and a not unfrequent excuse for the neglect is, that scouring of ditches to any extent incurs considerable labor and expense. No doubt they do, and no wonder, since so much work has to be done, when it is done. Were the ditches scoured out when they actually required it—nay, every year, if that is found necessary for the welfare of stock, fences, or drains—so little expense would be incurred at one time as to remove every complaint against the labor as a burden; but much better, in every case where it can be done, to incur the expense at once of converting an open ditch into a covered drain, than grudge the expense of keeping it in a proper state.

(797.) Should the fall from the mouth of the main drain to a river be too small, and there be risk, at times, of the overflowings of the river sending back-water into the drain, the drain should be carried down as far by the side of the river as will secure a sufficient fall for the outlet. Rather be at the expense of carrying the drain *under* a mill-course or rivulet, than permit back-water to enter it.

(798.) A *spirit level*, such as fig. 180, I have found a very convenient instrument for ascertaining such a point, and generally for taking levels in fields. It is furnished with eye sights *a b*, and

* See Prize Essays of the Highland and Agricultural Society, vol. xii, Plate I, for figure of this tile.

when in use is placed into a framing of brass, which operates as a spring to adjust it to the level position *d*, by the action of the large-headed brass screw *c*. A stud is affixed to the framing, and pushed firmly into a gimlet hole in the top of the short rod *e*, which is pushed or driven into the ground at the spot whence the level is desired to be ascertained. I need scarcely mention that the hight of the eye-sight from the ground is deducted from the hight of observation, and which quantity is easily obtained by having the rod marked off in inches and feet; but I may mention that you should use this instrument in all cases of draining on level ground, even where you are confident that you know the fall of the ground, for the eye is a very deceitful monitor for informing you of the levelness of ground. In one case of my own, I was pretty sure by the eye that the outlet to a division of drains in a field should fall, at some yards off, into an open ditch, which constantly contained spring water. The contractor of the drains was of the same opinion. On testing with the spirit level, however, we found that the bottom of the outlet would have been 8 inches below the bottom of the ditch instead of above it. As, in this particular case, it would have occasioned a cutting of 200 yards to get a proper fall for the outlet in another direction, I caused a narrow well to be sunk on the spot, 8 feet deep, to the stratum of gravel below, and, on being filled with stones, the gravel absorbed all the water from the drains. Such a spirit level, well finished, costs 15s. When not in use, the framing is concealed, and the spirit-tube protected by a movable cover; and the whole instrument, being only 8 inches in length, 1¾ inches deep, and 1 inch broad, and light withal, can be easily carried in the pocket, while its rod may be used as a staff.

(799.) It may happen that, through the undulatory nature of the ground, more than one outlet will be required to clear a field of water, that is, one division of drains may be let more easily out in one place, and another division more easily in another. In such a case, it should be well considered whether both outlets should be joined together or carried away separately, the latter being the less objectionable mode.

(800.) The *cutting* of the main drain should be *entirely finished* before the tiles are laid in it; and immediately after it is finished, it should be measured with the drain-gauge, fig. 155, to ascertain if it contains the specified dimensions and fall.

(801.) While the earth is throwing out toward the narrowest side of the head-ridge, that is, next the fence, the carts should be laying down the tiles and soles along the open side next the field; or they can be laid down before the drain is begun to be opened, and after its line of direction has been fixed. To be certain that the number of tiles and soles are laid down, they should each be placed end to end respectively along the whole line, the soles nearest the drain lying against the tiles, among which, of course, broken ones are not counted, though a sole fractured in two will lay down well enough in a good bed between two whole ones. The tiles with the opening in the side, along with its conjunctive small tile, fig. 179, should be laid down at the distances determined on for the small drains to enter the main drain. These preliminary arrangements should be carefully attended to, or much inconvenience may be occasioned in carrying tiles and soles to and fro to the person who lays them in. It is necessary beforehand to instruct the plowman who is to lay them down out of the cart, of the plan, as some mistake will inevitably ensue, if he is merely told to lay down the tiles and soles by the drain; for few plowmen reflect on the consequences of what they are doing. If, by inadvertence to these minutiæ of practice, more or fewer tiles or soles are laid down than re quired, part of the yoking of a pair of horses lost in laying them down, and part of another is also lost in leading them away to another place, while the unused tiles are in danger of being broken by frequent lifting, and all this waste of time arising from want of forethought.

(802.) The *person* intrusted with the laying of the soles and tiles into the drains, should be one who has been long accustomed to that kind of work, and otherwise a good workman, possessing judgment and common sense. If he is not a hired servant, he should be paid by day's wages, that he may have no temptation to execute the work ill; and to enable him to do it well let him take even more time than you imagine is necessary, especially at first, provided he executes what he does to the satisfaction of his employer. You will soon be able to ascertain how much work of this kind a man should do in a day, according to the circumstances of the case, and you can then judge whether he has been putting off his time, and admonish and encourage him accordingly.

(803.) This person should remain constantly at the bottom of the drains; and, to enable him to do so, he should have an *assistant*, to hand him the materials from the ground. The best assistant he can have, in my opinion, is a female field-worker. Such a one not only receives cheaper wages, but is dexterous in handling light materials, and quite able to lift tiles. soles, and turf easily.

(804.) Immediately before proceeding to *lay the sole-tiles*, the man should remove any wet, sludgy matter from the bottom of the drain with a scoop, fig. 171, and dry earth and small stones can be removed with a narrow draw-hoe, as in fig. 181, with a 2-feet handle *b*, and mouth *a* 3

Fig. 181.

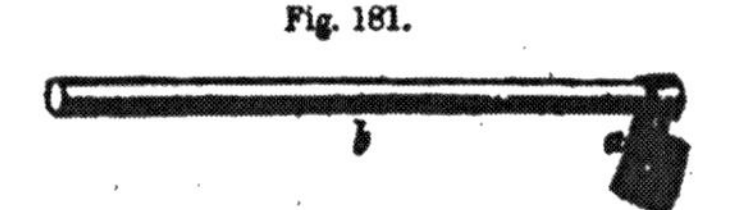

THE NARROW DRAW-HOE FOR DRAINS.

Fig. 182.

THE TROWEL FOR DRAINS.

inches in width, cost 1s. The sole is firmly laid and imbedded a little into the earth. Should it ride upon any point, such as a small stone or hard lump of earth, that should be removed; and a very convenient instrument for the purpose, and otherwise making the bed for the soles, is a mason's narrow trowel, as in fig. 182, 7 inches long in the blade *a*, 5 inches in the handle *c*, and $1\frac{1}{2}$ inches at *b*, and if of cast-steel will cost 2s., of common 1s. 3d. If a single sole has been determined to be laid on the ground, as n fig. 178, or a double one side by side, as in fig. 160, he lays them accord ingly. After laying 3 soles in length, he examines to see if they are straight in the face, and neither rise nor fall more than the fall of the drain. As a safe guide to him, in cases where the fall is not decidedly cognizable by the senses, a mason's plumb-level, such as fig. 183, will be found a convenient instrument. A mark at which the plummet-line *d f* will subtend an angle with the plumb-line *d e*, equal to the angle of the fall of the drain, should be made on the top of the opening *e*, which, in this case, may be supposed to be where the plumb *f* at present hangs; by which arrangement it is demonstrable that the angle *e d f* is always equal to the angle *b a c*, which is the angle of inclination of the fall. The breadth of the sole or soles, as the case may be, should occupy the exact width of the drain, and in the case of a main drain the soles are each 10 inches broad.

(805.) After 3 soles are thus placed, 2 tiles are set upon them, as represented in fig. 178, that is, the tiles *a* and *b* are so placed as that their joinings shall meet on the intermediate spaces *between* the joinings of the soles *c*, and this is done for the obvious reason that, should any commotion disturb one of the soles, neither of the tiles, partially standing upon

Fig. 183.

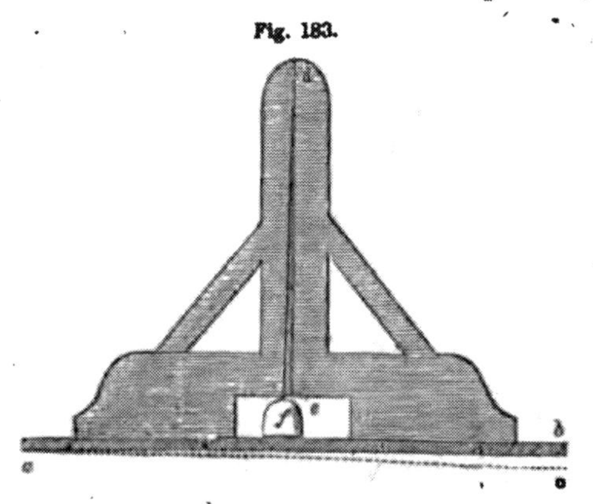

THE DRAINER'S PLUMB-LEVEL.

it, should be disturbed. In ordinary cases of water in a main drain, a tile of 4 inches wide and 5 inches high inside is a good size; and from this size they vary to 5¾ inches in width and 6½ inches in hight. Although the size of the tile varies, that of the main drain sole is always the same, that is, 10 inches wide and 12 inches long. Taking the useful tile of 4 inches in width and 5 inches in hight, its thickness being ¾ inch, there will be a space left in each side of 2¼ inches.

(806.) The *covering*, of whatever substance, should be laid in a row or in heaps along the line of the tiles. Turf is the best covering, and it is put over the tiles saddlewise. If the turf were cut 12 inches broad and 18 inches in length, it would just lap over the size of tile mentioned above, and rest its end upon the sole on both sides; and, if it be from 2 to 2½ inches in thickness, the small space left on each side between the turf and the walls of the drain would be filled up. It is as easy to cut the turfs of the exact sizes required as any other, by rutting off the swarded ground in regular breadths; and it would be as easy to cast the turfs in regular oblongs, as in the irregular pieces usually raised according to the whim of the spadesman. The old *flauchter-spade* of Scotland has long been used to cast turf, but it is a rude instrument at best, and not nearly so good as the common spade for a neat job. When cast, the turfs should be laid one above another in neat bundles of 3 or 4 turfs, which can be easily taken up, and placed safely into the cart, and not thrown singly in, to the risk of their being torn, broken, or put out of shape on being doubled up. They should be as carefully taken out of the cart, in the same bundles they were put in, and if not used immediately, should be put in large bundles to keep them supple and moist; but not so kept a long time, in case of their heating and fermenting. If used in summer, in very dry weather, some water should be thrown upon them to keep them moist, but water in winter might injure their texture by frost. If, on the other hand, they are used immediately, they should be laid down along the outside of the tiles, not in a single row like them, because taking them from the cart and lifting them up again is apt to tear them, but in the small bundles, which are placed as far asunder as the space the turfs would occupy were they laid singly along in line. Judging from the usual treatment which it receives, turf seems to be very little valued, being crumpled up, thrown down, and kicked about, until it becomes much broken and bruised, when it is not nearly so fit for a *covering for tile* as when raised from the ground. I need scarcely add that smooth turf is much better than tufty or heathery clods. Good turf is an expensive article, and not to be obtained everywhere. A man will cast from 4 to 6 cart-loads, of 1 ton each, per day, according to the smoothness and softness of the ground. Its usual thickness is about 3 inches, when 1 square yard will weigh about 54 lbs., and of course 1 ton will cover about 40 square yards, or 40 roods of 6 yards with turfs of 1 foot by 1½ feet. In the country carriage is the heaviest charge against turf; in towns it is charged from 8s. to 20s. a ton, and 8s. per square yard is charged for casting, carting, and laying turf for greens and borders.

(807.) On being handed to him, the man lays the turf, grass side down, over the tiles in a firm manner, taking care to cause the joinings of the turf to meet as near the middle of the tiles as practicable, and not over the joinings. Were the turf cut of the same breadth as that of the tiles used, the covering of a drain would proceed not only rapidly, but neatly and satisfactorily. He takes care not to displace the tiles in the least when the turf is being put over; and to secure the tiles in their respective places, he puts earth firmly between the covering and the sides of the drain as high as the turf over the tile. This earth is obtained from the soil that was thrown out; and if the subsoil is a strong clay, the surface soil is the best, but a porous subsoil answers the purpose. When all these things which I have described have been done, the drain will appear like the small drain, fig. 184, where the sole and the tile set upon it, the turf wrapped round the top of the tile, and the stuffing of earth on each side of the tile may all be easily observed.

(808.) The preparations for the junction with the small drains should be made during the completion of the main drain, for if the main tiles are taken up when the small drains are forming, in order to accommodate the small tiles, they will run the risk of being displaced, and of otherwise disturbing the current of water when it is to run in it. Whichever plan is adopted for letting in the small tiles, and be it ever remembered that the tiles with the open side are the best for the purpose, the man should never forget to make the openings at the stated distances the small drains should enter, and for this purpose he should be provided with a 6-feet rod, marked off in feet and inches, to measure the distances as near as he can, in regard to the fitting of the tiles. The covering of turf should, of course, not be put over the openings left for the small tiles, but the openings should not be left wholly unprotected after the main drain is finished, in case any thing should thrust earth or any other substance into the tile-duct, that might close up or otherwise injure the drain. A bundle of straw, or rather a turf, until the small drains are connected with them, will be sufficient to protect the openings against injury of this kind.

(809.) The *mouth of the main drain* at its outlet, whether in a ditch or river, should be protected with masonry, and dry masonry will do. The last sole, which should be of stone, should project as far beyond the mouth as to throw the water either directly upon the bottom, or upon masonry built up by the side of the ditch. The masonry should be founded below the bottom of the ditch, and built in a perpendicular recess in its side, with the outer face sloping in a line with the slope of the ditch. The sloping face can be made either straight, which will allow the water to slip down into the ditch, or like steps of a stair, over which the water will descend with broken force. It would be proper to have an iron grating on the end of the outlet, to prevent vermin creeping up the drain; not that they can injure tiles while alive, but in creeping too far up, they may die, and cause for a time a stagnation of water above them in the drain

(810.) If the ground fall uniformly toward the main drain over the whole field, the small drains should be proceeded with immediately after the main drain is finished; but should any hollow ground occur in the field too deep for its waters to find their way direct to the main drain, then a *sub-main drain* should be made along the lowest part of the hollow, to receive all the drainage of the ground around it, in order to transmit it to the main drain. The size of sub-main drains is determined by the extent of drainage they have to effect, and should they have as much to do as the main, they should have the same capacity, but if not, they should have less.

(811.) Sub-main drains are made in all respects in the same manner as main drains; but there may be this peculiarity in regard to them, that they will most probably have to receive small drains on both sides, on account of the position they may occupy in the area of a field, when they will require just double the number of tiles with openings in the side than the main. In order to avoid the interference of sediment from opposite small drains, these should not enter the sub-main directly opposite to each other, nor should their ends enter at right angles, but at an acute angle.

(812.) The sub-main drain should be as far below the small drains as the main itself, when it receives the small drains directly, and for the same reasons; and the main should be as far below the sub-main as the latter is below the small drains. The simple way to effect both these purposes is, to make the main drain deeper after its junction with the sub-main.

(813.) There is nothing now to prevent you proceeding with the *small drains*. In a field having a uniform surface, there is no difficulty or irregularity of work to be encountered in bringing the drains directly down the inclined ground into the main drain. Where sub-mains are employed in particular hollows, the ground comprehending the drainage belonging to each hollow should be distinctly marked off from the rest, that no confusion in the direction of the other small drains may ensue in the execution of the work. These markings should be made in the *water-shed of the ground*, from which the fall tends toward each sub-main, if more than one is required, and it may also tend toward the main drain. The markings can be made with pins driven in the lines determined by the water-shed.

(814.) In commencing the small drains from a fence on one side of the field, supposing that the ridges are 15 feet wide, and keeping in mind that, if the soil of the field is not strong clay, the drains need not be formed in the open furrows, it is requisite to measure the distance of the first drain from the fence, whatever that fence may be, at 16 feet. This space of 16 feet gives 2 feet for the fence-side, 14 feet from the fence-side to the drain, and one foot beyond the drain for the open furrow of the 15-feet ridge. Keeping the distance of every other drain from each other at the breadth of one ridge of 15 feet, or at any other multiple of that breadth, it is clear that every drain will fall within one foot of an open furrow. If the subsoil draws slowly, the drains should *not exceed* 15 *feet asunder*, and the *depth*, I should say, *not less than* 3 *feet*.

(815.) I know it is a common impression among farmers, that if a subsoil cannot draw water, there is no use of making drains in it, and this opinion I conceive to be quite correct in regard to pure clay subsoils, which cannot draw water at all. But the view I take of the matter is this, that pure clay subsoils are very limited in extent, and that many clays which *seem* quite impervious may draw water notwithstanding. Admitting that the subsoil draws water at all, which is the supposition in the present case, it is clear that the larger the area is extended for drawing it, the more water will be drawn into the drain. Now, a large area can only be secured by making drains deep and close together; and in the case supposed above, it appears to me that 3 feet in depth, with 15 feet asunder, will not give a greater area than is requisite for drawing water out of such ground. When, on the other hand, the subsoil is free, and discharges water as freely, so large an area is not required to dry the subsoil, and drains of less depth and at greater distance will answer the same purpose as in the other case, such as 30 inches in depth and 30 feet asunder. You must endeavor to make the depths and distances of the small drains suit the nature of the subsoil, for it is impossible for me to lay down here any absolute rule in a matter which admits of such diversity of character.

(816.) Small drains, as well as mains and sub-mains, should be completely cast out, gauged, and examined for the fall, before being attempted to be filled up; and the materials for doing so should be laid down beside them, as well as in the case of mains. The tiles for small drains are smaller than for mains and sub-mains, being 3 inches wide and 4 inches high, inside measurement, which may be considered a large tile in places where these of 2¾ inches wide by 3½ inches high are used; but so small ones are not made everywhere. There is this consideration in regard to the size of tiles which should be kept in view, that a substantial tile will have the chance of lasting much longer than a slight one, and the probability is, that the larger ones are the more substantial, which, however, may not actually be the case, but it is proper to examine whether they are heavy and firm, before you purchase your tiles. Be guided in your choice of them more on account of substantiality than cheapness, which, as I have said before, is quite of secondary consideration when brought into comparison with durability. Soles will also be required for small drains, for don't give credence to the absurd assumption that clay will retain its hardness at the bottom of a drain, because it happened to be hard when first laid open to the day. Soles for small drains are of different breadths, being 5 inches at one place and 7 inches at another; the former, 5 inches, I should conceive, *too narrow* for most purposes; for take even the narrowest tiles that are made, 2¾ inches inside—these are moulded at ⅝ inch thick, and allowing them to shrink ⅛ in the kiln, their thickness will be 1¼ inches; the outside breadth of the tile being thus 4 inches, leaves only 1¼ inches to divide between the two sides of the tile on a 5-inch sole, or just ½ an inch on each side, a small enough space, certainly. But as most soles for small drains are made of the same breadth, take a 3-inch tile, and it will be found by the same mode of calculation that only ¼ inch on each side of a 5-inch sole will be left, which is a much too narrow space to afford perfect steadiness to the tile. I would prefer the 7-inch soles as made in Ayrshire, and, of course, the breadth of the bottom of the drain should also be 7 inches. In other respects, the filling of the small drains is conducted in the same manner as the mains and sub-mains, and they are finished as represented by fig. 184.

Fig. 184.

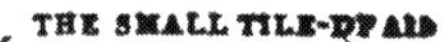

THE SMALL TILE-DRAIN.

(817.) While casting out the bottom of the end of each small drain, care should be taken in communicating it with the main or sub-main with which it is to be connected, that no displacement of tiles takes place in either; and when the bottom is cleared out, the turf or small bundle of straw left in the openings of the sides of the tiles is removed, and the opening examined, and any extraneous matter that may have got into the tiles removed. The places for the entrance of the small drain tiles having been prepared while constructing the main and sub-mains, there will be no difficulty of effecting the junction between the respective sorts of drains. Thus one small drain after another is finished, until the field, having been begun at one side, is furnished with drains by the time the other is reached. The small drain connecting the tops of all the small drains along the upper end of the field, should not be neglected. (767.)

(818.) The next procedure is the *filling up of the drains with the earth that was thrown out of them*, which is returned either with the spade or the plow, or both. When drains are furnished with stones, the plow may be used from the first, giving it as much *land* for the first bout or two as it can work with. If the earth has been thrown out on both sides, a strong furrow on each side of the top of the drain will fill in a considerable quantity of the earth; but, as the earth is generally thrown out on one side of the drain, and the plow can only advance the earth toward the drain while going in one direction, that is, going every other landing *empty*, or without a furrow, a more expeditious mode of leveling the ground, which, in the considerable labor of returning the earth into all the small drains of a field, is a matter of some importance, is to cleave down the mound of earth thrown out, and then take in a breadth of land on both sides of the drain, and gather it up twice or thrice toward the middle of the drain, which will constitute a prepared feering, after which the harrows will make the ground sufficiently level. This species of work, however, is only required when much earth has been thrown out, and thrown a distance from the drain, in deep draining; but in thorough-draining, what is accomplished by the plow is done with much less trouble. When the plow alone is used for this purpose, the first two furrows are taken round the mouth of the drain, and fall into it with considerable force; and, where tiles alone are used, such a fall of earth may be apt to break or displace them; and even the steadiest horses, which should only be employed at this work, run the risk of slipping in a hind foot into the drain, which, in attempting to recover, may be overstrained; and such an accident, trifling as it may seem, may be attended with serious injury to the animal. The safest mode, therefore, both for horses and tiles, is, in all cases, to put the first portion of earth into the drain with the spade, and this provision can be made in the agreement with the contractor; and there is this advantage attending the use of the spade that a better choice can be made, if desired, of the earth to be returned, the surface earth may first be put in before the poorer subsoil. (770.)

(819.) In regard to the quality of the earth which is employed to fill up the drains, some considerations are requisite. All *deep* drains, whether furnished with stones or tiles, should receive their supply of water from *below*, and not immediately from above through the soil; and all drains that receive their supply of water in this manner should be denominated deep drains, in reference to the nature of their functions, whatever may be their respective depths.

(820.) Were drains *entirely* filled with loose mould, or other loose materials, it is evident that the rain, percolating directly through them, will arrive in the drain loaded with as many of the impurities that the soil may contain as it could carry along with it in its downward course; but a primary object with drainers is to prevent impurities getting into the ducts of drains, because in time they might either collect in quantities in the ducts, or fill up the interstices between the stones; and the smaller the stones were broken, their upper stratum at least would the more easily be rendered inoperative as a drain. To prevent one and all of these mis chances, the practicable way is to return the clayey subsoil into the drain, where it will again soon consolidate, and resist the direct gravity of rain.

(821.) Keeping these distinctions in view, and applying them first to the case of strong clay soil, such as in the Carse of Gowrie, which does not draw water at all, were they filled up above the tiles with pure clay, the ultimate effect would be that the duct would remain open, but no water would ever enter it. To make them draw at all, there must loose materials be put above the tiles within 2 or 3 inches of the plane upon which the sole of the plow moves; and to obtain the greatest depth of loose materials for such drains, they should be made in the open furrows. As they cannot draw but through the loose materials, and are, in fact, covered ditches, they must receive their supply of water like any other ditch, from above; but here the analogy ceases, for instead of receiving their water direct from the top like a ditch, they should receive it by percolation through the plowed soil, and when the water has descended through the soil, deprived of most of its impurities, it meets the retentive subsoil across the whole area of the ridge, upon which it moves under the arable soil until it meets with the loose materials in the drains, by which it is taken down into the ducts to be conveyed away. The loose materials may be gravel, sand, peaty earth, scoriæ from furnaces, refuse tanners' bark, and such like.

(822.) In a subsoil that draws only a little water, were the clayey subsoil returned immediately above the *tiles*, it would have the effect of counteracting the purpose for which the drains were made, because it would curtail the drawing surface to only the hight of the tiles themselves. The method, therefore, to fill such drains is to put loose materials immediately above the tiles, to a hight not so far as in the case of pure clay drains, but to within ½ a foot of the plane of the plow's sole-shoe. Were the drains in such a subsoil, however, filled with *stones*, the case would be different, for these would secure a sufficient drawing surface, and the clayey subsoil may be returned immediately on their top with perfect propriety.

(823.) In the case of a free drawing subsoil to the bottom of the drain, the most retentive portion of the earth may be returned immediately above even *tiles*, for such a subsoil would still draw the moisture toward them; and were *stones* used, there would be left ample room for drawing with the most retentive part of the earth returned above them. But should the part of the drain occupied by the tiles or stones be of strong, impervious clay, although the soil above it be of the best drawing materials, as much of the loose subsoil should be placed above the tiles or stones as would give an easy access to the water, and all the space above that may consist of the strongest part of the clay.

(824.) The general rule, then, for filling the drains with the earth that has been thrown out of them is, that, with the exception of strong clay soils—the drains in which should be filled with porous materials, that the water on the surface may descend through them into the duct below, and be thence carried away—that, with this exception, every kind of drain should be filled near its top with the strongest soil afforded by the drain, in order to *prevent the descent of the water into the drain by the top*, but rather that the water shall seek its way through the plowed ground, and thence by the porous materials above the duct, and under the clay put in above them into the duct at the bottom. Through such a channel of filtration the water will have every chance of entering the duct in a comparatively pure state.

(825.) But the best mode, in my opinion, of draining land of any that has yet been described, has to be brought before your notice, and that is by the *union of stones and tiles* in the same drain. The method is represented in fig. 185, where a tile *a* rests on a sole; small stones are packed around the tile by the hand until they cover it as at *b*; the remaining small stones *c* are put in by any of the methods described above, but especially by the drain-screen; a covering is either put above them or small stones beaten down, as in the case of the stone drain, with the beater, and the earth returned upon them in either of the modes just described. The width of the bottom is 7 inches, the width of a good sole, width of the top 12 inches, depth $2\frac{1}{2}$ feet, composed of 18 inches of earth, and 12 inches to the top of the covering of the stones. This drain is constructed very similarly to the tile drains that have been described, by first laying the sole at the bottom and the tile upon it; but instead of covering the tile immediately with turf, small stones are packed by the hand on both sides until they cover its top. As these stones should be packed in as the laying of the tiles proceeds, they should be laid down in heaps, immediately after the tiles have been laid down, as near the drain as possible, so as they may be filled in baskets by the assistant, and handed down in them to the man in the drain. Two baskets are required for this purpose, one to be filled by the assistant when the other is emptying by the drainer. In filling up with stones afterward to their proper hight as to *c*, I would be afraid of using the drain-screen at first, in case the fall of the first stones upon those which were laid in by the hand, as at *b*, should by any chance fracture the *tiles* below them. I would rather fill up a few yards of the drain as high as required by the hand, and then use the drain-screen to let fall the stones upon the end of those previously filled in, from which they could be shoveled (fig. 38) or raked (fig. 174) down gently upon the stones over the tiles; or the stones could be filled in by the hand at first so high, while laying those at *b*, as to remove all danger from those falling in small quantities from the screen. The filling in from the drain-screen and carts should not be proceeded with until as much of the drain has been laid with tiles and packed in with stones by the hand, as to employ *at least* 2 single horse carts for one yoking, and should the weather seem favorable, not until that number of horses can be employed a whole day, because otherwise the time of the horses would be wasted. If the draining is of such an extent as to keep a pair of horses thus constantly employed, so much the better. In such a case, other hands than those employed in cutting the drain and laying the tiles should be employed in filling in the upper layer of stones with the screen, and beating down the small riddlings as a covering upon them. On the earth being returned into the drain the operation is completed.

Fig. 185.

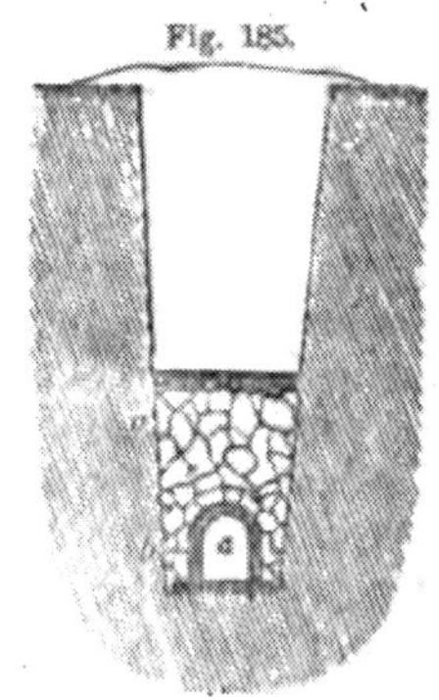

THE TILE AND STONE DRAIN.

(826.) This construction of drain is declared by every writer on and practitioner of draining, to be the *ne plus ultra* of the art, though I believe very few farmers have adopted it, not because there can the slightest objection be urged against it, but because, in cases where stones have to be quarried and broken, it is an expensive mode, and in other cases stones cannot be obtained at all. The last reason is a very good one, but that of the expense must fall to the ground where there is abundance of stones, as the advantage derived from their use along with the tile will be more than counterbalanced by the additional cost. The durability and efficiency of such a drain is undoubted. It is a perfect piece of work, inasmuch as the duct formed of the tile and sole presents the smoothest passage imaginable for carrying off water, and it is proof against the efforts of vermin. while the stones not only secure the duct in its place, but impart durability to the whole structure, which at the same time presents an extensive area to the subsoil. What other property that a good drain should have does this one not possess?

(827.) It may be satisfactory to you to have a general idea of thorough draining a field by a sketch of a ground plan, which is represented in fig. 186, where *a b* is the main drain formed in the lowest head-ridge; and if the field were of a uniform surface, the drains would run parallel to one another from the top to the bottom into the main drain, as those do from *a* to *c*, connected as they should be at the top with the drain *d e* running along the upper head-ridge. But as there may be inequalities in the ground, a very irregular surface cannot be drained in this manner, and must therefore be provided with sub-main drains, as *f g* and *h i*, which are each connected with a system of drains belonging to itself, and which may differ in character from each other, as *f g* with a large double set *k l* in

Fig. 186.

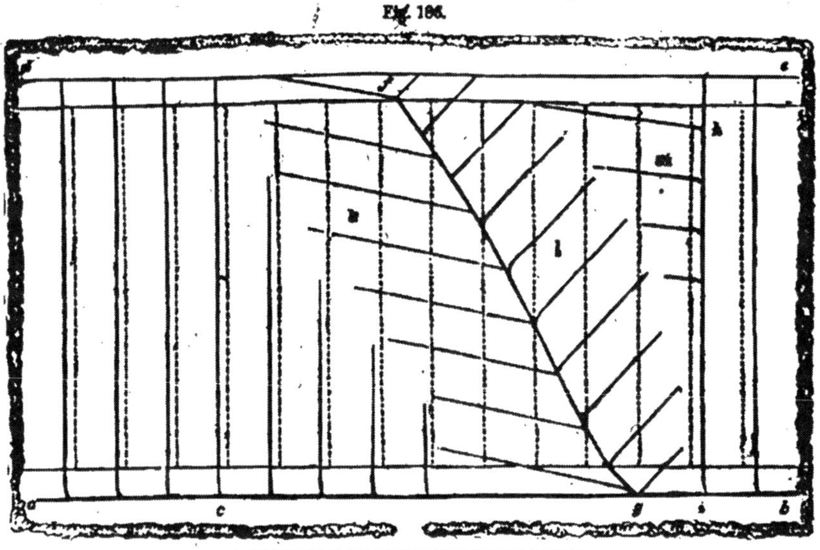

A PLAN OF A THOROUGH-DRAINED FIELD.

connection with it, and *h i* with only a small single set *m*; the sub-main *f g* is supposed to run up the lowest part of a pretty deep hollow in the ground, and the drains *k* and *l* on either side of it are made to run down the faces of the acclivities as nearly at right angles to the sub-main as the nature of the inclination of the ground will allow, so as always to preserve the natural tendency of water to find its way down the hollow. There is also a supposed fall of the ground from the hight above *l* toward *h*, which causes the drain at *m* to run down and fall into what would be a common drain *h i*, were it not, from this circumstance, obliged to be converted into a sub-main. The sub-main *f g* may be made as large as the main drain *a b*, as both have much to do; but the sub-main *h i* may be made comparatively smaller, and not larger, from the top of the field, than a common drain, until it reaches the point *h*, where the collateral drains begin to join it. The main drain should be made larger below *g* to *i* than above it, and still larger from *i* to *b*, which is its outlet. It will be observed that all the common drains *a* and *c*, and at *l* and *m*, have their ends curved, those at *k* not requiring that assistance, as they enter more obliquely into the main, from the position of the slope of the ground. The dotted lines represent the upper and lower head-ridges, and the open furrows of the ridges of the field; and it will be observed that the drains are not made to run in the open furrows—that is, the black lines in conjunction with the dotted, but along the furrow-brows of the ridges. This is done with the view of not confounding the open furrows and drains in the figure; but it is a plan which may be followed with propriety in subsoils otherwise than of strong clay; that is, of a light loam resting on a rather retentive subsoil; the water falling upon which should not be drained away by the small drains receiving it through their tops, but rather by the absorption of the water toward them from below the plowed soil, as far as the subsoil is porous. A hollow, such as that occupied by the sub-main drain *f g*, also indicates that the soil is a loam, and not strong clay. Although the ridges are supposed to be 15 feet wide, and they have been set off here at quarters of an inch, they bear no true relation to the size of the field; so that this diagram should not be considered as showing the relative proportions of the distances betwixt the drains and the size of the field.

(828.) The period of the rotation of cropping at which draining should be executed, requires consideration; but I believe it is now generally allowed to be best performed when the ground is in grass, and before the grass is plowed up. There are several advantages attending this period of cropping over every other. 1. Turf can be obtained at hand for covering the tiles; and although one year's grass may not afford very good turf for the purpose, yet if the turfs are carefully raised by the spade, and as carefully laid aside until used—not heaped upon one another to run the risk of rotting, but set down in a row with the grass-side up—and as carefully handled when about to be used, it will answer very well. In 2 or 3 years old grass, the turf is better; and in old pasture or meadow ground it is as good as can be procured elsewhere. At whatever age the turf is used, it should not be too rough or too thick, as it will not clap so closely over the tile in either state as it should. Sheep are the best stock for eating down the forage, and preparing the turf for this purpose. 2. Another advantage which grass-land possesses is the firm surface which it presents to cartage of materials, whether stones or tiles. If the stones are put in with the screen, the cart and barrow will pass lightly along the side of the drain; and if tiles are used, the grass forms clean ground for them to be laid down upon. 3. In grass, the filling in of the earth with the spade makes very neat work.

(829.) When it is determined to drain the land while in grass, the *season* of the year in which the drains should be opened is thereby in a great degree determined. It would scarcely be prudent to sacrifice the pasturage in summer, and no stock should be allowed to roam about a field that is in the act of being drained, not only on account of the possibility of their injuring themselves by slipping into the drains, but of injuring the drains by breaking down their edges, fracturing the tiles, or displacing the stones. It is therefore expedient to take the use of the summer's grass; but that the operation may commence soon in autumn, the grass should be by that time eaten down bare by an extra quantity of stock. These preliminary arrangements, then, being made, and the materials laid down as long as the weather is dry and the ground hard, the draining operations may be carried on through the winter, and as far into spring as to give time for the land to be plowed for the reception of the seed. Whether there are one or more sets of men engaged in cutting the drains, they should all work in the same field at the same time, as it invariably entails loss of time to drive materials with horses over different fields. With concentrated work, one field is drained after another, and this regularity of the order permits the eating down of the grass in succession, as regularly as the draining proceeds, so that none of the aftermath is sacrificed.

(830.) The *next* important point for consideration is, whether the great outlay upon land occasioned by draining can be compensated for by increase of produce? for if no increase of produce, adequate to repay the large outlay, can be guarantied, draining will not be persevered in. No one, beforehand, can give such a guaranty; but the experience of enterprising drainers, who have vested their capital in the experiment, has proved that draining—that is, effectual or thorough-draining, by whatever means that object is attained—not only compensates for the outlay incurred, but also improves the quality of the land and everything that grows upon it. Examples of amelioration, as well as of profit, effected by draining, inspiring confidence and stimulating imitation, I shall adduce.

(831.) The existence of moisture in the soil being most easily detected by its injurious effects on the crops usually grown upon it, the benefits of draining are also first indicated by the crops. On drained land, the straw of white crops shoots up steadily from a vigorous braird, strong long, and at the same time so stiff as not to be easily lodged with wind or rain. The grain is plump, large, bright colored, and thin skinned. The crop ripens uniformly, is bulky and prolific, more quickly won for stacking in harvest, more easily threshed, winnowed, and cleaned, and produces fewer small and light grains. The straw also makes better fodder for live-stock. Clover, in such land, becomes rank, long, and juicy, and the flowers are large and of bright color. The hay from it wons easily and weighs heavy to its bulk. Pasture-grass shoots out in every direction, covering the ground with a thick sward, and produces fat and milk of the finest quality. Turnips become large, plump, as if fully grown, juicy, and with a smooth and oily skin. Potatoes push out long and strong stems, with enlarged tubers, having skins easily peeled off, and a mealy substance when boiled. Live-stock of every kind thrive, become good tempered, are easily fattened, and of fine quality. Land is less occupied with weeds, the increased luxuriance of all the crops checking their growth. Summer fallow is more easily cleaned, and much less work is required to put the land in proper trim for the manure and seed; and all sorts of manures incorporate more quickly and thoroughly with the soil.

(832.) Thorough-drained land is easily worked with all the common implements. Being all alike dry, its texture becomes uniform, and, in consequence, the plow passes through it with uniform freedom; and even where pretty large sized stones are found, the plow can easily dislodge them; and moving in freer soil, it is able to raise a deeper furrow-slice; and the furrow-slice, on its part, though heavy, crumbles down and yields to the pressure and friction of the plow, forming a friable, mellow, rich-looking mould. The harrows, instead of being held back at times, and starting forward, and oscillating sideways, swim smoothly along, raking the soil into a uniform surface and entirely obliterating foot-marks. The roller compresses and renders the surface of the soil smooth, but leaves what is below in a mellow state for the roots of plants to expand in. All the implements are much easier drawn and held; and hence, all the operations can be executed with less labor, and of course more economically and satisfactorily on drained than undrained land.* All these effects of draining I have observed from my own experience.

(833.) "It is gratifying," says Mr. James Black, in reference to the effects of the Elkington mode of draining on the estate of Spottiswoode in Berwickshire, "to be enabled to state that the general result of the operations has been such as to bear out the calculations of the engineer, and to justify the most sanguine hopes that could have been formed of a valuable improvement. Bursts and springs, which formerly disfigured entire fields, and which rendered tillage precarious and unprofitable, are now not to be seen; and swamps, which were not only useless in themselves, but which injured all the land aronnd them, have been totally removed. The consequence is, that tillage can now in those parts be carried on without interruption, and with nothing beyond the ordinary expenditure of labor and manure; and a sward of the best grasses, raised and continued on spots which formerly only produced the coarsest and least valued herbage."

(834.) But draining has been found beneficial not only to the soil itself, to the processes of laboring it, to the climate in reference to crops, and to the growth of trees, but also to the health of the laboring population. Dr. Charles Wilson, Kelso, when comparing the health of the laboring population of the district of Kelso in two decennial periods, from 1777 to 1787 and from 1829 to 1839, came to this conclusion in regard to the effect of draining, that "our attention is here justly attracted by the extraordinary preponderance of cases of ague in the first decennium, where they present an average of $\frac{1}{4}$ of all cases of disease coming under treatment; and a closer examination of the separate years shows this proportion rising more than once to even as high as $\frac{1}{5}$; while, in the second decennium, the average proportion is only $\frac{6}{100}$ of the general mass of disease. Ague, then, as is well known to the older inhabitants of the district, was *at one time regularly endemic among us;* affecting every year a varying, but always a considerable portion of population, and occasionally, in seasons of unusual coldness and moisture, spreading itself extensively as an endemic, and showing its ordinary tendency, under such circumstances, of passing into a continued and more dangerous type. Ague was not usually in itself a disease of great fatality, the deaths recorded at the Dispensary having been only 1.81 per cent. of the cases treated—a sum which denotes its absolute mortality, while its relative mortality was 0.26, when viewed in connection with that from all other diseases. Still, if we keep in view how frequently it was known to degenerate into fevers of a worse form, and how often it terminated in jaundice, 'obstruction of the viscera of the abdomen,' and consequent dropsies; or even if we take into consideration the frequency of its recurrence, and the lengthened periods during which it racked its victims, we shall see much reason to be thankful that a plague so universal and so pernicious has been almost wholly rooted out from among us. Those who recollect what has been stated of the former swampy nature of the soil in our vicinity, and of the *extensive means which have been adopted for its drainage,* will, of course, have no difficulty in understanding why ague was once so prevalent, and under what agency it should now have disappeared; and will gratefully acknowledge the *twofold value of those improvements which have at once rendered our homes more salubrious and our fields more fruitful.*"†

(835.) But the most palpable advantage of draining land is the *profit* which it returns to the farmer. A few authenticated instances of the profits actually derived from draining will suffice to convince any occupier of land

* See papers by me on this subject in the Quarterly Journal of Agriculture, vols. vi. and viii.
† Quarterly Journal of Agriculture, vol. xii.

of the benefits to be derived from it. "I am clearly of opinion," says Mr North Dalrymple of Cleland, Lanarkshire, "that well authenticated facts on economical draining, accompanied with details of the expenses, value of succeeding crops, and of the land before and after draining, will be the means of stimulating both landlords and tenants to pursue the most important, judicious, and *remunerating* of all land improvements. The statements below will prove the advantages of furrow-draining; and as to the *profits* to be derived from it, they are *great*, and a farmer has only to drain a 5-acre field to have ocular proof upon the point."*

(836.) Without entering into all the minutiæ of the statements given by Mr. Dalrymple, it will suffice here to exhibit the general results. 1. One field containing 54 Scots acres cost £303 7s. to drain, or £5 12s.† per acre. The wheat off a part of it was sold for £11, and the turnips off the remainder for £25 13s. 4d. per acre. The soil was a stiff chattery clay, and let in grass for 20s. an acre; but in 1836, after having been drained, it kept 5 Cheviot ewes, with their lambs, upon the acre. 2. Another field of 18 acres cost £5 9s. the acre to drain. The wheat off a part of it fetched £13, the potatoes off another part £15 15s., and the turnips off the remainder £21 per acre. The land was formerly occupied with whins and rushes, and let for 12s. the acre; but when let for pasture after being drained, Mr. Dalrymple expected to get 50s. an acre for it. It may be mentioned, that the drains made by Mr. Dalrymple were narrow ones, 30 inches in depth, filled 18 inches high with stones or scoriæ from a furnace, and connected with main drains, 36 inches deep, furnished with tiles and soles.‡

(837.) Mr. James Howden, Wintonhill, near Tranent, in East-Lothian asserts from his experience, that although drains should cost as much as £7 the acre, yet on damp, heavy land thorough-draining will repay from 15 to 20 per cent. on the outlay.||

(838.) A farmer in Lanarkshire, who thorough-drained one-half of a 4-acre field, and left the other half undrained, in 1838, planted the whole field with potatoes, and from the drained portion realized £45, while the undrained only realized £13 the Scotch acre.§

(839.) A very successful instance of drainage is related to have taken place on the estate of Teddesley Hay, near the River Penk, in Staffordshire, belonging to Lord Hatherton, under the direction of his agent, Mr Bright. The soil is represented of a light nature, resting on a subsoil of stiff clay. The results are these:

Quantity of land drained.			Value of the land in its original state.				Cost of draining.			Value of the land in its present state.			
			Per acre.	Annual value.						Per acre.	Annual value.		
A.	R.	P.	s.	£	s.	D.	£	s.	D.	s.	£	s.	D.
78	1	36	10	39	4	3	262	15	0	27	105	18	9
19	1	32	10	9	14	6	74	9	8	35	34	0	9
38	0	3	16	30	8	3	52	14	2	40	76	0	9
82	2	2	15	61	17	8	346	16	4	30	123	15	4
30	3	24	10	15	9	0	121	5	8	35	54	1	6
81	1	34	8	32	11	8	153	16	4	22	89	12	2
36	3	16	10	18	8	6	142	8	0	30	55	5	6
33	[illegible]	0	8	13	4	0	80	5	2	26	49	18	0
10	2	33					90	8	0	50	26	15	3
10	0	8								21	10	11	0
9	0	0	12	5	8	0	76	9	8	30	13	10	0
15	0	11	16	12	1	0	41	9	4	33	24	17	3
21	2	10	15	16	3	5	66	0	0	30	32	6	10
467	0	9		254	10	9	1508	17	4		689	13	1

[† Calculate the pound at $4 80, and the shilling at 22 cents. *Ed. Farm. Lib.*]

* Quarterly Journal of Agriculture, vol. viii. ‡ Ibid., vol. viii. || Ibid., vol. viii. § Ibid., vol. x.

Here is an increase of £435 2s. 4d. a year by draining, with an expenditure of only £1,508 17s. 4d., or 29 per cent. on the capital expended.*

(840.) Mr. George Bell's experience of the good effects of thorough-draining on turnips gave, in the instance of Aberdeenshire yellow bullock turnips, which were raised from bone-dust in 1838, a crop of 16 tons 16 cwt. on 2 acres; whereas the same extent of undrained land only produced 6 tons 4 cwt. the acre. In 1839 the produce of potatoes on drained land was 175 cwt. the acre, whereas that on undrained land was only 70 cwt.

(841.) Besides the methods of draining land with stones and tiles as have just been described to you at much length, there are other methods which deserve your attention, because they may be practiced in particular situations in an economical manner. I do not believe that the methods I am about to mention are so effectual as those I have described, still I have no doubt any one of them may be made effectual in situations where the materials recommended are abundant. Besides, it is well to have a choice of methods of performing the same operation, that your judgment may be exercised in adopting the one most advisable in the circumstances in which you are placed; while, at the same time, you should never lose sight of this maxim in Agriculture, that that operation is most economically executed, or, at all events, affords most satisfaction in the end, which is executed in the most efficient manner, both as regards materials and workmanship, and the maxim applies to no operation so strongly as to draining, because of its permanent nature. The methods of draining which I am about to mention apply to every species of soil, from light loam, the heaviest clay, to bog.

(842.) The first method which I shall notice is applicable to a *tilly subsoil which draws water* a little, situate in a locality in which *flat stones are plentiful and sufficiently cheap.* Suppose a piece of land, containing 2 ridges of 15 feet in width, which had been gathered up from the flat, and in this form of plowing, as you have already learned, there is an open furrow on each side of a ridge (650). The drains are made in this manner: Gather up the land twice, by splitting out a feering in the crown of each ridge, and do it with a strong furrow. Should the 4-horse plow have been used for the purpose, the open furrow will be left 16 inches wide at bottom, and if the furrow have been turned over 12 inches in depth, and the furrow-slice laid over at the usual angle of 45°, the tops of the furrow-slices on the furrow-brow will be 32 inches apart, as from *a* to *a*, fig. 187. After this plowing the spade takes out a trench from the bottom of the open-furrow 8 inches wide at top *e*, 16 inches deep by *f*, and 3 inches wide at bottom at *g*. The depth of the drain will thus be 32 inches in all below the crowns of the gathered up ridges. The drain is filled by two flags *h h* being set up against its sides and meeting in the bottom at *g*; and they are kept asunder by a large stone of any shape, as a wedge, but large enough to be prevented by *h h* descending farther than to leave a conduit *g* for the water. The remainder of the drain is filled to *e* with small riddled stones with Mr. Robertson's drain-screen or with clean gravel. The stones are covered over with turf and earth like any other drain, or with small stones beaten down firmly. The expense of this method of draining is small: the spade-work may be executed at 1d. the rood of 6 yards, and of an imperial acre, containing 161⅓ of such roods, the cutting will cost 13s. 5½d. The flags, at 1 inch thick and 6 inches broad, will make 15 tons per acre, at 4d. the ton, will cost 5s. more. The broken stones, to fill 9 cubic feet in the rood of 6 yards, at 2½d. per rood, will cost £1 10s. 3½d. more; making in all about £2 8s. 8d. the acre, exclusive of carriage and plowing, which, though estimated, will yet make this a cheap mode of draining land so closely as 15 feet apart.†

Fig. 187.

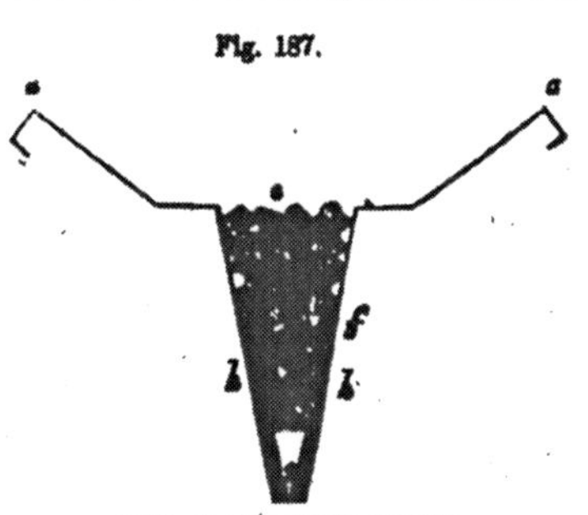

THE FLAT STONE DRAIN.

(843.) The draining of *mossy, light soils* where *peat is plentiful* may be effected in this way. The peats are made somewhat of the shape of drain-tiles but more massive, as may be seen in fig. 188. They are laid in the drain one *a* like a tile-sole, and another inverted upon it, as *b*, like a drain-tile, leaving a round opening between them for the passage of the water. These peats are cut out with a spade-tool, contrived some years ago by Mr. Hugh Calderwood, Blacklyres, Ayrshire. The spade is easily worked, and forms a peat with one cut, without any waste of materials; that is, the exterior semi-circle *b* is cut out of the interior semi-circle of *a*. A man can cut out from 2,000 to 3,000 peats a day with such a spade. The peats are dried in the sun in summer, with their hollow part upon the ground, and are stacked until used; and those used in drains have been found to remain quite hard. The invention of this spade, of which a figure, 189, is here given, tends to make the draining of Moorish soils more practicable than heretofore, and it may be done at 1-5 or 1-4 of the expense of ordinary drain-tiles. The frequent want of clay in upland moory districts renders the manufacture of drain-tiles on the spot impracticable, and their carriage from a distance a serious expense.‡

Fig. 188.

THE PEAT-TILE FOR DRAINS.

* On Land Drainage, &c   † Quarterly Journal of Agriculture, vol. vii.   ‡ Ibid., vol. vii.

(844.) Sir Joseph Banks alludes to the filling up of drains in bogs, which had been executed at great expense at Woburn, by the growth of the marestail *(Equisetum palustre)*. On examining the plant, Sir Joseph found "its stem under ground a yard or more in length, and in size like a packthread; from this a root of twice the size of the stem runs horizontally in the ground, taking its origin from a lower root, which strikes down perpendicularly to a depth I have not hitherto been able to trace, as thick as a small finger."* I have frequently met with the roots or stems of the marestail under ground, which, on being bisected by the drains, poured out a constant run of water for some time, but, when fairly emptied of it, and no longer receiving support by a due supply of moisture from above, they withered away. Although there is no doubt that, in the case mentioned by Sir Joseph Banks, the roots of the marestail penetrated deeper than the drains, yet the circumstance of their sending upward shoots which grew "along the openings left for the passage of water," proves that sufficient moisture had been left near the surface of the bog, notwithstanding the draining, to support the plant in life; in short, that the bog had been insufficiently drained, otherwise the rivation of support by moisture to the stem at the surface would inevitably have destroyed the vitality of the roots below.

Fig. 189.

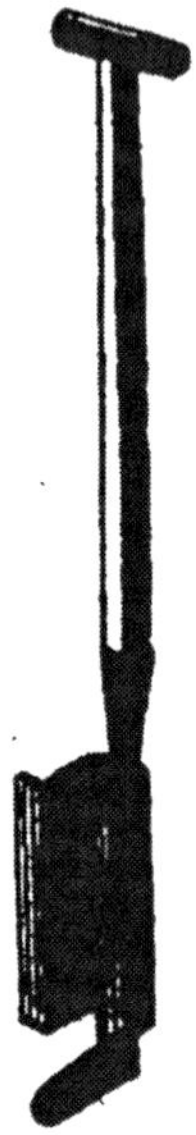

THE CALDERWOOD PEAT-TILE SPADE.

(845.) A plan similar to that described in (843) may be practiced on *strong clay land*. The open furrow is formed in the same manner with the plow, and, being left 16 inches in width, the spade work is conducted in this manner: Leave a scarsement of 1 inch on each side of the open furrow left by the plow, as seen below *a a*, fig. 190, and cut out the earth, 14 inches wide, perpendicularly, and 10 inches deep, as at *b b*. Then cast out from the bottom of this cut, with a spade 3 or 4 inches wide, a cut 5 inches or more in depth *c*, leaving a scarsement of 5 inches on each side of the bottom of the former cut *b b*. The bottom of the small cut will be found to be 32 inches below the crowns of the ridges, when twice gathered up with a strong furrow. The drain is filled up in this way: Take flagstones of 2 or 3 inches in thickness, as *d*, and place them across the opening of *c* upon the 5-inch scarsements, left by the narrow spade; they need not be dressed at the joints, as one stone can overlap the edges of the two adjoining, and they thus form the top of a conduit of pure clay in which the water may flow. As the water is made to flow immediately upon the clay, it is clear that this form of drain cannot be regarded as a permanent one; though a flag or tile sole laid on the bottom of the cut *c* would render it much more durable. The cutting of this form of drain, the workmen having to shift from one tool to another, will cost 1½d. the rood of 6 yards, which, at 15 feet apart, make 20s. 2d. the acre. The flags for covers will be 12 tons at 4d. per ton, 4s. more, in all 24s. 2d., but with 10 tons of soles the cost will be 3s. 4d. more, or 27s. 6d. the acre, exclusive of the carriage of stones and the labor of the plow. After the *joinings* of the flags are covered over with turf, the earth may be returned into the drain with the plow, but with precaution, and probably with the previous assistance of the spade; but, after all, the probability is that *flat* stones cannot be easily obtained in the neighborhood of strong clay, though this *form* of drain may be adopted in any subsoil where flat stones are abundant.†

Fig. 190.

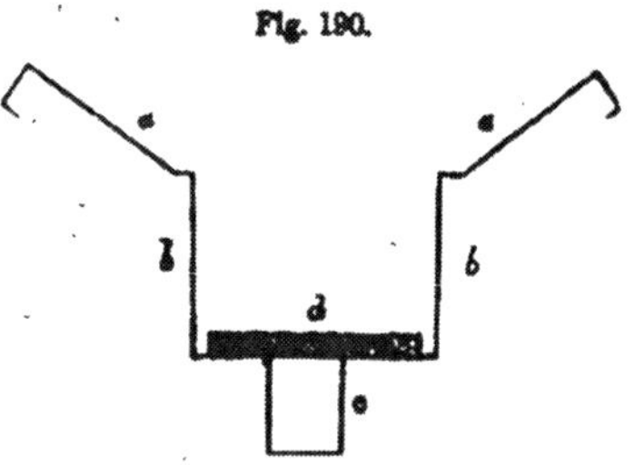

THE CLAY-LAND SHOULDER DRAIN.

(846.) A somewhat analogous mode to this last of draining heavy clay land is with the *wedge* or *plug*. As this mode of draining requires a very peculiar form of tools, they will be described as required for use.

(847.) The first remarkable implement used in this operation is the *bitting-iron*, represented by fig. 191, where *a* is the mouth, 1½ inches wide; *b* the bit, 6 inches in length; *c* the width of the bit, 4½ inches.—The bit is worked out of the body of the instrument, and laid with the best tempered steel; *e* is the tramp of the implement, placed 18 inches *d* from the mouth *a*; it would perhaps strengthen the power of the implement to have the tramp on the same side as the bit *c*; and *f* is the helve, which is of the length of that of a common spade.

Fig. 191.

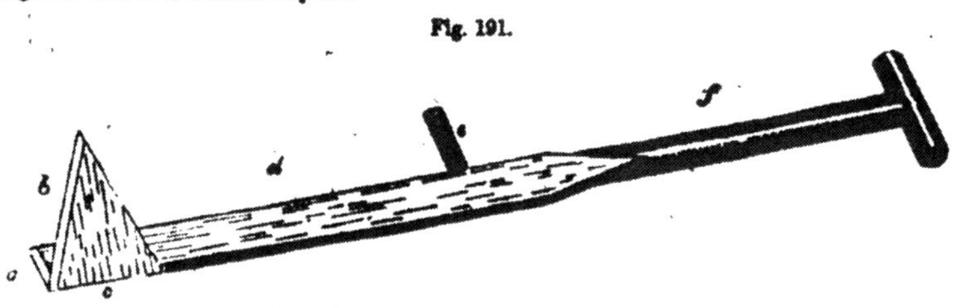

THE BITTING-IRON IN PLUG-DRAINING.

(848.) This species of draining is represented by those who have practiced it to be applicable to all soils that have a various and uncertain depth of vegetable mould, incumbent on a subsoil of

* Communications to the Board of Agriculture, vol. ii. † Quarterly Journal of Agriculture, vol. vii.

tenacious clay, exceedingly impervious to water, and never dry but by evaporation. It is, however, more suitable to pasture than to arable land, although it will suit all heavy soils that are far removed from stones, or where tiles cannot be conveniently made; but I would remark, on this last observation, that I would advise you to prefer *tile* draining to this mode, even although the tiles be very dear, either from distance of carriage or difficulty of manufacture on the spot.

(849.) The first process is to remove the surface turf, 12 inches in width and 6½ inches in depth, with the common spade, and to place it on the right hand side of the workman, with the grass side uppermost. A cut is then made in the clay, on each side of the drain, with an edging-iron, the circular-mouthed spade, as fig. 150; but a common spade will answer the purpose well enough, and it requires some skill and dexterity to remove this second cut properly. The first cut having been made with the spade 12 inches wide, the second should be made at such an angle down both sides of the drain, 9 inches deep, as that the breadth at the bottom shall be the exact width of the top of the plug *h*, fig. 192—that is, 4 inches wide. Carelessness in expert, or blundering in inexperienced workmen, in this part of the operation, has caused this kind of drain to fail.—The bitting-iron then completes the cuttings, by taking out the last cut 9 inches deep, and 1¾ inches wide at the bottom. This instrument is used in this manner: The workman gives its shaft such an angle with the ground line that, when pushed down to the requisite depth, it continues the cut made by the spade or edging-iron used previously, on the right hand side of the drain; and he does exactly the same on the opposite side of the drain, using his foot in both cases on the tramp *e*, fig. 191. On being forced down on the second side of the drain, the clay, that is now separated all round by the bit *b*, leans against the stem of the iron, and is easily lifted out, so that each bitful of the clay taken out by this instrument will have the form of an oblique parallelopipedon. If this part of the operation is performed inaccurately, the drain cannot succeed, because the angle and depth made by this instrument are of the utmost consequence in forming the bed which is to be occupied by the plug. Considerable accuracy of hand and eye is requisite; which, indeed, cannot well be acquired by workmen without much experience, but both may be soon acquired. The clay from the last two cuttings should be placed on the left hand side of the workmen, or the opposite side to that on which the upper turf was laid; and, from the last cutting being uppermost, it will come readily to hand when first returned into the drain. Any loose soil that may happen to remain at the bottom should be carefully taken out by a scoop spade, such as fig. 152, so as to leave the drain perfectly clean before the farther operations are effected.

(850.) The next implement used is the *suter* or *plug*, fig. 192, which consists of three or more pieces of wood *i*, 8½ inches in hight, 6 inches in length, 4 inches wide at the top *h*, and 1¾ inches

Fig. 192.

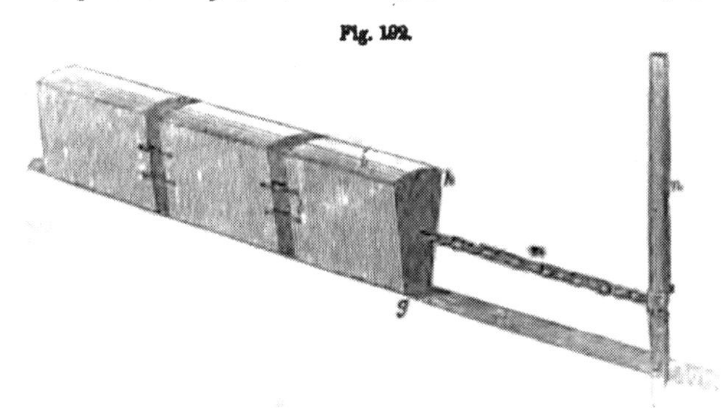

THE SUTERS OR PLUGS IN PLUG-DRAINING.

wide at the bottom *g*, joined together by means of iron links *l* sunk into the sides which allow them to pass in a cut with a slight curve. A single suter of 18 or 24 inches long would answer the same or perhaps better purpose. These dimensions are the guage of the opening of the drain described above.

(851.) The next step in the process is the placing of the plugs on their narrow edges in the bottom of the drain, which they will exactly fit, if the drains have been properly cut out. The most important part of the process is now to be done. The clay that was last taken out with the bitting-iron is well rammed down upon the plugs, the pieces of the clay being perfectly incorporated into one mass; then the next portion that was cut is returned, and equally well rammed down; and lastly, the turf is placed in the order it was taken out, and fixed in its original position. The whole earth and the turf are rammed down to the full length of the plug, with a rammer made for the purpose, or with such a one as is represented by fig. 175. The operation of ramming being finished, the lever *n*, fig. 192, is then struck into the bottom of the drain, and the plugs drawn forward to within 8 inches of their entire length by the power of the lever on the chain *m*, which is hooked to a staple in the end of the nearest plug. The work of ramming proceeds thus step by step until the whole drain is completed.

(852.) The finished drain is represented in section in fig. 193, where *o* is the duct left in the clay by the plugs, 8½ inches high; *p* is the clay that was rammed down above the plugs, 9 inches deep; and *r* is the returned turf, with the grassy side uppermost, 6½ inches, which again makes the surface smooth, making a drain of 2 feet in depth. These two figures are drawn to the scale of ½ of an inch to 2 inches.

(853.) Some particulars in the conducting of the work should be attended to. 1. Care should be taken to return *all* the earth that was cast out of the drain. This is a criterion of good work;

and for this purpose, the ramming being the most laborious part of the operation, the workmen are apt to execute it in an inefficient manner, and should therefore be strictly superintended in its execution. Four men and a boy are the best number of people for carrying on the work expeditiously; and only stout people should be employed, as the ramming is really a laborious process. 2. As few main-drains should be made as possible and the open ends of all should be protected against the inroads of vermin; or, what is a better finish, the lowest end of a plug drain should be furnished with tile and sole or stone. The main drains, of course, should be made larger than the ordinary drains, and they will have to be provided with proportionally larger plugs. The drains should be at a distance from each other, in proportion to the drawing nature of the subsoil. 3. No stock whatever should be allowed to enter the field while under this treatment, and even not until the earth over the drains has again become somewhat firm. After the drains of a field are all finished, the ground should be rolled with a heavy roller. 4. This sort of drain should not be made in frosty, snowy, or very rainy weather, as the earth to be rammed in will then be either too hard, crumbly, or too soft. A strict superintendence of the work when it is going on is the only guaranty for efficiency of work; for, as to the expedient of imposing fines upon poor workmen, they cannot be exacted without hardship, and perhaps injustice.

Fig. 193.

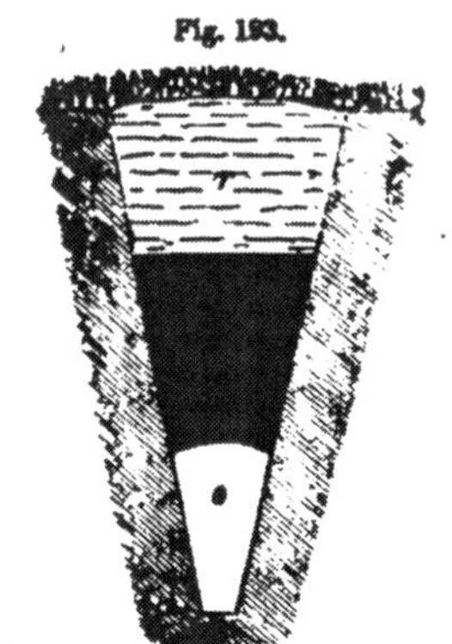

THE SECTION OF A PLUG DRAIN.

(854.) Mr. W. S. Evans, of Selkirk House, near Cheltenham, Gloucestershire, executed 300 miles of this kind of drain in 4 years, and is well pleased with its effects upon the land that has been subjected to it.

(855.) It is not so inexpensive a mode of draining as it at first sight appears, costing 1¾d. the lineal yard, or £5 0s. 10d. the acre, according to Mr. Evans's experience; but, according to another account, the expense is 4d. the rood of 6 yards, or £2 13s. 9¾d. the acre.

(856.) The principle of this mode of draining is said to have succeeded well on the tops of the Gloucestershire hills, where the bottoms of the drains descend to and are cut through rock, and where the bitting-iron and plug have been laid aside for the pick-ax, the channel formed by which is covered with flat stones, and the whole covered with clay rammed down as before described. *This* may be a permanent mode of draining; but, in plug-draining in clay, I have no doubt that water will have the effect of softening the sides of the duct, and causing the rammed wedge of clay above to slip downward; and, should the water ever reach to the wedge, the latter will inevitably crumble down, and either entirely fill up, or form a dam across the duct.*

(857.) An imperfect form of wedge-draining is practiced in some parts of England on strong clay soils, under the name of *sod*-draining. It is executed by removing the upper turf with the common spade, and laying it aside, for the purpose of making it the wedge at a subsequent part of the operation; and, if the turf is tough, so much the better for the durability of the sod-drain. Another spit is made with the narrow spade, fig. 170, and the last or undermost one is taken out with the narrowest spade, represented in fig. 194, which is only 2¼ inches wide at the mouth: and, as its entire narrowness cannot allow a man's foot being used upon it in the usual manner, a stud or spur is placed in front at the bottom of the helve, upon which the workman's heel is pressed, and pushes down the spade and cuts out the spit. The depth may be to any desired extent. The upper turf is then put in and trampled or beaten down into the narrow drain, in which it becomes wedged against the small shoulder left on each side of the drain, before it can reach the narrow channel formed by the last-mentioned spade, fig. 194; and the channel below the turf, being left open, constitutes the duct for the water. It will readily be perceived that this is a temporary form of drain under any circumstances, though it may last some time in grass land, but it seems quite unsuited for arable ground, which is more liable to be affected by dashes of rain than grass land; and, in any situation, the clay in contact with water will run the risk of being so much softened as to endanger the existence of the duct.

Fig. 194.

THE NARROWEST SPADE FOR SOD-DRAINS.

(858.) Another method of draining is performed on strong clay land by the *mole-plow*. This implement is almost unknown in Scotland, its use being confined to some parts of England, particularly in those parts where grass land on a clay subsoil abounds. It was, I believe, first introduced to the notice of Scottish agriculturists by the Duke of Hamilton, who caused it to be exhibited publicly on the occasion of the Highland and Agricultural Show at Glasgow in 1838. The day after the Show, I saw it exhibited in operation on a farm in the neighborhood of Glasgow, of strong clay land, for it seems to be best suited to operate in that kind of soil. Its object is to make a small opening in the soil at a given distance from the surface, in the form of a mole-run, to act as a duct for the water that may find its way into it; hence its name of *mole*-plow. It makes the pipe or opening in the soil by means of an iron-pointed cone, drawn through the soil by the application of a force considerably greater than that applied to a common plow.

(859.) The mole-plow as a draining machine can never be of much utility in a country like Scotland, whose alluvial formations, though not deficient in extent, are characterized more by the abundance of their stony matters than by their clays as occupying the place of subsoils; and it is only in the few patches of *carse land* that such clays occur as can be brought under the action

* Quarterly Journal of Agriculture, vols. iv. and xi.

of the mole plow. In all those subsoils where bowlders occur, whether large or small, the mole plow is so inapplicable, its usefulness is limited to such subsoils as consist of pure alluvial clays. In England, and where extensive flat districts of country occur, there the alluvium may be found which are the proper sphere of action for the mole-plow.

(860.) This plow is of extremely simple construction, as will appear from fig. 195, which is a view of it in perspective. It consists of a beam of oak or ash wood 6½ feet in length, and meas-

Fig. 195.

THE MOLE-PLOW.

uring 6 by 5 inches from the butt-end forward to 4 inches square at the bridle *b*. As the beam when in operation lies close upon the ground, and is indeed the only means of regulating the depth at which the conduit is to be formed, the lower side is sheathed all over with a plate of iron about ¼ inch thick. This plate at the proper place (4 feet 4 inches, or thereby, from the point of the beam) is perforated for the coulter-box; its fore-end is worked into an eye, which serves as a bridle, and is altogether strongly bolted to the beam. At the distance of a foot behind the coulter-box, a strong stub of wood is mortised into the beam at *c*, standing at the rake and spread which is to be given to the handles. Another plate of iron, of about 3 feet in length and ½ inch thick, is applied on the upper side of the beam; the coulter-box is also formed through this plate, and the hind-part is kneed at *c*, to fit upon and support the stub, to which, as well as to the beam, the plate is firmly bolted. The two stilts or handles *c f* are simply bolted to the stub, which last is of such breadth as to admit of several bolt-holes, by which the *hight* of the handles can be adjusted. That which may be termed the head of the plow is a malleable iron plate of about 2 feet in length; that part of it which passes through the beam, and is there fastened by means of wedges like the common coulter, is 7 inches broad and ¾ inch thick. The part *d*, below the beam that performs the operation of a coulter, is 9 inches broad, ¾ inch thick in the back edge, and thinned off to a knife-edge in the front. The share or *mole* is a solid of malleable iron, welded or riveted to the head; its length in the sole is about 15 inches, and in its cross section (which is a triangle with curved sides and considerably blunted on the angles) it measures about 3 inches broad at the sole, and 3½ inches in hight. A cylinder is, however, a better form than a triangle; but in either case the fore-part of the share is worked into a conical form, the apex being in the line of the sole, or nearly so. This, while it enables the share to penetrate the earth more freely, prevents a tendency to rising out of the ground. The tendency to rise is, however, not so great as may be supposed, for the center of motion in this implement being very low, not less than 12 inches under the surface of the ground, and the draft being applied horizontally, there is a strong tendency in the point of the beam and of the share, as in all similar cases of oblique draft, to sink into the ground (630)–(634); the effect of which, if not properly balanced by the effects of *form* in the parts, will give the mole-plow much unnecessary resistance.

(861.) In working this plow, the draft-chain is attached to the bridle-eye at *b*, and it is usually drawn by two horses walking in a circular course, giving motion to a portable horse capstan, that is constructed on a small platform movable on low carriage-wheels, and which is moored by anchors at convenient reaches of 50 to 60 yards. The mechanical advantage yielded by the horse capstan gives out a power of about 10 to 1, or, deducting friction, equal to a force of about 14 horses.

(862.) When the plow is entered into the soil and moved forward, the broad coulter cuts the soil with its sharp edge, and the sock makes its way through the clay subsoil by compressing it on all sides; and the tenacity of the clay keeps not only the pipe thus formed open, but the slit which is made by the broad coulter permits the water that is in the soil to find its way directly into the pipe. The plow is found to work with the greatest steadiness at 15 inches below the surface. The upper turf is sometimes laid over beforehand by the common plow, when the mole-plow is made to pass along the bottom of its furrow, and the furrow-slice or turf is again carefully replaced. This is the preferable mode of working this plow, as it serves to preserve the slit made by the coulter longer open than when it terminates at the surface of the turf, where, of course, it is liable to be soon closed up; but the least trouble is incurred when the plow is made to pass through the turf unplowed.

(863.) To work the whole apparatus efficiently, 2 horses and 3 men are required; and if the common estimate of 10s. a day for 2 horses and 1 man is taken, but which is too high, as you shall have occasion afterward to learn, and 3s. 6d. for the other 2 men, an acre of ground can be mole-drained for 13s. 6d., exclusive of the first cost and tear and wear of this apparatus, the cost of which cannot be less than £50. At this rate, this is the cheapest of all the modes of draining that you have yet heard of.

(864.) If the mole-plow is put in motion in soft clay, the slit made by the broad coulter will not remain open even for a single day; and, though it may again open in severe drouth, it will close again whenever the clay becomes moist. This plow seems fitted for action only in pure clay subsoils, and, when such are found under old grass, it may partially drain the ground with comparative economy; and, the process being really economical, it may be repeated in the course of years in the same ground. In my estimation, this mode of draining cannot bear a comparison for efficacy to tile-draining, although it is employed in some parts of England, where its effects are highly spoken of.*

(865.) It has lately been proposed by Mr. Scot, of Craigmoy, Stewartry of Kirkcudbright, to substitute tubes of larch wood for drain-tiles, in situations where larch is plentiful, and consequently cheap, and drain-tiles dear; and he considers that they would be equally efficient with tiles in many situations, and especially in mossy soils. Were larch tubes confined to draining mossy soils, I conceive they would answer the purpose well, not only on account of their length maintaining their original position in the drain, but on account of the durable nature of larch where water is constantly present, as is instanced in cases of great antiquity, such as the piles of larch upon which the city of Venice is founded. The larch tree that is felled in winter, and allowed to dry with the bark on, is much more durable and useful for every purpose, and infinitely more free of splits and cracks than that which is cut down in sap, and immediately deprived of its bark for tan.

(866.) The tube finished, fig. 196, presents a square of 4 inches outside, with a clear water-way of 2 inches. To those who wish to know how they are made, I refer to Mr. Scot's published statement;† but in doing this I must remark that the cost of these tubes will exceed that of clay tiles. For, take the cost of drain-tiles at 30s. per 1000, including carriage, that will be 1¼ farthings the lineal foot. Now, a lineal foot of larch tube contains say 1 superficial foot of timber at 1 inch thick, which will cost for carriage and sawing the timber 1 farthing; the fitting, boring and pins will cost other 2 farthings; and the timber, at 6d. the cubic foot, will increase the cost 2 farthings more, which altogether make the tube more than 3 times dearer than tiles; and, if the cost of the timber is thrown into the bargain, still they will be double the price of tiles.

Fig. 196.

THE LARCH DRAIN-TUBE.

(867.) The recommendation of wooden tubes for the purpose of draining land reminds me of many expedients which are practiced to fill drains, among which are brushwood, thorns, trees, and even straw-ropes. With the exception of the trunks of small trees, which, when judiciously laid down in drains, may last a considerable time, it is not to be imagined that brushwood of any kind can be durable. Hence, drains filled with them soon fall in. It could only be dire necessity that would induce any man to fill drains with straw twisted into ropes; and it could only have been the same cause, in situations where stones were scarce, and at a time when drain-tiles were little known in Scotland, such as was the case during the late war, that could have tempted farmers to fill drains with thorns. No doubt, the astringent nature of thorn-wood and bark may preserve their substance from decay under ground for a considerable time, but the sinking of holes in such drains, as I have seen, were infallible symptoms of decay. Only conceive what a mess such a drain must be that is "filled up to the hight of 8 or 10 inches either with brushwood, stripped of the leaves—oak, ash, or willow twigs being the best—and covered with long wheat straw, twisted into bands, which are put in with the hand, and afterward forced down with the spade; care being taken," the only case of it evinced in the whole operation, "that none of the loose mould is allowed to go along with them. The trench is then entirely filled up with earth, the first layer of which is closely trampled down, and the remainder thrown in loosely."‡ And yet such is the practice in several of the south-eastern and midland counties of England.

(868.) Of the *durability of common brick* when used in drains, there is a remarkable instance mentioned by Mr. George Guthrie, factor to the Earl of Stair, on Culhorn, Wigtonshire. In the execution of modern draining on that estate, some brick-drains, on being intersected, emitted water very freely. According to documents which refer to these drains, it appears that they had been formed by the celebrated Marshal, Earl Stair, *upward of a hundred years ago*. They were found between the vegetable mould and the clay upon which it rested, between the "wet and the dry," as the country phrase has it, and about 31 inches below the surface. They presented two forms—one consisting of 2 bricks set asunder on edge, and the other 2 laid lengthways across them, leaving between them an opening of 4 inches square for water, but having no soles. The bricks had not sunk in the least through the sandy clay bottom upon which they rested, as they were 3 inches broad. The other form was of 2 bricks laid side by side, as a sole, with 2 others built on bed on each other at both sides, upon the solid ground, and covered with flat stones, the building being packed on each side of the drain with broken bricks.‖

(869.) Various attempts have been made to *lessen the cost of cutting drains*. One of these is to cut the drains narrower than they used to be, for the obvious reason that the drawing power of drains lies more in their depth than breadth; and the cubical contents of drains, of any given length, have in consequence been much decreased, and the cost of digging them of course much lessened.

* See the Agricultural Surveys of Middlesex and Essex.

† Prize Essays of the Highland and Agricultural Society, vol. xiv., where the machinery for making these tubes is figured and minutely described.

‡ British Husbandry, vol. i.

‖ Prize Essays of the Highland and Agricultural Society, vol. xiv.

(870.) An attempt has been made with this view by Mr. Peter McEwan, Blackdub, Stirlingshire. His invention consists of the application of the plow in casting out the contents of drains, and it certainly displays much mechanical ingenuity, and really performs the work with a considerable degree of perfection. This application of the plow, however, is by no means adapted to every species of subsoil—the most common one, of a tilly clay, containing small stones and occasional bowlders, presenting insuperable difficulties to its progress; while in pure unctuous clay it cuts its way with ease, and lays aside the tenacious furrow-slice with a considerable degree of regularity. The instrument has thus a limited application, but a greater objection exists against it, inasmuch as it requires an inordinate amount of power to set it in motion, consisting of that of 12 horses. This circumstance alone still more limits its application, for there are comparatively few farms which employ 6 pairs of horses at work; and besides, it is almost impossible to yoke 12 horses together, so as to derive the amount of labor from them, as when yoked in pairs. It is truly distressing to see the horses with this plow, as I once had the opportunity of witnessing in a field, of favorable subsoil, too, in the neighborhood of Glasgow in 1838, on the occasion of the Highland and Agricultural Society's Show.

(871.) Mr. Smith, Deanston, has given a description of Mr. McEwan's draining-plow, which it is not necessary to particularize farther than that the horses go in two divisions, one on each side of the line of draught, yoked to a strong master-tree 10 feet long, arranged so as to have 4 abreast, when 8 horses are used, and 6 abreast when 12 horses are yoked.

(872.) With regard to the state of the work left by this plow, men follow with spades, and take out a bed for tiles or broken stones, and correct any deviation from a uniform fall in the bottom, occasioned by unevenness of ground. The tiles or stones are then put in the usual manner, and the earth is returned into the drain by the plow.

(873.) This drain-plow is made of two sizes, one weighing 5 cwt., costing £11, the other weighing 4 cwt., and costing £8 8s., and the bars or swingle-trees, necessary to accompany each plow, are 2 six-horse, 4 three-horse, and a strong chain, the whole costing £4 4s.

(874.) With regard to the length of drain cut by this plow, Mr. Smith estimates the time spent at 2 miles per hour for 8 hours; and allowing ¼ of it to be lost in turnings, the actual quantity of work done in 8 hours he takes at 3,126 roods of 6 yards, or about 19¼ acres, at 15 feet asunder, the drains being cut from 18 to 22 inches in depth. This quantity of work is corroborated by Mr. John Glen, Hilton, Clackmannanshire, who states that "we drain 400 Scotch chains in 9 hours," going down hill with the furrow, and up empty.

(875.) The rate of walking taken by Mr. Smith, at 2 miles the hour, is too great, as the distance traveled in plowing 1 imperial acre of ground, in the usual way, in a day of 10 hours, which constitutes a good rate of work, is 9¾ miles, or only 1742 yards per hour, including of course turnings. There is also a discrepancy in Mr. Glen's statement of draining nearly half the extent of his land in 9 hours, going half the time empty, with another statement where he says, "we have used Mr McEwan's drain-plow for the last 4 months, and have drained 837 chains, Scotch measure, with it;" that is, only 2 days' work in 4 months, with 6 horses, the drain being 18½ inches wide at top, and 8 inches wide at bottom, and from 15 to 17 inches in depth.

(876.) The cost of employing this drain-plow is thus given by Mr. Smith:

12 horses at 4s. a day each	£2	8	0
8 men at 2s. a day each	0	16	0
To cover interest of cost, and tear and wear of plow, say 1s. the hour	0	8	0
Total	£3	12	0

which is only 1½ farthings per rood of 6 yards.*

(877.) Other plows have been invented for making drains, which have attracted attention, and engaged the advocacy of friends in the immediate locality in which they originated, but seem never to have extended farther.

(878.) In 1832, Mr. Robert Green, a farmer in Cambridgeshire, published an account of a drain-plow of his invention. It cuts the ground 23 inches deep and 8 inches wide at top, and 2 inches at bottom, at three cuts; the first being 9 inches, the second 8 inches, and the third 6 inches deep. It is said to take the earth out clean, leaving none to shovel out. It cuts about 500 or 600 poles or rods of the above dimensions, at three times. It requires 4 horses at the first time, and 6 horses the other two, and 2 men and a boy to work them, at £1 10s. a day. The price of this plow is quoted at £15.†

(879.) In 1833, Mr. Thomas Law Hodges published an account of a drain-plow invented by Mr. John Pearson, Frotterden, near Cranbrook in Kent. The drain is taken out by it at three turns. Men follow with narrow scoops, and throw out all the loose earth clean, which finishes the drain at 26 inches deep, at an expense of 1d. the rod.

(880.) Both these implements are best adapted to strong clay subsoils, and best for plug-draining, especially Pearson's, when, after its operation, a long, narrow plug or slide of wood is used, the clay being rammed down upon which, it is then drawn forward by means of a windlass and rope. This plow is estimated to cost £9 5s., but with spades, scoops and rammers, it costs £18.‡

(881.) It may be asserted, without much fear of contradiction, that improvements by thorough-draining will never become general, or be made permanent, unless the assistance of the landlord be obtained. When left altogether to the tenant, want of capital and the shortness of the lease will tend at all times to limit the extent of improvements, and will seldom be made permanent, because the true interest of the tenant is to execute the work only in such a manner as will secure his own temporary purpose. To the proprietor, among the many inducements to improve his estate by draining, the greatest, at least the most satisfactory, is, that it yields an immediate and large return. If he has no spare money, he has only to borrow it at 4 per cent. and lend it,

* Smith's Remarks on Thorough-Draining.
† Green on Underdraining Wet and Cold Land.
‡ Hodges on the Use and Advantages of Pearson's Draining-plow.

out at 6 per cent., a per centage which no tenant will refuse to pay, and upon a security, too undoubted—that of his own property. No one will deny that a proprietor is as justly entitled to receive a fair return for money laid out in the improvement of his estate, as he is for that laid out on the original purchase of it. Hence, I would assume, as a general principle, that for every penny laid out by a proprietor upon ameliorations of any kind, he shall have an assurance of a return, either immediate or prospective—immediate, in the form of interest or of additional rent; prospective, in the increased value of his property, by which, in after leases, it will yield such an increase of rent as will repay the present outlay. Again, as regards a tenant, I will assume that he will make every improvement the subject of a calculation of profit or loss for *one lease only*, and that he will not lay out any money *merely for the purpose* of making improvements to extend beyond that period. Let him even have an assurance of a renewal of his lease, still, before that takes place, a valuation will be made of his farm, and in that valuation will be included his improvements; so that, while he originally disbursed the whole expense of them, he will in reality have to pay for them again in the shape of additional rent.

(882.) [The mechanical principles of draining have been already so fully discussed that I need not detain you a moment with their examination; but this will be the best place for considering a most interesting subject connected with soil, upon which the whole *necessity* of draining depends, I refer to the manner in which an excess of water proves injurious to the fertility of the soil.

(883.) In considering this subject it will be advisable to examine into the effects of water, 1st upon the mechanical condition of the soil, and 2d, upon its chemical constituents, reserving the influence which it exerts directly upon vegetation to be discussed on some future occasion.

(884.) If you call to mind what I have said regarding the mechanical constitution of soil (424), you will at once perceive that a soil *in situ* might not inaptly be compared to a porous solid permeated by innumerable tortuous channels, these channels being formed by the interstitial spaces occurring between the various particles composing the soil.

(885.) If water is added gradually to soil, the first effect will be doubtless to fill these channels, but from the attraction which the various components of soil have for water, they speedily draw it into their pores, and thus empty the channels so that even after a considerable addition, the soil, *taken as a whole*, does not lose its porosity although each particle *has its individual pores* filled with water. This is the healthy condition of soil; it is what I shall call *moist*, in contradistinction to *wet*. Soil in this state can be crumbled down in the hands without making them muddy, although it feels distinctly *damp*, and will lose, when heated to 212° F., from 20 to 50 per cent. of water.

(886.) If now more water should be added, the channels will be again filled, and as the pores of each particle are already saturated with moisture, they can again be emptied only by one of the two following methods: 1st, either very gradually by evaporation from the surface, as in *undrained soil*, or, 2d, much more rapidly and effectually by the channels having communication with some larger channel in a relatively lower level, as is the case in *drained soil*. Soil in which all the interstices between its particles are more or less filled with water may be called *wet* soil, and all such land must be drained before it can be properly and advantageously cultivated.

(887.) You will thus perceive that water does no harm, in fact it is absolutely necessary in soil so long as it does not alter its mechanical condition; but whenever it fills up the interstitial channels it becomes injurious for the following reasons: 1st, it prevents the circulation of air through the soil, as this takes place entirely through the medium of these channels; 2d, it impoverishes the soil by permitting soluble matter to soak through; because until these channels are filled there is no flow of liquid in the soil, except a very gentle current from below upward, produced by capillary attraction toward the drier particles near the surface.*

(888.) Again, an excess of water acts most injuriously in soil by reducing its temperature. This is owing to the extremely slight conducting power of water for heat, as compared to earthy matter, assisted also by the cold produced by continued evaporation. According to some experiments which I performed, the diminution of heat produced in this way amounts, in summer, on an average to 6½ degrees of Fahrenheit, which, according to Sir John Leslie's mode of calculating elevation by the mean temperature, is equivalent to a difference of 1,950 feet. When we consider the effects of elevation upon the nature and amount of produce, we shall have good reason to see the baneful effects of such a change as this represents.

(889.) Besides the above injuries inflicted by an excess of water, there are numerous effects upon the chemical changes in the soil, and also upon the plants themselves, all of which must be considered in their proper place. I trust, however, that what I have advanced will serve to impress sufficiently on your minds the evident necessity of thorough-draining in all situations where the soil is *wet*.—H. R. M.]

(890.) After pointing out the effects of draining in ameliorating the soil and promoting a healthy condition of vegetation, Professor Johnston proceeds to show the effects of water upon clay soil. "I shall add one important remark,' he says, " which will readily suggest itself to the geologist who has studied the action of air and water on the various clay beds that occur here and there as members of the series of stratified rocks. There are *no clays* which do *not* gradually *soften* under the united influence of air and of running water. *It is false economy, therefore, to lay down tiles without soles, however hard and stiff the clay subsoil may appear to be.* In the course of 10 or 15 years, the stiffest clays will soften, so as to allow the tile to sink, and many very much sooner. The passage for the water is thus gradually removed; and when the tile has sunk a couple of inches, the whole must be taken up. Thousands of miles of drains have been thus laid down, both in the low country of Scotland and in the southern counties of England which have now become nearly useless; and yet the system still goes on. It would appear even

* See Prize Essay on this subject by me in the Prize Essays of the Highland and Agricultural Society vol. xiii.

as if the farmers and proprietors of each district, unwilling to believe in or to be benefited by the experience of others, were determined to prove the matter in their own case also, before they will consent to adopt that surer system which, though demanding a slightly greater outlay at first, will return upon the drainer with no after-calls for either time or capital. If my reader," continues the Professor, "lives in a district where this practice is now exploded, and if he be inclined to doubt if other counties be farther behind the advance of knowledge than his own, I would invite him to spend a week in crossing the county of Durham, where he may find opportunities not only of satisfying his own doubts, but of scattering here and there a few words of useful advice among the more intelligent of our practical farmers."*

(892.) As the preservation of the fall in a drain on nearly level ground is of great importance in drying it, it may be satisfactory to have a demonstration of the fact that the angle subtended by the plumb-line df, in fig. 183, is equal to the angle of inclination of the drain $b\ a\ c$. The rule is, as radius : $a\ b$: : sine of the angle, $b\ a\ c$: $b\ c$, the hight of the fall: Or, multiply the natural sine of the angle $b\ a\ c$ by the length of the fall $a\ b$, and the same result will be obtained.†

(893.) The Romans practiced draining both with open and covered drains, the former in clay and the latter in porous soils. The instructions given by Palladius for the formation of drains may be received with surprise by modern practicers of the art on account of their correctness, and when their great antiquity is held in remembrance. "If the land is wet," he says, "it may be dried by drains drawn from every part. Open drains are well known: covered drains are made in this manner: Ditches are made across the field 3 feet deep; afterward they are filled half-way up with small stones or gravel, and then filled to the surface with the earth that was thrown out. These covered drains are let to an open one to which they descend, so that the water is carried off, and destroys no part of the field. If stones cannot be got, branches, or straw, or any kind of twigs, may be used in their place."‡

[We have omitted some calculations and tabular statements, designed to illustrate the particular and comparative *cost* of the several kinds of drains, according to the various materials employed in their formation, and other circumstances affecting the question.

All the elements of these calculations differ so widely from such as would be brought into account in this country, that the publication of them here would afford no reliable or useful data for those who might be disposed to enter on a system of draining, more or less extensive. Having so thoroughly exemplified and explained the principles of draining as conducted in a country where they are best understood and recently most extensively carried out, every reader can best judge for himself as to the cost and the probability of his being adequately remunerated by the results. These questions, as they arise, must depend, in all cases, on the peculiar circumstances that belong to them. As we have before said, the portions of a farm which most generally invite the operation of draining, are low grounds, and usually the most fertile, when divested of surplus moisture; and if any reliable inference is to be drawn from the profits of this operation in England, such as have been detailed, the increased crops to be expected, would warrant a heavy outlay, to say nothing of the *sanatary* effect, which has also been referred to. The reasons here given for not copying all the details, relating merely to the *cost* of draining by the different methods, and with the materials employed in England, might seem to apply to many others which have been given in respect to other branches of this operation, one of the great sources of the increased productiveness of English Agriculture; but, in the first place, the cases are not exactly parallel; and besides, we are so fully impressed with the importance and the profit of *draining* rich spots of land in our country, and so well satisfied that in the old States this important item in the management of land is too much overlooked, that it was deemed better to illustrate the subject in all its aspects, even though many of the examples given for that purpose may not be exactly applicable to the situation and circumstances of American landholders. For even these examples may present encouragement to justify the operation, in numberless instances where it has been neglected; and at all events, they serve more fully to explain one of the great problems of the day, with which every enlightened agriculturist ought to be familiar in all its phases. *Ed. F. Lib.*]

* Johnston's Elements of Agricultural Chemistry.
† See the practical application of this rule on a large scale illustrated in Denton on Model Mapping.
‡ Dickson's Husbandry of the Ancients, vol. i.

26. YOKING AND HARNESSING THE PLOW, AND OF SWING-TREES.

"No wheels support the diving pointed share;
No groaning ox is doomed to labor there;
No helpmates teach the docile steed his road;
Alike unknown the plow-boy and the goad;
But, unassisted through each toilsome day,
With smiling brow the plowman cleaves his way."
BLOOMFIELD.

(894.) HAVING inspected the varieties of soil within the sphere of your observation, and been told of the various modes in which the land may be stirred by the plow in winter, it will be proper for you to know the simple and efficient method by which horses are attached to and driven in the plow in Scotland, before the winter-plowing of the soil is begun, and to enable you to conceive the process more vividly, you will find a pretty accurate representation of a plow at work in Plate XIII.*

(895.) The first thing that will strike you is the extreme simplicity of the whole arrangement of the horses, harness, plow, and man, impressing you with the satisfactory feeling that no part of it can go wrong, and afford you a happy illustration of a complicated arrangement performing complicated work by a simple action. On examining particulars, you will find the *collar*, better seen in fig. 197, around the horse's neck, serving as a padding to preserve his shoulders from injury while pressing forward to the draught. Embracing a groove in the anterior part of the collar, are the *haims*, composed of two pieces of wood, curved toward their lower extremities, which are hooked and attached together by means of a small chain, and their upper extremities held tight by means of a leather strap and buckle; and they are moreover provided on each side with an iron hook, to which the object of draught is attached. The horse is yoked to the swing-trees by light chains called *trace-chains*, which are linked on one end to the hooks of the haims, and hooked at the other into the eyes of the swing-trees. A *back-band* of leather put across the back, near the

[* How little soever may be doing for disseminating in America among the rising generation of agriculturists a knowledge of the *principles* of their art, it may be said to the credit of the profession, that in the lightness and perfection of our gearing, and in the manner of attaching and using the motive power employed in the field of Agriculture, we are not behind, if, indeed, we are not in advance, of countries the most highly improved. Yankee ingenuity, in unrestricted play, under our free Government, has effected wonders in the form and structure and economy of agricultural implements and machinery, for saving cost and labor, except, perhaps, such as are of a costly nature, involving too much expense for common use. It is not to be maintained that our materials, especially of leather, or our workmanship in harness manufacture and saddlery, are by any means as perfect as in England; but so little have we to learn from them about "*yoking and harnessing* the plow," that we might have ventured to omit this chapter except that we choose, as well to gratify the curiosity as to instruct the minds of young and inquiring readers. England and Scotland are admitted to be in the van of all European nations in the march of improvement, especially in scientific Agriculture; and he would be deemed but a careless observer who should visit either the one or the other and come back without being able to tell if there were anything peculiar in their mode of harnessing their teams to agricultural implements. The next best thing to seeing for one's self is to be, as herein, authentically informed. The same reasons, without feeling the necessity of repeating them, will prompt us to give many other items that *might* be omitted were we to study nothing but *practical* usefulness and saving to our publishers.
Ed. Farm. Lib.]

groins of the horse, supports the trace-chains by means of simple hooks. The *bridle* has blinders, and while the horse is in draught it is customary to hang the *bearing-reins* over the tops of the haims. In some parts of the country there are no *blinders;* and there is no doubt that many horses so brought up will work very well without them. But in cases of horses of so timid a nature as to be easily frightened at distant objects, and those of so careless a disposition as to look much about them, they are useful in keeping the attention of the horse to his work. You observe there are two horses, the draught of the common plow requiring that number, which are yoked by the trace-chains to the *swing-trees*, which, on being hooked to the draught-swivel of the bridle of the plow, enable the horses to exercise their united strength on that single point; and being yoked abreast, they are enabled to exert their united strength much more effectually than if yoked atrip—that is, one before the other. The two horses are kept together either by a *leather strap*, buckled at each end to the bridle-ring, or by a *short rein* of rope passed from the bridle-ring to the shoulder of each horse, where it is fastened to the end of the trace-chain with a knot. The strap prevents the horse separating beyond its length, but allows their heads to move about loosely; the short reins prevent them not only separating, but keep their heads steady; and on this account, horses fastened with reins can be turned round more quickly and simultaneously than with the strap. The plowman guides the horses with *plow-reins*, made of rein-rope, which pass from both stilts to the bridle-ring of each horse, along the outermost side of the horse, threading in their way a ring on the back-band and sometimes another on the haims. The reins are looped at the end next the plowman, and conveniently placed for him under the ends of pieces of hard leather screwed to the foremost end of the helves; or small rings are sometimes put there to fasten the reins to. In many places, only one rein is attached to the near-side horse, and in others the horses are guided solely by the voice. It is perfectly obvious that the plowman must have a better and quicker command over his horses with a double than a single rein, and very much more so than by the voice alone.

(896.) Thus harnessed, each horse has not much weight to bear, nor is its harness costly, though made of the strongest harness leather, as this statement will show:

	Weight.		Value.		
Collar	15 lbs.		£1	0	0
Haims, when covered with plate-iron, and with a strap	7 "		0	5	6
Bridle	4½ "		0	10	0
Back-band	3½ "		0	8	0
Chains	8 "	at 7d. per lb.	0	4	8
Total	38 lbs.	and for each horse	£2	8	2

When compared with the weight of English harness, these are little more than feather-weight.

(897.) The *collars* are differently mounted in the *cape* in different parts of the country. The use of the cape is to prevent rain falling upon the top of the shoulder, and getting between the collar and shoulder, where, in draught, it would heat and blister the skin. In the Lothians, the cape of the form of fig. 197 is both neat and convenient. In Forfarshire, and somewhat more northerly, it is of the form of fig. 198, which lies flatter and comes farther back than the former; and it is certainly a complete protection from rain; but it makes the collar rather heavy, and its own weight is apt to loosen the sewing of white sheep-skin with which it is attached to the body of the collar. Fig. 199 is a form of cape common in England, which answers no purpose of protection from rain, but rather to catch the wind, and thereby obstruct the progress of the horse. Such a cape is frequently ornamented with flaring-colored red worsted fringe round the edge, or with large tassels from the corner and middle, or even with bells.

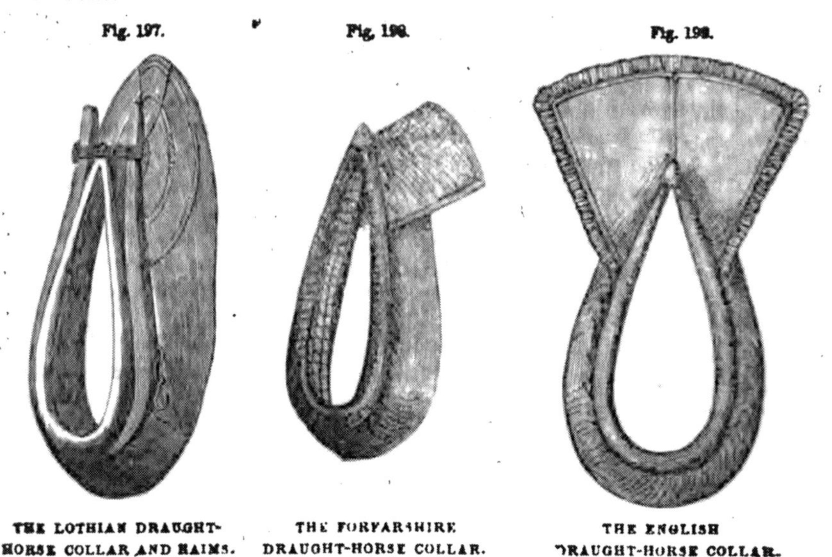

Fig. 197. THE LOTHIAN DRAUGHT-HORSE COLLAR AND HAIMS.

Fig. 198. THE FORFARSHIRE DRAUGHT-HORSE COLLAR.

Fig. 199. THE ENGLISH DRAUGHT-HORSE COLLAR.

(898.) With regard to ornamenting farm harness, it never appears, in my estimation, to greater advantage than when quite plain, and of good materials and excellent workmanship. Brass or plated buckles and brow-bands, worsted rosettes, and broad bands of leather tattooed with fillagree sewing, serve only to load and cover the horses when at work, and display a wasteful and vulgar taste in the owner. Whatever temptation there may be in towns to show off the grandeur of teams, you should shun such display of weakness in the country.

(899.) The English farmer is not unfrequently recommended by writers on Agriculture to adopt the 2-horse plan of working the plow; but the recommendation is never accompanied with such a description of the plow as any farmer could understand it who had never seen a plow with 2 horses at work; and it is not enough to tell people to adopt this or that plan, without putting it in their power to understand what is recommended.—To enable the English farmer, who may never have chanced to see a two-horse plow at work, and to facilitate the understanding of its arrangements by those who may have seen, but not have paid sufficient attention to it, the figure on Plate XIII. has been executed with a regard to show the just proportions of the various parts of the plow and the harness. The plow has been sufficiently well explained already, and keeping in mind the relative proportions of its parts, those of the horse and harness may be ascertained from this plate; for, so practically correct are those proportions, that any one desirous of mounting a plow in a similar manner may easily do so from this figure before them.

(900.) Although the reins alone are sufficient to guide the horses in the direction they should go—and I have seen a plowman both deaf and dumb manage a pair of horses with uncommon dexterity—yet the voice is a ready assistance to the hands, the intonations of which horses obey with celerity, and the modulations of which they understand, whether expressive of displeasure or otherwise. Indeed, in some of the midland counties of Scotland, it is no uncommon occurrence to observe the plowmen guiding their horses, both in the field and on the road, with nothing but the voice; but the practice is not commendable, inasmuch as those accustomed to it fall into the practice of constantly roaring to their horses, which at length become regardless of the noise, especially at the plow; and on the road the driver has no command over them, in any case even of the slightest emergency, when he is obliged to hurry and seize the bridle of the horse nearest to him at the time; and should one or both horses evince restiveness, when he can only have the command of one by the bridle, he runs the risk of being overcome by the other or by the cart.

(901.) The language addressed to horses varies as much as even the dialects are observed to do in different parts of the country. One word, *Wo*, to stop, seems, however, to be in general use. The motions required to be performed by the horse at work are—to go forward, to go backward, to go from you, and to come toward you; and the cessation of all these, namely, to stop or stand still.

To lessen or cease motion.—The word *Wo* is the common one for a cessation of motion; and it is also used to the making any sort of motion slower; and it also means to be careful, or cautious, or not be afraid, when it is pronounced with some duration, such as *Wo-o-o.* In some parts, as in Forfarshire, *Stand* has a similar signification; but, to stand without any movement at all, the word *Still* is there employed. In England, *Wo* is to stop.

To go forward.—The name of the leader is usually pronounced, as also the well-known *Chuck, Chuck,* made with the tongue at the side of the mouth, while impelling the breath.

To step backward.—*Back* is the only word I can remember to have heard for this motion.

To come toward you.—*Hie* is used in all the border counties of England and Scotland; *Hie here, Come ather,* are common in the midland counties of Scotland. In towns one hears frequently *Wynd* and *Vane.* In the west of England *Wo-ę* is used.

To go from you.—*Hup* is the counterpart to hie in the southern counties, while *haud aff* is the language of the midland counties; and, in towns, *Haap* is used where wynd is heard, and *Hip* bears a similar relation to vane. In the west of England *Gee agen* is used.

In all these cases, the speaker is supposed to be on what is called the *near-side* of the horse—that is, on the horse's left side. As a single word is more convenient to use than a sentence, I shall employ the simple and easily pronounced words *hup* and *hie*, when having occasion to describe any piece of work in which horses are employed.

(902.) [The *swingle* or *swing-trees, whipple-trees, draught-bars,* or simply *bars*—for by all these names are they known—are those bars by which horses are yoked to the plow, harrows, and other implements. In the plow-yoke, a set of swing-trees consists of 3, as represented in fig. 260, where *a* points out the bridle of the plow, *b b* the main swing-tree attached immediately to the bridle, *c c* the furrow or off-side little swing-tree, and *d d* the land or nigh-side little tree, arranged in the position in which they are employed in working. The length of the main-tree, between the points of attachment for the small trees, is generally 3½ feet, but this may be varied more or less; the length of the little trees is usually 3 feet between the points of attachment of the trace-chain, but this also is subject to variation.

(903.) Swing-trees are for the most part made of wood, oak or ash being most generally used; but the former, if sound English oak, is by much the most durable, though good Scotch ash is the strongest, so long as it remains sound, but it is liable, by long exposure, to a species of decay resembling dry-rot. As it is always of importance to know the why and wherefore of everything, I shall here point out how it may be known when a swing-tree is of a proper degree of strength. A swing-tree, when in the yoke, undergoes a strain similar in practice to that of a beam supported at both ends and loaded at the middle; and the strength of beams or of swing-trees in this state is proportional to their breadths multiplied into the square of their depths and divided by their lengths. It is to be understood that the *depth* here expressed is that dimension of the swing-tree that lies in the direction of the strain, or what in the language of agricultural mechanics is called the *breadth* of the swing-tree. To apply the above expression to practice—suppose a swing-tree

Fig. 200.

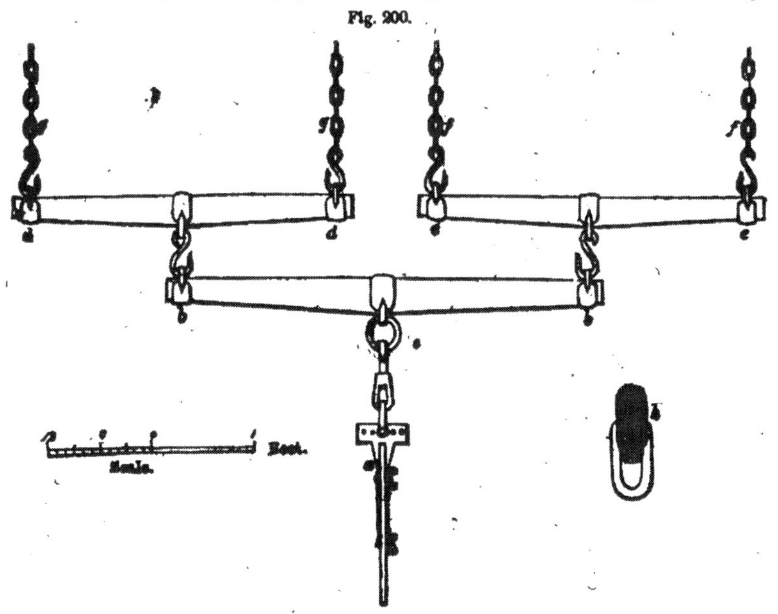

THE SWING-TREES FOR TWO HORSES.

of 3 feet in length between the points of attachment for the draught, that its breadth is $1\frac{1}{2}$ inches and depth 3 inches, and another of the same breadth and depth, but whose length is 6 feet, then in the case of the first we have $\frac{1\cdot5 \times 3 \times 3}{3 \text{ feet}} = 4\cdot5$; and in the second we have $\frac{1\cdot5 \times 3 \times 3}{6} = 2\cdot25$ —the strength of these two being as 2 to 1; and, to make the 6-feet swing-tree of equal strength with the other, the *breadth* must be increased directly as the length—that is to say, doubled—or the depth increased, so that its *square* shall be double that of the former. Hence a swing-tree of 6 feet long, and having a breadth of $1\frac{1}{2}$ inches and depth $4\frac{1}{4}$ inches, will be equal in strength to the 3-feet swing-tree with a breadth of $1\frac{1}{2}$ and depth of 3 inches; but, the depth remaining equal, the breadth is required to be *doubled*, or made 3 inches for the 6-feet swing-tree.

(904.) To find the absolute strength of a bar or beam, situate as above described, we have this rule: Multiply the breadth in inches by the square of the depth in inches, divide the product by the length in feet, and multiply the quotient by the constant 660 if for oak, or by 740 if for ash—the product will be the force in pounds that would break the swing-tree or the beam.* Here, then, taking the former dimensions as of a small swing-tree, $\frac{1\cdot5 \times 3^2 \times 740}{3} = 3{,}333$ lbs. the absolute force that would break the tree; but, taking into account the defect that all woods are liable to break from crossing the fibres and other contingent defects, we may allow $\frac{1}{2}$ to go for security against such contingencies, leaving a disposable strength equal to 1,666 lbs. It has been shown (634) that the usual force exerted by a horse in the plow does not exceed 168 lbs., but it occasionally rises to 300 lbs., and on accidental occasions even to 600 lbs.; but this is not much beyond $\frac{1}{3}$ of the disposable strength of the 3-feet swing-tree when its breadth and depth are $1\frac{1}{2}$ and 3 inches. The depth of such trees may therefore be safely reduced to $2\frac{1}{2}$ inches, and still retain a sufficient degree of strength to resist any possible force that can come upon it. In the large swing-tree the same rule applies; suppose its length between the points of attachment to be 3 feet 9 inches, its breadth $1\frac{3}{4}$ inches, and depth $3\frac{1}{2}$ inches, the material being ash as before; then $\frac{1\cdot75 \times 3\cdot5^2 \times 740}{3\cdot75} = 4{,}230$ lbs.; reducing this $\frac{1}{2}$ for security, there remain 2,115 lbs., but the greatest force that may be calculated upon from 2 horses is 1,200 lbs.; we have, therefore, nearly double security in this size of large swing-tree.

(905.) In proportioning the strength of swing-trees to any particular draught, let the greatest possible amount of force be calculated that can be applied to each end of the tree, the sum of these will be the opposing force as applied at the middle, and this may be taken as above (904) at 600 lbs. for each horse; but, for security, let it be 3 times or 1,800 lbs. each horse H, any number of horses being = H; and having fixed upon a breadth B for the tree, and L the length, C being the constant as before, then the depth D will be found thus: $\frac{L \times m H}{B \times C} = D^{\frac{1}{2}}$; or, in words, multi-

* Tredgold's Carpentry, art. 118.

ply the length into as many times 1,800 lbs. as there are to be horses applied to the tree, divide the product by the *constant* (740 for ash, or 660 for oak) multiplied into the breadth, the quotient will be the square of the depth, and the square-root of this will be the depth of the swing-tree, with ample allowance for assurance strength. In all cases, the depth at the ends may be reduced to ⅔ of that of the middle.

(906.) Wooden swing-trees ought always to be fitted up with clasp and eye mounting of the best wrought-iron, from 2 to 2½ inches broad, about 3-16 inch thick in the middle parts, and worked off to a thin edge at the sides; the part forming the eye may range from ½ inch diameter in the center eye of the large tree to ⅜ inch in the end clasps of the small trees; and they are applied to the wood in a hot state, which, by cooling, makes them take a firm seat. In the main tree, the middle clasp has usually a ring or a link *e* welded into it, by which the set is attached to the hook of the plow's bridle; the two end clasps have their eyes on the opposite edge of the swing-trees, with sufficient opening in the eyes to receive the S hooks of the small tree. The small are trees furnished with the S hooks, by which they are appended to the ends of the main trees; and end clasps are adapted to receive the hooks of the trace-chains *f f, g g*, a small part only of which are shown in the figure. The detached figure *h* is a transverse section of a tree showing the form of the clasps, the scale of which is double the size of the principal figure in the cut.

(907.) Though wood has hitherto been the material chiefly used for swing-trees, there have been some successful trials of malleable iron for the purpose. These have been variously constructed, in some cases entirely of sheet-iron turned round into a form somewhat resembling the wooden trees; but, in this form, either the iron must be thin, or the bar must be inconveniently heavy; if the former, durability becomes limited, by reason of the oxidation of the iron acting over a large surface, and soon destroying the fabric. Another method has been to form a diamond-shaped truss of solid iron rods, the diamond being very much elongated—its length being 3 feet, and its breadth about 4 inches, with a stretcher between the obtuse angles. A third has been tried, consisting of a straight welded tube of malleable iron, about 3 feet long and ¾ inch diameter. In this tube, acting as a strut, a tension-rod, also of malleable iron, is applied with a deflection of 4 inches, the extremities of the tension-rod being brought into contact by welding or riveting with the ends of the tubular strut, and eyes formed at the ends and middle for the attachment of the hooks and chains. A tree thus formed is sufficiently strong for every purpose to which it is applied, while its weight does not exceed 7 lbs.; and the weight of a wooden tree, with its mounting, frequently weighs 8 lbs. The price of a set of common wooden trees, with the iron mounting, is 12s., and of the iron trees 18s.

(908.) The foregoing remarks apply, so far as arrangement goes, to the common 2-horse swing-trees; but the various modes of applying horse-power, both as regards number and position of the horses, require farther illustration. The next I shall notice, therefore, is the 3-*horse yoke*, of which there are various modes—the simplest of which is, first, a pair, working in the common trees, fig. 200; and, for the third horse, a light chain is attached by a shackle to the middle of the main bar *b b*. To this chain the third horse is yoked, taking his place in front of the other two, in unicorn fashion. This yoke is defective, inasmuch as there are no means of equalizing the draught of the third horse.

(909.) Perhaps the most perfect method of yoking a 3-horse team, whether abreast or unicorn-fashion, is that by the compensation levers, fig. 201—a statical combination, which is at once cor-

Fig. 201.

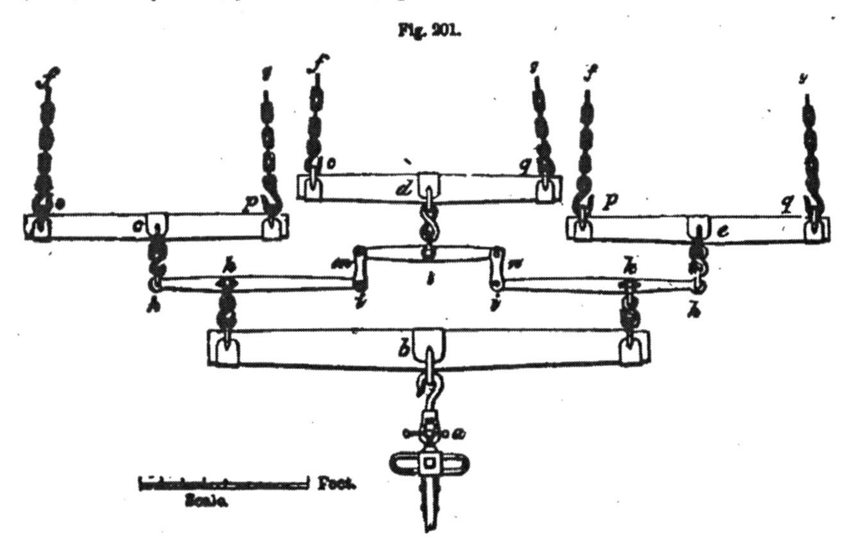

THE SWING-TREES FOR THREE HORSES.

rect in its equalization, scientific in its principles, and elegant in its arrangement, and I have to regret my inability to single out the person who first applied it. The apparatus in the figure is represented as applied to the subsoil-plow—*a* being the bridle of that plow; *b* is a main swing-tree, 5 feet in length, and of strength proportioned to the draught of 3 horses; and *c d e* are three small common trees, one for each horse. The trace-chains are here broken off at *f*, *g*, respectively, but are to be conceived as extending forward to the shoulders of the horses. Between the main swing-tree and the three small ones the compensating apparatus is placed, as in the figure, consisting of three levers, usually constructed of iron. Two of these, *h i* and *h i*, are levers of the first order, but with unequal arms, the fulcrum *k* being fixed at ⅓ of the entire length from the outward end of each; the arms of these levers are therefore in the proportion of 2 to 1, and the entire length of each between the points of attachment is 27 inches. A connecting lever *l*, of equal arms, and 20 inches in length, is jointed to the longer arms *i i* of the former, by means of the double short links *m*, *n*. The two levers *h i*, *h i*, are hooked by means of their shackles at *k* to the main swing-tree *b*; and the three small swing-trees *c*, *d*, *e*, are hooked to the compensation lever at *h*, *h* and *l*. From the mechanical arrangement of these levers, if the whole resistance at *a* be taken at 600 lbs., *k* and *k* will each require an exertion of 300 lbs. to overcome the resistance. But these two forces fall to be subdivided in the proportion of the arms of the levers *h i*; ⅔ of each, or 200 lbs., being allotted to the arms *h*, and the remaining ⅓, 100 lbs., to the arms *i*, which brings the system to an equilibrium. The two forces *i*, *i*, being conjoined by means of the connecting levers *m*, *n*, their union produces a force of 200 lbs., thus equalizing the three ultimate forces *h l h* to 200 lbs. each, and these three combined are equal to the whole resistance *a*; and the 3 horses that are yoked to the swing-tree *c*, *d*, *e*, are subjected to equal exertion, whatever may be the amount of resistance at *a* which has to be overcome.

(910.) The judicious farmer will frequently see the propriety of lightening the labor of some individual horse; and this is easily accomplished by the compensation apparatus. For this purpose, one or more holes are perforated in the levers *h i*, on each side of the true fulcrum *k*, to receive the bolt of the small shackles *k*. By shifting the shackle and bolt, the relation of the forces *h* and *i* are changed, and that in any proportion that may be desired; but it is necessary to observe that the *distance* of the additional holes, on either side of the central hole or fulcrum of equilibrium in the system, should be in the same proportion as the length of the arms in which the holes are perforated. Thus, if the distance between those in the short arm is half an inch, those in the longer arm should be an inch. By such arrangement, every increase to the exertion of the power, whether on the long or the short arm, would be equal.

(911.) The same principle of compensation has been applied to various ways of yoking, one of which is a complicated form of that just described. The main swing-tree and the compensation levers are the same, except that they may be a few inches shorter in all the arms, and the middle one of the three small swing-trees also shorter. The yoking is performed in this manner: The nigh trace-chain of the nigh horse is hooked to the end *o* of the swing-tree *c*, and his off-side trace-chain to the end *o* of the swing-tree *d*. The middle horse has his nigh-side chain hooked to the end *p* of the swing-tree *c*, while his off-side chain goes to the end *p* of the swing-tree *e*; and the off-side horse has his nigh-side chain attached to the end *q* of the middle swing-tree *d*, and his off-side to *q* of the swing-tree *e*. This system of yoking is complicated, and though in principle it equalizes the forces so long as all the horses keep equally ahead, yet it is in some degree faulty. Whenever the middle horse gets either behind or before his proper station—or out of that position which keeps all the swing-trees parallel to each other—the outside horses have a larger share of the draught upon one shoulder than upon the other; and, as this produces an unnecessary fatigue to the animal, it should be avoided. Such irregularity cannot occur with the simple mode of giving each horse his own swing-tree.

(912.) A modification of this compensation yoke has been contrived, as I am informed, by Mr. Bauchop, Bogend, Stirlingshire. The compensation levers are formed of wood, and in place of the connecting levers *l*, fig. 201, a chain, 2 feet in length, connects the ends *i i* of the levers *h i*; and in the bight of the chain, as at *k*, a pulley and strap are placed, to which a *soam* chain is hooked; the pulley from it oscillating in the bight of the chain serves the same purpose as the connecting lever *l*. In this mode of yoking, the horses work in unicorn-team, the middle horse pulling by the soam-chain.

(913.) *In the yoking of 4 horses*, various modes are also adopted. The old and simple method is for the plow-horses to draw by a set of common swing-trees, fig. 200; and to the center of the main swing-tree at *e* a soam-chain is hooked by means of a shackle or otherwise. The leading-horses are thus yoked by a second set of common swing-trees to the end of the soam. This is now seldom employed, but an improved method of applying the soam has been adopted in its place, which is represented by fig. 202, where *a* is the bridle of the plow, with its swivel-hook.—A pulley *b* of cast-iron, 6 inches diameter, mounted in an iron frame, of which an edge-view is given at *m*, is attached to the hook of the bridle. A link-chain *c* is rove through the frame of the pulley; and to one end of it, the short end, is hooked the main swing-tree *d* of a set of common trees for the plow-horses. The other end of the chain passes forward to a sufficient distance to allow the leading horses room to work; and to it is hooked the second set of common swing-trees at *e* for the leaders. In the figure, a part of the chain, from *f* to *g*, is broken off; but the full length is about 11 feet. In this yoke, the trace-chains of the nigh-side hind horse are hooked to the swing-trees at *h h*, and those of the off-side horse at *i i*, the leaders being yoked at *k k* and *l l* respectively. In this arrangement, the balance of forces is perfectly preserved; for the hind horses and the leaders, as they pull at opposing ends of the chain passing round a pulley, which must inevitably be always in equilibrium, each pair of horses has an equal share of the draught; and, from the principles of the common swing-trees through which each pair acts, the individual horses must have an equally perfect division of the labor, unless this equilibrium has been removed for the purpose of easing a weaker horse. In order to prevent either the hind horses or the leaders from slipping too much ahead, it is common to apply a light check-chain *o*, of about 15 inches long, connecting the two parts of the main-chain, so as to allow only a short oscillation round the pulley, which is limited by the check-chain. When this is adopted, care should be taken never to allow the check-chain to remain upon the stretch; for, if it do so, the advantage of equalization in the yoke is lost, and it becomes no better than the simple soam. In all cases of using a chain, that part of it which passes forward between the hind horses must be borne up by means of attachment to their back bands or suspended from their collars.

Fig. 202.

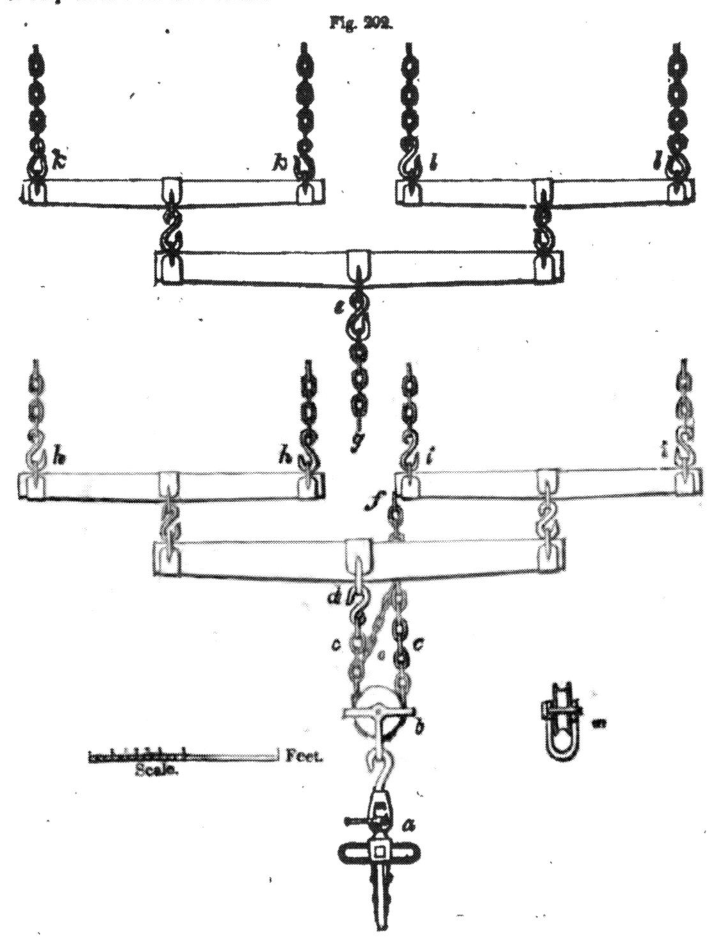

THE SWING-TREES FOR FOUR HORSES.

(914.) Mr. Stirling, of Glenbervie, Stirlingshire, recommends a method of yoking a team of four horses in pairs, the arrangements of which are represented in fig. 203; *a* is part of a main swing-tree of the common length, *b* a small swing-tree about 4 inches longer than the usual length, but both mounted in the usual form, except that, at each end of the small swing-trees, cast-iron pulleys *c c*, of 3 or 4 inches diameter, and set in an iron frame, are hooked on to the eyes of the swing-tree. The common trace-chains are rove through the frames of these pulleys, as in the figure; the ends *d d* of the chains are prolonged forward to the proper length for the nigh hind horse, and the ends *e e* are extended to the nigh leader. At the opposite end of the main swing-tree, which, in this figure, is cut off, the same arrangement is repeated for the off-side horses. The principle of action in this yoke is simple and effective, though different in effect from the former. There the two hind horses are equalized through the medium of their set of common swing-trees. The leading horses are alike equalized by their set, and thus the two pairs balance each other through the medium of the soam. Here, on the other hand, the two nigh-side horses have their forces equalized through the trace-chains, which are common to both, by passing over the pulleys *c c*; and the same holds in respect to the two off-sides. The couple of nigh-side and of off-side horses, again, are equalized through the medium of the one set of swing-trees. In both, therefore, the principle of equalization is complete, but there is a trifling difference in their economy. In the yoke, fig. 202, which I call the *cross balance yoke*, the soam chain and pulley are the only articles required in addition to the every-day gear. In that of fig. 203, which I call the *running balance yoke*, there is first the set of swing-trees, which, as they have to resist the force of 4 horses, must in all their parts be made stronger than the common set, agreeably to the rules before laid down; and to which are added the 4 pulleys, all of which are applicable only to this yoke. The trace-chain, though not necessarily stronger than those for common use, is required about three times longer than single horse-chains—that is to say, four horses will require the chains of six; but the chains of the leaders are more conveniently supported when they pass along the sides of the hind horses, and it is free of the set of swing-trees which dangle behind the leaders, of the method fig. 202.

(915.) In cases where 6, 8, and even 12 horses are required, such as for trenching, subsoil-plowing, and especially draining with the plow, the yoking is accomplished by modifications and ex-

Fig. 203.

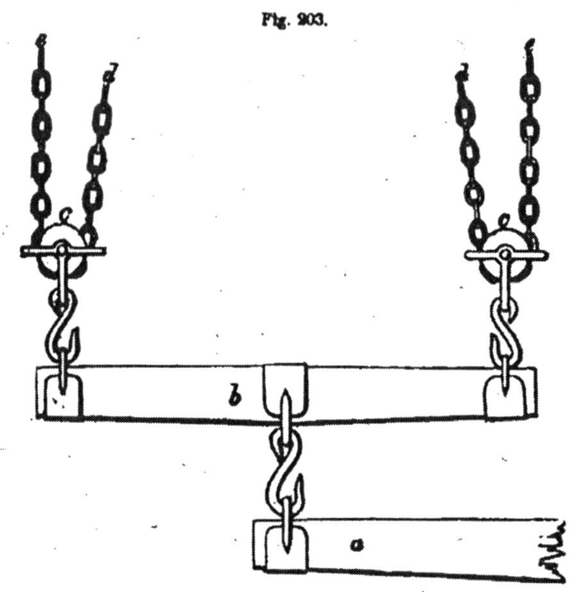

THE SWING-TREES ALSO FOR FOUR HORSES.

tension of the principles here laid down; for example, a team of 6 can be very conveniently applied with equalized effect by employing the compensation levers of fig. 202, along with 3 single swing-trees with pulleys at each end and running trace-chains, as in fig. 203.—J. S.]

---

27. PLOWING STUBBLE AND LEA GROUND.

"'T is time to clear your plowshare in the glebe."
GRAHAM.

(916.) When you take an extensive glance over the fields immediately after harvest, when the crop has been gathered into the stack-yard, you perceive that a large proportion of them are in stubble, while others are occupied by grass, turnips, and young wheat. On examining the stubbled fields particularly, you will observe young grass among the stubble in some fields, and nothing but stubble in others. You could not, of yourself, discover at once that these various states of the fields bear a certain proportion to one another, though they really do; and the cause of their being in those proportions is that they are cultivated under what is termed a "regular rotation of crops," which, when followed out, necessarily causes every field, in its turn, to carry the same series of crops. The numbers composing the series depend on the nature of the soil, and it shall be my duty to make you acquainted with them in due time. Meantime, suffice it to intimate that, when the stubble is in that state, the beginning of the agricultural year is arrived, when certain parts of it must undergo a change and be transformed into those which follow the ones you find them in.—Now, that part of the stubble-land which is devoid of any crop is the first to undergo a change, and it is effected by the plow, not at random, but by the application of those principles which have already been explained to you when we considered the "various modes of plowing land into ridges," from pages 289 to 302, where, as you may remember, the mode of plowing was said to be determined by the nature of the soil and subsoil. The stubble-land is generally all plowed before the lea is commenced with, and that part which is to bear the potato-crop next spring is first plowed, then that for the turnip-crop, and last of all for the bare fallow, when there is any

(917.) On *clay soil*, you will find the stubbled ridges of a rounded form, having been at least twice gathered up, fig. 139; and the way to keep them in a dry state during winter, on a considerable declivity, is to cleave them down without a gore-furrow, fig. 140, and without a mould-furrow, fig. 134, or to cleave them down with gore-furrows, fig. 141, and mould-furrows, when clay-land is flat. On *less strong soil*, casting with a gore-furrow (657) will preserve land dry whether it be flat or on a declivity.—On *light loams*, casting without gore-furrows, fig. 135, will serve the purpose. And on *sandy* and *gravelly soils*, crown and furrow (653) is the most appropriate mode of plowing stubble. It is rare that stubble-land is subjected to any other mode of plowing in winter. Snow should never be plowed in, nor the ground turned over when affected by frost, nor should strong clay soil be stirred when very wet, as it is apt to become very hard in spring, and of course more difficult to work.

(918.) In every variety of soil, plowed in the forms just described for winter, care should be taken to have plenty of channels, or *gaws* or *grips*, as they are usually termed in Scotland, cut in the hollowest places, so as surface-water may have them at every point by which to escape into the nearest open ditch. The gaws are first drawn by the plow laying them open like a feering—taking, in all cases, the hollowest parts of the ground, whether these may happen to cross the ridges or go along the open furrows; and they are immediately afterward cleared out with the spade of the *loose earth* left by the plow, and cast abroad over the surface. The fall in the gaws is made to tend toward a point or points best adapted to carry off surface-water by the shortest route, and do the least injury to the soil. The ends of the open furrows which terminate at the furrow along the side of the lowest head-ridge, as well as this furrow itself, should be cleared out with the spade, and cuts made at the hollowest places across the head-ridge into the ditch. This precaution of gaw-cutting should never be neglected in winter in any kind of soil, the stronger soils requiring more gaws than the lighter; for, as there is no foreseeing the injuries which a single deluge of rain may commit, it is never neglected by the provident farmer, though many small farmers, to their own loss, pay little heed to the necessity of its observance.

(919.) With regard to the plowing of *lea ground*, the most usual form in *strong soil* is to cast with a gore-furrow, fig. 136; and, on *less strong soil*, the same form of plowing without a gore-furrow; while, on the lightest soils of all, the crown and furrow is in most common use (653). Gathering up is a rare form of furrow for lea, though it is occasionally practiced on strong soil after gathered up or cast ridges, when it is a rather difficult operation to plow the furrow-brows and open furrows as they should be. The oldest lea is first plowed, that the slices may have time to mellow by exposure to the winter air; and that which is on the strongest land is for the same reason plowed before that on light. Lea should never be plowed in frosty weather, that is, as long as the ground is at all affected by frost, nor when there is rime on the grass, nor when the ground is very soft with rain; because, when ice or rime is plowed down, the non-conducting property of grass and earth, in regard to heat and cold, preserves the ice in an unaltered state so long as to chill the ground to a late period of the season; and, when the ground is too soft, the horses not only cut it into pieces with their feet, but the furrow-slice is apt to be squeezed out of its proper shape by the mould-board. Nor should lea be plowed when hard with drouth, as the plow in that case will take too shallow a furrow-slice, and raise the ground in broad, thin slabs, instead of proper furrow-slices. A semi-moist state of the ground in fresh weather is that which should be chosen for plowing lea. Gaws should not be neglected to be cut after lea-plowing, especially in the fields first plowed, and in strong land, always whether early or late plowed.

(920.) It is a slovenly though too common a practice to allow the head-ridges to remain unplowed for a considerable time after the rest of the field has been finished plowing, and the neglect is most frequently observed on stubble ground. The reasoning on the matter is that, as all the draughts cannot be employed on the head-ridges, it is a pity to break their number in beginning another entire field; and this reason would be a good one in summer, when there is little chance of bad weather occurring, but in winter it has no force at all, for the gaw-cuts cannot be properly executed until the field is entirely plowed; and to leave a plowed field to the risk of injury from wet weather, even for a day longer than you can help, shows little regard to future consequences, which may turn out far more serious than the beginning to plow a new field without all the draughts. No doubt, when land has been thorough-drained, there is less dread of ill consequences from the neglect of gaw-cutting; but, even in the most favored circumstances of drained land, I think it imprudent to leave isolated hollows in fields—and such are to be found in numbers on every farm—without the means of getting rid of any torrent of water that may fall at an unexpected time. Let, therefore, as many draughts remain in the field as will plow both head-ridges during the next day at longest; and if they can be finished in one yoking, so much the better.

(921.) With regard to the mode of plowing head-ridges for a winter furrow, some consideration is requisite. In stubble, should the former furrow have been cast with or without a gore-furrow, then, on reversing the casting, a ridge will be left on each side of the field, which will be most conveniently plowed along with the head-ridges by the plow going round parallel to all the fences of the field, and laying the furrow-slices toward them. The same plan could be adopted in plowing lea in the same circumstances. Should the furrow given to the stubble have been a cleaving down with or without gore-furrows, then the head-ridges should be cloven down with a gore-furrow along the ends of the ridges, and mould-furrows along their own crowns. On the ridges having been crown and furrowed, the head-ridges may be gathered up in early and late lea-plowing and in stubble; they may be cloven down without a gore-furrow along the ends of the ridges, especially in the upper head-ridge; and the half ridge left on each side of the field may be plowed by going the half of every bout empty.—But a better plan would be, *only if the ridges of the field are short*, to plow half of each head-ridge toward the ends of the ridges, going the round of the field, and passing up and down upon the half-ridge on each side empty, and then to plow those half-ridges with the other half of the head-ridges in a circuit, laying the furrow-slice still toward the ridges; all which will have the effect of casting the head-ridges toward the ends of the ridges.—When the ridges have been plowed in a completed form, a convenient mode of plowing the head-ridges on strong land is to gather them up, first making an open feering along the crowns.

(922.) Whatever mode of plowing the land is subjected to, you should take special care that it be plowed for a winter furrow in the best manner. The furrow-slice should be of the requisite depth, whether of 5 inches on the oldest lea, or 7 inches on the most friable ground; and it should also be of the requisite breadth of 9 inches in the former case, and of 10 in the latter; but as plowmen incline to hold a shallower furrow than it should be, to make the labor easier to themselves, there is less likelihood of their making a narrower furrow than it should be—a shallow and broad furrow conferring both ease on themselves, and getting over the ground quickly. A proper furrow-slice in land not in grass, or, as it is termed, in *red* land, should never be less than 9 inches in breadth and 6 inches in depth on the strongest soil, and 10 inches in breadth and 7 inches in depth on lighter soils. On grass-land of strong soil, or on land of any texture that has lain long in grass, 9 inches of breadth and 5 inches of depth is as large a furrow-slice as may possibly be obtained; but on lighter soil, with comparatively young grass, a furrow-slice of 10 inches by 6, and even 7, is easily turned over. At all seasons, but especially for a winter furrow, you should endeavor to establish for yourself a character for deep and correct plowing.

(923.) *Correct plowing* possesses these characteristics: The furrow-slices should be quite straight; for a plowman that cannot hold a straight furrow is unworthy of his charge. The furrow-slices should be quite parallel in length, and this property shows that they have been turned over of a uniform thickness, for thick and thin slices lying together present irregularly horizontal lines. The furrow-slices should be of the same hight, which shows that they have been cut of the same breadth, for slices of different breadths, laid together at whatever angle, present unequal vertical lines. The furrow-slices should present to the eye a similar form of crest and equal surface; because where one furrow-slice exhibits a narrower surface than it should have, it has been covered with a broader slice than it should be; and where it displays a broader surface than it should, it is so exposed by a narrower slice than it should be lying upon it. The furrow-slices should have their back and face parallel, and to discover this property requires rather minute examination after the land has been plowed; but it is easily ascertained at the time of plowing. The ground, on being plowed, should feel equally firm under the foot at all places, for slices in a more upright position than they should be, not only feel hard and unsteady, but will allow the seed-corn to fall down between them and become buried. Furrow-slices in too flat a state always yield considerably to the pressure of the foot; and they are then too much drawn, and afford insufficient mould for the seed. Furrow-slices should lie over at the same angle, and it is demonstrable that the largest extent of surface exposed to the action of the air is when they are laid over at an angle of 45°, thus presenting crests in the best possible position for the action of the harrows. Crowns of ridges formed by the meeting of opposite furrow-slices, should neither be elevated nor depressed in regard to the rest of the ridge, although plowmen often commit the error of raising the crowns too high into a crest, the fault being easily committed by not giving the feered furrow-slices sufficient room to meet, and thereby pressing them upon one another. The furrow-brows should have slices uniform with the rest of the ridge, but plowmen are very apt to miscalculate the width of the slices near the sides of the ridges, for if the specific number of furrow-slices into which the whole ridge should be plowed are too narrow, the last slice of the furrow-brow will be too broad, and will therefore lie over too flat; and should this too broad space be divided into two furrows, each slice will be too narrow and stand too upright. When the furrow-brows are ill made, the mould-furrows cannot be proportionately plowed out; because, if the space between the furrow-brows is too wide, the mould-furrows must be made too deep to fill up all the space, and *vice versa*. If the furrow-slices are laid too flat, the mould-furrows will be apt to throw too much earth upon their edges next the open-furrow, and there make them too high. When the furrow-brows of adjoining ridges are not plowed alike, one side of the open-furrow will require a deeper mould-furrow than the other.

(924.) You thus see that many particulars have to be attended to in plowing land into a ridge of the most perfect form. Plowmen differ much in bestowing attention on these particulars; some can never make a good crown, others a good furrow-brow and open-furrow, while others will make them all in a passable, but still objectionable manner. This last class of plowmen, however, is preferable to the other, because the injurious effects of the bad plowing of the former are obvious; whereas the effects of mediocre compared with first-rate plowing are not easy to ascertain, though no doubt the difference of their effects must be considerable in many respects. "It is well known," observes Sir John Sinclair, "that the horses of a good plowman suffer less from the work than those intrusted to an awkward and unskillful hand, and that a material difference will be found in the crops of those ridges tilled by a bad plowman, when compared to any part of the field where the operation has been judiciously performed."* Marshall contends that want of good tillage incurs a loss of as much as ¼ of the crops throughout the kingdom,† which may be an approximation to the truth in his day; but plowing is certainly now better performed in Scotland than it was, though it must be owned that by far the greatest part of the process is yet of a mediocre description, and the reasons for the mediocrity of the work are not difficult to find.

(925.) *Plowmen* cannot learn their profession at a very early age, when every profession ought to be acquired to attain a high degree of perfection in it, because plowing requires a considerable amount of physical power, even from the most expert plowmen, and it exacts the greatest exertion of strength by comparison from the youngest in years and the least initiated in the art; and after young men possess sufficient strength to hold the plow, they are left to acquire a knowledge of plowing more through sheer experience than by any tuition given them by those who are better acquainted with the art; and as excellence acquired in it cannot be bequeathed to the rising generation, its knowledge must be acquired *ab initio* by every generation. For example, to teach *boys* to plow it has been recommended "to put a cross-bar between the cheeks of the bridle, so as to keep them precisely at the same distance from each other, and then setting up a pole at the end of a furrow, exactly measured to the same line as that from which he starts, fixes his eye steadily upon it, and carries the plow in a direction precisely to that point."‡ To do all this implies that the *boy* has sufficient strength to hold a plow, which if he have, he will have come the length of a stout lad; and to "fix the eye steadily" upon a pole at a distance, while holding the plow with a staggering gait, and unable for want of breath to speak even a word to the horses, far less to guide them with the reins, is much beyond the power of any *lad*, instead of a *boy;* for it would require a very expert plowman to do that, for all that is nothing short of feering, and none but the expertest of the plowmen on a farm is intrusted to feer land; and, besides, no single pole always before the spectator can possibly guide any one in a straight line, for he may imagine he is moving by it in a straight line, while all the while

* Sinclair's Code of Agriculture. † Marshall's Gloucestershire, vol. i.

‡ British Husbandry, vol. ii.

ne may be deviating very widely from it. The truth is, the young man who is desirous of becoming a plowman in a short time should be taught day by day by an experienced plowman to temper the irons, and guide his plow according to his strength and talents. Very few young men have or are permitted to have such opportunities of learning, and the consequence is, that, as my observation confirms, the best plowmen are generally those who have been taught directly by their fathers, and work constantly upon their fathers' farms.

(926.) Were all the particulars of good plowing mentioned above (923) constantly attended to, there would be no *high crowned* ridges as at *a*, fig. 204, by bringing the two feering or the two open furrows too close to-

Fig. 204.

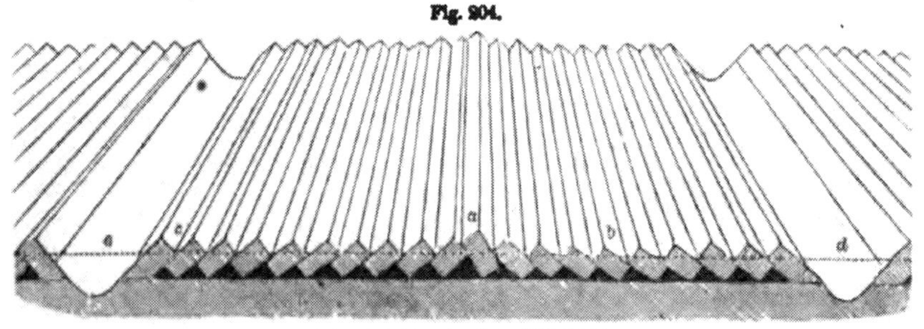

AN EXAMPLE OF BAD PLOWING.

gether, thereby causing the corn sown upon it to slip down both sides, and leave a space bare of seed on the best land of the ridge. There would be no *lean* flanks as at *b*, by making the furrow-slices there broader than they should be, with a view to plowing the ridge as fast as possible, and thereby constituting a hollow which becomes a receptacle for surface-water that sours the land; or when the soil is strong, it becomes so consolidated, that it is almost sure to resist the action of the harrows, especially when passed across the ridge; or in light soil it is filled up with the loose soil drawn by the harrows from the surrounding hights. There would be no *proud furrow-brows* as at *c*, by setting up the furrow-slices there more upright than they should be, to the risk of being drawn wholly into the open furrows when the harrows catch them too forcibly on leaving the ridge when cross-harrowed. And there would be no *unequal* open-furrows, as at *d*, by turning over a flatter mould-furrow on the one side than the other, which cannot fail to retain the greater quantity of seed. To extend this lengthened catalogue of ills accompanying bad plowing, I may mention that every sort of crop grows unequally on an ill-plowed ridge, because the soil is more kindly on the better plowed parts; but the evils of bad plowing are not confined to the season in which it is performed, as it renders land unequal when broken up again, and the thinner and harder portions cannot yield so abundantly as the deeper and more kindly. The line *d e*, fig. 204, shows the position of the surface before the land was plowed, and the furrow-slices, in relation to that line, show the unequal manner in which the ridge had been plowed.

(927.) It seems to be a prevalent opinion among agricultural writers,* that land when plowed receives a curvature of surface; whereas, correct plowing, that is, making the furrow-slices on the same ridge all alike, cannot possibly give the surface any other *form* than it had before it was plowed. If the former surface were curved, then the newly plowed sur-

* Low's Elements of Practical Agriculture, and British Husbandry, vol. ii.

face would also be curved; but if it were flat, the new surface will be flat also. No doubt, in gathering up a ridge, the earth displaced by the plow occupies a smaller area than it did before, but as the displacement only elevates it above its former level, the act of elevating it does not necessarily impart any curvature to it. It is quite true, however, that a ridge on being cross-harrowed, becomes curved, inasmuch as it becomes highest at the crown, because the harrows, in crossing, have a tendency to draw the soil toward the open sides of the ridge, that is, into the open furrows, where the least resistance is presented, and which will alter the uniformity of surface left by the plow; but this effect has no connection with the plowing. Seeing this external effect produced without knowing its cause, it is equally true that most plowmen endeavor to give the ridge a curvature, and this they accomplish by what I would designate bad plowing; that is, they give a slight cresting to the crown, which they support with a bout or two of well-proportioned furrow-slices; they then plow the flanks with narrow and rather deep slices set up a little high, to maintain the curvature, for about four bouts more, giving the last of these bouts rather less depth and hight than the rest, and the remaining three bouts next the furrow are gradually flattened toward the open furrows, which are endeavored to be finished off to the desired curved form by the mould-furrows. This artfulness produces a ridge of pleasing enough curvature, though it is exercised by the plowman with no intention to deceive; he, on the contrary, conceives all the while that he is displaying great skill in his art by so doing, and if he is not instructed better he will continue to practice it as an accomplishment. Such a device, however, sacrifices correct plowing to a fancied superiority of external appearance, as much as the crested furrow formerly spoken of (590), fig. 109. A thoroughly good plowman, and I have known a few, but only a few, of such valuable men, avoids so objectionable a practice, and plows always a true sound furrow, making it larger or smaller as the particular state of the work may require.

(928.) Without putting much value on the information, it may serve as a fact to refer to, in case it should be wanted, to state the weight of earth turned over in plowing. If 10 inches are taken as a fair breadth for a furrow-slice, there will be 18 such slices across a ridge of 15 feet in breadth; and, taking 7 inches as a proper depth for such a furrow-slice, a cross section of the slice will have 70 square inches. A cubic foot of earth is thus turned over in every 24¾ inches and a little more of length of such a slice; and taking 2.7 as the specific gravity of ordinary soil, every 24¾ inches and a fraction more of such a slice will weigh 12 stones 1 lb. imperial.*

(929.) The usual *speed* of horses at the plow may be ascertained in this way. A ridge of 5 yards in breadth will require a length of 968 yards to contain an imperial acre; and to plow which at 9 bouts, of 10-inch breadth of furrow-slice, counting no stoppages, will make the horses walk 9⅞ miles, which in 10 hours gives a speed of 1742⅖ yards per hour. But as ridges are not made of 968 yards in length, and as horses cannot draw a plow that distance without being affected in their wind, and as allowance must be made for time lost in turning at the ends of the ridges, as well as for affording rest to the horses, that speed will have to be considerably increased to do that quantity of work in the time. By experiment it has been found that 1 hour 19 minutes, out of 8 hours, are lost by turnings while plowing an acre on ridges of 274 yards in length, with an 8-inch furrow-slice.† Hence, in plowing an acre on ridges of 250 yards in length, which is the length of ridge I recommended as the best for horses in draught, when speaking of inclosures (456), in 10 hours, with a 10-inch furrow-slice, the time lost by turnings is 1 hour 22 minutes. I presume that the experiment alluded to does not include the necessary stoppages for rest to the horses, but which should be included; for however easy the length of ridge may be made for draught, horses cannot go on walking in the plow for 5 hours together (one yoking) without taking occasional rests. Now 250 yards of length of ridge give nearly 4 ridges to the acre, or 36 bouts; and allowing a rest of one minute in every other bout, 18 minutes will have to be added to the 1 hour 22 minutes lost, or very nearly 1⅔ hours of loss of time, out of the 10 hours, for turnings and rest. Thus 18,000 yards will be plowed in 8⅓ hours, or at the rate of 1 mile 422 yards per hour. I think this result is near the truth in regard to the plowing of lea in spring; it is too little in plowing red land in summer, and perhaps too much in plowing stubble land in winter; but, as lea plowing is the criterion by which all others are estimated, this result may be taken as a near approximation to the truth.

(930.) The comparative time lost in turning at the ends of long and short ridges may be seen from the following table, constructed from data furnished by the experiment above alluded to:

* Probably meaning 14 lb. to the stone.—*Ed. F. L.*

† Sinclair's Code of Agriculture.

Length of ridge.	Breadth of furrow-slice.	Time lost in turning.	Time devoted to plowing.	Hours of work.
Yards.	Inches.	H. M.	H. M.	H.
78	10	5 11	4 4	10
149	..	2 44	7 16	..
200	..	2 1	7 59	..
212	..	1 56½	8 3½	..
274	..	1 28	8 32	..

Thus it appears that a ridge of no more than 78 yards in length requires 5 hours 11 minutes of time to turn at the landings, to plow an acre in 10 hours, with a 10-inch furrow-slice; whereas a ridge of 274 yards in length only requires 1 hour 28 minutes for the same purpose, making a difference of 3 hours 43 minutes in favor of the long ridge in regard to saving of time. Consequently, in the case of the shortest ridge, only 4 hours 49 minutes out of the 10 can be appropriated to plowing, whereas in that of the long ridge, 8 hours 32 minutes may be devoted to the purpose. Hence so very short ridges require double the time of long ones to plow, and are thus a decided loss to the farmer. This is a subject well worth your experimenting on, by ascertaining the time usually taken in plowing and turning and resting on ridges of different lengths, in the different seasons, and in different soils. A watch with a good seconds-hand to mark the time will be required, and the observations should be made unknown to the plowmen, at their usual rate of work; for if you be constantly in the presence of the men, more than the usual work will be done, and less than the usual rests taken.

(931.) There is another circumstance, on some farms, which also greatly affects the speed of horses at work. I mean the *great steepness of the ground*; and it is not unusual to see the ridges traversing such steeps straight up and down. Ridges in such a position are laborious to plow to cart upon, to manure, and for every operation connected with farming. The water runs down the furrows when the land is under the plow, and carries to the bottom of the declivity the finest portion of the soil. In such a position a ridge of 250 yards is much too long to plow without breathing the horses. But although the general rule of making the ridges run N. and S. is the correct one, yet in such a situation as a steep acclivity, they should be made to slope along the face of the hill instead of running right up and down the acclivity, and the slope will not only be easier to labor in every respect, but the soil will be saved being washed so much away in the furrows; but the direction of the slope should not be made at random: it should go away to the right hand in looking up the acclivity, because the plow will then lay the furrow-slice down the hill when it is in the act of climbing the steep, and on coming down the hill the horses will be the better able to lay the slice even against the inclination of the ground. What the exact length of the ridges on such an acclivity should be, even with the assistance of the slope, I cannot positively say, but should imagine that 100 or 150 yards would be sufficient for the horses; but, at all events, there can be no doubt that it would be much better for the labor of the farm, as well as for the soil, that there should be 2 fields 100 yards broad each, one higher up than the other, than that the whole ground should be in one field 200 yards in breadth. I have all along been referring to very steep ascents.

(932.) There is still another arrangement of ridges which may materially affect the time required to labor them; I mean that where, by reason of irregularities in the fences or surface of the ground, ridges from opposite directions meet in a common line in the same field; and the question is, Whether the ridges should meet in an imaginary line or at a common head-ridge? Professor Low, when alluding to such an arrangement of ridges, says, that "the part where the opposite sets of furrows meet, may be made an open furrow, *or* a raised-up ridge or head-land, as circumstances may require."* When ridges meet from opposite directions, it is clear that they cannot be plowed at the same time without the risk of the horses encountering one another even upon a head-ridge; and where there is no head-ridge, should one set of ridges be plowed before the other, in the plowing of the second set, the end of the plowed land of the first will be completely trampled down. At the least, therefore, there should be one head-ridge betwixt two sets of ridges, that one set may be plowed before the other. But the most independent way in all respects with such a form of surface, is to treat it as if each set of ridges belonged to separate fields, and let each have a head-ridge of its own.

(933.) When *horses are driven in the plow beyond their step*, they draw very unequally together, and, of course, the plow is then held unsteadily. In that case, the plow has a tendency to take too much land; to obviate which the plowman leans the plow over to the left, in which position it raises a thin broad furrow-slice, and lays it over at too low an angle. On the other hand, when the plowman allows the horses to move at too slow a pace, he is apt to forget what he is about, and the furrow-slices most probably will then be made both too narrow and too shallow, and though they may be laid over at the proper angle, and the work appear externally well enough executed, yet there will be a want of mould in the plowed soil.

(934.) [The whole value of plowing, scientifically speaking, depends upon its having the effect of loosening the texture of the soil, and thus permitting a free circulation of air and moisture through its interstices, for the double purpose of increasing the rapidity of the disintegration of its stony portions, and of re-reducing to powder what had formerly been pulverized, but which, from the joint action of pressure, and the binding effect of root-fibres, had become agglutinated together.

(935.) Sufficient has already been said to draw your attention to the point of pulverizing the soil; in it lies one of the most important secrets of good farming. However well you may manure your land, however thoroughly you may drain it, you will never obtain the crops it is capable of yielding, unless you pulverize it; nay, so important did Jethro Tull think this, that he felt firmly persuaded that if you pulverized your soil well, you need not manure at all. I need hardly

* Low's Elements of Practical Agriculture.

tell you that we shall prove hereafter Jethro Tull to have carried his conclusions too far; but still so direct and unqualified a statement, from such a writer, should have its full influence upon all who wish to learn thoroughly the art of Agriculture. Always bear in mind that the *impalpable powder* is the active part of soil, and that no other portion has any *direct* influence upon vegetation, and you will then, at all times, be sufficiently impressed with the necessity of thorough-plowing, harrowing, &c.; indeed, you may rest assured that, except upon some few very light sands, you cannot pulverize the soil too much—economy alone must fix the limit of this useful operation.

(936.) But were I to stop here, you might naturally suppose that any season of the year would do equally well for plowing, provided it was before seed-time, and that the fixing of the time was regulated entirely with a view to economise labor. It is certainly true that, to a considerable extent, the time of plowing may be varied; but you may rest assured that, as a general rule, the sooner you plow after the removal of the crop, the better condition will your soil be in at the commencement of spring.

(937.) Several chemical processes of considerable consequence as respects the fertility of soil, occur after it has been plowed, which either take place very slowly, or not at all, while it lies unstirred; and, moreover, some of these take place to the greatest advantage during winter.

(938.) This is especially the case with the disintegration of mineral masses, nothing tending so powerfully to reduce even the hardest stones to powder as sudden changes of temperature, combined with the presence of much moisture. During rain or thaw after snow all the clods of earth and the pores of the more loosely aggregated stones become filled with water, which, of course, freezes, if the temperature is sufficiently reduced; and from its expansion during solidification, a peculiar property possessed in a marked degree by water, the particles of earth or stone, as the case may be, are pushed so far asunder that, when the thaw returns, it crumbles into fragments, which are again and again acted on until reduced to the state of soil.

(939.) This crumbling by frost is of the greatest importance in the case of stiff clays, for two reasons: 1st, because they are thus reduced much more easy to work; and, 2d, which is of far greater consequence, they are enabled to give up their alkalies more readily to water; and clayey minerals are fortunately the quickest to disintegrate, or rather to *decompose* by the action of the weather; and hence, every means that facilitates that process is valuable, because as we have already seen that those most valuable ingredients of soil, potass and soda, are of no use to plants unless they are soluble in water, and that they do not obtain this property until the mineral with which they have been associated becomes completely decomposed.—H. R. M.]

(940.) [In the previous remarks on the plow were embraced its construction, its principles of action, the principles on which its draught is exerted, and the resistance which it presents to the draught, as also some remarks on the system of plowing that each of the three leading varieties of plows have given rise to; and on this last branch of the subject I feel constrained to offer some farther remarks.

(941.) In treating of the form of the furrow-slice I have sufficiently evinced the preference that I give to the rectangular slice; and this I do on the broad principle that *deep plowing* ought to be the *rule*, and any other practice the *exception*. The exception may apply in a variety of cases, so well known to practical farmers that it would be presumptuous in me to point them out; but our "Book of the Farm" being peculiarly addressed to young farmers, the pointing out of a few of these cases of exception becomes more in place.

(942.) Shallow plowing, then, may be admissible in the case of a field that has been depastured with sheep, and to be simply turned up for a seed-furrow. The reason usually assigned for this, that the droppings of the sheep forming only a top-dressing, has given rise to the notion that a deep furrow would bury the manure to a depth at which its beneficial effects could not be reached by the plants of the crop that may be sown upon this field. While I allow that this is an admissible case, the *rationale* of the reasons assigned for it by practical men, may, on very fair grounds, be called in question. Thus, it is well known that the roots of vegetables in general push themselves out in pursuit of their nutriment, and with an instinctive perseverance they will pass over or through media which afford little or no nutriment, in order to reach a medium in which they can luxuriate at will. With the larger vegetable productions this is remarkably the case; and though, among those plants which the farmer cultivates, the necessity of hunting, as it were, for food cannot occur to a great extent, yet we are well aware that the roots of the cereal grasses may extend from 6 to 12 or more inches; and there is good reason to believe that their length depends upon the depth of the penetrable soil, and that the luxuriance of growth in the plant will in general be proportioned to that depth, soil and climate being the same.

(943.) Another case of exception to deep plowing, is in some of the courses of fallow plowing, where a deep furrow might be injurious; these occurring in the later courses. And a third is that of a seed-furrow, though in many cases this last is of doubtful recommendation.

(944.) In some of the clay districts, a system of shallow and narrow plowing is practiced, under the impression that the exposure of the soil, thus cut up in thin slices, tends more to its amelioration than a system of deep and broad plowing could effect. This supposition may, to a certain extent, be true, as a certain portion of the soil thus treated will undergo a stage of improvement; but allowing that it does so, the improvement is but a half measure. Soils of this kind are frequently deep, and, though apparently poor, they afford the stamina out of which may be formed the best artificial soils—the clay loam—which may be brought about by the due application of manure, and a proper, well directed, and continued system of plowing. On lands of this kind, the system of deep plowing will be always attended with beneficial effects; and instead of the apparently thin and hungry soil which the shallow system is more likely to perpetuate, the result might be a deep and strong clay loam. To effect this, however, there must be no sparing of expense or of labor, the draining must be efficient, and the manuring, especially with those substances that will tend to sharpen and yield porosity to the clay, must be abundantly supplied.

(945.) The most extensive suite of cases where a departure from the rule is admissible, are those lands where a naturally thin soil rests on a subsoil of sand or gravel variously impregnated with oxides of iron. To plow deep at once in such situations would run the risk of serious injury to the sparing quantity of soil naturally existing. But it is to be observed of soils of this kind, that the subsoil has always a tendency to *pan*, and if such do exist, the deep system should again come into requisition in the form of subsoil plowing, which, by destroying the pan—that frequent cause of sterility in soils of this kind—opens a way to the amelioration of both soil and subsoil.

(946.) There appears, in short, every reason for inculcating the system of deep plowing, not only where existing circumstances admit of its adoption, but where its ultimate effects are likely to induce a gradual improvement on the soil and all its products, admitting always that a variation in depth is proper and necessary under the varying circumstances of crops and seasons.

(947.) Though the Scotch swing plow has afforded the principal subject of what has been here given on this implement, it must not be lost sight of that numerous varieties of this important implement are to be found in other parts of the kingdom, many of which possess a high degree of excellence; and England is especially remarkable for these varieties. It has been already noticed (487) that the germ of improvement in the Scotch implements appears to have been obtained through England; but, like many other importations from that quarter, the necessities arising from circumstances of climate, of soil, and, perhaps not the least important, the paucity of pecuniary means, obliged the Scottish agriculturist to husband all his resources, and to call forth all his energies, in making the best and most economical use of his new acquisitions, so as in the end to outstrip his more favored brethren of the south. This will be found to have occurred, not only in the plow, but in the introduction of the turnip, of bone manure, and many other similar acquisitions.

(948.) It is remarkable, too, that the decided step taken in Scotland in regard to the rapid extension of the use of the improved plow, was long in retracing its steps back to England, and that the retrograde movement was gradual from the northern counties southward. In nearly half a century, this retrogression appears to have made very slow progress; and, like many other improvements which linger until some master-mind takes them in hand, the extension of the use of an improved plow met with little encouragement. In due time this subject was taken up by the Messrs. Ransome, of Ipswich; and, through their exertions, such changes have been produced in the plow as place the English agriculturist in possession of a command of these implements in such a variety of forms that no other country can boast of from the hands of one maker.

(949.) The numerous varieties (amounting to at least 100) of the plows constructed by the Messrs. Ransome seem to be chiefly adapted to the soils of England, and to the practice of her agriculturists; for we do not find that, when brought into Scotland, and placed in direct competition with the Scotch plow, that they ever gain a preference. There can be no question, however, that some of the varieties of these plows perform well, exhibiting work, when conducted by a skillful hand, that for its usefulness may compare with that of any implement now employed. The system of plowing in England being generally of the shallow character, and the modern plow for the most part of a light construction, adapted to the practice, it has been found that these plows were unable to resist the force required where the deep-plowing system is followed, as in many parts of Scotland. But a more serious objection to the introduction of these plows into Scotland lies in the frequent application to them of wheels. No plowman who has been able to wield the swing-plow will ever suffer himself to be incommoded with the addition of wheels to his plow (for he will always consider wheels an inconvenience), and this he does not from a conviction that wheels increase the labor of his horses, but because to himself they appear a source of annoyance; and here it may be farther remarked, as regards wheel-plows, that, since the wheels must always have a tendency to increase the draught (636), and on that account are objectionable, so also, if a plow can be wielded with equal and perhaps better effect without wheels than with them, the excuse that a wheel-plow may be wielded by a man of inferior qualifications is of small value. Any man may be trained to handle a plow, though every man will not be equally successful; and since in the whole of Scotland not a wheel-plow is to be found, except as a curiosity, while her plowing is at least not inferior to that of any part of the kingdom, and as the chances are surely equal that the plowmen are not all equally good, it is evident that plowing can be satisfactorily performed without wheels. If plowing can thus be performed over one part of the kingdom with an implement of the simplest form, and in a satisfactory and economical manner, there can be no necessity for using a more complicated and more expensive machine to perform the same work in another part of the kingdom, where it is at least not *better* done or done at less expense.

(950.) Having adverted to the plows of England, and particularly to those of the Messrs. Ransome, and though still impressed with the opinion that the simple Scotch swing-plow is preferable as an implement to the wheel-plows in their most improved form, and perhaps even to the *swing*-plows of England, it is proper to describe generally at least one example out of the many.

(951.) The example chosen for the purpose of illustration is Ransome's F F or Bedfordshire plow, represented in fig. 205. This plow has attained a high character for its general usefulness, and may be considered as the most perfect of modern English plows.

(952.) Without going into such a minute detail of its parts as has been done in the case of the three leading Scotch plows, I cannot avoid giving a short description of it. Like the greater part of modern English plows, it is constructed partly of wood and partly of iron. The body is of cast-iron, and is ingeniously formed for the attachment of the beam and handles. These are simply bolted to the body—a practice which leaves the parts in possession of their full strength instead of being weakened by the mortising, as is the case in the joining of the beam and left handle of the Scotch wooden plow. The body-frame rises from the middle of the sole-bar to the full hight of the upper edge of the beam, and the two are bolted together—the body being applied to the land-side of the beam, where its upper edge is seen at *a b*. The land-sides of the beam and of the body lie, therefore, in one plane behind; but the beam, at the point, lies 1¼ inches to the right of the body-plane. The handles are bolted to a vertical flange that projects from the hind part of the body; and, as an additional security to the right handle, a bar of cast-iron, extending from *a* to *c*, is laid upon the right-hand side of the beam, as seen in the figure, and fixed by the same bolts that connect the beam and the body. The hind part of this bar is also formed to embrace the root of the right handle, and is also secured by bolts. A farther security is effected by the application of the iron stay-bar *d e*, tying the beam and handles together. When the land-side of the plow is

Fig. 205.

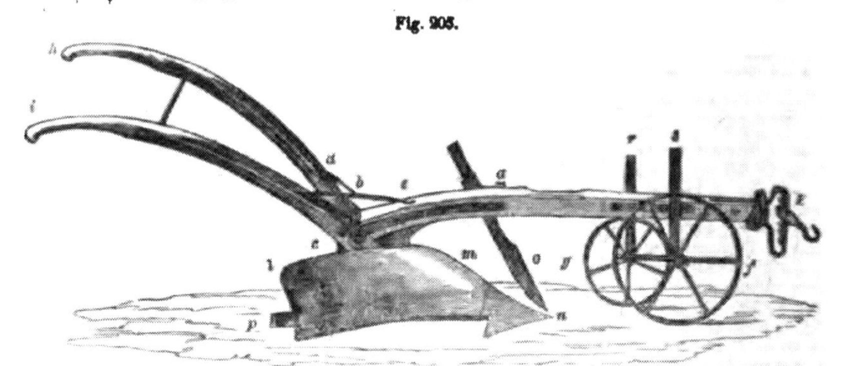

RANSOME'S F F, OR BEDFORDSHIRE PLOW.

vertical, the left handle *h* stands 10 inches to the left of the line of the land-side, and the width between the handles *h i* is 26 inches. Following the same rule as has been adopted in giving the details of the Scotch plows, we have the zero of the FF. figs. 121 and 122, Plate XII., at 27 inches behind the point of the share *n*, and the heel *p*, 11 inches behind the zero. The extremity of the handles is 4 feet 3 inches behind, and the point of the beam 4 feet 5 inches before the zero, making the horizontal length of the plow only 8 feet 8 inches. The hight of the handles above the base-line is 2 feet 10 inches; the hight of the body, at the junction *b* of the beam and handles, is 14 inches; the hight of the beam at the coulter-box is 17 inches, and at the point 16 inches; these hights being all as measured to the upper edge of the beam. The bridle *k* of this plow is similar to many others of this much-varied member, but differs in the material of which it is composed, being formed of cast-iron. In fig. 148, the plow is represented with two wheels *f* and *g*, which are mounted on stems *r s*; these move vertically in separate boxes, one on each side of the beam, and are held in position by clamp-screws. The larger wheel *f* runs in the furrow, and bears against the *land*, thus regulating the landing of the plow, while the wheel *g* runs *upon the land*, and regulates the earthing of the plow.

(953.) The active parts of the plow are also peculiar. The mould-board *l m* has been already given in detail in Plate XII., figs. 121 and 122; it is only necessary to add that its form indicates a medium of the convex and concave surfaces, and that its hight points out its inapplicability to deep plowing. The share *n*, as in a large proportion of all plows made by the Messrs. Ransome, is of cast-iron, but is very judiciously hardened at the point and along the edge of the feather on the lower surface only, which has the effect of throwing all the wear of the metal on the upper surface, thereby keeping the edge sharp below so long as the share lasts. This share has great breadth, being seldom under 7 inches at the broadest part. In the clays and chalky soils of England, the cast-iron share is both convenient and economical, though it is doubtful how they might answer in gravelly and stony soils, especially in the latter. The price of a cast-iron share is about 1s. 2d., and an allowance is made for them when worn out and returned of about 6d.; their duration may be taken at from 4 to 14 days, depending much on the texture of the soil. The coulter *o* in this plow, as in most others of the class, is fixed in a metal box bolted in the land-side of the beam. The land-side plane of the coulter, therefore, instead of crossing that of the plow's body, as in the Lanarkshire and other Scotch plows, is parallel to it, and stands altogether to landward of the land-side plane of the plow from top to bottom, and in some cases forms a small angle from the point landward. The body is frequently left entirely open on the land-side, except in so far as it is covered by the land-side flange of the sole-shoe, which is not more than 2 inches high, and it stands 1½ inches to landward of the body-frame, which last arrangement virtually brings the land-side plane of the body and of the fore part of the beam to coincide.

(954.) Having in a general way described the construction of the frame-work and the acting parts of this plow, there remains for me to say a few words on the wheels with which it is furnished. I have already (636-638) adverted to wheels, as they appear to me to affect the draught of plows, and have expressed myself in sufficiently distinct language to show that, in my opinion, they must in all cases be injurious, and tend to increase the resistance of the plow to which they are appended, whether they be applied within the body, or under the front, or any other part of the beam. That wheels may be of advantage for the working of a plow in the hands of an unskillful plowman may be true; but if this advantage is acquired by a certain additional expenditure of horse power, which, however much the proprietor of the team may blind himself to, will ultimately, though probably unheeded, tell on his profit and loss account, there will be no gain, but an ultimate loss. It must be admitted, even by the advocates of the wheel-plow, that though they may be handled with perfect regularity in plowing *along* ridges, whether the holder be an experienced plowman or not, yet in *cross* plowing they cannot by any means be brought so handily to follow the undulations of the surface. In leaving one ridge, the share will pass too shallow, and, in entering on the brow of the next, it will go too deep, or at least deeper than the average of the plowing. There is also the element of time, which in all farming operations is an important one; and here wheel-plows are found to come short by about 25 per cent. as compared with swing-plows. Mr. Pusey, in his paper on the draught of plows, incidentally observes: While the work of our plowing teams is at best but ¾ of an acre upon strong ground (and sometimes as much as 1 acre upon the lightest), the daily task performed by two Scotch horses upon strong land is 1¼ acres.[*] This deficiency of effect cannot be attributed to want of power in the horses, for English horses are at least not inferior to those employed in Scotland for agricultural purposes; neither can it be from unskillfulness in the plowmen, for even the most skillful seem to come short in this respect, by not being able to plow more than ¾ of an acre in a day, while with the swing-plow almost any plowman will turn over his acre a day. From the remarks of the same writer,[†] it is to be inferred that a Scotch swing-plow was incapable of being drawn through a certain clay soil by two horses, while the wheel-plows were found to perform the work with tolerable ease, though still a heavy draught. There may be such cases; but from the conditions of this particular case, where the draught that baffled the horses in the swing plow seems not to have exceeded 52 stones, there is an ambiguity in the matter that leads to doubts of the accuracy on the part of the observers of the experiment. We know well that in working the Scotch swing-plow in an 8 or 9-inch furrow on stiff land, the draught is not unfrequently as high as 7 cwt. or 56 stones; but two good horses never shrink from the task; and how a less draught, whatever be the soil, should have baffled the exertions of two good horses in a swing-plow, even in the Oxford clay, requires some farther investigation to be satisfactory.

(955.) Under all the circumstances, then—whether we take expenditure of horse power compared with the small saving in the pay of an inferior workman; the disadvantages attending the more complicated operation of plowing, compared with the celerity with which the swing-plow can be made to accommodate itself to all irregularities of ground; the loss of time, which is equivalent with capital, in plowing a given surface, when compared with the extent turned over by the swing-plow; and the probability that even the solitary instance of an apparent superiority in a wheel-plow may rest upon some oversight in observation—all seem to conspire to produce a conviction that a superiority exists in the swing-plow which is in some measure due to its deserving that appellation from an absence of wheels. And certainly, whatever be the merits of the modern improvements on English plows, they may be ascribed to any other cause than their possessing wheels, in whatever position they may be placed in the plow.

(956.) The plow under consideration is furnished with two wheels, see again fig. 205. The land-side wheel is 12 inches diameter, with a rim not exceeding 1 inch in breadth. The only purpose to which this wheel is applied is to regulate the depth of the furrow, for which purpose it runs upon the solid land. The furrow-wheel is 18 inches in diameter, with a breadth of rim equal to the former; its object is two-fold, serving in some measure to regulate the depth, by running on the bottom of the previously formed furrow, but its chief duty is by bearing against the edge of the furrow-slice that is about to be raised, to regulate the *breadth* of the slice, at the desired hight, by means of pinching-screws.

(957.) It is evident that both wheels perform a duty that either of them alone could do with perhaps equal effect, namely, the regulation of the depth; but the furrow-wheel performs a second office, regulating the breadth, which it can also do without interfering with its other duty. It would appear, therefore, that the land-side wheel may be set aside without impairing the efficiency of the plow; and we find, accordingly, that these plows are frequently used with only one wheel, which in itself performs both duties.

(958.) The consideration of these wheels, and their effects on the plow, suggests a farther objection to their utility in respect of the increased resistance they produce to the draught. If these wheels are to produce any effect at all, the plow-irons and yoke must be set so as to give the plow a bias both to *earth* and *land*. If the plow has not this, then whether it swim evenly, or have a bias, *from* both earth and land, in either case the wheels are ineffective, as they will neither bear upon the sole nor the edge of the furrow, but let the plow have the bias as proposed to both *earth* and *land*, the wheels will then both bear, and exert their efforts by reaction to counteract the tendency of the plow; on the one hand to sink deeper in the furrow, and on the other to cut a broader slice; and since "action and reaction are alike and in opposite directions," these antagonist forces will be in constant operation to a greater or less amount. Such effects will thereby increase the friction and consequent resistance in proportion to the amount of bias which has been given to the plow; and hence the conclusion is strengthened, that in all cases wheels are incumbrances and sources of increased resistance to the plow.

(959.) Among the numerous makers of plows in England whose works have come under our observation, besides those of Messrs. Ransome, I cannot omit to notice the names of Hart, of King, of Parker, and of Crosskill, all of whom take a high standing as plow-makers after the English fashions, and many of their productions are mounted with wheels. It is unnecessary to repeat any of the observations on that head, but should the preceding remarks come under the observation of any of the makers referred to, or of any other person who may take an interest in the subject, it will be gratifying to find that they endeavor to show how wheel-plows can be rendered more advantageous than swing-plows, and in doing so either practically or demonstratively, in a satisfactory manner, I shall be open to conviction, and ready to yield up that system which appears to me at present as the only tenable one; but it would, of course, be still more satisfactory to learn that these very humble efforts shall be of any use in satisfying those who take the trouble to inquire, that an extended application of the swing-plow practice might either be of individual or general importance.

(960.) Among the other numerous varieties, I cannot pass over the *two-furrow plow*, which, though seldom, if ever, seen in the hands of a Scotch farmer, is now rather extensively employed in some of the eastern counties of England, but more especially in Lincolnshire. These plows are constructed of a very effective and convenient form by Ransome and others, and are held to be very economical in point of draught, a pair of good horses being capable of working a two-furrow plow, or in cases of heavier soil three horses; the saving of labor in the one case being

* Journal of the Royal Agricultural Society of England, vol. i. † Ibid., vol. i.

one-half, in the other one-fourth. They are also mounted with wheels, and in the districts where they are employed, and the plowman accustomed to the implement, they make very fair work, the two furrows being in general laid very nearly alike. It must be conceded, however, that in the districts where these plows are used, the work is done with a very shallow furrow, seldom exceeding 3 or 4 inches, which may allow of 2 horses taking the draught. Where the deep-plowing system is fallowed, a two-furrow plow could not be drawn by fewer than 4 horses, which, as it would afford no saving, but rather the contrary, can never be expedient, or in any way advantageous; for though it may be urged, that when a light furrow only was required the two-furrow plow might offer some advantage, yet if it could not be applicable in every case, the inference is, that two sets of plows, double and single furrow, must be retained—a practice which cannot, under any circumstances, be recommended. The conclusion to be drawn from these remarks is, that though expedients, such as the two-furrow plow, may be very advantageously employed under a particular climate and soil, the practice cannot be held up as one of general application, or that could be rendered economical and advantageous under all circumstances.—J. S.]

(961.) This seems to me a befitting place to say a few words on *plowing matches*. I believe it admits of no doubt that, since the institution of plowing matches throughout the country, the character of our farm-servants as plowmen has risen to considerable celebrity, not but that individual plowmen could have been found before the practice of matches existed as dexterous as any of the present day, but the general diffusion of good plowing must be obvious to every one who has been in the habit of observing the plowed surface of the country. This improvement is not to be ascribed to the institution of plowing matches alone, because superior construction of implements, better kept, better matched, and superior race of horses; and superior judgment and taste in field labor in the farmer himself are too important elements in influencing the conduct of plowmen to be overlooked in a consideration of this question.

(962.) But be the primary motive for improvement in the most important branch of field labor as it may, there cannot be a doubt that a properly regulated emulation among workmen of any class, proves a strong incentive to the production of superior workmanship, and the more generally the inducement is extended, the improvement arising from it may be expected to be the more generally diffused; and on this account the *plow medals* of the Highland and Agricultural Society of Scotland being open for competition to all parts of Scotland every year, have perhaps excited a spirit of emulation among plowmen, by rewarding those who excel, beyond anything to be seen in any other country. Wherever 15 plows can be gathered together for competition at any time and place, there the plowman who obtains the first premium offered by those interested in the exhibition, is entitled to receive, over and above, the Society's plow medal of silver, bearing a suitable inscription, with the gainer's name. About 40 applications are made for the medals every year, so that at least 600 plowmen annually compete for them; but the actual number far exceeds that number; as, in many instances, matches comprehend from 40 to 70 plows, instead of the minimum number of 15. The matches are usually occasioned by the welcome which his neighbors are desirous of giving an incoming tenant to his farm, and its heartiness is shown in the extent of the assistance they give him in plowing a field or fields at a time when he has not yet collected a working stock sufficient for the purpose.

(963.) Plowing matches are generally very fairly conducted in Scotland. They usually take place on lea ground, the plowing of which is considered the best test of a plowman's skill, though I hold that drilling is much more difficult to execute correctly. The best part of the field is usually selected for the purpose, if there be such, and the same extent of ground, usually from 2 to 4 ridges, according to the length, is allotted to each portion of ground to be plowed. A pin, bearing a number, is pushed into the ground at the end of each lot, of which there are as many marked off as there are plows entered in competition. Numbers corresponding to those on the pins are drawn by the competing plowmen, who take possession of the lots as they are drawn. Ample time is allowed to finish the lot, and in this part of the arrangements I am of opinion that too much time is usually allowed, to the annoyance of the spectators. Although shortness of time in executing the same extent of work is not to be compared to excellency of execution, yet it should enter as an important element into the decision of the question of excellence. Every competitor is obliged to feer his own lot, guide his own horses, and do every other thing connected with the work, such as assorting his horses, and trimming his plow-irons, without the least assistance.

(964.) The judges, who have been brought from a distance, and have no personal interest in the exhibition, are requested to inspect the ground after all the plows have been removed, having been kept away from the scene during the time the plows were engaged. Now, this appears to me a very objectionable part of the arrangements, and it is made on the plea, that were the judges to see the plows at work, some particular ones might be recognized by them as belonging to friends, and their minds might thereby be biased by the circumstance. Such a plea pays but a poor compliment to the integrity of the judge; and any farmer who accepts of the responsible and honored office of judge, who would allow himself to be influenced by so pitiful a consideration, would deserve not only not to be employed in a similar arbitration again, but to be scouted out of society. One consequence of the exaction of this rule is, that the spectators evince impatience—the spectators, not the plowmen who have been competitors, for they are busily and happily occupied at the time in replenishing the inner man with rations of cheese and bread and ale provided to them by the possessor of the field who is to enjoy the profits of their labor—while the judges are taking no more than the proper time for deciding the plowing of, it may be, a large extent of ground. The judges ought, therefore, to be present during the whole time devoted to the competition, when they could calmly and certainly ascertain the nature and depth of the furrow-slices, and have leisure to mature their thoughts on points which may turn the scale against first impressions. That the bare inspection of the finished surface cannot inform them, in a satisfactory manner, whether the land has been correctly plowed or not, which can only be done by comparison of the soles of the furrows while the land is plowing, I shall endeavor to make clear to you by figures in a supposed case.

(965.) You have seen the action of different plows, which may be all employed in the same match; and you have seen that the East-Lothian form of plow lays over a slice of one form, as in fig. 108, and that the Lanarkshire plow lays over a slice of another form, as in fig. 109, and paragraph (591) acquaints you, that the latter form of slice, namely, that with the high crest and serrated furrow-sole, contains 1·7 less earth than the other. Now, were the surface work only to be judged of, which must be the case when judges are prohibited seeing the work done in the course of execution, the serrated extent of the furrow-sole cannot be ascertained by removing portions of the plowed ground here and there, so well as by constant inspection. As equal plowing consists in turning over equal portions of soil in the same extent of ground, other things being equal, a comparison of the quantity of earth turned over by these two plows may be made in this way: Suppose a space of 1 square yard, turned over by each of the two kinds of plows specified, taking a furrow-slice in both cases of 10 inches in breadth and 7 inches in depth, and taking the specific gravity of soil at 2·7, the weight of earth turned over by the East-Lothian would be 63 stones, while the Lanarkshire plow would only turn over 54 stones, making a difference of 9 stones of 14 lbs. in the small area of one square yard. In these circumstances, is it fair to say, that the horses yoked to the East-Lothian plow have done no more work than those yoked to the Lanarkshire, or that the crop for which the land has been plowed will receive the same quantity of loosened mould to grow in in both these cases? The prohibitory rule against the judges making their inspection during the plowing has been relaxed in several instances; but I fear more from the circumstance of the spectators losing their patience, while waiting for the decision after the excitement of the competition is over, than from regard to the justness of the principle. Thus far is the obvious view of the question regarding the mode in which plowing-matches are usually conducted; but in what follows will be found a more important view as affecting the integrity of good plowing.

(966.) [The primary objects of the institution of plowing-matches must have been to produce the best examples of plowmanship—and by the best, must be understood that kind of plowing which shall not only *appear* to be well done, but must be thoroughly and essentially well done. In other words, the award should be given to the plow that produces not only work of a proper surface finish, but which will exhibit, along with the first, the property of having turned up the greatest quantity of soil and in the best manner. That this combination of qualities has ceased to be the object of reward, is now sufficiently apparent to any one that will examine for himself the productions and rewards of recent plowing matches, and the causes of such dereliction are these:

(967.) The introduction by Wilkie of the Lanarkshire plow gave rise, as is supposed, to the high-crested furrow-slice. It cannot be denied that the plows made on that principle produce work on lea land that is highly satisfactory to the eye of a plowman, or to any person, indeed, whose eye can appreciate regularity of form; and, as there are many minds who can dwell with pleasure on the beauty of form, but who do not combine with that idea its adaptation to usefulness, it is no wonder that plows which could thus affect the mind through the sense of sight, should become favorites. While the crested system of plowing kept within bounds, it was all very good but in course of time the taste for this practice became excessive; and losing sight of the useful, a depraved taste, of its kind, sacrificed utility to the beautiful, in so far as plowing can be said to produce that impression. This taste came gradually to spread itself over certain districts, and plow-makers came to vie with each other in producing machines that should excel in that particular point of cresting. A keen spirit of emulation among plowmen kept up the taste among their own class, and very frequently the sons of farmers became successful competitors in the matches, which circumstance gave the taste a higher step in the social scale. Thus, by degrees, the taste for this mode of plowing spread wider and wider, until, in certain districts, it came to pervade all classes of agriculturists. At plowing matches in those districts, the criterion of good plowing became generally to be taken entirely from the appearance of the surface: furrow-slices possessing the highest degree of parallelism, *exposing* faces of equal breadth, and, above all, a high crest, carried off the palm of victory. I have seen a quorum of plow judges "plodding their weary way" for two hours together over a field, measuring the breadth of faces, and scanning the parallelism of slices, but who never seemed to consider the underground work of any importance in enabling them to come to a decision. Under such a system, it is not surprising that plowmen devote their energies to produce work that might satisfy this depraved taste, and that plow-makers find it their interest to minister to those desires, by going more and more into that construction of parts of the plow that would yield the so much desired results. Thus have those valuable institutions of plowing matches, in the districts alluded to, been unwittingly brought to engender a practice which, though beautiful as an object of sight, and, when within due bounds, also of utility, has induced a deterioration in the really useful effects of the plow.

(968.) But it is not yet too late to retrieve what has been lost. Let the Highland and Agricultural Society of Scotland, and all local Agricultural Associations, take up the subject, and institute a code of rules by which the judges of plowing matches shall be guided in delivering their awards. Let these rules direct attention to what is truly beneficial to the land, as well as what may be satisfactory to sight in plowing. When such rules shall have been promulgated from competent authority, we may hope to see plowing matches restored to their pristine integrity—doing good to all who are concerned in them, and restoring that confidence in their usefulness which is at present on the wane, but distrust in which has only arisen from an accidental misdirection of their main objects.

(969.) In connection with that part of the subject which has given rise to the foregoing remarks on plowing matches, it is not a little curious to find, that instead of the high-crested furrow being a modern innovation, it is as old as the days of Blith in 1652; and he, like the moderns, had entertained the same false notions of its advantages. In his curious work, under the section "How to plow as it may yeild most mould," he, in his quaint style says: "As for your ordinary seasons of plowing, your land being in good tillage, any well ordered and truly compassed plow will do, you may help yourself sufficiently in the making of your irons, if you would have the edge of your lying furrow lye up higher, which will yeeld most mould, then set your share-phin the shallower, and yet your ·ew the broader, and hold it the more ashore, the plowman going upon the land, and it will lay it with a sharp edge, which is a gallant posture for almost any land, especially for the lay turf beyond compare."*

(970.) The setting of the share-phin (feather) as here described, is precisely what is done in the modern plows to make them produce the high-crested furrow (590). Blith seems to consider that holding the plow "ashore" (to landward), aids the effect; it will make a slice thinner at one edge, but not more acute in the crest.—J. S.]

---

28. TRENCH AND SUBSOIL PLOWING, AND MOOR-BAND PAN.

> "If deep you wish to go, or if the soil
> Be stiff and hard, or not yet cleared of stones,
> The Scottish plow, drawn by a team four strong,
> Your purpose best will suit;——"
>
> GRAHAM.

(971.) Trenching of land with the spade has been a favorite operation in gardening for many ages; and since the plow became the substitute for the spade in field culture, it has been employed for the same purpose, of deepening the friable portion of the soil, and affording to the roots of plants a wider range in which to search for food. It is highly probable, however, that the plow could not have closely imitated the trenching of ground with the spade until after the introduction of the mould-board, which, comparatively speaking, is of very recent date, the ancient plow retaining its primitive simplicity of form until within a few centuries. Indeed, until the mould-board was added, it was scarcely in the power of the plow to trench the soil, that is, to reverse the position of the furrow-slice and mix the upper and lower soils together. When it was added, may now be difficult to ascertain; but fully two centuries ago, Hartlib, in his Legacie, intimates the practice of very deep plowing, with the mould-board in use, when he says: "There is an ingenious yeoman in Kent who hath two plows fastened together very finely, by the which he ploweth two furrows at once, *one under the other*, and so stirreth up the land 12 or 14 inches deep, which in deep land is good." This is essentially trench-plowing.

(972.) Within a very recent date, it has been recommended to plow land as deep as trenching, but so as to retain the stirred soil below the surface. Mr. Smith, Deanston, by the invention of his subsoil-plow, has been the means of directing the attention of agriculturists to this peculiar and apparently new process, which has obtained the appellation of *sub*soil-plowing. A figure and description of his *subsoil-plow* is given below. After the introduction of the mould-board, subsoil-plowing could not have been practiced; but prior to that improvement it is not improbable that the process was known and practiced, and so long ago even as by the Romans. It is uncertain what was the depth of the furrow usually made by the Roman plow, some commentators supposing, from a particular phrase used by Pliny, that it was as much as 9 inches, but at all events he designates a depth of furrow of 3 inches as a mere scarification of the soil. There is no doubt, however, from a passage of Columella, that the Roman farmers occasionally gave a deep furrow to good deep land, when he says: "Nor ought we to content ourselves with viewing the surface, but the *quality of the matter below* should be diligently inquired into, whether or no it is of earth. It is sufficient for corn if the land is equally good 2 feet deep." If they imagined that corn received benefit from the soil at the

* Blith's Improver Improved, p. 216, edition 1652.

distance of 9 feet below the surface, they would consider it as an advantage to plow as deep as their cattle were capable of, and their plow could go.* As the Roman plow had no mould-board, any *deep*-plowing effected by it would partake much more of the character of a subsoil than of a trench-plowing.

(973.) The effect of subsoil-plowing being merely to stir the subsoil without affecting its relative position, the best way of performing the operation is, as I conceive, in the following manner: and it may be executed either in winter or in summer according as it is made to form a part of the spring or summer's operations. It is best executed *across the ridges;* let, therefore, a feering of 30 yards in width be taken across them with the common plow from the upper fence of the field; and this is most easily effected by opening out feering furrow-slices parallel with and close to the fence, if it be straight, and another at 30 paces distant, and let the subsoil-plow follow in both the open feerings. The plow then closes the feerings, and so plows from one feering to another until the open furrow is formed in the middle of the feered space between them, followed implicitly all the time by the subsoil plow, which is held by one man, and the horses are driven by another. Feering after feering is thus made and plowed with the common plow, and followed by the subsoil until the whole field is gone over, with the exception of about the breadth of a ridge at each side of the field, upon which the horses had turned, and the neglect of which is probably of no great importance. Fig. 206 is given as

Fig. 206.

THE TRENCH OR SUBSOIL PLOWING.

a representation of the operation, where the plows and horses appear in black, and where the common plow with 2 horses precedes the subsoil one with 4. The depth taken by the plow is the usual one of 7 inches in stubble, which is seen as the upper furrow, succeeded by the subsoil-plow, which takes usually 9 inches in such a position, and whose furrow is seen in section below that of the other plow, making both furrows 16 inches deep. Care should be specially taken not to allow the subsoil-plow to approach within 2 inches of the covering of any drain, otherwise the drain will be torn up and materially injured. The drains in the figure are supposed to be 36 inches deep, filled 12 inches with tile and sole and small stones, and placed in every open furrow at 15 feet asunder, the curved form of the ground between them representing the ridges. This figure is not meant to give the exactly relative proportions of the different objects composing it.

(974.) The immediate effect of subsoil-plowing being to deepen the friable portion of the soil, it is evident that where the subsoil-plowed soil rests upon impervious or even retentive matter, that the operation will increase the depth, and, of course, the capacity of the soil for holding water. and on this account, in so far as respects itself, the operation after wet weather would do more injury than good to the crops growing upon it.—This is a very important fact in regard to the effects of subsoil-plowing.

* Dickson's Husbandry of the Ancients, vol. i.

considered in itself, and demands your serious consideration, because a misconception and disbelief of it continues to exist in some parts of the country, especially in England, and injury may thereby be inflicted on land which will require a considerable time to recover. But if injurious effects accompany subsoil-plowing, when it occasions excess of water, it is evident that were drains formed to give the water an opportunity to escape, it would do no injury to plow land to any depth. The misconception which I have alluded to as existing may not be easily dispelled, as the unusual depth to which subsoil-plowing is executed operates in the first instance as a drier of the *surface* of the ground, even when there has been no previous draining, and it also renders drained ground drier; and these immediate effects are regarded by improvers of land as all that are required to be effected by the operation, and, consequently, when it is easily ascertained that subsoil-plowing is a much cheaper operation than draining and seems to be equally efficacious, they are content to abide by it alone. I have no doubt that much of the land that has been subsoil-plowed in England has been so in consequence of the adoption of this opinion by farmers, upon whose attention the great comparative economy attending the process was so earnestly pressed some years ago by people of influence. They, however, who understood the nature and capability of subsoil-plowing, and Mr. Smith himself at their head, both published and publicly stated that to employ the subsoil-plow upon land having a retentive subsoil, without draining it in the first instance, would only aggravate the evil they wished to avoid. It is true that the subsoil-plow might penetrate through the retentive matter to an open substratum through which the water would escape; but the chance of meeting with such a rare arrangement of strata, forming the exception to the general structure of clayey subsoil, cannot afford a sufficient excuse for an indiscriminate use of the subsoil-plow.

(975.) It should, therefore, be laid down as a general rule, that no land ought to be subsoil-plowed unless it has been previously drained; for, where the subsoil is so porous naturally as not to require draining, neither will it require subsoil-plowing. After being thoroughly drained, any sort of land may be subsoil-plowed with safety—that is, no harm will accrue from it; but all sorts of land will not derive equal advantage from the operation. Taking it, therefore, for granted that draining should precede the subsoil-plow, the interesting inquiry arises—In what quality of subsoil does subsoil-plowing confer the greatest (if any) benefit to land?—the correct answer to which can alone determine the extent to which this operation should be carried. In the first place, in pure plastic clay, any opening made by the subsoil-plow passing through it would probably soon collapse together behind the implement. Through such a clay, in a dry state, the operation would be performed with great difficulty, if not prove impracticable. In what is usually called *till*—that is, clay containing sand veins, small stones, or small boulders—the subsoil-plow will pass sufficiently well, though slowly, and it will displace even pretty large stones, and the clay be afterward kept open for a time. Hardened masses of gravelly clay may be entirely broken up by this operation. I believe experience has established the effect of the subsoil-plow in these respects. It thus appears that the sphere of the subsoil-plow, as an operation of permanent utility, is limited to the breaking up of hard, gravelly subsoils; because it is scarcely supposable that it can keep pure clay always open, and it certainly admits of doubt that it will keep a tilly bottom constantly open, as experience has proved that percolation of water through a somewhat porous clay renders it more firm, by the well-known fact that such a soil, returned above a drain, soon becomes as firm as any other part of the field.

Hard, chalky, gravelly matter, and moor-band pan, are the only subsoils on which one would feel confident that subsoil-plowing would confer permanent benefit. I say *permanent* benefit; for I believe it is acknowledged that the process confers an immediate benefit in almost every case in which it has been tried; and on this account its keener advocates have claimed for it much of, if not the entire, advantages derived from its precursor, thorough-draining; and there seems some ground for the claim, inasmuch as subsoil-plowing is executed so soon after thorough-draining that it would be impossible to assert the superior claims of draining, were it not for its occurrence being more common without than with subsoil-plowing; whereas, when the latter is taken by itself, it cuts but a sorry figure. In vindication of his own invention, Mr. Smith endeavors to explain in general terms *why* indurated subsoil, when drained, should preserve the friability imparted to it by subsoil-plowing. "When drains have been some time executed," he says, "innumerable small fissures will be found in the subsoil, extending from drain to drain; these are caused by the contraction of the substance of the soil arising from its drier state. The contraction being greatest in the stiffest clays, the operation of the subsoil-plow admitting the air to a greater depth, the fissures take place under its operations, and generally reach to the level of the bottom of the drains." This is a natural enough explanation of the almost immediate effect of draining wet subsoils, and also of the almost immediate extension of its effects by a subsequent subsoil-plowing. "These fissures," he continues, "will get more or less silted or glutted up, from time to time, by the minute alluvial particles carried down and left in filtration by the rain-water," which is also a natural effect; but he adds, "the constant expansion and contraction of the unremoved subsoil, by the alternations of wet and dry, has a perpetual tendency to renew them;" and it is this effect which I question, because, before such expansion and contraction can be kept up, it must be assumed that the subsoil, after being subsoil-plowed, has no tendency to consolidate into its original state, whether of strong clay or any other substance; and yet the conviction, as I conceive, of all drainers must be, that every sort of subsoil, except hard rock, consolidates, however well it may have been stirred, though perhaps not to the degree of impermeability it may have possessed before; but, at all events, the more friable it becomes in its condition, it will be the less affected by the "alternations of wet and dry."*

(976.) It is allowed by all who have used the subsoil-plow that it requires much greater exertion from the horses to work it than the common plow, and that horses do not work well together in it for some time. With regard to the quantity of ground which a plow will subsoil in a day, in a long day in summer, 1 imperial acre may be calculated on being accomplished in favorable circumstances; but should obstructions occur, such as large boulder stones, ¾ of an acre is a very good day's work; and in winter ¾ is a good day's work without obstructions.

(977.) The great force required to work even the lightest of the Deanston subsoil-plows, which weighs 18½ stones—the heaviest weighing 28½ stones of 14 lbs.—and the insuperable bar which this circumstance places against its employment on farms working less than 3 pairs of horses, have induced people to contrive what they designate subsoil-plows, to be used as a substitute for the Deanston one; but, in every modification which I have seen proposed, their effect is quite different from what Mr. Smith proposes that his should produce. The Deanston subsoil-plow not only

* Smith's Remarks on Thorough Draining.

penetrates the subsoil to a determinate depth, but, by the simple contrivance of the feather, the subsoil is not only stirred, but pushed a little aside, and thereby partially mixed with the portion adjoining it. In doing this, there is little doubt that it is the action of the feather which causes the principal weight of the draught complained of in this plow. To avoid this redundancy of draught, as it is supposed to be, a feather is discarded, and simply a large tine bent forward at the point, as in the case of Gabell's subsoil-plow,* or a small scarifier, as in the case of the Charlbury one,† is substituted as its principal feature—a furrow being opened by the common plow preceding, as in the case of the subsoil one. But, as the subsoil plow makes a demand on the horses of the farm for a common plow as well as itself, it is proposed in the Charlbury one to have a small plow attached to the beam to lay over the furrow-slice, in order to prepare it for the tine or scarifier to pass along. But although a lightness of draught has been attained in both these instances, and a pair of horses have been said to be able to work Gabell's to the depth of 18 inches, it requires 4 horses to work the Charlbury to 12 inches—which, as a saving of labor, is of no great importance, except as regards the employment of the common plow in preparing the way; for the difference in the nature of the work performed by them differs so widely from that of the Deanston plow, that they cannot be said to be its substitutes, inasmuch as they only make ruts in the subsoil at the distance of the breadth of a furrow-slice from one another—namely, 8 inches at the least, most probably 9 inches, and not improbably 10 inches—and which, of course, leave ribs of hard land standing untouched in the subsoil. It is acknowledged by the proposers of these substitutes that the Deanston is more efficient than either; and, as Mr. Smith's opinion is that the heaviest of his plows does the most satisfactory work, it is clear that they can never be a substitute for his, provided he is correct in his views regarding the utility of *thorough* subsoil-plowing, which these substitutes certainly do not even profess to perform. The conclusion I would draw from the inefficiency of these modifications and substitutes for the Deanston subsoil-plow is, that none of them are likely to be used on farms employing less than 3 pairs of horses, such employment being the only object their contrivers had in offering them to public notice.

(978.) A modification has been proposed and practiced in the use of Mr. Smith's own subsoil-plow, which is, that instead of passing it in every furrow of the preceding small plow, it should pass in every other furrow.—The advantages said to be derived from this plan are, that it is cheaper, speedier, and the subsoil is not so much broken, though broken enough to allow the water to escape to the drains; but it is obvious that it is this very defective mode of operation which constitutes the great objection to the English substitution of the Deanston plow.

(979.) Instances are not wanting to prove that benefits have been derived from the conjoint operations of thorough-draining and subsoil-plowing;‡ but, as in almost every recorded case the combined effects of both operations are reported, it is impossible to ascertain the advantages derived from each. In a memorandum which I made some time ago, but now forget where the circumstance happened, I find it stated that a field of lea was drained in 1838, and the crop of oats from it in 1839 did not exceed 13 bushels per acre imperial; whereas, after it was subsoil-plowed, another crop of oats in 1840 gave 32 bushels per acre. It is well understood that the first crop of oats from lea newly thorough-drained never yields an increase; but, in this instance, I have no doubt that the second crop of oats

* Journal of the Royal Agricultural Society of England, vol. ii. † Ibid., vol. i.
‡ Journal of the Royal Agricultural Society of England, vols. i. ii. and iii.

would have been better than the first, without the use of the subsoil-plow, especially if the lea were old. In Mr. Laing's experience on Campend, Mid-Lothian, he has "found land to be more thoroughly dried after subsoil-plowing (especially when there was any approach to clay in the subsoil), with a drain in every alternate furrow, than with a drain in every furrow without it; in fact, on a stiff clay subsoil I have seen drains of little service, the water for some time standing on the top of them till evaporated, while in the very next field, which had been subsoil-plowed, there was an immense flow of water in every drain, and not a drop to be seen on the surface." Such an effect is not at all surprising, as it is well understood that subsoil-plowing greatly assists thorough-draining at first in drying land; and that effect appears the more striking when the land has not been so well drained as it should have been, which it certainly would not be with a drain in every alternate furrow, when it is so strong as that mentioned by Mr. Laing, where he says of a field of 10 acres that was subsoil-plowed in November, 1836—a wet season, certainly—that it "was at the time, and during the whole operation, so saturated with rain that the horses' feet sunk in the unplowed ground from 4 to 6 inches; which showed, though there was a drain in every alternate furrow, they had not drawn the water from the stiff, retentive subsoil. This circumstance," continues Mr. Laing, "convinced me the more of the necessity of persevering in subsoil-plowing, which *alone* enabled me to accomplish my object of thoroughly drying the soil."* The conclusion come to is scarcely a fair inference from the premises, which should rather, in the first instance, have brought conviction, from the nature and wet state of the soil, of the necessity of persevering in thorough-draining, by making a drain in every, instead of every alternate, furrow; and it was after *thorough*-draining had failed, that the drainer would be entitled to say that "subsoil-plowing *is an indispensable* accompaniment to furrow-draining," "or that *it alone* enabled him to accomplish his object of thoroughly drying the soil." I cannot refrain from making a passing remark here, that there is a strong propensity in farmers generally to laud the good properties of its auxiliaries at the expense of thorough-draining; and I can only account for the prevalence of the feeling, from the *fact* being well known to them that it is *cheaper* to subsoil-plow land than to thorough-drain it, the amount of labor to put it in the condition of being thoroughly dry depending upon its nature.

(980.) Mr. Melvin, Ratho Mains, Mid-Lothian, says what will readily be believed in Scotland, that "I have never seen *any* benefit from the use of the subsoil-plow upon damp-bottomed land that had not been drained;" and after a fair trial in a particular field of deep, soft, damp soil, of both operations conjointly, he expresses himself in terms which place the art of subsoil plowing in nearly its proper position. "Much, no doubt, of the improvement in the condition of this field, is to be attributed to draining, still, the quick absorption of the water in the furrows between the drains (the land being cast), the decided improvement of the drier part, and the uniformly equal crop, sufficiently attest the merits of subsoil-plowing." I have said that, in these remarks, Mr. Melvin has placed subsoil-plowing in *nearly*, but as I conceive, not altogether its proper position; because the field was drained in the alternate furrows, and the drier part of it was not drained at all. Now, had every furrow been drained, would not the water have been quickly absorbed; and had the drier part been drained, would it have required subsoil plowing at all?

(981.) With regard to the expense of subsoil-plowing, it may be fairly

* Prize Essays of the Highland and Agricultural Society, vol. xii.

taken at the cost of 3 pairs of horses and 3 men, and wear and tear of implements per day for every ¾ of an acre imperial plowed; and Mr. Pusey instances a case of a farm of not 600 acres of cold clay, the subsoil-plowing of which was estimated to cost £1,300, but how this sum was made out does not appear.* The returns of the grain crops seem to imply an increase of 25 per cent. at most, and in regard to green crops, an instance is given of a yield of turnips off peaty soil, resting on stiff clay and hard sand and gravel at Drayton in Staffordshire, belonging to Sir Robert Peel, of "four times the quantity in weight ever produced in the same field at any previous time," the large crop alluded to being 27 tons per acre, including tops.†

(982.) With regard to *trenching* the ground, it has long been practiced by gardeners with the spade, and its object is to bury the exhausted soil on the surface with all its seeds of weeds and eggs of insects, and bring up to the surface a comparatively fresh and unexhausted soil, not so rich in manure as the one buried down, but more capable, by its fresh properties, to make a better use of the manure put into it. Trenching with the spade is also practiced on farms on a large scale. From experience in both ways, I can maintain that it is cheaper to trench rough, stony ground with the spade than with the plow, giving consideration to the state of the soil when left by the two implements. The plow with 4 horses will turn over and rip up a strong furrow, and where there are no stones and roots, it will answer the purpose well enough; but where stones, though small but numerous, and if large, are encountered, the furrow becomes very uneven and unequal, the horses jaded, the men fatigued, the implement broken, and the work very imperfectly done. It is the same case with the roots of trees, and even of bushes, against which, when the plow comes, horses pull with vehemence, so as either to injure themselves, or break their tackling. At such work I had two valuable horses so much injured in their wind as to become unfit for ordinary farm work; and finding so I abandoned the plow for this purpose altogether. The same work, on the other hand, can be much better done with the spade, and when it is undertaken by a contractor who remains constantly with the spadesmen, it will be your own fault in superintendence, if the work be ill executed.

(983.) I have found this plan succeed in making good trenching. Let the ground to be trenched be laid off in lots with pins; and let the lots contain equal areas of five yards in breadth. The trench to be 14 inches in perpendicular depth in the solid ground on the average over the lot, the surface being left even with the general inclination of the field that is to be. The 14 inches out of the solid will give a depth of 16 inches in the trenched part of the ground. The contractors should be obliged to remove all stones, large and small, all roots, large and small, and every other thing that it likely to obstruct the future course of the plow, and lay them upon the surface of the trenched ground; and should large boulders be found a little below the surface, these must either be blown to pieces by gunpowder, and the fragments left on the surface, or farther sunk in the earth so as to be out of the reach of the plow in future, according as you find that you may have use for the stones for drains or foundations of fence dykes. The trenching is begun at the utmost limit of the rough ground, by each man rutting some breadths of 12 or 15 inches wide across his lot, and making a trench of the required depth of 14 inches, gauged by a stick kept constantly in his possession to guide him in the depth, that he may not have the plea of ignorance to urge in extenuation of his cupidity. The upper turf or spading is put on its back in the bottom of the trench; the

* Journal of the Royal Agricultural Society of England, vol. i. † Ibid., vol. iii.

soil is then dug and thrown upon it, care being taken to make the new ground level and even; and, lastly, shoveling the loose earth over the surface, leaving no inequalities in the bottom of the new trench. After one set of allotted spaces has been trenched in this way, another is ready marked off by the contractor for the men to enter upon as they finish their lots, and the second set should be marked off either along one end or one side of the field, whichever is found most convenient for the future operations of removing the trenched-up materials to their destination, that a whole piece of ground may be cleared for future operations without interfering with the progress of the trenching, the workmen employed in which should be called upon to do nothing else than their appointed tasks.

(984.) Ground that has lain in this rough state for years will no doubt require draining, and should be drained before or after being trenched, according to circumstances. It should be examined beforehand, by pits sunk here and there, whether the subsoil will afford a sufficient quantity of stones to thorough-drain the ground. If it is supposed or certain that it will, the ground should first be trenched to obtain the stones, and they being on the spot, the drains will be easily filled with them. If the stones be only to that amount as to form an ordinary covering to tiles, then tiles and soles should be used as the principal materials, and, in this case as well as the other, the ground should first be trenched. But if stones are plentiful near at hand, though not in the particular field under improvement, from a quarry hard by, for instance, then the drains should be opened and filled to the requisite depth before the surface of the old ground is broken up, that the cartage of the stones may first be borne by it; and the trenchers in that case should be obliged to cover the stones of the drains with turf, and level the ground over them as they proceed with the trenching.

(985.) The expense of trenching rough ground at 14 inches deep—and it should never be shallower, in order to insure a good plow-furrow ever after—is from 10d. to 1s. per pole, according to the roughness of the ground. I have had very rough ground, consisting of large roots of trees in scattered wood, with brushwood of birch, alder, whin, and broom, and containing as many stones as would have half-drained the ground, trenched 14 inches deep for 1s. per fall, Scotch measure, which is equal to £6 13s. 3d. the imperial acre, or rather more than 9½d. per pole, a large sum, undoubtedly, independent of draining, clearing away rubbish, and other horse and manual labor; but then the ground was rendered at once from a state of wilderness to one in which manure could be applied and covered in with an ordinary furrow-slice of mould. If this is not the *cheapest* mode, in a pecuniary point of view, of rendering ground available to cultivation, it is at all events the most pleasant to the feelings in the doing, and the most satisfactory when done.

(986.) But there is a mode of trenching ground which is best done with the plow, its object being to imitate the work of the spade by descending deeper than the ordinary depth of furrow, and of commixing part of the subsoil with the surface soil, which has been probably rendered effete by overcropping. Ground can be trenched with the plow in two ways, either with a large-sized common plow drawn by 4 horses in one of the ways pointed out before, fig. 202, or with one plow going before and turning over an ordinary furrow-slice, and another following in the same furrow drawn by 2 or more, usually with 3 horses, or both plows drawn by 3 horses each. It is best performed across the ridges. In either of the above ways the same effect is produced in similar soil, breaking up indurated gravel, deepening thin clays, ameliorating stiff clays by exposure to the air, and mixing old and new soils together, the ultimate effect on all being to *deepen* that portion of the soil which is used by the cultivated crops.

(987.) In one respect trenching has the same effect as subsoil-plowing, namely, the stirring of the ground to the same depth, the first plow turning over a furrow of 7 inches in depth, and the second going 8 or 9 inches deeper, making in all a furrow of 15 or 16 inches in depth; but in another respect the two operations leave the soil in very different states—the subsoil-plow stirs the soil to the depth named, but brings none of it to the surface, while the trench-plow does not altogether bring that which was undermost to the surface, but commixes the under and upper soils together. This latter practice has long been known in the midland counties of England, but the former has only been presented to the notice of the Scottish agriculturist since 1829.

(988.) It has been made a question, which is the better mode, if both are not alike, of making the soil fertile? the advocates of subsoil-plowing alleging that it is better to ameliorate the subsoil while under the soil by the admission to it of air and moisture; while those of the trench-plow answer that, if the object of both operations is to ameliorate the subsoil, it will become sooner so by being brought to the surface in contact with atmospheric air and moisture. But, say the promoters of subsoil-plowing, there are subsoils of so pernicious a nature, having the salts of iron and of magnesia in them, that the upper soil would be much injured by its admixture with such substances. No doubt, answer the trench-plowers, if the subsoil that contained these noxious ingredients in a large proportion were brought up in quantity, when compared with the bulk of the upper soil, injury would be done to it for a time, but they say it is not the abuse but the proper use of trench-plowing which they advocate; and of *such a subsoil*, they would use the discretion to bring up only a little at a time, which they have it in their power to do, until they accomplish their end, namely, that of ameliorating the whole depth of subsoil. But they maintain that by far the greatest proportion of subsoils do not contain those noxious ingredients; and, besides, the very best and quickest way of getting rid of even these is to bring them at once to the surface, for any of the acids, or the salts of iron, are easily neutralized by the action of lime, which is always applied to the surface; and those of magnesia are most easily reduced on free exposure to the air. And, moreover, they ask, If subsoils shall be ameliorated by air and moisture when stirred by a subsoil, why should they not also be ameliorated when stirred by a trench-plow? And they urge farther, that trenching may be practiced more safely without previous thorough-draining, than subsoil-plowing.

(989.) I have no hesitation in expressing my preference of trench to subsoil-plowing; and I cannot see a single instance, with the sole exception of turning up a very bad subsoil in large quantity, there is any advantage attending subsoil that cannot be enjoyed by trench-plowing; and for this single drawback of a very bad subsoil, trenching has the advantage of being performed in perfect safety, where subsoil-plowing could not be without previous draining. Mr. Melvin, Ratho Mains, mentions an instance of a field containing both damp and dry ground; the dry was trench-plowed in the autumn of 1836, an inch or two of the sandy gravel being brought up, and "was decidedly increased in fertility," both in the turnip and barley crops which followed.* I trench-plowed a field of 25 acres of deep black mould which had been worn out, with a 4-horse plow, taking and clearing a furrow from 14 to 16 inches deep in the solid land, and bringing up almost in every part a portion of the tilly subsoil, which was only

* Prize Essays of the Highland and Agricultural Society, vol. xii

drained to the extent of a few roods put in the face of a slope exhibiting spouts of water. The turnips that followed were excellent; the barley yielded upward of 50 bushels per acre imperial, and the year after a part was measured off and fenced, containing 6 acres, to stand for hay, which yielded of good hay 1999 stones of 22 lb. Another field, the year after, that was not drained, suffered injury after trench-plowing; but that was in consequence of having been caught with a premature fall of rain in the autumn before the trenched land could be ridged up, and it lay in the trenched furrow all winter. It is stated that Mr. Scott, Craiglockart, Mid-Lothian, "trench-plowed in the winter of 1833–4, with one common plow following another, a field of 20 acres, every two alternate ridges, and he has never observed on any of the crops the slightest difference."* This is, as I conceive, an unsatisfactory mode of testing the value of any sort of plowing land, as it is possible that the untrenched ridges derived a certain and it might be a sufficient advantage, in regard to drying, from the adjoining trenched ridges.

(990.) But while giving a preference to trench-plowing over subsoil, I am of opinion that it should not be generally attempted under any circumstances, however favorable, without previous thorough-draining, any more than subsoil-plowing, but when so drained there is no mode of management, in my opinion, that will render land so soon amenable to the means of putting it in a high degree of fertility as trench-plowing. Mr. Smith himself acknowledges the necessity of trench-plowing land in a rotation or so after the subsoil has been subsoil-plowed, in order to insure to it the greatest degree of fertility.† The experience in trench-plowing after thorough-draining of the Marquis of Tweeddale at Yester, East Lothian, may with great confidence be adduced in favor of the system. I have seen a field on Yester farm under the operation of draining which did not carry a single useful pasture plant, but which afterward admitted of the turnips being drilled across the face of inclining ground, and of presenting to sheep in winter as dry a bed as they could desire; and no farther gone than the spring of 1841, after the Swedish turnip-seed had been sown, a field was trench-plowed with 3 powerful horses in each plow, bringing up white and yellow tilly subsoil as unpromising in appearance as possible. The weather being very dry, this till became so hard that part of the field had to be rolled four times before they were reduced to powder, and after all the operations, there was apparently no sap left in the ground. White turnips were sown, came away, one-half being eaten off by sheep; and when the land was plowed up in spring 1842, it turned up to appearance a fine rich dark mould, rising in friable clods, and not a particle of till to be seen. No one need be afraid to bring up subsoil of any kind on thorough-drained land after the experience at Yester.

(991.) The advocates of subsoil-plowing seem to lay great stress on the laying of ground quite flat after that operation has followed thorough-draining, and of showing no open furrows in the field; because a uniform surface is the best for absorbing the rain, and transmitting it in the purest state to the drains. All this, however, is not peculiar to subsoil-plowing, for trenched land can be so treated if desired. But as to dispensing with open-furrows, the plan savors more of conceit than of possessing real utility. There is no way that has been contrived of plowing land so conveniently as in ridges, a portion of ground being allotted to each plowman, who is responsible for his own work; and the operations of sowing and reaping are easily marked off in equal distances to the work-people;

* Prize Essays of the Highland and Agricultural Society, vol. xii.
† Smith's Remarks on Thorough Draining.

and if in conducting all these operations few open-furrows seem desirable, there is the mode of plowing by two-out-and-two-in, fig. 138, which only leaves one open-furrow in every four ridges, and the ground as flat as you please. But the truth is, that a field cannot be plowed without making an open-furrow, but with either one plow making a feering in the middle and turning over the whole ground; or, if more than one plow is employed, they must follow one another in adjoining furrows—a plan inimical to good plowing, inasmuch as no plowman can hold so steady a furrow as when following up his own method of plowing, and few plows are exactly of the same guage on the furrow-sole; or the land must be plowed with a turn-wrest plow, beginning at one end and finishing at the other of the field plowing the whole of it itself, or followed by others of the same sort; but where such plows are used for such a purpose, other common plows must be provided on the same farm, as land for turnips or potatoes cannot be drilled up with the turn-wrest plows, as you will learn by-and-by.

(992.) The nature of *moor-band pan* is given below, and as to its destruction, although I have not had much experience of its obduracy, any case within my experience not exceeding 2 or 3 inches in thickness, which were easily ripped up with the 4-horse plow, and as easily mouldered down to dust on exposure to the winter's frost; yet there are places, such as in Aberdeenshire and Morayshire, where it is so deep and hard that extraordinary means are required to break it up. A remarkable and extensive band of this substance was encountered by Mr. Roderick Gray, Peterhead, when improving a part of the property of the Governors of the Merchants' Maiden Hospital of Edinburgh in that neighborhood. The moory surface was plowed with 4 horses. "At first the plow ran upon the pan, which it seemed impossible to penetrate; various trials were made, and the plan which ultimately succeeded was to have 4 men employed at the plow, and these were engaged as follows: One with a pick and spade made a hole when necessary, until it reached below the pan, and entered the plow at this hole; another held the plow; the third held down the beam, and kept the plow below the pan; and the fourth took care of the horses. In this way the upper stratum and pan were broken, and afterward they were brought into a sort of mould by the grubber and harrows."* However obdurate this substance may be to break up, it will yield to the air and moulder down into an innocuous powder of sand and gravel; but I should suppose that, after the plow was fairly entered below the crust it would not require to be held down.

(993.) [In describing the simple construction of the subsoil-plow, I shall not go to any length into its history. The implement, as now used, is generally know as Smith's subsoil-plow, having been brought into the present form by Mr. Smith, Deanston Works, who, in the year 1829, exhibited this plow at the Highland and Agricultural Society's Show at Dumfries, and obtained a premium from the Society for his invention and application of this useful implement.†

(994.) There is no doubt that plows, acting on the principle of Mr. Smith's, penetrating into, breaking, and stirring up the subsoil, without bringing it to the surface or mixing it in the first instance with the incumbent soil, have been long known. Mr. Holt, in his View of the Agriculture of the County of Lancaster, rendered in 1794 to the Board of Agriculture, when treating of the plows of that county, says: "Another instrument has been lately introduced, which Mr. Ecleston with propriety calls the *Miner*, which is a plow-share fixed in a strong beam, without mould-boards, and drawn by four or more horses, and follows in the furrow the plow (the common plow) has just made; and, without turning up the substratum, penetrates into and loosens from 8 to 12 inches deeper than the plow has before gone; which operation, besides draining the land, causes the water to carry along with it any vitriolic or other noxious matter by the substratum thus loosened. The roots of plants may penetrate deeper; and, in course of time, that which is but a barren substance may become fertile soil." This is truly the subsoil-plow of Mr. Smith, invented, laid aside, and forgotten for a period of 35 years.

(995.) Recent experience points out the reason why the earlier introduction of the subsoil-plow did not meet with the success which has attended Mr. Smith's, which, from the above description appears to be the same implement, for they appear in the essential parts to be almost exactly

* Prize Essays of the Highland and Agricultural Society, vol. viii.

† Ibid., vol. viii.

alike; yet the one has been lost sight of, while the other has come into all that notice which it deserves. The reason is now obvious: Without the necessary improvement of thorough-draining, subsoil-plowing is thrown away; and though thorough, or at least furrow draining, has been practiced in England for a long period,* the idea of combining the two seems not to have occurred to the agriculturists of that day. To Mr. Smith, therefore, is still due the merit of having brought these two powerful auxiliaries of Agriculture into effective coöperation.

(996.) Since its first appearance in 1829, Mr. Smith's plow has undergone various slight alterations, not affecting, however, its essential character, but chiefly in lightening its construction. The implement at first was made of enormous weight, sometimes so much as 5 cwt., but a few years' experience served to show that all its objects could be achieved with a plow of little more than half that weight; they are accordingly now generally made from 2 to 3 cwt. Fig. 207 represents

Fig. 207.

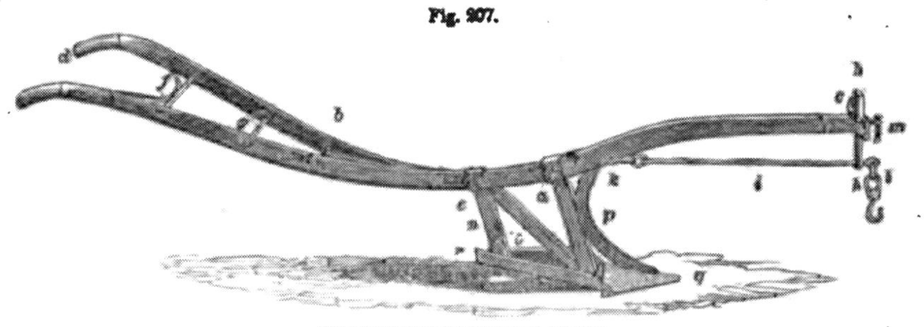

THE DEANSTON SUBSOIL-PLOW.

one of the modifications of the subsoil-plow as now manufactured by James Slight & Co., Edinburgh. It retains all the acting parts of Mr. Smith's without material change, except in weight, though in other respects it deviates slightly from the original. The beam, which is from 3 to 3½ inches deep at the fore sheath or slot, *a*, and 1 to 1½ inches thick, extends from *b* to *c*, a length of 7 feet 4 inches; at *c*, the point, it is diminished to 2½ by ¾ inches, and at *b* to about the same dimensions. The two handles, extending from *d* to *e*, are 6 feet 9 inches in length. They are thinned off at *e*, and bolted, one on each side, to the beam; the depth of the handles is 2 to 2½ inches, and are ⅝ to ¾ inch thick, worked into sockets at *d* in the usual manner for the reception of a wooden helve. The beam and handles are farther connected by stretcher tubes and bolts, the latter passing through all three at *b*, and binding them firmly together; the handles are also farther supported by the stretcher-bolts and bow, *f* and *g*. The beam is mounted at *c* with the bridle, which is at least 2 inches by ¾ inch, bolted on the point of the beam, being first formed into an oblong loop of 8 inches in length, standing at right angles to the beam, and having the opening vertical. To the front part of the loop is fitted a stout clasp, the two arms of which embrace the loop above and below, and admit of the slot *h h* to pass at once through them and the loop. The clasp and slot together have a motion along the loop right and left, and the slot itself has a motion vertically. The chain-bar *i*, is attached to the beam at *k*, and passes through an eye in the lower end of the slot *h*; to the chain-bar is then attached the draught-hook *l*, to which the yoke is applied. The motion above described of the slot *h*, and consequently of the chain-bar and draught-hook, afford ready means of adjusting the earthing and landing of the plow, and the position is retained by means of the pinching-screw *m*, which, by being screwed into the clasp, acts against the outside of the loop, drawing the slot and the loop into firm contact. The body consists of the two slots *a* and *n*, the first about 3 inches broad, the last about 2½ inches, and each ¾ inch thick; they are welded to a sole-bar 2 inches square, and 30 inches long, flush on the land-side. The head of the slots is worked into a kneed palm, which is strongly bolted to the beam, and the diagonal brace *o* is fitted in to resist the strain that tends to derange the form of the body. The coulter-bar *p* is 3 inches broad, ¾ inch thick at the back in the upper parts, becoming thinner downward, and is finished with a blunt edge and point; it is simply held in its place by being tongued into the beam, the fore-slot and the share. The share *q* is made after the same form as that of the common plow, having a feather to the furrow-side, and is spear-pointed. The length of the share is from 14 to 16 inches, and the breadth over the feather about 6 inches. It is fitted upon the prolongation of the sole-bar, and its socket is usually furnished with a short ear, by which it is fixed to the sole-bar to prevent its falling off, as the fixture of the coulter depends upon the share keeping its place. The feather *r* is a thin-edged bar, 3 inches deep and about ½ inch thick, thinned off on the upper edge; it is tapered off at the fore end where it joins the share, and is held in contact by being notched into it; but its chief supports are two palms, by which it is bolted to the sole-bar; and a sole-shoe of cast-iron, having a flange rising 6 inches on the land-side, completes the subsoil-plow, which, with the exception of the sole-shoe, is constructed entirely of malleable iron. The length of the plow over all is about 13 feet; the length of the sole 3 feet 3 inches; the hight of the handles 3 feet 6 inches; and at the point of the beam 2 feet 4 inches.—J. S.]

(997.) [Not much need be said regarding the efficacy of subsoil-plowing. After what I have stated of the immense value of a mixture of impalpable matter, and larger particles, in the form of a porous mass, I need scarcely say that anything capable of increasing the depth to which this porosity extends, must of necessity be advantageous. This, however, does not show any differ-

* See Sir James Graham's observations on the subject in vol. i. of the Journal of the Royal Agricultural Society of England.

ence between subsoil and trench plowing—in my opinion, the latter is the best in most instances, and this for the following reasons: All subsoils require ameliorating by exposure to air, before they are capable of acting beneficially to plants. This is owing to certain chemical changes which are produced by the joint action of air and water, and it is very evident that all these must take place much more rapidly when the subsoil, as in trench-plowing, is laid *upon the surface* of the field, and freely exposed throughout the winter, than when the air is merely admitted mor freely by the subsoil being *broken up* while it still remains *under the surface*. It may be averre that the trench-plow does not go so deep into the soil as the subsoil-plow; but still I cannot hel thinking that notwithstanding this disadvantage (if any such exists), it is in most cases the mo advisable of the two methods, if employed for *deepening* the soil. Not so, however, if used t assist in *draining the subsoil*. To prove its value for this purpose, I would earnestly direct you attention to the following valuable remarks of Professor Johnston: "The subsoil-plow is an *aux iliary* to the drain; in very stiff clay subsoils it is most advantageous in loosening the under lay ers of clay, and allowing the water to find a ready escape downward, and to either side until it reach the drains. It is well known that if a piece of stiff clay be cut into the shape of a brick, and then allowed to dry, it will contract and harden—cut up *while wet*, it will only be divided into so many pieces, each of which will harden when dry, or the whole of which will again attach themselves, and stick together if exposed to pressure. But tear it asunder *when dry*, and it will fall into many pieces, will more or less crumble, and will readily admit the air into its inner parts. So it is with a clay subsoil. After the land is provided with drains, the subsoil being very retentive, the subsoil-plow is used to open it up—to *let out the water*, and to let in the air. If this is not done, the stiff under-clay will contract and *bake as it dries*, but it will neither sufficiently admit the air"—nor let out the water—"nor open a free passage for the roots. But let this operation be performed when the clay is still too wet, a good effect will follow in the first instance; but after a while the cut clay will again *cohere*, and the farmer will pronounce subsoiling to be a useless expense *upon his land*. Defer the use of the subsoil plow till the clay is dry—it will then *tear* and *break* instead of *cutting*, and the openness will remain. Once give the air free access, and it, after a time, so modifies the drained clay, that it has no longer an equal tendency to cohere. Mr. Smith of Deanston very judiciously recommends that the subsoil-plow should *never* be used till at least *a year* after the land has been thoroughly drained. To attain those benefits which attend the adoption of improved methods of culture, . . . let the practical man make his trial *in the ways and with the precautions recommended by the author of the method*, before he pronounce its condemnation."* Thus you perceive that subsoil-plowing, *when properly performed*, will always be found useful in assisting the action of *drains*, but cannot be considered equal to deep or trench plowing, if an alteration is desired in the *depth of the soil*.

(998.) Another alleged advantage of subsoiling is the breaking in pieces the *moor-band pan*. I will therefore now say a few words respecting this enemy to good farmers. This ferruginous deposit which so frequently occurs in particular localities between the soil and subsoil is extremely hard and compact, and almost completely impermeable to water. Very much has been written concerning this substance, by persons who have but little knowledge of chemistry, and in their endeavors to prove the manner in which the deposit had been produced, and likewise the cause of its injurious action upon vegetation when newly brought to the surface, have made so many chemical errors that the whole subject appears at first sight wrapped in doubt, whereas, we believe that for all practical purposes its nature is already sufficiently well known.

(999.) Moor-band pan belongs to a class of bodies known to chemists under the name of *ochrey deposits*. These deposits, which so frequently occur in the beds of chalybeate springs, were carefully examined by Berzelius in 1832, and were found to consist of the two oxides of iron i chemical combination with two new organic acids, which he denominated the *crenic* and *apocrenic acids*. Feeling certain from various circumstances, that moor-band pan belonged to this class, I undertook an analysis to ascertain whether it contained these acids, and find that in each of two specimens of pan, sent to me for the purpose, there exists a large proportion of *crenic acid*, in one *apocrenic* also, and in the other *humic acid*; there can therefore be no longer any doubt about the composition of this substance; and instead of attempting to prove its injurious effects by relating the difference between the protoxide and peroxide of iron, and the fact of the peroxide being generally combined with water forming *hydrate*, none of which facts throw the least light upon the subject, we can readily explain all by reference to the chemical properties of the compounds of these two acids with iron. It is well known that iron *in solution* acts injuriously upon vegetation; and Berzelius has shown that the *crenate* and *apocrenate* of the protoxide of iron are both *soluble in water*; and that the same salts of the peroxide, although of themselves insoluble, are easily rendered so by ammonia, which substance is always produced in fertile soil; it follows, therefore, that moor-band pan must continue injurious to vegetation so long as the *crenates* and *apocrenates* of iron remain undecomposed. In the course of time, various chemical changes are effected by the joint action of air and moisture, which decompose these compounds, and give rise to new ones having no injurious effect upon vegetation.—H. R. M.]

* Johnston's Elements of Agricultural Chemistry.

29. DRAWING AND STORING TURNIPS, MANGEL-WURZEL, CABBAGE, CARROTS AND PARSNIPS.*

"Beneath dread Winter's level sheets of snow,
The sweet nutritious *Turnip* deigns to grow."
BLOOMFIELD.

(1000.) The *treatment of live-stock* receives early attention among the farm operations of winter; and whether they or land get the precedence depends entirely on the circumstance of the harvest having been completed late or early. If the harvest have been got through early, there is ample time to plow a large portion of stubble-land, in preparation of green crops in spring, before winter quarters are required to be provided for stock; but should it occupy all hands until a late period—that is, until the pas-

[* There are those who think, or affect to think, that nothing of practical Agriculture is to be learned, forsooth, by *reading*; while another class would maintain that it is all foolishness to read anything relating to the subject, except what may have a direct, immediate bearing on the objects and course of culture in which the reader is himself personally engaged. Such men treat as ridiculous the idea that a farmer should indulge in any curiosity about the crops which serve to make up the wealth, and the course of husbandry that constitutes the field practice of other States of their own country, much more those of foreign countries. Now we confess not to be, or to feel any ambition to be, a member of either of these classes. We confess to being well persuaded that *to books* we owe the creation and the spread of knowledge; and that the farmer who entertains right notions of his own respectability, and true position in society, will surely desire to be familiar with the natural and commercial history of *all the great staples* that serve to employ the industry of mankind, not only in his own but in all countries. There is no reason, that is not insulting and derogatory, why the agriculturist should not possess as general information as he of any other business or pursuit; and, if even there were, it does not follow that he should not possess information as to all the great branches and products of his own profession, even though some of them may never come within the range of his own cultivation, and may even be forbidden, by considerations of climate or other circumstances, from being produced in his own State or country.

Take, for example, this chapter on *Turnips*, and the uses made of them in England. Every one at all acquainted with agricultural literature knows of how comparatively recent date is the introduction of them, especially in a *field* crop in England, even as late as the end of the seventeenth century; yet he who has any pretensions to familiarity with the industrial resources of the nation from which we sprung, and with which we have the most extensive and important relations, must know that the spread of this single root has had an influence beyond calculation, on the wealth and industry and power of Great Britain. With no great degree of exaggeration it has been said that her national power has its *root in the turnip!* Now suppose an American traveler, especially an agricultural one, through England, to return without having made any observation as to the culture and uses of a crop thus influencing the destiny of a great nation! What would be said of such a dolt? Since, then, all cannot travel who may desire it, is it not the great province of *letters*—of *books*—to take the next place, and do the office of traveling? Too well do we know that there are those who will argue that because *we* do not and cannot rely on turnips, as the English do, to rear and fatten sheep and cattle—that climate, dearness of labor, want of capital to manure the land sufficiently, and want of the moisture indispensable to bring forward this crop, as well as the possession of Indian corn and other crops better adapted to our climate and purposes, all forbid its culture to an extent sufficient to render it a great national object, all going to interdict turnip culture; *therefore* it is out of place to admit a full exposition of this branch of English husbandry into an American work devoted to American Agriculture. The same cavilers might object that because turnips may never be with us a great *staple crop*, therefore it was superfluous and ill judged to occupy, in the last number, the little space which served to inform the reader that a good crop of English turnips, say twenty tons to the acre,

tures have failed to supply stock with the requisite quantity of food—provision for their support should be made in the steading in preference to plowing land. The usual occurrence is, that the harvest is entirely completed before the failure of the pasture; and, accordingly, I have described the methods of plowing the land before taking up the subject of winter treatment of live-stock; and in doing so, have included the plowing of *lea* after that of stubble-ground, in order to keep all the particulars of winter-plowing together, although the usual occurrence is, that the live-stock are snugly housed in the steading, and the stubble nearly all turned over, before the plowing of lea is commenced, unless there happen to be an old piece of lea to plow on strong land, in which case it should be turned over before the setting in of the winter's frost.

(1001.) Sheep always occupying the fields, according to the practice of this country, the only varieties of stock requiring accommodation in the steading in winter are cattle and horses. The horses consist chiefly of those employed in draught, which have their stable always at hand, and any young horses besides that are reared on the farm. Of the cattle, the cows are housed in the byre at night for some time before the rest of the

would extract and carry off 4,500 pounds of starch and sugar, 540 pounds of gluten, and 45 of oil. We are altogether—with all due deference to such wise and practical men—of a different opinion; and the question is, to which of the classes we have named does the reader belong? In *this* matter we go in for the "largest liberty." We would open wide to the mind of the young agriculturist the *whole field of practical Agriculture in every country*. If he is denied the pleasure and benefit of traveling, in person, to enlarge his mind by enlarged observation, there is the greater reason that he should *travel in books*. For what else was that enlightened observer. Mr. COLMAN, invited to visit and give us his "Personal Observations of European Agriculture," we would like to know? And what reader of any taste or ambition for knowledge would not rejoice as he reads them, to have been his *compagnon de voyage?* These are the views under which we have published, and shall continue to publish, much that may never be put in practice, or that is not practicable in our country exactly in the way, either in detail or extent, that it is done in others. We even think that although a man might not make a pound or a barrel more, for example, of apples, or beets, or barley, or sugar, yet that if our country schools were patronized and conducted as they ought to be, and provided with masters as enlightened, because as *well paid and as much honored*, as *professors of the military art*, not a boy would leave school without some knowledge of the native country and the constituent qualities and habits of these and of all other plants and animals. A little insight gained when young, would plant in him an appetite that would prompt to farther inquiry all his life, and the love of reading and research would take the place of sensuality and dissipation that idleness and even leisure moments always engender in men whose vacant minds are insensible to all thirst for knowledge or intellectual recreation. After all, there is one consideration from which we may take some comfort, even under the mortification of differing from wiser heads, whom we would fain not only please but oblige in all things. No one is obliged to read that for which he has no taste, and from which he may think no advantage *(money?)* is to be derived; and so, in this case, he who has no curiosity to know how the great branches of turnip and sheep husbandry are connected and carried on, where both constitute great items in the resources of our great and powerful mother country, may pass over all that follows on these subjects, to THE MONTHLY JOURNAL OF AGRICULTURE, where fifty pages of matter await him, in such variety as that he must be hard to please if he cannot find something sufficiently practical and enlightened, were it *even on sheep*, in the able letters of our friend Mr. RANDALL, which promise to make readers the least familiar *au courant* of that subject.

After all, we should be perfectly willing to leave it to the decision of any enlightened and liberal minded reader to say whether even this chapter on turnips and sheep feeding does not possess much intrinsic value and convey information both interesting and practically useful for all agricultural inquirers, except those great would-be monopolists of knowledge, who conceit themselves to be, in the agricultural, what the sun is in the natural world, the great fountain of light from which alone the least ray of information is to be derived.

With these explanations we shall be content for the future. *Ed. Farm. Lib.*

cattle are brought into the steading, in case the coldness of the autumnal dews and frosts should injure their milking properties; so that it is only the younger and feeding cattle that have to be accommodated, and of these the feeding are generally housed before the younger stock, which usually get leave to wander about the fields as long as they can pick up any food. I am only here describing what is the common practice, without remarking whether it is a good or bad one, as the whole subject of the treatment of cattle will very soon engage our attention.

(1002.) By the time the cattle are ready to occupy the steading, *turnips* should be provided for them as their ordinary food, and the supply at all times sufficient; and it should be provided in this way: The lambs of last spring, and the ewes which have been drafted from the flock as being too old or otherwise unfit to breed from any longer, are fed on turnips on the ground in winter, to be sold off fat in spring. The portion of the turnip-ground allotted sheep is prepared for their reception in a peculiar manner, by being *drawn or stripped*, that is, a certain proportion of the turnips is left on the ground for the use of the sheep, and the other is carried away to the steading to be consumed by the cattle. The reason for stripping turnips is to supply food to the sheep in the most convenient form, and, at the same time, enrich the ground for the succeeding crops by their dung, which is applied in such quantity as to prevent the ground being manured beyond what would be proper for the perfect development of the future crops; for it has been found that, were an entire good crop of turnips consumed on the ground, the yield of *corn* would be scanty and ill-filled. The usual proportion drawn, if a good crop, is $\frac{1}{2}$, but should the soil be in low condition, $\frac{1}{3}$ only is taken away, and should it be in fine condition, $\frac{2}{3}$ or even $\frac{3}{4}$ may be drawn; but, on the other hand, the quantity drawn is dependent upon the bulk of the crop. If the crop is very large, and the ground in very fine condition, $\frac{2}{3}$ may be drawn, but it is rarely the case that the soil is so rich and the crop so large as to make $\frac{1}{2}$ too great a proportion to be left to be consumed. If the crop is poor, $\frac{1}{3}$ only should be drawn, and a very poor crop should be wholly eaten on, whatever condition the soil may be in. There is another consideration which materially affects the quantity to be left on the ground, which is the occurrence of a poor crop of turnips over the whole farm. Hitherto I have only been speaking of that part of the crop of turnips which is to be appropriated to the use of the sheep, but when the entire crop is bad, that is, insufficient to maintain all the stock fully, then the proportion to be consumed by the sheep and cattle respectively, should be determined at the commencement and maintained throughout the season, that neither class of stock may receive undue advantage. In such a case, it is evident that neither the sheep nor cattle can be fattened on turnips; and other expedients must be resorted to to fatten them, such as either the sheep or cattle should get as many turnips as will feed them, and the other be fed on extraneous matter, or both classes of stock be left in lean condition. When foreign matters for feeding—such as oil-cake—can be procured, the cattle should get the largest quantity of them, and the sheep the largest portion of the turnips; because oil-cake can be more easily administered at the steading than turnips, and sheep, saving the trouble of manuring the ground afterward, can more easily be supplied with turnips. Thus, then, considerations of the state of soil and crop are required to determine the proportion of the turnip crop that should be drawn; but the standard proportion is $\frac{1}{2}$, and when that is deviated from it should only be from very urgent circumstances, such as those alluded to above.

(1003.) Fig. 208 shows how turnips are stripped in the various proportions noticed above. When ½ is drawn, it can be done in various ways, but each not alike beneficial to the land; for example, it can be done by leaving 2 drills *a* and taking away 2 drills *b*; or by taking away 3 drills *e* and leaving 3 drills *f*; or by taking away 6 drills *i* and leaving 6 drills *h*;

Fig. 208.

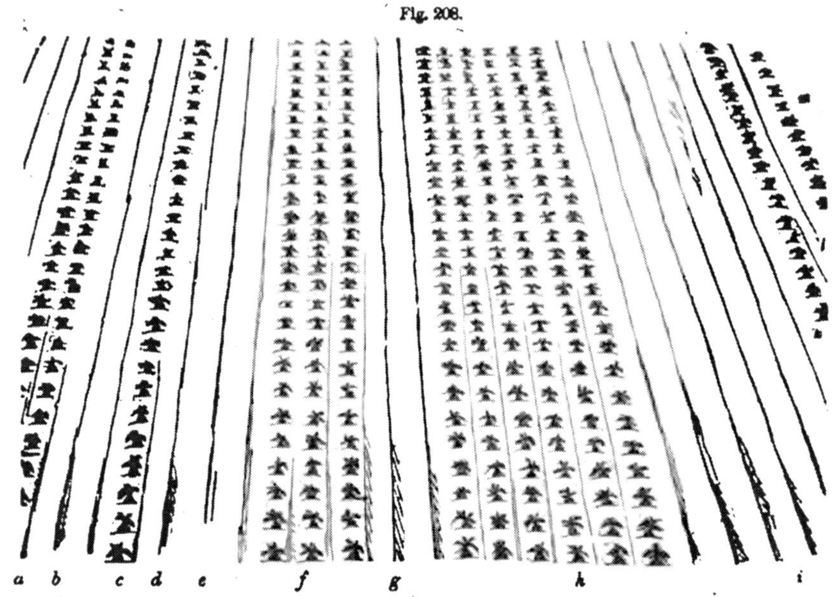

THE METHODS OF STRIPPING THE GROUND OF TURNIPS IN ANY GIVEN PROPORTIONS

or by taking away 1 drill *l* and leaving 1 drill *k*; and so on in every other proportion. Though the same result is attained in all these different ways, in so far as the turnips are concerned, there are cogent reasons against them all except the one which leaves 2 drills *a* and takes away 2 drills *b*; because, when 1 drill only is left, as at *l*, the sheep have not room to stand and lie down with ease between *k* and *m*, without interfering with the turnips, and, beside, sufficient room is not left for horses and cart to pass along *l*, without injuring the turnips on either side of the horses' feet or the cart-wheels; whereas, when 2 or more drills are pulled, as at *b*, and only 2 left, as at *a*, the sheep have room to stand and eat on either side of the turnips, and the cart can pass easily along *b* without injuring the turnips; that is, the horse walks up the center hollow of the drills, and a wheel occupies a hollow on each side. Again, when 3 drills are left, as at *f*, and 3 taken away, as at *e*, the sheep injure the turnips of the two outside rows to reach the middle one at *f*; and much more will they injure those at *h*, when 6 drills are left; and there is, besides, this serious objection to this latter mode, that when practiced on light soils it is observed that the succeeding grain crop is never so good on the ground that has been cleared as where the turnips are left. When other proportions are determined on, ⅓ may be easily left, by pulling 2 drills, as at *b*, and leaving 1, as at *c*; or ¼ may be left, by pulling 3 drills, as at *e*, and leaving 1, as at *c*; or ⅗ may be left, by pulling 2, as at *g*, and leaving 3, as at *f*. There are thus various ways in which the same and different proportions of turnips may be pulled and left on the ground; but in whatever proportion they may be taken, the rule of leaving 2 empty drills for the horses and carts to pass along without injury to the turnips, should never be violated.

(808)

(1004.) But the convenience and propriety of the plan of leaving 2 and taking 2 drills, when the ½ of the crop is to be eaten on, will be best appreciated in witnessing the mode of doing it, as shown in fig. 209, where the drills are represented on a larger scale than in the preceding figure.—One field-worker, being a woman, clears the 2 drills at *a*, and another simultaneously the other 2 at *b*; and in clearing these 4 drills, the turnips are thrown into heaps at regular distances, as at *c* and *d*, among the standing turnips of the 2 drills *e* and *f*, to the right of one woman and to the left of the other; and thus every alternate 2 drills left unpulled become the receptacle of the turnips pulled by every 2 women. The cart then passes along *a* or *b*, without touching the turnips either in *e* or *g* on the one hand, or in *f* and *h* on the other, and it clears away the heaps in the line of *c d*. In the cut the turnips are represented thinner on the ground

Fig. 209.

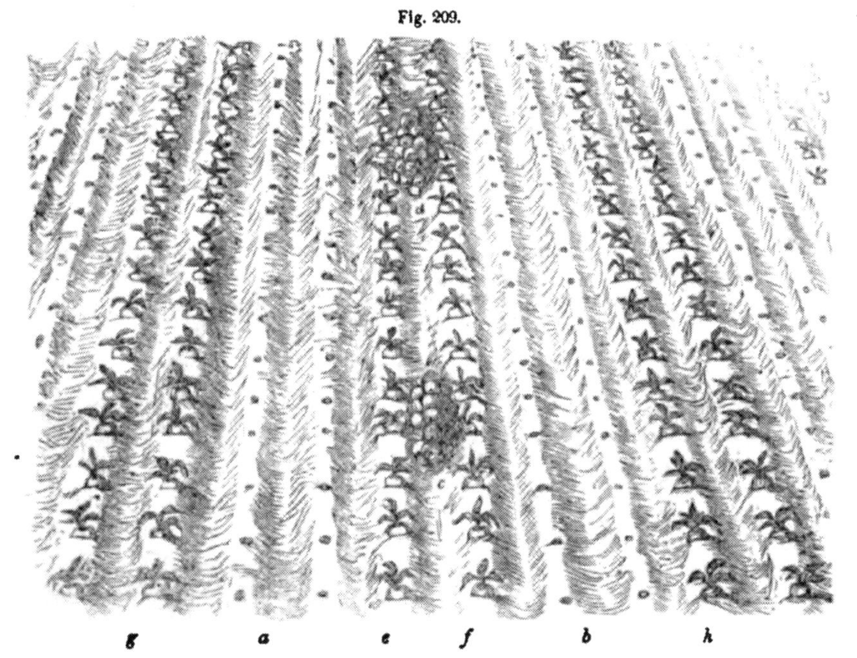

THE METHOD OF PULLING TURNIPS IN PREPARATION FOR STORING.

than they usually are, but the size of the bulb in proportion to the width of the drills is preserved both in the drills and in the heaps. The seats of the pulled turnips are shown upon the bared drills.

(1005.) The usual state in which turnips are thus placed in these temporary heaps, *c* and *d*, is with their tops on, but the tails are generally taken away. The most cleanly state, however, for the turnips themselves, and the most nutritious for cattle, is to deprive them of both *tops* and *tails*.—Many, and indeed I may say most farmers are impressed with the idea that tops of turnips make good feeding at the beginning of the season and especially for young beasts. The notion is quite a mistaken one, in regard to the feeding qualities of tops at any season, for there is really no *such* property in them. No doubt at that season they contain a large quantity of watery juice, which makes cattle devour them with avidity on coming into the steading off bare pasture, and they will even be eaten off before the turnips themselves are touched, when both are presented together; but observation and experience confirm me in the opinion that the time bestowed by cattle in consuming the turnip-tops is worse than so much valuable time thrown away; inasmuch as, in their cleanest state, tops are apt to produce a looseness in the bowels, arising partly, perhaps, from the sudden change of food from grass to such a succulent vegetable; and the complaint is much aggravated by the dirty, wetted, or frosted state in which they are usually given to beasts. This looseness never fails to bring down the condition of cattle so much that a considerable part of the winter passes away before they entirely recover from the shock which their system has thus received. Like my neighbors, I was impressed with the economic idea of using turnip-tops—and I believe it is solely as regards economy, rather than a conviction of their utility, that prompts farmers to continue their use—but their weakening effects upon cattle, especially young ones, caused me to desist from their use; and fortunate was the resolution, for ever after their abandonment my cattle throve better and the tops, after all, were not thrown away, as they served to assist the manuring of the field on which they had grown. I have no hesitation, therefore, in recommending you to deprive the turnips of both tops and tails before carrying them to the steading for the use of cattle. Sheep are not so easily injured by them as cattle, on account, perhaps, of their costive habit; and perhaps in spring, when turnips are naturally less juicy, tops might be of service to them as a gentle aperient, but then, when they might be most useful, they are the most scanty and fibrous.

(1006.) The tops and tails of turnips are easily removed by means of a very simple instrument. Figs. 210 and 211 represent these instruments, fig. 210 being formed from a portion of an old scythe reaping-hook, with a piece of the point broken off. This is a light instrument, and answers the purpose pretty well; but fig. 211 is still better. It is made of the point of a worn patent scythe, the very point being broken off, and the iron back to which the blade is riveted is driven into a helve, provided with a ferule around the end next the blade. This is rather heavier than the other instrument, and on that account removes the top more easily.

Fig. 210.

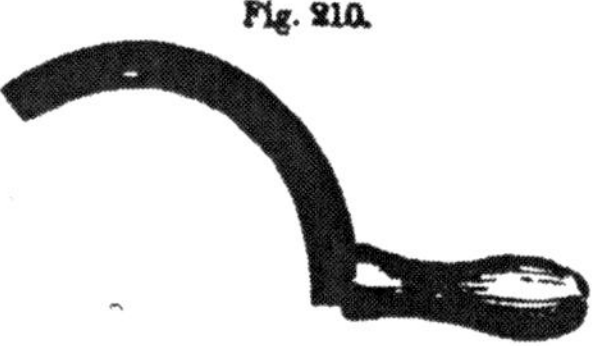

INSTRUMENT FOR TOPPING AND TAILING TURNIPS, MADE OF PART OF AN OLD SCYTHE REAPING-HOOK.

Fig. 211.

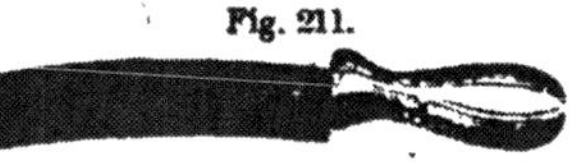

ANOTHER INSTRUMENT FOR THE SAME PURPOSE, MADE OF A PIECE OF OLD PATENT SCYTHE.

(1007.) The mode of using these instruments in the removal of the tops and tails of turnips is this. The field-worker moves along between the two drills of turnips which are to be drawn, as from *a*, fig. 209, and pulling a turnip with the left hand by the top from either drill, holds the bulb in a horizontal position, as represented in fig. 212, over and between the drills *e* and *f*, fig. 209, and with the hook or knife described above (1006), first takes off the root at *b* with a small stroke, and then cuts off the top at *a*, between the turnip and the hand, with a sharper one, on which the turnip falls down into the heap *c* or *d*, whichever is forming at the time. Thus, pulling one or two turnips from one drill, and then as many from the other, the two drills are cleared to the extent desired. Another field-worker acts as a companion to this one, by going up *b*, pulling the turnips from the drills on either side of her, and dropping them, topped and tailed, into the same heaps as her companion. The tops are scattered over the cleared ground. A left and a right-handed field-worker get on best together at this work.

(1008.) Due care is requisite, on removing the tops and tails, that none of the bulb be cut by the instrument, as the juice of the turnip will exude through the incision. Of course, when turnips are to be consumed immediately, this precaution is less necessary; but the habit of slicing off a part or hacking the skin of the bulb indicates carelessness, and should be avoided at all times.

Fig. 212.

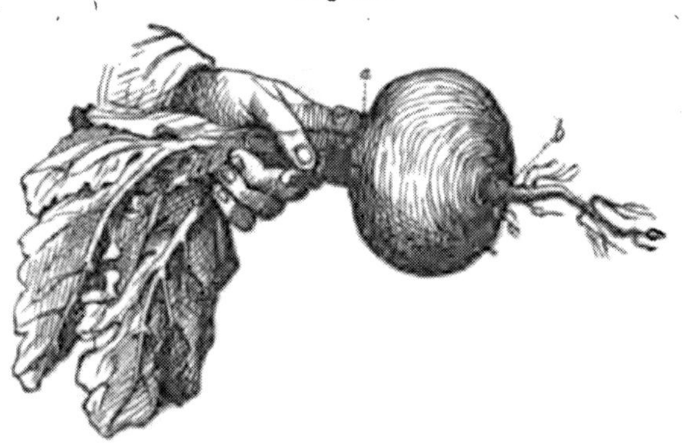

MODE OF TOPPING AND TAILING TURNIPS.

(1009.) When $\frac{2}{3}$ of the turnips are drawn and $\frac{1}{3}$ left, the field-worker goes up at *b*, fig. 208, and, pulling the 2 drills there, drops the prepared turnips between *c* and *d*, beyond the drill *c* that is left. When $\frac{3}{4}$ are pulled, as at *e*, and $\frac{1}{4}$ left on the ground, as at *c*, the turnips may still be dropped in the same place between *c* and *d*, the field-worker pulling all the 3 drills herself, and the horse walking along from *e* when taking them away. When 3 drills are pulled, as at *e*, and 3 left, as at *f*, which is not so good a plan of leaving the $\frac{1}{3}$ as the 2 and 2 I have described before (1003), the same field-worker pulls all the 3 drills, and drops the turnips along the outside row next herself of those that are left. When $\frac{3}{5}$ are left, as at *f*, and $\frac{2}{5}$ pulled, as at *g*, the field-worker goes up, pulling the 2 drills there, and dropping the turnips between the two rows next her of *f*. When 6 drills are pulled, as at *i*, which is not a good plan for leaving the $\frac{1}{2}$, 3 women work abreast, each pulling 2 drills, and all three drop the turnips into the same heap before the woman in the middle. This plan has the sole advantage of collecting a large quantity of turnips in one place and causing little carting upon the land. When the field is intended to be entirely cleared of turnips, the clearance is begun at the side nearest the gate, and carried regularly on from top to bottom of the field, the nearest part of the crop being taken when the weather is least favorable and the farthest when most so.

(1010.) These last remarks remind me of mentioning that when a field is begun to be stripped for sheep, that part should be chosen which will afford them shelter whenever the weather proves inauspicious. A plantation, a good hedge, a bank sloping to the south, or one in the direction opposite to that from which winds most prevail in the locality, or any marked inequality in the form of the ground, will afford shelter to sheep in case of necessity. On the sheep clearing this part first, it will always be ready for a place of refuge should it be required for protection against a storm. The utility of such shelter you shall be made acquainted with very soon.

(1011.) On removing prepared turnips from the land, the carts should be filled by the field-workers, as many being employed as to keep the carts going, that is, to have one filled by the time another approaches the place of work in the field. If there are more field-workers than will be required to do this, the remainder should be employed in topping and tailing. The topped and tailed turnips should be thrown into the carts by the hand, and not pricked by means of forks or graips; the cart should be placed alongside the drill near two or more heaps; and the carter should manage the horses and assist in the filling, until the turnips rise so high in the cart as to require from him a little adjustment in heaping, to prevent their falling off in the journey.

(1012.) As it is scarcely probable that your field-workers will be so numerous as to top and tail and assist in filling at the same time, so as to keep even 2 carts at work, it will be necessary for them to begin the pulling so much sooner, whether one yoking, or a whole day, or two days, but so much sooner, according to the bulk of the crop, as to keep the carts going when they begin to drive away the turnips; for it at all times implies bad management to let horses wait longer in the field than the time occupied in filling the cart. And yet how common an occurrence it is to see horses waiting until the turnips are pulled and tailed and thrown into the cart by perhaps only 2 women, the carter building them up not as fast as he can get them, but as slow as he can induce the women to give them! The driving away should not commence at all until there is sufficient quantity of turnips prepared to employ at least two carts one yoking; and, on the other hand, care should be taken not to allow more turnips than will employ that number of carts for that time to lie upon the ground before being carried away, in case frost or rain should prevent the carts entering the field as long as to endanger the quality of the turnip.

(1013.) Dry weather snould be chosen for the pulling of turnips, not only for the sake of cleanliness to the turnips themselves, but for the sake of the land, which should be cut up and poached by cart-wheels and horses' feet as little as possible; because, when land is much cut up in carrying away turnips, sheep have a very uncomfortable lair, the ruts forming ready receptacles for water, and are not soon emptied. No doubt thorough-draining assists to make land proof against such a condition; but let the land be ever so well drained, its nature cannot thereby be entirely changed—clay will always have a tendency to retain water on its surface and soil everything that touches it, when wetted by recent rains; and deep loam and black mould will still be penetrated by horses' hoofs, and rise in large masses with wheels immediately after rain. No turnips should therefore be led off fields consisting of these sorts of soils, however well drained, immediately after or during severe rain; nor should they be pulled at all, until the ground has again become consolidated.

(1014.) In commencing the pulling of turnips, one of the fields intended to support sheep should first be taken, in order to prepare space for them; and this is done while all the stock are engaged on pasture, which should not be bared too much, in case the sheep that are to be fed off on turnips fall off in condition upon it.

(1015.) Should the weather prove unfavorable at the beginning of the season—that is, too wet or too frosty—there should no more turnips be pulled and carried than will suffice for the daily consumption of the cattle in the steading; but whenever the ground is dry and firm and the air fresh, no opportunity should be neglected except from other more important operations—such as the wheat-seed—of storing as large a quantity as the time will permit, to be used when the weather proves interruptive to field operations. This is a very important matter, and, as I conceive, much neglected by most farmers, who too frequently place their cattle from hand to mouth for food. A very common practice is to employ one or two carts an afternoon's yoking, to bring in as many turnips as will serve the cattle for two or three days at most, and these are brought in with the tops on, after much time has been spent in the field in waiting for the pulling and tailing of the turnips. This slovenly mode of providing provender for cattle should be abandoned. It should be considered a work of the first importance in winter to provide cattle with turnips in the very best condition, independent of the vicissitudes of the weather; and this can only be done by storing a considerable quantity of them in good weather, to be used when the weather changes to a worse state. When a store is once made, the mind becomes easy under the certainty of having, let the weather prove ever so unpropitious, plenty of good food provided at home for the cattle, and having such a provision does not prevent you taking supplies from the field as long as the weather permits the ground to be carted upon with impunity, to be immediately consumed or to augment the store. How much better for all parties—for yourself, for men, horses and cattle—to be always provided with plenty of turnips, instead of being obliged to go to the field for every day's supply, and perhaps under the most uncomfortable circumstances! I believe few farmers would refuse their assent to this truth; and yet, how many violate it in their own practice! The excuses usually made for pursuing the ordinary practice are, that there is no time to store turnips when the potato-land should be plowed up and sown with wheat; that the beasts are yet doing well enough upon the pasture; and that it is a pity to pull the turnips while they continue to grow. It is proper to bestow all the time required to plow and sow the potato-land; and, after a late harvest, these may have to be done after the pasture has failed; but such an occurrence as the last being the exception to the usual condition of the crops and seasons, ought not to be adduced as an excuse applicable at all times; and as to the other excuses, founded upon the growing state of the turnips and the rough state of the pastures, they are of no force when adduced in compensation for the risk of loss likely to be incurred by a low condition in the stock. Rather than incur such a risk, give up the rough pasture to the sheep, or delay the working and sowing of the potato-land, or sacrifice a portion of the weight of a small part of the turnip crop by pulling it before reaching entire maturity. As for sheep, they are never at a loss for food, being constantly surrounded with turnips as long as the ground is bare.

(1016.) The *storing* of turnips is very well done in this way: Let a piece of lea ground, convenient of access to carts, be chosen near the steading for the site of the store, and if that be in an adjoining field, on a 15-feet ridge, so much the better, provided the ridge runs N. and S. Fig. 213 represents the form of the turnip-store. The cart with the topped and tailed turnips is backed to the spot of the ridge chosen to begin the store, and there emptied of its contents. The ridge being 15 feet wide, the store should not exceed 10 feet wide at the bottom, to allow a space of at least 2½ feet on each side toward the open furrow of the ridge, for the fall and conveyance of water. The turnips may be piled up to the hight of 4 feet, but will not easily lie to 5 feet on that width of base. In this way, the store may be formed of any length; but it is more desirable to make two or three stores on adjoining ridges than a very long one on the same ridge, as its farthest end may be too far removed for using a wheel-barrow to remove the stored turnips. Assorted straw, that is, drawn out lengthwise, is put from 4 to 6 inches thick above the turnips for thatch, and kept down by means of straw-ropes, arranged lozenge-shaped, and fastened to pegs driven in a slanting direction in the ground, along the base of the straw, as may be distinctly seen in the figure. Or a spading of earth, taken out of the furrow, may be placed upon the ends of the ropes to keep them down. The straw is not intended to keep out either rain or air—for both are requisite to preserve the turnips fresh—but to protect them from frost,

Fig. 213.

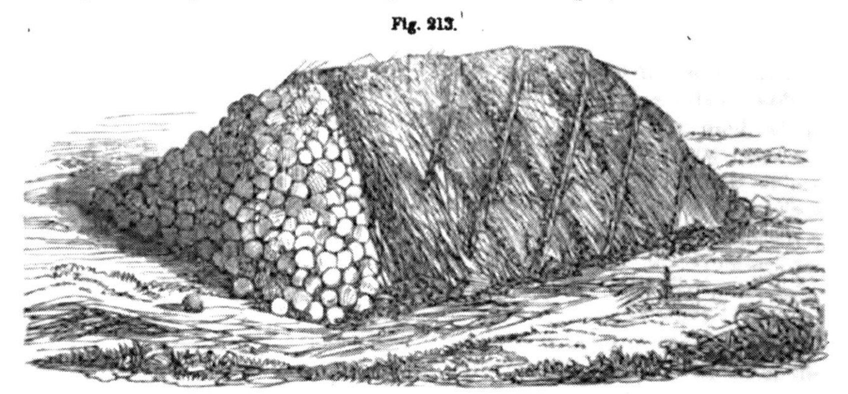

THE TRIANGULAR TURNIP-STORE.

which causes rottenness, and from drouth, which shrivels turnips. To avoid frost the end, and not the side, of the store should be presented to the north, whence frost may be expected to come. If the ground chosen is so flat, and the open furrows are so nearly on a level with the ridges as that a dash of rain would overflow the bottom of the store, a furrow-slice should, in that case, be taken out of the open furrows of the ridges with the plow, or a gaw-cut made with the spade, and the earth used to keep down the ropes.

(1017.) When the turnips are to be used from the store, the straw on the south end is removed, as seen in fig. 213, and a cart, or the cattle-man's capacious, light wheel-barrow, backed to it; and, after the requisite quantity for the day has been taken out, it should be replaced over the mouth of the store.

(1018.) Some people evince a desire to place the turnip-store in the stack-yard, on account, perhaps, of the straw; but there is not likely to be sufficient room, especially at the beginning of winter, for the turning of carts in an ordinary-sized stack-yard. I have seen turnips stored up between two stacks, in the early part of the season, but only as a temporary expedient, when there was a scarcity of straw.

(1019.) This is not the only form of store that will preserve turnips fresh and good for a considerable time. I have seen turnips heaped about 3 feet in hight, quite flat on the top, and covered with loose straw, keep very well. Other plans have been devised and tried, such as to pull them from the field in which they have grown, and set them upright with their tops on in another field, in a furrow made with the plow, and then cover the bulbs with the next furrow-slice; and another plan is to pull the turnips as in the former case, and carry them to a bare or lea field, and set them upright beside one another, as close as they can stand, with their tops and roots on. No doubt both these latter plans will keep turnips fresh enough, and an area of 1 acre will, by these methods, contain the growth of 4 or 5 acres of the field in which they had grown; but turnips are certainly not so secure from frost in those positions as in a store; and after the trouble of lifting and carrying them has been incurred, it would be as easy to take them to a proper store at once, where they would be near at hand, and save the farther trouble of bringing them again from the second field. And even if they were so set in a field adjoining the steading, they would occupy a much larger space than any store. Objectionable as these plans are, compared to triangular or flat-topped stores, they are better than storing turnips in houses, where they never fail to sprout on the top and become rotten at the bottom of the bin. Piling them against a high wall, and thatching them like a to-fall, preserves them very little better than in an outhouse. Stored in close houses, turnips never fail to rot at the bottom of the heap; and the heat engendered thereby not only endangers the rest of the heap, but superinduces on its surface a premature vegetation, very exhausting to the substance of the bulbs. Turnips put into pits dug in the ground, and covered with earth, have failed to be preserved. A plan has been recommended to drive stakes 2½ feet high into the ground, and wattle them together with brushwood, making an inclosure of 3 sides, in the interior of which the turnips are packed, and piled up to a point, and thatched, like the store in fig. 213; and the turnips are represented as keeping fresh in such a structure until June; and one advantage attending this plan is said to be, that "where room is rather limited in the rick-yard, one pile of this description will contain 3 times as much as one of those placed on the ground of a triangular shape; and the saving of thatch is also considerable."* But, as it appears to me, the providing of stakes and the trouble of wattling around an inclosure will far more than counterbalance any advantage of space or saving of straw for thatch, compared with the mode I have described in fig. 213; and besides all these inconveniences attending the plan, there is no necessity whatever for having a turnip-store in a rick-yard.

(1020.) With regard to *storing mangel-wurzel*, this plan seems unexceptionable. "It should be stored early in November. The best and cheapest mode is to build it up against some high wall, contiguous to your beasts' sheds, not more than 7 or 8 feet deep, carried up square to a certain hight, and then tapering in a roof to the top of the wall; protect the sides with thatched hurdles, leaving an interval between the roots and the hurdles, which fill up with dry stubble (straw); cover the roof with about 1 foot of the same, and then thatch it, so as to conduct all moisture well over the hurdles placed as a protection to the sides. In pulling the plants, care should be taken that as little injury be inflicted upon them as possible. Cleansing with a knife should on no account be permitted, and it is safer to leave some of the leaf on, than, by cutting it too close to impair the crown of the root. The drier the season is for storing the better; although I have never found the roots decayed in the heap by the earth, which, in wet weather, has been brought from the field adhering to them."† *Carrots* may be stored exactly in the same manner, and so may *parsnips*. *Cabbages* are stored by being soughed into the soil, or hung up by the stems, with the heads downward, in a shed. As cabbages are very exhausting to the soil, the plants should be pulled up by the roots when they are gathered, and the stems not merely cut over with a hook or knife, because they will sprout again.

(1021.) All these modes of storing turnips apply to all the varieties of the root usually cultivated, and which are much more numerous than necessary. Mr. Lawson enumerates and describes no fewer than 46 varieties cultivated in the field; namely, 11 of Swedes, 17 of yellow, and 18 of white,‡ the color names being derived as much from the color of the flesh as that of the skin. One kind from each of these classes seems al-

* Journal of the Royal Agricultural Society of England, vol. ii. † Ibid. vol. ii.
‡ Lawson's Agriculturists' Manual, and Supplement.

most requisite to be cultivated on every farm, although the yellow is omitted in some districts, and the Swedes in others. Where Swedes are omitted, they have never been cultivated, and where the yellow is the favorite, the Swedes are unknown; for where they are known their culture is never relinquished, and their extension is treading hard upon the yellow, and even curtailing the boundary of the white. The white varieties come earliest into use, and will always be esteemed on account of their rapid growth and early maturity, though unable to withstand the severest effects of frost. It is they which first support both cattle and sheep, being ready for use as soon as the pasture fails; and in storing them, only such a quantity should be prepared as will last to the end of the year. The yellows then follow, and last for about 2 months, that is, to the end of February or thereabout; and the same rule of storing a quantity for a specified time is followed in regard to them as with the whites. Then the Swedes finish the course, and should last until the grass is able to support the cattle, that is, to the end of May, or beginning of June, to which time they will continue fresh in store, if stored in proper time and in the manner recommended above; and the *most* proper time for storing them is before any vegetation makes its appearance on them in spring, which is generally about the end of March or beginning of April.

(1022.) Of all the 18 varieties of white turnips, I should say that the White Globe (*Brassica rapa, depressa, alba*, of De Candolle) *a*, fig. 214.

Fig. 214.

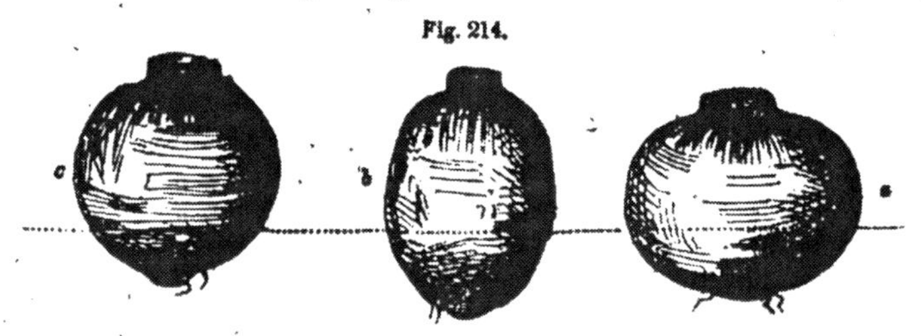

THE WHITE GLOBE TURNIP. THE PURPLE TOP SWEDISH TURNIP. THE ABERDEENSHIRE YELLOW BULLOCK TURNIP.

is the best for early maturity, sweetness, juiciness, size of root, weight of crop, and elegance of form. Its form is nearly globular, as its name indicates; skin smooth, somewhat oily, fine, and perfectly white; neck of the top and tap-root small; leaves long (frequently 18 inches), upright, and luxuriant. Though the root does not feel particularly heavy in the hand, it does not emit a hollow sound when struck; its flesh is somewhat firm, fine-grained, though distinctly exhibiting fibres radiating from the center; the juice easily exudes, and the rind is thin. Its specific gravity was determined by Dr. Skene Keith at 0·840; and its nutritive properties by Sir Humphry Davy, at 42 parts in 1,000; of which were, of mucilage 7, of sugar 34, and of albumen or gluten 1.* Mr. Sinclair mentions this remarkable fact in regard to the white turnip, that the quantity of nutritive matter contained in different roots of the same variety varies according to the size and texture of their substances. Thus, a root of the white-loaf turnip, measuring 7 inches in diameter, afforded only 72½ grains; while the same quantity of a root which measured only 4 inches, afforded 80 grains." So he forms this important conclusion, that "the middle-sized roots of the common turnip are therefore the most nutritious."†

(1023.) I suspect that our crops of white globe turnip ordinarily consist

* Davy's Agricultural Chemistry, edition of 1839.
† Sinclair's Hortus Gramineus Woburnensis, edition of 1824.

of middle-sized bulbs, or they contain many blanks, as the following statement tends to show. Taking the distance between the turnips at 9 inches —being that at which white turnips are usually thinned out—and taking the distance between the drills at 27 inches—the usual one—these distances embrace an area of 243 square inches of ground for each turnip. On each turnip occupying that area, there should be 25,813 turnips per acre imperial; and taking 30 tons per acre as a fair crop, each turnip will weigh nearly 1 lb. 1 oz.! Now, in an ordinary crop of white-globe turnips it is not beyond the truth to take them at 6 inches in diameter overhead; and having the specific gravity of white turnip as mentioned above, a 6-inch turnip should weigh 6 lbs., and the crop of course be, per acre, 69 tons 2 cwt., instead of 30 tons. The inevitable conclusion is, either that blanks, to the enormous extent of being able to contain 39 tons 2 cwt. of turnips per acre, occur in the ordinary crops of white globes—that is, the number on the acre is only 9,445 turnips, instead of 25,813; or the average distance between the turnips may be 20 inches instead of 9. When actual results fall so very far short of anticipation, the important and interesting inquiry arises, Whether the great deficiency is occasioned by the death of plants after the singling process has been completed? or the average size and weight of each turnip are much less than we imagine? or the distance left by the singling is greater than we desire?—or from all these causes combined? From whichever of these causes, singly or combined, the result arises, it is worthy of serious investigation by the farmer; for the bulk of the crop may really depend more on these less obvious circumstances than on the mode of culture. Let us see.

(1024.) Weights and sizes of turnips have already been ascertained with sufficient accuracy. The white globes exhibited at the Show of the Highland and Agricultural Society at Inverness in October, 1839, gave a girth varying 28½ to 34 inches, and a weight varying still more—from 8 lbs. to 15½ lbs. each root; so that 3 roots of the same girth of 30½ inches, varied in weight respectively 8 lbs., 9¾ lbs., and 14½ lbs.* After the statement of these facts, our surprise at realization falling so far short of expectation may be moderated; for we see crops, of apparently the same bulk, weigh differently; and turnips growing on the same field exhibit different fattening properties; and different localities produce turnips of different bulk. Whence arise these various results? These weights are by no means the utmost to which this turnip attains, examples occurring in some seasons of weights from 18 lbs. to 23 lbs.;† and I have pulled one from among Swedes, weighing 29 lbs., including the top.‡ From 30 to 40 tons per imperial acre is a good crop of this kind of turnip.

(1025.) Of the yellow turnip, Mr. Lawson has described 17 varieties, of which perhaps the greatest favorite is the green-top Aberdeen Yellow Bullock (*Brassica rapa, depressa, flavescens*, of De Candolle). This is a good turnip, of the form of an oblate spheroid, as seen at *c*, fig. 214; color of the skin below the ground, as well as of the flesh, a deep yellow orange, and that of the top bright green. The leaves are about 1 foot long, dark green, rather soft, spreading over the bulb, and collected into a small girth at the top of the turnip; the tap-root small. Its specific gravity, as determined by Dr. Keith, is 0·940; and its nutritive property, according to Sinclair, is 44 in 1,000 parts, of which 4¾ are of mucilage, 37¾ of sugar, and 1½ of bitter extract or saline matters. This root feels firm and heavy

* Quarterly Journal of Agriculture, vol. x.

† Lawson's Agriculturist's Manual.

‡ The Norwich Mercury, of July, 1841, makes mention of a turnip—a white one, we presume—exhibited at Fakenham market, and sent from Van Diemen's Land in strong brine, which weighed 84 lbs., having a girth of 5 feet 2 inches. It is said to have weighed 32 lbs. when pulled.

in the hand, with a skin smooth and fine, flesh firm, but not so juicy, nor the rind so thin as the globe.

(1026.) Selected specimens exhibit a circumference of the larger diameter of from 27 to 30 inches, which vary in weight from 6 lbs. to 8½ lbs. each, but specimens may be found weighing as much as from 9 to 11 lbs., and those of the same diameter sometimes show a difference of 1 lb in weight: yellow turnips seldom yield so heavy a crop as either the globe or Swedes, 30 tons the imperial acre being a good crop; but their power of fattening is greater than that of white turnips. In some parts of the kingdom, they are grown in preference to Swedes, especially where light soils predominate; but from my own experience in raising Swedes on the driest gravelly soil, of a superior description to the yellow, I believe that if Swedes always received the sort of culture they require, they would in every soil exceed the yellows in weight and nutrition; and a strong proof of the soundness of this opinion may be found in the rapid inroads which they have of late years made, and are making, upon the confines of the yellows.

(1027.) Of the 18 varieties of the Swedish turnip described by Mr. Lawson, the Purple-Top (*Brassica campestris, napo-brassica, rutabaga*, of De Candolle,) has long obtained the preference, and certainly if weight of crop, nutritious property, and durability of texture are valuable properties in a turnip, none can exceed the Swedes. They are of an oblong form, as seen at *b*, in fig. 214, having the color under ground and of the flesh a deep yellow orange, and the upper part above the ground a dusky purple. The leaves are about 1 foot long, standing nearly upright, of a bluish green color, and growing out of a firm conical base, which forms the neck of the bulb. The skin is somewhat rough, the rind thicker than either of the two former sorts of turnip, and the flesh firm. This turnip feels heavy and very hard in the hand. According to Dr. Keith, the specific gravity of the yellow Swede is 1,035, and of the white 1,022, and Sir Humphry Davy estimates its nutritive property at 64 in 1,000 parts, of which 9 are starch, 51 sugar, 2 gluten, and 2 extract. Dr. Keith states that he found the Swedish turnip heaviest in April, at the shooting out of the new leaves, and that after its flower stem is fairly shot in June, the specific gravity of the root decreases to 0·94, that exactly of the yellow turnip. This fact shows the relative values of those turnips, and also of the time in spring, namely, before April, for storing the Swedes, after which they should not remain in the ground in a growing state. As Sir Humphry experimented on Swedish turnips grown in the neighborhood of London, where they are confessedly inferior to those in the northern counties, his results as to their nutritive properties may be considered below the true mark.

(1028.) Picked specimens have exhibited a girth of from 25 to 28 inches, varying in weight from 7 lbs. to 9½ lbs., but the weight of this, like all other turnips, is not in proportion to the bulk, as a 25-inch one gave a weight of 9½ lbs., while one that measured 26 inches only weighed 7 lbs. It is not an uncommon thing, however, to see them from 8 lbs. to 10½ lbs. A crop of from 16 to 20 tons may be obtained by very ordinary culture, but in the neighborhood of large towns, such as Edinburgh, from 28 to 34 tons are obtained on the imperial acre I have heard of 50 or 60 tons boasted of, but suspect that such weights had been calculated for the whole field from very limited and selected spots; nevertheless, a large and equal crop will sometimes be obtained, under favorable circumstances, for I remember seeing a crop of 50 acres within the policy of Wedderburn, Berwickshire, in 1815, then farmed by Mr. Joseph Tod, Whitelaw, on traversing which I could not detect a single turnip of less apparent size than a man's head. The crop was in no part weighed, but it was let to be consumed by cattle and sheep, the half being eaten off by wether sheep at 6d. a head per week, and realized £21 per imperial acre! Taking a man's head at 7 inches in diameter, and the specific gravity of a Swedish turnip at 1·035, the weight of each turnip should have been 11½ lbs., and taking 19,360 turnips per acre, at 12 inches apart, the distance at which Swedish turnips are singled, and 27 inches wide in the drill, the weight of the crop should have been 99 tons 7 cwt. Taking the calculation in another form, let us see the result of £21 at 6d. a head per week. That gives 32 sheep to the acre, and taking Mr. Curwen's estimate of a sheep eating 24 lbs. a day, exclusive of 4 lbs. a day of waste, for 180 days, or half a year, the weight of crop by this method should have been 61 tons 14 cwt. Statements, however, regarding the quantity of turnips eaten by sheep are various. One given by Sir John Sinclair is a consumption of 21 acres of 44 tons each, by 300 sheep in 180 days, or half a year, which gives 38 lbs. a day for each sheep.* If we take this allowance of 38 lbs. the crop mentioned above should have weighed 85 tons 1 cwt. to have paid £21 per acre! The usual allowance is 16 young sheep to the ordinary acre of 30 tons, which is 23¼ lbs. a day to each, and 10 old sheep, which is 37¼ lbs. to each, and both are probably near the truth; but the exact consumption of food by live-stock is a subject worthy of experimental investigation.

(1029.) The proportion of the top to the root is less in the Swedish than in other sorts of turnips, as evinced in the experiments of Mr. Isaac Everett, South Creake, Norfolk, which, on a crop of Swedes grown at 18 inches, and 27 inches apart in the rows, of an average of 17 tons 9 cwt., gave 3 tons 3 cwt. of tops, on the 15th December, after which they were not worth weighing; and while mentioning this experiment, I may advert to a fact derivable from it, that tops are lighter in a crop raised on drills than one in rows on the flat surface; that is, while, in the above case, 28 tons 8 cwt. of topped and tailed turnips afforded only 5 tons 10 cwt. of tops from the drilled land, those from the rows on the flat surface yielded 6 tons 16 cwt. from a crop very little heavier, namely, 28 tons 16 cwt.

(1030.) The yellow turnip will continue fresh in the store until late in spring, but the Swedes have a superiority in this respect to all other turnips. The most remarkable instance I remember of Swedes keeping in the store, in a fresh state, was in Berwickshire, on the farm of Whitsome Hill, when in the possession of Mr. George Brown, where a field of 25 acres of excellent Swedes was pulled, rooted, and topped, and stored in the manner already described, in fine dry weather in November. This extensive storing was undertaken to have the field sown with wheat. The store was opened in February, and the cattle partook of the turnips and continued to like them until the middle of June, when they were sold fat, the turnips being then only a little sprouted, and somewhat shriveled, but exceedingly sweet to the taste. There is a property possessed by the Swedish turnip which stamps a great value upon it as a root for feeding stock, which is, the larger it grows the greater quantity of nutritive matter it contains. According to Sinclair, 1,728 grains of large-sized Swedes contained 110 grains of nutritive matter, whereas small-sized ones only yielded 99 grains,† affording a sufficient stimulus to the farmer to raise this valuable root to the largest size attainable.

(1031.) A comparative view of the specific gravity and nutritive proper-

* Sinclair's Account of the Husbandry of Scotland, vol. ii., Appendix.

† Sinclair's Hortus Gramineus Woburnensis.

ties of the turnips just described may prove to you both an interesting statement and a memorandum of facts, as far as at present known.

Specific gravity of yellow Swedish turnip in December .... 1·033
It is heaviest in April, about the shooting of the new leaves, and in June, after the development of the flower stalk, it is .... 0·940
Specific gravity of white ditto .... 1·022
.. .. yellow bullock .... 0·940
.. .. white globe .... 0·849
.. .. carrot .... *0·018

Grains.	Mucilage, or Starch.	Saccharine matter.	Gluten.	Extract or saline matter.	Total soluble or nutritive matter.
1,000 of Swedish turnip contain	9	51	2	2	64
.. yellow bullock	4¼	37¾		1¼	44
.. white globe	7	34	1		42
.. mangel-wurzel	13	119	4		136
.. orange-globe	25¾	106¾	1¼	less than 1	135¼
.. sugar-beet	17¾	126¾	1¼	1	146¾
.. field-cabbage	41	24	8		73
.. carrot (red)	3	95		½	98½
.. carrot (white)	2	98		1	105
.. kohl-rabi					60
.. parsnip	9	90			99†

(1032.) A summary of the foregoing results may here be given, thus: A 7-inch diameter of white turnip affords 72½ grains, whereas a 4-inch turnip yields 80 grains of nutritive matter in the same bulk. On the contrary, a large Swedish turnip affords 110 grains, and a small one only 99 grains of nutritive matter in the same bulk. Swedish turnip is superior to cabbage in nutritive matter in the proportion of 110 to 107½; the white turnip inferior in the proportion of 80 to 107½; and carrots superior in the proportion of 187 to 107½. A good crop of Swedes weighs from 30 to 35 tons, of yellow from 30 to 32 tons, and of white globe from 30 to 40 tons the imperial acre. A bushel of turnips weighs from 3 stones to 3 stones 3 lbs. of 14 lbs. to the stone, that is, from 42 to 45 lbs. the bushel. A young sheep eats about 18 lbs., and an old one about 24 lbs. of turnips every day;‡ or, by another authority, a young sheep eats 23 lbs., and an old one 37 lbs. a day. The usual allowance to an ordinary crop of turnips is 20 young Black-faced, or 16 Leicester, and 16 old Black-faced and 10 Leicester, or 1 three-year-old ox to the imperial acre; that is, a young Black-faced will consume about 126 lbs., an old one 168 lbs.; a young Leicester 161 lbs., an old one 259 lbs., and an ox about 1 ton of turnips every week. For sheep a crop of turnips of 30 tons will be required, and one of 26 tons will suffice for an ox during 180 days. In making this last estimate, the state of the crop should be taken into consideration, a crop of small yellow or white turnips, if regular, takes longer time to consume in proportion to the bulk than a crop of larger turnips, but a crop of large Swedish turnips, though apparently thin on the ground, takes a much longer time to be consumed than a thicker crop of small roots. There is *no certainty* in these calculations; at the same time, they are perhaps near enough the truth to enable you to lay on to turnips such a lot of sheep or cattle as will about consume a crop in a given time.

(1033.) The prices of turnips depend almost entirely on the demand of the locality. In the neighborhood of towns they are always high priced, where an ordinary crop of white will fetch £10, of yellow £12, and of Swedes £16 an imperial acre. They are chiefly purchased by milkmen, or cow-feeders, as they are usually called in Scotland. In the country,

* Keith's Agricultural Report of Aberdeenshire.
† Journal of the Royal Agricultural Society of England, vol. ii.; Sinclair's Hortus Gramineus Woburnensis; Davy's Agricultural Chemistry. ‡ Curwen's Agricultural Hints.

about £5 10s. for white, and £8 for Swedish turnips, to be carried off the land, are given; and, when consumed on the ground by sheep, £3 to £5 an acre are considered a fair price; and, when on the premises by cattle, £5 for white, and from £5 to £7 per acre for Swedes, with straw. A fairer plan for both the raiser and consumer of turnips is to let them by the week at so much a head of stock put on to consume them. At the usual price of 3d. per head per week for young sheep, for the ordinary period of 26 weeks, makes a cost for keep of 6s. 6d. for the season; and, if it take 16 sheep to consume an acre, the turnips will realize about £5 5s. per acre For old sheep, 6d. per head per week is given, which is just double the cost for the season of the other—namely, 13s.—which, for 10 sheep, will realize £6 10s. per acre. For cattle, 5s. per head per week are given, with straw; and, if an ox take 26 weeks to eat an acre, the turnips will realize £6 10s. Thus, an acre of turnips that will support 10 old sheep for the season is worth more than one that will support 16 young sheep; but why old sheep should cost more to keep them than young does not appear; it would be fairer for the owner of the sheep to make the rate of keep exactly proportionate between the young and old. In plentiful years 2d., and in scarce years 4d. per head are given for young sheep, and other stock in proportion.

(1034.) These three kinds of turnip seem to possess all the properties desiderated by the farmer, and more than these, in my opinion, need not be cultivated; for although, in peculiar seasons, it is possible that, in a particular locality, some other variety may attain greater perfection and prolificacy, yet I believe that, in the long run, these will bear comparison with any variety that has yet been introduced into cultivation, provided they are of pure kinds.

(1035.) There are one or two hybrids of turnips worth mentioning, and which are so named, although it is probable that most of the varieties of turnips in use are natural hybrids. One is called Dale's Hybrid, being a cross betwixt the green-topped Swede and the globe, but whether the white or green-topped globe I do not know. It possesses more of the properties of the yellow turnip than of either of its progenitors; and it has the advantage of arriving sooner at maturity, and may therefore be sown later than the ordinary yellow turnip. The other hybrid is called the Lawtown Hybrid, being a cross between the green-topped Swede and the green-topped globe, the result of which is a heart-shaped, white-fleshed, green-topped turnip, considerably harder than the globes, with its leaves set on like those of the Swedes. The results of these two crosses are—a yellow turnip, Dale's, which arrives sooner at maturity than the older varieties, and a white globe, the Lawtown, which is more hardy than any other variety of white.

(1036.) With regard to the crop afforded by these hybrids, in an experiment made, in 1835, by Mr. John Gow, Fettercairn, Kincardineshire, the Dale attained to 28 inches in girth, and yielded 23 tons, and the Lawtown to 32 inches in girth, and yielded 27 tons the imperial acre.*

(1037.) Although storing is the proper method of securing turnips for use during a storm of rain or snow, when the turnip-field should not be entered by a cart, yet it is necessary that you should be provided with expedients for obtaining food for your cattle should you be overtaken by a storm, with a scantiness of provision in hand. As both rain and snow exhibit prognostics of their approach, and should these indicate a serious fall or storm, send all the field-workers and plowmen to the turnip-field, and

* Lawson's Agriculturist's Manual.

pull the turnips in the form in which the land is in the course of being stripped; and, removing only the tails, throw the turnips into heaps of from 3 to 6 cart-loads each, according to the thickness of the crop, taking care to place the tops of the uppermost turnips on the heap upon the outside, in order to protect the bulbs from frost, should it come suddenly and unaccompanied with snow. To these heaps rain will do no harm, and hey will serve to point out where they are, should snow cover them and the ground. As the turnips gathered in frost or snow should be immediately consumed and not stored, they may be thrown into the cart with a fork or graip, and their tops taken off at the steading, where this process can be done in the severest weather, when women could not stand out in the field to do it.

(1038.) I have given fig. 215 to show you what I conceive to be an ill-formed turnip, and also one which stands so much out of the ground as to

Fig. 215.

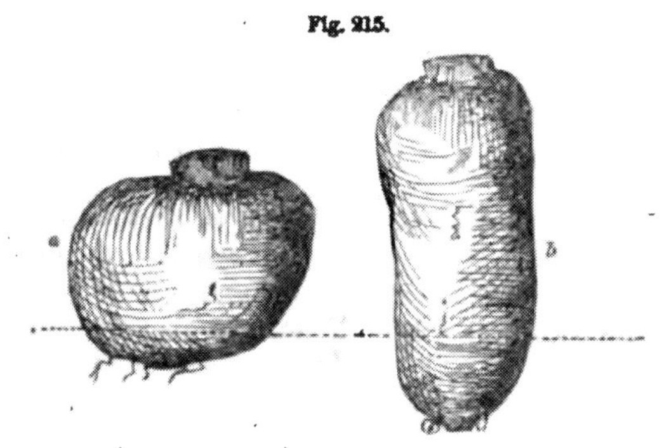

AN ILL-SHAPED TURNIP. THE TANKARD TURNIP.

be liable to injury from frost—where *a* is an ill-formed turnip, inasmuch as the upper part of it around the top being hollow, rain, snow or rime may lodge there, and find their way into the heart, and corrupt it, as is actually found to take place. All white turnips, when allowed to remain on the ground after they have attained maturity, become soft and spongy, of inferior quality in the heart, and susceptible of putrefaction, which frequently overtakes them in sudden changes from frost to thaw, and reduces them into a saponaceous pulp. This fact supplies a strong reason for storing white turnips after they come to maturity, which state is indicated by the leaves losing their green color and becoming flaccid. There are some sorts of white turnips that become spongy in the heart early in the season, and among these I would pronounce the Tankard-shaped, such as is represented by *b* in fig. 215; as are also a flat-shaped red-topped, and a small flattish white turnip, both much cultivated among small farmers, because, being small, they are supposed to require little manure to bring them to maturity, and this class of people are apt to spread manure as thin as possible upon the land, to make it go the farther. I need scarcely tell you that thrift attends the cultivation of only the best varieties of turnip. The dotted line in figs. 214 and 215 represents the surface of the ground, by which will be seen the relative depths to which these kinds of turnips descend into the soil when growing.

(1039.) I think it useful to give you a tabular view of the number of turnips there should be on an imperial acre, at given distances between the drills, and between the plants in the drills, and of the weight of the crop at specified weights of each turnip, that you may compare actual receipts with defined data, and endeavor to ascertain whether differences in the crop in those respects arise from deficiency of weight in the turnip itself, or too much thinning out of the plants. The distance between the drills is taken at the usual width of 27 inches; the distance between the plants is what is allowed to the different sorts of turnips; and the imperial acre contains 6,272,640 square inches. On altering the width between the drills, a calculation from these data can easuy be made of what ought to be the weight of crop at these given weights of turnips. On comparing a usual crop of 20 tons of Swedes with these data, and keeping in view the distance of 12 inches aimed at between the plants, the inevitable conclusion is that the average weight of each turnip in that crop must be less than 3 lbs., or the distance between the turnips greater than 12 inches.— In the one case your skill in raising a crop is almost rendered abortive, and in the other your negligence in wasting space by too much thinning out appears conspicuous. An amendment in both particulars is therefore requisite, and fortunately is attainable in both; for, as you perceive that but a slight difference in either of these particulars makes a great difference in the weight of the crop, your endeavor should be both to make the turnip heavy, and to maintain the desired distance between the plants inviolate. For example, 5-lb. turnips, at 9 inches asunder, give a crop of 57 tons, 12 cwt.; whereas the same weight of turnip at 11 inches apart gives a crop of 10 tons less. Now, how easy is it for careless people to thin out the plants to 11 instead of 9 inches; and yet, by so doing, a difference of no less than 10 tons, or 18 per cent. on a crop, is sacrificed. And again, a difference of only 1 lb. on the turnip—from 4 lbs. to 5 lbs.—at 9 inches asunder, makes a difference of 11 tons, or 25 per cent. per acre on the crop! So that a difference of only 1 lb. in each turnip, and 2 inches in the distance between them, makes the enormous difference of 21 tons on the whole crop! Who will say, after this, that these particulars do not require the most serious consideration in the treatment of the turnip crop?

Usual distance between the drills.	Usual distances between the plants.	Area occupied by each plant.	Number of turnips there should be per imperial acre.	Weight of each turnip.	Weight which the crop should be per imp. acre.
Inches.	Inches.	Square inches.		Lbs.	Tons. Cwt.
				1	11 10
				2	23 0
				3	34 11
27	9 between the plants of white turnips.	243	25,813	4	46 0
				5	57 12
				6	69 2
				7	85 2
				8	92 0
				1	10 7
				2	20 14
				3	31 1
27	10 between the plants of yellow turnips.	270	23,232	4	41 8
				5	51 15
				6	62 2
				7	72 9
				8	82 16
				1	9 8
				2	18 16
				3	28 5
27	11	297	21,120	4	37 13
				5	47 3
				6	56 11
				7	66 0
				8	75 8
				1	8 13
				2	17 6
				3	25 19
27	12 between the plants of Swedes.	324	19,360	4	4 12
				5	43 5
				6	51 18
				7	60 11
				8	69 4

(1040.) On comparing the amount of what the crop should be with instances given in the newspapers of what are considered great crops, it will be seen that these, after all, are no more than what they should be; and they are only the result of what might be expected to be attained by combined skill and care in culture. In the instances adduced in the Mark-Lane Express in 1840, crops were considered heavy which ranged from 40 to 60 tons per acre; and the Leinster Express of the same year mentions turnips having been raised on Lord Charleville's property in Ireland to a still greater amount—namely, of yellow Aberdeen, 49 tons, 13 cwt.; of yellow Tancred, 60 tons, 10 cwt.; of Swedish, 60 tons, 10 cwt.; and of white Tancred, 79 tons, 18 cwt. Such statements prove one of two things—either that large crops of turnips are more easily raised than farmers deem practicable, or great errors have been committed in making out these results. It is quite possible for great errors to be committed in making returns from any other mode of ascertaining the amount of a crop of turnips, than by topping and tailing a whole field, and weighing every cart-load separately. For example: Suppose that 1 square yard is measured in a field of turnips in this way—that is, if the distance of 1 yard is measured *from a turnip* (see fig. 208) along a drill, then the yard will embrace 5 turnips of white and 4 of Swedes; whereas, if the measurement is begun *between two turnips*, the same measure will only embrace 4 turnips of white and 3 of Swedes—making, in the former case, a difference in amount of 1 turnip out of every 5, and, in the latter, 1 out of every 4; and, if the weight of a statute acre has been calculated on such-like data, the crop will, in the case of the white turnips, be returned 1·5, and in that of the Swedes ¼ beyond the truth. Again, if the yard is placed *across two drills*, their produce will be included within the yard, the distance between them being only 27 inches; but, if the yard be placed across one drill only, then its produce alone will be included, as the yard will not reach to the drill on either side; and, if the produce of the whole field is calculated on such data, the result, in the latter mode of measurement, will just give half the amount of the other. Such ways of ascertaining the weight of a crop, when thus plainly stated, appear ridiculous enough; but it is an error which country people, who are not aware of the effects of the powers of numbers when squared, are very liable to fall into. The part of the field, too, from which the data are taken, may make a very great difference in the result over the whole; as, even on true turnip-soil, how different will the size and number of turnips be on a rising knoll and a hollow! The difference is not very obvious on looking upon the tops alone, but it is made very apparent after sheep have eaten off the leaves, and just begun to break upon the bulbs. The plan, too, of filling one cart-load or so and weighing it, and filling the rest of the cart-loads to a similar extent, without weighing them, is a fallacious one, when the fact is, as shown above, that turnips grown on the same field differ in weight, and therefore a few more or less in a small cart-load will make a considerable difference in the amount over the whole field. I question much whether any person ever weighed every cart-load of turnips as they were brought out of a field, or ever measured many places of the same field, to ascertain the number and weight of turnips in them; and, unless some plan approaching to either or both of these is adopted, the results obtained will never prove satisfactory. When the trouble of weighing every cart-load is wished to be avoided, the smallest and the largest and the middle-sized turnips should be pulled, topped and tailed, and chosen from every part of the field where a difference of size and number is found to occur, such as in hollows, on knolls, on sloping and level ground, at the top and bottom of the field; and each turnip weighed, and the tops weighed too, separately if desired, and then the average weight of the turnip may be relied on. A convenient machine for such a purpose is one of Salter's spring steel-yards, with a basin suspended from it by chains, in which a turnip may be placed with ease and celerity. Besides doing this, the distance from center to center of the tops of the turnips, before they are pulled, should be measured and noted down, and the average distance from turnip to turnip would then be ascertained. Having thus obtained correct data of the weight and number of turnips within the given limits of a field, the amount of the crop would then be so ascertained as to insure confidence in the result. The average girth of the turnips could be ascertained at the same time if desired; but this is not an essential element in determining the weight of the crop.

(1041.) It may prove interesting to you to know the periods at which the various kinds of turnips in culture were introduced. According to the name given to the plant in this country, the Swedes are natives of Sweden; the Italian name *Navoni de Laponia* intimates an origin in Lapland; and the French names, *Chou de Lapone, Chou de Swede*, would indicate an uncertain origin. Sir John Sinclair says: "I am informed that the Swedes were first introduced into Scotland in *anno* 1781–2, on the recommendation of Mr. Knox, a native of East Lothian, who had settled at Gottenburg, whence he sent some of the seeds to Dr. Hamilton."* There is no doubt they were first introduced into Scotland from Sweden, but I believe their introduction was prior to the date mentioned. The late Mr. Airth, Mains of Dunn, Forfarshire, informed me that his father was the first farmer who cultivated Swedes in Scotland, from seeds sent him by his eldest son then settled in Gottenburg, when my informant, the youngest son of a large family, was a boy of about 10 years of age. This would make the date of their introduction 1777; and this date is corroborated by the silence preserved by Mr. Wight regarding the culture of such a crop by Mr. Airth's father, when he undertook the survey of the state of husbandry in Scotland, in 1773, at the request of the Commissioners of the Annexed Estates, and when he would not have failed to report so remarkable a circumstance as the culture of the Swede. Mr. Airth sowed the first portion of seed he received in beds in the garden, and transplanted the plants in rows in the field, and thus succeeded in raising good crops for some years, before sowing the seed directly in the fields. I have not been able to trace the history of the yellow turnip; but it is probable that it originated, as is supposed by Professor Low, in a cross between a white and the Swede;† and, as its name implies, this may have been in Aberdeenshire. All the white varieties of field turnips obtained at first the name of the "Norfolk whites," from the circumstance of their having been first cultivated in that county, to any extent, by Lord Townshend, who, on coming home from being ambassador to the States-General, in 1730, paid great attention to their culture, and for which good service he obtained the appellation of "Turnip Townshend." It is rather remarkable that no turnips should have been raised in this country in the fields until the end of the 17th century, when it was lauded as a field-root as long ago as Columella, and in his time even the Gauls fed their cattle on them in winter. The Romans were so well acquainted with turnips that Pliny mentions having raised them 40 lbs. weight.‡ Turnips were cultivated in the gardens in England in the time of Henry VIII.‖ Dale's hybrid originated in a few ounces of a hybridal seed being sent, in 1822 or 1823, by the late Mr. Sherriff, of Bastleridge, Berwickshire, to Mr. Robert Dale, Libberton West Mains, near Edinburgh, who, by repeated selection and impregnation, brought it to what it is, a good yellow turnip, and now pretty extensively cultivated.§ The Lawtown hybrid originated about 8 or 10 years ago, by Capt. Wright, of Lawtown, in Forfarshire, crossing the green-topped white with the green-topped Swede, to harden the white; which object proved successful, but its culture has not been pushed. By sowing the Swede beside the white Lawtown, the latter has been converted into a yellow turnip, possessing the properties of the Swede; and, were the cross still farther pushed, I have no doubt that a distinct variety of the Swede would be obtained. A variety of Swedes was brought into notice, about 4 or 5 years ago, by Mr. Laing, Duddo, Northum

* Sinclair's Account of the Husbandry of Scotland, vol. i., *note*.

† Low's Elements of Practical Agriculture.

‡ Dickson's Husbandry of the Ancients, vol. ii.

‖ Phillips's History of Cultivated Vegetables, vol. ii.

§ Lawson's Agriculturist's Manual.

berland, who found it among his ordinary Swedes, and observed it by its remarkably elegant form of leaf, which is much notched near the base. It is getting into use, and possesses the valuable property of resisting the effects of spring for at least a fortnight longer than the common varieties, as I had a favorable opportunity of observing in Berwickshire late in spring 1841, and on this account may be stored and kept in a fresh state to a very late period of the season.

(1042.) As *cabbages* are considered good food for cows giving milk, it may be desirable to say a few words as to their use. The varieties of cabbage most suited for field culture are the Drum-head (*Brassica oleracea, capitata, depressa*), and the great round Scotch or white Strasburg, from which the German sour-krout is chiefly made (*Brassica oleracea, capitata, spherica alba* of De Candolle). Of these two, the drum-head is the most productive, and the Scotch stands the winter best. It is alleged by Sinclair that, for the purposes of the dairy, 1 acre of cabbages is worth 3 of turnips; but wherein this advantage consists is not stated, which it ought to have been, as he mentions that the nutritive matter contained in Swedish turnips is superior to that in the cabbage, in the ratio of 110 to 107½. There is no doubt, however, that the taste of milk is less tainted by cabbages than turnips, and I believe more milk may be derived from them; but there is considerable difference of opinion with respect to the effects of cabbage on butter and milk, and there is no doubt that a decayed leaf or two in a head of cabbage will impart both to butter and milk a strong, disagreeable taste. "This," says Sinclair, "I have long had an opportunity of proving."* If planted in drills usually made for turnips, these two kinds of cabbage will require to be placed in good soil, 18 inches asunder at least, which will give 12,907 plants to the acre, and, at 24 inches apart, the number of plants will be 9,834; and if they at all attain to the weight that cabbages sometimes do—that is, from 18 lbs. to 23 lbs. each—the lowest number, 18, will give a crop of 78 tons; but the usual crop is from 35 to 40 tons per acre. Their uses are to feed milch cows, to fatten oxen, and sheep are very fond of them. It is questionable how far their culture should be preferred to turnips, excepting on soil too strong for turnips, as they require a fine, deep, strong soil, and a large quantity of manure—means too valuable to be expended on cabbages, as an economical crop, in Scotland. I have no experience of the cabbage as a food for milch cows or feeding cattle, but know they are much relished by ewes at the season of lambing.

(1043.) The turnip-rooted cabbage (*Brassica campestris, napo-brassica, communis* of De Candolle) is little known in English culture, though it is cultivated in the fields in France. Its root is either white or red, and its neck and petioles greenish or purplish. It has a woody, short stem, produced by the formation and decay of the leaves; and, as new leaves are formed by the central bud of the stem, the lower leaves drop off, and thus the top of the bulb assumes the appearance of a stem; and Dr. Neill observes it has a root under ground as sweet as a Swedish turnip. In both these respects it is very similar to our Swedish turnip, but whether it could be made to assume the form of, or has given origin to, that valuable root, I must leave to be determined by the botanical physiologist.

(1044.) The cow-cabbage or Cesarean kale (*Brassica oleracea, acephala, arborescens* of De Candolle), which created such a noise a few years ago, deserves only a passing notice. "This plant," says Don, "is almost similar in habit to the palm kale, but the stem rises to the hight of from 10 to 16 feet, the leaves are not so puckered nor rolled inward at the edges, nor do they hang down so much. The stem is naked and simple, crowned by a head of leaves like a palm-tree. Sixty plants of this variety are said to afford sufficient provender for one cow for a year; and, as the side leaves are only to be used, it lasts four years without fresh planting. In La Vendee it is said to attain the hight of 12 or 16 feet. In Jersey this plant is sufficiently hardy, and where it grows from 4 to 12 feet. The little farmers there feed their cows with the leaves, plucking them from the stem as they grow, leaving the crown at the top. The stems, being strong, are also used by them for roofing small out-houses. When the gathering of the leaves is finished, at the end of the year, the terminating bud or crown is boiled, and is said to be particularly sweet. It is not sufficiently hardy to stand the climate of Britain, unless planted in a very sheltered situation."†

(1045.) There is still another variety of the cabbage tribe which deserves notice—the turnip-stemmed cabbage or kohl-rabi (*Brassica oleracea, caulo-rapa, alba* of De Candolle). The varieties of this plant are numerous, but the best suited for field culture are the large red and green sorts. It is a native of Germany, where it is much cultivated, as it also is in the low countries and the north of France, where it is chiefly given to milch cows, for which it is well adapted on account of its possessing little of that acridity which is found in the turnip to affect butter and milk. It is taken up before the frost sets in, and stored like potatoes or turnips for winter use. Its habits and produce are similar to the Swedish turnip, the part of the plant resembling which is a swollen bulb at the top of the stem, which, when divested of leaves, may readily be mistaken for a Swedish turnip. Hares are so fond of it, that on farms where they abound, its culture is found to be impracticable. Sir Thomas Tyrwhitt first introduced it into England from Germany.‡

(1046.) Although the parsnip (*Pastinacea sativa edulis* of De Candolle) is too tender a root for general cultivation in this country, it deserves notice on account of its fattening properties when given to all domesticated animals. "The parsnip," says Don, "has been partially introduced of late years as a field-plant, and is nearly equal to the carrot in its product of saccharine and nutritive matter. Its culture as a field-plant has chiefly been confined to the island of Jersey, where it attains a large size, and is much esteemed for fattening cattle and pigs. It is considered rather more hardy than the carrot, and its produce is said to be greater. . . . The variety best suited for the field is the *large Jersey*. . . . In the fattening of cattle, it is found equal, if not superior, to the carrot, performing the business with as much expedition, and affording meat of exquisite flavor, and a highly juicy quality. The animals eat it with much greediness. It is reckoned that 30 perches, where the crop is good, will be sufficient to fatten an ox 3 or 4 years old, when perfectly lean, in the course of 3 months. They are given in the proportion of about 30 lbs weight

* Sinclair's Hortus Gramineus Woburnensis.
† Don's General Dictionary of Botany and Gardening, vol. i.
‡ Sinclair's Hortus Gramineus Woburnensis; and Lawson's Agriculturist's Manual.

morning, noon and night, the large ones being split in 3 or 4 pieces, and a little hay supplied in the intervals of those periods. And when given to milch cows with a little hay, in the winter season, the butter is found to be of as fine a color and as excellent a flavor as when feeding in the best pastures. Indeed, the result of experiment has shown that not only in neat cattle, but in the fattening of hogs and poultry, the animals become fat much sooner, and are more healthy, than when fed with any other root or vegetable; and that, besides, the meat is more sweet and delicious. The parsnip-leaves being more bulky than those of carrots, may be mown off before taking up the roots, and given to cows, oxen, or horses, by whom they will be greedily eaten."* The leaves may be greedily eaten when no other green food is presented as a choice to cattle, but I have no doubt that cattle will make very little progress toward condition when using them. The weight of the largest parsnips grown in gardens in Scotland, varies from 10 ounces to 2 lbs. each.†

(1047.) The carrot (*Daucus carrota sativa* of De Candolle) is raised in the fields in several parts of this country. The varieties most suited for field-culture are the large orange, Altringham, long red, and green-top white. In giving a detailed statement of the general treatment of the carrot, Mr. Burrows says, in regard to their use in winter: "I take up, in the last week of October, with 3-pronged graips, a sufficient quantity to have a store to last me out any considerable frost or snow that may happen in the winter months. The rest of the crop I leave in the ground, preferring them fresh out of the earth for both horses and bullocks. The carrots keep best in the ground, nor can the severest frosts do them any material injury. The first week in March, it is necessary to have the remaining part of the crop taken up, and the land cleared for barley. The carrots can either be laid in a heap, with a small quantity of straw covered over them, or they may be laid in some empty outhouse or barn, in heaps of many hundred bushels, provided they are put together dry. This latter circumstance it is indispensable to attend to; for if laid together in large heaps when wet, they will certainly sustain much injury. Such as I want to keep for the use of my horses until the months of May and June, in drawing over the heaps (which is necessary to be done the latter end of April, when the carrot begins to sprout at the crown very fast), I throw aside the healthy and most perfect roots, and have their crowns cut completely off and laid by themselves. By this means, carrots may be kept the month of June out in a high state of perfection."‡ When the ground is desired to be cleared for wheat, carrots should be taken up in autumn, and stored in the manner described for mangel-wurzel (1020), in a *dry* state, though with fewer precautions against the frost. Arthur Young gives the average produce of an acre of carrots in Suffolk at 350 bushels; but Mr. Burrow's crops averaged upward of 800 bushels, which, taking the bushel at 42 lbs., will make the former crop 6 tons 11 cwt., and the latter 15 tons exactly. In the fields in Scotland, the Altringham carrot has been grown to 1¼ lbs. and in gardens to 2¼ lbs.; and a crop of the large orange-carrot, manured with night-soil, has been raised by Mr. Spiers, of Calcreuch, at the rate of 9 tons the acre—probably the Scotch acre—which is equal to 7 tons 1 cwt. the imperial.§

(1048.) Varieties of the common potato (*Solanum tuberosum*) are also used in the feeding of cattle, but as the crop is of more importance as human food, I shall reserve the description of storing them until the proper season, in autumn, when they are removed from the ground. Meantime, I may mention that the varieties raised exclusively for cattle are the common yam, red yam, and ox-noble.

---

30. FEEDING SHEEP ON TURNIPS IN WINTER.

"Now, shepherds, to your helpless charge be kind,
Baffle the raging year, and fill their pens
With food at will; lodge them below the storm,
And watch them strict; for, from the bellowing East,
In this dire season, oft the whirlwind's wing
Sweeps up the burden of whole wintry plains
At one wide waft, and o'er the hapless flocks,
Hid in the hollow of two neighboring hills,
The billowy tempest whelms."

THOMSON.

(1049.) Having prepared room on the turnip land for the sheep intended to be fattened upon turnips, by removing the proportion of the crop in the manner described above, that is, by drawing 2 rows and leaving 2 rows alternately, and having prepared that part of the field to be first occupied by the sheep, which will afford them shelter in case of need, the first thing to be afterward done is to carry on carts the articles to the field requisite to form a temporary inclosure to confine the sheep within the ground allotted them. It is the duty of the shepherd to erect temporary inclosures,

---

* Don's General Dictionary of Botany and Gardening, vol. i.
‡ Communications to the Board of Agriculture vol. vii.
† Lawson's Agriculturist's Manual.
§ Lawson's Agriculturist's Manual.

and as, in doing this, he requires but little assistance from other laborers, he bestows as much time daily upon it until finished as his avocations will allow.

(1050.) There are two means usually employed to inclose sheep upon turnips, namely, by *hurdles* made of wood, and by *nets* of twine. Of these I shall first speak of the *hurdle* or *flake*. Fig. 216 represents 2 hurdles set

Fig. 216.

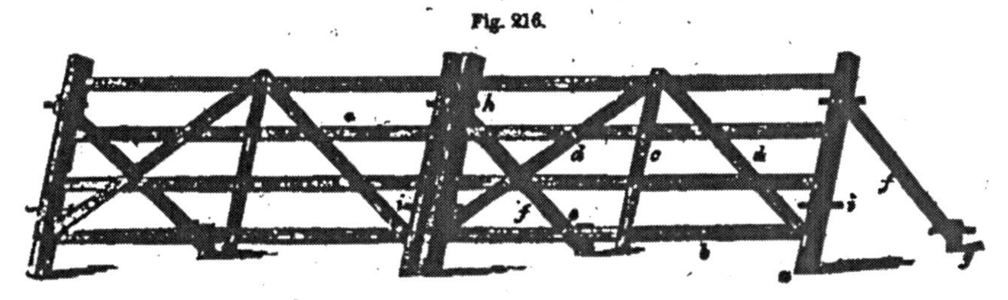

HURDLES OR FLAKES SET FOR CONFINING SHEEP ON TURNIPS.

as they should be. The mode of setting them is this; but in doing it, the shepherd requires the assistance of another person,—a field-worker will serve the purpose: The flakes are set down with the lower ends of their posts in the line of the intended fence. The first flake is then raised up by its upper rail, and the ends of the posts are sunk a little into the ground with a spade, to give them a firm hold. The second flake is then raised up and let into the ground in the same way, both being held in that position by the assistant. One end of a stay *f* is then placed between the flakes near the tops of their posts, and these and the stay are made fast together by the insertion, through the holes in them, of the peg *h*. The peg *i* is then inserted through near the bottom of the same posts. The flakes are then inclined backward away from the ground fenced, until their upper rail shall be 3 feet 9 inches above the ground. The stake *e* is driven into the ground by the wooden mallet, fig. 218, at such a point as, where the stay *f* is stretched out from the flakes at the above inclination, that a peg shall fasten stake and stay together, as seen at *g*. After the first two flakes are thus set, the operation is easier for the next, as flake is raised after flake, and fastened to the last standing one in the manner described, until the entire line is completed.

(1051.) Various objections can be urged against the use of flakes, the first being the inconvenience of carrying them from one part of a field to another in carts, and of their liability to breakage in consequence; as also the shepherd himself cannot set them up well and speedily without assistance, and even with that they require a good deal of time in setting up. They are also easily upset by a high wind blowing behind them; and, when in use they require almost constant repair and replacing of pegs, stays and stakes; though, when repaired and set carefully by at the end of the season, they will last several years. The mode of making flakes, and their price, are mentioned below.

(1052.) The other method of inclosing sheep on turnips is with nets made of twine of the requisite strength. These nets having square meshes when stretched upon the stakes, usually extend to 50 yards in length, and stand $3\frac{1}{2}$ feet in hight. They are furnished with a rope along both sides passing through the outer meshes, which are called the "top" and "bottom rope" as the position of either may be at the time. These ropes are wound round the stakes by a peculiar sort of knot called the "shepherd's knot." The stakes are best formed of thinnings of ash-trees that have been planted very thick together and grown up long and small, and they should be 3 inches in diameter and 4 feet 9 inches long; allowing 9 inches of a hold in the ground, 3 inches between the ground and the bottom of

Fig. 217.

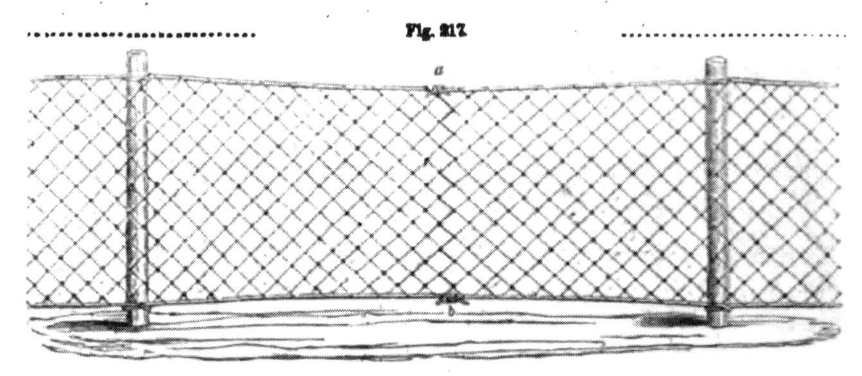

SHEEP NET SET FOR CONFINING SHEEP ON TURNIPS.

the net, and 3 inches from the top of the net to the top of the stake; or they may be made of larch woodings, 4 inches in diameter and 4 feet 9 inches long; but every kind of wood of which they may be made should be seasoned with the bark on before being cut into stakes. They are pointed at one end with the ax, and that end should be chosen to be pointed which will make the stake stand in the same position as when it was growing in the tree, for its bark, it has been found, is then in the best state for repelling rain.

(1053.) A net is set in this way: If the ground is in its usual soft state the stakes may simply be driven into the ground with a hard-wood *mallet*

Fig. 218. Fig. 219.

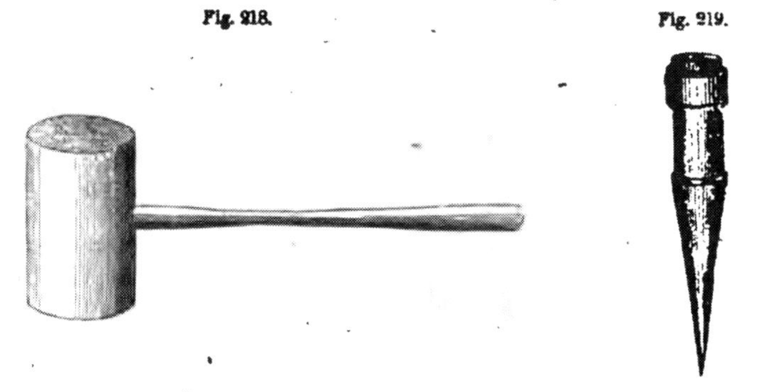

THE SHEPHERD'S WOOD MALLET. THE DRIVER.

fig. 218, in the line fixed on for setting the net, at distances of 3 paces asunder. The wood of the apple-tree makes the best mallet, as not being apt to split. Should the soil be thin and the subsoil moderately hard, a hole sufficiently large for a stake may be made in the subsoil with the tramp-pick, fig. 37; but should the subsoil be so very hard as to require a larger hole to be made than what can easily be formed by the tramp-pick, or should the ground be so dry and hard as to require the use of any instrument at all, the most efficient one for the purpose is one called a *driver*, fig. 219. It is formed of a piece of pointed hard-wood, strongly shod with iron, and its upper end is protected by a strong ferule of iron to prevent its splitting by the strokes of the mallet. The stakes being thus driven so that their tops may not be less than 4 feet high, along as many sides of the inclosure as are required at the place to form a complete fence.

(1054.) A net is set in this manner: Being in a bundle, having been rolled up on the arms and fastened together by the spare ends of the top and bottom ropes, these are unloosened and tied to the stake that has been driven close to the fence, whatever that may be, and then the net is run out in hand toward the right as far as it will extend in a loose manner, on the side of the stakes facing the ground the sheep are to occupy. On coming to the next stake from the commencement, the bottom rope gets a turn to the left round the stake, and the top rope above it a similar turn round the same stake, so as to keep the leading coil of the rope uppermost. The bottom rope is then fastened with the shepherd's knot to the stake, 3 inches from the ground, and the top rope is fastened with a similar knot near the top of the stake, stretching the net even and upward, and in this way the net is fastened to one stake after another until the whole of it is *set up*, as it is called, care being taken to make the top of the net run uniformly throughout its entire length.

(1055.) The shepherd's knot is made in this way: Let *a*, fig. 220, be the continuation of the rope which is fastened to the first stake, then press the second stake with the hand toward *a* or the fastened end, and at the same time tighten the turn round the stake with the other hand by taking a hold of the loose end of the rope *d*, and moving it so as to cause it to pass under *a* at *c*, and screwing it round the stake to *b*, where the elastic force of the stake will secure it tight under *a* at *b* when the stake is let go. The bottom rope is fastened first, to keep the net at the proper distance from the ground, and then the top rope is fastened to the same stake in the same manner. Proceed in this manner at each successive stake until the whole net is set up. A net may be thus set up either toward the right or the left as the starting point may be situate, but in proceeding in either direction care must be taken to pass the top and bottom ropes round the stakes, so as the leading coil of the rope is always uppermost toward the direction in which the net is to be set up. Thus, in fig. 220, the rope *d* was uppermost until it was passed under *a*, because the setting of the net in this case is from right to left, and it continues to be uppermost until it reach the next stake to the left. If both the cord and stake are dry, the knot may slip as soon as made, but if the part of the stake at *b* where the knot is fastened is wetted a little, it will make the rope keep its hold until the cord has acquired the set of the knot. With a new rope that is greasy, and a smooth stake, it is difficult for the knot to retain its hold even with the assistance of water.

Fig. 220.

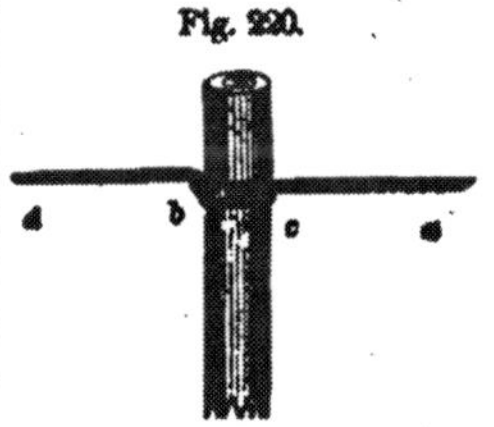

THE SHEPHERD'S KNOT, BY WHICH A NET IS FASTENED TO A STAKE.

(1056.) There are some precautions required in setting a net beside this of the ropes. If the net is new, the cords may be stretched as tight as you please, because they will stretch considerably; but if old, the least damp or rain afterward will stretch them so as to cause them to break. If the net is at all in a damp state, it should be set very tight, because rain cannot make it tighter; and if not set very tight, the first dry weather will so slacken the cords as to loosen all the knots, and make the net slip down the stakes; but even if it should not be slackened to that extent, it will be so slackened as to shake about with the wind, and bag down and touch the ground. Such an occurrence will create the trouble to the shepherd of re-setting the whole net, and the best way of avoiding this trouble is to have the nets in a dry state when they are set. In wet weather shepherds take the opportunity of a dry moment of setting a dry net in anticipation along a new break of turnips, and they also hang up wet nets to dry on the outside of the stakes away from the sheep. Nets should never be wound up in a wet state, even for a short time, as they will soon mould and rot.

(1057.) On commencing the setting of another net, its top and bottom ropes are fastened to those of the last net, and the ends of the nets themselves are brought together by interlacing the meshes of both with a piece of string, as at *a*, fig. 217. Here the knots in the top and bottom ropes are seen, and the twine interlacing the meshes are made to appear stronger than that of the net only to let it be perceived. Thus one net is set after another, until the whole intended area is inclosed. Where there is a turn in the line of nets in going from one side of the inclosure to another, as seen on the right side of fig. 226, if there is much of the net left at the turn, it should be brought down the next side; in which case the stake at the corner should be driven very securely down, as there will be a considerable strain upon it from the nets pulling from different directions, and this will especially be the case in damp weather. But the safer and perhaps better plan is to take a fresh net at the turn and fasten it to a stake, and run on the other net in its own line until it is suspended either in setting or coiling it around the top of a stake. All surplus ends of nets should be carefully hung upon the back of the stakes when wet, to dry and get the air. Part of the nets will thus cross ridges, and part will run along a ridge. Where they cross ridges that have been but once gathered-up, or plowed crown-and-furrow, the bottom of the nets will be nearly close to the open-furrows, but where they cross a gaw-cut in rather strong land, a stake or two should be made to lie upon the bottom rope to keep it down, for some sheep have a trick of creeping under the net, when they find a suitable opening; and where nets cross ridges which have been twice gathered-up, one stake should be driven at one side of the open-furrow, and another at the crown of the ridge, and the bottom rope will then run nearly parallel to the surface of the ground.

(1058.) In setting nets, in whatever position, care should be taken to keep each side of the inclosure in the same plane—that is, each side exactly in a straight line, and the surface of its nets perpendicular; and the different lines should meet at right angles to one another, so that every break of turnips occupied by the sheep should either be a rectangle or a square; because the strain upon the ends will then be equalized over the entire cords and stakes of each side, and no undue pressure exerted on any one stake. A shepherd who knows his business so as to pay attention to these particulars, will preserve his nets and stakes a much longer time in a serviceable state, than one who is ignorant or careless about them.

(1059.) The shepherd should always be provided with net-twine to mend any holes that may be made in the nets; and where they happen to be set across hare-roads, the hares will invariably keep their runs open; which being the case, it is much better to allow them to remain open, than in filling them up to have them cut through daily.

(1060.) When flakes or nets have been set round the first break, the ground may be considered in a proper state for the reception of sheep; and the ground should be so prepared before the grass fails, that the sheep to be fattened may not in any degree lose the condition they have acquired on the grass; for you should always bear in mind that it is much easier to improve the condition of lean sheep that have never been fatter than to regain the condition of those that have lost it. Much rather leave pastures a little rough, than risk the condition of sheep for the sake of eating it down. The rough pasture will be serviceable to the portion of the sheep-stock that are not to be fattened, such as ewes in lamb and aged tups. Let sheep, therefore, intended to be fattened, be put on turnips as early as will maintain the condition they have acquired on the grass. By a *break* of turnips is meant that part of the crop which is being consumed by the sheep.

(1061.) As the tops of white turnips are long and luxuriant at the commencement of the season, the first break or inclosure should be made smaller than those which succeed, that the sheep may not have too many tops at first on a change of food from grass to turnips, and which they will readily eat to excess, on account of their freshness and juiciness. Let the sheep fill themselves with turnips pretty well before taking them to the next break. The second break may be a little larger than the first, and the third may be of the proper size—that is, contain a week's consumption of food. These considerations will cause the shepherd some trouble for two or three weeks in the beginning of the season; but they are trifling compared with the advantage derived from it by the sheep. Rather let him have the assistance of a field-worker to shift the nets, than neglect the precautions. When the tops wither in the course of the season, and one night of sharp frost may effect that, or after the sheep have been accustomed to the turnip, the danger is over. The danger to be apprehended is diarrhœa or severe looseness of the bowels, which is an unnatural state in regard to sheep, and they soon become emaciated by it; many sink under it, and none recover from such a relaxation of their system until after a considerable lapse of time.

(1062.) Another precaution to be used on this head is, to avoid putting sheep on turnips for the first time in the early part of the day when they are hungry. The danger may be apprehended with tops in a dry state, but when they are wet by rain or snow, or half-melted rime, they are most likely to do the harm. The afternoon, then, when they are full of grass, should be chosen to put the sheep on turnips, and they will immediately begin to pick the tops, but will not have time to injure themselves. Should the weather prove wet at first, and the ground be either somewhat too clayey or soft, and the sheep thereby find an uncomfortable lair, it would be advisable to allow them to rest in an adjoining grass field for a few nights until the ground becomes consolidated (which will soon take place) by their constant and repeated tramplings.

(1063.) Sheep when put on turnips are selected for the purpose. Ewes being at this season with young, whether as a flying or standing flock, are never, in Scotland, put on turnips in winter, but continue to occupy the pastures, part of which, if left on purpose in a rough state, will suffice to support them as long as the ground is free of snow. The reason why *great ewes*, as ewes in lamb are called, are never put on turnips, is the chance of getting too fat, which if they do, they will produce small lambs and run great risk of being attacked by inflammatory complaints at the lambing time. Tups are most frequently put on turnips, especially tup hoggs, but they are never folded in the same part of the field as the feeding sheep, having a snug corner somewhere to themselves, or else the turnips are led to them in a sheltered part of a grass field. Young sheep, that is, lambs of the same year, are always put on turnips, whether with the view of feeding them fat at once, or enlarging the size of their bone. Every year a certain number of old ewes, unfit for farther breeding from want of teeth, or means of supplying milk, are drafted out of the standing flock to make room for the same number of young females into the ewe flock, and are fattened off upon turnips. It sometimes happens that the castrated male lambs of last year, instead of being sold, have been grazed during the summer, and are fattened off the second season on turnips. All these classes of sheep, of different ages, may be mixed together and occupy the same break of turnips. It is seldom that the last class, namely, the lambs of last year, are kept on to the second year, but the draft ewes are always fed along with the young sheep, and they prove useful in breaking the turnips and eating up the picked shells. A mixture of old and young sheep are less useful to one another when turnips are cut by machines.

(1064.) Since I have had occasion to mention some of the classes of sheep, it may not be out of place here to make you acquainted with the technical names which they receive in respect of age and sex, and which I shall always employ when speaking of them in future. A new-born sheep is called a *lamb*, and retains that name until it is weaned from its mother and able to support itself. The name is modified according to the sex and condition of the animal: when a female, it is a *ewe-lamb*, when a male, a *tup-lamb*, and the last name is changed to *hogg-lamb* when the creature undergoes emasculation. After a lamb has been weaned, until the first fleece is shorn from its back, it receives the name of *hogg*, which cognomen, like that of lamb, is modified according to the sex and condition of the animal; namely, a female is called a *ewe-hogg*, a male a *tup-hogg*, and a castrated male a *wether-hogg*. After the first fleece has been shorn, another change is made in the nomenclature; the ewe-hogg then becomes a *gimmer*, the tup-hogg a *dinmont* or *shearling-tup*, and the wether-hogg a *dinmont*, and these names are retained until the fleece is shorn the second time. After this operation another change is effected in all the names, the gimmer being then called a *ewe* if she is *in lamb*, but if she has failed being in lamb she is said to be a *tup-eill gimmer* or *barren gimmer*, and if she has never been put to the ram she gets the name of *yield gimmer*. If a ewe who has borne lambs fails again to be in lamb, she is called a *tup-eill ewe* or *barren ewe*. After the ewe has ceased to give milk, or become dry, she is said to be a *yield ewe*. The shearling-tup is called a *2-shear tup* when the fleece has been taken off him the second time, and the dinmont commonly a *wether*, but more correctly a *2-shear wether*. After a ewe has been shorn three times she is called a *twinter ewe*, that is, a *two-winter ewe*; a tup that has been so treated is called a *3-shear tup*; and a wether still a *wether*, or more correctly a *3-shear wether*, which is an uncommon name among Leicester sheep, as the castrated sheep of that breed are rarely kept to so great an age. A ewe that has been four times shorn gets the name of a *three-winter ewe* or *aged ewe*; a tup is called an *aged tup*, a name which he retains ever after, whatever his age, but they are seldom, except for special reasons, kept beyond this age; and the wether is now a *wether* properly so called. A *tup* and *ram* are synonymous terms. A ewe when she is removed from the breeding flock is called a *draft ewe*, whatever her age may be, and gimmers that are put aside as unfit for breeding from are called *draft gimmers*, and the lambs, dinmonts, or wethers, that are drafted out of the fat stock are called the *sheddings*, or *tails*, or *drafts*. In England a somewhat different nomenclature prevails. There sheep bear the name of *lamb* until 8 months old, after which they are called *ewe* and *wether teggs* until once clipped. Gimmers are called *theaves* until they bear the first lamb, when they are named *ewes of 4-teeth*, next year *ewes of 6-teeth*, and the year after *full-mouthed ewes*. Dinmonts are called *shear hoggs* until clipped, when they are 2-*shear wethers*, and ever after they are called *wethers*.

(1065.) When sheep are on turnips they are invariably supplied with dry fodder, hay or straw—hay being the most nutritious, though most expensive; but sweet, fresh oat-straw answers the purpose very well. The fodder is supplied to them in racks. There are various forms of straw racks for sheep—some being placed so high that sheep can with difficulty reach the fodder, and others are mounted high on wheels. The form represented in fig. 221 I have found convenient, containing as much straw at

Fig. 221.

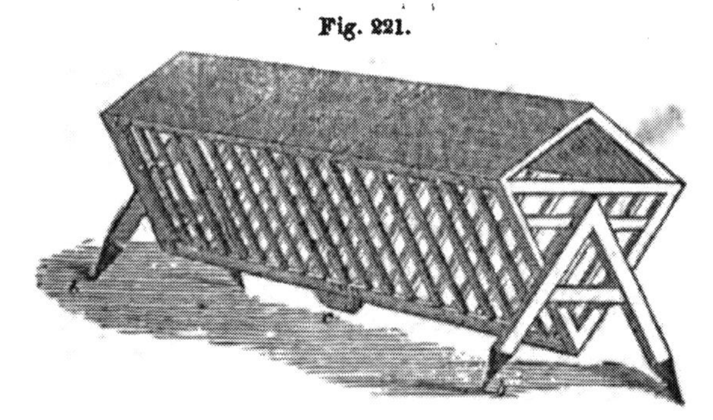

THE SHEEP STRAW OR HAY RACK.

a time as should be given, admitting the straw easily into it, being easily moved about, of easy access to the sheep, and being so near the ground as to form an excellent shelter. It is made of wood, is 9 feet in length, 4½ feet in hight, and 3 feet in width, having a sparred rack with a double face below, which is covered with an angled roof of boards to throw off the rain. The rack is supported on 2 triangular-shaped tressels *b*, shod with iron at the points, which are pushed into the ground, and act as stays against the effects of the wind from either side. The billet *c*, fixed on the under or acute edge of the rack, rests upon the ground, and, in common with the feet, supports it from bending down in the middle. The lid *a* is opened on hinges when the fodder is put into the rack. There should at least be 2 such racks in use; because, when set at an angle to each other against the weather point, the space embraced between them forms an excellent shelter for a considerable number of sheep. (Fig. 226.) Such a rack is easily moved about by 2 persons, and their position should be changed according to a change of wind indicative of storm.

(1066.) It is the duty of the shepherd to supply these racks with fodder, and one or all of them may require replenishment daily. This he effects by carrying a bundle of fodder at any time he visits the sheep. When carts are removing turnips direct from the field, they carry out the bundles; but it is the duty of the shepherd to have them ready for the carters in the straw-barn or hay-house. For shelter alone the racks should be kept full of fodder. Fodder is required more at one time than another; in keen, sharp weather, the sheep eat it greedily, and when turnips are frozen they will have recourse to it to satisfy hunger, and after eating succulent tops they like dry fodder. In rainy, or in soft, muggy weather, sheep eat fodder with little relish; but it has been remarked that they eat it steadily and late, and seek shelter near the racks prior to a coming storm of wind and rain or snow; in fine weather, on the other hand, they select a lair in the more exposed part of their break.

(1067.) Until of late years sheep were allowed to help themselves to turnips in the early part of the season; and in consuming them the tops were first eaten, and then the bulbs were scooped out as far as the ground would permit. When a large proportion of the turnips of the break were thus eaten, the *shells*, as the bottom part fast in the ground is called, were picked out of the ground with an instrument made for the purpose. Its name is a *turnip-picker*, and the mode of using it may be seen in fig. 222. Its handle is 4 feet long, and blade 10 inches, including the eye for the hand e. By its mode of action, you will see that the tap-root of the turnip is cut through and the shell separated from the ground at one stroke. A very common form of these pickers is with the mouth cleft in two, between which cleft the tap-root is embraced, and the shell and root are pulled up together. It is found, however, that the tap-root contains an acrid juice detrimental to the stomach of sheep, so that the better plan is to cut it off and leave it on the ground to rot. The best form of blade may be seen in fig. 223, and fig. 224 shows the objectionable form of the same instrument.

Fig. 222.

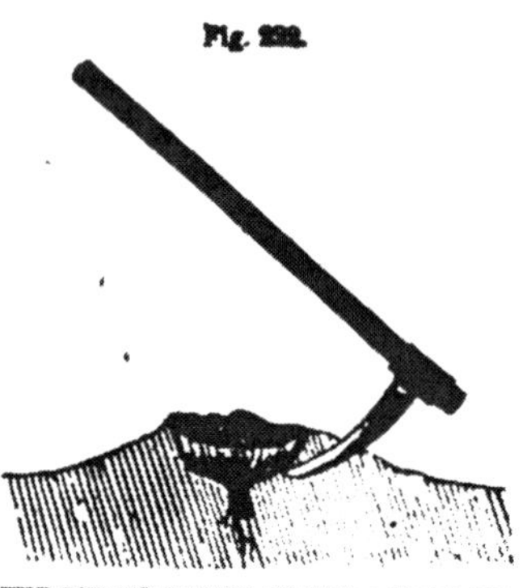

THE TURNIP-PICKER IN THE ACT OF CUTTING OFF THE TAP-ROOT AND PULLING UP THE SHELL OF THE TURNIP.

Fig. 223. Fig. 224.

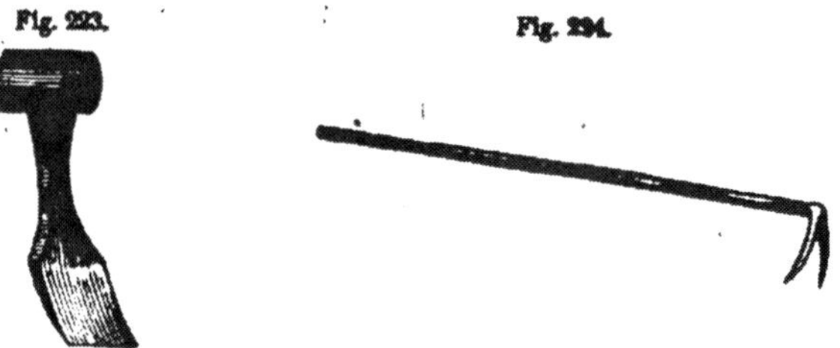

THE BEST FORM OF TURNIP-PICKER. OBJECTIONABLE FORM OF TURNIP-PICKER.

(1068.) Only half of the ground occupied by the shells should be picked up at one time, by removing every alternate double row of them, in order to make the sheep spread over a greater space while consuming them.—When the ground is dry, the shells should be pretty clean eaten up before a new break of turnips is formed; but, a few being left, the sheep will come over the ground again and eat them up, though in a shriveled state, especially in frost, when they are sweeter and softer than turnips.

(1069.) But the more recent and better plan of serving turnips to sheep—and it should be universally adopted—is to cut them into small pieces with a turnip-slicer into troughs conveniently placed for use, while at the same time the sheep have liberty to eat the turnips themselves. A convenient and expeditious form of turnip-slicer is described below at fig. 252, which description you should peruse at once; and a simple form of turnip-

Fig. 225.

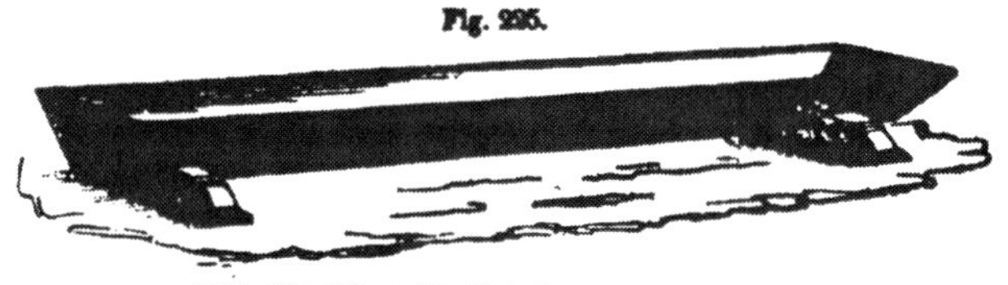

THE TURNIP-TROUGH FOR SHEEP-FEEDING.

trough is here represented by fig. 225. It is 8 feet long, and made acute at the bottom, for the more easy seizure of the pieces of turnip by the mouths of the sheep, by nailing two boards together upon the two triangular-shaped ends, and placing it upon billets for feet. The troughs are set in a line along the outside 2 rows of turnips about to be pulled. The turnip-cutter is wheeled to each trough successively by the field-worker, who works the handle, and its hopper is filled by another worker, who tops and tails the turnips. The sheep range themselves on either side of the trough.

(1070.) I have constructed fig. 226 to give you a bird's-eye view of the manner in which a turnip-field should be fitted up for sheep. There are,

Fig. 226.

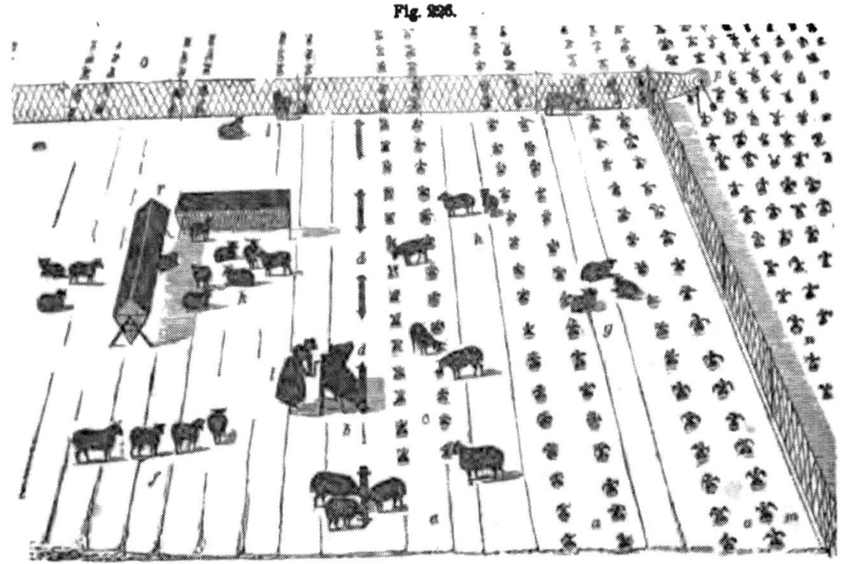

THE MODE OF OCCUPYING TURNIP-LAND WITH SHEEP.

in the first place, the turnips themselves *a*, of which half have been drawn by pulling 2 drills and leaving 2 alternately. The ground upon which they are growing is represented partly bare, because they are supposed to have been pulled up in the progress of the turnip-cutter advancing from one side of the break to the other; and it constitutes the break. As matters are represented, the turnip-slicer *b* is proceeding up beside the 2 drills *c*, and depositing the cut turnips into one of the small troughs *d*, out of another of which some of the sheep are eating, while others are helping themselves from the bulbs in the drills *c*. The sheep are represented scattered over the ground as they are usually seen, some following one another in a string *f* toward the place where their food is preparing for them, while others, *g*, are lying resting, regardless of food. Some, *h*, are standing, as if meditating what next to do, and others, *i*, examining the structure of the nets. Some nibble at the dry fodder in the racks *r*, while *k*, a group, lie under their shelter. The field-worker *l* is slicing the turnips with the machine. Such are the usual occupations of sheep when they have abundance of food at their command. The nets *m* are represented as inclosing two sides of the break, the other two sides being supposed to be composed of the fences of the field, and not represented. The turnips *n*, to the right of the nets, appear undrawn, while those *o*, above the nets, are stripped, indicating that the progress of the breaks at this time is upward toward the top of the field, in a line with the direction of the drills, and, of course, with that of the ridges; and this part of the plan is not a matter of indifference, because the breaks should so succeed one another in their passage across the field, as that the land, when cleared of turnips, may be plowed from end to end and ridged up, if desired. In a large field, which engages the sheep for a considerable part of the season, the land is plowed as each stretch of breaks is cleared, in order to preserve the virtue of the manure; and this is of more importance in a large than in a small field, over which a large number of sheep will soon pass. In plowing up land, however, with this intent, care should be taken not to deprive the sheep of any natural shelter they have enjoyed; and to secure this to them as long as practicable, the breaks should be so arranged as to make those first formed along the lowest and most sheltered part of the field, so that the sheep could resort at the bottom of the set of breaks they are occupying, after the first set had been given up and plowed to the top of the field, and so on in succession. Such an arrangement requires more consideration than at first sight may appear, and its neglect may much inconvenience the sheep for want of shelter; and shelter to sheep in winter does not merely imply protection from unusual inclemency of the weather for a night or two, but also preservation of the fleece, and comfort to the flock throughout the season. The remainder of the net along the upper part of the break is represented coiled round the top of a stake at *p*, and there also the mallet and driver await their use.

(1071.) I have already stated that *tups* or *rams* are fed on turnips in a separate division from the feeding sheep. Some apportion them in a space in the same, while others give them a break in another field; but I would prefer giving tups turnips in a small grass paddock, and cutting them with the small lever turnip-slicer represented in fig. 246, and described minutely below. Where the lot of tups is large, say 40 or 50, it may create, it is true, more trouble to fetch their turnips to them than to inclose them on turnips; but this consideration should be always borne in mind, in regard to tups, that whenever they and female sheep become aware of the presence of each other in the same field, and even in contiguous fields, neither party will rest to feed. The air will carry the scent of their bodies to each other, and, whenever any of the females show a tendency toward coming into season, the scent of the males confirms it, and, becoming restless themselves, they have a tendency to render the rest of the flock so also. And if tups are in a separate fold by themselves, away from the rest of the sheep, they cause as much trouble to the shepherd in visiting them there as a larger flock; whereas, were they near home in a grass paddock, he could visit them frequently in going and coming to his house at his hours of repast.

(1072.) Sheep are sometimes assisted in their feeding on turnips with other substances, such as oil-cake and corn. Either of them is administered in a covered box, to protect it from injury from weather. Such a box is represented in fig. 227, the construction of which requires no explanation. I have never had any experience of feeding sheep on oil-cake or corn, having mostly farmed turnip-land, upon which sheep never failed to become abundantly fat without any adventitious aid. On deaf and clay soils, however, oil-cake may prove beneficial; and it may be presented in these boxes to sheep on grass in winter as their entire food. Oil-cake has the effect of keeping the dung of sheep in a moist state. It is supplied them in a bruised state, partly in powder, and partly in bits, as it falls from the oil-cake crusher—a convenient machine, the construction and operation of which will be described when treating of the feeding of cattle. I believe there is little use of measuring the quantity of oil-cake to sheep, even when on turnips, as they will eat it when inclined, and some sheep eat it more heartily than others. The discriminating choice of food manifested by sheep is a valuable hint, in fattening them, to supply them with

Fig. 287.

THE OIL-CAKE OR CORN BOX FOR FEEDING SHEEP.

different kinds of food, such as oil-cake, corn, hay, straw, and turnips, at one and the same time, that every sheep may take his choice daily; but, in case such a mode of feeding may be costly, it is worth while to try experiments on the subject, in order to ascertain whether, when a number of articles are presented at the same time to sheep for their choice, less of the most costly kind is not proportionally consumed than when supplied separately. On this principle, corn may be put in one box, and oil-cake in another, and so of other substances; and, although it is an indubitable fact that sheep will feed quite fat upon turnips alone with fodder on turnip-soil, yet they may become sooner ripe upon mixed than simple food; and the time thus gained may more than compensate (or at least compensate) for the cost of the various materials employed in feeding.

(1073.) *Salt* has been frequently given to sheep on turnips, but with what advantage I have never satisfactorily learned. I have given them it, and the eagerness with which they followed the shepherd when he came at the stated hour to lay down small quantities here and there over the break, upon flat stones, and the relish with which they enjoyed it, was very remarkable; yet the great desire for it continued but a short time, and then every day they took so little that it appeared as if they were trifling with it; and hence I could perceive no benefit they derived from its use. Perhaps the cultivator who paid the greatest attention to the use of salt to animals was the late Mr. Curwen, of Workington Hall, Cumberland, who used to give from 2 to 4 ounces per week to sheep, if fed on dry pastures; but, if feeding on turnips or rape, they were supplied without stint. "It is, in fact, *indisputably proved*," says Mr. Cuthbert W. Johnson, "that if *sheep are allowed free access to salt, they will never be subject to the disease called the rot*. Is not this a fact worthy of a farmer's earliest, most zealous attention? Some recent experiments also lead me even to hope that I shall one day or other be able to prove it to be a *cure* for this devastating disease. I have room but for one fact: Mr. Rusher, of Stanley, in Gloucestershire, in the autumn of 1828, purchased, for a mere trifle, 20 sheep, *decidedly rotten*, and gave each of them, for some weeks, 1 ounce of salt every morning. 2 only died during the winter; the surviving 18 were *cured*, and have now, says my informant. lambs by their sides."*

(1074.) There are some inconveniences attending the feeding of sheep on turnips *in winter*, which necessarily you should be made aware of. A heavy rain may fall for some days, and render the land quite soft and poachy, though it had been previously thoroughly drained, or even naturally dry. As the wet will, in such a case, soon subside, the removal of the

* Johnson's Observations on the Employment of Salt.

sheep for a night and day to an old grass-field will give the land time to become firm; and a small quantity of oil-cake will suffice to support the sheep all the time they will be in the grass-field. A very heavy rain may fall in a day, and inundate the lower end of the field with water, which may take some days to subside. The best way of preventing the sheep approaching the inundated part is to fence it off with a net. A fall of snow, accompanied with wind, may cover the sheltered part of the field, and leave the turnips bare only in the most exposed. In this case, the sheep must feed in the exposed part, and the racks placed there for shelter. But the snow may fall heavily, and lie deep over the whole field, and cover every turnip out of reach. Two expedients only present themselves in such a case; the one is to cast the snow from the drills containing the turnips, and pile it upon those which have been stripped. This task cannot be performed by the shepherd alone, or by the field-workers. The plowmen must bring their stable-shovels, fig. 149 or fig. 176, and clear the turnips; but, in doing this in severe frost, too many turnips should not be exposed at one time, in case they become frosted, which they are apt to be when exposed suddenly to frost from under snow. The advantage of casting the snow is that it gives the sheep an immediate access to the turnips; but a disadvantage attends it when the snow lies for a considerable time, all the manure being left by the sheep in the channels cut out of the snow, and, of course, none in those parts upon which the snow has been piled. The best plan to pursue at first, under the circumstances, is, in my opinion, to adopt the other expedient alluded to above—namely, to give the sheep oil-cake in the troughs, fig. 227, for a time, in a sheltered place of the field, until it is seen whether the snow is likely soon to disappear; and, should it lie longer than afford time to consume the turnips, then the first expedient of casting off the snow may be resorted to at once, and its disadvantages submitted to. In the great fall of snow, in spring 1823, my turnip-field was covered over 4 feet deep. Having no oil-cake, and finding it impossible to remove the sheep, the snow was cast in trenches, in which they soon learned to accommodate one another, and all throve apace. A fresh fall of snow a few days after came from the opposite quarter, and covered up the trenches, which had to be cleared out again. The snow continued upon the ground until the end of April, and as there was no time after that to put manure on the land which had been covered with piled-up snow—and, indeed, its soft state rendered the operation impracticable—the succeeding crop of barley grew in strips corresponding to the trenches. Even a supply of oil-cake would not, in this case, have superseded the trenching of the snow, to get the turnips eaten in time for the barley-seed.

(1075.) While young sheep and tups are thus provided with turnips during winter, the *ewes in lamb* find food on the older grass, which, for their sakes, should not be eaten too bare in autumn. Where pastures are very bare, or when snow covers the ground, they should either have a few turnips thrown down to them upon the snow, or, what is better, clover-hay given them in a sheltered situation. The best hay for this purpose is of broad or red clover, and next meadow-hay; but as you can only give the kind you happen to have, much rather give them turnips than hay that has been heated or wetted, or is moulded, as in either of those states it has a strong tendency to engender diseases in sheep, such as consumption of the lungs and rot of the liver; and in regard to great ewes, it is apt to make them cast lamb. If turnips cannot be had, and the hay bad, give them sheaves of oats, or clean oats in troughs, or oil-cake; but whatever extraneous food is given, do not supply it in such quantity as to fatten the ewes, but only to keep them in fair condition. In the severe snow-storm of 1823, I put my ewes into an old Scots-fir plantation, into which only a small quantity of snow had penetrated, and there supplied them with hay laid on the snow around each tree. A precaution is requisite in using a Scots-fir plantation in snow for sheep; its branches intercepting the snow in its fall to the ground are apt to be broken by its weight, and fall upon the sheep and kill them; and in my case, a ewe was killed on the spot by this cause. The branches should therefore be cleared of the snow around where the sheep are to lodge by shaking them with poles or long forks, assisted by ladders if the case require it. In driving ewes heavy with lamb, through deep snow to a place of shelter, plenty of time should be given them to creep along, in case they should overreach themselves, and the exertion thereby cause them to cast lamb.

(1076.) In some parts of Scotland, and more generally in England, rape as well as turnips is grown for winter food for sheep. The rape (*Brassica rapus oleifera* of De Candolle) cultivated in this country, is distinguished from the colsat of the Continent by the smoothness of its leaves. It has been cultivated for the fattening of sheep in winter from time immemorial. The green leaves, as food for sheep, are scarcely surpassed by any other vegetable, in so far as respects its nutritious properties; but in quantity it is inferior both to turnips and cabbages. Its haulm may be used as hay with nearly as much avidity as cut straw.* The consumption of rape by sheep should be conducted in exactly the same manner as that of turnips. In England, that intended for sheep is sown broadcast and very thick, in which state it is certainly very suitable for them. In Scotland, it is raised in drills like turnips; and although not so conveniently placed for sheep as the broadcast, the top leaves being somewhat beyond their reach from the bottom of the drill, yet this form permits every cleansing process of the land during summer, and thus renders the culture of rape as ameliorating a crop for land as any other green crops raised for the purpose. It is acknowledged on all hands that, for raising seed for oil, the drill form of culture is far the best.

(1077.) Every kind of sheep, of whatever breed, when kept in the low country, should be treated in winter in the way described above, though the remarks there are meant to apply to the peculiar management of Leicester sheep, which is the usual breed cultivated where sheep form an integral part of the mixed husbandry. Where a Leicester flock is so kept, the ewes are regarded as a *standing flock*; that is, they have themselves been bred upon the farm upon which they are supported, and are used as breeders, until considered no longer profitable, when they are fed off. But on many lowland farms, the mixed husbandry is only practiced to a partial extent, no flock of ewes being kept for breeding, and only wethers intended to fatten on turnips are bought in on purpose. Some farmers, instead of wethers, buy old ewes, dinmonts, or lambs. When wethers are bought, the breeds generally selected for the purpose are Cheviots and Black-faced from the mountains, where they are bred, and where large standing flocks of ewes are kept for the purpose of supplying the demand for lambs. Turnip-sheep are thus easily obtained at fairs in autumn; but where certain stocks have acquired a good name, purchasers go to the spot and buy them direct from the breeders.

(1078.) Sheep on turnips have little shelter afforded them but what the fences of the field can give. In some cases, this is quite sufficient; but in others, it is inadequate. Of late years, the subject of shelter has attracted attention, and artificial means have been suggested, consisting of various devices involving different degrees of cost, not merely for protection against sudden outbreaks of weather, but with the view of gradually improving the condition of sheep, both in carcass and wool. It is a natural expectation that a fat carcass should produce more wool, and constant shelter improve its quality.

(1079.) One plan for shelter and comfort, a slight remove from the usual practice, was first tried by Mr. Hunter of Tynefield, in East-Lothian, in 1809, by littering the break occupied by the sheep in the field with straw, and supplying them with turnips upon it. In this way he littered 300 sheep upon 25 acres of turnips, which afforded 36 tons the acre, with the straw of 60 acres of wheat, weighing 1 ton the acre imperial. The sheep were thus treated 5 months on the ground, and fetched 2s. a head more than those treated in the usual manner. This increase of price is an advantage; but it is not all advantage, as the trouble of leading, at intervals, 60 tons of straw to the field; of leading the same, in the shape of manure, from that field to another; and of carrying the turnips from the drills to the fold, should be deducted from it. When turnips are laid upon

* Don's General Dictionary of Botany and Gardening, vol. i.

straw, sheep cannot bite them easily; and this is an objection to laying down whole turnips to sheep on grass, instead of cutting them with a turnip-slicer; and among damp litter, sheep almost invariably contract foot-rot, as 7 of Mr. Hunter's did.*

(1080.) Another plan of affording shelter to sheep on turnips is that of movable sheds to lie in. Fig. 228 gives a floor-plan of such a shed, 15 feet long, 7 feet wide, with an opening of ½ of an

Fig. 228.

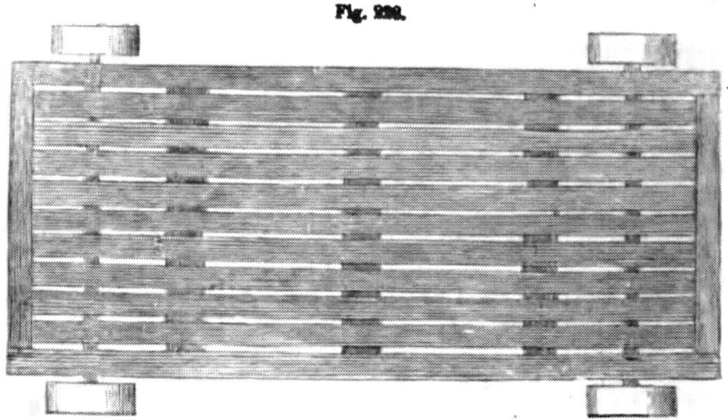

THE FLOOR OF A MOVABLE SHED FOR SHEEP ON TURNIPS.

inch between the floor-deals. The floor-frame rests on 2 axles of iron supported upon 4 iron wheels, 1 foot diameter, which raise it 6 inches above the ground. Fig. 229 gives a side elevation of the shed, with the form of the roof, made of deals, lapping over each other, and elevated 5 feet above the floor; and fig. 230 is an end elevation of the same. One side and both ends, when the shed is in use, could be boarded in the quarter from which the wind comes; and if the boards are fastened dead, the shed should be wheeled round to suit the wind; but if boarding is considered too expensive a mode of fitting up such sheds, hurdles clad with thin slabs, or wattled with straw or willow against the ends and side, might answer the same purpose. A horse is required to wheel such a shed to any distance. A shed of the above dimensions might accommodate about a score of sheep, and its cost is said to be £4. But should this construction be considered too unwieldy, the shed could be made of two pieces of half the size, which would easily be moved about by people, and when placed together on end, would form an entire shed of the proper dimensions. Thus, fig. 231 represents two short floors placed together on 8 wheels; and fig. 232 a side-elevation and roof of two half-sheds, mounted on wheels set together. The scale attached to fig. 232 gives the relative proportions of every part. The cost of 2 half-sheds will of course be more than a whole one. Whether any one will incur the cost of sheltering sheep on turnips

Fig. 229.

THE SIDE-ELEVATION OF A MOVABLE SHED FOR SHEEP ON TURNIPS.

Fig. 230.

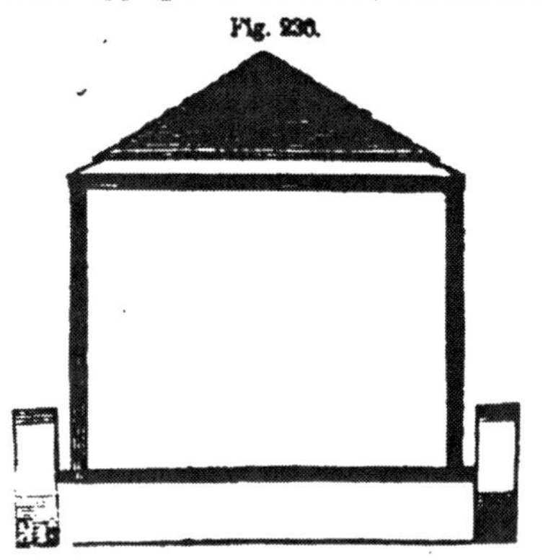

THE END ELEVATION OF A MOVABLE SHED FOR SHEEP ON TURNIPS.

* Sinclair's Account of the Husbandry of Scotland, vol. ii. Appendix.

sheds is, I conceive, questionable; and it might be some time ere sheep would be in- to enter them.*

Fig. 231.

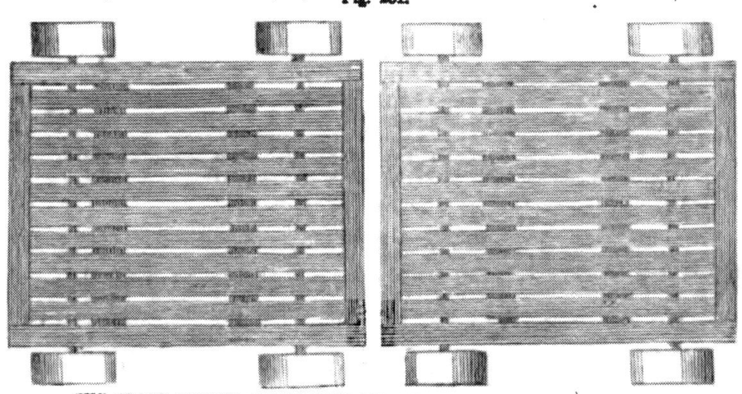

THE FLOOR OF TWO SHORT MOVABLE SHEDS FOR SHEEP ON TURNIPS.

(1081.) A third plan is to erect sheds and courts at the steading, to be littered when required and the sheep daily supplied with cut turnips. This plan, as I conceive, would afford more shelter and protection than by putting down litter, or erecting movable sheds in the field. I re-

Fig. 132.

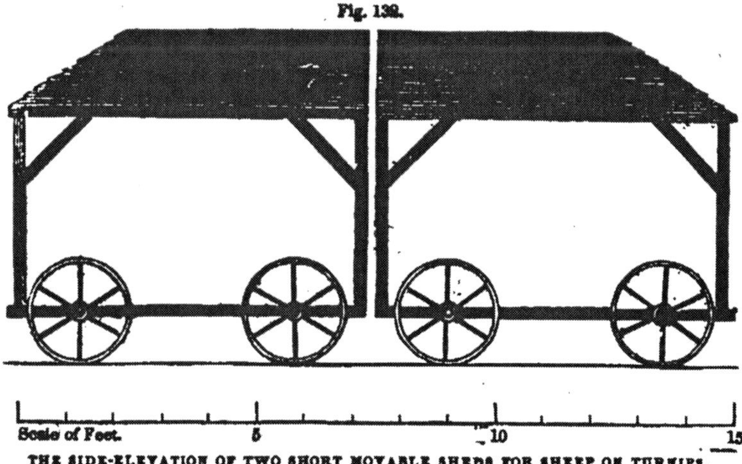

THE SIDE-ELEVATION OF TWO SHORT MOVABLE SHEDS FOR SHEEP ON TURNIPS.

member of seeing, more than 20 years ago, the courts and sheds erected at his steading by the late Mr. Webster of Balruddery, Forfarshire; so that the recent practice and suggestions on the subject by English sheep-feeders possess at least no novelty. The results of Mr. Webster's experiments, I believe, were not very encouraging. Mr. Childers, M. P. for Malton, fed 40 Leicester wether-hoggs on turnips, 20 in the field and 20 in a shed. The shed consisted of a thatched erection of rough deals, having a floor of slabs raised 18 inches above the ground, with a small court belonging to it. The boarded floor was swept every day, and fresh straw put over the court after every shower of rain. The sheep were divided into as equal lots as could be drawn, the score to be fed in the shed weighing 183 stones 3 lbs., and those in the field 184 stones 4 lbs. Each lot got as many *cut* turnips as they could eat, which amounted to 27 stones every day; 10 lbs. of linseed cake, or ½ lb. to each sheep, per day; ¼ pint of barley to each sheep; and a little hay, and a constant supply of salt. They were fed from 1st January to 1st April; and, on the fourth week, the hoggs in the shed ate 3 stones fewer turnips every day; in the ninth week, 2 stones still fewer and of linseed-cake 3 lbs. per day. The results were these:

Date.	20 shed-hoggs.		Increase.		20 field-hoggs.		Increase.	
	Sts.	lbs.	Sts.	lbs.	Sts.	lbs.	Sts.	lbs.
January 1	183	3			184	4		
February 1	205	0	21	11	199	8	15	4
March 1	215	10	10	10	208	2	8	8
April 1	239	[illegible]	23	11	220	12	12	10
Total increase..........			56	6			36	8

* Quarterly Journal of Agriculture, vol. xi.

(889)

"Consequently," says Mr. Childers, "the sheep in the shed, though they consumed nearly 1-5 less food, have made ⅓ greater progress."* Thus, in 4 months, the shed-fed hoggs gained about 1 stone a head more than those in the field, and were worth 8s. a head more. This experiment of shed-feeding corroborates the ordinary experience in the progress of fattening sheep: namely, that the greatest progress is made at the beginning and end of the season. In the beginning, the fat is laid on in the inside, to fill up; and at the end it is laid on on the outside, after the acquirement of muscle in the intermediate period.

(1082.) Lord Western pursues the plan of shed-feeding his Anglo-Merino sheep, to the extent of confining them all the year round. His folding-yards are spacious, and surrounded by sheds, which are only 10 feet wide, and 6 or 7 feet high—built, in the cheapest manner, of timber that would otherwise be burnt. The yards are well littered, and to a considerable depth, and they never heat. After three years' experience, his lordship is "decidedly of opinion that the fatting stock thrive quicker, and the sheep with their lambs also do better than out of doors." Turnips, cabbages and salt constitute their food.†

(1083.) Similar experiments have been tried in Scotland with success. Mr. Wilkin, Tinwald Downs, Dumfriesshire, fed 20 cross-bred Cheviot and Leicester hoggs in courts and sheds, on turnips, grass and oil-cake, and their increased value over others in the field was estimated at from 22s. to 25s.; and Mr. John MacBryde, Belkar, fed both Leicester and Cheviot wethers in *stalls* on turnips, rice, sago, sugar, and linseed-oil, and realized 7s. a head more than from those fed in the field.‡ But, in estimating the advantages derived from shed-feeding, the trouble occasioned in bringing the turnips from, and taking the manure to the field, should always be borne in mind. But should the plan leave no profit, yet if it improve the quality of the wool in its most essential particulars, it is worthy of consideration.

(1084.) Sheep are not fed on turnips on every kind of farm. Carse-farms are unsuited to this kind of stock; and, where turnips can be raised on them, cattle would be more conveniently fed. There being, however, abundance of straw on clay-farms, sheep might be fed in small courts and sheds at the steading on oil-cake, or any other succedaneum for turnips.

(1085.) On farms in the neighborhood of *large* towns, whence a supply of manure is obtained at all times, turnips are not eaten off with sheep; but on those near small towns they are so employed to manure the land. They are bought in for the purpose, and consist of Cheviot or Black faced wethers, or Leicester hoggs, or draft-ewes, which, if young, feed more quickly than wethers of the same age.

(1086.) On dairy-farms there is as little use for sheep as near towns, except a few wethers to eat off part of the turnips that may have been raised with bone-dust, or any other specific manure, in lieu of farm-yard dung.

(1087.) On pastoral farms, sheep are not fattened on turnips; but their treatment in winter possesses exciting interest. There are *two kinds of pastoral* farms, and, as this is the first opportunity I have had of considering the peculiarities of their management, I shall here make some general remarks on their constitution and fitness for rearing sheep.

(1088.) The first thing that strikes you on examining a pastoral country is the *entire want of shelter*. After being accustomed to see inclosed and protected fields in the low country, the winding valleys and round-backed hills of a pastoral one appear, by comparison, naked and bleak.—You are not surprised to find bare mountain-tops and exposed slopes in an alpine country, because you scarcely conceive it practicable for man to inclose and shelter elevated mountains; but among green hills and narrow glens, where no natural obstacles to the formation of shelter seem to exist, but, on the contrary, whose beautiful outlines indicate sites for plantations that would delight the eye of taste, independent of their utility as shelter to their owner's habitation; and he, having experienced their utility in that respect, could not refuse similar comfort to the dumb and patient creatures dependent on his bounty. Hence, the hurricane that a planter arrests in its progress toward his own dwelling, ceases at the same time to annoy the peace of his flocks and herds. The chief difficulty of forming shelter by planting is the expense of inclosing it; for, as to the value of trees from a nursery, it is a trifle compared to the advantage derived from the shelter where they grow; and yet, in a mountainous country, there is no want of materials for inclosing, no want of rock to produce stones for building rough but substantial stone-dykes; labor is but required to remove and put them together; and, as a simple means of their removal, it is surprising what a quantity a couple of men will quarry, and a couple of single-horse carts will convey, in the course of a summer. The carriage, too, in every instance, could be made down-hill, fresh rock being accessible at a higher elevation as the building proceeds upward.

(1089.) Suppose a hill-farm containing 4 square miles, or 2,560 acres, were inclosed with a ring-fence of planting of at least 60 yards in width, the ground occupied by it will amount to 174 acres. A 6-feet stone wall round the inside of the planting will extend to 13,600 roods of 6 yards, which, at 6s. 6d. per rood, will cost £612. But the sheltered 2,386 acres will be worth more to the tenant, and of course to the landlord, than the entire 2,560 acres unsheltered would ever have been, while the proprietor will have the value of the wood for the cost of fencing. Besides, it should be borne in mind by the proprietor that planting, as a ring-fence to one farm, shelters one side of 4 other farms of the same size, which is an inducement to extend the benefits of shelter; and these, moreover, can be afforded on a large scale at a cheaper cost than on a small—so much so that, were neighboring proprietors to undertake simultaneously the sheltering of their farms on a systematic plan, not only would warmth be imparted over a wide extent of country, but efficient fencing would be accomplished along march fences at half the cost to each proprietor.

(1090.) Low pastoral farms should be stocked with *Cheviot*, and high with the more hardy *Black-faced*, sometimes called the Heath and Mountain Sheep; and, although the general treatment of both breeds is nearly alike, yet their respective farms are laid out in a different manner. A Cheviot sheep-farm contains from 500 to 2,000 sheep; that which maintains from 500 to 1,000

* Journal of the Royal Agricultural Society of England, vol. I.

† Mark-Lane Express, 16th Dec. 1839. ‡ Quarterly Journal of Agriculture, vol. xi.

is perhaps the highest rented, being within the reach of the capital of many farmers; and one that maintains from 1,000 to 2,000 is perhaps the most pleasant to possess, and, if it have arable land attached to it, will afford pretty good employment to the farmer, though, with good shepherds under him, and no arable farm, he could manage the concerns of 6,000 sheep as easily as those of 500. A shepherd to every 600 Cheviot sheep is considered a fair allowance, where the ground is not very difficult to traverse, and it may be held as a fair stent to put 1,000 sheep on every 1,200 acres imperial."*

(1091.) Every Cheviot sheep-farm should have arable land within it, to supply turnips and hay to the stock, and provision to the people who inhabit it. It is true that all the necessaries, as well as the luxuries, of life may be purchased; but no dweller in the country will hesitate a moment to choose the alternative of raising the necessaries of life and having them at command, to going perhaps many miles to purchase the most trivial article of domestic use. It is not easy to determine the proportion which arable land should bear to pastoral, to supply the requisite articles of provision; but perhaps 2 acres arable to every 20 breeding-ewes the pasture maintains may supply all necessaries. Taking this ratio as a basis of calculation, a pastoral farm maintaining 1,000 ewes, a medium number, would require 100 acres of arable land, which would be labored by 2 pair of horses, on a 4-course shift; because, *pasture* not being required on the *arable* portion of the farm, new grass will be its substitute. The farm will thus be divided into 25 acres of green crops, 25 acres of corn after them, 25 acres of sown grasses, and 25 acres of oats after the grass.—Manure will be required for 25 acres of green crop, which will partly be supplied by the 100 acres of straw, by bone-dust, and by sheep on turnips after bone-dust. To render the straw into manure there are 4 horses; cows of the farmer, the shepherd, and plowmen; with perhaps a few stirks, the offspring of the cows, and a young colt or two, in the farm-yard. The arable land should have a ring-fence of thorn, if the situation will admit of growth or of stone.

(1092.) The *steading for such a farm* may be of the form of fig. 28, containing a 4-horse threshing-mill, driven by water if possible, by horses by necessity; a corn-barn, straw-barn, chaff-house, stable, byre, cart-shed, wool-room, and implement-room for the shepherd's stores.

(1093.) The pasture division of the farm should be subdivided into different lots, varying in number and dimension according to the age and kind of the stock to be reared upon each. The nature of the land determines the age and kind of stock to be reared upon it; for it is found that some land will not suit breeding-ewes, and others are unsuitable for hoggs. If the pasture consist chiefly of soft, rough land, hoggs are best adapted for it; but, if short and bare, ewes will thrive best upon it. That farm is best which contains both conditions of pasture, to maintain both breeding and rearing stock. In subdividing a farm into lots, each should, as much as possible, contain within itself the same quality of pasture, whether rough or short; for, should fine and coarse grass be included within the same lot, the stock will remain almost constantly upon the fine, to the risk of even reducing their condition. To the extent of 1-5 of coarse to fine may be permitted within the same lot, without apprehending much detriment to stock. Should a large space of inferior soil lie contiguous to what is much better, they should be divided by a fence, and, if requisite, a different breed of sheep reared upon each. By these arrangements not only a greater number of sheep may be maintained upon a farm, but the larger number will always be in better condition.†

(1094.) The draining of pastoral farms is an operation of great importance, as a superior class of plants will thereby be encouraged to grow in places occupied by coarse herbage, nourished by superabundant and stagnant water. A plan of laying out hill-drains may be seen in fig. 145.—Their collected waters may be conveyed away to a contiguous rivulet or hollow in open main-drains, like that in fig. 146. A spouty swamp, of whatever extent, and wherever occurring, should be drained by coupled stone-drains, like fig. 159, cut to the bottom of under water; and the ordinary drains for conveying the water in the branches should be formed with a cover, like fig. 147. The arable portion of the farm should, of course, be drained by parallel drains, as represented in fig. 186, of the form of coupled drains, like fig. 159; and, if tiles are near as well as stones, like fig 185. One means of keeping part of the surface dry is to have the channel of every rivulet, however tiny, that runs through the farm, scoured every year in those parts where accumulated gravel causes the water, in rainy weather or at the breaking up of a storm, to overflow its banks; because the overflowed water, acting as a sort of irrigation, sets up a fresh vegetation, which is eagerly devoured by sheep in spring, to the risk of their health; and the sand carried by it is left on the grass on the subsidence of the water, much to the injury of the teeth and stomachs of the sheep. The confinement of water within its channels also prevents it leaving the land, where inundated, unduly wet.

(1095.) In recommending a connection of arable with a pasture farm, my object is simply to *insure* an abundant supply of provision for sheep in winter. Were our winters so mild as to allow sheep to range over the hills in plenty and safety, no such connection need be formed—or, at least, to a greater extent than would supply provisions to its inhabitants, when situated far from a market. But when we are aware that severe storms at times almost overwhelm a whole flock, and protracted snows and frosts debar the use of the ground for weeks together, it is necessary that provision be made for the support of stock in those calamitous circumstances; and, surely there is no better or more legitimate mode of supporting them than of raising provision for them upon their own ground. I am quite aware of the folly of trusting to corn in a high district for rent, and am also aware that stock alone must provide that; and I have seen too many instances of failure in trusting to corn and neglecting stock; nevertheless, it cannot be denied that the *more stock are provided with food and shelter in winter*, the less loss will be incurred during the most inclement season. Let one instance, out of many that could be adduced, suffice to show the com-

* Little's Practical Observations on Mountain Sheep.

† A Lammermuir Farmer's Treatise on Sheep in High Districts. The Lammermuir Farmer was the late Mr. John Fairbairn, Hallyburton, a man of good sense and an excellent farmer, and whose acquaintance I was happy to cultivate.

parative immunity from loss enjoyed by food and shelter being provided for sheep in winter. In the wet and cold winters of 1816 and 1818, the *extra*—that is, the more than usual—loss of sheep and lambs on the farm of Crosscleuch, Selkirkshire, was as follows:

In 1816,	200 lambs, at 8s. each	£80	
	40 old sheep, at 20s. each	40	
			£120
In 1818,	200 lambs, at 8s. each	£80	
	30 old sheep, at 20s. each	30	
			£110
	Value of total *extra* loss		£230

whereas, on the farm of Bowerhope, belonging to the same farmer and on which ½ more sheep are kept, the *extra* loss in those years was as follows:

In 1816,	70 lambs, at 8s. each	£28	
	10 old sheep, at 20s. each	10	
			£38
In 1818,	50 lambs, at 8s. each	£20	
	8 old sheep, at 20s. each	8	
			£28
	Value of total *extra* loss		£66
	Deduct loss on Crosscleuch		230
	Value saved on farm of Bowerhope		£164*

(1096.) Food and shelter being both necessary for the proper treatment of sheep in winter on pastoral farms, the means of supplying them demand the most serious attention of the store-farmer. During winter, sheep occupy the lower part of the farm. Hoggs are netted on turnips in the early part of the season, and ewes and other sheep subsist on the grass as long as it is green. The division allotted to green crop in the arable part of the farm contains 25 acres, and, allowing 3 acres for potatoes for the use of the farmer and his people, there remain 22 acres for turnips; and as land among the hills is generally dry, turnips grow well upon it; so that 30 double-horse cart-loads to the acre, of 15 cwt. each, may be calculated on for a crop. It is judiciously recommended by Mr. Fairbairn to carry off, about the end of October or beginning of November, if the weather is *open*—that is, fresh—before the grass fails, 4-5 of the turnips, and store them in heaps as in fig. 213, and as described in (1016); and allow the *ewe-hoggs*, retained to maintain the number of the ewe-flock after the draft-ewes have been disposed of, to eat the remaining 1-5 off the ground, with whatever small turnips left when the others were pulled; and, to strip the land in that proportion, 1 drill should be left and 4 carried off. This is, as I conceive, an excellent suggestion for adoption on every hill-farm, especially as it secures the turnips from frost, and, at the same time, gives the entire command of them whenever they are required in a storm.

(1097.) It is found that *hoggs* fall off in condition on turnips in spring, in a high district, if confined exclusively upon turnip-land; not certainly for want of food, but probably from too much exposure to cold from want of shelter. They are, therefore, always removed from the turnips in the afternoon to their pasture, where they remain all night, and again brought back to the turnips in the following morning. It is obvious that this necessary treatment, under the circumstances, deprives the land of much of the manure derivable from the turnips; and hence farm-dung should be put on the land before the sowing of the following grain crop, where the previous turnips had been raised with bone-dust. The hoggs continue their daily visit to the land until all the turnips are consumed; which, amounting in all to 4½ acres, may last, under the peculiar treatment, 17 score of hoggs—the number kept for refreshing the ewe-stock—about 6 or 7 weeks. After the land has been cleared of the turnips, the hoggs should be daily supplied from the store on their pasture with 1 double cart-load to every 8 scores, which will be consumed in about 4 hours; and, after that, they depend on the grass for the remainder of the day. Round turnips, having no hold of the ground, give way to the upward bite of the sheep with the lower jaw teeth, and prove troublesome to them when laid down upon grass. When taken out of a store, they should therefore always be cut with a slicer. Hoggs are treated in this way until March, or longer if the weather is bleak; and the advantages of it are that they are maintained in their condition, and become proof against the many diseases which poverty engenders; and their fleece weighs 1 lb. more at clipping-time. The cost of 8 acres of turnips given to hoggs, valued at £3 an acre in a high district, is 17d. each, which is so far counterbalanced by the additional pound of wool which the cost insures, and which is worth from 10d. to 1s. per lb. The balance of 5d. to 7d. a head, which is the true cost of the keep of the sheep, is a trifle compared to the advantage of bringing them through the winter in a healthy state and in fair condition.

(1098.) As to the older sheep, they must partly depend, in frost and snow, upon the 14 acres of turnips yet in store, and upon hay, and, of course, upon pasture in fresh weather. The hay is obtained from the 25 acres of new grass, which may be all made into hay; but allowing 5 acres for cutting-grass given in suppers to horses and cows, there remain 20 acres for hay, which, at 120 hay-stones (of 22 lbs. to the stone) per acre, give 2,400 hay-stones, or 3,771 stones imperial. The 1,000 ewes will eat 1½ lbs. and the hoggs ½ lb. each every day, besides the two cart-loads of turnips among the lot. At this rate, the hay will last 31 days, which is a shorter time than many storms continue; but if the whole 25 acres of new grass were made into hay, it would last 40 days. But the rule should be to begin with a full hand of hay at the commencement of farming,

* Napier's Treatise on Practical Store-Farming.

and preserve what may be left over in a favorable season, and mix it with the new of the following season, for any subsequent unusual continuance of storm.

(1099.) But in storm, their provender cannot be given to sheep upon snow, safely and conveniently, as ground-drift may blow and cover both; and no place is so suitable for the purpose as a *stell*, a term, according to Dr. Jamieson, literally signifying a covert or shelter. There still many store-farmers skeptical of the utility of stells, if we may judge of their opinions their practice; but I presume no great sagacity is required to discover the fact, that stock much more comfortably lodged in a drifting storm within a high inclosure than upon an open heath. A stell may be formed of planting or high stone-wall. Either will afford shelter; but the former most, though most costly, as it should be fenced by a stone-wall. Of this class I conceive the form represented by fig. 233 a good one, and which may be characterized an *outside* stell. It

Fig. 233.

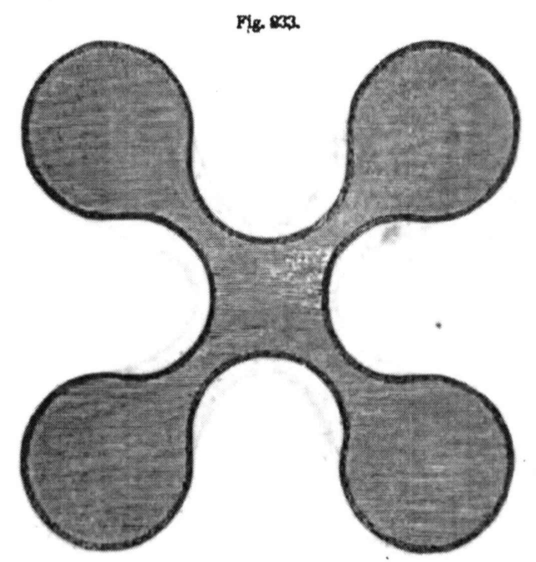

THE OUTSIDE STELL SHELTERED BY PLANTING

has been erected by Dr. Howison, of Crossburn House, Lanarkshire, and proved for 30 years. The circumscribing strong black line is a stone-wall 6 feet high; the dark ground within is covered with trees. Its 4 rounded projections shelter a corresponding number of recesses embraced between them, so that, let the wind blow from whatever quarter it may, two of the recesses will always be sheltered from the storm. The size of this stell is regulated by the number of the sheep kept; but this rule may be remembered in regard to its accommodation for stock, that each recess occupies about ⅛ part of the space comprehended between the extremities of the 4 projections; so that, in a stell covering 4 acres—which is perhaps the *least* size they should be—every recess will contain ½ an acre. "But, indeed," as Dr. Howison observes, and which observation applies to the general benefit derived from every species of shelter, "were it not from motives of economy, I know no other circumstance that should set bounds to the size of the stells; as a small addition of walls adds so greatly to the number of the trees, that they become the more valuable as a plantation; and the droppings of the sheep or cattle increase the value of the pasture to a considerable distance around in a tenfold degree."*

(1100.) As a modified improvement of this form of stell, Dr. Howison proposes the one in fig. 234, which consists in giving shelter in its interior as well as on the outside, and may therefore be denominated a *double* stell. This form has never yet been tried; but if made on an adequate scale, I have no doubt of its efficiency. Instead of one opening into its interior at *a*, I think it should have one at the head of each recess at *b*, for the purpose of facilitating the shifting of the sheep from the outside into the interior chambers *c*, on a dangerous change of the wind. The hay-stack will be conveniently placed in the center of the stell at *d*. This stell should scarcely occupy less ground than 7 acres to be really useful; thus, 1 acre in each of the projections, making 4 acres, divided into ½ an acre for each interior chamber *c*, and ½ an acre of wood around it; and each recess *b*, with the one at *a*, occupying ¾ of an acre, make other 3 acres, or 7 acres in all. But it would be better to occupy even more ground. The dark line circumscribing both the interior and exterior is a wall-fence, which would no doubt make this form of stell somewhat expensive, but it would have the great advantage of accommodating a large proportion of the flock for a long time at one place. Stells of this construction, besides affording shelter, would form embellishments to a pastoral country, and might, moreover, make a fence betwixt one farm and another. For instance, if it were desired to divide a 4-square-mile farm into one of 2 square miles, which had been fenced with a ring-fence planting, a few of these stells, placed in a row down the middle of the farm, with a single dyke from stell to stell, would not only divide the

* Prize Essays of the Highland and Agricultural Society, vol. xii.

large farm into two small ones, but provide stells for both; and being double, half the number of ordinary ones would suffice.

(1101.) In making stells of planting, I think it would be desirable to have the outside row of such trees as do not project branches from their tops: branches, in such a place, only serve to drop water upon the sheep lying in the outside recesses or inside chambers; and the dropping is so far injurious to the sheep as to chill them with cold, or entangle their wool with icicles, before they get up at daybreak to shake themselves free of the wet. This form of tree is found in the spruce, which affords, moreover, excellent shelter by its evergreen leaves and closeness of sprays, descending to the very ground. It should be employed to back the inside as well as the outside walls; and the space between them to be filled with Scots fir, larch, or such hard-wood trees as will grow at the elevation. It must, however, be borne in mind, that as every soil does not suit

Fig. 234.

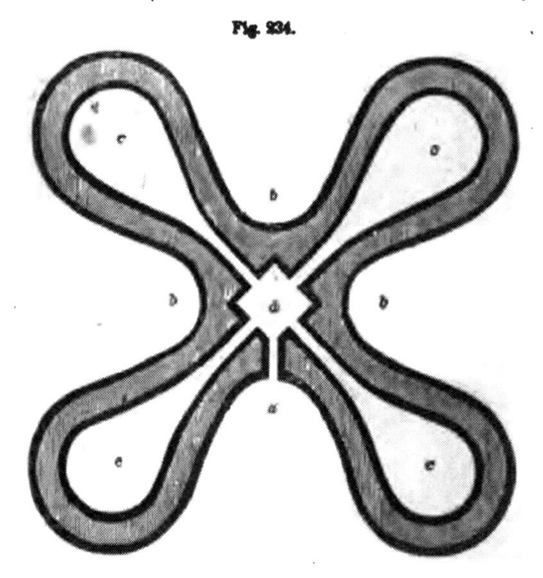

THE DOUBLE STELL SHELTERED BY PLANTING.

spruce, it is impossible to follow this rule implicitly. Larch grows best among the debris of rocks and on the sides of ravines; Scots fir on thin dry soils, however near the rock they may be; and the spruce in deep, moist soils.

(1102.) With regard to the number of stells or stone fences on a farm, Lord Napier recommends the establishment of what he calls a "system of stells," which would place one in the "particular haunt" of every division of the flock. In this view, he considers that 24 stells would be required on a farm maintaining 1,000 sheep; that is, 1 to little more than every 40 sheep.* However desirable it may be to afford full protection and shelter to stock, it is possible to overdo the thing—that is, incur more trouble and expense than necessary in accomplishing the object. On a farm where the practice is for the whole hirsel to graze together, it will almost be impracticable to divide them into lots of 40, one lot for each stell; and even if the division were accomplished, it would be with great waste of time, much bodily fatigue to the shepherd and his dog, and considerable heating to the sheep. I rather agree in opinion with Mr. William Hogg, shepherd at Stobohope, that stells should easily contain 200 sheep, or even 300 should be put into one on emergency; because, in the bustle necessarily occasioned by the dread of a coming storm, a large lot of 200 could easily be shed off from the rest, and accommodated in the recesses of a stell like fig. 233, which are accessible from all quarters; and 5 such stells would accommodate the whole hirsel of 1,000 sheep.

(1103.) Suppose, then, that 5 such stells were erected at convenient places, not near any natural means of shelter, such as a crag, ravine, or deep hollow, but on an open rising plain, over which the drift sweeps unobstructed, and on which, of course, it remains in less quantity than on any other place. With a stack of hay inside and a store of turnips outside, everything would be ready for the emergency. On a sudden blast coming, the whole hirsel might be safely lodged for the night in the leeward outside recesses of even one or two of the stells, and, should prognostics threaten a *lying storm*, next day, all the stells could be inhabited in a short time. Such a stell as fig. 234, filled outside and in, could hold the whole hirsel at one time. Lord Napier recommends a stack of hay to be placed close to the outside of every small circular stell; but these, I conceive, would be a great means of arresting the drift which would otherwise pass on.

(1104.) Mr. Fairbairn recommends a form of stell something like fig. 233, without the planting, having 4 concave-sides, and a wall running out from each salient angle, as in fig. 235; each stell

* Napier's Treatise on Practical Store-Farming.

to occupy ½ an acre of ground, to be fenced with a stone-wall 6 feet high, if done by the landlord; but if by the tenant, 3 feet of the wall to be built with stone, and coped other 3 feet with turf, which last construction, if done by contract, would not cost more than 2s. per rood of 6 yards.

Fig. 235.

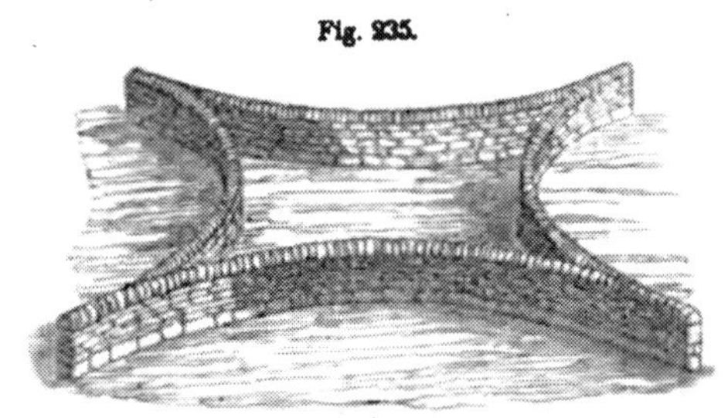

THE FORM OF STELL RECOMMENDED BY MR. FAIRBAIRN.

An objection to this form of stell without a planting is, when the wind strikes into any of the recesses, it is arrested in its progress by coming against the perpendicular face of the wall, from which it strikes upward, and then throws down the snow immediately beyond it; where, in this particular form, the drift would be deposited in the inside of the stell; and hence it is, I presume, that Mr. Fairbairn objects to sheep being lodged in the inside of a stell.* This form, though affording more shelter, seems open to the same objections as may be urged against the forms of the ancient stells, *a*, *b*, or *c*, fig. 236, the remains of many of which may be observed among the hills, and might yet screen sheep from a boisterous blast in summer.

Fig. 236.

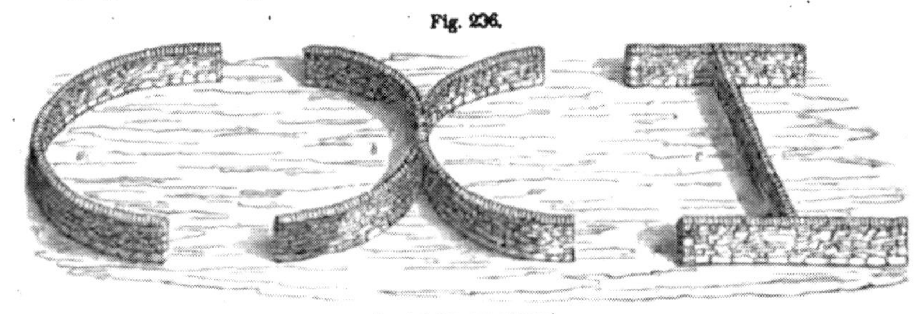

THE ANCIENT STELLS.

(1105.) There is much difference of opinion regarding the utility of *sheep-cots* on a store-farm.—These are rudely-formed houses in which sheep are put under cover in wet weather, especially at lambing time. Lord Napier recommends one to be erected beside every stell, to contain the hay in winter, if necessary; and Mr. Little even advises them to be built to contain the whole hirsel of sheep in wet weather. It seems a chimerical project to house a large flock of sheep for days, and perhaps weeks; and, if even practicable, it could not be done but at great cost. I agree with those who object to sheep-cots in high farms, because, when inhabited in winter, even for one night, by as many sheep as would fill them, an unnatural hight of temperature is occasioned.—Cots may be serviceable at night when a ewe becomes sick at lambing, or when a lamb has to be mothered upon a ewe that has lost her own lamb, because, these cases being few at a time, the cot never becomes overheated.

(1106.) In an unsheltered store-farm it is found requisite to have 2 paddocks, and the number is sufficient to contain all the invalid sheep, tups, and twin lambs, until strong enough to join the hirsel. Hay should be stacked within, and the turnips stored around the outside walls, or in the planting of the stells. Tups may graze with the hirsel in the early part of the summer; but, as no ordinary dyke will confine them in autumn, they should be penned in one of the stells, on hay or turnips, until put to the ewes.

(1107.) Where a rivulet passes through an important part of the farm, it will be advisable to throw *bridges* across it at convenient places for sheep to pass along without danger, either to better pasture or better shelter on the opposite bank. Bridges are best constructed of stone, and, though rough, if put together on correct principles, will be strong; but, if stones cannot be found fit for arches, they may do for buttresses, and, across these, trees may be laid close, and held together by transverse pieces 6 feet long, which, when covered with tough turf, will form a broad and safe roadway.

(1108.) These are all the remarks that occur to me in reference to the management of a low pastoral farm in winter; and, although many of them are equally applicable to a *high store-farm*, yet their circumstances are so far different as to warrant modifications of management. There is one circumstance which obviously renders modifications in management necessary, and that is the difference in habit betwixt the *Black-faced* and Cheviot breeds of sheep, the former being the best suited for a high farm. Some of the hill-farms extend to the highest points of our mountain ranges, to 4,000 feet above the level of the sea, and embracing many thousand acres; and, as land at that

* A Lammermuir Farmer's Treatise on Sheep in High Districts

elevation cannot be expected to yield much nutritious vegetation, many acres in some places are required to support a single sheep, so that a farm containing 1,000 sheep may require from 2,000 to 5,000 acres; but there are few hill-farmers who possess only 1,000 sheep. The circumstance of elevation and seclusion from roads also imposes modifications in the feeding from that pursued in the lower country. The store-farmers of the lower country sell what lambs they can spare, after retaining as many as will keep their ewe-stock fresh. They thus dispose of all their wether-boggs, the smaller ewe-boggs, and draft-ewes—which, if parted with at an early age, say 3 years, become more easily fattened on turnips in the low country than wethers of the same age. Suppose that 1,000 ewes wean 1,000 lambs, 500 of these will be wether and 500 ewe-hoggs; of which latter 17 score, or 340, will be retained, and the remaining 160 disposed of. It is the practice of the hill store-farmer, on the other hand, to purchase these lambs—rear them until fit, as wethers, to go to the low country to be fed fat on turnips—and, being a purchaser of lambs, keeps fewer breeding-ewes than wethers.

(1109.) It seems impracticable to have arable land on a hill-farm—at least, hill-farmers are unwilling to admit that turnips are the best food for their stock in winter. Whatever may prompt them to object to any arable culture on their farms, it would require very cogent reasons to prove that Black-faced sheep would not thrive well on turnips in the hills, if these could be raised in sufficient quantity upon the spot. Doubtless on many farms, far removed from the great thoroughfares of the country, it would be very difficult to bring even a favorable spot into culture, and especially to raise green crops upon them as they should be; but, on the other hand, there are many glens among the hills, not far removed from tolerable roads, in which culture might be practiced to great advantage, the produce of which would assist to maintain the condition of the flock through a stormy period of 6 weeks or 2 months.

(1110.) As a corroborative proof of the utility of some culture on hill-farms, it is the practice of many hill-farmers to take either turnips or a rough grazing for their stock in the lower part of the country, as nearly adjacent to their homes as food can be procured; and many lowland-farmers who possess hill-farms besides, bring down their young sheep to the low country in winter, and put them on turnips. When turnips, however, are taken for this purpose, a considerable expense is incurred, and a rough pasture, though less efficacious than turnips, may bring the stock through the dreary part of winter tolerably well; but the conveniences of home are wanting here, and when snow falls deep, and covers the ground for weeks together, little provision has been made to get at the turnips in the fields; and then whins and bushes afford the only food where there is no hay, but where there is, it is of course given them: but then, in this case, there was no use of incurring the expense and enduring the fatigue of the flock going from home, when hay could be given them in their own haunts. Hence the necessity, wherever turnips are, of storing a large proportion to be used in emergencies. Where a Scots-fir plantation is near a haunt of sheep, they need not starve; for a daily supply of branches, fresh cut from the trees, will not only support them, but make them thrive as heartily as upon hay alone; and, if a small quantity of hay is given along with the fir-leaves, they will thrive better than on hay alone.*

(1111.) One inducement may make some hill-farmers send their stock to a lower country in winter—namely, the want of adequate shelter at home. Their *hills* are bare of wood, the few trees being confined to the glens, and of course sheep can find no shelter in their usual grounds; and it is surprising how susceptible of cold even Black-faced sheep are when the atmosphere is becoming moist. They will cover down, creep into corners and beside the smallest bushes for shelter, or stand hanging their heads and grinding their teeth, having no appetite for food. If a piercing blast of wind follows such a cold day, the chances are that not a few of them perish in the night; and, if thick snow-drift comes on, they drive before it, apparently regardless of consequences, and get into some hollow, where they are overwhelmed. Thus the utility of stells becomes apparent, and many hearty wishes are no doubt expressed for them by the farmer and his shepherd, when too late to save the flock.

(1112.) Much diversity of opinion exists in regard to the best form of stell for high pastures, where wood seldom grows. At such a hight the spruce will not thrive; and the larch, being a deciduous tree, affords but little shelter with its spear-pointed top. There is no tree but the evergreen Scots-fir fit for the purpose; and, when surrounding a circular stell, such as is represented by fig. 237, it affords very acceptable shelter to a large number of sheep. In reference to this particular form of stell, it consists of 2 concentric circles of wall, represented by the dark lines in the figure, inclosing a planting of Scots-fir, and having a circular space *a* in the center for sheep, which can be made as large as to contain any number. This may be denominated an *inside* stell in contradistinction to that in fig. 233, and has been proved efficient by the experience of Dr. Howison. Its entrance, however, is erroneously made wider at the mouth than next the interior circle *a*, which has the effect of increasing the velocity of the wind into the circle, or of squeezing the sheep when they enter the passage in numbers. Were the passage parallel it would be better, but if wider at the inner end it would be of still better construction.

(1113.) But, where trees cannot be planted with any prospect of success, stells may be formed without them, and indeed usually are; and, of all the forms that have been tried, the *circular* has obtained the preference on hill-farms; but the difficulty of determining their size as the best is still a matter of dispute among hill-farmers. Lord Napier thinks that 7 yards diameter is a good size, and that the largest should not exceed 10 yards, inside measure; while Mr. William Hogg approves of 18 yards. I am inclined to agree with Mr. Hogg. In the first place, the circular form is better than a square, a parallelogram, or a cross; because the wind striking against a curved surface, on coming from any quarter, is divided into two columns, each weaker than the undivided mass; whereas, on striking against a straight surface, though its velocity is somewhat checked, it is still undivided, and its force still great, when it rebounds upward with increased force, and, curling over the top of the wall, throws down the snow a few yards only beyond it—that is, into the interior of the figure. Any one who has noticed the position of drifts of snow on each side

* Little's Practical Observations on Mountain Sheep.

of a straight stone-dyke, will remember that the leeward-side of the dyke is completely filled up, and that on the windward a hollow is left often clear to the ground between the snow and the dyke. Every form of stell, therefore, that presents a straight face to the drift, will be filled up be-

Fig. 237.

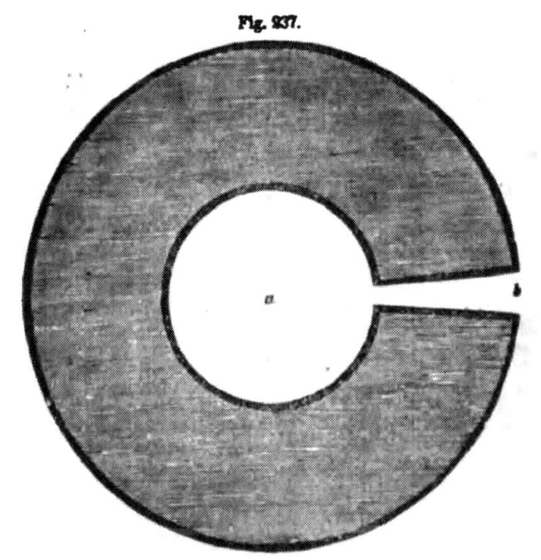

THE INSIDE CIRCULAR STELL, SURROUNDED WITH PLANTING.

hind the front wall, and can be no protection to sheep against being blown over with snow Of two curves, that which has the larger diameter will, it is obvious, divide the drift the farther asunder. A stell of small diameter, such as 7 yards, dividing a mass of drift, the current of air immediately over the stell is suddenly cut in two, but to so small a hight that the snow from the air above falls between them into the stell. When, on the other hand, a stell of large diameter, as of 18 yards, divides a column of air, this is so much deflected on each side that the current above the stell is widely divided to a considerable hight, and, long before it regains its former state, it has passed over the stell, where it deposits its snow; and hence, near such a stell the snow is found to accumulate in a triangular shape, with its apex away from the stell quite to leeward of the most distant part of the wall, and of course leaves the interior free of snow. Fig. 238 repre-

Fig. 238.

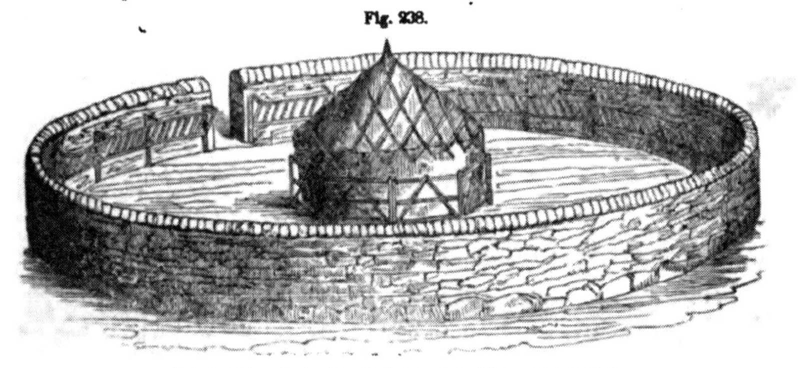

THE CIRCULAR STELL FITTED UP WITH HAY-RACKS.

sents one of 18 yards diameter inside, surrounded by a wall of 6 feet high, the first 3 feet of which may be of stone, and the other 3 feet of turf, and will cost 2s. 4d. per rood of 6 yards, if erected by the tenant, but if by the landlord, and wholly of stone and coped, will cost 7s. per rood: this size gives 9½ roods, which at 7s. makes its cost £3 5s. 4d., including the quarrying and carriage of the stones—a trifling outlay compared to the permanent advantage derived from it on a hill-farm.—The opening into the stell should be from the side toward the rising ground, and its width 3 feet, and of the whole hight of the wall, as seen in the figure; or it is sometimes a square of 3 or 4 feet, on a level with the ground, in which case the stell is entered by stile-steps over the wall. Such a structure should supersede the use of every antiquated form that has been tried, such as the single crescent *a*, fig. 236, double crescent *b*, or double T *c*; and it will easily contain 10 score of sheep for weeks, and even 15 or 16 score may be put into it for a night without being too much crowded together.

(1114.) Stells should be fitted up with *hay-racks* all round the inside, as in fig. 238, not in the expensive form of the circle, but of a many-sided regular polygon. It is a bad plan to make sheep eat hay by rotation, as recommended by Lord Napier and Mr. Little, but condemned by Mr. Fairbairn, for the timid and weak will be kept constantly back, and suffer much privation for days at a time. Let all have room and liberty to eat at one time, and as often as they choose. The hay-stack should be built in the center of the stell, as in fig. 238, where it should be placed on a basement of stone, raised 6 inches above the ground. A small stack, 5 yards in diameter at the base, 6 feet high in the stem, with a top of 6 feet in hight, will contain about 450 hay-stones of hay, which will last 200 sheep 33 days, about the average duration of a long storm; but upon the same base a much greater quantity of hay could be built. The interior circumference of the stell measures 160 feet round the hay-racks, and if 8 or 9 six-feet flakes were put round the stack, at once to protect the hay and serve as additional hay-racks, that would give 47 feet more, which together afford 1 foot of standing room to each of 200 sheep at one time, and supersede the objectionable plan of feeding them in rotation.

(1115.) Stells form an excellent and indispensable shelter for sheep in a snow-storm, when deprived of their pasture; but it has occurred to me that, in want of stone-stells, very good stells or chambers might be made of snow of any form or size desired. Even around the space occupied by sheep, after a heavy fall of snow, a stell might be constructed of the snow itself, taken from its interior and piled into walls as wide and high as required. Such a construction would remain as long as the storm endured, a new storm could be made available for repairs, and even after the ground was again clear, the snow-walls would remain as screens for some time after. A small drain or two, in case of a thaw, would convey away the water as the snow melted. As long as the ground continues green, natural shelter is as requisite as stells; these consist of rocks, crags, braes, bushes, heather, and such like. To render these as available to sheep as practicable, the ground should be cleared of all obstructions around them, and bushes planted in places most suited to their growth, such as the whin *(Ulex europæa)*, in poor thin clay, and it is a favorite food of sheep in winter; the broom *(Genista scoparia)*, on rich light soil; the juniper *(Juniperus communis)*, in sandy soil; the common elder *(Sambucus nigra)*, in any soil, and it grows well in exposed, windy situations; the mountain ash *(Pyrus aucuparia)*, a hardy grower in any soil; and the birch when bushy *(Betula alba)*, grows in any soil, and forms excellent clumps or hedges for shelter, as well as the hazel *(Corylus avellana)*, and the common heaths *(Erica vulgaris* and *tetralix)*, when they get leave to grow in patches to their natural hight in peaty earth.

(1116.) Since *hay* is the principal food given to sheep in snow or in black frost, it is matter of importance to procure them this valuable provender in the best state and of the best description. It has long been known that irrigation promotes, in an extraordinary degree, the growth of the natural grasses; and perhaps there are few localities which possess greater facilities for irrigation, though on a limited scale, than the Highland glens of Scotland. Rivulets meander there through haughs of richest alluvium, which bear the finest description of natural pasture plants, and yet irrigation is entirely neglected in those regions. Were the rivulets in winter subdivided into irrigating rills, the produce of these haughs might be multiplied many fold. It is not my purpose *here* to describe the management of irrigating meadows—that I will do ere all the winter operations terminate; nor is it my intention to describe the best mode of converting natural grass into hay, for that will form part of our occupation in the summer season: all that is requisite to be said in this place on the subject of irrigated meadows is, that, as they might be formed with great advantage to stock in many places where they are at present neglected, I cannot too earnestly draw the attention of hill-farmers to their utility; and although the localities in which they can be constructed are limited in extent, they will not be the less valuable on that account. One obstruction to their formation is the necessary fencing required around them, to prevent the trespass of stock while the grass is growing for hay. Besides places for irrigation, there are rough patches of pasture frequently found in the hills, probably stimulated to growth by latent water performing a sort of under-irrigation to the roots of the plants, which should be mown for hay; and to save farther trouble, *this* hay should be ricked on the spot, and surrounded by small hurdles, through which the sheep could feed in frosty weather from the rick, and keep themselves in fair condition. They would assemble round the stacks at stated hours, and, after filling themselves with dry food, again wander over, it may be, the bare but green sward for the remainder of the day, until severe black frost make them frequent the stacks; and when snow comes, the stells would be their place of refuge and support. As the hay in the stacks is eaten in, the flakes should be drawn closer around them, to allow it to approach again within reach of the sheep.

(1117.) [*Sheep-flakes or hurdles.*—Flakes are constructed in two different forms. The one represented by fig. 216 is the strongest and most durable, but is also the most expensive in first cost. The figure exhibits 2 flakes joined and supported, in the way they are placed, to form a fence. Each flake of this construction, with its fixtures, consists of 14 pieces, viz. 2 side-posts *a*, 4 rails *b*, and 3 braces *c d d*, which go to form the single flake; and 1 stay *f*, 1 stake *g* or *e*, and 3 pegs *h* or *i*, which are required for the fixing up of each flake. The scantling of the parts are the side-posts 4½ feet long, 4 inches by 2 inches. The rails 9 feet long, 3½ inches broad by 1 inch thick. The braces, 2 diagonals 5 feet 2 inches long, 2½ inches broad by ¾ inch thick, and 1 upright 4 feet long, and of like breadth and thickness. The stay is 4½ feet long, 4 inches broad, and 2 inches thick, and bored at both ends for the pegs; the stake 1½ feet long, pointed and bored. The pegs 1 foot long, 1½ inch diameter.

(1118.) The preparation of the parts consists in mortising the side-posts, the mortises being usually left round in the ends, and they are bored at equal distances from the joining and stay pegs. The ends of the rails are roughly rounded on the edges, which completes the preparation of the parts; and when the flake is completed, its dimensions are 9 feet in length, and 3 feet 4 inches in breadth over the rails; the bottom rail being 9 inches from the foot of the post, and the upper rail 5 inches from the head.

(1119.) The other form of flake, which is by far more extensively employed, though by no means the best, consists of the same parts, except that it has always five rails, and the only material difference in the scantling is, that the rails are all 1¼ inches square. An essential difference also occurs in the preparation or manufacture of this kind of flake. The ends of the rails are all turned round by machinery, and the side-posts are bored for their reception, as well as for the pegs by like machinery. The five rails in the flake are divided in hight as follows: The bottom rail 9 inches from the foot of the posts; the spaces *between* the first and second and the second and third rails, are each 7 inches, and the two upper spaces are respectively 8 and 9 inches, leaving, as before, 5 inches of the post above the upper rail.

(1120.) Flakes of this last description are extensively manufactured in Perthshire, where young larches are abundant, for of that wood they are generally made. Their price, when sold in retail by fifties or hundreds, is 1s. 9d. to 2s. each flake, including all the parts, sold in pieces; the expense of putting the parts together is usually 2d. each flake, including nails. The bar-flake first described is not generally to be found in the market, and is chiefly made to order; the price about 2s. 6d. each flake, with fixtures.—J. S.]

(1121.) As hurdles in England are somewhat differently put together, as well as made of a different sort of wood; and as the folding of sheep on turnips is differently managed in that country from what is given above, it seems proper to adve. for a little to both these subjects; and first as to the structure of hurdles.

(1122.) Where the common crack-willow (*Salix fragilis*) will grow, every farmer may have poles enough every year for making 2 or 3 dozen hurdles to keep up his stock. To establish a plantation, large cuttings 9 or 10 feet long should be pushed, not driven, into moist soil, and on being fenced from cattle, will soon shoot both in the roots and head, the latter being fit to be cut every seventh year. Where soil for a willow-plantation does not naturally exist, the farmer can buy his hurdles ready-made at 16s. the dozen; when made at home they cost 4d. each, and when the shepherd makes them they cost only his time. Hurdle-makers go the round of the country and make at 4d. and mend at 2d. each, finding their own tools.

(1123.) "A hurdle-maker's tools," says Mr. Main, "are a hand-saw, light hatchet, draw-shave, flamard, a center-bit and stock, a tomahawk, and gimlet. He has also a rending-frame, which is a common tressel *a*, fig. 239, on which 2 strong poles, *b*, are laid, leaning and connected by a

Fig. 239.

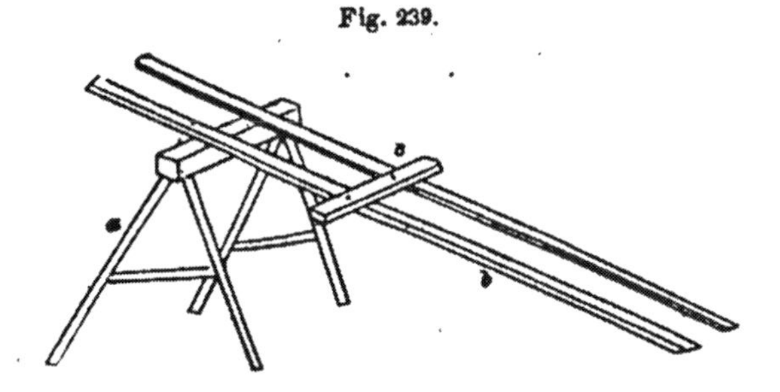

THE RENDING-FRAME IN HURDLE-MAKING.

piece, *c*, called a bridge. Besides this, he has a contrivance for shaving the poles, &c. In doing this, 2 auger-holes are bored in a post, to admit 2 stout square iron stubs, having ears to assist in withdrawing them when done with. The stubs project from the surface of the post about 6 inches, let in about 3 feet from the ground, and 8 inches from each other, though not exactly horizontally, the one nearest the workman being higher than the other, as seen at *a*, fig. 240. The

Fig. 240.

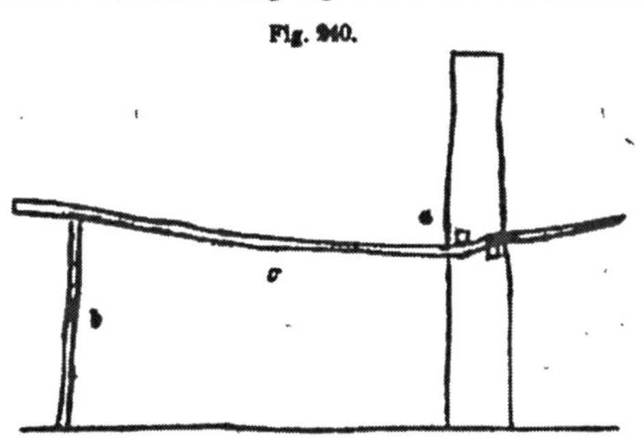

THE POSITION OF THE STUBS AND STANDARD IN HURDLE-MAKING.

use of these stubs is to hold the poles while they are shaved; being at the same time supported by a standard, *b*, about 3 feet from the post, and having a sharp short spike on the top to steady the pole *c* under the action of the draw-shave. In the same post a square staple, *a*, fig. 241, is driven, to hold the feet of the heads while they are mertised, assisted by a low stool, *b*.

(1124.) "All these things being ready, the poles are prepared for the different purposes to which they can be converted. The butt-end of the pole is first sawed off; 4½-feet lengths make a pair

Fig. 241.

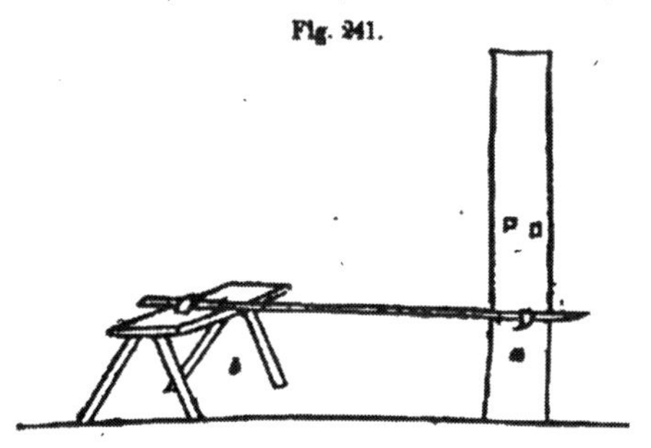

THE SQUARE STAPLE AND STOOL IN HURDLE-MAKING.

of heads, *a*, fig. 242; 9-feet lengths make a pair of slots, *b*; 5-feet lengths make a pair of stay-slots, *c*: and 3½-feet lengths make a pair of uprights, *d*.

Fig. 242.

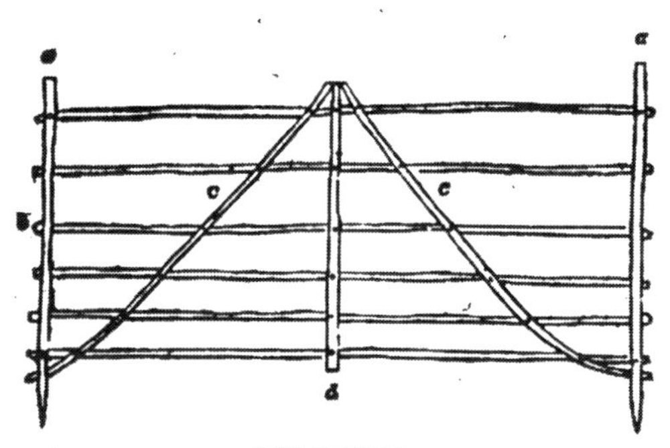

THE HURDLE.

(1125.) "The next proceeding is rending the different pieces, which is done at the rending-frame, fig. 239. The piece is put over the bridge *c*, with the butt-end upward. The flamard *a*, fig. 243—an edge-tool of iron, with a wooden handle—is placed across the pith, and driven down with a wooden baton *b*. When entered down 1 or 2 feet, the pole is brought up to bear upon the bridge, and at the same time on the under side of the top of the tressel. The pole being kept down by the left hand, while the flamard is guided by the right, by bending and turning the pole, the cleavage is performed from end to end with great exactness. They next undergo a little chopping or hewing with the hatchet, to cut off the knobs on the outside, keeping the inside as square as possible. The next operation is shaving off the bark and all irregularities, and giving each member of the hurdle its proper form.

Fig. 243.

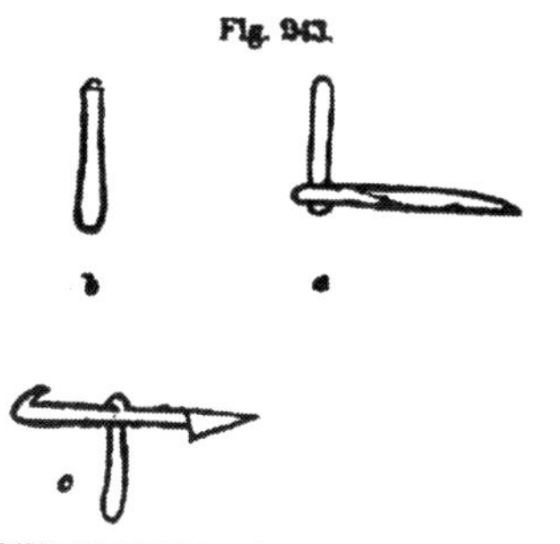

THE FLAMARD, BATON, AND TOMAHAWK, IN HURDLE-MAKING.

(1126.) "The maker next proceeds to form the hurdle; 4 low stumps are driven into the ground to mark the length, and 4 other to mark the distance between the upper and lower slots; a pair of heads, one at each end, are laid down in their right position, the flat or pith side upward; the 6 slots are then laid at due distances upon the heads, and the latter are *scribed* to the size of each slot, to regulate the mortises. The hurdle-maker uses no foot-rule in his operations, he having rods cut to the different lengths of the respective pieces; and the entire distances between the slots are arranged by the eye, the lower ones being gradually closer together, as seen in fig. 242; and the strongest pair of slots are usually chosen for the highest and lowest of the hurdle. One of the heads is then placed on the staple *a*, fig. 241, and resting on the top of the mortising-stool *b*, to which it is fixed in an opening by a wedge. The center-bit and stock drills out a hole at each end of the mortises, and also one for the diagonal-brace slot, about 2 inches below the lowest slot, and a little out of the line of the mortises above. It will be observed that mortises made by a center-bit leave an intermediate piece between the apertures, which is taken out by the tomahawk *c*, fig. 243, a tool made for the purpose. One end is a sharp stout-pointed knife, which cuts each side of the middle piece left in the mortise, and the other end hooks out the piece not dislodged by the knife. The mortising, which with a mallet and chisel would take up 1 hour, is done with the center-bit and hawk in 5 minutes. This head is now hammered on to the slots, and the other head is prepared and hammered on in the same way. The top and bottom slots are next nailed to the heads, and then the upright slot exactly in the middle. The 2 stay-slots are cut with a bend at the bottom, and rather sharply pointed; the points are driven through these oblique mortises, and their heads brought up to bear on the top of the upright, and nailed to each slot from top to bottom. The hurdle is then raised on its feet, and the nails clenched, which finishes the business. The gimlet is used for every nail, and a small block of wood placed under each slot while the nail is driven. The nails used are of the best iron, and what are called *fine-drawn*—not square, but rather flattened, to facilitate clenching, on which much of the strength of the hurdle depends; the head of the nail is somewhat large. Their price is 6d. per lb.; 100 poles at 18s. make 36 hurdles, which, including nails and workmanship, cost £1 11s. 6d., or 10s. 6d. per dozen. Although the horizontal slots are cut 9 feet long, the hurdle, when finished, is only somewhat more than 8 feet, the slot-ends going through the heads 1 or 2 inches; 2 hurdles to 1 rod of 16 feet, or 8 to 1 chain of 22 yards, are the usual allowance."

(1127.) A larger kind of hurdle, called *park hurdles*, worth 2s. each, is made for subdividing meadows or pastures, and are a sufficient fence for cattle. From all this it is obvious that when the small hurdles are used for sheep, the larger class must be obtained to fence cattle; whereas the Scotch flake described above (1117), and in fig. 216, answer both purposes at once, and are therefore more economical.

(1128.) "The hurdles being carted to the field," continues Mr. Main, "are laid down flat, end to end, with their heads next to, but clear of, the line in which they are to be set. A right-handed man generally works with the row of hurdles on his left. Having made a hole in the hedge, or close to the dyke, for the foot of the first hurdle, with the *fold-pitcher*, fig. 244—which is a large iron dibber, 4 feet long, having a well-pointed, flattened bit, in shape similar to the feet of the hurdles—he marks on the ground the place where the other foot is to be inserted, and there with his dibber he makes the second hole, which, like all the others, is made 9 inches deep. With the left hand the hurdle is put into its place, and held upright while lightly pressed down by the left foot on the lowest slot. This being done, the third hole is made opposite to, and about 6 inches from, the last. The dibber is then put out of hand, by being stuck in the ground near where the next hole is to be made; the second hurdle is next placed in position, one foot on the open hole, and the other foot marks the place for the next hole, and so on throughout the whole row. When the place of the second foot of a hurdle is marked on the ground, the hurdle itself is moved out of the way by the left hand, while the hole is made by both hands. When the whole row is set, it is usual to go back over it, giving each head a slight rap with the dibber, to regulate their hight, and give them a firmer hold of the ground. To secure the hurdles steady against the rubbing of the sheep, couplings, or, as they are commonly called, *copses*, are put over the heads of each pair where they meet, which is a sufficient security. These couplings are made of the twigs of willow, holly, beech, or any other tough shoots of trees, wound in a wreath of about 5 inches diameter.

Fig. 244.

THE FOLD-PITCHER IN HURDLE-SETTING.

(1129.) "The number of hurdles required for feeding sheep on turnips is one row the whole length of the ridges of an inclosed field, and as many more as will reach twice across 2 eight-step lands or ridges, or 4 four-step lands—that is, 48 feet, or 3 or 4 ridges of 15 feet. This number, whatever it may be, is sufficient for a whole quadrangular field, whatever number of acres it may contain. The daily portions are given more or less in length, according to the number of the flock. Two of these portions are first set, the sheep being let in on the first or corner piece. Next day they are turned into the second piece, and the cross-hurdles that inclosed them in the first are carried forward, and set to form the third piece. These removes are continued daily till the bottom of the field is reached; both the cross-rows are then to spare, and are carried and set to begin a new long-row, close to the off-side of a furrow, and the daily folding carried back over 2 or 4 lands as at first. It is always proper to begin at the top of a field, if there be any difference of the level, in order that the flock may have the driest lair to retire to in wet weather.

(1130.) "When there is a mixed flock—that is, couples, fattening and store sheep—two folds or pens are always being fed off at the same time, which only require an extra cross-row of hurdles. The couples have the fresh pens, while the lambs are allowed to roam over the unfolded turnips, by placing the feet of the hurdles, here and there, far enough apart, or by lamb-hurdles made with open panels for the purpose. The fattening sheep follow the couples, and have the bulbs picked up for them by a boy. The stores follow behind and eat up the shells."* It is never the practice in Scotland to put ewes with their lambs upon turnips, as new grass is considered much better for them, but the only ewe and lamb that can be seen on turnips in winter are of the peculiar breed of Dorsetshire. The *store*-sheep in Scotland—that is, the ewe-hoggs—are always fed as fully as the wether-hoggs which are intended to be fattened. In England the entire turnip-stock—ewes, lambs and wethers—are all intended for the butcher, and even, if possible, sold before the turnips are ended. The whole have hay or trough-meat, either in the field or in the sheep-house, on wet or stormy nights. An acre of good turnips maintains 5 score of sheep for 1 week.

(1131.) *Nets*, by which sheep are confined on turnips in winter, are made of good hempen twine, and the finer the quality of the material, and superior the workmanship bestowed on the manufacture of the twine, the longer will nets last. Being, however, necessarily much exposed to the weather, they soon decay, and, if guided carelessly, can scarcely be trusted more than a season. No treatment destroys them so rapidly as laying them by for the season in a damp state; and if rolled up wet, even for a few days, they become mildewed, after being affected with which nothing can prevent them rotting. They should never be laid by either damp or dirty, but washed and thoroughly dried in the open air before being rolled up and stowed away. It is alleged by shepherds that nets decay faster in drouth and exposure to dews and light in summer, than in

* Quarterly Journal of Agriculture, vol. ii.

winter. Several expedients have been tried to preserve nets from decay—among others, tanning, in imitation of fishermen; but, however well that process may suit nets used in the sea, it makes them too hard for the shepherd's use in tying the knots around the stakes. Perhaps a steeping in Kyan's solution might render them durable, and preserve their pliability, at the same time. The Company's charge is 5s. per cwt. for nets and cordage. It should be kept in mind that nets made of twine bleached by acids, or other chemical process, should not be submitted to Kyan's solution.

(1132.) *Sheep-nets* are wrought by hand, at least I have never heard of machinery being yet applied to their manufacture. They are simply made of *dead netting*, as it is technically called, which consists of plain work in regular rows, and is wrought by women as well as men. A shepherd ought to know how to make nets as well as mend them, which he will not do well unless he understand, in the first place, how to make them.

(1133.) All the instruments required in this sort of net-making are a *needle* and *spool*. "Needles are of two kinds—those made alike at each end with open forks, and those made with an eye and tongue at one end and a fork at the other. In both needles the twine is wound on them nearly in the same manner—namely, by passing it alternately between the fork at each end, in the first case, or between the fork at the lower end and round the tongue at the upper end, in the second case; so that the turns of the string may lie parallel to the length of the needle, and be kept on by the tongue and fork. The tongue and eye needle is preferable both for making and mending nets, inasmuch as it is not so liable to be hitched into the adjoining meshes in working; but some netters prefer the other kind, as being capable of holding more twine in proportion to their size." An 8-inch needle does for making nets, but a 4-inch one is more convenient for mending them.—*Spools*, being made as broad as the length of the side of the mesh, are of different breadths. They "consist of a flat piece of wood of any given width—of *stout* wood, so as not to warp—with a portion cut away at one end, to admit the finger and thumb of the left hand to grasp it conveniently. The twine in netting embraces the spool across the width; and, each time that a loop is pulled *taut*, half a mesh is completed. Large meshes may be made on small spools, by giving the twine two or more turns round them, as occasion may require." "In charging your needle, take the twine from the *inside* of the ball. This prevents tangling, which is at once recommendation enough. When you charge the needle with *double* twine, draw from 2 separate balls."* It is almost impossible to describe the art of netting by words, so as to render it intelligible, and I shall not therefore attempt it; but it may be learned from any shepherd. In joining the ends of *twine* together—which, in mending, is necessary to be done—the *bend* or *weaver's* knot is used; and in joining top and bottom ropes together, in setting nets, the *reef*-knot is best, as the tighter it is drawn the firmer it holds.

(1134.) Sheep-nets run about 50 yards in length when set, and weigh about 14 lbs. Hogg-nets stand 3½ feet in hight, and dinmonts 3 feet 3 inches, and both are set 3 inches above the ground. Stakes, to have a hold of 9 inches of the ground, bear the net 3 inches from the ground, and be 3 inches above the net-cord, should be 4¼ feet in length for the dinmont, and 4 feet 9 inches for the hogg-net. The mesh of the hogg-net is 3½ inches in the side, and of the dinmont 4½ inches; the former requires 9½ meshes in the hight, the latter 8½. The twine for the hogg-net is rather smaller than that for the dinmont, but the *top* and *bottom rope* of both are alike strong. A hogg-net costs 12s., or under 3d. per yard; a dinmont 10s., or under 2½d. per yard, on the Border, as at Berwick-upon-Tweed and Coldstream; but they are now sold in the prison of Edinburgh, being the work of the prisoners, at 7s. 6d., or under 2d. per yard; while in London the charge is 4½d. per yard.

(1135.) It is generally imagined that nets are not suitable for confining Black-faced sheep on turnips, chiefly because they are liable to be entangled in them by their horns; but this objection against the use of nets, is not insuperable, as the following circumstance will show. A farmer, a very extensive feeder of Black-faced sheep, on seeing my Leicester hoggs on turnips confined by nets, expressed a willingness to try the same method of confining his own sheep, adducing the great expense of hurdles as a reason for desiring a change. After getting a pattern net from me to stand 4 feet high, he got others made like it; and so successful was his experiment the first season, that he ever after inclosed a large proportion of his Black-faced sheep by nets. There occurred a few cases of entanglement for some days at first, but as his shepherd was constantly employed among his large flock, and having none else to attend to, no harm arose either to sheep or net, and in a short time the sheep became aware of the trap and avoided it. They never attempted to overleap the nets, though they would never have hesitated to do so over a much higher wall.

(1136.) [*Turnip-Slicers for Sheep.*—Machines for slicing roots, and particularly for the turnip, are constructed in a great variety of forms, but may be classed under two leading groups—those that cut the turnip simply into circular disks, as generally adopted for the feeding of cattle, and those that cut at one operation into oblong rectangular pieces or parallelopipedons, commonly practiced for feeding sheep; forming a somewhat more complicated class of machine. This last class, as coming first in the order of application, I shall first describe. Turnip-slicers for sheep may be again subdivided into lever and revolving machines; and of the many varieties under these forms, there are the stationary, the portable, the wheelbarrow, and what may be called the locomotive machine. This last being rendered so by its attachment to a cart, and by its own motion thus communicated, performs the operation of slicing while it travels over the field.

(1137.) The first introduction of the turnip-slicer is, like many other equally useful inventions, lost in obscurity, but it is most probable that, like the cultivation of the root itself, it originated in England; and it is likewise probable that the first attempt was the simple *chopper* still used to chop turnip for cattle. It appears uncertain whether the lever or the revolving slicer came first into use, as does also the time of their introduction. But we have an authentic record of a premium having been offered in 1806 by the Board of Trustees for the Encouragement of Arts and Manufactures in Scotland, for a revolving turnip-slicer. This was awarded to John Blaikie car-

* Bathurst's Notes on Nets.

penter to the late Lord Polwarth, then Mr. Scott of Harden, which is believed to have been the earliest application of that form of the machine in Scotland.

(1138.) *Lever Turnip-Slicer for Sheep.*—The first of the sheep turnip-cutters that I shall notice, is one of the lever form, but in its mechanical construction may be very aptly called the gridiron turnip-cutter, and is represented in an entire form in fig. 245, which is a perspective view

Fig 245.

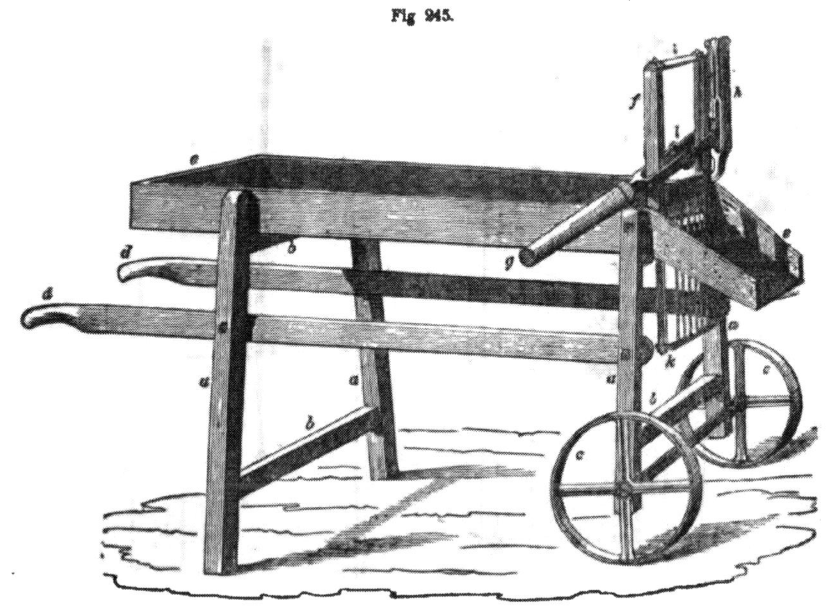

THE LEVER TURNIP-SLICER FOR SHEEP.

of the machine. It consists of a wooden frame supporting a trough, together with the cutting apparatus. The frame is formed of the four posts *a, a, a, a,* which are 2½ inches square. The front pair stand 15 inches in width, over all at top, the hind pair 19 inches; in both they spread a little below, and are separated to a distance of about 34 inches. Each pair is connected by cross-rails *b, b.* and they are connected longitudinally by the bars *d, d,* 4½ feet long, which form also the handles of the wheelbarrow; being bolted to the posts at a suitable hight for that purpose; their scantling is 2 by 1½ inches. A pair of wheels, *c c,* of cast-iron, 9 to 12 inches diameter, fitted to an iron axle, which is bolted to the front posts, gives it the conveniency of a wheelbarrow. The trough *e,* into which the turnips are laid for cutting, is 4 inches deep, and 3½ feet long, besides the sloping continuation of it in front of the cutters, for throwing off the sliced turnips. The cutting apparatus consists of a grooved frame of iron *f,* in which the compound cutter moves up and down by means of the lever handle *g.* A forked support, *h,* is bolted by a palm to the farther side of the wooden frame, and at the extremity, *i,* of the fork, a swing link is jointed. The lower end of the link is jointed to the extremity of the lever, which is likewise forked, forming its fulcrum; and the gridiron-cutter, *k l,* is also jointed by its top-bar to the lever at *l.* While the point *l,* therefore, of the cutter moves in a parallel line by its confinement in the grooves of the frame *f,* the fulcrum is allowed to vibrate on the joint *i* of the swing link—thus allowing an easy vertical motion to the cutter through the full range of its stroke. For the better illustration of the cutting apparatus the following figures are given on a larger scale. Fig. 246 is a front view of the cutter-frame, and fig. 247 a horizontal section of the same, including that of the grooved frame *f.* In fig. 247, *a b c* is a section of the grooved frame, with the cutter-frame set in the grooves. The grooves are ½ inch wide, and the cheeks of the frame 1½ inches by ½ inch, making the parts *a b c* 1½ inches square; *a d a* is the bottom bar of the cutting-frame, 1 inch broad, and ½ inch thick, kneed at the ends to receive the lower ends of the cutting-frame: *e* is the edge view of the slicing-knife, as fixed in the cutter-frame, 4 inches broad by ½ inch thick, and *f f f f f f* are the vertical or cross cutting-knives, also as seen from above. In fig. 246, *d* again marks the bottom bar of the cutter-frame, *e* is the slicing-knife, and *f f f f f f* the shanks of the cross cutting-knives—these are rivetted at top into *e,* and at bottom into *d*; *g g* are the side-bars of the cutter-frame, ¾ inch by ½ inch, into which the knife *e* is riveted, and to which the bar *d* is attached by screw-nuts. The top bar *h* welded to *g g* swells out in the middle, where it is perforated for the joint-bolt of the lever, as seen at *l* in fig. 245, and forms as a whole the gridiron-cutter.

(1139.) Figs. 248 and 249 are views of the knives on a still larger scale. In the first fig., 248, together with the portion *f* broken off, *a a b* is a cross section of the slicing-knife, and *c d e f* a cross cutting-knife, with its shank; here *a d* is the cutting edge, *c* being the body, and *e f* the shank of the knife. The length of the cutting edge *a d* may vary from ½ to 1 inch, according to the practice of the feeder, the shank *e f* being about ½ inch broad, and the whole ½ inch thick, except the cutting edge, that alone being sharpened and steeled, as well as the edge of the slicing-knife.

The second fig., 249, is a section of a cross cutting-knife on the line *x x* of fig. 248, together with a part of the shank *c*.

(1140.) The whole length of the cutter-frame, fig. 246, is about 20 inches, apportioned thus: from the bottom bar, *d*, to the edge of the slicing-knife, 12 inches; breadth of the knife, as before, 4 inches; and from the back of the knife to the top of the frame, 4 inches. The width of the frame

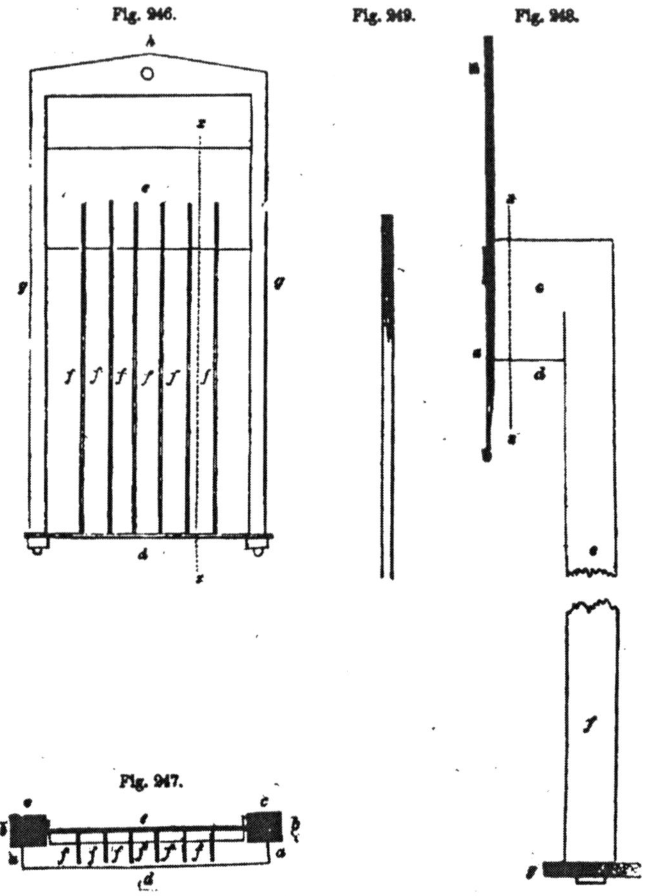

THE FRONT AND SIDE VIEWS OF THE CUTTERS OF THE LEVER TURNIP-SLICER.

over all may be 9 inches, and the cross-cutters set at from 1 to 1½ inches apart. The grooved frame must of course be constructed to receive and admit of the range of stroke of the cutter-frame. It is to be remarked that the slicing and cross-cutting is performed with this machine by one operation, the slicing edge being only ½ inch in advance of the cross-cutters, as at *b* in fig. 248.

(1141.) In operating with this machine, the trough is filled with turnips, and the operator lays hold of the lever *g*, fig. 245, with the right hand, while in his left he holds a short baton. Having raised the lever, and with it the cutter-frame, he pushes a turnip with the baton against the gridiron, and bringing down the lever, the knives cut off a slice, and divide it into oblong pieces; these may partly remain between the backs of the cross-cutters, until the succeeding stroke is effected, when the several portions of this slice will discharge those of the first, and so on.

(1142.) The principle of the gridiron-slicer is not confined to this particular mode of construction, nor even to a reciprocating action. Its application to a revolving disk machine was brought forward some years ago by Mr. Hay; but owing probably to its greater expense, has been but partially adopted.

(1143.) The gridiron has also been applied in combination with a revolving crank motion; the gridiron reciprocating, and that in a horizontal position. This modification appears to have originated in Roxburghshire, and appears to possess some advantages, the chief of which is, that, as the roots lie directly *upon* the gridiron, they are more likely to be regularly sliced than in those machines machines where the roots lie only *against* the cutters, as in the common vertical disk machines. This machine is essentially a gridiron turnip-slicer, with a reciprocating motion, derived from a rotary motion. The latter is produced by turning a winch handle, the axle of which carries a fly-wheel and two crank levers, or, more properly, the crank of the winch serves for both; the throw of the handle being 15 inches, while that of the crank is only 7 inches from the axis. A connecting-rod on each side of the machine connects the cranks with the gridiron-cutter, producing the reciprocating motion, which is in the horizontal direction; and to render the motion as easy as possible, the frame of the gridiron moves upon slide-rods. From the circumstance of the motion of the cutter being horizontal, and the turnips lying directly upon the grid iron, it can be easily constructed to cut both ways, that is, with the *out* as well as the *in* stroke; the gridiron for this purpose being furnished at both ends with the slicing-knife and the cross-cutting-knives, as described in (1138) fig. 246. The machinery here described is mounted on a wooden frame, 4 feet 6 inches long, 22 inches wide, and 34 inches high, and over the gridiron is placed a square hopper of wood or of sheet-iron, into which the turnips are thrown by an assistant, the machine being driven by a man. A bar of division is placed across in the middle of the hopper, serving as the point of resistance against which the turnips are pressed while the slice is being made; and as the turnips lie on the bars of the gridiron with their full weight, they will for the most part be in a position to secure a slice of uniform thickness being removed. A slight modification has been made on this machine by placing the gridiron on radius bars, making the cutter move in an arc of about two feet radius, instead of moving in a slide as above described. The radius bars produce a lighter motion, but have no effect on the cutting principle.

(1144.) With a view to economy, the regular slicing of turnips is of more importance than many farmers are aware of. When a part of the turnip is cut into very thin, and even into fragments of slices, a very considerable proportion of it goes to waste. In choosing a turnip-slicer, therefore, one of its points should always be, that it should cut as far as possible to a uniform size, whatever that size may be, and not pass a large proportion of the sliced turnips in thin-edged slices, or thin and small fragments of slices.

(1145.) *Wheel Turnip-slicer for Sheep.*—This machine, alluded to in (1136), has, since its introduction, undergone many modifications. From being made entirely of wood, it came to be made entirely of iron; but this last being less convenient for moving about, has induced the more general introduction of a disk of cast-iron, carrying the cutters, mounted on a wooden frame, which is generally again mounted on wheels like a wheelbarrow. Fig. 250 is a perspective of this ma-

Fig. 250.

THE WHEELBARROW TURNIP-SLICER FOR SHEEP.

chine; the wooden frame, which is 36 inches long and 15 inches wide over the posts at top, but spreads a little wider below, is formed with four posts, *a a a a*, one of which is only partially seen in the figure; they are 2½ inches square, and stand about 32 inches in hight. The posts are connected on the sides by top-rails *b b*, and two brace-rails *c c* below, one of which serves to support the spout *d*, which discharges the sliced turnips. The sides of the frame thus formed are connected by cross-rails above and below, *e e e*, and is there furnished with the handle-bars *f f*, bolted to the posts, and projecting a convenient length beyond them at one end. The barrow-wheels *g g*, of 12 inches diameter, are fitted to an iron axle, which is bolted to the posts in front. The hopper *h* is fixed upon the top-rail by means of a cast-iron sole bolted upon the rail, and is farther supported by a wooden bracket at each side, as seen at *i*, and by the iron stay *k*. The slicing-wheel *l* is a disk of cast-iron, carrying three sets of cutters. The disk is mounted on an axle passing through its center, where it is fixed, and which is supported on bearings placed on the top-rails, and when worked, it is turned by the winch-handle *m*, fixed upon the axle. Fig. 251 is a section

Fig. 251.

THE SECTION OF THE DISK AND HOPPER OF THE WHEELBARROW TURNIP-SLICER.

of part of this machine, cutting it through the hopper, and the disk, &c., to exhibit some of the parts more in detail. *a a* are parts of two of the posts, *b b* the top-rails, and *c* one of the end-rails of the frame, covered by the boarding *d* of the spout. *f* is one of the pillow-block bearings of the axle, the other being kept out of view by the hopper, and the winch-handle is applied at *e*. *g g* is the disk shown also in section. The sole of the hopper is represented at *h*; it has a flange between *h* and *b*, by which it is bolted to the top-rail *b*; and the sole itself is a cylindrico-concave plate of 12 inches in length at the bottom of the concavity, 9 inches in breadth, and is placed at an angle of 45°. It is also furnished with a flange at each side, whereby the sides *i* of the hopper are attached to the sole. *k* is the foot of one of the brackets referred to in fig. 250, rising in the position of the dotted lines, for supporting the hopper; and *l* is a light tie-bar, cut by the section, which is applied also to bind the sides of the hopper.

(1146.) The disk or wheel *g g* is a plate of cast iron, 32 inches in diameter and ½ inch thick, encircled by a heavy ring of the same metal, to give it momentum when in action. The face of the disk is divided into three segmental compartments around a plain and central portion, which is 9 inches diameter. This central part lies in the general plane of the disk, while the segmental portions diverge from the plane in the direction of the circle, causing them to take the form of portions of three separate helical or spiral surfaces of 9 inches in breadth. Their divergence from the plane of the disk does not, however, exceed ¾ of an inch at the termination of a segment, or such other space as may be determined upon for the thickness of the slices. By this construction, three slits are formed in the disk, passing obliquely through, one at the termination of each segment; and the steel slicing-knife, 12 inches in length and 1½ inches in breadth, is fixed by bolts, so as to form the entering edge of each segment, as seen in fig. 250—the flat face of the knife lying in the general plane of the disk.—The terminal edge of each segment lies exactly behind the leading edge of the next, so that, when the slicing-knife is affixed to a leading edge, the edge of the knife covers 1½ inches of the length of the preceding segment. Into the border, which is three

Fig. 252.

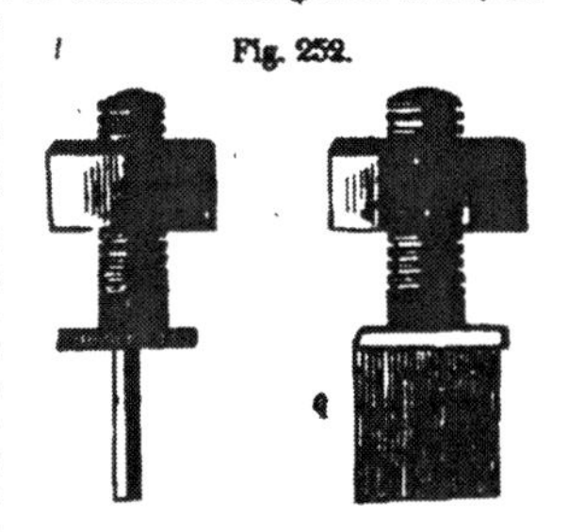

THE CROSS-CUTTERS OF THE DISK OF THE WHEELBARROW TURNIP-SLICER.

covered by the slicing-knife, are placed from 6 to 10 lancet shaft-cutters, their length being just equal to the width of the slit, and their distance apart proportioned to the number employed, or the breadth at which the turnips are required to be cross-cut The cross-cutters are formed as represented in fig. 252, where *a* is an edge, and *b* a side view of a cutter, with its tail and screw-nut, by which it is fixed into the disk.

(1147.) It will be seen that the action of those compound cutters is very similar to that of the gridiron, the slicing and cross-cutting knives acting together, though the slicing-knife is here also about ¼ inch in advance of the cross-cutters; and from the construction of the disk, and arrangement of the feeding-hoppers, the turnip is applied with great regularity, and in close contact with the spiral surface of the segments of the disk. The slope of the sole-plate in the hoppers gives the turnip a constant tendency to keep in contact with the surface and the cutters, thereby securing regular and good performance by the machine.

(1148.) The wheel turnip-slicer has been applied in a variety of forms, such as cutting on a horizontal direction, the turnips being placed in a hopper right above the disk; and both vertically and horizontally it has been adopted on the locomotive principle, attached in various modes to a cart. Perhaps the most successful of those modes is that produced at the late Show of Implements at Edinburgh, under the auspices of the Highland and Agricultural Society of Scotland, by Mr. Kirkwood, Tranent. It is a common slicing disk, mounted on a carriage with two wheels, from the axle of which, and by their own resistance, motion is communicated to the disk by means of a beveled gearing. The carriage is simply hooked on to a cart which conveys the turnips.—The cutting process can be stopped at pleasure by means of the common clutch and lever; and the whole machine, being constructed of iron, will be very durable.

(1149.) *Cylinder Turnip-Slicers.*—Turnip-slicers for sheep have been also constructed in a variety of forms with the cutters set in the surface of a cylinder, and been in use for many years.—In Roxburghshire it has been long and successfully employed in the locomotive principle—not driven by any machinery from the cart to which it is attached, but, being simply hooked to the cart, is drawn forward; and the machine being of some weight, and moving upon wheels of 3 feet or more in diameter, armed with spikes on their tires to prevent them sliding over the surface of the ground, these give motion to the cutting cylinder; while a boy, sitting on the cart which contains the turnips that are to be cut, throws them into the hopper of the machine, from which they are dropped over the surface of the grass on which the sheep are feeding.

(1150.) A modification of the cylinder-slicer was patented in 1839 by Mr. Gardner, Banbury; the principle of the patent lies in the form and arrangement of the cutters, which are set in three divisions upon the surface of the cylinder. The arrangement of the cutters is peculiar, and difficult to describe without the aid of a figure. The cylinder on which the cutters are placed is 15 inches in diameter and 12 inches long. Its periphery is divided into 3 compartments, each forming a portion of a spiral, so that the commencement of one and the termination of the next leaves a slit across the periphery, corresponding in some degree with that described on the disk of the wheel turnip-cutter. The original cylindrical slicers had the slicing-knife extending in an unbroken edge across the surface of the cylinder, and the cross-cutters placed under and behind it. The improvement on which the patent is based may be described as cutting the slicing-knife into a number of sections, say of 1 inch each in length. The two extreme sections remain in the original position on the cylinder. The section next to that on each side is removed backward upon the surface of the cylinder, say 1½ inches, and there fixed. The section on each side next to those is in like manner set back, and so on till the whole are placed on the surface of the cylinder. By this arrangement, the slicing-cutters form two converging lines, *en echellen*, and this is repeated three times on the periphery of the cylinder. The cross-cutters are formed by a part of the slicing-cutter being bent to a right angle with the former.

(1151.) This cylinder machine, by reason of the cutters acting in succession from their position, *en echellen*, works with great ease and cuts regularly, but, withal, makes the slices too small, and has a tendency to produce waste, though this fault could be easily rectified by enlarging the sections of the knife, and lengthening the cross-cutters. The cylinder is mounted in a wooden frame, and the hopper is so arranged that the turnip, while being cut, tends always to apply itself to the surface of the cylinder.—J. S.]

(1152.) There is a mode of preserving corn for sheep on turnips which has been tried with success in Fife. It consists of a box like a hay-rack, as in fig. 253, in which the corn is at all times kept closely shut up, except when sheep wish to eat it, and then they get to it by a simple contrivance. The box *a b* contains the corn, into which it is poured through the small hinged lid *y*. The cover *c d*, concealing the corn, is also hinged, and when elevated the sheep have access to the corn. Its elevation is effected by the pressure of the sheep's fore-feet upon the platform *e f*, which, moving as a lever, acts upon the lower ends of the upright rods *g* and *h*, raises them up, and elevates the cover *c d*, under which their heads then find admittance into the box. A similar apparatus gives them access to the other side of the box. The whole machine can be moved about to convenient places by means of 4 wheels. The construction of the interior of the box being somewhat peculiar, another, fig. 254, is given as a vertical section of it, where *b* is the hinged lid by which the corn is put into the box, whence it is at once received into the hopper *d*, the bottom of which being open, and brought near that of the box, a small space only is left for the corn to pass into the box, the hopper forming the corn-store; *a* is the cover of the box raised on its hinges by the rod *f*, acted upon by the platform *e f*, fig. 253; and, when in this position, the sheep put their heads below *a* at *c*, and eat the corn at *d*. Machines of similar construction to this have also been devised to serve poultry with corn at will.* It is a safer receptacle for corn in the field than the open oil-cake trough, fig. 230; but animals require to be made acquainted with it before they will use it with confidence.

(1153.) It is not my purpose to dilate fully on the *diseases* of animals, the symptoms and treatment of which you will find satisfactorily described in the published works of veterinarians; but,

* Prize Essays of the Highland and Agricultural Society, vol. vii.

nevertheless, it is necessary you should know something of the various diseases animals are liable to, when subjected to the usual treatment of the farm. Were you not warned of the consequences of this, you would not know how to check the progress of disease, but allow it to proceed, until

Fig. 253.

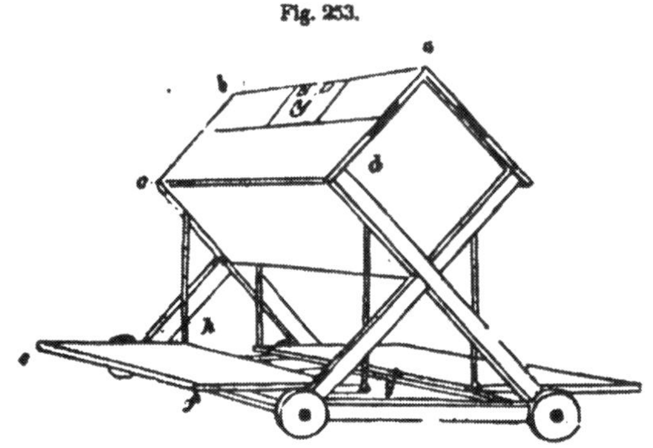

THE CORN-BOX FOR SHEEP ON TURNIPS.

Fig. 254.

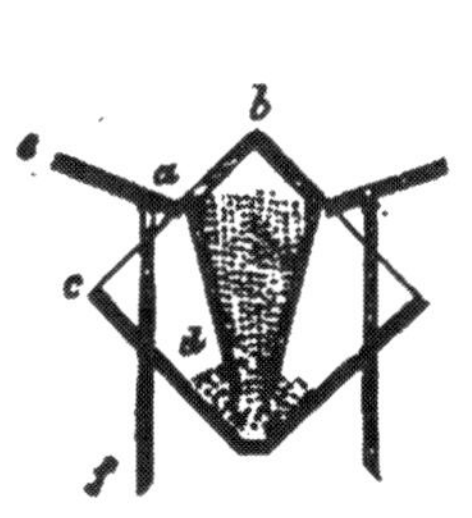

THE VERTICAL SECTION OF THE INTERIOR OF THE CORN-BOX.

the life of the animal were endangered. It is, however, not desirable that you should consider yourself as a veterinarian, because, not being a professional man, your practical knowledge will necessarily be confined to the cases arising from the casualties of your own stock, and hence your experience will never enable you to become so well acquainted with any disease, nor so many, as the professional man, while you would rely so much upon your own knowledge, crude as it must be, as to undertake the treatment of every case of illness that occurred on your own farm; and thus be prompted to try experiments which may prove dangerous to the safety of the animal, but it is very desirable, because much conducive to your own interest that you should be acquainted with the most easily recognized symptoms of the commonest diseases incidental to domesticated animals, and with the general principles of their treatment. If you but knew how to distinguish between local and general affections, and to apply the proper preliminary treatment, you would place the afflicted animal in such a state of safety until the arrival of the veterinary surgeon, as the disease might be easily overcome by him, and your animal restored to health in a short time. Farther than this you have no right to aspire as an amateur veterinarian; for it cannot be too strongly impressed upon you that, before you can be competent to learn the art of healing, you must have an accurate knowledge of the anatomical structure and the physiology of the domestic animals. At the same time, as a branch of general knowledge, veterinary science ought to have your regard, and more especially as your profession places you in a position to occupy the field which affords the most numerous, varied, and interesting cases for veterinary practice. And besides, you should have "some insight," as Professor Dick suggests, "into a subject with which all who have any pretension to a knowledge of horse-flesh *ought to have some acquaintance*. And if you bear in mind that, in a compendious view of the principles upon which alone the diseases of domestic animals can be properly treated, you will find an antidote to the quackery by which many valuable animals are sacrificed, and serious expense and vexation occasioned."*

(1154.) On this suggestion, I would notice the complaints of sheep, with the view of letting you see their interesting nature, and of the expediency of your becoming acquainted with their principles. The first which presents itself on sheep, in the low country in winter, is *purging*, occasioned by eating too heartily of the tops, when first confined on turnips. At first, the complaint is not alarming, and the physicing may do good ultimately; but should it increase, or continue beyond the existence of the exciting cause, it may pass into diarrhœa, causing prostration of strength, and at last terminate in dysentery. When the purging is moderate, the pain is inconsiderable; but when aggravated, the mucous membrane, which is the *seat* of the disease, acquires a tendency to inflammation, and griping and colicy pains are the consequence. The disease should not be thought lightly of, but speedily checked. When the green food, as in this case, is obviously at fault, the sheep should be removed to dry pasture until the symptoms disappear. One year, I remember, the white-turnip tops grew so luxuriantly that when Leicester hoggs were put on in October, they were very soon seized with purging, and the symptoms were much aggravated by alternate falls of rain and raw frosts. The sheep were removed to a rough, moory pasture, which had been reserved for the ewes; and while there, I caused the field-workers to switch off the turnip tops with sickles, and thus got rid of the cause of complaint. In a short time, the hoggs were restored to the turnips, and throve apace; though the wool behind was much injured by the fæcal discharge. And this is one of the losses incurred by such a complaint; and at a season, too, when it would be improper to clip the soiled wool away, to the risk of making the sheep too bare below to lie with comfort upon the cold ground.

(1155.) Sheep are sometimes infested with a species of *louse*, the *Trichodectes sphærocephalus*, characterized by Mr. Denny as having the head nearly orbicular, the clypeus rugulose and ciliated with stiff hairs, and the third joint of the antennæ longest and clavate.

(1156.) This animal is perhaps induced to make its appearance by an increase of condition after a considerable period of poverty. It is seldom seen on Leicester sheep, because, perhaps, they are seldom in the state to induce it; but hill-sheep are not unfrequently infested by it, and when

* Dick's Manual of Veterinary Science, Preface.

so, it is amazing what numbers of the vermin may be seen upon a single sheep; its powers of reproduction seeming prodigious. It lodges chiefly upon and below the neck, where it is most effectually destroyed by mercurial ointment, which should not, however, be applied, *in quantity*, in very cold or in very wet weather; and in these circumstances, tobacco-juice and spirit of tar may be safely used. A quart bottle of decoction of tobacco-leaf, containing a wine glass of spirit of tar, is a useful lotion, for many purposes, for a shepherd to have constantly in his possession. Professor Dick says that, in slight visitations of the louse, a single dressing of olive-oil will cause its disappearance.

(1157.) Another disease to which sheep are subject on passing from a state of poverty to improved condition is *scab*, and hoggs are most susceptible of it. This disease indicates its existence by causing sheep to appear uneasy and wander about without any apparent object; to draw out locks of wool with its mouth from the affected parts, as the disease increases; and, lastly, to rub its sides and buttocks against every prominent object it can find, such as a stone, a tree, a gate-post, the nets, and such like. Mr. Youatt says that it arises from an insect, a species of *acarus*;* but whether this be the case or not, one remedy is efficacious, namely, mercurial ointment; a weak one of 1 part of the ointment with 5 of lard for the first stage, and the other a stronger, of 1 part of ointment and 3 of lard, for an aggravated case. The ichorous matter from the pustules adheres to and dries upon the wool, and gets the name of *scurf*, which should first be washed off with soap and water before applying the ointment. The scab is a very infectious disease, the whole flock soon becoming contaminated; but the infection seems to spread not so much by direct contact as by touching the objects the animals infected have rubbed against. Its direct effects are deterioration of condition, arising from a restlessness preventing the animal feeding, and loss of wool, large portions not only falling off, but the remainder of the broken fleece becoming almost valueless; and its indirect effects are propagation of the disease constitutionally, and hence the loss to the owner in having a scabbed flock, for no one will purchase from one to breed from that is known to be, or to have been, affected by scab. With regard to the very existence of this disease, it is held disgraceful to a shepherd not to be able to detect its existence at a very early stage, and more so to allow it to make head in his flock, however unobservant he may have been of its outbreak. When it breaks out in a standing flock, it must have been latent in its constitution or in the ground, when the shepherd took charge of it, for some shepherds have only the skill to suppress, not eradicate it; but it is his duty to examine every sheep of his new charge, and of every one newly purchased; before they are allowed to take to their hirsel, and also to make inquiry regarding the previous state of the ground.

(1158.) On soft ground sheep are liable to be affected with *foot-rot*, when on turnips. The first symptom is a slight lameness in one of the fore-feet, then in both, and at length the sheep is obliged to go down, and even creep on its knees, to get to its food. The hoof, in every case, first becomes softened, when it grows misshaped, occasioning an undue pressure on a particular part; this sets up inflammation, and causes a slight separation of the hoof from the coronet; then ulcers are formed below where the hoof is worn away, and then at length comes a discharge of fetid matter. If neglected, the hoof will slough off, and the whole foot rot off; which would be a distressing termination with even only one sheep, but the alarming thing is, that the whole flock may be similarly affected, and this circumstance has led to the belief that the disease is very contagious. There is, however, much difference of opinion among store-farmers and shepherds on this point, though the opinion of contagion preponderates. For my part, I never believed it to be so, and there never would have been such a belief, had the disease been confined to a few sheep at a time; but though numbers are affected at one time, the fact can be explained from the circumstance of all the sheep being similarly situate; and as it is the nature of the situation which is the cause of the disease, the wonder is that any escape affection, rather than that so many are affected. The first treatment for cure is to wash the foot clean with soap and water, then pare away all superfluous hoof, dressing the diseased surface with some caustic, the spirit of tar and blue vitriol being most in vogue, but Professor Dick recommends butter of antimony as the best; the affected part being bound round with a rag, to prevent dirt getting into it again; and removing the sheep to harder ground, upon bare pasture, and there supplying them with cut turnips. The cure indicates the prevention of the disease, which is careful examination of every hoof before putting sheep upon red land, and paring away all extraneous horn; and should their turnips for the season be upon soft, moist ground, let them be entirely sliced, and let the sheep be confined upon a small break at a time, and thus supersede the necessity of their walking almost at all upon it for food. I may mention that sheep accustomed to hard ground, when brought upon that which is comparatively much softer, are most liable to foot-rot, and hence the necessity of frequent inspection of the hoof when on soft ground; and as some farms contain a large proportion of this state of land, frequent inspection should constitute a prominent duty of the shepherd.

(1159.) Erysipelatous complaints occur in winter among sheep. "*Wildfire*, it is said," remarks Professor Dick, "generally shows itself at the beginning of winter, and first attacks the breast and belly. The skin inflames and rises into blisters, containing a reddish fluid, which escapes and forms a dark scab. The animal sometimes fevers. Venesection (blood-letting) should be used, the skin should be washed with a solution of sugar of lead, or with lime-water, and physic given, such as salts and sulphur; afterward a few doses of nitre."†

(1160.) There is, perhaps, no circumstance upon which an argument could be better founded in favor of arable land being attached to a hill-farm, for the purpose of raising food to be consumed in stormy weather, than on the fatality of the disease commonly called *braxy*. It affects young sheep, and chiefly those of the Black-faced breed, which subsist upon the most elevated pasture. Indigestion is its primary cause, exciting constipation, which sets up acute inflammation of the bowels, and death ensues. The indigestion is occasioned by a sudden change from succulent to dry food, and the suddenness of the change is imposed by the sudden occurrence of frost and snow, the latter concealing the green herbage which the sheep have been eating; and obliging

* Youatt on Sheep. † Dick's Manual of Veterinary Science.

them to subsist upon the tops of old heather, and twigs and leaves of bushes that overtop the snow. By this account of the origin of the disease, it is obvious that, were stells provided for shelter, and turnips for food, the braxy would never affect young hill-sheep, at least under the circumstances which usually give rise to it. The Ettrick Shepherd thus describes its symptoms: "The loss of cud is the first token. As the distemper advances, the agony which the animal is suffering becomes more and more visible. When it stands, it brings all its four feet into the compass of a foot; and sometimes it continues to rise and lie down alternately every two or three minutes. The eyes are heavy and dull, and deeply expressive of its distress. The ears hang down, and, when more narrowly inspected, the mouth and tongue are dry and parched, and the white of the eye inflamed. . . . The belly is prodigiously swelled, even so much that it sometimes bursts. All the different apartments of the stomach are inflamed in some degree."* Violent inflammation succeeds, with a tendency to mortification and sinking, so that, after speedy death, the touch of the viscera, and even of the carcass, is intolerable. Its effects are so sudden, that a hogg apparently well in the evening will be found dead in the morning. Cure thus seems almost unavailable, and yet it may be effected, provided the symptoms of the disease are observed in time; when, if blood is drawn freely from any part of the body, such as by notches made across the under side of the tail, from the vein under the eye, and that behind the fore-arm, and a dose of salts administered in warm water, the animal will most probably recover.† But the grand object is *prevention* of the disease by a timely supply of succulent food; and if turnips cannot be obtained, it may be worth the store-master's consideration whether oil-cake should not be given to the sheep along with hay, during a storm. The laxative property of oil-cake is well established, and its carriage to the remotest hill-farm comparatively easy. Mr. Fairbairn recommends salt to be given to young sheep, when shifted suddenly from fresh to dry food; and no doubt, as a condiment in support of the healthy action of the stomach, it would prove useful; and more especially in the case of cattle and sheep, the structure of whose digestive organs renders them peculiarly liable to the effects of indigestion; and on this account it would be a valuable assistant to the more nutritious oil-cake. And instead of entirely acquiescing in the Ettrick Shepherd's recommendation "to pasture the young and old of the flocks all together," as has been done in Peeblesshire, to the eradication, it is said, of the braxy—as being in many cases impracticable and attended with no profit, Mr. Fairbairn rather observes, "Let the pasture for a hirsel, as was observed before, be as nearly as possible of one soil. To overlook this is a mighty error, and the surest means of making the flock unequal. The heath should also be regularly burned, and the sheep never allowed to pasture *long upon soft grass*." And as a last resource in an attempt to eradicate the disease everywhere, be would have the sheep put on turnips, as "an infallible antidote against the progress of the malady;" and which he has "invariably found gives a settling stroke to the disease."‡ This last remedy doubtless being effective, I would recommend its adoption rather as a preventive than a cure of the disease.

(1161.) The Ettrick Shepherd mentions the existence of 4 kinds of braxy—namely, the *bowel sickness*, the *sickness in the flesh and blood*, the *dry braxy*, and the *water braxy*—all originating in the same cause, producing modified effects—namely, a sudden change of food from succulent to dry, inducing constipation of the bowels and consequent inflammation; and they are all a class of d'seases allied in their nature to hoven in cattle, and flatulent colic or botts in horses.

---

31. DRIVING AND SLAUGHTERING SHEEP.

"Pierced by Roderick's ready blade,
Patient the sickening victim eyed
The life-blood ebb in crimson tide
Down his clogged beard and shaggy limb,
Till darkness glazed his eye-balls dim."

SCOTT.

(1162.) ALTHOUGH it is unusual for farmers who possess a *standing* flock —and most farmers who practice the mixed husbandry have one—to dispose of their fat sheep in winter—that is, before the turnips are all consumed—yet as farmers, who, having no standing flock, purchase a flying one every year, of sheep in forward condition, and in such numbers as to consume the turnips allotted to them in a short time, do dispose of their fat sheep in winter, it is proper that you should be made acquainted here with the *driving of sheep upon roads*, and the general practice of the *mutton-trade*. The sheep most forward in condition in autumn are yeld ewes and wethers, the tup-eill ewes being already fat and sold.

(1163.) Sheep are purchased from farmers both by *dealers* and *butchers*.

* Hogg's Shepherd's Guide. † The Mountain Shepherd's Manual.
‡ A Lammermuir Farmer's Treatise on Sheep in High Districts

Dealers buy from farmers in wholesale, and sell to butchers in retail; so they constitute a sort of middle-men; but, unlike most middle-men, their avocation is fully as useful to both parties as to themselves, inasmuch as they purchase at once the whole disposable stock of the farmer, and they assort that stock, and present it at the markets which the different classes of their customers, the butchers, are in the habit of frequenting, in the most suitable form. They thus act the part which the wool-staplers do, in assorting the different qualities of wool between the grower and the manufacturer. They buy either at fairs, or on the farmer's own premises. In the former case they pay ready money, and lift the stock immediately; in the latter they pay at the time the stock is lifted by agreement. In lifting their bargains, they appoint one time among all the places they have purchased, to make up their entire drove; for it is less costly for their people to drive a large one than a small. Dealers chiefly buy at the country fairs, where they have ample choice, and only purchase on the farmer's premises when stock happens to be scarce, and prices likely to advance. Butchers purchase chiefly in the market-towns in which they reside, though they also attend fairs, and pick up a few fat lots which will not bear the long journeys of the dealers; and in this case they pay ready money and lift immediately, as dealers do. But when they purchase on the farmer's premises, they usually lift so many at a time, according to agreement, and pay only for what they lift. Every farmer should avoid this practice, as every time the butcher comes for his lot the sheep have to be gathered, and the whole handled, that he may take away only those which suit his present purpose; and this commotion is made most probably every week, the whole stock being disturbed by the shouting of men and the barking of dogs, among whom those of the butcher are not the least noisy or the least active. Farmers take their stock either to fairs or market-towns, and there meet the respective sorts of purchasers, the dealers never appearing as purchasers in towns, the butchers there ruling paramount.

(1164.) When a dealer purchases on the farmer's premises, he lifts his lot *at any time of day* that best suits his own arrangements. He begins to lift the first lot in the more distant part of the country, and, proceeding on the road in the direction of their destination, he lifts lot after lot until the whole are gathered, to the amount of many hundreds. In this way he may lift a lot in the forenoon on one farm, and another in the afternoon on another; and this is a much more satisfactory way for the farmer to dispose of his stock than the one he allows the butcher to adopt. But when a farmer is to drive his own sheep to market, he starts them at a time when the journey will do them the least injury. Sheep should not begin their journey either when too full or too hungry; in the former state they are apt to purge on the road, in the latter they will lose strength at once.—The sheep selected for market are the best conditioned at the time; and, to ascertain this, it is necessary to handle the whole lot, and shed the fattest from the rest; and this is best done about midday, before the sheep feed again in the afternoon. The selected ones are put into a field by themselves, where they remain until the time appointed them to start. If there be rough pasture to give them, they should be allowed to use it, and get quit of some of the turnips in them. If there is no such pasture, a few cut turnips on a lea-field will answer. Here all their hoofs should be carefully examined, and every unnecessary appendage removed, though the *firm* portion of the horn should not be touched. Every clotted piece of wool should also be removed with the shears. The sheep should also be marked with *keil*, or *ruddle*, as it is called in England—the ochrey-red ironstone of mineralogists, which occurs in abundance near Platte, in Bohemia.* The keil-mark is put on the wool, and on any part of the body you choose—the purpose being to identify your own sheep in case of any being lost in the fair. The parts usually chosen for marking Leicester sheep are top of shoulder, back, rump, far and near ribs. The mark is made in this way: Take hold of a small tuft of wool at any of the above parts with the right-hand fingers, and seize it between the fore and middle fingers of the left hand, with the palm upward; then color it with the keil, which requires to be wetted, if the wool itself is not damp. Short-wooled sheep are usually marked on the head, neck, face and rump, or with a bar across the shoulders, and generally too much keil is put upon them. The sheep being thus prepared should have food early in the morning, and be started on their journey about midday—the season, you will remember, being winter. Let them walk gently away; and, as the road is new to them, they will go too fast at first—to prevent which, the drover should go before them, and let his dog bring up the rear. In a short distance they will assume the proper speed, about 1 mile the hour. Should the road they travel be a green one, the sheep will proceed nibbling their way onward at the grass, along both sides; but if a turnpike, especially a narrow one, the drover will require all his attention in meeting and being passed by every class of vehicle, to avoid injury to his charge. In this part of their business, drovers generally make too much ado, both themselves and their dogs; and the consequence is that the sheep are driven more from side to side of the road than is requisite. On meeting a carriage, it would be much better for the sheep were the drover to go forward, instead of sending his dog, and point off, with his stick, the leading sheep to the nearest side of the road; and the rest will follow, as a matter of course, while the dog walks behind the flock, and brings up the stragglers. Open gates to fields are sources of great annoyance to drovers, the stock invariably making an endeavor to go through them. On observing an open gate before, the drover should send his dog behind him over the fence, to be ready to meet the sheep at the gate. When the sheep incline to rest, let them lie down. Before night-fall the drover should inquire of lodging for them for the night, as in winter it is requisite to put them in a grass-field, and supply them with a few turnips or a little hay, the road-sides being bare at that season. If turnips or hay are laid down near the gate of the field they occupy, the sheep will be ready to take the road in the morning; but, before doing this, the drover should ascertain whether the road is infested with stray dogs; if which be the case, the sheep should be taken to the safest spot and watched. Many dogs that live in the neighborhood of drove-roads, and more especially village dogs, are in the habit of looking out for sheep to worry, at some distance from their homes. The chief precaution that can be used under such an apprehension is, for the drover to go frequently through the flock with a light, and be late in retiring to rest, and up again early in the morning. This apprehension regarding dogs is not solely in regard to the loss sustained by worrying, but, when sheep have been disturbed by them, they will not settle again upon the road. The first day's journey should be a short one, not exceeding 4 or 5 miles.—Upon drove-roads, farms will be found at stated distances with food and lodging for the drover and his flock, at a moderate charge. Allowing 8 miles a day for a winter-day's travel, and knowing the distance of your market by the destined route, the sheep should start in good time, allowance being made for unforeseen delays, and one day's rest near the market.

* Jameson's Mineralogy, vol. iii.

(1165.) The *farmers' drover* may either be his shepherd, or a professional drover hired for the occasion. The shepherd knowing the flock makes their best drover, if he can be spared so long from home. A hired drover gets 2s. 6d. a day of wages, besides traveling expenses, and he is intrusted with cash to pay all the necessary dues incidental to the road and markets, such as tolls, forage, ferries, and market custom. A drover of sheep should always be provided with a dog, as the numbers and nimbleness of sheep render it impossible for one man to guide a capricious flock along a road subject to many casualties; not a young dog, who is apt to work and bark a great deal more than necessary, much to the annoyance of the sheep, but a knowing, cautious tyke. The drover should have a walking-stick, a useful instrument at times in turning a sheep disposed to break off from the rest. A shepherd's plaid he will find to afford comfortable protection to his body from cold and wet, while the mode in which it is worn leaves his limbs free for motion. He should carry provision with him, such as bread, meat, cheese or butter, that he may take luncheon or dinner quietly beside his flock while resting in a sequestered part of the road, and he may slake his thirst in the first brook or spring he finds, or purchase a bottle of ale at a roadside ale-house. Though exposed all day to the air, and even though he feel cold, he should avoid drinking spirits, which only produce temporary warmth, and for a long time after induce chilliness and languor. Much rather let him drink ale or porter during the day, and reserve the allowance of spirits he gives himself until the evening, when he can enjoy it in warm toddy beside a comfortable fire, before retiring to rest for the night. The injunction to refrain from spirits during the day I know will sound odd to the ear of a Highland drover; but though a dram may do him good in his own mountain-air, and while taking *active* exercise, it does not follow that it will produce equally good effects on a drove-road in the low country in winter, in raw and foggy weather. I believe the use of raw spirits does more harm than good to all drovers who indulge in the practice. He should also have a good knife, by which to remove any portion of horn that may seem to annoy a sheep in its walk; and also a small bottle of a mixture of tobacco-liquor and spirit of tar, with a little rag and twine, to enable him to smear and bandage a sheep's foot, so as it may endure the journey. He should be able to draw a little blood from a sheep in case of sickness. Should a sheep fail on the road, he should be able to dispose of it to the best advantage; or, becoming ill, he should be able to judge whether a drink of gruel, or a handful of common salt in warm water, may not recover it so as to proceed; but, rather than a lame or jaded sheep should spoil the appearance of the flock, it should be disposed of before the flock is presented in the market.

(1166.) The many casualties incidental to sheep on travel, more especially in winter, require consideration from the farmer before undertaking to send his stock to a distant market-town, in preference to taking them to a fair, or accepting an offer for them at home. A long journey in winter will cost at least 1s. a head, and their jaded appearance may have the effect of lowering their market price 2s. or 3s. a head more. Under any circumstances, when you have determined on sending your sheep to a market-town, it is, I believe, the best plan, after the journey, to intrust them to a salesman, rather than stand at market with them yourself, as you cannot know the character of the buyers so well as he does, nor can you know what class of purchasers your lot may best suit. The convenience attending the employment of a salesman is now generally felt, because it not only saves the personal annoyance of attending a market, but your money is remitted to you through a bank in the course of the day. The only precaution requisite in the matter is to become acquainted with a salesman of judgment; for as to honesty, if he have not *that*, he is of course quite worthless. In attending country fairs, the case is otherwise; there being no salesman, you yourself must stand by your lot. Before attending the fair, you should make up your mind what to ask for your stock, in accordance with the current market prices; but, notwithstanding this, you may come away with more or less cash than you anticipated, because the actual state of *that* market will be regulated by the quality and quantity of the stock brought forward, and by the paucity or numbers of buyers who may appear. After your sheep are fairly placed, you should inquire of friends of the state of prices before you sell, and on doing this you will frequently find the market in a most perplexing state from various causes. Thus, there may be too many sheep for the buyers, when the market will be dull, and remain so all day. On the other hand, the stock may be scanty for the buyers, when a briskness may start in the morning and continue even till the whole stock are sold off. There may be briskness in the morning, the buyers purchasing; dullness at midday, buyers declining; and briskness again in the afternoon, buyers again purchasing. There may be excessive dullness in the morning, occasioned by the buyers lying off and beating down prices, and, finding they cannot succeed, buy briskly all the afternoon. There may be dullness in the morning, arising from the dealers finding the condition of the stock below their expectation. The markets are never better for the farmer than when they begin brisk early in the morning, and the stock are all sold off early. These are the vicissitudes of a market; they are interesting, demand attention, and are worth examination. You will frequently observe a trifling circumstance give a decided tone to a market. A dealer, for instance, who generally buys largely, and having bought for many years respectably in that particular fair, will mark the prices of the day by his purchases; so that other people, particularly sellers, observing the prices given by him, will sell briskly and with confidence. There is no use, at any time, of asking a much higher price than the intrinsic value of your stock, or than you will willingly take; for, although your stock may be in particularly fine condition, and of good quality, and therefore worth more than the *average* price of the market, still their value must conform to the *rate* of the market, be it high or low, and it is not in your power to control it; though, if prices dissatisfy you, you have it in your power to take your stock home again. There is a common saying applicable to all public markets, and is now received as a maxim, because indicating the truth, that "the first offer is the best," that is, the first offer from a *bona fide* buyer; for there are people to be found in all markets who, having no serious intention of buying at market price, make a point of offering considerably below it, with the view of catching a bargain from a greenhorn, or from one tried of standing longer in the fair, and they sometimes succeed in their wishes; but such people are easily discovered, and therefore cannot *deceive* any but inexperienced sellers.

(1167.) There are certain *rules* which, by tacit consent, govern the principles upon which all public markets of stock are conducted, and they are few and simple. There is a *custom* payable for all stock presented at fairs, exigible by the lord of the manor, or other recognized authority. After entering the field, your stock can take up any unoccupied position you choose, appointed for the particular kind of stock you have to show. No one, on pretence of purchasing, has a right to interfere with a lot which is under inspection by another party. Neither have you any right to show your lot to more than one party at a time, unless each party consent to it. When a bargain is made, there is no necessity for striking hands, or exchanging money, as an earnest of it. When a bargain is made, a time may be stipulated by the purchaser for lifting the stock; and, until they are delivered to him or his accredited agents, they continue at the risk of the seller. When counted over before the purchaser, the price becomes immediately due. When the money is paid, there is no obligation on the seller to give a discount off the price, or a *luck-penny*, as it is termed; but purchasers sometimes make offers in a way to humor the prejudices of the seller—that is, they offer the price demanded, on condition of getting back a certain sum, or amount of luck-penny, to bring the price down to their own ideas; in such a case, when such an offer is accepted, the seller must return the luck-penny conditioned for, when he receives the money. Some times, when parties cannot agree as to price, the offerer proposes to abide by the decision of a third party; but, in doing this, you virtually relinquish your power over your own stock. Sometimes bills and bank-postbills are tendered by dealers in part or entire payment of what they purchase; but it is in your power to refuse any form of cash but the legal tender of the country, such as Bank of England notes, or gold, or silver. If a bill of exchange or promissory note is proffered instead of ready money, you are quite entitled to refuse the bargain; for the usage of trade in a fair implies the condition of ready money;* or you may demand a higher price to cover the risk of the bill being dishonored. The notes of any bank you know to be good, you will, of course, not refuse. After the stock are delivered, they are at the risk of the purchaser. Some dealers' *top's-men*—that is, the men who take charge of their master's lots after delivery—demand a gratuity for their trouble, which you are at liberty to refuse. All these rules, in as far as relates to money and the delivery of stock, apply to the stock purchased by dealers on your own farm. When you *purchase* stock at a fair, people will be found on the ground willing to render your drover assistance in taking them out of it, and of setting them fairly on the road. Such people are useful on such occasions, as it may happen, especially in the case of sheep, that one or more may break away from their own flock and mix with another, when there may not only be difficulty in shedding them out, but those into whose lot yours have strayed may show unwillingness to have their stock disturbed for your sake, though it is in your power to follow your strayed stock, and claim it anywhere by the wool-mark.

(1168.) The way that fat is laid on sheep while on turnips, and the mode of judging of a fat sheep, are these: Hoggs, when put on turnips in winter, are generally lean; for although they had been in good condition as lambs when weaned from their mothers in summer, their growth in stature afterward is so rapid that their flesh is but little intermixed with fat. For the first few weeks on turnips, even in the most favorable circumstances as to quality of food, warmth of shelter, dryness of land, and pleasantness of weather, they make no apparent advancement in condition; nay, they rather seem to fall off, and look clapped in the wool, and indicate a tendency to delicacy, in consequence, I suppose, of the turnips operating medicinally on their constitution as an alterative, if not as a laxative; but immediately after that trying period of young sheep, especially trying in bad weather, is past, when the grass has completely passed through them, and the stomach and intestines have become accustomed to the more solid food of the turnip, their improvement is marked, the wool seeming longer and fuller, the carcass filled out, the eyes clear and full, and the gait firm and steady. They then thrive rapidly, and the more rapidly the drier the weather.

* The Farmer's Lawyer.

(1169.) The formation of fat in a sheep destined to be fattened, commences in the inside, the *net* of fat which envelops the intestines being first formed, and a little deposited around the kidneys. After that, fat is seen on the outside, and first upon the end of the rump at the tail-head, which continues to move on along the back, on both sides of the spine or back-bone to the bend of the ribs, to the neck. Then it is deposited between the muscles, parallel with the cellular tissue. Meanwhile, it is covering the lower round of the ribs descending to the flanks, until the two sides meet under the belly, whence it proceeds to the brisket or breast in front, and the shaw or cod behind, filling up the inside of the arm-pits and thighs. While all these depositions are proceeding on the outside, the progress in the inside is not checked, but rather increased, by the fattening disposition encouraged by the acquired condition; and hence, simultaneously, the kidneys become entirely covered, and the space between the intestines and lumbar region or loin gradually filled up by the net and kidney fat. By this time, the cellular spaces around each fibre of muscle is receiving its share, and when fat is deposited there in quantity, it gives to meat the term *marbled*. These inter-fibrous spaces are the last to receive a deposition of fat; but after this has begun, every other part simultaneously receives its due share, the back and kidneys receiving the most, so much so that the former literally becomes *nicked*, as it is termed; that is, the fat is felt through the skin to be divided into two portions, from the tail-head along the back to the top of the shoulder, the tail becoming thick and stiff, the top of the neck broad, the lower part of each side of the neck toward the breast full, and the hollows between the breast-bone and the inside of the fore-legs, and between the cod and the inside of the hind thighs, filled up. When all this has been accomplished, the sheep is said to be *fat* or *ripe*.

(1170.) When the body of a fat sheep is entirely overlaid with fat, it is then in the most valuable state as mutton; but few sheep lay on fat entirely over their body, one laying the largest proportion on the rump, another on the back; one on the ribs, another on the flanks; one on the parts adjoining the fore-quarter, another on those of the hind-quarter; one more on the inside, and another more on the outside. Taking so many parts, and combining any two or more of them together, you may expect to find, in a lot of fat sheep, a considerable variety of condition, and yet any one is as ripe in its way as any other.

(1171.) Taking these data for your guide, you will be able to detect, by handling, the state of a sheep in its progress toward ripeness. A *ripe* sheep, however, is easily known by the *eye*, by the fullness exhibited in all the external parts of the particular animal. It may exhibit wants in some parts when compared with others, but you easily see that these parts would never become so ripe as the others; and this arises from some *constitutional* defect in the animal itself, because, if this were not so, there is no reason why all the parts should not be alike ripe. Whence this defect arises remains to be considered afterward. The state of a sheep that is obviously not ripe cannot altogether be ascertained by the eye; it must be *handled*, that is, it must be subjected to the scrutiny of the hand. Now, even in so palpable an act as handling discretion is requisite. A full-looking sheep need hardly be handled on the rump, for he would not seem so full unless fat had been first deposited there. A thin-looking sheep, on the other hand, should be handled on the rump, and if there be no fat there, it is useless handling the rest of the body, for assuredly there will not be so much as to deserve the name of fat. But between these two extremes of condition there is every variety to be met with; and on that account examination by the hand is the rule, by the eye alone the exception; but the hand is much assisted by the eye, whose acuteness detects deficiencies and redundancies at once. In handling a sheep the points of the fingers are chiefly employed, and the accurate knowledge conveyed by them through practice of the true state of condition is truly surprising, and settles a conviction in the mind that some intimate relation exists between the external and internal state of an animal. And hence this practical maxim in the judging of stock of all kinds, that no animal will appear *ripe to the eye*, unless as much fat had previously been laid on in the inside as his constitutional habit will allow. The application of this rule is easy. Thus, when you find the rump nicked on handling, you may expect to find fat on the back; when you find the back nicked, you would expect the fat to have proceeded to the top of the shoulder and over the ribs; and when you find the top of the shoulder nicked, you would expect to find fat on the under side of the belly. To ascertain its existence below, you will have to *turn him up*, as it is termed; that is, the sheep is set upon his rump with his back down and his hind-feet pointing upward and outward. In this position you see whether the breast and thighs are filled up. Still, all these alone would not let you know the state of the inside of the sheep, which should, moreover, be looked for in the thickness of the flank; in the fullness of the breast, that is, the space in front from shoulder to shoulder toward the neck; in the stiffness and thickness of the root of the tail, and in the breadth of the back of the neck. All these latter parts, especially with the fullness of the inside of the thighs, indicate a fullness of fat in the inside; that is, largeness of the mass of fat on the kidneys, thickness of net, and thickness of layers between the abdominal muscles. Hence the whole object of feeding sheep on turnips seems to be to lay *fat* upon all the bundles of fleshy fibres, called muscles, that are capable of acquiring that substance; for as to bone and muscle, these increase in weight and extent independently of fat, and fat only increases their magnitude.

(1172.) I have spoken of the *turning up* of a fat sheep; it is done in this way: Standing on the *near* side of the sheep, that is, at *its* left side, put your left hand under its chin, and seize the wool there, if rough, or the skin, if otherwise; place your knees, still standing, against its ribs, then bowing forward a little, extend your right arm over the far loin of the sheep, and get a hold of its flank as far down as you can reach, and there seize a large and firm hold of wool and skin. By this, lift the sheep fairly off the ground, and turning its body toward you upon your left knee under its near ribs, place it upon its rump on the ground with its back to you, and its hind feet sticking up and away from you. This is an act which really requires strength, and if you cannot lift the sheep off the ground, you cannot turn it; but some people acquire a sleight in doing it, beyond their physical powers. I believe the art consists in jerking the sheep off its feet at once, before it suspects what you are going to do; for if you let it feel that you are about to lift it as a dead weight, the probability is, that you will not be able to make it lose hold of the ground, as it is surprising how dexterously sheep contrive, in the circumstances, to retain hold of the ground with the point of the hoof of the *near hind-foot*, which, if you cannot force away, you cannot turn the sheep. I remember seeing 4 shepherds defeated in the attempt to turn 5 dinmonts belonging to the late Mr. Edward Smith, Marledown, Northumberland. None of the shepherds, even the longest and strongest, could turn all the 5 sheep, and one of them, a short though stout man, could not turn one of them, they were so broad in the back, so round and heavy. The ability to turn a sheep is not to be regarded as a feat in a shepherd, but a necessary act in connection with many important operations, as you shall see afterward.

(1173.) Sheep are easily *slaughtered*, and the operation is unattended with cruelty. They require some preparation before being deprived of life, which consists of food being withheld from them for not less than 24 hours, according to the season. The reason for fasting sheep before slaughtering, is to give time for the paunch and intestines to empty themselves entirely of food, as it is found when an animal is killed with a full stomach, the meat is more liable to putrefy, and not so well flavored; and as ruminants always retain a large quantity of food in their intestines, it is reasonable they should fast somewhat longer to get quit of it than animals with single stomachs. Sheep are placed on their side on a stool, called a *killing stool*, to be slaughtered, and requiring no fastening with cords, are deprived of life by a thrust of a straight knife through the neck, between its bone and the windpipe, severing the carotid artery and jugular vein of both sides, from which the blood flows freely out, and the animal soon dies. The skin, as far as it is covered with wool, is taken off, leaving that on the legs and head, which are covered with hair, the legs being disjointed by the knee. The entrails are removed by an incision along the belly, after the carcass has been hung up by the tendons of the houghs. The net is carefully separated from the viscera, and rolled up by itself; but the kidney fat is not then extracted. The intestines are placed on the inner side of the skin until divided into the *pluck*, containing the heart, lungs, and liver; the *bag*, containing the stomach; and the *puddings*, consisting of the viscera or guts. The bag and guts are usually thrown away, that is, buried in the dunghill, unless when the bag is retained and cleaned for haggis. The pluck is either fried or made into haggis. The skin is hung over a rope or pole under cover, with the skin-side uppermost, to dry in an airy place.*

[* By the Abbé Cornea, a man of great learning and observation, it was remarked that the people of the United States possessed "*bacon*-stomachs." Their want of relish, not to say their distaste—more especially females—for *mutton*, is one of the obstacles to the extension of sheep husbandry in the United States. An opinion generally prevails—borrowed, as many of our opinions are, from England—that mutton does not attain perfection in juiciness and flavor under four or five years; but we are disposed to believe, with an experienced victualer of our acquaintance, and a good judge of mutton, alike in the field and the shambles, that this is a mistake, and that well-fed and fatted mutton is never better than when it gets its full growth, in its second year; nor can the farmer afford to keep it longer, unless the wool would pay for the keep. We have not the epicures and men of wealth who would pay the butcher the extra price, which he must have, that would enable him to pay a remunerating price to the grazier for keeping his sheep two or three years over.

The common mistake in the management of mutton in our country is that we *eat it exactly at the wrong time after it is killed.* It should be eaten, as a fried chicken should, immediately after being killed, and, if possible, before the meat has time to get cold; *or,* if not, then it should be kept a week or more—in the ice-house, if the weather require—until the time is just at hand when the fibre passes the state of toughness which it takes on at first, and reaches that incipient or preliminary point in its progress toward putrefaction when the fibres begin to give way and the meat becomes tender.

We were gratified lately to see, in a large *smokery* of Mr. Clements, in Philadelphia, quite a large number of *corned legs of mutton*, which had been sent in from Ohio. If mutton can be brought into vogue in that shape, it will be an additional inducement for rearing sheep in many situations removed from available markets.

Who need complain, at a watering-place, if he can secure for his breakfast or his dinner a good mutton-chop, such as is to be had at Caldwell's White Sulphur Springs, peppered, and broiled and served up hot, *with no gravy but its own?* Can anything be more *toothful*—more *wholesome?* *Ed. Farm. Lib.*]

(1174.) The carcass should hang 24 hours in a clean, cool, airy, dry apartment, before it is cut down. I say *cool* and *dry*, for if warm the meat will not become firm, and if damp a clamminess will cover it, and will never feel dry, and have a fresh, clean appearance. The carcass is divided in two by being sawed right down the back-bone. The kidney-fat is then taken out, being only attached to the peritoneum by the cellular membrane, and the kidney is extracted from the *suet*—the name given to sheep tallow in an independent state.

(1175.) In almost every town there is a different way of cutting up a carcass of mutton; and, it being here impossible to advert to them all, I shall select those of Edinburgh and London, and distinguish them as the Scotch and English modes. Although the English mode is upon the whole preferable, having been adopted to suit the tastes of a people long acquainted with domestic economy, it must nevertheless be admitted that meat is cut up in Scotland in a cleanly and workmanlike manner; but on the other hand, it will not be denied by those who have observed for themselves, that the beauty and cleanliness of meat, as exhibited in London, call forth the admiration of every connaisseur. The Scotch mode is represented in fig. 255, where, in the hind-quarter, *a* is the *jigot* and *b* the *loin*, and, in the fore-quarter, *c* the *back-ribs* and *d* the *breast*. It will be observed that the jigot is cut with a part of the haunch or rump, and the fore-quarter right through the shoulder into 2 pieces. The English mode is represented in fig. 256, where, in the fore-quarter, *a* is the shoulder, *b* and *b* the

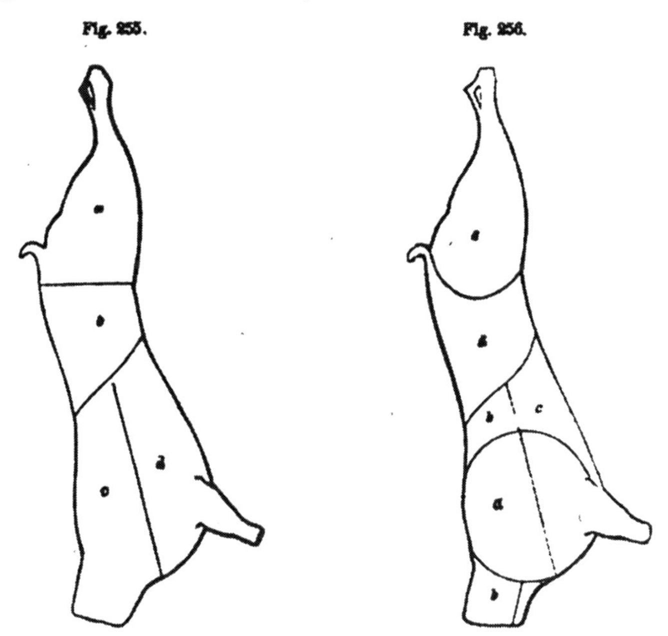

Fig. 255.

THE SCOTCH MODE OF CUTTING UP A CARCASS OF MUTTON.

Fig. 256.

THE ENGLISH MODE OF CUTTING UP A CARCASS OF MUTTON.

neck, and *c* the breast after the shoulder is removed; and, in the hind-quarter, *d* is the loin and *e* the leg. The leg here is cut short, without any of the haunch, like a ham; and the shoulder is preserved whole.

(1176.) The jigot *a*, fig. 255, is the handsomest and most valuable part of the carcass, and on that account fetches the highest price. It is either a roasting or a boiling piece. Of Black-faced mutton it makes a fine roast, and the piece of fat in it called the *Pope's eye* is considered a delicate *morceau* by epicures. A jigot of Leicester, Cheviot or South-Down mutton makes a beautiful "boiled leg of mutton," which is prized the more the fatter it is, as this part of the carcass is never overloaded with fat. The loin *b* is almost always roasted, the flap of the flank being skewered up and it is a juicy piece. For a small family, the Black-faced mutton is preferable; for a large, the South-Down and Cheviot. Many consider this piece of Leicester mutton roasted as too rich, and, when warm, this is probably the case; but a cold roast loin is an excellent summer dish.—The back-rib *c* is divided into two, and used for very different purposes. The fore-part, the neck, is boiled, and makes sweet barley-broth; and the meat, when well boiled, or rather the whole pottage simmered for a considerable time *beside* the fire, eats tenderly. The back-ribs make an excellent roast; indeed, there is not a sweeter or more varied one in the carcass, having both ribs and shoulder. The shoulder-blade eats best cold, and the ribs warm. The ribs make excellent chops. The Leicester and South-Downs afford the best mutton-chops. The breast *d* is mostly a roasting piece, consisting of rib and shoulder, and is particularly good when cold. When the piece is large, as of South-Down or Cheviot, the gristly part of the ribs may be divided from the true ribs, and helped separately. The breast is an excellent piece in Black-faced mutton, and suitable to small families; the shoulder being eaten cold, while the ribs and brisket are sweet and juicy when warm. This piece also boils well; or, when corned for 8 days, and served with onion sauce, with mashed turnip in it, there are few more savory dishes at a farmer's table. The shoulder *a*, fig. 256, is separated before being dressed, and makes an excellent roast for family use, and may be eaten warm or cold, or corned and dressed as the breast mentioned above. The shoulder is best from a large carcass of South-Down, Cheviot or Leicester, the Black-faced being too thin for the purpose; and it was, probably, because English mutton is usually large, that the practice of removing it originated. The neck-piece *b b* is partly laid bare by the removal of the shoulder, the fore-part being fitted for boiling and making into broth, and the best end for roasting or broiling into chops. On this account this is a good family piece, and in such request among the tradesmen of London that they prefer it to any part of the hind-quarter. Heavy mutton, such as the Leicester, South-Down and Cheviot, supply the most thrifty neck-piece. The breast *c* is much the same sort of piece as in the Scotch method, but the ribs are here left exposed at the part from which the shoulder has been removed, and constitute what are called the *spare ribs*, which may be roasted, or broiled, or corned. The back end of the breast makes a good roast for ordinary use. The flap of the loin left attached to this piece may be used in making broth. The loin *d* is a favorite roast in a family; and when cut double, forming the chine or saddle, it may grace the head of the table of any public dinner. Any of the kinds of mutton is large enough for a saddle; but the thicker the meat, of course the larger the slice. The leg *e* is cut short and roasted. When cut long, taking in the hook-bone, it is similar to a haunch of venison, and roasted accordingly. A fat Black-faced wether yields a good haunch.

(1177.) The different sorts of mutton in common use differ as well in quality as in quantity. The flesh of the Leicester is large, though not coarse-grained, of a lively red color, and the cellular tissue between the fibres contains a considerable quantity of fat. When cooked it is tender and juicy, yielding a red gravy, and having a sweet, rich taste; but the fat is rather too much and too rich for some people's tastes, and can be put aside; and it must be allowed that the lean of fat meat is far better than lean meat that has never been fat. Leicester sheep generally attain to heavy weights—hoggs reaching 18 lbs. or 20 lbs., and dinmonts 30 lbs. per quarter; but the 5 dinmonts which I mentioned before as having defeated the shepherds in turning up, were 55 lbs. a quarter overhead, when killed at Newcastle in November, a few weeks after they were shown.

(1178.) Cheviot mutton is smaller in the grain, not so bright of color, with less fat, less juice, not so tender and sweet, but the flavor is higher and the fat not so luscious. The weight attained by a hogg may be taken at 14 lbs. or 15 lbs., and by a wether at 22 lbs.; but Mr. Fairbairn mentions having fattened 5 wethers in 1818 which averaged 30 lbs. a quarter.*

(1179.) Black-faced mutton is still smaller in the grain, of a darker color, with still less fat, but more tender than the Cheviot, and having the highest flavor of all. The ordinary weight of a fat wether is about 18 lbs. or 20 lbs. a quarter; but I remember seeing a lot of 5-year-old Black-faced wethers, exhibited at the first Show of the Highland and Agricultural Society at Perth, belonging to Lord Panmure, that averaged 40 lbs. a quarter.

(1180.) The mutton of South-Downs is of medium fineness in grain, color pleasant red, fat well intermixed with the meat, juicy, tenderer than the Cheviot, and of pleasant though not of so high a flavor as the Black-faced. The ordinary weight may be from 16 lbs. to 22 lbs. a quarter, but 3 wethers exhibited by Mr. Grantham, at the Show of the Smithfield Club in 1835, weighed, on the average, 41½ lbs. a quarter.†

(1181.) Tup-mutton of any breed is always hard, of disagreeable flavor, and in autumn not eatable. The mutton of old ewes is dry, hard and tasteless, but of young well enough flavored, but still rather dry. Hogg-mutton is sweet, juicy and tender, but flavorless. And wether-mutton is the meat in perfection, according to its kind.

(1182.) The average quantity of fat afforded by each sheep of every class, sold in any given market in Scotland, is perhaps not great. In Glasgow, for example, where heavy animals of all sorts are generally sold, the fat afforded by all the sheep—consisting chiefly, I presume, of Cheviot and Black-faced—exclusive of lambs, amounting to 57,520 head, sold in 1822, was only, on the average, 4 lbs. 13 oz. per head.‡ From 8 lbs. to 12 lbs. is the ordinary quantity obtained from *Leicester* sheep slaughtered on farms of good land; and in Edinburgh I find that 7 lbs. is considered an average from Black-faced and Cheviot sheep, which shows that the quality of mutton sold there is better than that in Glasgow.

(1183.) As you may frequently hear it remarked in the course of your experience as a farmer, that 5-year-old mutton is the best, it is worth while considering whether the case can be so. Two subjects of inquiry immediately present themselves on hearing this remark: one, whether sheep require 5 years to put them in condition for use? and the other is, whether it is treating them properly to postpone putting them in condition for use until they shall attain the age of 5 years? If truth is implied in the first inquiry, then that breed of sheep must be very unprofitable which takes five years to attain its best state; but there is no breed of sheep in Great Britain which requires 5 years to bring it to perfection. Therefore, if truth is implied in the second inquiry, then it must be folly to restrain sheep coming to perfection until they have attained the age of

* A Lammermuir Farmer's Treatise on Sheep in High Districts. † Youatt on Sheep.
‡ Cleland's Account of the Highland and Agricultural Society's Show at Glasgow, in 1828.

5 years. It is not alleged by the lovers of 5-year-old mutton that it bestows profit on the farmer, for the allegation only insists on its being best at that age. But such an allegation involves one of two absurd conditions in Agriculture; namely, the keeping a breed of sheep that cannot, or the keeping of one that you should not allow to attain to perfection before it is 5 years. Either of these conditions makes it obvious that mutton cannot be in its *best* state at 5 years. The fact is, the idea of 5-year-old mutton being super-excellent, is founded on a prejudice, which probably arose from this circumstance: Before winter food which could maintain the condition on stock that had been acquired in summer, was discovered, sheep lost much of their summer-condition in winter, and of course an oscillation of condition occurred year after year until they attained the age of 5 years; when their teeth beginning to fail, would cause them to lose their condition the more rapidly. Hence, it was expedient to slaughter them not exceeding 5 years of age; and, no doubt, at that age mutton would be high flavored that had been exclusively fed on natural pasture and natural hay. But such treatment of sheep cannot now be justified on the principles of modern practice; because both reason and taste concur in mutton being at its *best* whenever sheep attain their *perfect state of growth and condition*, not their largest and heaviest; and as one breed attains its perfect state at an earlier age than another, its mutton attains its best before another breed attains *its* best state, although its sheep may be older; but taste alone prefers one mutton to another, even when both are in their best state, from some peculiar property. The Black-faced sheep, for instance, it is preferred by many, because of the flavor of its mutton; and this property it has most probably acquired from the heathy pasture upon which it is brought up. But if flavor alone is to decide the point, the Welch mutton is much the superior. So far as juiciness is concerned, a Leicester hogg has more of it than any Black-faced sheep; and the darkness of the flesh of the latter arises solely from the breed, as it seems to form the connecting link, in this country, between the sheep and the goat, the latter of which always has dark-colored flesh. Judged by the scale of perfection of growth, Leicester mutton is best in the dinmont; and as it may require five years to bring a Black-faced wether to that state when constantly confined upon the hills, Black-faced mutton may then be considered in its best state, *because* it is 5-year-old, but so far from being in the condition it would have attained had it been brought down to the low country when a lamb and fed upon the best food, it would still be lean, and, of course, not in a state of perfect growth; whereas, in the low country, it would attain perfection of growth at 3 years, and then its mutton must be at its *best*; for beyond that age—that is, if kept to 5 years on such food—it would become too fat, and lose much of its delicacy. The cry for 5-year-old mutton is thus based on very untenable grounds.

(1184). *Markets* for sheep are held in all large towns, and the butchers in the small ones supply themselves from the farmers. The Edinburgh weekly market, on Wednesday, supplies the Black-faced mutton in perfection, and the Cheviot is also very good. In Morpeth, on Wednesdays, are to be seen Leicesters in the highest state of condition, which are brought up with avidity for the colliers around Newcastle. In London, on Mondays, the South-Downs are seen in great perfection, this being the favorite mutton of the Capital.

(1185). A great trade in the transmission of live-stock and meat from the east coast of Scotland to London, has arisen since the establishment of steam navigation. From inquiry, I found that, in the year ending May 1837, there were shipped 4,221 old sheep, and 11,672 barrel-bulk of meat, chiefly mutton, which, at 2½ cwt. per barrel-bulk, give 29,175¼ cwt.* The meat is sent by butchers at the different shipping ports, and the live-stock by dealers, butchers and farmers. When you determine to send your stock to London, you should, in the first place, establish a correspondence with a live-stock salesman, who will pay all charges on board ship and at market, and remit the balance in course of post. The charges consist of freight, which for sheep is 3s. 6d. a herd, commission, hay or grass on board, dues and wharfage, hay or grass on shore, and driving to market. You will, of course, never ship meat, but you should, nevertheless, be well acquainted with all the pieces into which a carcass of beef or mutton is cut up, that you may know whether your stock is of the description to supply the most valuable pieces of meat; for, without this knowledge, unless, in short, you know the wants of a market, you cannot know whether you are supplying its requirements, or whether your stock ought to realize the top prices.

(1186.) On the supposition that you send sheep to London by steam on your own account, they should be of the following description, to command the best prices, and unless they are so you had much rather dispose of them at home. They should be ripe, compact, and of light weight; carrying a large proportion of lean on the back, loins, and shoulders, with a full, round leg and handsome carcass. Such, from 14 lbs. to 20 lbs. a quarter, will take readily, but they will draw the most money at from 16 lbs. to 18 lbs. The nearer in their form and quality they approach the South-Downs, the more likely to command top prices. True-bred Cheviots and the Black-faced Linton breed approach very near to the South-Down, and command as high a price. Half-breds, between Leicester tups and the above sorts of Cheviot and Black-faced ewes, form valuable sheep. The old Black-faced breed are too *thin*, and therefore styled *goaty* in Smithfield, and when only half-fat, or *half-meated*, as the condition is there termed, fetch middling prices, however good their flavor may be. Pure-bred Leicesters are too *fat*, unless sent young, and not exceeding 20 lbs. a quarter, but above that weight, fetch inferior prices, so much so that a difference of only 1d. per lb. may perhaps constitute all the difference between a profit and loss on their export. This last remark applies to every other breed, and shows the expediency of only exporting the best form of sheep.

(1187.) Never attempt to drive stock on foot on your own account to a distant market, when you have steam-conveyance to the place of destination. A simple comparison of the results of the two methods of traveling will show you at once the advantage of steam-conveyance. It has been ascertained that a journey of 400 miles on land causes a loss of 6 stones out of 40 stones, or 12 per cent.; whereas the loss by steam is only 2 stones out of 50. But besides this great difference in the loss itself, the state in which the remainder of the flesh is left, it is worth 6d. a stone less after land travel; and when stock are sent to graze in that state, they require a month to take with the pasture, whereas the steam-carried will thrive again at the end of a fortnight. Besides all these disadvantages of land travel, the juices of the meat of fat stock never recover their natural state, while, by being carried by steam, they do. Were heavy and high-conditioned stock to be traveled by land, they would inevitably sink under the attempt, while by steam any degree of condition may be conveyed

* See an article on the preparation of live-stock and meat for exportation by steam, in vol. viii. of the Quarterly Journal of Agriculture, drawn up by me on information derived from Mr. James Dickson, in Orkney, who has had great experience in every matter relating to meat and live-stock; and also from other sources.

with comparative ease. The *time*, too, spent on a land journey is of consideration, when a more expeditious mode of traveling is in your option.

(1188.) With regard to the relative weights of offal and meat afforded by sheep, there are recorded instances of their proportions, and of a fat South-Down wether they were these, namely:

Live weight, 13 *st.* 10 *lbs.*

Offal.	Lbs.	Oz.	Meat.	Lbs.	Oz.
Blood and entrails	13	0	Fore-quarter	29	0
Caul and loose fat	21	4	do.	28	12
Head and pluck	8	12	Hind-quarter	33	8
Pelt	15	12	do.	32	0
Total	58	12	Total	113	4*

I may mention that the *carcass* consists of the entire useable meat of the body, which, when sawed down the middle of the back-bone, is divided into two *sides*, which, when again divided by the 5th rib, makes the carcass to consist of 4 *quarters*. The remainder of the animal consists of offal—namely, of fat, entrails, head and skin. In purchasing fat live-stock, the butcher is supposed to pay the market value of the carcass, bone and meat to the farmer, reserving the offal to himself for his profit and risk. The relative proportions of mutton and offal have probably never been absolutely ascertained, as they must differ in different breeds of sheep; but there is little doubt that, in the Leicester breed, the meat bears a higher proportion to the offal than in any other breed. In the above case, the meat is about ⅔ and offal ⅓ of the whole weight; or, more nearly, the meat is as 123¼ : 182, and the offal as 58¾ : 182. And in the same breed it has been said that the proportion of bone is as low as 1 oz. to 1 lb. flesh; but I much doubt this, because Mr. Donovan found in a leg of mutton, which is the most fleshy part of the carcass in proportion to the bone in it, weighing 9¾ lbs., 16 oz. of bone; another of 9 lbs. 6 oz., 15 oz.; and a leg of small Scotch mutton, of only 6 lbs. weight, afforded 10¼ oz. of bone.

(1189.) There is a rule mentioned by Mr. Ellman, of Glynde, in Sussex, by which the age of mutton may be ascertained by certain marks on the carcass, and it is an infallible one. He says: "Observe the *color* of the *breast-bone* when a sheep is dressed—that is, where the breast-bone is separated—which, in a lamb, or before it is 1 year old, will be quite red; from 1 to 2 years old, the upper and lower bone will be changing to white, and a small circle of white will appear round the edges of the other bones, and the middle part of the breast-bone will yet continue red; at 3 years old, a very small streak of red will be seen in the middle of the 4 middle bones, and the others will be white; and at 4 years, all the breast-bone will be of a white or gristly color."†

(1190.) The experiments of Mr. Donovan prove that meat of all kinds loses a considerable proportion of weight on being cooked. His results on mutton were: The average loss on *boiling* legs of mutton is 10 per cent.; so that, if the butchers' price were 6d. per lb., the boiled mutton would cost 7½d. The average loss of *roasting* legs of mutton is 27 7-10 per cent.; so that, at the butcher's price of 6d. per lb., the roasted mutton would cost 8¼d. per lb. The average loss of roasting shoulders of mutton is 28 per cent.; and, were the butcher's price 5d. per lb., the roasted shoulder would cost 6 9-10d. per lb. The average loss, therefore, in boiling mutton is 10 per cent., and in roasting it 27 85-100 per cent. These results differ considerably from those obtained by Professor Wallace, who, in the case of *boiling* 100 lbs. of mutton, detected a loss of 21½ per cent., instead of 10 per cent.; and in that of *roasting* 100 lbs., the loss was 31½ instead of 28 per cent.— These discrepancies might, perhaps, be easily explained were we acquainted with every particular connected with both sets of experiments, such as the state of the meat before and after being cooked. In these respects, in his own experiments, Mr. Donovan says: "I used meat of sufficient but not unprofitable fatness, such as is preferred by families; the meat was, in all cases, a little rare at its center, and the results were determined with the utmost care."‡

(1191.) Good *ham* may be made of any part of a carcass of mutton, though the *leg* is preferred, and for this purpose it is cut in the English fashion. It should be rubbed all over with good Liverpool salt, and a little saltpetre, for 10 minutes, and then laid in a dish and covered with a cloth for 8 or 10 days. After that it should be rubbed again slightly for about 5 minutes, and then hung up in a *dry* place, say the roof of the kitchen, until used. Wether mutton is used for hams, because it is fat, and it may be cured any time from November to May; but tup mutton makes the largest and highest flavored ham, provided it be cured in spring, because it is out of season in autumn.

(1192.) There is an economical way of using fat mutton well adapted for the laboring people of a farm. The only time Scotch farm-servants indulge in butcher-meat is when a sheep *falls*, as it is termed—that is, when it is killed before being affected with an unwholesome disease, and the mutton is sold at a reduced price. Shred down the suet small, removing any flesh or cellular membrane adhering to it; then mix among it intimately ¼ oz. of salt and a tea-spoonful of pepper to every pound of suet; put the mixture into an earthen jar, and tie up tightly with bladder. One table-spoonful of seasoned suet will, at any time, make good barley-broth or potato soup for two persons. The lean of the mutton may be shred down small, and seasoned in a similar manner, and used when required; or it may be corned with salt, and used as a joint.

(1193.) Where Leicester sheep are bred, and the farmer kills his own mutton, suet will accumulate beyond what can be used for domestic purposes. As long as it is fresh it should be *rynded* or *rendered*, as it is termed—that is, prepared for preservation—because the fibrous and fleshy matter mixed with it soon promotes putrefaction. It should be cut in small pieces, removing only fleshy matter. It is then put in an earthern jar, which is placed within a pot containing warm

* Sussex Agricultural Report. † British Husbandry, vol. ii. ‡ Donovan's Domestic Economy, vol. ii

water, at the side of the fire, merely to simmer, and not to boil. As every portion put in is melted, another succeeds, until the whole is melted; and the melted mass should be very frequently stirred. Suet melts at from 98° to 104° Fahr. After being fused a considerable time, the membraneous matter comes to the top, and is taken off; and when obtained in quantity, and squeezed, this scum constitutes the *cracklings* which are sometimes used for feeding dogs. The purified suet may then be poured through a cullender, into a dish containing a little water, upon which it consolidates into a cake; and the cakes are either sold to the candle-makers or candles taken in exchange. "Many plans for purifying fats," says Ure, "have been proposed. One of the best is to mix 2 per cent. of strong sulphuric acid with a quantity of water, in which the tallow is beated for some time with much stirring, to allow the materials to cool, to take off the supernatant fat, and re-melt it with abundance of hot water. More tallow will thus be obtained, and that considerably whiter and harder than is usually procured by the melters."* Some people melt suet in a pot over the fire, where it is apt to be burnt; and some even fry it in a frying-pan, which may answer for culinary purposes, but cannot, of course, be disposed of to the candle-makers.

(1194.) Mutton-suet consists of about 77 parts of stearine and 23 of oleine in every 100 parts.—The former is solid, the latter fluid. The specific gravity of suet is 0·936. When a piece of solid suet is broken, innumerable minute granules separate from the mass; and these, when examined by the microscope, exhibit definite forms, being polyhedral, bounded within the limits of a sphere, or oblong, of very firm consistence, and, when measured, give dimensions varying in length from 1-400 to 1-900, and in breadth from 1-200 to 1-500 part of an inch.† The constituent parts of suet, according to Chevreul, are carbon 78·996, hydrogen 11·700, and oxygen 9·304.‡

(1195.) Fat is very generally distributed in the animal frame. It is "abundant under the skin, in what is called the cellular membrane, round the kidneys, in the folds of the omentum, at the base of the heart, in the mediastinum, the mesenteric web, as well as upon the surface of the intestines, and among many of the muscles. It varies in consistence, color and smell, according to the animal from which it is obtained. Thus, it is generally fluid in the cetaceous tribes, soft and rank-flavored in the carnivorous, solid and nearly scentless in the ruminants; usually white and copious in well-fed young animals, yellowish and more scanty in the old. Its consistence varies also according to the organ of its production, being firmer under the skin and in the neighborhood of the kidneys than among the movable viscera. Fat forms 1-20 of the weight of a healthy animal; but as taken out by the butcher it is not pure, for, being of a vesicular structure, it is always inclosed in membranes, mixed with blood, blood-vessels, lymphatics, &c."‖

(1196.) Sheep is one of the most useful, and therefore one of the most valuable, of our domestic animals; it not only supports our life by its nutritious flesh, but clothes our bodies with its comfortable wool. All writers on diet have agreed in describing mutton as the most valuable of the articles of human food. "Pork may be more stimulating, beef perhaps more nutritious, when the digestive powers are strong; but, while there is in mutton sufficient nutriment, there is also that degree of consistency and readiness of assimilation which renders it most congenial to the human stomach, most easy of digestion, and most contributable to health. . . . Of it, almost alone can it be said that it is our food in sickness as well as in health; its broth is the first thing that an invalid is permitted to taste, the first thing that he relishes, and is his natural preparation for a return to his common aliment.§ In the same circumstances, it appears that fresh mutton, broiled or boiled, takes 3 hours to digest; fresh mutton, roasted, 3¼ hours; and mutton-suet, boiled, 4½ hours.¶

(1197.) But the products of sheep are not merely useful to man, they also promote his luxuries. The skin of sheep is made into leather, and, when so manufactured with the fleece on, makes comfortable mats for the doors of our rooms, and rugs for our carriages. For this purpose the best skins are selected, and such as are covered with the longest and most beautiful fleece. Tanned sheep-skin is used in coarse book-binding. White sheep-skin, which is not tanned, but so manufactured by a peculiar process, is used as aprons by many classes of artisans, and, in Agriculture, as gloves in harvest; and, when cut into strips, as twine for sewing together the leathern coverings and stuffings of horse-collars. Morocco leather is made of sheep skins as well as of goats, and the bright-red color is given to it by cochineal. Russia leather is also made of sheep-skins, the peculiar odor of which repels insects from its vicinity, and resists the mould arising from damp—the odor being imparted to it in currying, by the empyreumatic oil of the bark of the birch tree. Besides soft leather, sheep-skins are made into a fine, flexible, thin substance, known by the name of parchment; and, though the skins of all animals might be converted into writing materials, only those of the sheep and she-goat are used for parchment. The finer quality of the substance called vellum is made of the skins of kids and dead-born lambs, and for the manufacture of which the town of Strasburgh has long been celebrated.

(1198.) Mutton-suet is used in the manufacture of common *candles*, with a proportion of ox-tallow. Minced suet, subjected to the action of high-pressure steam in a digester at 250° or 260° Fahr., becomes so hard as to be sonorous when struck, whiter, and capable, when made into candles, of giving very superior light. Stearic candles, the late invention of the celebrated Guy-Lussac, are manufactured solely from mutton-suet.

(1199.) Besides the fat, the intestines of sheep are manufactured into various articles of luxury and utility, which pass under the absurd name of catgut. "All the intestines of sheep," says Mr Youatt, "are composed of 4 coats or layers, as in the horse and cattle. The outer or *peritoneal* one is formed of that membrane by which every portion of the belly and its contents is invested, and confined in its natural and proper situation. It is highly smooth and polished, and it secretes a watery fluid which contributes to preserve that smoothness, and to prevent all friction and concussion during the different motions of the animals. The second is the *muscular* coat, by means

* Ure's Dictionary of the Arts, art. *Fat*.
† Raspail's Organic Chemistry.
‡ Liebig's Animal Chemistry.
‖ Ure's Dictionary of the Arts, art. *Fat*.
§ Edinburgh Encyclopædia, art. *Aliments*, as quoted by Youatt on Sheep.
¶ Combe on Digestion and Dietetics.

of which the contents of the intestines are gradually propelled from the stomach to the rectum, thence to be expelled when all the useful nutriment is extracted. The muscles, as in all the other intestines, are disposed in two layers, the fibres of the outer coat taking a longitudinal direction, and the inner layer being circular; an arrangement different from that of the muscles of the œsophagus, and in both beautifully adapted to the respective functions of the tube. The submucous coat comes next. It is composed of numerous glands, surrounded by cellular tissue, and by which the inner coat is lubricated, so that there may be no obstruction to the passage of the food. The *mucous* coat is the soft villous one lining the intestinal cavity. In its healthy state, it is always covered with mucus; and when the glands beneath are stimulated—as under the action of physic—the quantity of mucus is increased; it becomes of a more watery character; the contents of the intestines are softened and dissolved by it; and by means of the increased action of the muscular coat, which, as well as the mucous one, feels the stimulus of the physic, the fæces are hurried on more rapidly and discharged."* In the manufacture of some sorts of cords from the intestines of sheep, the outer peritoneal coat is taken off and manufactured into a thread to sew intestines, and make the cords of rackets and battledores. Future washings cleanse the guts, which are then twisted into different-sized cords for various purposes. Some of the best known of those purposes are whip-cords, hatters' cords for bow-strings, clock-makers' cord, bands for spinning-wheels (which have now almost become obsolete), and fiddle and harp strings. Of this last class of cords—the source of one of our highest pleasures—it has long been subject of regret that those manufactured in England should be so inferior in goodness and strength to those of Italy; and the reason assigned is that the sheep of Italy are both smaller and leaner than those of this country.—The difficulty lies, it seems, in making the treble strings from the fine peritoneal coat, their chief fault being weakness, whence the smaller ones are hardly able to bear the stretch required for the higher notes in concert-pitch—maintaining, at the same time, in their form and construction, that tenuity or smallness of diameter which is required to produce a brilliant and clear tone.†—However contemptible this subject may appear in the estimation of some, it is worth attending to by those interested in enhancing the profits of our native products, and more especially when it is considered that harp-strings sell as high as from 6d. to 2s. apiece.

(1200.) While adverting to the uses of the skin of the sheep, it may be useful to give an idea of its physical structure, a knowledge of which being requisite for an acquaintance of the rationale of its diseases. "It is composed of 3 textures. Externally is the *cuticle* or *scarf-skin*, which is thin, tough, devoid of feeling, and pierced by innumerable minute holes, through which pass the fibres of the wool and the insensible perspiration. It seems to be of a scaly texture; but this is not so evident in the sheep as in many other animals, on account of a peculiar substance, the *yolk*, which is placed on it to nourish and protect the roots of the wool. It is, however, plainly enough to be seen in the scab and other cutaneous eruptions to which the sheep is liable. Below this is the *rete mucosum*, a soft structure, its fibres having scarcely more consistence than mucilage, and being with great difficulty separated from the skin beneath. This seems to be placed as a defence to the termination of the blood-vessels and nerves of the skin, and these are in a manner enveloped and covered by it. The color of the skin, and probably that of the hair or wool also, is determined by the *rete mucosum*, or at least the hair and wool are of the same color as this substance. Beneath is the *cutis* or *true skin*, composed of innumerable minute fibres crossing each other in every direction—highly elastic, in order to fit closely to the parts beneath, and to yield to the various motions of the body; and dense and firm in its structure, that it may resist external injury. Blood-vessels and nerves, countless in number, pierce it, and appear on its surface under the form of *papillæ*, or minute eminences; while, through thousands of orifices, the exhalant absorbents pour out the superfluous or redundant fluid. The true skin is composed principally or almost entirely of gelatine; so that, although it may be dissolved by long continued boiling, it is insoluble in water at the common temperature. This organization seems to have been given to it not only for the sake of its preservation, while on the living animal, but that it may become afterward useful to man." It would appear that there are circumstances which materially limit the action of the power of excretion and absorption in the skin of the sheep. It is surrounded by a peculiar secretion, adhesive and impenetrable to moisture, the yolk, destined chiefly to preserve the wool in a soft, pliable and healthy state. On this account there can be little perspiration going forward from the skin, and hence few diseases are referable to change in that excretion. Also, there is little radiation of animal heat, both on account of the interposition of the yolk, and of the non-conducting power of the wool. The caloric disengaged from a sheep is only 1·7 part of that from man, though the weight of the animal is ½ of that of man—that is, only half the animal heat radiates from a sheep, from a given surface, that does from a man. This it is which enables the ewe and its lamb to endure the colds of spring without detriment; and also, when sheep are crowded together in an open fold, no unnatural or dangerous state of heat is thereby produced.‡

* Youatt on Sheep.

† Ure's Dictionary of the Arts, art. *Catgut*; also, *Leather*, *Parchment*.

‡ Youatt on Sheep.

32. REARING AND FEEDING CATTLE ON TURNIPS IN WINTER.

"The cattle from th' untasted fields return,
And ask, with meaning low, their wonted stalls,
Or ruminate in the contiguous shade."
THOMSON.

(1201.) THE first thing to be done with the courts in the steading, before being taken possession of by the cattle, is to have them *littered plentifully with straw*. The first littering should be abundant, as a thin layer of straw upon the bare ground makes an uncomfortable bed; whereas a thick one is not only comfortable in itself, but the lower part of it acts as a drainer to the heap of manure above it. There is more of comfort for cattle involved in this little affair than most farmers seem to be aware of; for it is obvious that the first layer of litter, when thin, will soon get trampled down, and in rainy weather soon become poached—that is, saturated with wet and pierced with holes by the cattle's feet—so that any *small* quantity of litter that is afterward laid upon it will but absorb the moisture below it, and never afford a *dry* lair to the cattle. On the other hand, when the first layer is thick, it is not poached even in wet weather, because it is with difficulty pierced through by feet, and it instantly drains the moisture that falls upon it, and of course keeps the bedding comparatively dry.

(1202.) There is, however, sometimes a difficulty of obtaining sufficient straw at this season, from various causes, among which may be mentioned a dislike in farmers to thresh a stack or two of the new crop at so early a period, even when there is no old straw or old stack of corn to thresh; but however recently formed the stacks may be, and inconvenient to thresh their produce at the time, it should be done rather than stint the cattle of bedding; and should bad weather immediately set in—an event not unlikely to happen—the cattle may be so chilled in their ill-littered quarters as not entirely to recover from it during the winter; and hence may arise a serious reduction of profit.

(1203.) It may happen, on the other hand, with plenty of old stacks, there may be want of water to drive the threshing-machine; and this is no uncommon predicament at the commencement of winter on many farms which depend upon surface-water for their supply; and a windmill is in no better plight in want of wind. Where such contingencies *may* happen, a sufficient quantity of litter should be provided for in good time, and there are various ways of doing this. Those who still use the flail may employ it at any season; and those having horse threshing-mills are equally independent. Access to bog-land gives the command of making coarse herbage into hay during summer; but in regard to the use of other products of bog-land for litter, precaution is requisite, for the turfy matter on the top on being used as a bottoming for courts, with the view of absorbing their moisture, will inevitably become as a sponge of water after the first fall of rain, and the cattle will soon render the whole bedding a poached mass. I once tried the experiment under the most favorable circumstances of getting the turf well dried, and yet could not get rid of the inconvenience of poaching until the courts were entirely cleared of their contents. Those who are annoyed with ferns in their pastures should cut them down and won them for litter, and a most excellent foundation they make for straw. Those who can cut grass, or gather dry leaves in woods, should do so in summer, or immediately after harvest, for a day or two, with the harvest people. By attending to one or all of these provident measures, a comfortable bed may be provided for cattle at the commencement of the season, under the most unfavorable circumstances in regard to a command of straw.

(1204.) Suppose, then, that all the courts and hammels are plentifully littered for the reception of the cattle, the next step is to arrange the different classes of cattle in their respective places. The different classes of cattle are cows, calves of the year, 1-year-olds, 2-year-olds, bulls, heifers in calf, and any extra cattle.

(1205.) *Cows* occupy the byre Q, fig. 3, Plate III. (42.) Each should always occupy the stall she has been accustomed to, and all will then go and come into their stalls without interfering with one another. They thus learn to become very quiet in the stall, both to the cattle-man who feeds them, and the dairy-maid who milks them. Each stall should have a manger *c*, fig. 10 (47), elevated 20 inches above the floor, lined with wood or stone, and having an edging of plank 8 inches in depth, to keep in the food. The usual plan is to place the mangers of byres on a level with the floor, down to which the cow has to stretch her neck to get to the turnip, or other food, and in doing this she is obliged to support herself almost wholly on one leg. This awkwardness of position is itself a certain proof that the animal is ill at ease while eating. There should be as much room behind the manger to the gutter as to allow the cow to lie at ease, whatever be her size, like a horse in a stall with a low hay-rack. Each stall should have a travis-board to separate it from the next (149). Some people are great advocates for double stalls, both in byres and stables, to hold a pair of animals each. In a byre, that plan is objectionable for several reasons; a cow is a capricious creature, and not always friendly to her neighbor, and one of them, in that case, must be bound to the stake on the same side as she is milked from; to avoid which inconvenience to the dairy-maid, the cow must be put aside nearer her neighbor in the same stall, which may prove unpleasant to both parties, or her neighbor in the adjoining stall be put aside nearer *her* neighbor, which may prove equally inconvenient. Neither is it a matter of indifference to the cow from which side she is milked, for many will not let down their milk if the milk-maid sits down to the unaccustomed side. The safest plan, therefore, in every respect, is for each cow to have her own stall. The floor of the stall should be causewayed only as far as shown at *m*, fig. 10, and the remainder at *f* should be of beaten earth, and this plan is intended to save the fore-knees of cows from injury. *Cattle* lie down and rise up by resting on the fore-knees, and when they have to do so on a hard pavement, injury will likely arise to the knees, if the pavement is not always covered with litter. I remember seeing a valuable Short-Horn cow, in Ireland, get injured in the knees from this cause; they swelled so much and continued so long in a tender state that she would not lie down at all; and all the while her owner was not aware of the cause until I suggested it. On the removal of the pavement, and proper treatment of the parts affected, they recovered. Cows are bound to the stake *h*, fig. 10, either by seal *c d*, fig. 11, or baikie, *e k g*, fig. 12, and *either* secures the animal sufficiently. The seal is made entirely of iron chain, and slides up and down the inclined stake *h* by means of the iron ring *d*; the baikie is made partly of wood *e*, and partly of rope *k* and *g*. Of the two modes of ligature I prefer the seal, because its construction permits the animal turning its head so much round as to be able to lick herself as far as the loin, whereas the baikie only admits of a constrained up and down motion along a perpendicular stake (48); and, besides, it is an impracticable mode of binding in connection with the use of a manger, because it prevents the animal stepping back to avoid it.

(1206.) *Calves of the year* should occupy court K, fig. 3, Plate III. (62.) In such receptacles they are put together male and female, strong and weak, but having plenty of trough room around two of the walls, they can all be amply provided with food at the same time, without the danger of the stronger buffeting about the weaker. The openings into the shed in which they take up their abode at night is at D, and in the center of the court stands the straw-rack *o*, formed like fig. 19 where straw is scarce, as on gravelly soils, or like figs. 20 and 21 where it is plenty. The troughs for the turnips are fitted up as in fig. 18, which is there represented as a short one, to show the finishings of the ends, but which, of course, may be extended to any length, as may be seen by *z* in K, fig. 4, Plate IV. There is a water-trough *w* in the same court, it being essential for young stock to have water at will, and especially when they do not get as many turnips as they can eat. When they do, cattle do not feel the want of water, the juice of the turnip supplying them with sufficient liquid. In the same Plate IV. may be seen the shed D, under the granary, connected with the court K, having a straw-rack *h'* fitted up at one end. The turnip-store for this court is at *g*; and *x* is the mouth of the liquid-manure drain, to carry off any superfluous water. In the calves of the year occupying this court K, where there is a good deal of traffic in going to and from the corn-barn C, the young creatures will become familiarized with the people, and have a chance of getting pickings of corn from the barn.

(1207.) The court I is fitted up precisely with the same conveniences of feeding-troughs *z*, water-trough *w*, straw-racks *h'* and *o*, and turnip-store *i*, as the other court for the 1-year-olds. It will be observed that the shed D, in both courts, has two entrances, which is the usual plan; but, in my opinion, the comfort of the cattle is more secured with only one entrance, inasmuch as all draft is prevented; and although the object of two entrances is a laudable one in affording a means of escape to a beast that may be ill-used by the rest, that advantage to one is dearly bought at the sacrifice of comfort to the others, and after all it is doubtful whether the contingency can be avoided in this way.

(1208.) As I have said before (62), I prefer hammels to large courts, for young beasts; because the heifers could be separated from the steers, and each of the classes subdivided to suit color, strength, age, temper, or any other point in which a few agree, and differ from the rest; and it is surprising how much better the same beasts look when assorted. In a large court, all are put together, and, if there be plenty of room for every one to do as it likes, no harm may accrue; but where too many are crowded together, which is almost always the case on farms where winterings are bought, some will be knocked about and kept back from their meat, and obliged to eat it at untimeous hours; and in either plight will be stinted in their growth and condition. Only one beast so used makes a serious drawback on the value of the lot, for it must be drafted from the rest and sold separately, at a reduced price, to the vexation of the owner, when too late to retrieve the loss. Now, no such occurrence can take place in hammels, where every difference in character, age, and strength of animals, can be nicely assorted; and this is the more requisite in beasts that have been bought in to be fed, than those brought up together at home.

(1209.) The *2-year-olds, intended to be fattened for the butcher*, occupy the hammels M, where are inner sheds at M, feeding-troughs *z*, liquid-manure drains *x*, in the courts, and where fodder is supplied in the inside of the sheds, in racks, in three of the corners, and the turnip-stores of which are at *e* and *f*. The sheds being 14 feet wide and 18 feet long, and the courts 30 feet long by 18 feet in width, each hammel will accommodate 4 steers, not merely at the beginning of the feeding season, but at its end, when they shall have attained the weight of at least 70 stones each imperial.

(1210.) Occasionally the cow stock requires to be renewed, one or two at time, by *young heifers*; and as these, when in calf, should not of course be fattened, they not be put in the hammels of the feeding-stock of their own age, namely, the 2-year-olds, but have hammels to themselves at N, which are fitted up in precisely the same manner as at M, with feeding-troughs *z*, straw-racks in the corner of the sheds, liquid-manure drains *x*, and turnip-stores *p* and *q*. Their size, inside the shed, is 17 feet long by 14 feet wide, and the court 20 feet long by 17 feet wide, so that each can accommodate 3 heifers in calf. The *old cows* which these heifers are intended to supersede have to be fattened, and they can be accommodated with one of the hammels at N.

(1211.) The *servants' cows* are accommodated in the byre Y, fitted up in the same manner as the other byre Q, having an outer court *v*, water-trough *w*, liquid-manure drain *x*, and turnip-store *h*.

(1212.) When *oxen are fattened in byres* instead of hammels, they are accommodated in the same manner as the cows are in either Q or Y; but instead of each having a stall, they are usually bound up in pairs in double stalls, with a partition in the turnip-trough, placed on the ground, and a travis between every pair. Stalls of this construction are often as narrow as 7 feet, but 8 feet is the more common width. I have already condemned the crowded state in which oxen, fed in byres, are usually placed (49), and shall not again advert to the subject here. When cattle are bound to the stake for the first time, for the season, they are apt to be restless until reconciled to their confinement, which they will be ere long, if provided with plenty of food.

(1213.) Bulls occupy the hammels X, which are fitted up with feeding-troughs *z*, water-troughs *w*, liquid-manure drains *x*, and racks in the corners of the sheds X. More than one bull-calf may be put together; but more than one bull that have served cows are never intrusted together.

(1214.) Having thus accommodated all the cattle, according to their kinds and ages, in their respective places in the steading, for the winter, let us now attend to the treatment which each class should daily receive during their confinement.

(1215.) And to begin with the *cows*. The first piece of work connected with the treatment of cows in winter, is to milk them at day-break, which cannot be at a very early hour this season. On farms on which cows are bred, they are heavy in calf in winter; so most of them will be dry, and those still yielding milk, being the latest to calve, will give but a scanty supply. It is, therefore, not as *milch*-cows they are treated at this season. After milking is finished by the dairy-maid, the usual practice is to give the cows, though heavy in calf, a feed of cold turnips, on an empty stomach, which I have always considered an injudicious practice; and its injudiciousness is evinced by the fact of the fœtus showing unequivocal symptoms of its existence in the womb, in the same manner as after a drink of cold water in the morning. I would, therefore, give them a mouthfull of fresh oat-straw, to prepare the stomach for the turnips. While amusing themselves with this fodder, the cattle-man, whose duty it is to take charge of all the cattle in the steading in winter, cleans out the byre of its litter and dung with the graip, fig. 257, and shovel (fig. 149), and wheelbarrow, and spreads it equally over the court, sweep the gutter and causeway clean with a birch or broom-besom. Having shut the byre-door and left the half-door into the court open for fresh air, the cattle-man leaves the cows until he has supplied the fattening and young beasts with turnips, which having done, he returns to the cow-byre, bringing litter-straw with him, and gives them their allowance of turnips for the first meal. Cows in calf never get as many turnips as they can eat, the object being not to fatten, but support them in a fair condition for calving; for were they fed fat, they would run the risk of life at calving through inflammation, and the calves would be small. It is not easy to specify the number or weight of turnips that should be given to cows; but I conceive that ½ of what a feeding ox would consume will suffice.

Fig. 257.

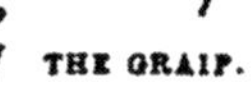

THE GRAIP.

(1216.) There are three ways of supplying cows with turnips, either through the openings of the wall at their heads, as at *o*, fig. 10, and through the door, fig. 9, from the store in the shed *s*, into the trough *c*; or with basketfulls, carried by the stall; or with barrow-loads, wheeled along a passage at their head, as described in (45), and emptied into the same trough *c* from the same store *s*, as seen in plan at *m*, fig. 4, Plate IV., by the back-door into the byre.

(1217.) With the willow-basket or *skull*, is the most common way of serving cows or cattle in byres with turnips. It is about 2 feet in diameter, with holes wrought into each side, under the rim, for handles, and costs about 1s. 6d.; but they are apt to become rotten or broken after the natural sap is dried out of the willows, which is generally in a few months' time, and then they become very brittle. In short, a skull seldom lasts more than a year or two; and as a number of them are required about a steading where a variety of beasts are fed on turnips, their cost, though individually trifling, becomes in the aggregate so considerable as to make its avoidance desirable. A basket of wire or small iron rods has been substituted in some places. A wire basket is represented by fig. 258, where the rim *a b c*, which forms its mouth, is a flat slip of iron about ¾ of an inch in breadth, and the keel or bottom *a d c* is of the same dimensions and materials. Holes are punched through them, at about 3 inches apart from each other. The small iron rods are inserted through them, receiving a bend to suit the form of the basket, and the ends of those attached to the rim *a b c* are shouldered below, and made fast with a counter-sink rivet above. The spaces left at the ends of the keel, under the rim, at *a* and *c*, form the handles. The cost is about 2s. 6d. each, and with due care—such as the replacement of a rod now and then, when broken—will last from 5 to 10 years. Were the keel made straight at *d*, the basket would stand steadier to be filled.*

Fig. 258.

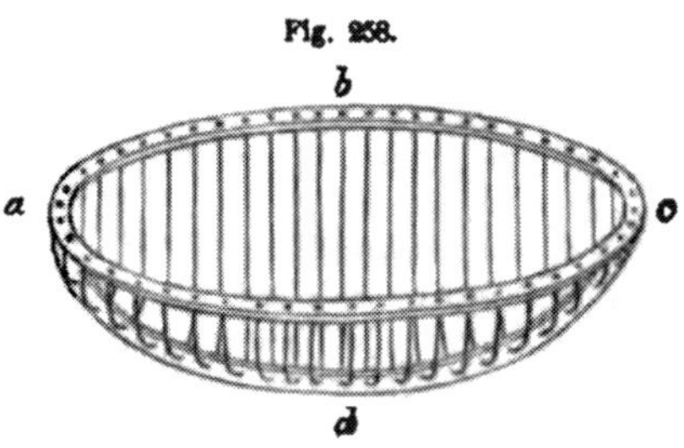

THE WIRE TURNIP-BASKET.

(1218.) Before the turnips are put into the troughs, the remains of the

* Quarterly Journal of Agriculture, vol. xi.

fodder given in the morning should be strewed down for litter, and the troughs cleaned out. The turnips should always be put into the troughs in a regular order, beginning at the same end of the byre, and finishing at the other, and after the turnips have been given, the cows should be permitted to eat them in quiet, for nothing irritates animals more than to be handled and worked about when feeding. The turnips consumed, and the stalls comfortably littered with straw, the cows will lie down and chew the cud until mid-day, when they should be turned into the court to enjoy the fresh air, lick themselves and one another, drink water from the trough, and bask in the sun. They should go out a while every day, in all weathers, until they calve, except perhaps in a very cold, wet day. One hour may be long enough at a time. In loosening cows from their stalls, a plan requires to be pursued to prevent confusion. In the first place, every cow, in the beginning of the season, should be put in the stall she has occupied since she first became an inmate of the byre; and she will always go to *it*, and no other, avoiding the least collision with the rest. In loosening them from the stalls, they should be so one by one, always beginning at the same end of the byre, and finishing at the other, and not indiscriminately. This will prevent collision on the floor and jamming in the doorway on going out—accidents injurious to animals with young. After their return, they should also be bound in the same regular order from one end of the byre to the other, and this will prevent any one being forgotten to be bound, and to remove every temptation from even a greedy cow running up into another one's stall for the sake of snatching a little of her food, no food should be lying in the troughs when they return to their stalls; and no food that they like—such as turnips, mangel-wurzel, and the like—should be given them immediately on returning to the byre, because the expectation of receiving it will not only render them impatient to leave the court, but make them restless in the stall until they receive it. This plan, contrary to usual practice, will, it is obvious, suppress all anxiety, and thereby prevent violation of discipline, and, of course, necessity for correction. When subjected to this regular form of discipline, they will soon obey it, and make no confusion, but conduct themselves peaceably. On their return to the byre, let a little fodder be given; and after a lapse of time, say at 3 P. M., give them their evening meal of turnips, after which they should be littered for the night.

(1219.) The treatment of *oxen* in a byre is different from that of cows; they get as many turnips as they can eat, and are not permitted to leave their stalls until sold off fat. As it is not usual for oxen to be fed in byres and hammels on the same farm, what I have to say in regard to feeding in the byre should be considered in lieu of the plan of what I shall have to say on feeding them in hammels. After the cow-byre doors have been opened, and the stalls cleared into the gutter of any dung that might annoy the dairy-maid, the cattle-man goes to the feeding-byre, and, first removing any fodder that may have been left from the previous night, and any refuse of turnips or other dirt, from the troughs, gives the cattle a feed of turnips at once. The quantity to be given at this time should be ⅓ of what they can eat during the day; for they should be fed 3 times a day—in the morning, at noon, and at sunset; and in distributing the food, the same regularity should be observed as in the case of the cows, that is, the same ox should always be the first supplied, and the same ox the last to receive his portion. When cattle find their food given them in regular order, they never become impatient for their turn. It is a good plan to begin serving at the farthest end of the byre, because then the cattle-man has no occasion to pass and disturb those which have been served; and in the case of what is called a *double-headed* byre, in which cattle stand on both sides, tail to tail, both sides should be served simultaneously by alternate beasts, thus still leaving those which have been served undisturbed. With the half-door left open for the admission of fresh air, and the expulsion of heated air through the ventilators (fig. 8), the cattle-man leaves them to enjoy their meal in quietness.

(1220.) Much has been said on the expediency of wisping and currying cows and fattening-beasts in the byre; and no doubt many satisfactory reasons could be urged in favor of the practice, when they are entirely confined. But, as it occurs to me, animals that are allowed to be at liberty at one part of the day do not require—or at least to a much smaller degree—any artificial dressing, inasmuch as they can dress their own skin, when at liberty, much better than any cattle-man. Nevertheless, where cattle are constantly confined in the byre, as is the case of all beasts fattened in a byre, it seems indispensable for their good health to rub their skin every day by some process; and I believe there are no better instruments for the purpose than a simple curry comb and a wisp of straw. In performing this operation, however, it should only be done when the cattle are not at food; and you should see to this, for there is a strong propensity in people who have charge of animals to dress and fondle with them when at food; from no desire, I am sure, of tormenting them, but the contrary. Still it is a habit which has a tendency to irritate all animals unnecessarily, and should be prevented; for any one may soon satisfy himself, from observation, that an animal is never more jealous of being approached than when eating his food—as witness the grumble from a dog or the scowl from a horse.

(1221.) Whenever the cattle have eaten their turnips the byre should be completely cleared of the dung and dirty litter with the graip, shovel, besom and barrow belonging to the byre. A fresh foddering and a fresh littering being given, they should be left to themselves to rest and chew the cud, until the next time of feeding, which should be about mid-day, when another ⅓ of turnips is given to each ox; after finishing which more fodder should be supplied, and what dung may prove annoying, drawn into the gutter. In the afternoon, before daylight goes, the dung should again be cleared out, and the last supply of turnips for the day, another ⅓, given to each ox; and before leaving them for the night, and after the turnips are eaten up, a fresh foddering should be given, and the litter shaken up and augmented where requisite. After eating a little fodder the cattle lie down and rest until visited at night.

(1222.) Where cattle are fattened in *hammels*, a somewhat different procedure is adopted. While the dairy-maid is milking the cows in the byre, the cattle-man cleans the troughs of the hammels with an old shovel, and gives the first supply of turnips for the day to the cattle; and in doing this he should adopt the same rule as to regularity as with the cows in the byre, always beginning with the same hammel

(1223.) It is now well understood that sliced turnips afford great facilities to cattle in filling their stomachs with food with the least trouble; and the instruments used for this purpose are much simpler than those which have been described for sheep in (1138) and (1145). Not an uncommon instrument for the purpose is an old sharp spade, with which turnips are broken in as many pieces as desired; but it is objectionable, in as far as it breaks them into unequal pieces, the round turnips rolling away from its strokes, and it scatters the hard ones in splinters. Much better instruments will be found in the two hand turnip-choppers described below, figs. 263 and 264. A single perpendicular stroke with either of these instruments cuts a turnip into a certain number of pieces; but in using fig. 263 a little dexterity is required to save its cutting edges from being injured against the bottom of the trough. The dexterity consists in first getting a hold of the turnip with the instrument by a gentle tap, and then lift up the turnip, striking it against the bottom of the trough with a smart stroke, when it will fall into pieces before the knives touch the trough; but the constant exercise of dexterity is scarcely to be looked for in an ordinary cattle-man, and, therefore, fig. 264 may be pronounced the more useful instrument of the two, for the studs serve to guard the cutting surface from injury. But where a cattle-man has charge of a large number of cattle receiving cut turnips, a more expeditious process of slicing them is required, and this will be obtained by the use of the lever turnip-slicer, described below in fig. 259. This machine is placed beside the turnip-store, where it slices the turnips into the skull placed under it, and, being light, can easily be carried from store to store, unless where the distance is great, when another machine should be provided. It will be observed that all these implements cut turnips into large pieces, which are sufficiently small for cattle, sheep requiring theirs cut into long narrow slips, to suit the form of their mouth.

(1224.) Cattle naturally feeling more appetized in the morning than during the day, their morning meal should be large, and while employed at it the cattle-man should furnish their racks with fresh oat-straw, to which they will repair from the turnips, and lie down in the open court or within the shed, according to the state of the weather, and chew their cud with composure. At mid-day their troughs should again be replenished with turnips, and again before daylight is gone. The quantity given at the evening meal partly depends upon the state of the moon; for cattle, as well as sheep, will always feed during the night in moonlight, a habit which I have frequently observed in both animals; and from this fact I conclude that if light were placed beside cattle in the byre, they would also feed during the long winter nights, and, of course, fatten quicker.* The last foddering of straw is given after the evening meal of turnips; and, during the day, whenever the shed or court requires litter, the refuse straw of the foddering may be spread abroad, and in rainy weather it should be brought direct from the straw-barn.

(1225.) The younger cattle in the courts next receive their turnips, and of these the calves should have the precedence, as they take longer time to finish their meal than their older compeers. They occupying the court K, fig. 4, Plate IV., the turnips are wheeled from the store *g* to the troughs, and there broken with one of the hand turnip-choppers, fig. 263 or 264, or sliced in the store with the lever-slicer, fig. 259. Their fodder is put both in the open straw-rack *o* and that under the shed at *h*′, and their litter strewed after the young beasts in the other large court have been served with turnips.

(1226.) Immediately after the calves, the year-olds in the court I are served with turnips, fodder and litter, in the same order. All young beasts should get as many turnips as they can eat; but should the crop prove insufficient for this, let the calves have their full share, and the year-olds rather put on short allowance; but in a case of this kind occurring, the most prudent plan, perhaps, would be to purchase oil-cake for the fattening beasts, to be given along with some turnips, and let all the young

* That highly-prized bird in France, the ortolan, feeds at dawn, and when confined for the purpose of being fattened, an artificial dawn is produced every three hours during the night by artificial light, when it eats its food, and thereby becomes much sooner fat.

beasts have their full share of turnips. To insure them still farther with this, the cows might also have oil-cake.

(1227.) The young heifers in the hammels N, and the bulls in the hammels X, next receive their turnips, and as neither of them get as many as they can eat, their proportion is divided into two small meals, one served after all the rest in the morning, and the other before the rest in the evening. Both these classes depending much upon fodder for food, it should be of the sweetest and freshest straw, and supplied at least 3 times a day, morning, noon and evening; and having water at command, and liberty to move about, they will maintain a fair condition. The heifers are supplied from their own turnip-stores *p* or *q*, and the bulls from that belonging to the servants' cow-byre *h*.

(1228.) With regard to the supply of turnips to the servants' cows, much depends on the terms of the agreement made with the servants. Where a specified number of cart-loads are given, the servant may choose to give them to his cow during the earlier part of the winter or not, because, when she is dry, it is not usual to give her turnips; but if in milk, the servant's family may give what they choose from their own store. On the other hand, if the farmer has agreed to treat the servants' cows in the same manner as his own, then the cattle-man takes charge of them in the manner I have already described (1215).

(1229.) From the beginning of the season until the end of the year, white turnips alone are used, after which, to the end of the winter season, the yellows are brought into requisition, or Swedes where these are not cultivated. When turnips are brought from the field in a very dirty state, which will inevitably be the case in wet weather from clayey soil, they ought to be washed in tubs of water, and when they are so as long as the earth is fresh, they will be the more easily cleansed; and this is not so troublesome and expensive a business as may at first sight appear. A large tub of water, placed at a store when about to be filled with turnips, a field-worker, taking a small fork, picks up a turnip with it, and dashing it about in the water for an instant, pulls it off against the edge of the store or barrow; and in this way cleanses a great number in a short time, much faster than the cattle-man can wheel them away and serve and break them to the beasts. A friend of mine used a very curious mode to wash turnips. Whenever any of the fields of his farm, along which was the lead that conducted the water from the dam to the threshing-mill, were in turnips, he filled the lead pretty full of water, by keeping down the sluice at the mill. He then topped and tailed the turnips in the field, and emptied them into the lead, from a cart when the distance to the turnips was considerable, and from a hand-barrow, carried by field-workers, when they were near. The sluice at the mill was then opened a little, and the gentle current thereby created in the water floated the turnips to the steading, where they were taken out and carried to the stores in barrows. When the turnips were very dirty, they were washed in the lead by a person pushing them about with a pole. That some provision for cleaning turnips is sometimes necessary, is obvious to me, for I have seen very fine cattle getting turnips to eat in such a state that the dirt actually bedaubed them to the very eyes, the tops being left on to make the matter worse. Surely no one will say that filth, in any shape, is beneficial to cattle; not that they dislike to lick earth, but then they do so only when they feel they require it to rectify acidity in the stomach.

(1230.) When turnips have not been stored, and are brought from the field as required, it is highly probable that they will be in a frozen state at times, when, even if broken by the instruments in use, they will be masticated by cattle with difficulty, besides the danger they run of being chilled by them; for cattle always have a staring coat after eating frozen turnips. This being the case, means should be taken to thaw them, and the most available is to put them in tubs of cold water for some hours before being given to the cattle. Such expedients, to avoid greater evils, of course always incur expense, and it will be much greater than the comparatively trifling one of storing the same quantity of turnips at the proper season, which, when done, every such petty source of vexation will be removed.

(1231.) It is supposed that a fattening ox, which will attain 70 stones imperial at the end of the season, consumes on an average, during the season, a double horse load of turnips per week, and, as carts are usually loaded in field work in winter, their weight may be estimated at about 12 cwt.; so each ox will consume about 1¾ cwt. or 14 stones a day, or 4⅔ stones of each 3 meals, and about 16 tons during the season of 26 weeks. The calves may consume ½ or 7 stones, and the 2-year-olds ¾, or 10½ stones a day. These comparative statements are given from no authenticated data, for I suppose that no comparative trials with different ages of cattle have ever been made, but only from what I imagine to be near the truth; and some such estimate, at the beginning of the season, is useful to be made, that you may know whether your turnips will answer the stock. It has been correctly ascertained, however, by Mr. Stephenson, Whitelaw, East Lothian, in a careful experiment of 17 weeks, that an ox, yielding under 30 stones of beef, consumes 1 cwt., or 8 stones every day;* and if cattle consume food somewhat in proportion to their live weight, in similar circumstances, as is believed, the above ratios may be pretty correct. And yet Mr. Boswell of Kingcausie's four 2-year-olds, fed entirely on turnips, and which increased in live weight, in four months, from 40 to 45 cwt., only consumed a little more than 27 tons of yellow bullock turnips, or 8½ stones each day.† So that Mr. Boswell's cattle, of from 45 to 50 stones each, consumed only a very few more turnips than Mr. Stephenson's, of 28 stones each. Such discrepancies show how little we can yet anticipate when we undertake to fatten cattle. But there is this that may be said in explanation of this difference, which is, however, merely conjectural, that Mr. Stephenson's lightest lot experimented on may have been West Highlanders, Mr. Boswell's Aberdeenshires, and my supposition is made in reference to well-bred Short-Horns. It will be observed that cows receiving ⅓ of oxen, namely, 4⅔ stones a day, each skullful will contain rather more than 32 lb.

(1232.) The most personally laborious part of the duty of a cattle-man in winter is *carrying straw in large bundles on his back to every part of the steading.* A convenient means of carrying it is with a soft rope about the thickness of a finger, and 3 yards in length, furnished at one end with an iron ring through which the other end slips easily along until it is tight enough to retain the bundle, when a simple loop-knot keeps good what it has got. Provided with 3 or 4 such ropes, he can bundle the straw at his leisure in the barn, and have them ready to lift when required. The iron ring permits the rope to free itself readily from the straw when the bundle is loosened.

(1233.) The *dress* of a cattle-man is worth attending to, in regard to its appropriateness for his business. Having so much straw to carry on his back, a bonnet or low-crowned hat is most convenient for him; but what is of more importance, when he has charge of a bull, is to have the color

* Prize Essays of the Highland and Agricultural Society, vol. xii.

† Ibid., vol. xi.

of his clothes of a sombre hue, free of all gaudy or strongly-contrasted colors, especially *red*, because that color from some cause is peculiarly offensive to bulls. It is with red cloth that the bulls in Spain are irritated at their celebrated bull-baits. Instances are in remembrance of bulls turning upon their keepers, not perhaps because they were habited in red clothes, but probably because there was some red color about them, or that they contrasted strongly with what their keepers usually wore. It was stated at the time, that the keeper of the celebrated bull Sirius, belonging to the late Mr. Robertson of Ladykirk, had on a red night-cap when he was killed by him. One day, when walking with a lady across a field, for a short cut to a road, my own bull, the one represented in the plate of the Short-Horn bull, than which a more gentle and generous creature of his kind never existed, made toward us, and seemed unusually excited. This conduct did not arise from the circumstance of a stranger being in the field, for many strangers, both male and female, visited him in the field. I could ascribe his extraordinary excitement to no other cause than to the red shawl worn by the lady; for when she left the field he resumed his wonted quietness of conduct. I remember observing him more than usually excited, on another occasion, in his hammel, when his keeper, an aged man who had attended him for years, was beside him on a Sunday afternoon, I ascribed his excited state to the new red night-cap, instead of the usual black hat, which his keeper wore on the occasion; and on my desiring him to throw it away, the animal became again quite quiet. Be the *rationale* of the thing what it may, it is prudential in a *cattle*-man to be always habited in a sober suit of clothes.

(1234.) *Regularity in regard to time* is the chief secret in the successful treatment of cattle. Cattle, dumb creatures though they be, soon understand your plans in regard to what affects themselves, and there is none with which they reconcile themselves more quickly than regularity in the time of feeding; and none on the violation of which they will more readily show their discontent. No cattle-man can keep regular time without a watch; and if he has not one of his own, lend him one that will keep time well. His day's work in winter may be divided thus: Let him be astir and have his breakfast over by daybreak, which cannot be very early at that season. The first thing he should do is to go to the cow-byre, and remove with the graip. into the gutter, any dung that would immediately interfere with the dairy-maid in milking the cows. She should be at the byre in time for this purpose. Leaving her there, he goes to the fattening beasts in the hammels, and first cleans out the same trough, always beginning at the same end, of all refuse, with his shovel; and immediately as he cleans one trough he replenishes it with turnips from the store at hand, and breaks them with any of the instruments used. He thus proceeds from one hammel to another until the six are gone over, or as many as are occupied. It is not an easy matter to say exactly how long time this should take in doing, but say half an hour, 30 minutes. He then proceeds immediately to the calves in the large court, cleans out the dirt from the troughs, replenishes them with fresh turnips from the store, and breaks them; and he does this, having long troughs and fewer turnips, say in 15 minutes. He next goes to the 2-year-old court, and does the same in it; and, having a few more turnips to wheel out of the store and break, he will take a little more time, say 20 minutes. The bulls in the hammels may take 10 minutes to clean out their troughs and supply them with their small quantity of turnips. And the same time, 10 minutes, may suffice to give the heifers a little fresh fodder, for they should not get cold turnips on empty stomachs, more than cows; with another 10 minutes to supplying the old cows, or extra beasts, with turnips. Having thus given all the cattle that are at liberty something to do for some time, he returns to the cow-byre with a bundle of fodder of fresh oat-straw, which he distributes among the cows, and which they pick during the time he is clearing the byre of litter and dung; and to do all this may require 30 minutes. Shutting the principal door of the byre, and leaving the half-door to the court open for air, he leaves the cows with the fodder, and cleans out the servants' cow-byre, and fodders the cows, which may take other 30 minutes. Taking then a bundle of litter, he goes again to the byre, and spreading any refuse fodder as litter, and cleaning out the troughs, he supplies the cows with their allowance of turnips, and shaking up the straw which he has just brought as litter, he leaves them again to eat and rest awhile. All this may require other 30 minutes; and 10 minutes may suffice to give the heifers their small quantity of turnips, and the old cows their fodder, and 10 minutes more to litter the servants' cows, the servants themselves having supplied the turnips as they choose. All the cattle having now been once fed, brings the time to 25 minutes past 10 A. M., if the operations began at 7 o'clock. The next step to be taken is to supply those which get as many turnips as they can eat, with fodder and litter, and for this purpose he takes the fodder fresh from the straw-barn, and fills all the straw-racks in the large courts, whether in the open air or under the shed. The old fodder should be pulled out before the fresh is put in; but this is seldom attended to. He then strews the open courts and sheds with litter where it is chiefly required; namely, along the side of the troughs where the beasts stand to eat the turnips, and where they have lain under the sheds. The hammels are then supplied with fodder and litter, the refuse fodder probably being sufficient for the latter purpose, as long as the weather is dry. The bulls and heifers should also have fodder and litter. All this business with the straw, and making it up into bundles for the afternoon, may take up 50 minutes, and bring the time to ¼ past 11 A. M. What with cleaning out the troughs and supplying the hammels and courts again with turnips of the midday meal, and letting out all the cows—including those of the servants—into their respective courts, 12 o'clock will have arrived, which is the hour of dinner for all the work-people. The people have an hour to themselves, to 1 P. M., to refresh and rest. At 1 P. M. the cattle-man resumes his labors by bunching up *windlings* of straw, which are small bundles having a twisted form, of 10 lbs. weight or more each, for each of the cows' supper, and also larger bundles in the ropes for fodder. Having prepared these just now or at any other leisure moment, he takes a bundle of fodder to the byre, supplies the troughs, and brings the cows in from the court, and ties them to their stakes. He does the same with the servants' cows. He then replenishes the straw-racks in the courts and hammels with what little fodder is required. He then litters the sheds comfortably for the night. He lays the windlings of straw in a corner of the servants' cow-byre for the night's suppering, and he does the same in the other byre; and the reason he does this in preference to letting them remain in the straw-barn is, to avoid the danger of taking a light into the straw-barn when the windlings are to be used. By the time all this business with the straw has been done, it is time to give the cows their second meal of turnips, so that they may have them eaten up before the milk-maid comes again at dusk to milk them. The feeding beasts in the hammels are then supplied with turnips broken for them, then the calves, then the young beasts in the other court, and then the bulls and heifers, in the same order as formerly. He then litters the servants' cows for the night, by which time it will be time for the other cows to be milked; immediately after which they are littered for the night, and the doors closed upon them, and thus the labors of the day are finished.

(1235.) In thus minutely detailing the duties of the cattle-man, my object has been to show you rather how the turnips and fodder should be distributed relatively than absolutely; but at whatever hour and minute the cattle-man finds, from experience, he can devote to each division of his work, you should see that he performs *the same operation at the same hour and minute every day.* By paying strict attention to time, the cattle will be ready for and expect their wonted meals at the appointed times, and will not complain until they arrive. Complaints from his stock should be distressing to a farmer's ears; for, he may depend upon it, they will not complain until they feel hunger; and if allowed to hunger, they are not only losing condition, but rendering themselves, by discontent, less capable of acquiring it, even should their food happen to be regularly given them for the future. Whenever, therefore, you hear petitioning and impatient lowings from cattle at any steading, you may safely conclude that matters there, in so far as regards the cattle, at least, are conducted in a very irregular manner. The rule, then, simply is, *Feed and fodder cattle at fixed times, and dispense their food and fodder in a fixed routine.* I had a striking instance of the bad effects of irregular attention to cattle. An old staid laborer who was appointed to take charge of the cattle, was quite able and very willing to undertake the task. He was allowed to take his own way at first; for I had observed that many laboring men display great ingenuity in arranging their work. Lowings from the stock were heard in all quarters, both in and out of doors; and they intimated that my ancient cattle-man was not endowed with the organ of order, while I observed that the poor creature himself was constantly in a state of perspiration. To put an end to this disorderly state of things, I apportioned his whole day's work by his own watch; and on his implicitly following the plan, he was not only soon able to satisfy the wants of every creature committed to his charge, but had abundant leisure besides to lend a hand at anything else that required temporary assistance. His heart overflowed with gratitude when he found he could easily make all the objects of his charge happy; and his kindness to them all was so sincere, that they would have done whatever he liked. A man better suited for this occupation I never saw.

(1236.) Now, you may consider that all these minute details regarding the treatment of cattle are frivolous and unnecessary. But the matter is really not so; and it is of importance for your own interests to tell you so, for you will admit that where a number of minutiæ have to be attended to, unless taken in some order, they are apt either to be forgotten altogether, or attended to in a hasty manner; and none of these conditions, you will also admit, are conducive to correct management. Observe, then, the number of minute things the cattle-man has to attend to. He has various classes of cattle under his charge—cows, fattening beasts, young steers, calves, heifers, bulls, and perhaps extra beasts besides; and he has to keep all these clean in their various places of abode, and supply them all with food and fodder 3 times in a short winter's day of 7 or 8 hours. Is it possible to attend to all these particulars, as they should be, without a matured plan of operations? The cattle-man requires a plan for his own sake, for were he to do one thing just when the idea struck him, his mind, being guided by no fixed rule, would be as apt to forget as to remember anything he had to do. And besides, the injurious effects which irregularity of attendance tends to produce upon the condition of animals, seem to render a plan of operations absolutely necessary to be adopted Before you can see the full force of this observation, you require to be told that food given to cattle in an irregular manner—such as too much at one time and too little at another, frequently one day and seldom in another—and the same with fodder and litter, thus surfeiting them at one time, hungering them at another, and keeping them neither clean nor dirty, never fails to prevent them acquiring that fine condition which better management insures. And still farther to show you its force, you may not be sensible of any deficiency of condition under the most irregular management, from the want of the means of comparing your beasts with others; but an appeal to figures will show you the risk of loss you are unconsciously incurring. Suppose you have 3 sets of beasts, of different ages, which should get as many turnips as they can eat, and each set to contain 20 beasts; that is, 60 beasts in all. Suppose, moreover, that, by irregular management, each of these beasts acquires only ½ lb. less of live weight every day than they would under proper management, this would make a loss of 30 lbs. a day of live weight, which, over 180 days, the duration of the fattening season, will make 5,400 lbs. of live weight, or (according to the common rules of computation) 3,240 lbs., or 231 stones of beef, which, at 6s. the stone (not a high price), show a deterioration of £69 6s. in the value of the whole herd at the end of the season. The question, then, resolves itself into this, Whether it is more for your interest to lose this sum annually, or make your cattle-man attend to your beasts according to a regular plan, any form of which it is in your own power to adopt and pursue?

(1237.) What I have narrated above applies to the ordinary mode of feeding cattle, but extraordinary means are sometimes applied to attain a particular object. You may have, for instance, a pair of very fine oxen which you are desirous of exhibiting at a particular show, not altogether for the sake of gaining the premium offered, but partly for the honor of carrying off the prize from contemporaries. In this case they should have a hammel comfortably fitted up for themselves; that is, possessing all the means of satisfying their wants, both of food and shelter. Your ingenuity should be taxed to devise means that will anticipate every desire; and this you will be the better able to do after you have determined on the sort of food you wish to support them upon. If, regardless of expense, you will present a choice of food, there should be a trough for sliced Swedish turnips—a manger for bean-meal—another for bruised oats—a third for broken oil-cake*—a rack for hay—and a trough for wa-

[* The parts of this disquisition on rearing and feeding cattle on turnips which may seem not to possess practical information or value for the American husbandman, may yet be read with entertainment as affording a familiar view of the details of the most important branch of agricultural economy in a country famed for its advancement in the arts as well of tillage as of manufacture.

Few problems can be of more consequence to a great number of farmers, than the cheapest and best mode of fattening stock. There is a sufficient number of our farmers engaged in it render it a matter of national importance. It opens, as observed by an able writer and practical farmer, a wide field for calculation and inquiry as to the cost per pound, of putting on weight according to the food the animal is fed on.

There is, no doubt, a vast difference in the action of food, in producing increased weight, depending upon its quality and description, the mode in which it is administered, the temper and breed of the animal, and, above all, whether the creature is placed in a cold, damp, and exposed situation, (to which many of our farmers give little or no attention,) or in nice, comfortable, dry and warm quarters.

On this subject we particularly recommend the reader to be on the look-out for, and to read

ter; for water at will I conceive essential when so much dry food is administered. Then there should be abundance of straw for litter and warmth, and a regular dressing of the skin every day, to keep it both clean and healthy, as *fat* oxen can reach but very few parts of their bodies with their tongue. So much for winter treatment. In summer they should get cut clover in lieu of the turnips and hay, and all the other auxiliaries to the dry materials and straw, as already stated. But all these will not avail to attain your object, if constant attention be not given, and everything conducted with the utmost regularity in regard to time. True, they get as much as they can eat, but then what they eat should be administered with judgment. It will not suffice to set an adequate portion of each sort of food daily before them, to be taken at will; one or more kinds should be given at stated times, that each may possess the freshness of novelty and variety—not all at one time, but every one at such a time as one or both the animals may incline most to have. All these considerations demand attention, and afford exercise to the judgment. Oxen, when thus fattened, cannot travel any distance on foot; they must be conveyed on carriages built for the purpose, and even on these, if the distance is great, they will fall off in condition, as the confinement in, and motion of, the carriage proving irksome, prevent animals taking their food so heartily as they would do at home. I knew a 3-year-old-off bull that lost 30 stones live weight on being carried partly by steam-ship and partly by railway to a show.

(1238.) The *names* given to cattle at their various ages are these: A new-born animal of the ox tribe is called a *calf*, a male being a *bull-calf*, and a female a *quey-calf*, *heifer-calf*, or *cow-calf*, and a castrated male gets the name of *stot-calf*, or more commonly, simply a *calf*. The term calf is applied to all young cattle until they attain a year old, when they are called *year-olds* or *yearlings*, saying *year-old bull*, *year-old quey* or heifer, *year-old stot*; *stirk* is applied to both a young ox and quey, and *stot* in some places means a bull of any age. In another year they are named *2-year-old bull*, *2-year-old quey* or *heifer*, *2-year-old stot* or *steer*. In England females are called *stirks* from calves to 2-year-old, and the males *steers*. The next year they are called *3-year-old bull*; females, in England, from 2 to 3-year-old, *heifers*, in Scotland *3-year-old queys*, and when they are kept for breeding, and bear a calf at that age, they get the name of *cows*, the same as in England, and the males *3-year-old stots* or *steers*. Next year the *bulls* are *aged*, the *cows* retain that name ever after, and the *stots* or *steers* are *oxen*, which they continue to be at any age they are kept. A cow or quey that has been served by the bull is said to be *bulled*, and are then *in calf*, and from that circumstance are called in England *in calvers*.

---

attentively, a Prize Essay on Fat and Muscle, that we shall give in an early number, perhaps in our next, which will be the last of the Second Volume.

An important point of the question, says the writer referred to, and one to which Mr. Stephens does not seem to have adverted, is the comparative *worth of the manure* from the various materials used for feeding. It seems to be admitted, that there is a well grounded preference for that from richly fed animals, and we doubt not that in the farther progress of the application of science to the subject, we shall have an accurately graduated scale of the intrinsic value of manures, from every kind of food, vegetable and animal—from dry wheat straw to oil-cake, and from fish to fat mutton.

The extent to which oil-cake is used as food for beasts in England is entirely justified on the ground of the additional value it imparts to the manure. This is one of the resources for feeding our animals and our land that American Farmers seem as yet not to have studied or understood with the care we may suppose it deserves from the extent to which it is practiced in other countries. *Ed. Farm. Lib.*]

'989\

A cow that has either *missed* being in calf, or has *slipped* calf, is said to be *eill;* and one that has gone dry of milk is called a *yeld-cow.* A cow giving milk is a *milch cow.* When 2 calves are born at one birth, they are *twins;* if three, *trins.* A twin bull and quey calf are called *free martins,* in which case the quey never produces young, but has no marks of a hybrid or mule. *Cattle, black cattle, horned-cattle,* and *neat-cattle,* are all generic names for the ox tribe, and the term *beast* is used as a synonym. An ox that has no horns is said to be *dodded* or *humbled.* An aged bull that is castrated is called a *segg;* and a quey that has had the ovaries obliterated, to prevent her breeding, is called a *spayed heifer* or *quey.*

(1239.) Cows are kept on every species of farm, though for very different purposes. On *carse* and *pastoral* farms they are merely useful in supplying milk to the farmer and his servants. On *dairy* farms, they afford butter and cheese for sale. On some *farms near large towns,* they chiefly supply milk for sale. And on *farms of mixed husbandry,* they are kept for the purpose of breeding young stock.

(1240.) On carse and pastoral farms, cows receive only a few turnips in winter, when they are dry, and are kept on from year to year; but where the farmer supplies milk to his work-people, as a part of wages, they are disposed of in the yeld state, and others in milk, or at the calving, bought in to supply their place, and these receive a large allowance of turnips, with perhaps a little hay. On these farms, little regard is paid to the breed of the cow, the fact of being a good milker being the only criterion of excellence.

(1241.) On true dairy farms, the winter season is not a favorable one for making butter and cheese for sale; for do what you like to neutralize the effect of the usual rooted green crops on these products, and especially butter, they remain unpalatable to the taste. The cows are therefore in calf during this season, and receive the treatment described above until the period of calving in spring.

(1242.) *In and near* large towns, the dairy-man must always have milk to supply his customers, and it is his interest to render the milk as palatable as possible. For the purpose of maintaining the supply, he buys cows at all seasons, just calved or about to calve. He disposes of the calves, without attempting to fatten them; and to render the milk he sells palatable, he cooks all the food partaken of by the cows. When the cows run dry, they are fattened for the butcher, and not allowed to breed again.

(1243.) The cows in the public dairies in Edinburgh are supported in winter on a variety of substances, namely, turnips, brewers' and distillers' grains called draff, dreg, malt comins, barley, oats, hay-seeds, chaff, cut hay. One or more of these substances, with turnips, are cooked together, and the usual process in doing this and administering the cooked food, is as follows:—Turnips, deprived of tops and tails, and washed clean, are put into the bottom of a boiler, and covered near its top with a quantity of malt comins, cut hay, hay-seeds, chaff, or barley, or more than one of these, as the articles can be procured. Water is then poured into the boiler sufficient to boil them, and a lid placed upon it. After being thoroughly boiled and simmered, the mess is put into tubs, when a little pounded rock-salt is strewed over it, and chopped into a mash with a spade. As much dreg is then poured upon the mash as to make it lukewarm, and of such a consistence as a cow may drink up. From 1 to 1½ stable-pailfulls of this mixture—from 40 to 60 pints imperial—according to the known appetite of the cow, is then poured into the trough belonging to each. The trough is afterward removed and cleaned, and the manger is ready for the reception of fodder—hay or straw. This mess is given 3 times a day, *after* the cows have been milked, for dairy-men understand that animals should not be disturbed while eating their food. The times of milking are 6 A. M., 12 noon, and 7 P. M. The sweet milk and cream obtained by these means, and received direct from the dairy, are pretty good. The former sells in Edinburgh at 1d., and the latter at 1s. the imperial pint. Dr. Cleland states the price of sweet milk in Glasgow at 1½d. the imperial pint.

(1244.) It will be observed that none of the articles usually given to cows are so expensive as oil-cake, cabbages, kohl-rabi, or cole-seed. These products were employed by the late Mr. Curwen in his experiments to ascertain the cost of raising milk for supplying the poor, and the results show they left him very little profit.*

(1245.) There is little milk in winter on a farm which supports cows for breeding stock, being only derived from one or two cows that are latest of calving in spring. All the spare milk may probably be eagerly bought by cotters who have no cows; but should that not be the case, a little butter may be made once in ten days or a fortnight, which if not palatable for the table may be used in making paste, and other culinary purposes. A little saltpetre, dissolved in water, certainly modifies the rank taste of turnips in both butter and milk.

(1246.) Cattle are fed on other substances than turnips, either with themselves or in conjunction with turnips. Oil-cake and potatoes are the most common substances used for this purpose. Linseed-oil and linseed have been recommended, and many are fed at distilleries on draff and *dreg,* as the refuse of distillation are termed; and these are sold to the farmers for the purpose of feeding. Oats, barley-meal, and bean-meal, have also been pressed into the service of feeding cattle.

(1247.) The *potatoes used in feeding cattle* are either the common kinds known in human food, or others raised on purpose, such as the yam and ox-noble; and they are given either alternately with turnips, or together. In feeding cattle with potatoes of any kind, and in any way,

* Curwen's Agricultural Hints

there is considerable risk of flatulency and choking. To prevent the latter, the potatoes should be smashed with a hammer, or with an instrument like a pavior's rammer, and though juice should come out in the operation, no loss is incurred, as it is considered of no service in feeding. To prevent flatulence from potatoes is no easy matter; but a friend of mine used a plan which completely answered the purpose, which was, mixing cut straw with the broken potatoes. The straw obliging the cattle to chew every mouthfull before being swallowed, may prevent such a large quantity of gas being generated in the paunch as bruised potatoes alone would do, and it is this gas which occasions that distressing complaint called *hoven*. A farm-steward, who had considerable experience in feeding cattle on potatoes on a led-farm, always placed as many potatoes, whole, before cattle as they could consume, and they never swelled on eating them, because, as he conjectured, and perhaps rightly, they do not eat them so greedily when in their power to take them at will, as when doled out in small quantities. This fact confirms the propriety of mixing cut straw among potatoes that are given in small quantities, in order to satisfying the appetite, and filling the paunch with unfermentable matter. The only precaution required in giving a full supply of potatoes is to give only a few and frequently at first, and gradually to increase the quantity.

(1248.) *Oil-cake* has been long and much employed in England for the feeding of cattle, and it is making its way in that respect into Scotland. It consists of the compressed husks of linseed after the oil has been expressed from it, and is formed into thin oblong cakes. The cakes are broken into pieces by a machine described in fig. 264. Cattle are never entirely fed on oil-cake, but in conjunction with other substances, as turnips, potatoes, cut hay, or cut straw. When given with cut hay or straw, an ox will eat from 7 to 9 lbs. of cake a-day, and the hay or straw induces rumination, which the cake itself is not likely to do. When given with other substances, as turnips or potatoes, 3 lbs. or 4 lbs. a-day will suffice. A *mixture* of oil-cake and cut meadow-hay

Fig. 259.

THE LEVER TURNIP-SLICER FOR CATTLE.

forms a very palatable and nutritious food for oxen, and is a favorite one in England. Oil-cake costs from £7 to £10 a ton.

(1249.) [*Turnip-slicers for Cattle.* In the description formerly given of machines for cutting turnips for sheep, that described in (1145) may be again adverted to, fig. 250, the wheel turnip-cutter. This machine is equally well adapted to slice for cattle as for sheep, and is frequently fitted up to slice for cattle only. More frequently it is finished as described with the cross-cutters, fig. 252, and when wanted to slice for cattle, the *cross-cutters* are removed. This is easily accomplished, by first lifting the slicing-knives from the disk of fig. 250, then unscrewing all the nuts of the cross-cutters, and removing them from their places. The slicing-knives are then again placed as before, and the machine is prepared to cut the turnips into plain slices. This machine costs from £4. 4s. to £5.

(1250.) *Lever Turnip-slicer.* One of the cheapest and most efficient turnip-slicers is represented in fig. 259. It was brought before the Highland and Agricultural Society of Scotland by Mr Wallace of Kirkconnell, as an improvement on a preëxisting machine of the same kind.* It has since undergone some farther improvements in the hands of James Slight and Company, Edinburgh; and for the purpose of regular and perfect slicing of turnips, it may be held as as the best and cheapest now employed. The machine as produced by Mr. Wallace, is represented in fig. 259; where *a b* is the stock or sole of the machine, about 34 inches long, 6 inches broad, and 2 inches in thickness. The sole is in 2 pieces, connected by an iron bar or strap *a c*, which is repeated on the opposite side, and the whole bolted together, as in the figure. The 2 pieces forming the sole are separated longitudinally from each other, so as, with the two side-straps of iron, to form a rectangular opening of 9 inches by 6 inches, bounded on the two ends by the parts of the sole, and on the two sides by the side-straps, which, to the extent of the opening, are thinned off to a sharp edge, and thus form the two exterior cutters *d, e*, as seen in fig. 260, which is a transverse section through the cradle of the machine. The sole is supported at a hight of 2 feet upon 4 legs, fig. 259, and the lever *d e* is jointed at *d* by means of a bolt passing through it and the ears of the side-straps, as seen at *d*. The lever is 4 feet in length, its breadth and thickness equal to that of the sole, but is reduced at the end *e* to a convenient size for the hand. Two cutter blocks *f* and *g* are appended to the sole by mortice and tenon, and farther secured by the bolts which pass through the side-strap at that place. Into these blocks the remaining cutters *h h* and *i i*, fig. 260, are inserted in corresponding pairs, and also secured by bolts; the cutters thus arranged form a cradle-shaped receptacle, into which the turnip is laid to be sliced. The lever *d e* is armed with a block of wood *m*, loosely fitted to the cradle; and its lower face is studded with iron knobs, the better to prevent the turnip sliding from under it. The transverse section *d e f f*, fig. 260, shows the position of six cutters *d e, h h*, and *i i*, as inserted in the wooden block *d e f f*; and *k l* is the lever, seen in section, with the block *m* attached.

Fig. 260

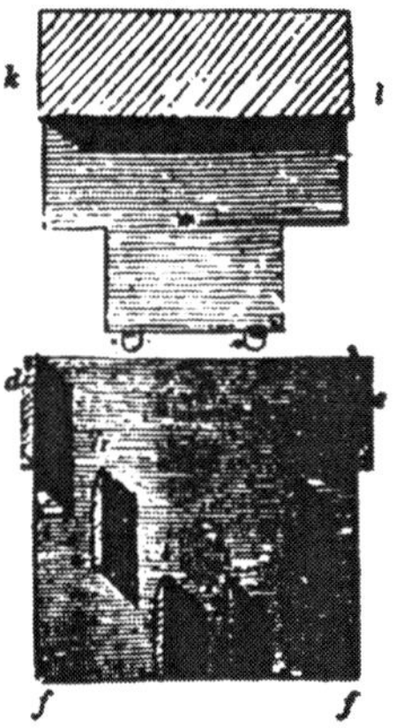

TRANSVERSE SECTION SHOWING THE POSITION OF THE CUTTERS.

(1251.) The late improvement by Messrs. Slight and Company consists in the application of cast-iron knife-blocks, which give greater strength to the machine, and a more ready and secure fixture of the interior knives, and of introducing 8 cutters instead of 6, which makes a more convenient size of slice. Fig. 261 is a section of the cradle, as it appears with the cast-iron knife-block; *a* is the body of the block which is attached to the sole through the medium of a flange behind, and fixed by bolts. The external cutters *b b* are a part of the side straps, as before; and the interior cutters are fixed in pairs, *c c, d d*, and *e e*, by their respective bolts passing through the cutters and the block.

(1252.) In using this machine the workman takes hold of the lever at *e*, fig. 259, with the right hand, and having raised it sufficiently high, he, with the left hand, throws a turnip into the cradle. The lever is now brought down by the right hand, which, with a moderate impetus, and by means of the block *m*, sends the turnip down upon the cutters, through the openings of which it passes while the cutters are dividing it, and the whole falls away in perfectly uniform slices. In most cases, it is found more convenient to have a boy to throw in the turnips, and this will somewhat expedite the work. One advantage of this turnip-slicer—and it is an important one—is that, with unerring certainty, it cuts every slice of uniform thickness; the slab-slices, indeed, may of course vary, but all are free of the smallest portion of waste. Its cheapness also is of importance, especially when it is considered that, in a given time, it will slice weight for weight of turnips with the most elaborate machine in use, the power applied being also equal. The price is 28s. to 30s. It is also extremely portable, and can be carried about by one person. An objection has been urged to this slicer, namely, that the turnips must all be put into it one by one; and it is perhaps unnecessary to remark that this objection applies to all turnip-slicers. For though the hopper of some may be capable of containing a number of turnips at one time, yet that number may be considered as having been deposited there individually. The price of the lever turnip-slicer for sheep, as in fig. 245, is £3 10s.

Fig. 261.

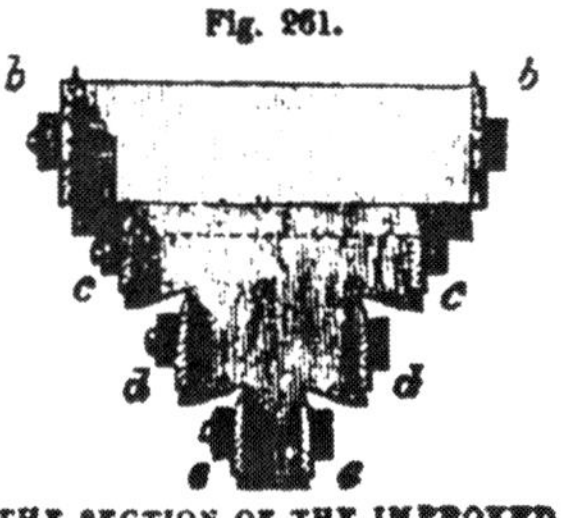

THE SECTION OF THE IMPROVED CRADLE.

(1253.) *Cross Turnip-cutter.* There is another very simple and useful turnip-cutter, which is frequently used when thin slicing is thought of less importance, but is more especially useful where the cooking system is adopted for either cows or horses, thin slicing being in such cases not called for. This instrument is represented in fig. 262. The cutting part of it consists of 2 steel-edged blades, 8 inches in length and 4 inches in depth. They are slit half-and-half at their middle point, so as to penetrate each other, standing at right angles, forming the *cross-cutter, a a a a*. They are then embraced in a four-split palm, and riveted. The palm terminates in a short shank *e*, which is again inserted into the hooped end of a wooden handle *b*, 3 feet in length, which is finished with a crosshead *c*. The price of this instrument is 8s. 6d. The mode of using it is obvious. It is held by the hand in a vertical position; and when placed upon a turnip, one

* Prize Essays of the Highland and Agricultural Society.

thrust downward cuts it into quarters. This instrument is also varied in its construction, being sometimes made with 3 and even with 4 blades, dividing the turnip into 6 or into 8 portions.

(1254.) Another individual form of the same species is represented by fig. 263. It has two blades *a a*; but they, instead of crossing, stand parallel to each other, and therefore divide the turnips into three portions, resembling slices, of considerable thickness, the middle one being 1½ inches thick. In the construction of this cutter, a blunted stud is formed at the extremities of each blade, and there projects below the cutting edge about ½ inch, serving as guards to save the cutting edges from receiving injury when they have passed through the turnip, or otherwise striking any hard surface. These guards, it may be remarked, would form a useful addition to all this class of cutters. The arms *b* of the blades rise to a hight of 9 inches, widening upward to 3½ inches, to give freedom to the middle slice to fall out. The two arms coalesce above, and are then formed into the socket *c*, to receive the handle, which—as in the cross-cutter, fig. 262—terminates in a crosshead.

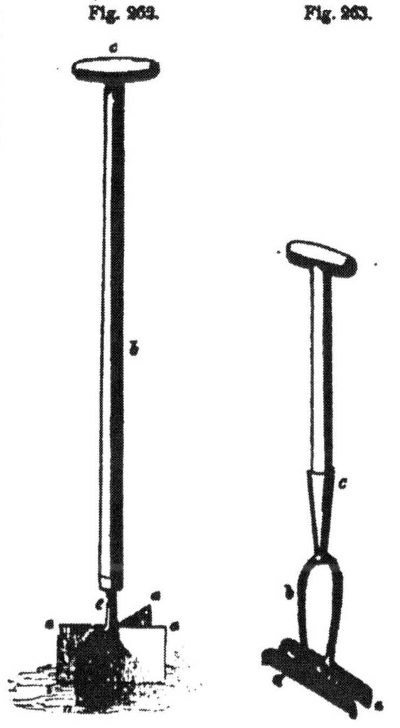

THE HAND TURNIP-CHOPPER, WITH CROSS BLADES. THE HAND TURNIP CHOPPER, WITH PARALLEL BLADES.

(1255.) *Oil-cake Breaker.*—Machines for preparing oil-cake for more easy mastication by cattle or sheep are made in a variety of forms. One of those forms is similar in principle to that of the early bone-crushing machines; namely, a revolving axle, armed with several series of teeth, which are so arranged as to pass in succession through the interstices of a line of strong teeth or prongs, against which the cakes lies, and is reduced to fragments by the successive action of the revolving teeth. Of this form there are various modifications, all serving the same purpose with nearly equal success.

(1256.) A different form of the machine, and which is held superior in the principle of its construction, is here exhibited in fig. 264, which is a view of the machine in perspective, wherein *a, a, a, a*, are the four posts of a wooden frame, on which the machinery is supported. The frame is 39 inches in length and 20 inches in width over the posts at top, the hight being 33 inches. *b b* are two top-rails, 34 inches in length, and the scantling of their timbers should not be less than 2½ inches square. The posts are supported toward the bottom by the four stay-rails *c, c, c*; and the top-rails are held in position by cross-rails *d*, one only of which is seen in the figure. Of the machinery, the acting parts consist of 2 rollers, studded all over with pyramidal knobs or teeth. These are arranged in zones upon each roller, and having a smooth space or zone between each of the knobbed zones; the knobs of the one roller corresponding to the smooth space in the other. The rollers *e* and *f* are constructed with an axle or shaft, that of the first *e* being 25 inches long, and of the second *f* 23 inches, and each 1½ inches square. Journals are formed upon these shafts, to run in the bearings which are placed on the top-rails *b b*, as afterward described. In this figure, *g g* are two pinching screws, which serve to regulate the distance at which the rollers are to work and, consequently, the degree of coarseness to which the cake is to be broken. The wheel *h*, of 20 inches diameter, is placed upon the shaft of the roller *e*, and the pinion *i*, of 3 inches diameter, with its shaft, and the winch-handle *k*, act upon the wheel *h*, giving a very considerable mechanical advantage to the power which is applied to the machine. The fly-wheel *l* is likewise placed upon the shaft of the pinion *i*, and is requisite in this machine to enable the power to overcome the unequal resistance of the work. On the farther end of the shaft of each of the rollers, there is mounted a wheel of 4½ inches diameter, for the purpose of carrying both rollers at the same speed. These wheels, one of which is seen at *m*, are formed with long teeth, to admit of the roller *f* approaching to or receding from the other, which is stationary *in place*. A feeding hopper *n* is placed over the line of division of the two rollers; it is 16½ inches long, 3 inches wide, and 14 inches deep. In forming the hopper, two upright pieces, 3 inches by 2 inches, are bolted to the inside of the top-rail, their position being between the shafts of the 2 rollers, and these form the ends of the hopper. They are then boarded on each side, which completes the machine. The hopper is here represented in section, the near portion of it being supposed entirely removed, in order to exhibit more distinctly the construction of the rollers.

(1257.) Fig 265 is a farther illustration of the construction of the rollers being a transverse section of the two, *a a* are the shafts, the shaded part *b* one of the plain disks which go to form the smooth zones on the body of the roller; it is 4 inches diameter and 1 inch thick; *c* is one of the knobbed disks, its body being of the same diameter and thickness as the former; but having the 4-sided pyramidal knobs set around it, the diameter, measuring to the apex of the knobs, is extended to 6 inches. One roller for the machine here described requires 5 plain and 6 knobbed disks, beginning and ending with a knobbed disk. In the other the arrangement is reversed, bringing out the alternation of the plain and knobbed zones before alluded to as more distinctly represented in fig. 266, which is a plan of part of the rollers, *c c* being two of the knobbed disks, and *b b b* three of the plain.

(1258.) *Fig.* 267 represents one of the bearings or plummer-blocks for the journals of the rollers. *a* is the bed of the plummer-block, *b* and *c* the brass bushes, and *d* the cover. The bush *b*, which corresponds to the roller *e*, fig. 264, is always stationary, while *c*, which is acted upon by the

Fig. 264.

THE OIL-CAKE BREAKER.

screw, is advanced toward, or withdrawn from, *b*, as the size to which the cake is to be broken may require. These plummer-blocks are bolted down to the top-rails of the frame, to which also the separate bearings of the pinion-shaft are likewise bolted.

Fig. 265.

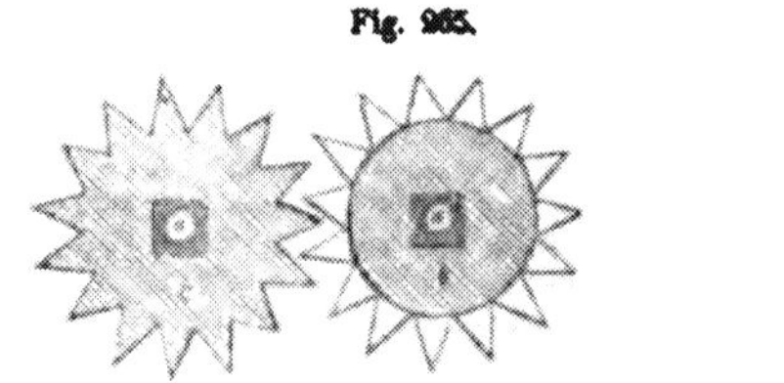

THE TRANSVERSE SECTION OF THE ROLLERS.

Fig. 266.

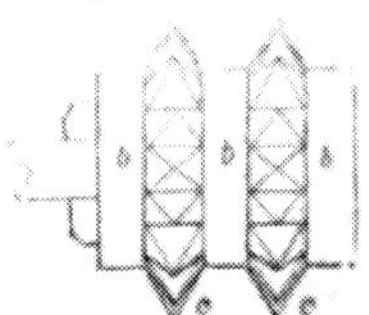

THE PLAN OF PART OF A ROLLER.

(1259.) It may be proper to remark here that the machine now described is of a good medium size, and with a man to drive and a boy to feed in the cakes, it will break about half a ton in an hour. The price is from £4 to £4 10s. The amount of its performance can be augmented or diminished to only a small extent, for as its feed is necessarily confined to one cake at a time, the only change that can be made on its production must depend upon the celerity of its motions. Hence it is one of those machines that cannot easily be adapted to *large* and to *small* establishments with any view, in this latter case, to amelioration of form; for the almost only means of doing so must be by giving it a quicker or slower motion, which can only affect the expense of construction to a very small amount, so small as hardly to be appreciable. In addition to what is

Fig. 267.

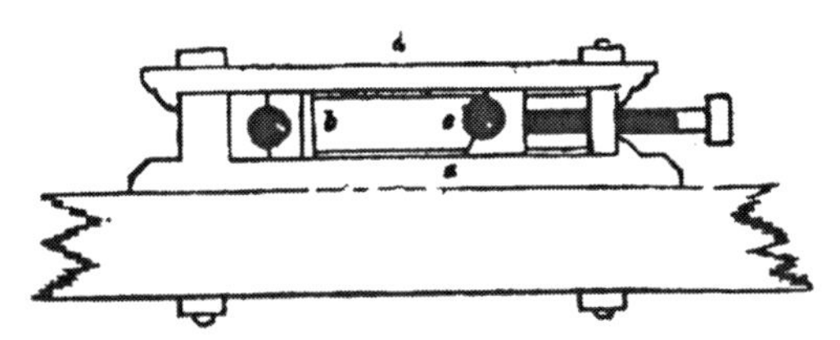

PART OF THE TOP-RAIL, WITH PLUMMER BLOCK.

shown of this machine in fig. 264, the rollers are frequently covered with a movable wooden case, which gives a more tidy appearance to it, and, moreover, it is always desirable that the frame below should contain a *shoot* formed of light boarding, that will receive the broken cake from the rollers, and deliver it at one side of the machine into a basket or other utensil in which it can be removed to the feeding stations.—J. S.]

(1260.) Mr. Brodie, Abbey Mains, East Lothian, made an experiment on feeding cattle, from October, 1836, to June, 1837, on different kinds of food. There were 4 lots of cattle, consisting of 5 each. The first lot was fed on turnips and straw, which, being the usual treatment, formed the standard of comparison. The second lot had half the weight of turnips and 30 lbs. of oil-cake a day. A third lot was fed on the last quantity of turnips and bean-meal and bruised oats. And the fourth had distillery grains and ground beans. The value of the cattle, when put up to feed, was £11 a piece, and they were of the Aberdeenshire polled breed. This is a summary of the cost of feeding:

		£	s.	d.
Lot 1.	White turnips at 8s. 4d., Swedes at 12s. 6d. per ton, cost	£53	9	10
	Average cost of each beast per week	0	6	3
Lot 2.	Turnips as above, oil-cake, £7 15s. per ton, cost	£48	16	0
	Average cost of each beast per week	0	5	9
Lot 3.	Turnips as above, bean-meal 5s., bruised oats 3s. 6d per bushel, cost	£58	8	1
	Average cost of each beast per week	0	6	8
Lot 4.	Turnips and bean-meal, as above, draff 4s. 6d. per quarter, dreg 2s. 6d. per puncheon, cost	£63	3	2
	Average cost of each beast per week	0	7	2

The ultimate results are as follows:

Lots.	Live weight.	Beef.		Tallow.		Hide.	
	sts.	sts.	lbs.	sts.	lbs.	sts	lbs.
1.	536	283	3	36	10	27	13
2.	552	295	10	41	6	29	6
3.	517	280	7	37	2	26	13
4.	545	280	0	36	11	25	7

"Upon the whole," concludes Mr. Brodie, "it is evident, by these experiments, that feeding with turnips as an auxiliary has been the most advantageous mode of using turnips, as, by the above statement, it is apparent that if the cattle of the first lot had only been allowed half the quantity of turnips which they consumed, and had got oil-cake in lieu of the other half, as was given to the second lot, the expense of their keep would have been lessened £4 13s., and from superior quality of beef, their value would have been increased £10, making together £14 13s."* Three remarks occur to me to make on the progress of this experiment; the first is, that if the cattle had been sold on the 7th April, 1837, when they were adjudged by competent farmers, they would not have repaid the feeder his expenses, as the prime cost of lot first, with the cost of feeding to that time, amounted to £95 1s. 8d., and they were only valued at £82; lot second cost £90 12s., and they were valued at £88 10s.; lot third cost £93 4s., and were valued at £77; and lot fourth cost £97 4s. 5d., and their value was only £81 10s. And this is almost always the result of feeding cattle, because *ripeness* only exhibits itself toward the *end of the feeding season*, and it is only after that state of condition is indicated that the quality of the meat improves so rapidly as to enhance its value so as to leave a profit. As with sheep so with cattle; with good beasts the inside is first filled up before the outside indicates fineness. Another remark is, that this result should be a useful hint to you to weigh well every consideration before disposing of your fattening beasts in the middle of the feeding season. The last remark I have to make is, that the cattle of lot first, continuing to receive the same sorts of food they had always been accustomed to, throve more rapidly at first than the beasts in the other lots, but afterward lost their advantage; thereby corroborating the usual experience of stock not gaining condition immediately on a change of food, even of a better kind, such as from turnips to grass.

(1261.) Mr. Mowbray of Cambus, in Clackmannanshire, made experiments in the winter of 1839-40 on feeding cattle with other than the ordinary produce of the farm, but as the cattle were

* Quarterly Journal of Agriculture, vol. viii.

not all sold at the same time I need not relate the details; and I mention the experiments for the sake of some of the conclusions that may be deduced from them. It would appear that cattle may be fed on turnips and hay as cheaply as on turnips and straw, for this reason, that when straw is given as fodder, more turnips are consumed, and, therefore, when turnips are scarce, hay may be used with advantage. It also appears that cattle may be fed cheaper on distillery refuse of draff and dreg than on turnips and straw, but then the food obtained from the distillery requires more time to bring cattle to the same condition, which in some circumstances may be an inconvenience.*

(1262.) *Linseed-oil* has been successfully employed to feed cattle by Mr. Curtis of West Rudham, in Norfolk. The mode of using the oil is this: First ascertain how much cut straw the oxen, intended to be fed, will consume a week, then sprinkle the oil, layer upon layer, on the cut straw, at the rate of 1 gallon per week per ox. The mixture, on being turned over frequently, is kept 2 days before being used, when a slight fermentation takes place, and then the oil will scarcely be discerned, having been entirely absorbed by the straw, which should of course be the best oat straw. This mixture, when compared with oil-cake, has stood its ground. The cost of the oil is not great, its average price being about 34s. the cwt. of 12¼ gallons, a gallon of fine oil weighing 9.3 lb., which makes the feeding of an ox cost only 2s. 10d. per week.†

(1263.) Mr. Curtis has fed cattle for upward of 20 years upon what he calls *green malt*, which consists of steeping light barley "for 48 hours in soft water, when the water is let off and the barley is thrown into a round heap, in a conical form, till it gets warm and begins to sprout freely. It is then spread out and turned over repeatedly as it grows. The only care required is, that the sprout or future blade does not get cut off, as the malt will then lose much of its nutritious quality." He finds this substance, which costs with its labor 1s. a stone, preferable to oats at 10d. in their natural state.‡

(1264.) A method of feeding cattle has been adopted by Mr. Warnes, Jr., Trimingham, Norfolk, which, in a manner, combines both the substances used by Mr. Curtis, and deserves attention. The substances consist of linseed-meal and crushed barley. The barley may either be used malted, that is, in a state of "green malt," as designated by Mr. Curtis, or crushed flat by bruising cylinders. Crushed oats, boiled peas and bean flour may all be substituted for the barley, and used with the linseed meal. The mode of making this compound is thus recommended by Mr. Warnes: "Put 168 lbs. of water into an iron cauldron or copper or boiler, and as soon as it boils, not before, stir in 21 lbs. of linseed meal; continue stirring it for 5 minutes; then let 63 lbs. of the crushed barley be sprinkled by the hand or one person upon the boiling mucilage, while another rapidly stirs and crams it in. After the whole has been carefully incorporated, which will not occupy more than 5 minutes, cover it closely down and throw the furnace door open. Should there be much fire, put it out. The mass will continue to simmer, from the heat of the cauldron, till the barley has entirely absorbed the mucilage. The work is then complete, and the food may be used on the following day. When removed into tubs, it must be rammed down to exclude the air, and to prevent its turning rancid. It will be seen that these proportions consist of 3 parts of barley to one of linseed, and of 2 parts of water to 1 of barley and linseed included. Also, that the weight of the whole is 13 stones when put into the cauldron; but after it has been made into compound and become cold, it will be found in general reduced to something less than 15 stones, which will afford 1 bullock for a fortnight 1 stone per day, containing 1½ lbs. of linseed. It will keep a long time if properly prepared. The consistency ought to be like that of clay when formed for bricks." In regard to the nutritive properties of this compound Mr. Warnes testifies thus: "The last of my experimental bullocks for 1841 was disposed of at Christmas at 8s. 6d. per stone. He weighed 60 stones 5 lbs., of 14 lbs. to the stone, and cost £7 17s. 6d. thirteen months previously; so that he paid £17 10s. for little more than one year's keeping. His common food was turnips or grass; 14 lbs. a day of barley or peas compound were given him for 48 weeks, and an unlimited quantity the last 5 weeks; when, considering the shortness of that time, his progress was perfectly astonishing, not only to myself, a constant observer, but to many graziers and butchers who had occasional opportunities of examining him. Altogether the weight of compound did not exceed 2 tons 4 cwts., at a cost of only £3 16s per ton."‖

(1265.) This successful result obtained by Mr. Warnes shows that cattle may be *profitably* fed on prepared food, though the results of several experiments which have been made by farmers in Scotland lead to an opposite conclusion; yet Mr. Warnes's statement contains no comparison, for it is quite possible that the nutritious materials employed by him, namely, linseed-meal and bruised barley, would have fed a bullock equally well in their naturally cold state as when cold, after being cooked warm. As to the expediency of cooking food for cattle, Mr. Warnes goes so far in opinion as to say that "neither oil nor linseed should be used in a *crude* state, but formed into mucilage by being boiled in water;" but this opinion was evidently given when the results obtained by Mr. Curtis on feeding cattle with linseed-oil in a crude state, were unknown to him; for although he admits "that linseed-oil will fatten bullocks, experience has placed beyond a doubt. Among the fattest beasts ever sent to the London market from Norfolk was a lot of Scotch heifers, grazed (?) on linseed-oil and hay." Yet he adds: "But the quantity given per day, the cost per head, or anything relative to profit or loss, I never heard." I should therefore like to see a comparison instituted between the nutritive properties of linseed-meal and bruised barley, or peas or bean-meal, in their ordinary state, and after they had been boiled and administered either in a hot or cold state, and also between the *profits* arising from both. Until this information is obtained, we may rest content with the results obtained by some very accurate experiments, conducted by eminent farmers, on the same food administered in a warm and in a cold state, and which go to prove that food is *unprofitably* administered to cattle in a *cooked* state. I shall now lay some statements corroborative of this conclusion before you.

* Prize Essays of the Highland and Agricultural Society, vol. xiv. † Ibid., vol. xiv. ‡ Ibid., vol. xiv.
‖ Warnes's Suggestions on Fattening Cattle.

(1266.) The first I shall notice, though not in detail, are the experiments of Mr. Walker, Ferrygate, East Lothian. He selected, in February, 1833, 6 heifers of a cross between country cows and a Short-Horn bull, that had been on turnips, and were advancing in condition, and divided them into 2 lots of 3 heifers each, and put one lot on raw food and the other on steamed, and fed them three times a day—at daybreak, noon and an hour before sunset. The food consisted of as many Swedes as they could eat, with 3 lbs. of bruised beans and 20 lbs. of potatoes, ½ stone of straw and 2 ounces of salt to each beast. The three ingredients were mixed together in a tub placed over a boiler of water, and cooked by steaming, and the bruised beans were given to the lot on raw food at noon, and one-half of the potatoes in the morning and another half in the afternoon. It was soon discovered that the lot on the cooked food consumed more turnips than the other, the consumption being exactly 37 cwt. 16 lbs., while, when eaten raw, it was only 25 cwts. 1 qr. 14 lbs., the difference being 55 lbs. every day, which continued during the progress of the experiment for 3 months.

(1267.) Steers were experimented on as well as heifers, there being 2 lots of 2 each. They also got as many Swedish turnips as they could eat, but had 30 lbs. of potatoes and 4½ lbs. bruised beans, 2 ounces of salt, and ½ stone of straw each, every day.

(1268.) The cost of feeding the heifers was as follows:

3 heifers on *steamed* food: Cwts. qrs. lbs.

					£	s.	d.
Consumed of Swedish turnips,	37	0	16, at 4d. per cwt.		£0	12	4½
.. .. Potatoes	3	3	0, at 1s. 3d.		0	4	8
.. .. Beans, 1 bushel	0	2	7		0	3	0
.. .. Salt					0	0	0¾
Coals and extra labor					0	2	0
Cost of 3 heifers 1 week, or 7s. 4½d. per week each					£1	2	1¼

3 heifers on *raw* food: Cwts. qrs. lbs.

				£	s.	d.
Consumed of Swedish turnips,	25	1	14, at 4d. per cwt.	£0	8	6½
.. .. Potatoes, beans and salt, as above				0	7	8¾
Cost of 3 heifers 1 week, or 5s. 5d. each per week				£0	16	3

2 stots on *steamed* food: Cwts. qrs. lbs.

				£	s.	d.
Consumed of Swedish turnips,	28	2	0, at 4d. per cwt.	£0	7	10
.. .. Potatoes	3	3	0, at 1s. 3d.	0	4	8
.. .. Beans	0	2	7	0	3	0
.. .. Salt				0	0	0½
Coals and extra labor				0	1	6
Cost of 2 stots for 1 week, or 8s. 6¼d. each per week				£0	16	0½

2 stots on *raw* food: Cwts. qrs. lbs.

				£	s.	d.
Consumed of Swedish turnips,	17	2	0, at 4d. per cwt.	£0	5	10
.. .. Potatoes, beans and salt, as above				0	7	8½
Cost of 2 stots for 1 week, or 6s. 9¼d. each per week				£0	13	6½

(1269.) The following table shows the progress of condition made by the heifers and stots:

Cattle.	Average live weight of 3 at commencement of feeding.	Average live weight of 3 at end of feeding.	Average increase of live weight in 3 months.	Average dead weight of beef.	Average weight of tallow.	Average weight of hide.	Average weight of offal.
	Sts.	Sts. lbs.	Sts. lbs.	Sts. lbs.	Sts. lbs.	Sts. lbs.	Sts. lbs.
Heifers on steamed food,	74	90 0	16 0	50 9	7 11	3 12	26 9
Heifers on raw food,	74	89 3	15 0	50 1	8 4	4 4	26 10
Stots on steamed food,	84	103 4	19 0	56 12	8 11	5 12	28 3
Stots on raw food.	90	106 5	15 0	58 6	8 8	5 4	30 4

(1270.) The comparative profits on cooked and raw food stand thus:

	£	s.	d.
Live weight of heifers, when put to feed on *steamed* food, 74 sts.=42 sts. 4 lbs. beef, at 5s. 6d. per stone, sinking offal	£11	12	7
Cost of keep 12 weeks 5 days, at 7s. 4½d. per week	4	19	0
Total cost	£16	11	7
Live weight of the same heifers, when finished feeding on steamed food, 90 sts.=50 sts. 9 lbs., at 6s. 6d. per stone, sinking offal	16	9	1½
Loss on steamed food on each heifer	£0	2	6½
Live weight of 1 heifer, when put to feed on *raw* food, 74 sts.=42 sts. 4 lbs. beef, at 5s. 6d. per stone, sinking offal	£11	12	7
Cost of keep 12 weeks 5 days, at 5s. 5d. per week	3	8	10½
Total cost	£15	1	5½
Live weight of the same heifer when finished feeding on raw food, 89 sts. 3 lbs.=50 sts. 1 lbs., at 6s. 6d. per stone, sinking offal	16	5	5½
Profit on raw food on each heifer	£1	4	0

	£	s.	d.
Live weight of 1 stot when put up to feed on *steamed* food, 84 sts.=50 sts. 4 lbs., at 5s. 6d. per stone, sinking offal	£13	4	6
Cost of keep 12 weeks 5 days, at 8s. 6¼d. per week	5	8	
Total cost	£18	12	[illegible]
Live weight of the same stot after being feed on steamed food, 104 sts. 7 lbs.=56 sts. 10 lbs., at 6s. 6d. per stone, sinking offal	18	8	7¾
Profit on each stot on steamed food	£0	3	8½
Live weight of 1 stot when put on *raw* food, 90 sts.=51 sts. 6 lbs., at 5s. 6d. per stone, sinking the offal	£14	2	10½
Cost of 12 weeks 5 days' keep, at 6s. 9½d. per week	4	6	1
Total cost	£18	8	11½
Live weight of the same stot after being fed on raw food, 106 sts. 7 lbs.=58 sts. 6 lbs., at 6s. 6d. per stone, sinking offal	18	19	9½
Profit on each stot on raw food	£0	10	10

(1271.) The facts, brought out in this experiment, are these: It appears that turnips lose weight on being steamed. For example, 5 tons 8 cwts. only weighed 4 tons 4 cwts. 3 qrs. 16 lbs. after being steamed, having lost 1 ton 3 cwts. 12 lbs. or 1-6 of weight; and they also lost 1-5 of bulk when pulled fresh in February; but on being pulled in April the loss of weight in steaming decreased to 1-6. Potatoes did not lose above 1-50 of their weight by steaming, and none of their bulk. The heifers on steamed food not only consumed a greater weight of fresh turnips, in the ratio of 37 to 25, but after allowing for the loss of steaming, they consumed more of the steamed turnips. Thus, after deducting 1-5 from 37 cwts. 16 lbs.—the weight lost in steaming them—the balance, 29 cwts. 2 qrs. 17 lbs., is more than the 25 cwts. 1 qr. 14 lbs. of raw turnips consumed, by 4 cwts. 1 qr. 3 lbs. All the cattle, both on the steamed and raw food, relished salt; so much so, that when it was withheld, they would not eat their food with the avidity they did when it was returned to them.

(1272.) Steamed food should always be given in a fresh state—that is, new made; for, if old, it becomes sour, when cattle will scarcely touch it, and the sourer it is they dislike it the more. "In short," says Mr. Walker, "the quantity they would consume might have been made to agree to the fresh or sour state of the food when presented to them. . . . We are quite aware that to have done a large quantity at one steaming would have lessened the expense of coal and labor, and, also, by getting sour before being used, saved a great quantity of food; but we are equally well aware that, by so doing, we never could have fattened our cattle on steamed food."

(1273.) An inspection of the above table will show that both heifers and stots increased more in live-weight on steamed than on raw food; the larger profit derived from the raw food arising solely from the extra expense incurred in cooking the food. It appears, however, that a greater increase of tallow is derived from raw food. The results appear nearly alike with heifers and stots of the same age; but if the stots were of a breed possessing less fattening properties than cross-bred heifers—and Mr. Walker does not mention their breed—then they would seem to acquire greater *weight* than heifers, which I believe is the usual experience. The conclusion come to by Mr. Walker is this: "We have no hesitation in saying that, in every respect, the advantage is in favor of feeding with raw food. But it is worthy of remark that the difference in the consumption of food arises on the turnips alone. We would therefore recommend every person wishing to feed cattle on steamed food to use potatoes, or any other food that would not lose bulk and weight in the steaming process, as there is no question but, in doing so, they would be brought much nearer to each other in the article of expense of keep. . . . Upon the whole, we freely give it as our opinion that steaming food for cattle will never be attended with beneficial results under any circumstances whatever, because it requires a more watchful and vigilant superintendence during the whole process than can ever be delegated to the common run of servants, to bring the cattle on steamed food even upon a footing of equality, far less a superiority, to those fed on raw food."*

(1274.) One of the stots that had been fed on raw and another on steamed food were kept and put to grass. In their external condition no one was capable of judging how they had been fed. They were put to excellent grass on the 20th May, and the stot on raw food gained condition until 20th July, when, perhaps, the pasture may have begun to fail. That on steamed food fell off to that time 3 stones live weight. On 20th August both were put on cut grass, and both improved, especially the one that had been on steamed food, until the 18th October, when both were put on turnips, on which both became alike by the 10th November, relatively to what they were at the beginning of the season; that is, the stot that had been on raw food increased from 103 to 120 stones, and the other from 106 to 118 stones, live weight.

(1275.) Similar results as to profit were obtained by the experiments of Mr. Howden, Lawhead, East Lothian. "To me," he says, "it has been most decidedly shown that preparing food in this way [by steaming] is anything but profitable. Local advantages—such as fuel and water being at hand—may enable some others to steam at less expense; but in such a situation as mine, I am satisfied that there will be an expense of 10s. a head upon cattle incurred by the practice. A single horse-load of coals, carriage included, costs me 10s.; and exactly 6 cart-loads were required and used in preparing the food for cattle, equal to 6s. 8d. each, and probably as much more would not be an over-estimate for the additional labor in the 3 months." A few facts, worthy of attention, have been brought to light by Mr. Howden's experiment. It seems that raw potatoes and water will make cattle fat—a point which has been questioned by some of our best farmers. Potatoes, beans and oats, taken together, will feed cheaper, in reference to time, than turnips or potatoes separately; and from this fact may be deduced these, namely, that potatoes, when used alone, to pay their expense, would require the beef fed by them to fetch 4d. per lb.;

* Prize Essays of the Highland and Agricultural Society, vol. x.

(518)

turnips alone, 3½d.; and potatoes and corn together, 3d., and at the same time yielding beef of finer quality. There is a curious fact to be observed in the table given by Mr. Howden. Of 6 heifers, 1 in a lot of 3 weighed 1022 lbs.; and another, in another lot of 3, weighed also 1022 lbs., on 5th March, when both were put up to be experimented on; and on the 5th June following both were of the same weight, namely, 1176 lbs., both showing exactly an increase of 154 lbs.; both being supplied with the same *weight* of food, namely, 140 lbs. of turnips, to the one given raw, to the other cooked. This is a remarkable coincidence; but here it ends, and the superiority of cooked food becomes apparent; for the beef of the heifer fed on raw turnips weighed 5 sts. 12 lbs. and its tallow 5 sts. 10 lbs.; whereas, the beef of the one fed on steamed turnips weighed 44 sts. 4 lbs., and its tallow 6 sts. 22 lbs. How is this to be accounted for? Partly, no doubt, in the cooking of the food; but partly, I should suppose, from the state of the animal indicated by its hide, the the thinner one of the heifer fed on steamed turnips weighing 3 sts. 10 lbs., showing a greater disposition to fatten—that is, to lay on more rapidly the valuable constituents of beef and tallow—than the thicker hide of the other heifer fed on raw turnips, which weighed 4 sts. 4 lbs. It is but justice, however, to the raw turnips to mention a fact to which Mr. Howden adverts. The turnips appropriated to the experiment were, it seems, stored against a wall, one store having a northern and the other a western aspect; but whether from aspect, or dampness, or other cause, those intended to be eaten *raw* had fermented in the store awhile before being observed, and thus, becoming unpalatable, of the 18 tons 15 cwts. stored, about 2¼ tons were left unconsumed; so that, in fact, the heifers upon raw turnips did not receive so much food, or in so palatable a state as those on the steamed. It seems steaming renders tainted turnips somewhat palatable, while it has a contrary effect on tainted potatoes, the cattle preferring these raw. Turnips require a longer time to steam, and according to Mr. Howden's experience, they lose ⅛ or 1-10 more of their weight than potatoes.* You may observe, from the state of the turnips in the store, the injudiciousness of storing them *against a wall*, as I have before observed (1019).

(1276.) Mr. Boswell of Kingcausie, in Kincardineshire, comes to the same conclusion in regard to the unprofitableness of feeding cattle on cooked food. He says, "It appears that it is not worth the trouble and expense of preparation to feed cattle on boiled or steamed food; as, although there is a saving in food, it is counterbalanced by the cost of fuel and labor, and could only be gone into profitably where food is very high in price and coal very low." His experiments were made on 10 dun Aberdeenshire horned cattle, very like one another, and their food consisted of the Aberdeen yellow bullock-turnips and Perthshire red potatoes. The 5 put on raw food weighed alive 228 stones 11 lbs., and the other 5 on cooked food 224 stones 6 lbs. imperial. When slaughtered, the butcher considered both beef and tallow "to be perfectly alike." Those fed on raw food cost £32 2s. 1d., and those on cooked £34 5s. 10d., leaving a balance of expense of £2 3s. 9d. in favor of the former. The opinions of feeders of cattle are not alike on all points. Thus, Mr. Boswell says, "The lot on raw consumed much more food than those on steamed," a fact directly the reverse of that stated by Mr. Walker in (1266). "Twice a week, on fixed days," he continues, "both lots got a small quantity of the tops of common heath, which acted in the way of preventing any scouring; in fact, turnip-cattle seem very fond of heather as a condiment." . . . "The dung of the steamed lot was from first to last in the best state, without the least appearance of purging, and was free of that abominable smell which is observed when cattle are fed on raw potatoes, or even when a portion of their food consists of that article. Another fact was observed, that after the steamed lot had taken to their food, they had their allowance finished sooner than the raw lot, and were therefore sooner able to lie down and ruminate." There is a curious fact mentioned by Mr. Boswell regarding a preference and dislike shown by cattle for turnips in different states. "When *raw* turnips and potatoes were put into the stall at the same time, the potatoes were always eaten up before a turnip was tasted; while, on the other hand, steamed turnips were eaten in preference to steamed potatoes."†

(1277.) Some curious and interesting facts have been arrived at by Mr. Stephenson, Whitelaw, East Lothian, in his experience of feeding cattle. They are detailed by him in a paper on feeding different lots of cattle, not with cooked and raw food, but with different sorts of food in a raw state. He divided a number of cattle into 3 lots, containing 6 in each lot, and fed one on oil cake, bruised beans, and bruised oats, in addition to whatever quantity of turnips they could eat, and potatoes for the last few days of the experiment; another lot received the same sort of food, with the exception of the oil-cake; and the third lot was fed entirely on turnips. The live weights of the lots varied considerably from 486 to 346¾ imperial stones. I need not detail the particulars of the experiment, which was conducted from November, 1834, to March, 1835, for 17 weeks, as they present nothing remarkable; but their results are worthy of your attention.

(1278.) Each beast in the lot that got oil-cake cost, in 17 weeks, £5 2s. 7d., or 6s. per week; in the lot fed on corn, £3 17s., or 4s. 6d. per week; and in that fed entirely on turnips, £1 18s. 7½d., or 2s. 3d. a week. Estimating the value of the fed beef at 6s. 6d. per imperial stone, there was a *loss* of 12½ per cent. sustained on the lot fed on oil-cake; a *gain* of 8½ per cent. on that fed on corn; and a *gain* of 22 per cent. on that fed entirely on turnips.

(1279.) This was the cost incurred for producing every 1 lb. of increase of live weight, the lot fed on oil-cake increasing from 486 to 594 stones; that on corn from 443 to 544 stones; and that on turnips from 346¾ to 395¼ stones.

The oil-cake cost 4 9-10 pence to produce 1 lb. of live weight.
.. corn .. 3 9-10 .. ..
.. turnips .. 4 4-10 .. ..

It thus appears that the joint agency of corn and turnips produces 1 lb. of live weight at the cheapest rate of the three modes adopted.

---

* Prize Essays of the Highland and Agricultural Society, vol. x.

† Ibid., vol. x.

(1280.) Another conclusion come to from the data supplied by this experiment is, that it took—

90 lbs. of turnips to produce 1 lb. of live weight.
40 lbs. of potatoes .. .. ..
8 7-10 lbs. of corn .. .. ..
21 8-10 lbs. of oil-cake .. .. ..

And the cost of doing this was as follows:

90 lbs. of turnips, at 4d. per cwt.......................... 3 2-10d. per 1 lb. of live weight
40 lbs. of potatoes, at 1s. 6d. per cwt .... 6 4-10d. .. ..
8 7-10 lbs. of corn at 3s. 3d. per bushel of 60 lbs....... 5 7-10d. .. ..
21 8-10 lbs. of oil-cake, at ¾d. per lb. or £7 per ton......16 3-10d. .. ..

Could these results be proved to be *absolutely* correct, there would be no difficulty of assigning the degree of profit to be derived from employing any of these substances in the feeding of cattle. Is not the inquiry, however, of as much importance, even in a national point of view, as to deserve investigation at some sacrifice of both cost and trouble?

(1281.) You should not suppose that cattle consume food of any sort in a uniform ratio; for see actual results. The lot that was fed entirely on turnips increased in the first 32 days of the experiment only 8 stones, whereas the same beasts in 46 days immediately preceding those on which the experiment began, increased 48½ stones; and in one 8 days of the 46 they consumed 160 8-10 lbs. each of white globe turnips every day, and increased 1 lb. of live weight for every 65 4-10 lbs. of turnips consumed. The 90 lbs. taken above as the quantity of turnips required to produce 1 lb. of live weight is therefore not absolute, but assumed as a medium quantity, for it will happen that 1,000 lbs. will not produce 1 lb. of live weight. What the circumstances are which regulate the tendencies of cattle to fatten, are yet unknown. The fact is, cattle consume very different quantities of turnips in different states of condition, consuming more when lean, in proportion to their weight, than when fat. A lean beast will eat twice, or perhaps thrice, as many turnips as a fat one, and will devour as much as ¼ part of his own weight every day, while a very fat one will not consume 1-10. I had a striking example of this one year, when I bought a very lean 2-year-old steer, a cross betwixt a Short-Horn bull and Angus cow, for £6 in April; and he was a large-boned, thriving creature, but his bones were cutting the skin. He was immediately put on Swedish turnips; and the few weeks he was on them, before being turned to grass, he could hardly be satisfied, eating three times as much as the fat beasts in the same hammel. He was grazed in summer, and fed off on turnips and sold in April following for 17 guineas. Some stots of Mr. Stephenson's, in November, eat 2 7-10 lbs. for every stone of live weight they weighed; the year after the quantity decreased to 1 9-10 lbs., and after the experiment was included, when their live weights were nearly doubled, they consumed only 1 5-10 lbs.

(1282.) The object which Mr. Stephenson had in conducting the experiment the results of which are narrated above, were fourfold: 1. To compare cattle fed partly on oil-cake with those which had none; 2. To compare those fed partly on corn with those which had none; and 3. To compare those fed solely on turnips with those which had different sorts of food. The results were, that oil-cake is an unprofitable food for cattle, that corn yields a small profit, that turnips are profitable, and that when potatoes can be sold at 1s. 6d. per cwt. they are also unprofitable. "When any other food than turnips," observes Mr. Stephenson, "is desired for feeding cattle, we would recommend bruised beans as being the most efficient and least expensive; on this account we would prefer bruised beans alone to distillery offal. As regards linseed-cake, or even potatoes, they are not to be compared to beans." . . . "We give it as our opinion, that whoever feeds cattle on *turnips alone* will have no reason, on the score of *profit*, to regret their not having employed more expensive auxiliaries to hasten the fattening process. This opinion has not been rashly adopted, but has been confirmed by a more extended and varied experience in the feeding of cattle than has fallen to the lot of most men." 4. Another object he had in these experiments was, to ascertain whether the opinion is correct or otherwise, that cattle consume food in proportion to their weights. On this subject Mr. Stephenson says "that cattle consume food something nearly in proportion to their weights, we have very little doubt, *provided they have previously been fed in the same manner, and are nearly alike in condition.* Age, sex, and kind have little influence in this respect, as the quantity of food consumed depends much on the length of time the beast has been fed, and the degree of maturity the animal has arrived at—hence the great difficulty of selecting animals to be experimented upon. To explain our meaning by an example, we would say that 2 cattle of the same weight, and which had been previously kept for a considerable time on similar food, would consume about the same quantity. But, on the contrary, should 2 beasts of the same weight be taken, the one fat and the other lean, the lean beast would perhaps eat twice, or perhaps thrice, as much as the fat one—more especially if the fat one had been for some time previously fed on the same food, as cattle eat gradually less food until they arrive at maturity, when they become stationary in their appetite." . . . "We shall conclude," he says, "by relating a singular fact," and a remarkable one it is, and worth remembering, "that *sheep* on turnips will consume nearly in proportion to *cattle*, weight for weight, that is, 10 sheep of 14 lbs. a quarter, or 40 stones in all, will eat nearly the same quantity of turnips as an ox of 40 stones; but turn the ox to grass, and 6 sheep will be found to consume an equal quantity. This great difference may perhaps," says Mr. Stephenson, and I think truly, "be accounted for by the practice of sheep cropping the grass much closer and oftener than cattle, and which, of course, prevents its growing so rapidly with them as with cattle."*

(1283.) Still another question remains to be considered in reference to the feeding of cattle in winter, which is, whether they thrive best in *hammels* or in *byres at the stake?* The determination of this question would settle the future construction of steadings; for, of course, if more profit were certainly yielded to the farmer to feed his cattle in hammels than in byres, not only would

* *Prize Essays of the Highland and Agricultural Society, vol. xii.*

no more byres be erected, but those in use converted into hammels; and this circumstance would so materially change the form of steadings, as to throw open the confined courts, embraced within quadrangles, to the influence of the sun, at the only season these receptacles are required, namely, in winter. Some facts have already been decided regarding the comparative effects of hammels and byres upon cattle. Cattle are much cleaner in their persons in hammels than in byres. No doubt they *can* be kept clean in byres, but not being so, there must be some difficulty incidental to byre-management, and it consists, I presume, in the cattle-man finding it more laborious to keep the beasts clean in a byre, than in hammels; otherwise the fact is not easily to be accounted for, for he takes no *special* care to keep beasts in hammels clean. Perhaps when cattle have liberty to lie down where they please they may choose the driest, because the most comfortable spot; whereas, in a byre, they must lie down upon what they cannot see behind them. There is another advantage derived from hammels; the hair of cattle never scalds off the skin, and never becomes short and smooth, but remains long and mossy, and all licked over, and washed clean by rain, until it is naturally *cast* in spring, and this advantage is felt by cattle when sent to market in winter, where they can withstand much more wet and cold than those which have been fed in byres. A third advantage is, that cattle from hammels can travel the road without injury to their feet, being accustomed to be so much upon their feet, and to move about. It has been alleged in favor of byres, that they accommodate more cattle on the same space of ground, and are less expensive to erect at first than hammels. That in a given space more beasts are accommodated in byres there is no doubt, and there is as little doubt that more beasts are put in a byre than should be; but I have great doubts that it will cost more money to accommodate a given number of cattle in the hammel than in the byre system; because hammels can be constructed in a temporary form of wood and straw, and make beasts very comfortable at a moderate charge, whereas byres cannot be formed in that fashion; and even in the more costly form of roofs and walls, the shedding of hammels requires, comparatively to a byre, but a small stretch of roof; and it is well known that it is the roof and not the bare masonry of the walls that constitute the most costly part of a steading. I have seen a set of hammels, having stone and lime walls, and feeding-troughs, and a temporary roof, erected for £1 for every beast it could accommodate, and no form of byre could be built at that cost. But all these advantages of hammels would be of trifling import, if it can be proved by experience that cattle afford larger profits on being fed in byres; and unless this superiority is established in regard to either, the other is undeserving of preference. How then, stands the fact? Has experiment ever tried the comparative effects of both on anything like fair terms? Mr. Boswell of Balmuto, in Fifeshire, and of Kingcausie, in Kincardineshire, has done it; and it shall now be my duty to make you acquainted with the results.

(1284.) To give as much variety to this experiment as the circumstances would admit, it was conducted both at Balmuto and Kingcausie, and the beasts selected for it were of different ages, namely, 2 and 3-year-olds. At Balmuto 4 three-year-olds were put in close byres, and 4 in open hammels, and the same number of 2-year-olds were accommodated in a similar manner at Kingcausie. Those at Kingcausie received turnips only, and of course straw, at Balmuto a few potatoes were given at the end of the season, in addition to the turnips. The season of experiment extended from 17th October, 1834 to 19th February, 1835. The results were these:—

	Sts.	lbs.	St.	lb.
The 4 hammel-fed 2-year-olds at Kingcausie gained of live weight			45	8
.. 4 .. .. 3-year-olds at Balmuto			45	0
			91	8
.. 4 byre-fed 2-year-olds at Kingcausie gained of live weight,	32	7		
.. 4 .. 3-year-olds at Balmuto	36	0		
			68	7
Gain of live weight by the hammel-fed,..............................			23	1

This is, however, not all gain, for the hammel-fed consumed more turnips, the Aberdeen yellow bullock, than the byre-fed.

	Tons.	cwts.	qrs.	lbs.
Those at Kingcausie consumed more by	1	7	2	6
And those at Balmuto	2	4	3	22
Total more consumed,	3	12	2	0

In a pecuniary point of view, the gain upon the hammel-fed was this:—23 stones 1 lb. live weight, = 13¾ stones beef, at 6s. per stone, gives £4 2s., from which deduct the value of the turnips, at 4d. per cwt., £1 4s. 2d., leaving a lance of £2 7s. 10d.

(1285.) It is a prevalent opinion among farmers, that young cattle do not lay on weight so fast, as old. But this experiment contradicts it; for the 2-year-olds in the hammels at Kingcausie gained 44 stones 22 lbs., on their united weights of 320 stones 7 lbs., in the same time that the 3-year-olds in the hammels at Balmuto, weighing together 350 stones, were of gaining 46 stones. Besides, the young beasts in the hammels at Kingcausie gained over those in the byre 12 stones 15 lbs., while the older cattle in the hammels at Balmuto gained over those in the byre only 10 stones. So that, in either way, the young cattle had the advantage over the older.

(1286.) Mr. Boswell observes that "hammels ought never to be used unless when the climate is good, and the accommodation of courts dry and well sheltered; and, above all, unless when there is a very large quantity of litter to keep the cattle constantly clean and dry." Shelter is essential for all sorts of stock in any situation, and the more exposed the general condition of the farm is, the more *need* there is of shelter; but be the situation what it may, it is, in my opinion, quite possible to render any hammel sheltered enough for stock, not by the distribution of planting, but by temporary erections against its weather-side; and these means will be the more effectual when the hammel is placed facing the meridian sun, which it should be in every case If these particulars are attended to, and a rain-water spout placed along the eave in front to prevent the rain from the roof falling into the court, and an open drain, with convenient gratings, connected with all the courts, is properly made, the quantity of straw required will not be inordinate as I have myself experienced when farming dry turnip-soil. Mr. Boswell's testimony in favor of hammels is most satisfactory: it is this. "From the result of my own experiment, as well as the unanimous opinion of every agriculturist with whom I have conversed on the subject, I feel convinced that there is no point more clearly established than that cattle improve quicker, or, in other words, *thrive better in open hammels* than in close byres."*

(1287.) I have dwelt the longer on the subject of feeding cattle, because of its great importance to the farmer, and because of the uncertainty sometimes attending its practice to a profitable issue; and there is no doubt that whether it leaves a profit or not depends entirely on the mode in which is prosecuted. Many are content to fatten their cattle in any way, or because others do so, provided they know they are not actually losing money by it, but if they do not make their cattle in the ripest state they are capable of being made, they are, in fact, losing part of their value. But how are they you may ask, best to be made ripe? There lies the difficulty of the case, and it must be attended with much difficulty before a man of the extensive experience in fattening cattle as Mr. Stephenson, would express himself in these words: "We have had great experience in feeding stock, and have conducted *numbers of experiments* on that subject with all possible care, both in weighing the cattle alive, and the whole food administered to them and in every experiment we made *we discovered something new*. But we have seen enough to convince us that, were *the art of feeding better understood, a great deal more beef and mutton might be produced from the same quantity of food* than is generally done." So far should such a declaration deter you from fattening cattle, it should rather be a proof of the wideness of the field that is still open for you to experiment in.

(1288.) There are but few *diseases* incidental to cattle in a state of confinement in winter, these being chiefly confined to the skin, such as the affection of *lice*, and to accidents in the administration of food, as *hoven* and *obstruction of the gullet* may be termed.

(1289.) *Lice.* When it is known that almost every species of quadruped found in the country, and in a state of nature, is inhabited by one or more pediculidæ, sometimes peculiar to one kind of animal, at other times ranging over many, it will not excite surprise that they should also occur on our domestic ox. Indeed, domestication and the consequences it entails, such as confinement, transition from a low to a higher condition, high feeding, and an occasional deviation from a strictly natural kind of food, seem peculiarly favorable to the increase of these parasites. Their occurrence is well known to the breeder of cattle, and to the feeder of fat cattle; and they are not unfrequently a source of no small annoyance to him. Unless when they prevailed to a great extent, they are probably not the cause of any positive evil to the animal, but, as their attacks are attended with loss of hair, an unhealthy appearance of the skin, and their presence is always more or less unsightly, and a source of personal annoyance to cattle, they may much impair the animal's look, which, when it is designed to be exhibited in the market, is a matter of no small consequence. As an acquaintance with the appearance and habits of these creatures must precede the discovery and application of any judicious method of removing or destroying them, I shall describe the species now which are the most common and noxious to the ox, and afterward to the other domestic animals of the farm. They may be divided into two sections, according to a peculiarity of structure, which determines the mode in which they attack an animal, namely, those provided with a mouth formed for sucking, and such as have a mouth with two jaws formed for gnawing. Of the former there are 3 species, which are very common, attacking the ox, the sow, and the ass.

(1290.) *Ox-louse (Hæmatopinus eurysternus)*, fig. 268. It is about 1 or 1½ lines in length, as seen by the line below the figure, the head somewhat triangular, and of a chesnut color, the eyes pale brown, antennæ pale ochre-yellow, thorax darker chesnut than the head, with a spiracle or breathing-hole on each side, and a deep furrow on each side anteriorly; the shape nearly square, the anterior line concave, abdomen broadly ovate, grayish-white, or very slightly tinged with yellow, with 4 longitudinous rows of dusky horny excrescences, with 2 black curved marks on the last segment; legs long and strong, particularly the 2 fore-pairs, the color chesnut; claws strong and black. This may be called the common louse that infests cattle. It is most apt to abound on them when tied to the stall for winter feeding; and a notion prevails in England that its increase is owing to the cattle feeding on straw. The fact probably is, that it becomes more plentiful when the animal is tied up, in consequence of its being then less able to rub and lick itself, and the creature is left to propagate, which it does with great rapidity, comparatively undisturbed. It generally concentrates its forces on the mane and shoulders. As the parasite is suctorial, if it is at all the means of causing the hair to fall off, it can only be by depriving it of the juices by which is nourished, which we can conceive to be the case when the sucker is inserted at the root of the hair; but it is more probable that the hair is rubbed off by the cattle themselves, or is shorn off by another louse to be just noticed. The egg or nit is pear-shaped, and may be seen attached to the hairs.

Fig. 268.

THE OX-LOUSE. HÆMATOPINUS EURYSTERNUS.

(1291.) *Ox-louse (Trichodectes scalaris)*, fig. 269.—This parasite is minute, the length seldom exceeding ½ a line. The head and thorax are of a light rust color, the former of a somewhat obcordate shape, with two dusky spots in front: the third joint of the antennæ longest, and spindle-shaped (in the horse-louse, *Trichodectes equi*, that joint is clavate); abdomen pale, tawny, pubes-

* Prize Essays of the Highland and Agricultural Society, vol. xi.

ceat, the first 6 segments with a transvere dusky or rust-colored stripe on the upper half, a narrow stripe of the same color along each side, and a large spot at the hinder extremity; legs, pale tawny. Plentiful on cattle; commonly found about the mane, forehead, and rump. near the tail-head. It is provided with strong mandibles, with two teeth at the apex, and by means of these it cuts the hairs near the roots with facility. Both these vermin are destroyed by the same means as the sheep-louse (1156).

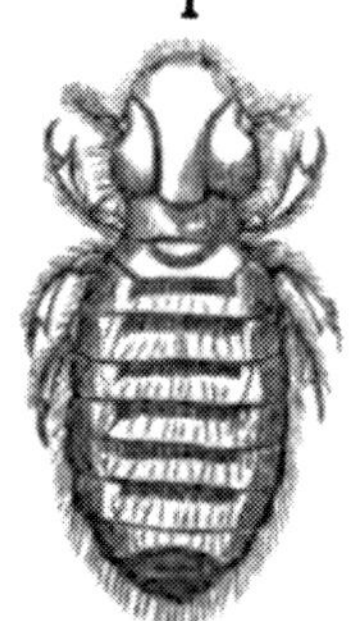

Fig. 296

THE OX-LOUSE, TRICHODECTES SCALARIS.

(1292.) *Choking.*—When cattle are feeding on turnips or potatoes, it occasionally happens that a piece larger than will enter the gullet easily, is attempted to be swallowed, and obstructed in its passage. The accident chiefly occurs to cattle receiving a limited supply of turnips, and young beasts are more subject to it than old. When a number of young beasts in the same court only get a specified quantity of turnips or potatoes once or twice a day, each becomes apprehensive, when the food is distributed. that will not get its own share, and therefore eats what it can with much apparent greediness, and not taking sufficient time to masticate, swallows its food hastily. A large piece of turnip, or a small potato, thus easily escapes beyond the power of the tongue, and, assisted as it is by the saliva, is sent to the top of the gullet, where it remains. Cattle that project their mouths forward in eating, are most liable to choke. When turnips are sliced and potatoes are broken, there is less danger of the accident occurring even among young cattle. The sight of the obstruction, its consequent effects, and remedial measures for its removal, are thus described by Professor Dick. "The obstruction usually occurs at the bottom of the pharynx and commencement of the gullet, not far from the lower part of the larynx, which we have seen mistaken for the foreign body. The accident is much more serious in ruminating animals than in others, as it immediately induces a suspension of that necessary process, and of indigestion, followed by a fermentation of the food, the evolution of gases, and all those frightful symptoms which will be noticed under the disease *hoven.* The difficulty in breathing, and the general uneasiness of the animal, usually direct at once to the nature of the accident, which examination brings under the cognizance of the eye and hand. *No time must be lost in endeavoring to afford relief*; and the *first* thing to be tried is, by gentle friction and pressure of the hand upward and downward, to see and rid the animal of the morsel. *Failing in this,* we mention first the great virtue we have frequently found in the use of mild lubricating fluids, such as warm water and oil, well boiled gruel, &c. The gruel is grateful to the animal, which frequently tries to gulp it, and often succeeds. Whether this is owing to the lubrication of the parts, or the natural action superinduced, it is unnecessary to inquire; but the fact we know, that a few pints of warm gruel have often proved successful in removing the obstruction. *If this remedy should be ineffectual,* the foreign body may perhaps be within the reach of the small hand which a kind dairy-maid may skillfully lend for the purpose. *If this good service cannot be procured,* the common probang must be used, the cup-end being employed. Other and more complicated instruments have been invented, acting upon various principles—some, for example, on that of bruising the obstructing body; and the use of these requires considerable skill. *Disappointed in all,* we must finally have resource to the knife."* You may try all these remedies, with the exception of the knife, with perfect confidence. The friction, the gruel, the hand, and the probang, I have successfully tried; but the use of the knife should be left to the practical skill of the veterinary surgeon.

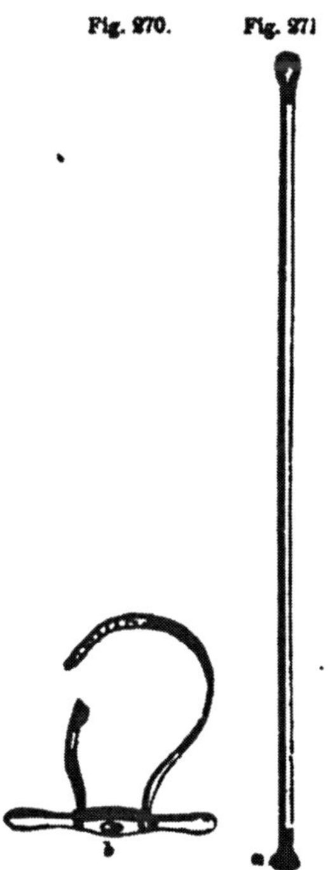

Fig. 270. THE MOUTH-PIECE FOR THE PROBANG.

Fig. 271. THE PROBANG.

(1293.) The common *probang* is represented in fig. 271, *a* being the cup-end, which is so formed that it may partially lay hold of the piece of turnip or potato, and not slip between it and the gullet, to the risk of rupturing the latter, and being of larger diameter than the usual state of the gullet, on pressing it forward distends the gullet, and makes room for the obstructing body to proceed to the stomach. Formerly the probang was covered with cane, but is now with leather, which is more pliable. It is used in this manner: Let the piece of wood, fig. 270, be placed over the opened mouth of the animal as a bit, and the straps of leather attached to it buckled tightly over the neck behind the horns, to keep the bit steady in its place. The use of the bit is, not only to keep the mouth open without trouble, but to prevent the animal injuring the probang with its teeth, and it offers the most direct passage for the probang toward the throat. Let a few men seize the animal on both sides by the horns or otherways, and let its mouth be held projecting forward in an easy position, but no fingers introduced into the nostrils to obstruct the breathing of the animal, nor the tongue forcibly pulled at the side of the mouth. Introduce now the cup-end *a* of the probang, fig. 271, through the round hole *b* of the mouth-piece, fig. 270, and push it gently toward the throat until you feel the piece of the turnip obstructing you; push then with a firm and persevering hand, cautioning the men, previous to the push, to hold on firmly, for the

* Dick's Manual of Veterinary Science.

passage of the instrument may give the animal a little pain, and cause it to wince and even start away. The obstruction will now most likely give way, especially if the operation has been performed before the parts around it began to swell; but if not, the probang must be used with still more force, while another person rubs with his hands up and down upon the distended throat of the beast. If these attempts fail, recourse must be had to the knife, and a veterinary surgeon sent for instantly.

(1294.) *Hoven.*—The hoven in cattle is the corresponding disease to the gripes or batts in horses. The direct cause of the symptoms is undue accumulation of gases in the paunch or large stomach, which, not finding a ready vent, causes great pain and uneasiness to the animal, and, if not removed in time, ruptures the paunch and death ensues. The cause of accumulation of the gases is indigestion. "The structure of the digestive organs of cattle," says Professor Dick, "renders them peculiarly liable to the complaint, while the sudden changes to which they are exposed in feeding prove exciting causes. Thus, it is often witnessed in animals removed from confinement and winter feeding to the luxuriance of the clover field; and in house-fed cattle, from the exhibition of rich food, such as peas-meal and beans, often supplied to enrich their milk. We have already mentioned that it sometimes proceeds from obstructed gullet. The symptoms bear so close a resemblance, both in their progress and termination in rupture and death, to those so fully described above, that we shall not repeat them. The treatment mostly corresponds, and it must be equally prompt. The mixture of the oils of linseed and turpentine is nearly a specific."* The recipe is, linseed-oil, raw, 1 lb.; oil of turpentine, from 2 to 3 oz.; laudanum, from 1 to 2 oz., for one dose. Or hartshorn, from ½ to 1 oz., in 2 pints imperial of tepid water. In cases of pressing urgency, from 1 to 2 oz. of tar may be added to ½ pint of spirits, and given diluted, with great prospect of advantage. These medicines are particularly effective in the early stage of the disease, and should therefore be tried on the first discovery of the animal being affected with it. Should they not give immediate relief, the probang may be introduced into the stomach, and be the means of conveying away the gas as fast as it is generated; and I have seen it successful when the complaint was produced both by potatoes and clover; but I never saw an instance of hoven from turnips, except from obstruction of the gullet. The trial of the probang is useful to show whether the complaint arises from obstruction or otherwise, for should it pass easily down the throat, and the complaint continue, of course the case is a decided one of hoven. Placing an instrument, such as in fig. 270, across the mouth, to keep it open, is an American cure which is said never to have failed. But the gas may be generated so rapidly that neither medicines nor the probang may be able to prevent or convey away, in which case the apparently desperate remedy of *paunching* must be had recourse to. "The place for puncturing the paunch," directs Professor Dick, "is on the left side, in the central point between the lateral processes of the lumbar vertebra, the spine of the ileum, and the last rib. Here the *trochar* may be introduced without fear. If air escape rapidly, all is well. The canula may remain in for a day or two, and, on withdrawal, little or no inconvenience will usually manifest itself. If no gas escape, we must enlarge the opening freely, till the hand can be introduced into the paunch, and its contents removed, as we have sometimes seen, in prodigious quantities. This done, we should close the wound in the divided paunch with 2 or 3 stitches of fine catgut, and carefully approximate and retain the sides of the external wound, and with rest, wait for a cure, which is often as complete as it is speedy."† To strengthen your confidence in the performance of this operation, I may quote a medical authority on its safe effects, in the human subject, even to the extent of exposing the intestines as they lay in the abdomen. "I should expect no immediately dangerous effects from opening the abdominal cavity. Dr. Blundell has stated that he has never, in his experiments upon the rabbit, observed any marked collapse when the peritoneum was laid open, although in full expectation of it. The great danger to be apprehended is from inflammation, and the surgeon, of course, will do all in his power to guard against it."‡ I once used the trochar with success in the case of a Skibo stot which had been put on potatoes from turnips, and as he was in very high condition, took a little blood from him, and he recovered very rapidly. In another year I lost a fine 1-year-old Short-Horn quey by hoven, occasioned by potatoes. Oil and turpentine were used, but as the complaint had remained too long, before it was noticed by myself, late at night, the medicine had no effect. The probang went down easily, proving there was no obstruction. The trochar was then thrust in, but soon proved ineffectual, and as I had not the courage to use the knife to enlarge the opening the trochar had made, and withdraw the contents of the paunch with the hand, the animal sank and was immediately slaughtered. The remedies cannot be too soon applied in the case of hoven.

Fig. 272.

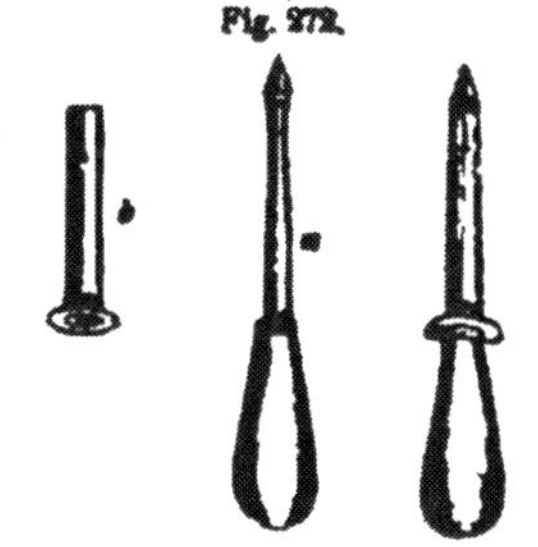

THE TROCHAR.

(1295.) The *trochar* is represented in fig. 272. It consists of a round rod of iron *a*, 5 inches in length, terminating at one end in a triangular, pyramidal-shaped point, and furnished with a wooden handle at the other. The rod is sheathed in a cylindrical cover or case *b*, called the canula, which is open at one end, permitting its point to project, and furnished at the other with a broad, circular flange. The canula is kept tight on the rod by means of a slit at its end nearest the point of the rod, which, being somewhat larger in diameter than its own body, expands the slit end of the canula until it meets the body when the slit collapses to its ordinary size, and the canula is kept secure behind the enlarged point as at *c*. On using the trochar, in the state as seen by *c*, it is forced with a thrust into the place pointed out above, through the skin into the paunch; and on withdrawing the rod by its

* Dick's Manual of Veterinary Science. † Ibid.

‡ Stephens on Obstructed and Inflamed Hernia.

handle—which is easily done, notwithstanding the contrivance to keep it on—the canula is left in the opening, to permit the gas to escape through its channel. On account of the distended state of the skin, the trochar may rebound from the throat; and in such an event, a considerable force must be used to penetrate the skin.

(1296.) The *fardlebound* of cattle and sheep is nothing more than a modification of the disease in horses called stomach-staggers, which is caused by an enormous distension of the stomach. "In this variety, it has been ascertained," says Professor Dick, "that the *maniplies* are most involved, its secretions are suspended, and its contents become dry, hard, and caked into one solid mass. Though the constipation is great, yet there is sometimes the appearance of a slight purging, which may deceive the practitioner."* The remedial measures are, first, to relieve the stomach by large drenches of warm water, by the use of the stomach-pump. Searching and stimulating laxatives are then given, assisted by clysters, and then cordials.

(1297.) *Warts* and *angle-berries* are not uncommon excrescences upon cattle. They are chiefly confined to the groin and belly. I have frequently removed them by ligature with waxed silk thread. Escharotics have great efficacy in removing them—such as alum, bluestone, corrosive sublimate.

(1298.) *Encysted tumors* sometimes appear on cattle, and may be removed by simple incision, having no decided root or adhesion. I had a 2-year-old Short-Horn quey that had a large one upon the front of a hind foot, immediately above the coronet, which was removed by simple incision by a veterinary surgeon. What the true cause of its appearance may have been, I cannot say; but the quey, when a calf, was seen to kick a straw-rack violently with the foot affected, and was lame in consequence for a few days; after which, a small swelling made its appearance upon the place, which, gradually enlarging, became the loose and unsightly tumor which was removed.

(1299.) A gray-colored scabby eruption, vulgarly called the *ticker*, sometimes comes out on young cattle on the naked skin around the eye-lids, and upon the nose between and above the nostrils. It is considered a sign of thriving, and no doubt it makes its appearance most likely on beasts that are improving from a low state of condition. It may be removed by a few applications of sulphur ointment.

(1300.) In winter, when cows are heavy in calf, some are troubled with a complaint commonly called a *coming down of the calf-bed*. A part of the womb is seen to protrude through the vaginal passage when the cow lies down, and disappears when she stands up again. It is supposed to originate after a very severe labor. Bandages have been recommended, but, in the case of the cow, they would be troublesome, and indeed are unnecessary; for if the litter is made firm and higher at the back than the front part of the stall, so as the hind-quarter of the cow shall be higher than the fore when lying, the protrusion will not occur. I had a cow that was troubled with this inconvenience every year, and as she had no case of severe labor in my possession, I did not know whether, in her case, it was occasioned by such a circumstance; but it seemed to give her no uneasiness, when the above preventive remedy was resorted to.

(1301.) It not unfrequently happens to cattle in large courts, and more especially to those in the court nearest the corn-barn, that an oat-chaff gets into one of their eyes in a windy day. An irritation immediately takes place, causing copious watering from the eye, and, if the chaff is not removed, a considerable inflammation and consequent pain soon ensue, depriving the sufferer of the desire for food. To have it removed, let the animal be firmly held by a number of men, and as beasts are particularly jealous of having anything done to their eyes, a young beast even will require a number of men to hold it fast. The fore-finger should then be gently introduced under the eye-lid, pushed in as far as it can go, and being moved round along the surface of the eye-ball, is brought round to its original position, and then carefully withdrawn and examined, to see if the chaff has been removed along with it, which it most likely will be; but if not, repeated attempts will succeed. A thin handkerchief around the finger will secure the extraction at the first attempt. Fine salt or snuff have been recommended to be blown into the eye when so affected, that the consequent increased discharge of tears may float away the irritating substance; but the assistance of the finger is less painful to the animal, and sooner over, and, as it is an operation I have frequently performed with undeviating success, I can attest its efficacy and safety.

---

33. DRIVING AND SLAUGHTERING CATTLE.

"Frisk, dance and leap, like full-fed beasts, and even
Turn up their wanton heels against the Heaven;
Not understanding that this pleasant life
Serves but to fit them for the butcher's knife."

FLAVEL.

(1302.) It is requisite that cattle which have been disposed of to the dealer or butcher, or are intended to be driven to market, should undergo a preparation for the journey. If they were immediately put to the road to travel, from feeding on grass or turnips, when their bowels are full of

* Dick's Manual of Veterinary Science.

indigested vegetable matter, a scouring might ensue, which would render them unfit to pursue their journey; and this complaint is the more likely to be brought on from the strong propensity which cattle have to take violent exercise on feeling themselves at liberty from a long confinement. They in fact become *light-headed* whenever they leave the hammel or byre, so much so that they actually "frisk, dance and leap," and their antics would be highly amusing, were it not for the apprehension they may hurt themselves against some opposing object, as they seem to regard nothing before them. I remember seeing a dodded Angus stot lĕt out of a byre running so recklessly about that at length he came at full speed with his head against the wall of the steading, and was instantly felled to the ground. Before any one could run to his assistance he sprang upon his feet and made off again at full speed, holding his head high and tail on end, as if he felt proud of having done a feat which no one else could imitate. With distended nostrils and heaving flanks he appeared dreadfully excited; but on being put into his byre he soon calmed down. On being let out for the first time cattle should be put awhile into a large court, or on a road well fenced with inclosures, and guarded by men, to romp about. Two or three times of such liberty will make them quiet; and, in the mean time, to lighten their weight of carcass, they should get hay for a large proportion of their food. These precautions are absolutely necessary for cattle confined in byres, otherwise accidents may befall them on the road, where they will at once break loose. Even at home serious accidents sometimes overtake them, such as the breaking down of a horn, casting off a hoof, spraining a tendon, bruising ribs, and heating the whole body violently; and, of course, when any such ill luck befalls, the animal affected must be left behind, and become a drawback upon the value of the rest, unless kept on for some time longer.

(1303.) Having been prepared for the road, the *drover*—who may be your own shepherd, or a hired professional drover—takes the road very slowly for the first two days, not exceeding 7 or 8 miles a day. At night, in winter, they should be put into an open court and supplied with hay and water and a very few turnips; for if the turnips are suddenly withdrawn from them, their bellies will become what is termed *clinged*, that is, shrunk up into smaller dimensions—a state very much against a favorable appearance in a market. After the first two days they may proceed faster, say 12 or 13 miles a day if very fat, and 15 if moderately so. When the journey is long and the beasts get faint in travel, they should get corn to support them. In frosty weather, when the roads become very hard, they are apt to become *shoulder-shaken*, an effect of founder; and if sleet falls during the day, and becomes frozen upon them at night, they may become so chilled as to refuse food, and shrink rapidly away. I had a lot of 12 Angus oxen so affected on their road to Glasgow, when overtaken in an unexpected storm in May, that I could scarcely recognize them in the market. Cattle should, if possible, arrive the day before in the neighborhood of a distant market, and be supplied with a good feed of turnips and hay, or grass, to make them look fresh and fill them up again; but if the fair is only a short distance, they can travel to it early in the morning.

(1304.) In driving cattle the drover should have *no dog*, which will only annoy them. He should walk either before or behind, as he sees them disposed to proceed too fast or loiter on the road; and in passing carriages the leading ox, after a little experience, will make way for the rest. In other respects their management on the road is much the same as that of sheep, though the rate of traveling is quicker. Accommodation will be found at night at stated distances along the road. On putting oxen in a ferry-boat the shipping of the first one only is attended with much trouble. A man on each side should take hold of a horn, or of a halter made of any piece of rope, should the beast be hornless, and other two men, one on each side, should push him up behind with a piece of rope held between them as a breaching, and conduct him along the plank into the boat, which, if it have low gunwales, a man will require to remain beside him until one or two more of the cattle follow their companion, which they will most readily do. In neglecting this precaution in small ferry-boats, I have seen the first beast leap into the water, and then it was difficult to prevent some of the rest doing the same thing from the quay.

(1305.) Whatever time a lot of cattle may take to go to a market, they should never be *overdriven*. There is great difference in management in this respect among drovers. Some like to proceed on the road quietly, slowly, but surely, and to enter the market in a placid, cool state. Others, again, drive smartly along for some distance and rest to cool awhile, when the beasts will probably get chilled and have a staring coat when they enter the market; while others like to enter the market with their beasts in an excited state, imagining them then to look gay; but distended nostrils, loose bowels and reeking bodies, the ordinary consequences of excitement, are no recommendations to a purchaser. Good judges are shy of purchasing cattle in a heated state, because they do not know how long they may have been in it, and, to cover any risk, will give £1 a head below what they would have bid for them in a cool state. Some drovers have a habit of thumping at the hindmost beast of the lot with his stick while on the road. This is a reprehensible practice, as the flesh, where thumped, will bear a red mark after the animal has been slaughtered, the mark getting the appropriate name of *blood-burn*, and the flesh so affected will not take on salt, and is apt to putrefy. A touch upon the shank, or any tendonous part, when correction is *necessary*, is all that is required; but the voice, in most cases, will answer as well. The flesh of overdriven cattle, when slaughtered, never becomes properly firm, and their tallow has a soft, melted appearance.

(1306.) A few large oxen in a lot look best in a market on a position rather above the eye of the spectator. When a large lot is nearly alike in size and appearance, they look best and most *level* on a flat piece of ground. Very large fat oxen never look better than on ground on the same level with the spectator. An ox, to look well, should hold his head in a line with the body, with lively ears, clear eye, dewy nose, a well-licked hide, and stand firmly on the ground on all his feet. These are all symptoms of high health and good condition. Whenever you see an ox shifting his standing from one foot to another, he is *foot-sore*, and has been far driven. When you observe him hanging his head and his eyes watering he feels ill at ease inwardly. When his coat stares he has been overheated some time, and got a subsequent *chill*. All these latter symptoms will be much aggravated in cattle that have been fed in a byre. You may discover when a beast has been fed at the stake with the seal or baikie, by observing a fretted and callous mark on the *top* of the neck immediately behind the ears; by the hoofs being rather overgrown at the points; by marks of dung, or at least much resting, upon the outside of the hams; and very frequently by the remains of lice upon the tail-head and top of shoulder, their scurf remaining, or the hair shorn off altogether.

(1307.) In all *customs relating to markets* it is the same with cattle as with sheep (1167) And an ox puts on fat precisely in the same manner as a sheep (1169).

(1308.) In *judging* cattle the procedure is somewhat different from that of sheep, inasmuch as the hair of cattle not hiding their form so effectually as wool does that of sheep, the *eye* is more used than the hand; indeed, in the case of ripe fed cattle, the eye alone is consulted. The hand as well as the eye is brought into use in judging of *lean* cattle to lay on to grass or to fatten on turnips; and when we come to consider that matter in summer and autumn, I shall let you know the use of the hand in determining the qualities essential to a good *lean* beast. Meantime our business is with fat beasts; and although judging them by the eye is not a difficult thing in itself, it is rather difficult to describe in words. With the assistance, however, of the accompanying figures, I hope you will obtain some useful hints toward acquiring a knowledge of the art. When you look at the *near* side of a *ripe* ox in profile—and this is the side usually chosen to begin with—whatever be its *size*, imagine its body to be embraced within a rectangled parallelogram, as in fig. 273; and if the ox is filled up in all

Fig. 273.

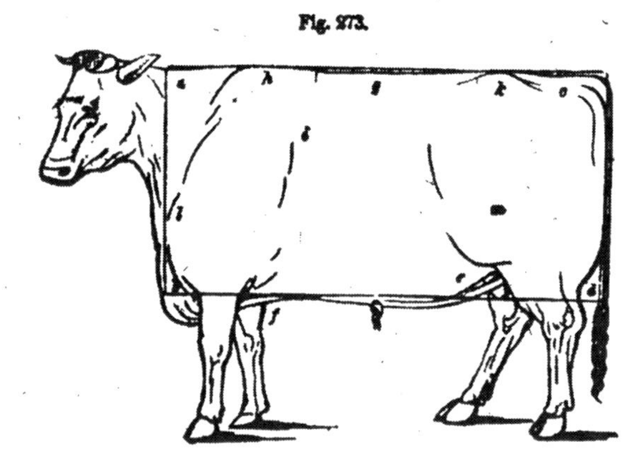

THE SIDE VIEW OF A WELL FILLED-UP FAT OX.

points, his carcass will occupy the parallelogram *a b c d* as fully as in the figure; but, in most cases, there will be deficiencies in various parts—not that *all* the deficiencies will occur in the same animal, but different ones in different animals. The flank *e*, for instance, may be shrunk up, and leave a space there to the line; or the brisket *f* may descend much farther down than is represented; or the rump *c* may be elevated much above the line of the back; or the middle of the back *g* may be much hollowed below the line; or the top of the shoulder *h* may be much elevated above it; or a large space may be left unfilled in the hams above *d*. Then a similar survey should be made behind the animal; the imaginary line should inscribe it also within it the perimeter of a rectangled parallelogram, though of different form from the other, as represented in fig. 274, where the breadth of the hook-bones, *a* and *b*, is maintained as low as the points *c* and *d*; and the *closing* between the legs at *e* is also well filled up. This figure gives a somewhat exaggerated view of the appearance of a fat ox behind; but still it gives the form of the *outline* which it should have. Then go in front of the ox, and there imagine the outline of the body at the shoulder, inscribed within a rectangled parallelogram *a b c d*, fig. 275, of exactly the same dimensions as the one in fig. 274. The shoulder, from *a* to *b*, is apparently of the same breadth as across the hook-bones, from *a* to *b*, fig. 274. The off-side of the animal may of course be expected to be similar in outline to the near side. Having thus obtained an idea of the outline which a fat ox should have, let us now attend to the filling up of the area of the parallelogram.

Fig. 274.

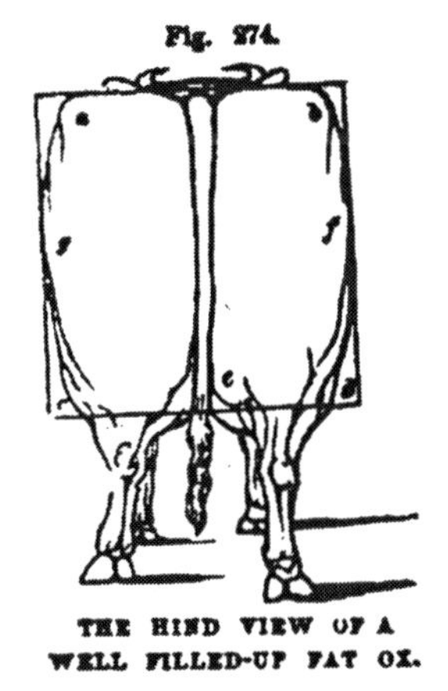

THE HIND VIEW OF A WELL FILLED-UP FAT OX.

Fig. 275.

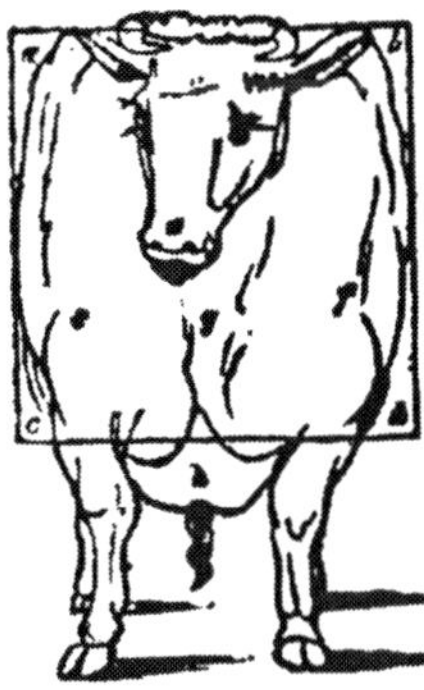

THE FRONT VIEW OF A WELL FILLED-UP FAT OX.

(1309.) On looking again at the near-side view, fig. 271, observe whether the ribs below and on each side of *g* are rounded, and nearly fill up the space between the more projecting points *h* and *k*, that is, between the shoulders and the hook. Observe also whether the shoulder *h* is flat, somewhat in the same plane as *g*, or more rounded and prominent; and whether the space behind the shoulder, at *i*, is hollow or filled up. Observe, again, whether the shoulder-point *l* is projecting and sharp, or rounded off; and whether the neck, between *a* and *l*, is flat and sunk, or sweeps finely in with the shoulder. Observe yet more, whether the muscles at *m* are thin and flat, or full and rounded; and whether the hook-bone *k* projects or sinks in, or appears to connect itself easily with the rump *c* on the one hand, and with the ribs *g* on the other. With all these alternative particulars before you, they should be arranged in the following manner, to constitute *points* in perfection.

(1310.) The line from the shoulder to the hook, from *h* to *k*, fig. 273, should be parallel to the back-bone. The space on each side of *g*, along the ribs from *g* to *h*, and along the loin from *g* to *k*, does not fall in with the line *h* and *k*, but should be a little nearer, and almost as high as the back-bone, with a rounding fall of the ribs down the side of the animal. The loin, from *k* to *g*, should be perfectly flat above, on the same level as the back-bone, and drop down on this side, in connection with the utmost rounding of the ribs. The point of the hook *k* should just be seen to project, and no more; and the space between it and the rump *c* should gradually sweep round to the narrower breadth of the pelvis, as seen from *a* to *c* or *b* to *c* in fig. 276. *i* is placed at the utmost bend of the ribs, along which a straight line should touch every point through *i*, from the front of the shoulder to the buttock. The triangular space comprehended within *a h l* should gradually taper from the shoulder-point to the head. A straight line from *l*, the shoulder-point, should touch every spot from it to *m*. The line of the back should be straight from *a* to *c*; the tail should drop perpendicularly from *c* to *d*; and the belly should sweep level, not high at *e* nor dropping at *f*. There are thus three straight lines along the side of a fat ox, from *a* to *c*, one through *i*, and from *l* to *m*. Proceeding behind the animal to fig. 274, the space between the hooks, from *a* to *b*, should be level, but a little rounded off at both ends, and the bone at the top of the tail only being allowed to project a *little* upward. The muscles on each side below the hooks, at *g* and *f*, when fuller than the hooks, is no *deformity* but should they be no fuller, they are right. The muscles at *c* and *d*, down the side of the hams, are allowed to sweep gradually toward the hock joints of the legs. The *closing* at *e* should be well filled up to furnish the rounds fully, but freely, for *packed* rounds prevent easy motion of the hind legs. Sometimes the tail is hid in a channel left by the muscles between *e* and *f*, but this is not usually the case. On going to the front view, fig. 275, the shoulder-top between *a* and *b* should be filled out with a natural round, and the muscles below it upon the shoulder-blades should always project farther than the breadth of the shoulder-top, and in this respect the fore-quarter differs from the hind, where the muscles below the hooks seldom project beyond them. The shoulder-points *e* and *f* should not be prominent, but round off with the muscles of the neck toward *g* where the round of the front of the neck falls from the head to the breast where the upper part of the brisket *h* meets it, and projecting a little in front, is rounded below and forms the lowest part of the body of a fat ox, and should be well filled out in breadth to spread the fore-legs asunder. The fore-legs are usually farther apart than the hind, but the hind at times, when the *shaw* or *cod* is large and fat, is as much and even more apart.

(1311.) The objectionable deviations from these points are as follows: In fig. 273, a hollow back at *g* is bad, showing weakness of the back-bone. A high shoulder-crest at *h* is always attended with a sharp thin shoulder, and has the effect of bringing the shoulder tops *a* and *b*, fig. 275, too close ogether. A long distance between *g* and *k* makes the loins hollow, and gives to a beast what is called a *washy* appearance, and is always attended with a liability to looseness in the bowels. This washiness is generally attended with an inordinate breadth of hooks, from *a* to *b*, fig. 274, and causes them to project much beyond the muscles below. A sharp projecting hook is always accompanied with flat ribs at *g*, fig. 273, and ribs when flat give the animal a hollow side, which bears little flesh, the viscera being thrown down into the cavity of the belly, which droops considerably below the line; but in the event of the muscles of the abdomen having a greater weight to bear, they become thicker and stronger, and, accordingly, the flesh there becomes less valuable, and it has also the effect of thinning away the thick flank *e*. Flatness of rib is also indicative of hollowness of the space behind the shoulder, so much so, indeed, that the animal seems as if it had been gripped in too firmly there. As the flesh is taken away from the shoulder-blade by a sharp shoulder and hollow ribs, so the shoulder-joint *l* projects the more, and causes a thinness of the neck between *a* and *l*. The rump-bone, at *c*, frequently rises upward, thereby spoiling the fine straight line of the back; and whenever this happens, the rump between *k* and *c* wants flesh and even becomes hollow, thereby much deteriorating the value of the hind-quarter. A projecting hook *k* also thins away the muscles about *m*, and behind it to the rounds; and this again is followed by an enlargement of the openings at the closing *e*, fig. 274. Whenever the shoulder becomes thin and narrow, when viewed in front, as in fig. 275, the shoulder-points *e* and *f* are wider than from *a* to *b*, and while this effect is produced above, the

Fig. 276.

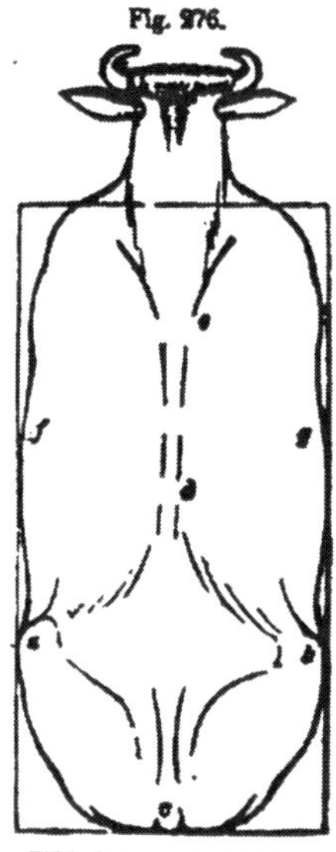

THE BACK VIEW OF A WELL-FILLED FAT OX.

brisket *h* below becomes less fat, and permits the fore-legs to stand nearer each other. A greatly commendatory point of a fat ox is a level broad back from rump to shoulder, because all the flesh seen from this position, as is endeavored to be represented by fig. 276, is of the most valuable description; where the triangular space included between *a*, *b*, *c*, is the rump the triangular space between *a*, *b*, *d*, the loin; and the space between *a* and *e*, deflecting on both sides toward *f* and *g*, the ribs, the value of all which parts are enhanced the more nearly they all are on a level with each other. All that I have endeavored to describe, in these paragraphs, of the points of a fat ox, can be judged of alone by the eye, and most judges never think of employing any other means; but the assistance derived from the hand is important, and in a beginner cannot be dispensed with.

(1312.) The first point usually *handled* is the end of the rump at the tail-head, at *c*, fig. 273, although any fat here is very obvious, and sometimes attains to an enormous size, amounting even to deformity. The hook-bone *k* gets a touch, and when well covered, is right; but should the bone be easily distinguished, the rump between *k* and *c* and the loin from *k* to *g* may be suspected, and, on handling these places, the probability is that they will both be hard, and deficient of flesh. To the hand, or rather to the points of the fingers of the right hand, when laid upon the ribs *g*, the flesh should feel soft and thick and the form be round when all is right, but if the ribs are flat the flesh will feel hard and thin, from want of fat. The skin, too, on a rounded rib, will feel soft and mobile, the hair deep and mossy, both indicative of a kindly disposition to lay on flesh. The hand then grasps the flank *e*, and finds it thick, when the existence of internal tallow is indicated. The cod is also fat and large, and on looking at it from behind seems to force the hind legs more asunder than they would naturally be. The palm of the hand laid along the line of the back from *c* to *h* will point out any objectionable hard piece on it, but if all is soft and pleasant, then the shoulder-top is good. A hollowness behind the shoulder at *i* is a very common occurrence; but when it is filled up with a layer of fat. the flesh of all the fore-quarter is thereby rendered very much more valuable. You would scarcely believe that such a difference could exist in the flesh between a lean and a fat shoulder. A high narrow shoulder is frequently attended with a ridged back-bone, and low-set narrow hooks, a form which gets the appropriate name of *razor-back*, with which will always be found a deficiency of flesh in all the upper part of the animal, where the best flesh always is. If the shoulder-point *l* is covered, and feels soft like the point of the hook-bone, it is good, and indicates a well-filled neck-vein, which runs from that point to the side of the head. The shoulder-point, however, is often bare and prominent. When the neck-vein is so firmly filled up as not to permit the points of the fingers into the inside of the shoulder-point, indicates a well tallowed animal; as also does the filling-up between the brisket and inside of the fore-legs, as well as a full, projecting, well-covered brisket in front. When the flesh comes down heavy upon the thighs, making a sort of double thigh, somewhat like the shape at *d* and *e*, fig. 274, it is called *lyary*, and indicates a tendency of the flesh to grow on the lower instead of the upper part of the body. These are all the *points* that require *touching when the hand is used;* and in a high-conditioned ox, they may be gone over very rapidly.

(1313.) Cattle are made to fast before being slaughtered, as well as sheep. The time they should stand depends on the state of the animal on its arrival at the shambles. If it has been driven a considerable distance in a proper manner, the bowels will be in a tolerably empty state, so that

12 hours may suffice, but if full and just off its food, 24 hours will be required. Beasts that have been overdriven, or much struck with sticks, or are in any degree infuriated—or *raised*, as it is termed—should not be immediately slaughtered, but allowed to stand on dry food, such as hay, until the symptoms disappear. While such precautions are certainly necessary to preserve meat in the best state, we can scarcely credit the loss there must be incurred every year in Smithfield market in London, by the injuries sustained by the animals being driven through the streets of an immensely peopled metropolis. The state of many of the animals in the market on a Monday morning is truly pitiable. "The loss to the grazier," says a writer who advocates the removal of the market to the suburbs, "is in the difference in value of his sheep or cattle, when they arrive in the neighborhood of the metropolis, and when offered for sale in Smithfield after intense suffering from hard blows, driving over the stones, from hunger, thirst, fright, and the compressed state in which they are constrained to be packed; the sheep and beasts the whole time, from their raised temperature, clouding the atmosphere of Smithfield with dense exhalations from their bodies. The London butcher, carrying on a respectable trade, will at all times, when he enters the market, reject such cattle or sheep as are what is termed in a *mess;* that is, depressed, after excitation by being overlaid or overdriven, or such as have been more than usually troublesome in getting into the market, and, consequently, will be in a more worried and exhausted condition. It is to be observed that all animals brought into Smithfield, especially on the Monday's market-day, are more or less in the condition above described." He goes on to state that "a calculation has been made, that 512,000 serious and extensive œdematous bruises are, in the course of one year, discovered on cattle after they are slaughtered. The pain these bruises must occasion to the cattle, and the loss to the butcher or the public, is exclusive of those parts of the animal which suffer most from the conduct of the drovers, namely, the head, especially the nasal organs, and concussions of the brain by blows on the horns, besides the more acute suffering from blows on the hocks." The beef consumed in London, in 1836, amounted to 9½ millions of stones, which, at 6s. a stone, gives a total value of £2,850,000; and if its deterioration is taken only at half a farthing per pound, the annual loss sustained by the bruises of cattle alone will amount to £69,270 16s.*

(1314.) Cattle are slaughtered in a different manner from sheep, and the mode differs in different countries. In the great abbatoirs at Montmartre, in Paris, they are slaughtered by bisecting the spinal cord of the cervical vertebræ; and this is accomplished by the driving of a sharp-pointed chisel between the second and third vertebra, with a smart stroke of a mallet, while the animal is standing, when it drops, and death or insensibility instantly ensues, and the blood is let out immediately by opening the blood-vessels of the neck. The plan pursued in this country is, first to bring the ox down on his knees, and place his under-jaw upon the ground, by means of ropes fastened to his head and passed through an iron ring in the floor of the slaughter-house. He is then stunned with a few blows from an iron ax made for the purpose, on the forehead, the bone of which is usually driven into the brain. The animal then falls on its side, and the blood let out by the neck. Of the two modes the French is apparently less cruel; for some oxen require many blows to make them fall; I once witnessed an ox receive nine blows before he fell. I have heard it alleged by butchers that the separation of the spinal cord pro-

* The Question of the Smithfield Market Fully Considered.

ducing a general nervous convulsion throughout the body, prevents the blood flowing so rapidly and entirely out of it as when the ox is stunned in the forehead. The skin is then taken off to the knees, when the legs are disjointed, and also off the head. The carcass is then hung up by the tendons of the hough, on a stretcher, by a block and tackle, worked by a small winch, which keeps good what rope it winds up by a wheel and rachet.

(1315.) Every farm on which sheep are killed should have a proper slaughter-house, such as is seen at fig. 1, Plate I. at *y*, the ground-plan of which may be seen in fig. 2, plate II., where it is represented to be 10 feet by 9 feet, a space too small when cattle are slaughtered, but it is very rare that an ox is slaughtered on a farm, except when one is so that is likely to be lost by some acute disease, in which contingency the animal is slaughtered in the straw-barn and hung up by the baluks; but it is quite easy, when the slaughter-house is fitting up, to have a block and tackle and small winch, erected at a convenient corner, to be used on such occasions. The floor should be laid with clean-droved pavement, and have a decided slope to the side at which the drain is made to take away the dirty water occasioned by cleaning, which it should always be thoroughly. The walls should also be plastered, and a ventilator placed on the roof, to maintain a draft of air. The site chosen for the slaughter-house should be in a cool place, away from the sun's influence in summer, for in a heated apartment meat never becomes firm. A locked closet is useful to hold the knives, steel, and stretchers, and the outer door provided with a good thumb-latch and lock and key; the key, of course, always to remain in the farm-house until needed.

(1316.) It is the shepherd's duty to act the part of butcher on a farm. He should learn to slaughter gently, dress the carcass neatly and cleanly, in as plain a manner as possible, without *flourishes*, as the figures incised on the membraneous skin are called, and separate the valuable from the worthless portion of the entrails, keeping the loose fat by itself, and hang up the skin in a suitable place to dry. It is his duty also to keep the slaughter-house neat and clean.

(1317.) After the carcass has hung 24 hours, it should be cut down by the back-bone, or chine, into two *sides*. This is done either with the saw or chopper; the saw making the neatest job in the hands of an inexperienced butcher, though it is the most laborious, and with the chopper is the quickest, but by no means the neatest plan, especially in the hands of a careless fellow. In London the chine is equally divided between both sides, while in Scotland one side of a carcass of beef has a great deal more bone than the other, all the spinous processes of the vertebra being left on it. The bony is called the *lying* side of the meat. In London the divided processes in the fore-quarter are broken in the middle when warm, and chopped back with the flat side of the chopper, and this has the effect of thickening the fore and middle ribs considerably when cut up. The London butcher also cuts the joint above the hind knee, and, by making some incisions with a sharp knife, cuts the tendons there, and drops the flesh of the hind-quarter on the flanks and loins, which causes it to cut up thicker than in the Scotch mode. In opening the hind-quarter he also cuts the aitch-bone or pelvis through the center, which makes the rump look better. Some butchers in the north country score the fat of the *closing* of the hind-quarter, which has the effect of making that part of both heifer and ox look like the udder of an old cow. There is far too much of this scoring practiced in Scotland, and ought to be abandoned, and let the pieces have more their natural appearance.

(1318.) In cutting up a carcass of beef the London butcher displays great expertness; he not only discriminates between the qualities of its different parts, but can cut out any piece to gratify the taste of his customers. In this way he makes the best use of the carcass, realizes the largest value for it, while he gratifies the taste of every grade of customers. A figure of the Scotch and English modes of cutting up a carcass of beef will at once show you their difference, and on being informed where the valuable pieces lie, you will be enabled to judge whether the oxen you are breeding or feeding possess the properties that will enable you to demand the highest price for them.

(1319.) The Scotch mode of cutting up a carcass of beef is represented

Fig. 277.

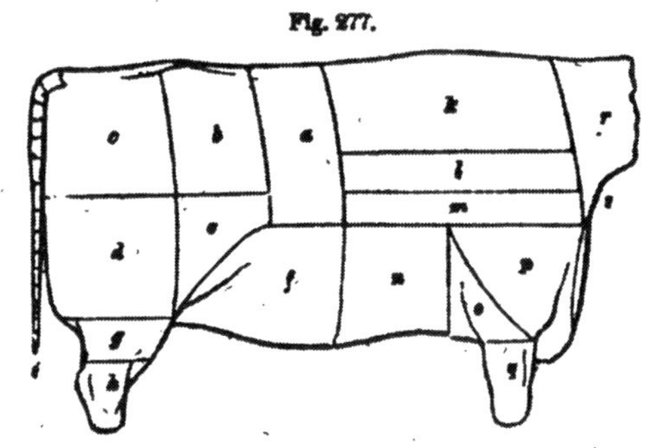

THE SCOTCH MODE OF CUTTING UP A CARCASS OF BEEF.

in fig. 277, and these are the names of the different pieces of meat:

In the hind-quarter.	*In the fore-quarter*
a, The sirloin, or back sey.	*k*, The spare rib or fore sey.
b, .. hook-bone.	*l*, .. runner. } large and small.
c, .. buttock, } the rump.	*m*, .. runner, } large and small.
d, .. large round } the rump.	*n*, .. nineholes.
e, .. thick flank.	*o*, .. brisket.
f, .. thin flank.	*p*, .. shoulder-lyar.
g, .. small round.	*q*, .. nap or shin.
h, .. hough.	*r*, .. neck.
i, .. tail.	*s*, .. sticking piece.

a the surloin is the principal roasting-piece, making a very handsome dish, and is a universal favorite. It consists of two portions, the Scotch and English sides, the former is the one above the lumbar bones, and is some what hard in ill-fed oxen; the latter consists of the muscles under these bones, and are generally covered with fine fat, and are exceedingly tender. The better the beast is fed the larger is the under muscle, better covered with fat, and more tender to eat; *b* the hook-bone and *c* the buttock are cut up for steaks, beef-steak pie, or minced collops, and both these, along with the sirloin, fetch the highest price; *d* is the large round, and *e* the small round, both well known as excellent pieces for salting and boiling, and are eaten cold with great relish; *h* the hough is peculiarly suited for boiling down for soup, having a large proportion of gelatinous matter. Brown soup is the principal dish made of the hough, but its decoction forms an excellent *stock* for various dishes, and will keep in a state of jelly for a considerable time. The synovial fat, skimmed off in boiling this piece, and poured upon oatmeal, seasoned with pepper and salt, constitutes the famous *fat brose* for which Scotland has long been celebrated. In the making, this brose should not be much stirred, and the oatmeal left among the fat gravy in small, dry lumps. It was of this piece that the old favorite soup of Scotland, called *skink*, was made. *e* is the thick and *f* the thin flank both excellent pieces for salting and boiling; *i* is the tail, and insignificant as it may seem, it makes a soup of very fine flavor. Hotel-keepers have a trick of seasoning brown soup, or rather beef-tea, with a few joints of tail, and passing it off for genuine ox-tail soup. These are all the pieces which constitute the hind-quarter, and it will be seen that they are valuable both for roasting and boiling, not containing a single coarse piece. In the fore-quarter is *k*, the spare rib or fore-sey, the six ribs of the back end of which make an excellent roast, and when taken from the side opposite to the *lying* one, being free of the bones of the spine, makes a large one; and it also makes excellent beef-steaks and beef-steak pie. *l* and *m* are the two runners, and *n* the nine-holes, make salting and boiling pieces; but of these, the nine-holes is much the best, as it consists of layers of fat and lean without any bone; whereas the fore parts of the runners have a piece of the shoulder-blade in them, and every piece connected with that bone is more or less coarse-grained; *o* the brisket eats very well boiled fresh in broth, and may be corned and eaten with boiled greens or carrots; *p* the shoulder-lyar is a coarse piece, and fit only for boiling fresh to make into broth or beef-tea; *q* the nap or shin, is analogous to the hough of the hind-leg, but not so rich and fine, there being much less gelatine in it; *r* the neck makes good broth, and the sticking-piece *s* is a great favorite with some epicures, on account of the pieces of rich fat in it. It makes an excellent stew, as also sweet barley broth, and the meat eats well when boiled in it. These are all the pieces of the fore-quarter, and it will be seen that they consist chiefly of boiling-pieces, and some of them none of the finest, the roasting-piece being confined to the six ribs of the spare-rib *k*, and the finest boiling-piece, corned, only to be found in the nine-holes *n*.

(1320.) In some of the largest towns of Scotland a difference of 1d. per lb. may be made between the roasting and boiling pieces, but in most towns, and in the country villages, all the pieces realize the same prices, and even the houghs and shins fetch 3d. per lb.

(1321.) In the English mode the pieces are cut up somewhat differently,

Fig. 278.

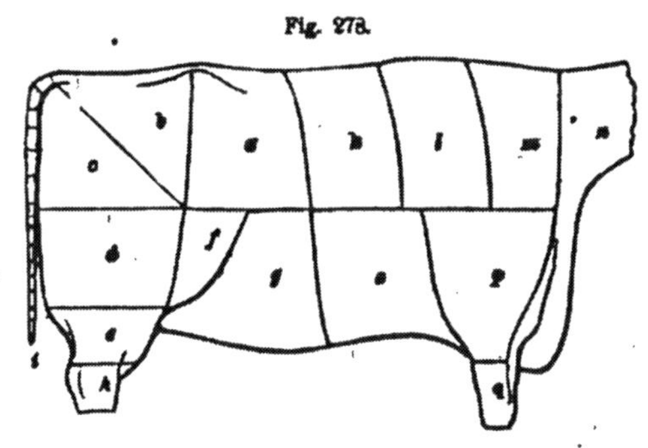

THE ENGLISH MODE OF CUTTING UP A CARCASS OF BEEF

especially in the fore-quarter. Fig. 278 shows this mode, and it consists of the following pieces:

In the hind-quarter.

a, The loin.
b, .. rump.
c, .. aitch-bone.
d, .. buttock.
e, .. hock.
f .. thick flank
g .. thin flank.
h, .. shin.
i, .. tail.

In the fore-quarter

k, The fore rib.
l, .. middle rib.
m, .. chuck rib.
n, .. clod. and sticking, and neck.
o, .. brisket.
p, .. leg-of-mutton piece.
q, .. shin.

1071)

a, the loin, is the principal roasting piece; *b*, the rump, is the favorite steak-piece; *c*, the aitch-bone, the favorite stew; *d*, the buttock, *f*, the thick flank, and *g*, the thin flank, are all excellent boiling pieces when corned; *e*, the hock, and *h*, the shin, make soup, and afford *stock* for various purposes in the culinary art; and *i* is the tail for ox-tail soup—a favorite English luncheon. In the curious case of assessing damages against the Bank of England for removing the famous Cock eating-house in Threadneedle-street, it was produced in evidence, that, in the 3 years 1837–8–9, there had been 13,359 ox-tails used for soup; and as 36 tails make 10 gallons of soup, there had been served up 59,369 basins, at 11d. the basin, making the large amount of £2,720 13s. 4d. for this article alone.* These are all the pieces in the hind-quarter, and it will be seen they are valuable of their respective kinds. In the fore-quarter, *k*, the fore rib, *l*, middle rib, and *m*, chuck rib, are all roasting pieces, not alike good; but in removing the part of the shoulder-blade in the middle rib, the spare-ribs below make a good broil or roast; *n*, the neck, makes soup, being used fresh, boiled, and the back end of the brisket *o* is boiled corned, or stewed; *p*, the leg-of-mutton piece, is coarse, but is as frequently stewed as boiled; *q*, the shin, is put to the same uses as the shin and hock of the hind-quarter. On comparing the two modes of cutting up, it will be observed that in the English there are more roasting pieces than in the Scotch, a large proportion of the fore-quarter being used in that way. The plan, too, of cutting the line between *b* and *c*, the rump and aitch-bone in the hind-quarter, lays open the steak-pieces to better advantage than in the Scotch buttock, *c*, fig. 277. Extending the comparison from one part of the carcass to the other, in both methods, it will be seen that the most valuable pieces—the roasting—occupy its upper, and the less valuable—the boiling—its lower part. Every beast, therefore, that lays on beef more upon the upper parts of its body is more valuable than one that lays the same quantity of flesh on its lower parts.

(1322.) The relative values of the pieces differ much more in London than in Scotland. The rumps, loin and fore ribs fetch the highest price; then come the thick flank, buttock and middle rib; then the aitch-bone, thin flank, chuck-rib, brisket and leg-of-mutton piece; then the clod, sticking and neck; and last of all the legs and shins. In actual pecuniary value, the last may bear a proportion of only one-fourth of the highest price.

(1323.) Of the qualities of beef obtained from different breeds of cattle, I believe there is no better meat than from the West Highland breed for fineness of grain, and cutting up into convenient pieces for family use. The Galloways and Angus, when fattened in the English pastures, are great favorites in the London market. The Short-Horns afford excellen steaks, being thick of flesh, and the slice deep, large and juicy, and their corned flanks and nineholes are always thick, juicy and well mixed. The Herefords are somewhat similar to the Short-Horns, and the Devons may perhaps be classed among the Galloways and Angus, while the Welsh cannot be compared to the West Highland. So that, taking the breeds of Scotland as suppliers of good beef, they seem to be more valuable for the table than those of England. Any beef that I have seen of Irish beasts is inferior, but the cattle derived from Britain, fed on the pastures of Ireland, afford excellent meat. Shetland beef is the finest grained of all, but the pieces are very small.

(1324.) In regard to the proportion of beef and tallow generally obtained from cattle, Dr. Cleland states that of 14,566 head of cattle sold in the Glasgow market in 1822, averaging exactly

* John Bull, 18th January, 1841.

44 stones imperial, each yielded 5½ stones, which is exactly ⅛ of the weight of beef.* From 4 of the oxen experimented on by Mr. Stephenson, and which were slaughtered at the same time, these results were obtained:

Cattle.	Live weight.	Beef.		Tallow.		Hide.		Other offals.	
Nos.	Sts.	Sts.	lbs.	Sts.	lbs.	Sts.	lbs.	Sts.	lbs.
1.	112	66	2	8	10	5	11	32	5
2.	100	58	6	7	7	4	12	29	3
3	108	62	3	9	0	4	12	26	13
4.	109	62	4	9	4	5	12	31	8

The proportion of tallow to beef is here nearer 1-7 than ⅛ over the whole beasts; and there is another result worth attending to, which is that of Nos. 2 and 3, which had the same weight of hide, namely, 4 sts. 12 lbs.; No. 3 must have had the finest skin and touch, for its superiority in every respect is apparent, both in weight of tallow, which was 9 sts. to 7 sts. 7 lbs.; in weight of beef, which was as 62 sts. 3 lbs. to 58 sts. 6 lbs.; and in lightness of other offals, which were as 26 sts. 13 lbs. is to 29 sts. 3 lbs. Besides difference in quality of the same weight of hide, a lighter hide, under similar circumstances, will produce the same results as those above. For example: of two of the heifers fed by Mr. Howden on raw and steamed food, which afforded the same live weight when put up to feed, namely, 1022 lbs., and the same live weight after the experiment was concluded, namely, 1176 lbs., one had a light hide, that is, 52 lbs. weight, and the other a heavier one, 60 lbs., and the light one was accompanied with 620 lbs. of beef and 96 lbs. of tallow, while the heavier hide was associated with only 572 lbs. of beef and 80 lbs. of tallow. All these proportions very nearly indicate the tallow at 1-7 of the beef; but sometimes the proportion of tallow is very much greater and much less. I had a young Short-Horn cow which slipped her second calf, and was fed when she became dry, which was in November, and in May following was sold to the late Mr. Robert Small, flesher, Dundee, when she yielded 72½ stones of beef and 27 stones of tallow, being in the ratio of 2¾ to 1. On the other hand, Lord Kintore's large ox that was exhibited at the Highland and Agricultural Society's Show at Aberdeen in 1834 only yielded 16 sts. 7 lbs. of tallow, from 173 sts. 4 lbs of beef; being in the ratio of 10½ to 1.† There are, perhaps, not sufficient data in existence to determine the true proportion of offal of all kinds to the beef of any given fat ox; but approximations have been made which may serve the purpose until the matter is investigated by direct experiment, under various circumstances. The dead weight bears to the live weight a ratio varying between .571 and .605 to 1; and on applying one or other multiple to the cases of the live weight, you will find a pretty correct approximation. The tallow is supposed to be 8-100 of the live weight, so that the multiple is the decimal of .08. The hide is supposed to be 5-100 of the live weight, so, to obtain its weight, a multiple of .05 should be used. The other offals are supposed to bear a ratio of about ¼ of the live weight, so that the multiple of .28 is as near as can be proposed under existing experience.‡

(1325.) *Beef* is the staple animal food of this country, and it is used in various states—fresh, salted, smoked, roasted, and boiled. When intended to be eaten fresh, "the *ribs* will keep the best, and with care will keep 5 or 6 days in summer, and in winter 10 days. The middle of the *loin* is the next best, and the *rump* the next. The *round* will not keep long, unless salted. The *brisket* is the worst, and will not keep longer than 3 days in summer, and a week in winter."‖ In cooking, a piece of beef, consisting of four of the largest ribs, and weighing 11 lbs. 1 oz., was subjected to roasting by Mr. Donovan, and it lost during the process 2 lbs. 6 oz., of which 10 oz. were fat, and 1 lb. 12 oz. water dissipated by evaporation. On dissection, the bone weighed 16 oz., so that the weight of meat fit for the table was only 7 lbs. 11 oz. out of 11 lbs. 1 oz. It appears that when the butchers' price of ribs is 8½d. per lb., the cost of the meat when duly roasted is 11½d. per lb., and the average loss arising from liquefaction of fat and evaporation of water is 18 per cent. With sirloins, at the price of 8½d. per lb., the meat cost, when roasted, 1s. 1 1-6d. per lb., at a loss of 20½ per cent. A loss of 18 per cent. was also sustained on boiling salted briskets; and on salted flanks at 6d. per lb. the meat cost 7½d. per lb., at a loss of 13 1-5 per cent.§ In regard to the power of the stomach to digest beef, that which is eaten boiled with salt only is digested in 2 hours 45 minutes. Beef, fresh, lean, and rarely roasted, and a beef-steak broiled, takes 3 hours to digest: that fresh and dry-roasted, and boiled, eaten with mustard, is digested in 3 hours and 30 minutes. Lean fresh beef fried takes 4 hours to digest, and old hard salted beef boiled does not digest in less than 4 hours 15 minutes. Fresh beef-suet boiled takes 5 hours 30 minutes to digest.¶

(1326.) The usual mode of preserving beef is by *salting*; and when intended to keep a long time, such as for the use of shipping, it is always salted with brine: but for family use it should be salted dry with good Liverpool salt, without saltpetre; for brine dispels the juice of the meat, and saltpetre only serves to make the meat dry, and give it a disagreeable and unnatural red color. Various experiments have been made to cure beef, with salt otherwise than by hand-rubbing, and in a short space of time; and also to preserve it from putrefaction by other means than salt. Those of Messrs. Payne and Ellmore, of London, consist in putting meat in a copper which is rendered air-tight, and an air-pump then creates a vacuum within it, thereby extracting all the air out of the meat; then brine is pumped in by pressure, which entering into every pore of the meat formerly occupied by the air, is said to place it in a state of preservation in a few minutes. M. Gannal, of France, preserved the carcass of an ox from putrefaction for 2 years by injecting

* Cleland's Account of the Highland and Agricultural Show at Glasgow in 1828

† Quarterly Journal of Agriculture, vol. vii.

‡ Prize Essays of the Highland and Agricultural Society, vol. xii.

‖ The Experienced Butcher. § Donovan's Domestic Economy vol. ii.

¶ Combe on Digestion and Dietetics.

4 lbs. of saline mixture into the carotid artery. Whether any such contrivance can be made available for family purposes, seems doubtful. Up to the 10th October, 1842, salted beef imported was subject to a duty of 12s. per cwt.; since that date both salted and fresh pay a duty of 8s. per cwt. from foreign countries, and 2s. from British possessions. Up to that period the importation of live stock was prohibited, by the same act; bulls and oxen can now be imported from foreign countries at £1 each, and from British possessions, 10s.; cows, 15s. and 7s. 6d.; calves, 10s. and 5s.; sheep, 3s. and 1s. 6d., and lambs, 2s. and 1s. each.* Salted beef cured with wood smoke is converted into a ham, and very highly relished. The tierce of salted beef for the navy contains 300 lbs., consisting of 38 pieces of 8 lbs. each.

(1327.) Cattle are useful to man in various other ways than affording food from their flesh, their offals of tallow, hides, and horns, forming extensive articles of commerce. Of the *hide*, the characteristics of a good one for strong purposes are strength in its middle or *butt*, as it is called, and light on the edges or *offal*. A bad hide is the opposite of this, thick in the edges and thin in the middle. A good hide has a firm texture, a bad one loose and soft. A hide improves as the summer advances, and it continues to improve after the new coat of hair in autumn until November or December, when the coat gets rough from the coldness of the season, and the hide is then in its best state. It is surprising how a hide improves in thickness after the cold weather has set in. The sort of food does not seem to affect the quality of the hide; but the better it is, and the better cattle have been fed, and the longer they have been well fed, even from a calf, the better the hide. From what has been said of the effect of weather upon the hide, it seems a natural conclusion that a hide is better from an ox that has been fed in the open air than from one fed in a byre. Dirt adhering to a hide injures it, particularly in byre-fed animals; and anything that punctures a hide, such as *warbles* arising from certain insects, is also injurious. The best hides are obtained from the West Highland breed of cattle. The Short-Horns produce the thinnest hides, the Aberdeenshire the next, and then the Angus. Of the same breed, the ox affords the strongest hide; but as hides are applied to various uses, the cows, provided it be large, may be as valuable as an ox's. The bull's hide is the least valuable. Hides are imported from Russia and South America; and the number imported in 1838 was 301,890. The duty on hides, by the new Tariff, is 6d. per cwt. for dried and 3d. for wet.

(1328.) Hides, when deprived of their hair, are converted into leather by infusion of the astringent property of bark. The old plan of tanning used to occupy a long time; but such was the value of the process, that the old tanners used to pride themselves in producing a substantial article. More recent discoveries have prompted tanners to hasten the process, much to the injury of the article produced. Strong infusions of bark make leather brittle; 100 lbs. of skin, quickly tanned in a strong infusion, produce 137 lbs. of leather; while a weak infusion produces only 117½, the additional 19½ lbs. serving only to deteriorate the leather, and cause it to contain much less textile animal solid. Leather thus highly charged with tannin is so spongy as to allow moisture to pass readily through its pores, to the great discomfort and danger of persons who wear shoes made of it. The proper mode of tanning lasts a year or a year and half, according to the quality of the leather wanted, and the nature of the hides. A perfect leather is recognized by its section, which should have a glistening marbled appearance, without any white streaks in the middle.† Leather is applied to many important purposes, being made into harness for agricultural and other uses. It is used to line the powder magazines of ships-of-war; to make carding-machines for cotton and other mills; belts to drive machinery; to make soles of shoes; and, when japanned, to cover carriages. Calves' leather is used in bookbinding. The duty on tanned hides is now fixed at £10 per £100 value. The hair taken off hides in tanning is employed to mix with plaster, and is surreptitiously put into a hair-mattresses. The duty imposed on foreign cattle hair is 6d. per cwt.

(1329.) "The principal substances of which *glue* is made," says Dr. Ure, "are the paring of ox and other thick hides, which form the strongest article; the refuse of the leather-dresser; both afford from 45 to 55 per cent. of glue. The tendons, and many other offals of slaughter-houses, also afford materials, though of an inferior quality, for the purpose. The refuse of tanneries—such as the ears of oxen, calves, sheep, &c.—are better articles; but parings of parchment, old gloves, and in fact animal skins in any form, uncombined with tanning, may be made into glue."‡

(1330.) *Ox-tallow* is of great importance in the arts. Candles and soap are made of it, and it enters largely into the dressing of leather and the use of machinery. Large quantities are annually imported from Russia. Of the exports from St. Petersburgh, consisting of 4½ millions of poods, at least 3½ millions are exported to this country, at the value of £2,306,150. at £35 per ton.‖ Of the quantity imported in 1837, 1,294,000 cwts. were retained for home consumption. Ox-tallow consists of 76 parts of stearine and 24 of oleine out of the 100. The duty on tallow by the new Tariff is 3s. 2d. per cwt. from foreign countries, and 3d. per cwt. from the Colonies.

(1331.) The *horns* of oxen and sheep are used for many purposes. "The horn consists of two parts: an outward horny case, and an inward conical-shaped substance, somewhat intermediate between indurated hair and bone," called the *flint* of the horn. "These two parts are separated by means of a blow on a block of wood. The horny exterior is then cut into three portions by means of a frame-saw. The lowest of these, next the root of the horn, after undergoing several processes by which it is rendered flat, is made into combs. The middle of the horn, after being flattened by heat and its transparency improved by oil, is split into thin layers, and forms a substitute for glass in lanterns of the commonest kind. The tip of the horn is used by the makers of knife-handles and of the tops of whips, and for other similar purposes. The interior, or core of the horn, is boiled down in water. A large quantity of fat rises to the surface; this is put aside and sold to the makers of yellow soap. The liquid itself is used as a kind of glue, and is purchased by the cloth-dresser for stiffening. The bony substance which remains behind is then sen

* The Act to amend the Laws relating to the Customs, Table A.

† Ure's Dictionary of the Arts, art. *Leather Tanning*.

‡ Ibid., art. *Glue*.

‖ McCulloch's Dictionary of Commerce, art. *Tallow*.

to the mill, and, being ground down, is sold to the farmers for manure. Besides these various purposes to which the different parts of the horn are applied, the clippings which arise in comb-making are sold to the farmers at about 1s. per bushel. The shavings which form the refuse of the lantern-makers are sold as manure."* Horn, as is well known, is easily rendered soft and pliant in warm water; and by this, and the property of adhesion like glue, large plates of horn can be made by cementing together the edges of small pieces rendered flat by a peculiar process, as a substitute for glass. For this purpose, the horns of goats and sheep are preferred, being whiter and more transparent than those of any other animal. Imitation of tortoise-shell can be given to horn by the use of various metallic solutions. Horn, also, when softened, can be imprinted with any pattern by means of dies.† The duty on horns is 1s. per ton, and on hoofs £1 per £100 value.

---

34. TREATMENT OF FARM-HORSES IN WINTER.

> "But loose betimes, and through the shallow pond
> Drive the tired team, and bed them snug and warm;
> And with no stinting hand their toil reward."
>
> GRAHAM.

(1332.) With the exception of a few weeks in summer, farm-horses occupy their stable all the year round. It is situate at O, fig. 3, Plate III., where it is seen with two doors and two windows in front, and surmounted with two ventilators on the roof. Its plan may be seen at O, fig. 4, Plate IV., where it is represented as containing 12 stalls. The fitting-up of the stable in all its particulars of stalls, floor, and accommodation, having already been fully dilated on when treating of the steading, from (23) to (34), more seems unnecessary to be said in regard to these particulars in this place.

(1333.) Farm-horses are under the immediate charge of the plowmen, one of whom works a pair, and keeps possession of them generally during the whole period of his engagement. This is a favorable arrangement for the horses, as they work much more steadily under the guidance of the same driver, than when changed into the hands of different; and it is also better for the plowman himself, as he will perform his work much more satisfactorily to himself, as well as his employer, with horses familiarized to him than strange ones. In fact, the man and his horses must become acquainted before they can understand each other; and when the peculiarities of each party are mutually understood, work becomes more easy, and of course greater attention can be bestowed upon it. Some horses show great attachment to their driver, and will do whatever he desires without hesitation; others show no particular regard; and the same difference may be remarked of plowmen toward their horses. Upon the whole, however, there seems to be a very good understanding in this country between the plowman and his horses; and, indeed, independently of this, I believe there are few masters disposed to allow their horses to be ill treated, because there is no occasion for it; for horses which have been brought up upon a farm, in going through the same routine of work every year, become so well acquainted with what they have to do in every department of work, that should a misunderstanding arise between them and their driver, you may safely conclude that the driver is in the wrong.

(1334.) The treatment which farm-horses usually receive in winter is this: The plowmen, when single, get up and breakfast before daybreak, and by that time go to the stable, where the first thing they do is to take

* Babbage on the Economy of Machinery and Manufactures. † Ure's Dictionary of the Arts, art. *Horn*.

out the horses to the water.* While the horses are out of the stable the rest of the men take the opportunity of cleansing away the dung and soiled litter made during the night, into the adjoining court-yard K, fig. 3, Plate III., with their shovels (fig. 149), wheelbarrow and besoms. While the horses are absent usually one of the plowmen supplies each corn-box with corn from the corn-chest. It is not an unusual practice to put the harness on while the horses are engaged with the corn; but this should by no means be allowed. Let the horses enjoy their food in peace, as many of them, from sanguine temperament or greed, cannot divest themselves of the feeling that they are about to be taken away from their corn, if worked about during the time of feeding. The harness can be quickly enough put on after the feed is eaten, as well as the horses curried and brushed and the mane and tail combed. A very common practice, however, is to dress the horses while eating, which should not be allowed. A better plan in all respects is to let the horses eat their corn undisturbed, and then dress and harness them afterward, and it has the advantage of allowing them a little time between eating their corn and going out to work, which, if of a violent nature, undertaken with a full stomach, may bring on an attack of *batts* or colic.† The plan which I have just described is intended to apply to single men who live together, and who have their own victuals to cook. But should the plowmen be married men the best arrangement for them is to go to the stable when they rise, water the horses, clean the stable, corn the horses, bind them up, and, shutting the door, leave them in quiet to eat their food as long as they themselves are in taking their breakfast, which by that time should be made ready by their wives. On returning to the stable after breakfast the horses should then be dressed, combed and harnessed, when they will come out quite fresh and clean to go to yoke and after their feed has been a little time in their stomachs.

[* It can nowhere, we trust, be unseasonable to exhort the farmer to give attention, habitually, to the *treatment* of all domestic animals, and more especially to those to whose labor he owes the fruits and profits of his industry. Common humanity, to say nothing of Christian benevolence, demands, and the more especially as all power of complaint is denied them, that every animal that works should be not only *suitably and well fed*, but that it should be saved from suffering by thirst; that its skin be kept clean and in good condition, which a few minutes' use of the curry-comb and brush will effect, and that its eyes and feet should be often examined and protected from injury; while a sore back or gall on any part of the body should be deemed a disgrace to master as well as man. Trite and commonplace as this word in their favor may appear, we cannot withhold it, in view of so much neglect in working animals as we see committed on many farms that fall under our observation; for whether we speak of it or not, we never visit or pass a farm without noticing these things. Those whose benevolence does not move them thus to care for the feelings of the humblest beast in their employment should yet be prompted to look after their comfort, as well in the stable as the field—whether in or out of harness—by the consideration that a sufficiency of substantial food, personal care and kind treatment always beget proportionate willingness and efficiency—as much or even more with the brute than with the human. For ourselves, we take the occasion to reiterate our persuasion that the almost universal preference for the horse over the ox and mule, as a beast of labor and burden, has its origin, in a great measure, in—[alas! not *bygone*] ages of war and barbarism, which made power and fleetness identical; and that with a vast saving of national wealth these more economical animals might and ought to be substituted for the short-lived and perishable horse. *Ed. Farm. Lib.*]

[† We well remember, when residing in Baltimore, to have been within an ace of killing one of the best geldings we ever straddled, by giving him a brisk eight-mile-an-hour canter immediately after he had swallowed his oats. As soon as he took the spur we knew there was something wrong; and, turning back at the end of the first mile, scarcely got him in his stall before he was prostrate and swollen like an inflated bladder. By dint of hard rubbing and the prompt use of enemas he was fortunately saved for some years more of gallant service. *Ed. Farm. Lib.*]

(1335.) Men and horses continue at work until 12 noon, when they come home, the horses to get a drink of water and a feed of corn and the men their dinner. Some keep the harness on the horses during this short interval, but it should be taken off to allow both horses and harness to cool, and at any rate the horses will be much more comfortable without it, and it can be taken off and put on again in a few seconds; and, besides, the oftener the men are exercised in this way they will become the more expert. When the work is in a distant field, rather than come home between yokings, it is the practice of some farmers to feed the horses in the field out of the nose-bags; and the men to take their dinners with them, or be carried to them in the field by their own people. This plan may do for a day or two in good weather on a particular occasion; but it is by no means a good one for the horses, for no mode so effectual for giving them a chill could be contrived than to cause them to stand on a head ridge for nearly an hour on a winter day, after working some hours. A smart walk home can do them no harm, and if time is pressing for the work to be done, let the horses remain a shorter time in the stable. The men themselves will feel infinitely more comfortable to get dinner at home. There is a practice in England connected with this subject that I think highly objectionable, which is, doing a day's work in one yoking. For a certain time horses, like men, will work with spirit, but if made to work beyond that time they not only lose strength, but their very spirit is wrung out of them, and in the latter part of the time will do their work in a careless manner. Horses thus kept for 7 or 8 hours upon the stretch at work must be injured in their constitution, or, if able to withstand it, it must be either at the expense of bad work executed at the latter part of the yoking, or of curtailment of hours of a full day's work, or of extraordinary feeding, either of which expedients is no compensation for bad management. Common sense tells a man that it is much better for a horse to be worked a few hours smartly, and have his hunger satisfied before feeling fatigue, when he will again be able to proceed with fresh vigor, than to be worked the same number of hours without feeding. I can see no possible objection that can be offered to horses receiving a little rest and food in the middle of a long day's work, but I see many and serious ones to their working all day long without rest and food. Suppose, then, that men and horses come home at midday, the usual dinner-hour of agricultural laborers, the first thing to be done is to give the horses a drink at the pond on the way to the stable; and there should then be no washing of legs. From the water the horses proceed to the stable, where the harness is taken off; and as the men then have nothing else to do, every man gets the corn from the steward at the corn-chest for his own horses in nose-bags, or in a small corn-trough or box which each man keeps for the purpose. Of these two sorts of things for carrying the corn in the stable I prefer the trough, as being most easily filled and emptied of corn. The horses are bound up, the stable door shut, and the men go to their own houses to dinner, which should be in readiness for them. After dinner they proceed to the stable, when the horses will be found to have finished their feed, and when a small quantity of fodder may be thrown before them fresh from the straw-barn, for at this time of year farm-horses get no hay. The men may have a few minutes to converse until 10 minutes to 1 P. M., when they should give the horses a slight wisp down, put on the harness, comb out their tails and manes, and be all ready to put on the bridles the moment 1 o'clock arrives, which is announced by the steward.*

[* In our country, in the towns generally, the "ten-hour system," as it is called, prevails, while in the country the practice is to be up by daylight, winter and summer, and to labor until dark.

(1336.) The afternoon yoking is short, not lasting longer than sunset, which at this season is before 4 P. M., when the horses are loosened out of yoke and brought home. After drinking again at the pond they are gently passed through it to wash off any mud from their legs and feet, which they can hardly escape collecting in winter. But in washing the men should be prohibited wetting their horses above the knees, which they are most ready to do should there be any mud upon the thighs and belly; and to render this prohibition effectual, I have expressly stated, when speaking of the construction of a horse-pond (125), that it should not be made deeper at any part than will take a horse to the knee. There is danger of contracting inflammation of bowels or colic in washing the bellies of horses in winter; and to treat mares in foal—which they will be at this time of year—in this way is little short of madness. If the feet and shanks are cleared of mud, that is all that is required in the way of washing in winter. On the horses entering the stable and having their harness taken off, they are well strapped down by the men with a wisp of straw. Usually two wisps are used, one in each hand; but I am sure the work is much better done with one, shifting the hand as occasion requires, and directing the attention to one place at a time. A couple of wisps may very properly be taken to rub down the legs and clean the pasterns, rendering them as dry as a moderate length of time will admit. All this is done not quite in the dark, for there is still a glimmer of twilight in the western horizon, but too much in the dark to allow its being well done. After the horses are rubbed down, the men proceed to the straw-barn and bundle each 4 windlings of fodder-straw, one to be given to each horse just now, and the other two to be put above the stalls across the small fillets *p*, fig. 7, which run along the stable for the purpose.

(1337.) When 8 P. M. arrives, the steward, provided with light in a lantern, summons the men to the stable to give the horses a grooming for the night and their suppers. The sound of a horn, or ringing of a bell, are the usual calls on the occasion, which the men are ready to obey. I may here remark in passing, that the sound of a horn is pleasing in a calm winter night, and I never hear it without its recalling to my mind the goatherds' horns in Switzerland, pouring out their mellow and prayer-like strains at sunset, the time for gathering the flocks together from the mountain sides, on their way to the folds in the neighboring village. Lights are placed at convenient distances in the stable to let the men see to groom the horses. The grooming consists first in currying the horse with the curry-comb, to free him of all dirt that may have adhered to the skin during the day, and which has now become dry and flies off. A wisping of straw removes the roughest of the dirt loosened by the curry-comb. The legs ought to be thoroughly wisped, not only to make them clean, but dry of any moisture that may have been left in the evening, and at this time the feet should be picked clear by the foot-picker of any dirt adhering around between the shoe and the foot. The brush is then used to remove the remaining and finer portions of dust, from which, in its turn, it is cleared by a few rasps of the curry-comb. The wisping and brushing, if done with some force

---

with the usual intervals for meals. But while mechanics in town, under a vigilant boss, are at work, they keep working all the time; not so with men hired by the day at the *public* expense to work *and to vote!* In the country, too, if a stage drives along the road, you will see a man stop his plow and a whole gang let fall their hoes or axes to look at the passer-by, as if they had never seen a stage or a horseman before.

In the South the slaves "knock off" for two hours at midday in very hot weather, but being almost universally *tasked*, where the work will admit of it, they generally get through in less than ten hours. *Ed. Farm. Lib.*]

and dexterity, with a combing of the tail and mane, should render the horse pretty clean; but there are more ways than one of grooming a horse, as may be witnessed by the skimming and careless way in which some plowmen do it. It is true that the rough coat of a farm-horse is not easily cleaned, and more especially in a work-stable where there is much dust floating about and no horse-clothes in use; but rough as it is, it may be *clean* though not *sleek*; and it is the duty of the steward to see that the grooming is done in an efficient manner. A slap of the hand upon the horse will soon let you know whether there is any loose dust in his hair. Attendance at this time will give you an insight into the manner in which farm-horses ought to be cleaned and generally treated in the stable. The straw of the bedding is then shaken up with a fork such as in fig. 279. This figure has rather longer prongs, and is too sharp for a *stable fork*, which is most handy for shaking up straw when about 5 feet in length, and least dangerous of injuring the legs of the horses by puncture when blunt. The united prongs terminate at their upper end in a sort of spike or tine, which is driven into a hooped ash shaft, as better seen in fig. 280, which is a steel-pronged fork of the form used in Lincolnshire, and is an excellent instrument for working among straw. This mode of mounting a fork is much better than with socket and nail, which are apt to become loose and catch the straw. The horses then get their feed of oats, after which the lights are removed and the stable-doors barred and locked by the steward, who is custodier of the key. In some stables a bed is provided for a lad, that he may be present to relieve any accident or illness that may befall any of the horses; but where the stalls are properly constructed, there is little chance of any horse strangling himself with the collar, or any becoming sick where a proper ventilation is established.

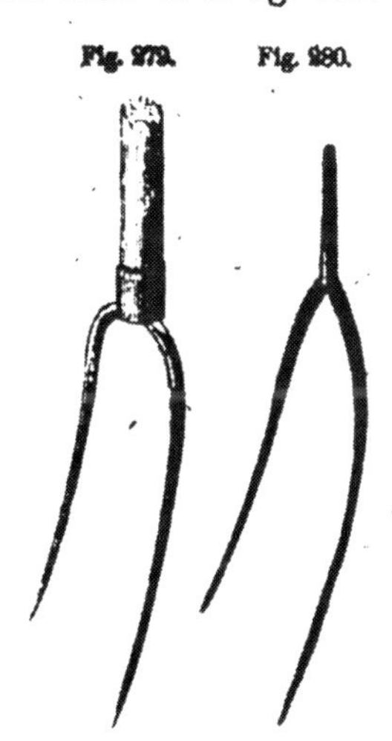

Fig. 279. THE COMMON FORK.

Fig. 280. THE LINCOLNSHIRE STEEL FORK.

(1338.) In winter it is usual to give farm-horses a mash, once, at least and sometimes thrice a week. The mash consists of either steamed potatoes, boiled barley or oats, mixed sometimes with bran, and sometimes seasoned with salt. The articles are prepared in the boiler *b′* in the boiling-house U, fig. 4, Plate IV., in the afternoon by the cattle-man or a field worker, or any other person appointed to do it, and put into tubs, into which it is carried to the stable by the men, and dealt out in the troughs used to carry the corn to the horses, with a shovel. It is warm enough when the hand can bear the heat. The quantity of corn put into the boiler is usually as much as that given raw, and in preparation swells out considerably, so that the mash acquires considerable bulk. The horses are exceedingly fnd of mash, and when the night arrives for its being dealt out, show unequivocal symptoms of impatience until they receive it. The quantity of raw oats given to farm-horses, when on full feed, is 3 lippies a day, by measure and not by weight; but taking horse-corn at almost the highest figure of 60 lbs. per firlot, each feed will weigh 3¾ lbs., the daily allowance amounting to 11¼ lbs.; but the lippy-measure, when the corn is dealt out, is most frequently not struck, but heaped, or at least hand-waved, so that the full allowance will weigh even more than this. As horses work only 7 or 8 hours a day in winter, their feeding is lessened to perhaps 2 full feeds a day or 7½ lbs., divided into three portions, namely, a full feed in the morning, ½ feed at midday, and ½ feed at night; and on the nights the mash is given, the evening ½-feed of oats is saved. One season, as a mash, I tried steamed potatoes, with salt alone, of which the horses were excessively fond, and received three times a week, and on which they became very sleek in the skin and fat, notwithstanding much heavy work; but in spring, when the long day field-work was resumed, they seemed to me to be all affected with shortness of wind. Have cooked potatoes necessarily this effect upon horses? I may mention that oats, when desired to be cooked, must be *boiled*, as steaming only burns the outside, and does not penetrate into the interior, having somewhat the effect of kiln-drying. Oats, in fact, and barley too, must be macerated to be cooked, and to do this effectually *warm* water must be used.

(1339.) I have often thought that the usual careless manner of placing the lights in the stable in the evening is highly dangerous to the safety of the building; and yet, in the most crowded and dirty stables, no accidents of fire almost ever happen. Sometimes the candle is stuck against a wall by a bit of its own melted grease; at other times it hangs by a string from the roof in an open lantern, set apparently on purpose to light straws. A good stable lantern is still a desideratum; and it should be made to hold a candle, and not an oil-lamp, as being the most cleanly mode of light for carrying about; and if the candle could be made to require no snuffing it would be perfect. A common tin lantern, with a horn glass, is what is commonly in use to carry the candle in the air; but when it gets blackened with smoke in the inside, it is of little use to give light outside. I have seen a globe lantern of glass made very strong for use on board of ship, but it has an oil-lamp. I observe Messrs. Palmer and Company, London, advertising a "weather *candle*-lamp for *safety*, and for use in *wind* and *rain*, and which requires *no snuffing*." The candle costs 8½d. per lb., and burns 5 hours for 1d. I have not seen this lamp, but judging from its figure, and if the glass is made strong enough, it would seem to answer the purpose, and is certainly not dear. If safe, it might be taken into the straw-barn, hammels, &c. at night.

(1340.) This is the usual routine of the treatment of farm-horses in winter, and when followed with some discernment in regard to the state of the weather, is capable of keeping them in health and condition. The horses are themselves the better of being out every day; but the species of work which they should do daily must be determined by the state of the weather and the soil. In very wet, frosty, or snowy weather, the soil cannot be touched; but then threshing and carrying corn to market may be conducted to advantage, and the dung from the courts may be taken out to the fields in which it is proposed to make dung-hills. This latter piece of work is best done when the ground is frozen hard. When heavy snow falls, nothing can be done out of doors with horses, except threshing when the machine is impelled with horse-power. In a very rainy day, the horses should not go out, as everything about them, as well as the men, become soaked; and before both or either can be again made comfortable, the germs of serious disease may be laid in both. When it is fair above, on the other hand, however cold the air or wet the soil, some of the sorts of out-door work mentioned above may be done by the horses; and it is better for them to work only one yoking a day than to stand idle in the stable. Work-horses soon show symptoms of impatience when confined in the stable even for a day, on Sundays, for example; and when the confinement is much prolonged, they even become troublesome. When such occasions happen, which they do in continued snow-storms, with the ground covered deep, the horses should be ridden out for some time every day, and groomed as carefully as when at work. Exercise is necessary to prevent thickening of the heels, a shot of grease, or a common cold. Fat horses, when unaccustomed to exercise, are liable to molten grease.

(1341.) It is an advisable plan for a farmer to breed his own horses; and, on a farm which employs 6 pairs, two mares might easily bear foals every year, and perform their share of the work at the same time, without injury to themselves. The advantage of breeding working stock at home is, that, having been born and brought up upon the ground, they not only become naturalized to the products of its particular soil, and thrive the better upon them, but also become familiarized with every person and every field upon it, and are broken into work without trouble or risk. The two mares should work together, and be driven by a steady plowman; and their work should almost always be confined to plowing, particularly in winter and spring, when they are big with young, for the shaking in the shafts of a cart is nothing in their favor. In driving home turnips, and leading out dung in winter, over most probably not the smoothest of roads, mares in foal should not be employed, their driver rather plowing with them, when that operation can be performed, or assisting the other men at their carts with manual labor.

(1342.) Supposing, then, that one or two mares bear foals every year, the young horses, their produce, consisting of foals, year-olds, and two-year-olds, should be accommodated in the steading N, figs. 3 and 4, Plates III. and IV., according to age, where there are more than one of the same age, the older being apt to knock about the younger; but where one only of every age is brought up, they may be placed together for the sake of companionship, as horses are very social animals, and they learn to accommodate themselves to one another's tempers. Where blood foals are bred as well as draft, they should have separate hammels, the latter being too rough and overbearing, but the bloods generally contrive to obtain the mastery. Young horses never receive any grooming, and are even seldom handled; but they should all be accustomed to be led in a halter from their youngest period.

(1343.) The food usually given to young horses in winter is oat-straw for fodder, and a few oats; and where they are wintered among the young cattle in the large court K, they have the chance of a few pickings of corn from the corn-barn, or the refuse of hay from the litter of the work-horse stable, and then they seldom get corn. The fact is, young horses are generally unjustly dealt with; they are too much stinted of nourishing food, and the consequences established by the treatment are a smallness of bone which deprives them of requisite strength for their work, and a dullness of spirits which renders their work a burden to them. I speak of what I have seen of the way in which a large proportion of the farm-horses of this country are brought up when young. Their treatment seems to be derived from the opinion that very little nourishing meat should be given to young horses. Instead of this, they should receive a stated allowance of corn—and if bruised, so much the better—according to their ages; and when a mash is given to the work-horses, the young ones should always have a share. Should a mash be grudged as being too extravagant for young horses, they should get Swedish turnips or potatoes every day; for some moist food is requisite with dry fodder and corn.

(1344.) The names usually given to the different states of the horse are these: The new-born young is called a *foal*, a male one being a *colt foal* and a female a *filly-foal*. After being weaned, the foals are called *colt* or *filly*, according to the sex, which the male retains until broken in for work, when he is a *gelding* or *horse*, which he retains all his life; and the filly is then changed into *mare.* When the colt is not castrated, he is an *entire colt;* which name he retains until he serves mares, when he is a *stallion* or *entire horse.* A mare, when served, is said to be *covered by* or *stinted to* a particular stallion; and after she has borne a foal, then she is a *brood-mare,* until she ceases to bear, when she is a *barren mare* or *eill mare;* and when dry of milk, she is said to be *yeld.* A mare, while bearing a foal, is said to be *in foal.**

(1345.) You have seen that though cattle gain weight when fed on cooked food, compared to others fed on the same substances in a raw state; yet the expense of cooking counterbalances any advantage gained in weight, and it is therefore inexpedient to undertake the trouble of it. These results might have been anticipated from the peculiar functions of the stomach of the ox; for he *chews the cud,* that is to say, he masticates the food, as he takes it into his mouth in a very imperfect manner, rendering it only so small as to be able to swallow it with some degree of force, in which state it reaches the paunch or first stomach; where, if it decomposes immediately and generates gas, it produces the disease of the *hoven,* which has been spoken of already (1294). But should it not decompose—which is the usual condition of the food—it is again brought up to the mouth, and undergoes a thorough mastication, after which it is swallowed and finds its way to the stomach, which contains the gastric juice, there to be digested for the purpose of being assimilated into the system. Now, all that we can do for the ox in cooking his food, is to save him the trouble of chewing the cud, and to put the food into that state in which it is at once fit to be acted upon by the gastric juice. In doing this, we attempt to imitate and enter into competition with a complicated natural process, and, as might be expected in the circumstances, exhibit our inferiority. In the state, however, in which cooked food is presented to the ox, chewing the cud is not altogether saved him, as the straw which he chooses to eat undergoes that operation, and therefore assists in keeping that important function in exercise. It is doubtful that the ox would retain his wonted good health, were we able entirely to suspend the action of that function in him; and it is therefore questionable policy to attempt it to a farther degree than to reduce his food so small as to render it fit to enter the paunch, with still less mastication than he would have to give it in its ordinary state.

(1346.) The case, however, of the horse is very different. His is a single or simple stomach, which must be filled at once with well-masticated food, before the gastric juice can act upon it in a proper manner; and should any food which enters it in an insufficiently masticated state, escape beyond the influence of the juice into the bowels, it may decompose there, generate gas, and produce the analogous disease of *hoven* in cattle, namely, flatulent colic or *batts.* To render food in such a state at first as shall save the horses the trouble of mastication, is, therefore, to do him a good service; and hence, cooked food is in a proper state for feeding a horse, and it has also been proved to be economical. Still, the cooking will be carried to an injurious degree, if it shall, by dint of ease of deglutition, prevent the flow of the sufficient quantity of saliva into the stomach which is necessary to complete digestion—"the quantity of which," says Professor Dick, "is almost incredible to those who have not had an opportunity of ascertaining it, but which the following fact will testify. A black horse had received a wound in the parotid duct, which became fistulous. When his jaws were in motion in the act of eating hay, I had the curiosity to collect in a glass measure the quantity which flowed during 1 minute, by a stop-watch; and it amounted to nearly 2 drachms more than 2 oz. in that time. Now, if we calculate that the parotid gland on the opposite cheek poured into the mouth the same quantity in the same time, and allow that the sublingual and sub-maxillary gland, on each side combined, pour into the mouth a quantity equal to the two parotids, we then have no less than 8 oz. of saliva passing into the mouth of a horse in one minute, for the purpose of softening the food and preparing it for digestion."† Yet it is impossible for any horse to swallow food in the most favorable state it can be made for swallowing, without moving his jaws to a certain degree, and this insures a certain quantity of saliva entering his stomach.

(1347.) But more than this, cooked food may be presented in too nutritious a state for the stomach; and there may be, on the other hand, too little nutriment in the food given; for, "the digestive organs of the horse, like those of the ox," says Professor Dick, "are very capacious, and are evidently intended to take in a large proportion of matter containing a small proportion of nutriment; and if the food upon which they are made to live is of too rich a quality, there is, by the excitement produced, an increase of the peristaltic motion, in order to throw off the superabundant quantity which has been taken into the stomach and bowels. It is necessary to give, therefore, a certain quantity of *bulk, to separate, perhaps, the particles of nutritious matter,* that the bowels may be enabled to act upon it properly. A horse could not live so well on oats, if fed entirely upon them, as when a portion of fodder is given; with them a certain quantity is required. But this may be carried too far, and the animal may have his bowels *loaded* with too large a quantity of unnutritious food;" as witness the nature of the *steep* before alluded to; "and nothing less than such a mass as will render him incapable to perform any active exertion, will be sufficient to afford him even a scanty degree of nourishment. A horse living on straw in a straw-yard be-

[* Some use the foolish expression, *out* of such a horse or stallion, instead of *by* the stallion and *out* of the mare; and, again, they say "My horse was *sired* by Messenger," instead of *got* by Messenger. As John Randolph once observed, they might as well say he was *damn'd* by a particular mare. There are proprieties and technicalities belonging to every sort of business, which should be learned at least by those who follow, or are immediately interested in it.

Ed. Farm. Lib.]

† Quarterly Journal of Agriculture, vol. iii.

comes pot-bellied. Hence it is, that a proper arrangement in the properties and proportions of his food becomes a matter of important consideration."* These and the preceding remarks comprehend all the rationale of feeding both cattle and horses, and, if carefully considered, may conduct you to adopt such an appropriate mixture of materials in your possession as may serve to maintain the strength, good health and condition of your horses, on the one hand, and to do so economically on the other. Meantime, I shall enumerate a few of the attempts that have hitherto been made of making mixtures of food for horses, with the view of ascertaining whether cooked or raw food, in a prepared or natural state, maintains horses in the best order.

(1348.) The most careful set of experiments that have yet been recorded in supporting *farm-horses* on *boiled* and *raw* grain, and on raw grain *prepared* and in a *natural* state, was made by Mr. James Cowie, Halkerton Mains, Kincardineshire. He subjected no fewer than 12 horses to the experiment, dividing them into 3 sets of 4 each, and keeping each set on a separate fare. The horses were weighed on 1st March, when the experiment began, and their weights varied from 9 cwts. 3 qrs. to 12 cwts. 1 qr. 4 lbs.; and they were again weighed on 1st May, at the end of the experiment, and their weights then ranged from 9 cwts. 2 qrs. 23 lbs. to 12 cwts. 1 qr. 14 lbs. Thus the range of weight did vary much at both the periods, though the individual weights did. Their ages ranged from 4 to 12 years.

(1349.) The facts brought out in this experiment were, that the horses fed on *unbruised raw* and on *boiled* grain, gave results so very nearly alike that it seems inexpedient to incur the expense of *cooking* food for horses, as that costs about 1½d. on two feeds for each horse. This is a rather remarkable result, for one should have expected that the *boiled* grain would have had the advantage. *Bruised raw* grain seems the most nourishing, and, in not requiring cooking, of course the most economical, mode of feeding work-horses. For, all the horses that had been on *boiled* and *unbruised raw* grain lost 70 lbs. each; and that amount of loss in an animal of 10 cwts. or 12 cwts. is considerable; whereas those which had been on *bruised* grain, though given raw, either gained weight or lost none. And as to the economy of using grain in this state, *besides the cooking*, it is alleged that boiled whole grain passes through the horse undigested as well as raw grain when whole, and that the quantity which thus escapes is equal to 1-6 of what a horse consumes; whereas, the grain that is *bruised* undergoes a considerable degree of digestion at least, before passing away. If the loss is taken at 1-6 on a horse which gets 12 lbs. daily of oats whole, a yearly saving might be effected of about 2 quarters of corn, by giving him 10 lbs. of bruised instead.

(1350.) Many economical forms of mixtures have been recommended for farm-horses, and these are among them:

10 lbs. of chaffed straw, at £1 per ton	1d.
10 lbs. of oats, at 3s. per bushel	9
16 lbs. of turnips, at 10s. per ton	1
Expense of cutting and chaffing	0½
Cost of one horse each day	11½d.

16 lbs. of hay, at 3s. 6d. per cwt.	6d.
5 lbs. of oats, at 3s. per bushel	4½
16 lbs. of turnips, at 10s. per ton	1
Cost of one horse each day	11½d.

28 lbs. of steamed turnips	3½d.
7 lbs. of coals, at 1s. per bushel	1
Expenses of steaming	0½
16 lbs. of straw, at £1 per ton	1½
Cost of one horse each day	6½d.

This last mixture, containing no corn of any kind, is said to "succeed remarkably well, and although the horses perspired considerably while at work, they kept their condition exceedingly well," and has been adopted by some farmers in the South of England, and by Mr. Karkeek, the veterinary surgeon, as having been "highly recommended by several practical farmers."† No doubt, horses can live upon turnips, as well as upon grass, without corn, and they may be said to work upon them; but I quite agree with Mr. Stewart, when he observes: "What the owner might call *work* is not known. In this country, grass alone will not produce workable horses," and the same may more truly be said of turnips and straw. "If food is not given," continues Mr. Stewart, "work cannot be taken. Every man who has a horse has it in his power to starve the animal; but that, I should think, can afford little matter for exultation."‡ Turnips are frequently given to farm-horses in the evening in lieu of a feed of corn, and even in lieu of a hot mash at night; and horses are very fond of Swedish turnips, which, on being washed, are generally set before them whole, unless some of the men take the trouble of cutting them into slices with their knives; but the best way would be to have them sliced on purpose by Wallace's turnip-slicer, fig. 259, which has been already described. Potatoes are given to horses in a raw state, in the same manner as turnips, and they seemed to be relished by them, but not so fondly, in so far as I have observed, as Swedish turnips. But of the sorts of food of the root kind, there is none which gives horses so much delight as the carrot. It is a pity that this root can only be cultivated successfully on very light soil, otherwise it would be worth while to raise as many, at least, as would support the horses, in conjunction with corn, all winter. Stewart says that "for slow-working horses carrots may supply the place of corn quite well, at least for those employed on the farm."‖ They would get fat enough on 70 lbs. of carrots a day, but would want stamina without corn.§ Carrots

* Quarterly Journal of Agriculture, vol. iii.
† Prize Essays of the Highland and Agricultural Society, vol. xiv.
‡ Stewart's Stable Economy. ‖ Ibid.
§ An error has crept into (1031), where the specific gravity of the carrot is stated at 0·018 instead of 0·610

are easily and successfully grown in the Island of Guernsey; but they are not given to horses on account of an allegation that "when on this food their *eyes* are injured." The same writer mentions a similar effect produced by the parsnip at a certain season of the year. "To horses," he says, "parsnips are frequently given, and have the property of making them sleek and fat; but in working they are observed to sweat profusely. If new, and cut sufficiently small, no other ill effect results, except, indeed, at one period of the year, toward the close of February, when the root begins to shoot; if then given, both horses and horned cattle are subject, on this food, to an inflammation in the eye, and epiphora or watery eye; in some subjects, perhaps, producing blindness."* Horses are very fond of bread; a piece of bread, and especially oat-cake, will take a horse in the field when a feed of corn cannot. It is quite common in Holland to see travelers at a village inn, take a black loaf and slice it down with a bread-knife in a trough for their horse. Upon the principle of economy, M. Longchamp has proposed to feed the cavalry of France with a bread composed of ¾ of of boiled potatoes and ¼ oatmeal, properly baked in an oven. The usual allowance of oats for a horse, at 10 lbs. costs 13 sous; but 10 lbs. of this bread will only cost 5 sous

(1351.) But independent of all succedanea, which may be given to horses at times as a treat, and as affording a beneficial change of food, there should be a regular feed prepared for farm-horses, which should be administered every day, and any deviation from which should be regarded as a relish or treat. There are two formulæ which I shall give, which have been found to make excellent prepared food for farm-horses, and they may be prepared without much trouble, provided the proper apparatus is erected for the purpose. The first is given in quantity of each day for one horse:

In the morning....	3½ lbs. of oat and bean meal, 11 lbs. of chopped straw,	14½ lbs.
At midday........	3 lbs. of oat and bean meal, 12 lbs. of chopped straw,	15 ..
At night..........	1½ lbs. of oat and bean meal, 11 lbs. of steamed potatoes. 2 lbs. of chopped straw,	14½ .. 44 lbs.

This quantity is quite sufficient for the strongest farm-horses, and less will be consumed by ordinary ones, but that can be regulated according to circumstances, by withdrawing a little meal and straw, still retaining the proportions. The usual allowance of oats, as you have seen, weighs 11½ lbs. a day, when the grain is of the finest quality; but as horses seldom receive the finest oats, and are usually supplied with what are called common oats, which do not weigh so heavy, the usual allowance may be taken at 10 lbs.; and when hay is given to the horses in spring they eat at least 1½ stones of 22 lbs.=33 lbs. every day. This mixture, on the other hand, contains no hay, and only 8 lbs. of oats and bean-meal, or 6 lbs. of barley-meal instead, if more convenient to be given, and 11 lbs. of steamed potatoes, which cannot be estimated at much value on a farm beyond the cost of steaming.† The value of the ordinary and the prepared food can easily be estimated, and it will be found that the prepared is the cheapest, and at the same time better for the horses' health, and equally well for them as to condition and spirit. The mixture is made in this way: The meal and chopped straw are put and mixed together in a tub, and a little salt sprinkled over it. The steamed potatoes are then poured hot into the tub over the straw, and the whole is formed into a mash with a shovel, and let stand awhile to acquire an equal temperature throughout, and to swell the meal into a pulpy state with the potatoes, before being divided out to the horses.

(1352.) A formula is given by Prof. Low, consisting of chopped straw, chopped hay, bruised or coarsely-ground grain and steamed potatoes by weight, in equal parts, with 2 oz. of salt; and of this from 30 lbs. to 35 lbs., or 32½ lbs. on an average, to be given to a horse every day.‡ This mixture, including hay, will be more expensive than the above; and I am doubtful that 35 lbs. of it will satisfy a farm-horse on active work in spring, when he can eat 33 lbs. of unchopped hay a day, without corn.

(1353.) It appears at first sight somewhat surprising that the idea of preparing food for farm-horses should only have been recently acted on; but I have no doubt that the practice of the turf and of the road, of maintaining horses on large quantities of oats and dry rye-grass hay, has had a powerful influence in retaining it on farms. But now that a more natural treatment has been adopted by the owners of horses on fast work, farmers, having now the example of post-horses standing their work well on prepared food, should easily be persuaded that, on slow work, the same sort of food should have even a more salutary effect on their horses. How prevalent was the notion, at one time, that horses could not be expected to do work at all unless there was *hard meat* in them! "This is a very silly and erroneous idea, if we inquire into it," as Prof. Dick truly observes; "for whatever may be the consistency of the food when taken into the stomach, it must, before the body can possibly derive any substantial support or benefit from it, be converted into *chyme*—a pultacious mass; and this, as it passes onward from the stomach into the intestinal canal, is rendered still more fluid by the admixture of the secretions from the stomach, the liver and the pancreas, when it becomes of a milky appearance, and is called *chyle*. It is then taken into the system by the lacteals, and in this *fluid*, this *soft* state—*and in this state only*—mixes with the blood and passes through the circulating vessels for the nourishment of the system."‖ Actuated by these rational principles, Mr. John Croall, a large coach proprietor in Edinburgh, now supports his coach-horses on 8 lbs. of chopped hay and 16 lbs. of bruised oats; so does Mr. Isaac Scott, a Postmaster, who gives 10 lbs. or 12 lbs. of chopped hay and 16 lbs. of bruised oats to large horses; and to carry the principle still farther into practice, Captain Cheyne found his post-

* Quayle's Agriculture of the Channel Islands.
† Quarterly Journal of Agriculture, vol. iv.
‡ Low's Elements of Practical Agriculture.
‖ Quarterly Journal of Agriculture, vol. iii.

horses work well on the following mixture, the proportions of which are given for each horse every day; and this constitutes the second of the formulæ alluded to above:

In the day .. { 8 lbs. of bruised oats. 3 lbs. of bruised beans. 4 lbs. of chopped straw. } 15 lbs.

At night .. { 22 lbs. of steamed potatoes. 1½ lbs. of fine barley-dust. 2 lbs. of chopped straw. 2 ounces of salt. } 25½ lbs.

Estimating the barley-dust at 10d. per stone; chopped straw 6d. per stone; potatoes, steamed, at 7s. 6d. per cwt.; and the oats and beans at ordinary prices, the cost of supper was 6d., and for daily food 1s. with cooking, in all 1s. 6d. a horse each day.*

(1354.) [*Hay and Straw-Cutters.*—Machines for chopping hay and straw form now an important article in the class of implements for preparing food for horses and cattle. In England the straw-cutter or chaff-cutter is held, very properly, in high estimation by the farmer, and its value, in an economical point of view, seems to be fully appreciated by all. In Scotland, with all its boasted economy in the various walks of Agriculture, the straw-cutter is but partially employed, and it is chiefly among those farmers who, to a well-established experience superadd scientific skill, that the employment of the straw-cutter, together with the other members of the class of food-preparing machines, are brought to bear upon the establishment in a systematic reform. There can be no doubt that ere long the food-preparing system will become as universal among farmers as the threshing-machine is already, and straw-cutters, corn-bruisers and steaming-apparatus will be seen in every well-regulated steading. To the full development of such a system there exists one especial obstacle, which is that defect, in the minds of many men, which prevents their forming a systematic arrangement of any given subject, and from being indifferently qualified to draw conclusions from a series of facts, which individually appear isolated and loosely connected, but which, in the aggregate, are capable of bringing out important results. For example, there are many individuals who may have procured the requisite machinery to have enabled them to follow the system here alluded to, but, owing to the absence of properly organized methods of procedure in the different processes, and losing sight of the advantages to be derived from a proper combination of effects, by viewing only the results in detail, the well-intended trial ends in disappointment, and the machinery set aside as unprofitable; whereas, under proper direction, it would certainly have achieved the object. In using machinery of this kind it should always be borne in mind that the more constantly and regularly it is kept in operation, so much the more productive it will be in saving expense to its proprietor, provided such machinery be of a kind that can be rendered available as a means of saving expense; and, from the nature of things, no machine will be continued in any practice after it has been ascertained to possess only negative properties.

(1355.) Straw-cutters are of very various construction, rising in the perfection and complication of parts, from the simple knife, jointed at one end to a table and wielded by the right hand, as a lever of the second order, chopping the straw or hay that is presented to it by the left. From this simple and primitive form they rise in gradation to a class of elaborate machines, too numerous to be described individually, but out of which the following varieties are selected as appearing most worthy of attention. I shall pass over some of the early machines, which, however ingenious, were unnecessarily complicated; such as those which enjoyed the advantage of a revolving web to carry forward the substance to be cut, and having also the means of moving the substance, not uniformly, but by starts, the progressive action being performed in the interval of the strokes of the cutter, the substance, at the same time being alternately compressed and relaxed, that is, compressed while the knife is cutting, and relaxed during the progressive stage. Such mechanical appliances are now, for the most part, laid aside, and the machine is proportionally simplified.†

(1356.) The straw-cutting machines now in general use may be arranged under three varieties, and in the order of seniority stand as follows: 1. Those having the cutting-knife or knives attached on the disk of a fly-wheel. 2. Those having the knives placed upon the periphery of a skeleton cylinder; and 3. Those having numerous knives set round the surface of a small solid cylinder. This last being the simplest form of the modern machine I shall place it first in the order of description.

(1357.) *The Canadian Straw-Cutter.*—This machine, as the name implies, is an importation from Canada, a description of it having been sent thence by Mr. Fergusson of Woodhull, now of Fergus, Upper Canada, to the Highland and Agricultural Society, in whose Transactions it was first published;‡ but the present figure is taken from the machine as made by James Slight and Company, who have greatly improved the construction of the cutting cylinder. Fig. 281 is a view in perspective of this machine. It consists, first, of a wooden frame, of which *a a a a* are the four posts, 2½ inches square, the front pair 43 inches in hight, and the back pair 36 inches. These are connected by two side-rails, one of which is seen at *b*, and a cross-rail, *c*, which last serves also to support the bottom of the feeding-spout. These rails are 2½ inches deep by 1½

[† Not to have it said that this great work had been impaired in any of its essential parts, and especially in such engravings as are employed to illustrate it, we give all the cuts of machinery, though well aware that some of them have been well supplanted by more economical inventions of our own ingenious countrymen. The presentation of them will be acceptable, too, as we may suppose, to the manufacturers and venders of implements, patrons of this work.

Ed. Farm. Lib.]

* Quarterly Journal of Agriculture, vol. iii.

‡ Prize Essays of the Highland and Agricultural Society, vol. xii.

inches in thickness. The posts are farther connected by four light stay-rails below; and the frame, when thus joined, measures 15 inches in width at the front, where the rollers are applied, 23 inches in width behind, and 40 inches in length at bottom, but only 36 inches at the top-rail

Fig. 281.

THE CANADIAN STRAW-CUTTER.

measured over all. The feeding-spout, *d*, is 40 inches in length, 9 inches in width within, at the feeding end, and 18 inches behind; the depth is 4 to 6 inches.

(1358.) The acting parts of this straw-cutter consist of the cutting-cylinder *e*, which is 9 inches in length and 6½ inches in diameter to the edge of the cutters. It is armed with 24 cutters or knives; its axle runs in plummer-blocks, bolted upon the posts, and carries likewise the wheel *f* of 9½ inches in diameter. The pressure cylinder *g* is a plain cylinder of hard wood, beech or elm, turned true upon an iron axle, which runs in plummer-blocks similar to the former. The length of the pressure cylinder is 9 inches, and its diameter 7 to 8 inches; it carries no wheel, but revolves by similar contact with the cutting-cylinder. The pressure-cylinder is furnished with a pair of adjusting screws at *h h*, which act upon the plummer-blocks of the cylinder, and afford the means of regulating the pressure of the one cylinder upon the other. The shaft *i*, which has also its plummer-blocks, carries at one end a pinion of 3¼ inches in diameter, which acts upon the wheel *f*, while, at the other end, it carries the fly-wheel *l* of 34 inches diameter and 60 lbs. weight. The winch-handle *m* is also attached to the shaft *i*, and serves to put the machine in motion.

(1359.) Fig. 282 is a transverse section of the cutting-cylinder, showing the position of the cutters and their insertion into grooves which are planed out of the solid cast-iron forming the body of the cylinder; *a* is the axle, and *b* the body of the cylinder, which is 4 inches in diameter, and has 24 cutters inserted in its periphery. Fig. 283 is a longitudinal section of the same, for the purpose of exhibiting the manner in which the cutters are secured in their places. *a* is the axle, and *b* the body as before, *c c* being two opposite cutters. The body is 8 inches in length but is furnished with two caps *d d*, which make it up to 9 inches. The caps are cupped out, so that their edges *e e e e* embrace the ends of the body, and at the same time enter into notches cut in each end of the cutters, as seen in fig. 284, which is a cutter detached, and drawn to a larger scale exhibiting the notch *a a* into which the edges of the cupped ends enter. By this arrangement, the numerous cutters are all held firmly in their grooves: for so soon as the caps are applied, and fixed by the keys *f f*, fig. 283, being driven through the axle, the caps are pressed home upon the body and the cutters. On the other hand, when it is found requisite to remove a cutter, for sharpening or other purposes, it is only necessary to drive out one of the keys *f*, to withdraw the cap, and the cutters can be lifted out of their grooves without trouble.

(1360.) As this machine acts entirely by direct pressure, it will readily be observed that, in working it, the straw being laid in a trough *d*, fig. 281, and brought in contact with the cutting-cylinder and its antagonist, the hay or straw will be continuously drawn forward by means of the two cylinders; and when it has reached the *line of centers* of the two, it will be cut through by the direct pressure of the cutting-edges of the one against the resisting surface of the other

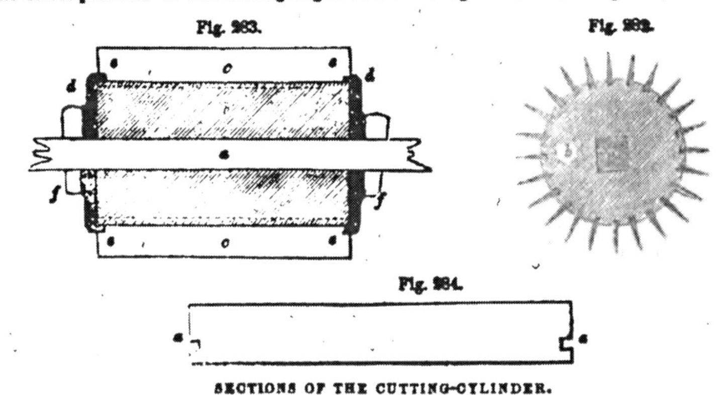

SECTIONS OF THE CUTTING-CYLINDER.

cylinder, and the process goes on with great rapidity. The straw is cut in lengths of about ½ inch, and though it passes in a thin layer, yet the rapidity of its motion is such, that when driven by hand, at the ordinary rate of 44 turns of the handle per minute, the number of cuts made by the

Fig. 285.

THE CYLINDER STRAW-CUTTER WITH STRAIGHT KNIVES.

cutting-cylinder in that time is 360; and the quantity, compared by weight, will be three times, nearly, what any other straw-cutter will produce, requiring the same force to work it, that is to say, a man's power. There is one objection to this machine, which is, the wearing out of the resisting cylinder but this balanced by the excess of work performed, and by the circumstance that the wearing cylinder can be removed at an expense not exceeding 2s., and it will last from 3 to 6 months. The price of the Canadian straw-cutter is £6,10s.

(1361.) *Cylinder Straw-cutter*, so named here from its having the knives or cutters (generally two, but sometimes four) placed on the periphery of a skeleton cylinder, the knife lying nearly in the plane of revolution. Besides the cutting cylinder, they necessarily have a pair of feeding-rollers, which bring forward the substance to be cut, and also, from the velocity of their motion, regulate the length of the cut. Two forms of the machine exist, the essential difference of which is that, in the one, the cutters are placed upon the cylinder with a large angle of obliquity to the axis, generally about 35°, and are therefore bent and twisted until their edges form an oblique section of the cylinder, while the box, or the orifice through which the substance is protruded for being cut, lies parallel to the axis of the feeding-rollers. In the other variety, the knives are placed parallel to the axis of the cylinder, and therefore straight in the edge, while the cutting-box is elongated into a nozzle, and is twisted to an angle of 15° with the axis of the feeding-rollers. To this form of the machine I shall at present chiefly confine myself.

(1362.) The cylinder straw-cutter with straight knives, as constructed by James Slight and Co., Edinburgh, at prices from £7 10s. to £8 10s., is represented by fig. 285, being a view in perspective of the machine, while fig. 286 is a section of the principal parts; and in the two figures the same letters refer to the corresponding parts of each. The machine is made entirely of iron, chiefly cast-iron. The two side-frames *a a*, are connected together, at a width of from 12 to 15 inches, by the stretcher-bolts *b b*, two of which are seen in the right-hand side of the figure, and a third below on the left; a fourth is formed of the bed-plate *c*, which is bolted to a projecting bracket, and carries the cheeks or frame *d* of the feeding-rollers *e* and *f*. The lower roller *e* carries upon its axle the driving-wheel *g*, and also the feeding-wheel, indistinctly seen in the figure, but which works into its equal wheel *i*, fitted upon the axle of the upper roller *f*. In the machine, when adapted for hand power, the rollers vary from 5 to 8 inches in length, and are 3¼ inches in diameter, and fluted. In the apex of the side-frames, bearings are formed for the axle of the cutter-wheels *k*, which form the skeleton cylinder, and whose axle carries also the driving-pinion *l*, acting upon the wheel *g*. The cutter-wheels are 11 inches diameter, and are set at from 10 to 13 inches wide. Intermediate between the feeding-rollers and the cutter-wheels is placed the cutting-box or nozzle *m*, bolted to the roller-frame in the position represented in fig. 286. On the farther end of the cutter-wheel axle the fly-wheel *n*, of 4 feet diameter, is fixed; and on the near end of

Fig. 286.

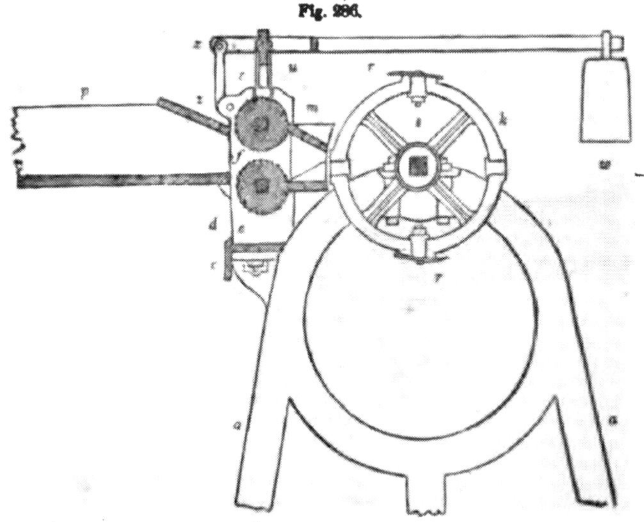

A TRANSVERSE SECTION, SHOWING THE RELATION OF THE PRINCIPAL PARTS.

the same, the winch-handle *o*, by which the machine is worked. The feeding-trough *p* is 4 feet in length, from 5 to 8 inches wide at the feeding end, and 18 inches behind. The depth is 6 inches, and the trough is formed of ¾ inch deal. It is hooked to the roller-frame at the mouth, and is supported by the jointed foot *q*. The cutters *r r*, from 10 to 13 inches in length, and 3½ to 4 inches in breadth, are made of the finest steel, backed with iron. The cutters are fixed upon the cylinders, each with two screw-bolts, as seen at *s*, passing through the ring of the wheel, and they are placed slightly eccentric to it; the cutting-edge being about ¼ inch more distant from the center than the back. To secure the regular feed of the rollers, the lower one turns in fixed bearings; but the other is at liberty to rise and fall in the fork *t* of the roller-frame. In order farther to secure a uniform pressure on this roller, a bridge *u* is inserted on the fork *t*, resting on both journals of the roller. A compensation lever *v*, has its forked fulcra in *x* through the strap *x x*, which is hooked on to pins in the roller-frame; and it thus bears upon the bridge *u* at both sides by means of the forked end, as seen at *a'* in fig. 285. A weight *w* is appended to the extremity of the lever, which, thus arranged, keeps a uniform pressure on the upper roller, while it is always at liberty to rise or fall according to the thickness of the feed which the rollers are receiving.

(1363.) Fig. 287 is a direct front view of the cutting-box detached, on a scale of 2 inches to a foot. *a a* are the ears by which it is bolted to the roller-frame; *b c* and *d e* are the upper and lower extreme edges of the nozzle, while *f g h i* is the base, as applied to the roller-frame. The angle *b g f* or *e i k* is the obliquity which is given to the nozzle, and is about 15° with the horizontal axis. Fig. 288 is a plan of the same, in which the same letters of reference apply to the corresponding parts. *f g* is the base, lying on the same plane with the ears *a a*. The curve line *b c* is the contour of the upper edge of the nozzle, and the dotted line *b e* represents the lower edge *d e* of fig. 287.

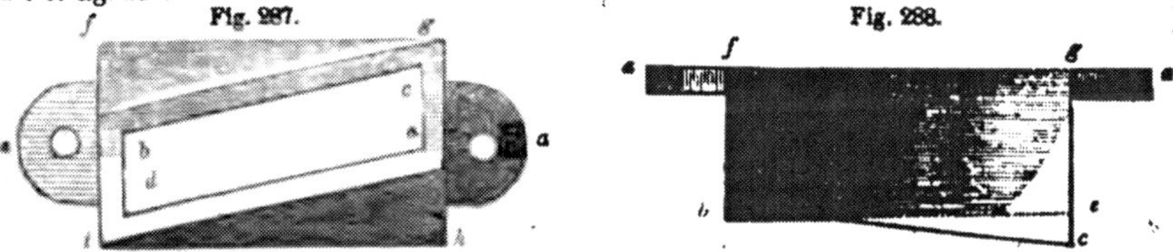

Fig. 287. Fig. 288.

THE CUTTING-BOX DETACHED.

(1364.) The obliquity of the nozzle, as here represented, serves the purpose of causing the knife to make a progressive cut, approaching to the effect of clipping; thereby preventing the shock that would otherwise arise, were the nozzle and the edge of the knife parallel at the instant when the cut commences. The action of the machine would perhaps be improved by an increased obliquity of nozzle, as the stroke of the knives would thus be still less felt; but this could not easily be done unless the knife were placed oblique also in the opposite direction. But this involves a difficulty which the present machine was intended to obviate, namely, the preserving a straight-edged knife, which, in the hands of an unskilled workman, is much easier adjusted than a twisted one. And farther, a moderate degree of obliquity is preferable to an excess; in the former, the intervals of the cuts are so considerable as to allow the fly-wheel to exert its natural effect of storing, as it were, a quantity of momentum or force, which it freely gives out to meet the resistance of the next cut; when, on the contrary, the obliquity is very great, as is frequently the case in the machines with twisted knives, and more especially when there are four cutters applied to the cylinder, there may, in such cases, be no interval of action in the cutters, for before one has completed its cut, the next has commenced, and the advantages of a fly-wheel are, in such cases, nearly, if not altogether, lost; hence we find that some of the machines with twisted knives are extremely heavy to work.

(1365.) There is, however, another cause of increased labor in the working of some straw-cutters, on the principle now before us. In many of the cylinder machines, the cutter-wheels are of a diameter so large as to render the operation of cutting with them one of great labor. This arises from the circumstance of the action of the machine being a combination of the effects of the lever. The winch-handle is a lever, say of 12 inches radius, and the radius of the cutter-wheel, measured to the edge of the cutter, may be taken at 6 inches; here there is a mechanical advantage of two to one in favor of the power. Let the cutter-wheel be increased to 16 inches, or a radius of 8 inches, the proportion of the leverage is now materially changed, and the mechanical advantage in favor of the power is only 1½ to 1. The power, therefore, suppose it exerts in the first case a force of 30 lbs., it will require, in the second case, a force of 40 lbs., to overcome the same resistance, and so on in that proportion. But the fly-wheel, under the like circumstance, loses part of its effect, though not to the same amount; for, suppose its radius to the center of gyration to be 2 feet, the mechanical advantage of its momentum would be, in the first case, as 4 to 1, and in the second, as 3 to 1. In the construction of all cylinder straw-cutters, therefore, it is of importance to make the cutters of small diameter, that is to say, never to exceed 11 inches. It were, perhaps, better that they should be less than this, but on no account should they exceed it, especially for hand-machines. In the case of steam or water power, any small increase of resistance is less important, provided that a countervailing object is to be attained by it, such as a machine already made, or the like.

(1366.) While on the subject of *fly-wheels*, it may not be out of place to make a few remarks, pointing out where those auxiliaries to machinery may with propriety be applied, and where they ought not. In the first place, it may be asserted that in no case can a fly-wheel act as a *generator* of power; and under a false impression of this supposed function of fly-wheels, numerous instances occur of their misapplication, or at least a misconception of their effects; and, secondly, the only available function of fly-wheels is their capability of acting as *reservoirs* of that power or force that is communicated to them while in motion. Thus a comparatively small force applied to a heavy fly-wheel for a few seconds, will, on the principle of its absorbing and partially retaining that power and force, accumulate a momentum that may, through the agency of mechanical means, be discharged on a particular point, and produce an instantaneous effect that the first mover never could accomplish without such means. This is finely exemplified in the machine for punching and cutting thick iron plates and bars; and the principle applies in all cases where fly-wheels can be employed with advantage. The principle of action is this: Fly-wheels may be employed with advantage in every case where the intensity of either the *power* or the *resistance* is variable; and where both are variable, it becomes still more necessary. On the other hand, where both the power and the resistance are uniform, a fly-wheel may be held as an incumbrance, and can only act as a load upon the first mover. In the steam-engine, for example, of any form in which a crank is used to communicate motion to machinery, the fly-wheel is indispensably necessary; and such is the requisite governing power of fly to keep up steady motion, that its momentum is sufficient to compensate for considerable variation in the resistance of the machine or machinery upon which it operates. With water-wheels, however, the power is perfectly uniform, and is the resistance is also uniform, as in grist-mills and even threshing-mills, fly-wheels would be worse than useless; but where the resistance is intermitting, such as rolling and tilt-mills, and punching-machines, a heavy fly becomes necessary, in which the power of the first mover can be accumulated for a short period and which will be expended during the succeeding short period that the driven machine is in action. In every case where manual labor is applied, fly-wheels are useful if not essential. This arises from the power itself being variable; for the power of a man working a winch varies according to the different positions which the winch occupies in the course of its revolution, and has been ascertained to range in the proportion of 30 lbs. and 60 lbs. The rationale of this is, that when he is in a position that restricts his exertion to 30 lbs., he might not be able to overcome the resistance unless at a very slow rate; but in the position where he can exert a force of 60 lbs., he can do more than overcome the resistance. And here it is that the fly-wheel comes to his aid: for suppose the resistance requires an actual force of 45 lbs., while he is putting forth 60 lbs., there is a surplus of 15 lbs. This last quantity the ever-ready fly-wheel, whose velocity, from its inertia, he is not able greatly to increase; but it takes up, with a small increase of velocity, the surplus force exerted by the hand, and this is stored up in the mass of the wheel, to be delivered out again at the next weak point in the revolution of the winch. Hence, a nearly equable force is produced to act upon the machine to which the power of the man is directed. If this power is directed upon an intermitting machine, such as a straw-cutter, the demands upon the fly-wheel are very much increased; but as the point of the machine to which the power is applied moves at a much slower velocity than the center gyration in the fly, and as the intermissions of the resistance are not likely to coincide exactly with the increments or decrements of the power, there will be a mutual compensation going on among the forces to bring out a uniform result. There is a possibility that a coincidence of the above circumstances may occur; hence, it is sometimes of consequence to observe the placing of the winch, so as to counteract any defect of compensation.

(1367.) The power of horses to impel the machinery being of nearly uniform intensity, requires no regulator in itself; but it comes under the general law, if the resistance is intermitting. Thus, in the threshing-machine, which is slightly variable in resistance, it would, if worked with horse-power, be considerably improved by the addition of a well proportioned fly-wheel, of which more in another place. In various other machines worked with horse-power, where the resistance is frequently intermitting, such as blowing bellows, pumping, and the like, a fly is indispensable; while in a malt or other mill, whose resistance is uniform, the fly would be an incumbrance. Steam-power applied to a threshing-machine requires, as already observed, no additional fly-wheel; but that of the steam-engine for such purpose, should be above the standard allowed for ordinary purposes.

(1368.) The general theory of the application of fly-wheels may be repeated in a few words. They are usefully employed in all cases of intermitting resistance and of variable force, whether in the first mover or in the resistance. Where the motion is uniform and not intermitting, the first mover being also uniform, the fly-wheel is in almost every case unnecessary, and frequently an obstruction.

(1369.) The determination of the weight of a fly-wheel for any given purpose, is a problem not very definite in its results; but approximations to it have been made by men of eminence. Among these, we find Tredgold stating a rule for the fly-wheels of steam-engines,* which, for practical purposes is convenient, and comes near to the general practice; though this is to be taken with considerable latitude, seeing that the practice of engineers differs considerably on this point; and the rule, though it applies to heavy fly wheels with tolerable exactness, does not agree with practice in the case of fly-wheels for the hand, and other small machines. But the following approximation will be a tolerable guide in practice for the weight of small fly-wheels.

(1370.) Taking the average force that a man will exert in turning a winch of 12 inches radius at 23 lbs., when he turns it 45 times per minute, the rule will be:

RULE.—Multiply 20 times the force in pounds exerted on the winch by its radius in feet, and divide this product by the cube of the radius of the fly-wheel in feet, multiplied into the number of the revolutions per minute; the result will be the area of a section of the rim of the fly in square inches.

EXAMPLE.—The force applied to the winch being 23 lbs., its radius 1 foot, and the revolutions per minute 45; required the section of the rim of a fly-wheel whose radius is 1½ feet, or 3 feet diameter.

$$\frac{20 \times 23 \times 1}{1{\cdot}5^3 \times 45} = \frac{460}{151} = 3 \text{ inches, the area of section of the rim nearly.}$$

(1371.) Though this formula will serve for small fly-wheels, whose velocities range from 40 to 80 revolutions per minute, it becomes necessary, in order to make it agree with practice, to change the constant. Thus, for velocities ranging from 80 to 150, the number 10 will be substituted for 20; and from 150 to 300, the number 5. In this last case, the fly-wheel cannot exceed 2 feet diameter; and in the former, it is restricted to 5 feet.

(1372.) *The Disk Straw-cutter.*—Of the disk straw-cutter the varieties are very numerous, and they form a very important order of this machine; being that, also, which is for the most part employed in England, it is the most numerous of the class. The principal feature, the cutting-knife, fixed upon the fly-wheel, is invariable, except that it sometimes carries one, at other times two knives. The machinery or details are exceedingly varied. In some, it is adapted to cut of various lengths by means of ratchet-wheels and lever-catches applied to the motion of the feeding-rollers, and at the same time to move the substance forward only in the intervals of the stroke of the knife; in others, the latter qualification only is attended to; in a third, a continous motion of the substance is deemed sufficient; and these varieties of motion are produced by other and various arrangements of spur, bevel, and screw geerings.

(1373.) The machine selected for illustration is one in which two knives are employed, and which gives to the substance to be cut a continuous motion forward. The figure here representing this, is taken from a machine manufactured by John Anderson and Son, founders, Leith Walk.

* Tredgold on Steam-Engines.

at a price of £10 10s. Fig. 289 is a view of this machine in perspective. The chief parts of the frame-work are of cast-iron, consisting of a frame *a a* on each side of the machine, which are supported transversely by the truss *b*. The front part of the side-frames extend upward and form

THE DISK STRAW-CUTTER WITH CONVEX KNIVES.

the feeding-roller frame. The cutting-plate is attached in front of the latter portion of the frame work, and is dressed truly off for the passage of the knife over its face. The feeding-trough *e* is connected in the fore-part to the roller-frame, and along its bottom to the upper edge of the side frames. The back end of the trough is supported in a light wooden-frame. The principal shaft *f* is supported on two projecting brackets *g* and *h*, and upon it is mounted the single-thread screw *i*, and the fly-wheel *k*; on the extreme end also of the shaft, the winch-handle *l* is attached. A bracket carries one end of a small shaft, on which the screw-wheel *n*, of 21 teeth, is mounted, and is turned by means of the screw, when the fly-wheel is put in motion. On the opposite end of the small shaft *n*, a spur-wheel is also placed, and acts upon another of equal diameter placed on the axle of the lower feeding-roller. This last, as well as the upper roller, are furnished with the usual long-toothed pinions, for admitting of the rise and fall of the upper roller. The upper roller is supported in a light frame that rises and falls in a slide of the roller-frame, and this is acted upon by a lever and weight, of which the hook only is seen in the figure at *o*. The cutting-knife *p* is 18 inches in length, and 4 inches in breadth. It is firmly bolted upon the arm of the fly-wheel, and its cutting-edge, which is convex, is so formed that every successive point, in passing the edge of the cutting-plate, forms equal angles with the edge of that plate. In many of the disk machines, the cutting-edge of the knife is concave, formed on the same principle of equal angles, and, in effect, is the better of the two.

(1374.) The dimensions of the principal parts of this machine are as follows. Width of the frames 14 inches; length of cast-iron frames 30 inches, and hight 3 feet; length of feeding-rollers 12 inches, and their diameter 4½ inches; length of feeding-trough 5 feet, and width 12 inches. The fly-wheel is 4 feet 3 inches diameter, and the hight to its center is 3 feet. From the entire weight of the fly-wheel being supported at one angle of the frame, the spreading brackets *q r* are attached, to give the machine stability.

(1375.) *The Steaming Apparatus.*—The means employed for cooking food for horses and cattle, are either boiling or steaming. In the first, an open vessel is of course employed, in which the roots or other substances are placed, with a sufficient quantity of water. This method has been found inconvenient in many respects; and when the establishment is extensive, the vessel is required to be incommodiously large, and is withal not economical.

(1376.) Steaming in a separate vessel has been adopted in preference to the former method and has been followed in a variety of forms, but these may be ranked under two distinct kinds. The first is an open vessel, a boiler generally of cast-iron, having a channel or groove of 1 inch wide and 2 inches deep formed round its brim. The vessel is placed over a furnace properly constructed, and is partly filled with water. The groove is also filled with water. A sheet-iron cylindrical pan, of 3 to 4 feet in depth, and of a diameter suited to pass into the groove of the water-vessel (which is generally about 3 feet diameter), is also provided. The pan has a perforated bottom to admit steam freely from the lower vessel. It is also furnished with an iron bow by which it can be suspended, and by which it can be conveniently tilted while suspended. This is the steaming-pan; and for the purpose of moving it to and from the boiler, a crane, mounted with wheel and pinion and a chain, completes the apparatus. To put this in operation, the pan is filled with the substances to be steamed, and covered over either with a deal-cover or with old canvas bags. It is then placed upon the boiler by means of the crane, and the fire being pretty strongly urged till the water in the boiler gives off its steam, which, passing up through the bottom of the pan, and acting upon the contents, produces in a few hours all the results of boiling. The water in the groove of the boiler serves as a sealing to prevent the escape of steam without passing through the pan. But notwithstanding this, it is evident that the steam can hardly ever reach the temperature of 212°; and hence, this apparatus is always found to be very tardy in its effects. When the contents of the pan have been found sufficiently done, the whole is removed from the boiler by means of the crane, and tilted into a large trough to be thoroughly mixed, and thence served out to the stock. A general complaint has been urged against this construction of apparatus, arising from the slowness of the process of cooking by it, and consequent expense of fuel. Boilers of the form here described are not well calculated to absorb the maximum of caloric that may be afforded by a given quantity of fuel, neither is the apparatus generally the best adaptation for the application of

Fig. 290.

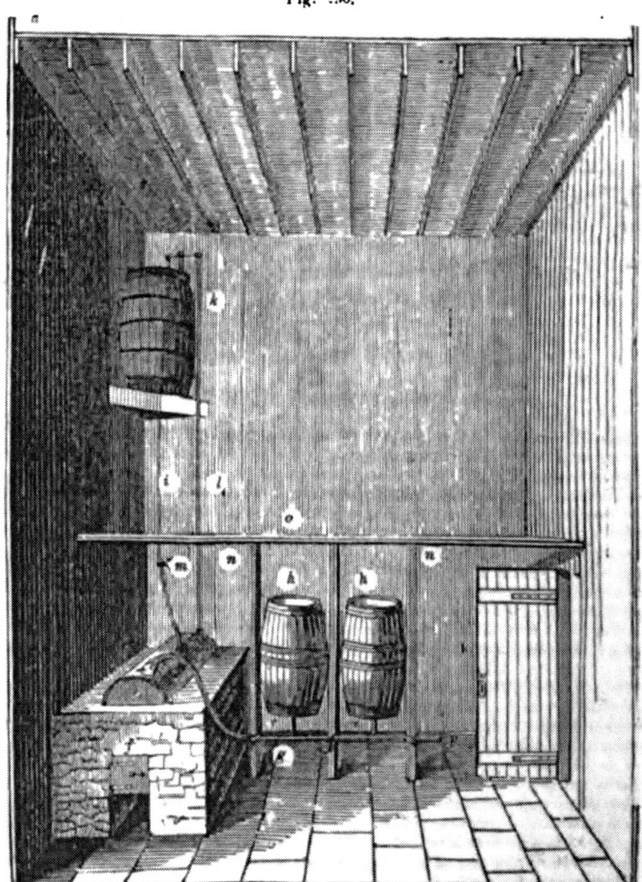

THE CLOSED-BOILER STEAMING APPARATUS.

steam to the substances upon which the steam has to act. Such boilers, as already observed, can never produce steam of a higher temperature than 212°. If they did, the shallow water-luting formed by the marginal groove, would be at once thrown out by the steam-pressure; for it is well known that the addition of 1° to the temperature of the steam increases its elasticity equal to the resistance of a column of water about 7 inches high. A groove, therefore, of 7 inches in depth would be required to resist the pressure, which would even then be only 1·5 lb of pressure on the square inch. Under such circumstances, the temperature in the steaming-pan will always be under 212°. Hence the tedious nature of the process by using this apparatus.

(1377.) The apparatus which deserves the precedence of the above mode is here represented in fig. 290. The principle of its construction is that of a closed boiler, in which the steam is produced under a small pressure of 3 to 4 lbs. on the inch. It is then delivered through a pipe to one or more separate vessels containing the substances that are to be cooked; and these vessels are so arranged as to be readily engaged or disengaged with the conducting steam-pipe. The outline *a b c d* of the figure represents a section of the steaming-house, with the apparatus in due order of arrangement, and of the extent that may be capable of supplying an establishment of from 10 to 16 horses. The boiler *e* is of a cylindrical form, 20 inches in diameter and 4 feet in length. It is set in brick-work *f*, over a furnace of 14 inches in width, with fire-grate and furnace-door. The brick building requires to be 6 feet 6 inches in length, 4 feet 6 inches in breadth, and the hight about 3 feet 6 inches. The furnace is built with a circulating flue, passing first to the farther end of the boiler, then, turning to right or left according as the chimney may be situated, returns to the front of the boiler, and terminates in the chimney on the side opposite to the first turning. The flues should be not less in width at the upper part than one-fourth the diameter of the boiler; and their hight will be about one-third the diameter. The steam-pipe is attached to the boiler at its crown, takes a swan-neck bend downward to within 12 inches of the floor at *g*, and terminates at *p*: it is furnished with as many branch nozzles as there are intended to be steaming-vessels. The steam-pipe may be either cast-iron or lead, and 2 inches diameter in the bore. The receptacles or steaming-vessels *h h* are usually casks of from 50 to 100 gallons contents. They are mounted with 2 iron gudgeons or pivots, placed a little above mid-hight; they are besides furnished with a *false* bottom, supported about 3 inches above the true bottom; the former being perforated with a plentiful number of holes, to pass the steam which is introduced between the two bottoms. The connection between the steam-pipe and the receptacle may be either by a stop-cock and coupling-screw—which is the most perfect connection—or it may be by the simple insertion of the one nozzle within the other, in the form of a spigot and faucet. In this latter case, the nozzle that leads from the steam-pipe is stopped with a wooden plug, when the receptacle is disengaged. Besides the steam-pipe, the boiler is furnished with a pipe *i*, placed in connection with a cistern of water *k*, the pipe entering into it by the bottom, and its orifice closed by a valve opening upward, the lower extremity of the pipe passing within the boiler to within 3 inches of its bottom. A slender rod *l* passes also into the boiler through a small stuffing-box; and to its lower end, within the boiler, is appended a float, which rests upon the surface of the water within the boiler. The upper end of this rod is jointed to a small lever which has its fulcrum supported on the edge of the cistern a little above *k*; the opposite end of the lever being jointed to a similar but shorter rod, rising from the valve in the bottom of the cistern. This forms the feeding apparatus of the boiler, and is so adjusted by weights that when the water in the boiler is at a proper hight, the float is buoyed up so as to shut the valve in the cistern, preventing any farther supply of water to pass into the boiler, until, by evaporation, the surface of the water has fallen so far as to leave the float unsupported, to such extent as to form a counterpoise to the valve, which will then open, and admit water to descend into the boiler, until it has again elevated the float to that extent that will shut the valve in the cistern. By this arrangement, it will be perceived that the water in the boiler will be kept nearly at a uniform hight; but to accomplish all this, the cistern must be placed at a certain fixed hight above the water in the boiler, and this hight is regulated by the laws which govern the expansive power of steam. This law, without going into its mathematical details at present, in so far as regards this point, may be stated in round numbers as follows: That the hight of the surface of the water in the cistern must be raised above the surface of that in the boiler, 3 feet for every pound-weight of pressure that the steam will exert on a square inch of surface in the boiler. Thus, if it is estimated to work with steam of 1 lb. on the inch, the cistern must be raised 3 feet: if 2 inches, 6 feet; 3 inches, 9 feet; and so on. If the steam is by any chance raised higher than the hight of the cistern provides for, the whole of the water in the boiler may be forced up through the pipe into the cistern, or until the lower orifice of the pipe, within the boiler, is exposed to the steam, which will then also be ejected through the pipe; and the boiler may be left dry. Such an accident, however, cannot occur to the extent here described, if the feeding-apparatus is in proper working order; and its occurrence to any extent is sufficiently guarded against by a safety-valve.

(1378.) The safety-valve of the steam-boiler is usually a conical metal valve, and always opening outward; it ought always to be of a diameter large in proportion to the size of the boiler and steam-pipe, so as to insure the free egress of any rapid generation of steam. For a boiler of the size under consideration it should be 2 inches in diameter on its under surface—that being the surface acted upon—this gives an area of fully 3 square inches: and if loaded directly, or without the intervention of a lever, for steam of a pressure of 1 lb. on the inch, it will require 3 lbs.; if 2 lbs. on the inch, 6 lbs.; if 3 lbs. on the inch, 9 lbs., and so on. With these adjustments, the steam, should it rise above the proposed pressure, will, instead of forcing the water through the feed-pipe, raise the safety-valve, and escape into the atmosphere until the pressure is reduced to the intended equilibrium.

(1379.) Another precautionary measure in the use of the steam-boiler is the guage-cock, of which there are usually two, but sometimes one, a two-way cock; they are the common stop-cock, with a lengthened tail passing downward, the one having its tail terminating about 1½ inches below the proper water level in the boiler, the other terminating 1½ inches above that level, which allows a range of 3 inches for the surface of the water to rise or fall. The first, or water-cock, then, when opened, will throw out water by the pressure of the steam upon its surface, until the surface has sunk 1½ inches below its *proper* level, when steam will be discharged, thus indicating the water in the boiler to be too low, and that measures should be taken to increase the supply. When the second or steam-cock is opened, it will always discharge steam alone, unless the water shall have risen so high as to come above its orifice, in which case the cock will discharge *water*, indi eating a too large supply of water to the boiler, and that it should be reduced; for this purpose, the feed-pipe *i* is provided with a stop-cock *m*, whereby the admission of water can be entirely prevented at the pleasure of the attendant.

(1380.) The foregoing description refers to a steaming-apparatus of the best description, and implies that the water-cistern can be supplied either from a fountain-head, or that water can be pumped up to the cistern. But there may be cases where neither of these are easily attainable. Under such circumstances the feed-pipe may rise to the hight of 4 or 4½ feet, and be surmounted by a funnel, and under it a stop-cock. In this case, also, a float with a wire stem, rising through a stuffing-box on the top of the boiler, must be employed—the stem may rise a few inches above the stuffing-box, in front of a graduated scale—having the zero in its middle point. When the water is at the proper hight in the boiler, the top of the stem should point at zero, and any rise or fall in the water will be indicated accordingly by the position of the stem. To supply a boiler mounted after this fashion, the first thing to be attended to, before setting the fire, is to fill up the boiler, through the funnel, to the proper level, which will be indicated by the float pointing to zero; but it should be raised, in this case, two or three inches higher. In this stage, the gauge-cocks are non-effective; but when the steam has been got up, they, as well as the float, must be consulted frequently; and should the water, by evaporation, fall so low as 3 inches below zero, a supply must be introduced through the funnel. To effect a supply, in these circumstances, the steam must be allowed to fall rather low, and the funnel being filled, and the stop-cock opened, the water in the former will sink down through the tube, provided the steam be sufficiently low to admit its entrance, but the first portion of water that can be thus thrown in will go far to effect this, by sinking the temperature. The sinking of the temperature by the addition of a large quantity of cold water, is the objection to this mode of feeding; but this is obviated to some extent from the circumstance, that unless the steaming receptacles are large or numerous, the first charge of water will generally serve to cook the mess, when a fresh charge can be put in for the next.

(1381.) In using this steaming-apparatus, it has been noticed that the casks are furnished with gudgeons, which play in the posts *n n*; these are kept in position by the collar-beam *o* to which they are attached; the casks being at liberty to be tilted upon these gudgeons. They are charged when in the upright position, and the connection being formed with the steam-pipe, as described, they are covered at top with a close lid or a thick cloth, and the process goes on. When the substances are sufficiently cooked, the couplings *r r* are disengaged, the upper part of the cask is swung forward, and their contents discharged into a trough which is brought in front of them for that purpose.

(1382.) The connections with the steam-pipe are sometimes, for cheapness, formed by a *sliding tube* of copper or brass, about 4 inches in length, which, after the nozzle of the cask and that projecting from the steam-pipe are brought directly opposite to each other, is slid over the junction, and as a moderate degree of tightness only is requisite in such joints, a strip of sacking wrapped round the ends of the slider is found sufficient. On breaking the connection, and opening the exit nozzles, the steam will of course flow out, but this is checked by a wooden plug, or even a potato or slice of turnip, thrust into the orifice, may be sufficient. It is advisable, however, that a main stop-cock should be placed in the steam-pipe anywhere between the boiler and the first receptacle.

(1383.) The most perfect connection between the steam-pipe and the receptacles is a *stop-cock and coupling-screw*. These should be of 1¼ inches bore, they are more certain in their effect, and more convenient in their application, though attended with more expense in the first cost of the apparatus. In this case no main-cock is required. The extremity of the steam-pipe should, in all cases, be closed by a small stop-cock, for the purpose of draining off any water that may collect in the pipe from condensation. A precaution to the same effect is requisite, in the bottom of each cask, to draw off the water that condenses abundantly in it; or a few small perforations in the bottom will effect the purpose.

(1384.) It must be remarked, in regard to steaming, that in those establishments where grain of any kind is given in food in a cooked state, dry grain cannot be cooked, or at least boiled to softness in dry steam, the only effect produced being a species of parching; and if steam of high temperature is employed, the parching is increased nearly to carbonization. If it is wished, therefore, to boil grain by steam, it must be done by one of the two following methods: The grain must either be soaked in water for a few hours, and then exposed to the direct action of the steam in the receptacle; or it may be put into the receptacle with as much water as will cover it, and then by attaching the receptacle to the steam-pipe, by the coupling stop-cock, or in the absence of stop-cocks, by passing a bent leaden pipe from the steam-pipe, over the upper edge of the receptacle and descending again inside, to the space between the false and the true bottoms; the steam discharged thus, by either method, will shortly raise the temperature of the water to the boiling point, and produced the desired effect.*

(1385.) The time required to prepare food in this way varies considerably, according to the state of the apparatus, and the principle of its construction. With the apparatus just described, potatoes can be steamed in casks of from 32 to 50 gallons contents, in 30 to 45 minutes. In casks extending to 80 gallons, an hour or more may be required. Turnips require considerably longer time to become fully ready, especially if subjected to the process in thick masses, the time may be stated at double that of potatoes. When the apparatus is ill constructed, the time, in some cases, required to cook turnips, extends to 5 hours. And, with reference to the apparatus first described (1376), the time is seldom under 5 hours.

(1386.) The prices of steaming-apparatus vary accordingly to quality and extent; but, on an average, the open boiler and pan apparatus, including a power-crane, will range from £7 to £10; and the other, fig. 290, the price ranges from £8 to £16. The expense of building the furnace, and supplying mixing-troughs, will add about £2 10s. to each.

* See an article by me in the Quarterly Journal of Agriculture, vol. vi.

(1387.) *Corn-Bruisers.* In following up an economical system of feeding, the bruising of all grain so applied forms an important branch of the system, and, as might be expected, numerous are the varieties of machines applied to the purpose. These naturally arrange themselves under *three* distinct kinds. 1st. Machines which act on a principle that partakes of cutting and bruising, by means of grooved metal cylinders, and is applied to those chiefly driven by the hand. 2d. Machines adapted to bruise only by means of smooth cylinders; this is applied exclusively to those driven by steam, or other agency more powerful than the human hand. And, 3d, Breaking or grinding by the common grain millstones, and of course, only worked by power.

(1388.) That variety of the first division which I shall particularly notice, is represented in perspective in fig. 291. It is constructed almost entirely of cast-iron, except the hopper and discharging-spout; but its frame or standard may with propriety be formed of hard-wood, when circumstances render the adoption of that material desirable. In the figure, *a a a* is the frame-work, consisting of two separate sides, connected by two stretcher-bolts, the screw-nuts of which are only seen near to *a* and *a* below. A case *b b*, formed of cast-iron plates, is bolted on the projecting ears at the top of the frame, and contains the bruising cylinders. The cylinders are 4 inches in diameter, and 6 inches in length, of cast-iron or of steel. They have an axle of malleable iron passing through them, having turned journals, which run in bearings formed on the cast-iron side-plates of the case, the bearings being accurately bored out to fit the journals. The spur-wheels *c* and *d* are fitted upon the axle of the cylinders, *c* having 14 teeth, and *d* 24 teeth. The cylinder corresponding to *d* is perfectly smooth, while that of *c* is grooved at a pitch of ¼ inch, and about 3-32 inch deep worked to sharp edges. The grooves lie obliquely on the face of the cylinder, being at an angle of 10° with the axle. The winch-handle *e* is attached to the axle of the roller *c*, whose bearings are permanent, while those of *d* are movable, being formed in separate plates, and fitted to slide to a small extent in a seat, for the adjustment of the cylinder to any desired grist.

Fig. 291.

THE HAND CORN-BRUISER.

This adjustment is effected by means of the screws *f*, which act upon the sliding-plates of the bearings. *g* is one of the bearings of a feeding-roller, placed also within the case; it is turned by means of a toothed-wheel fitted upon the farther end of its axle, and which is driven by another wheel of 24 teeth on the axle of the cylinder *d*. The fly-wheel *h* is fitted upon the axle of the cylinder *c*, and is 3½ feet in diameter; *i* is the feeding-hopper, attached to the top of the case by two small hooks; and *k* is a wooden spout to convey the bruised grain from the case.

(1389.) Fig 292 is a section of the case, and the cylinders, detached from the frame. *b b* are the two ends of the case cut by the section; *c* is the grooved cylinder, *d* the smooth, and *l* is the feeding-roller, it is 3½ inches diameter, and has cylindrical grooves formed on its surface to convey the grain; *o* is a cover of cast-iron fixed upon the top of the case; it has two rounds ears *n n*, with eye-holes which serve to steady the hopper, and to which it is screwed by the hooks already mentioned. A hopper-shaped opening *m* is formed in the cover; it is six inches long, 3 inches wide at top, and 1 inch at bottom, and the edges fit closely upon the feeding-rollers. Two plate-iron sliders are fitted upon the surface of this little hopper, which serve to enlarge or contract the opening longitudinally, and are fixed by screw-bolts in each plate; the head of one of the bolts is seen at *o*. *p p p p* are ears by which the case is bolted together, and *q q* are prolongations of the side-plate of the case; *r r* are additional plates of sheet-iron, to prevent the grain from being thrown over the cylinders unbruised.

Fig. 292.

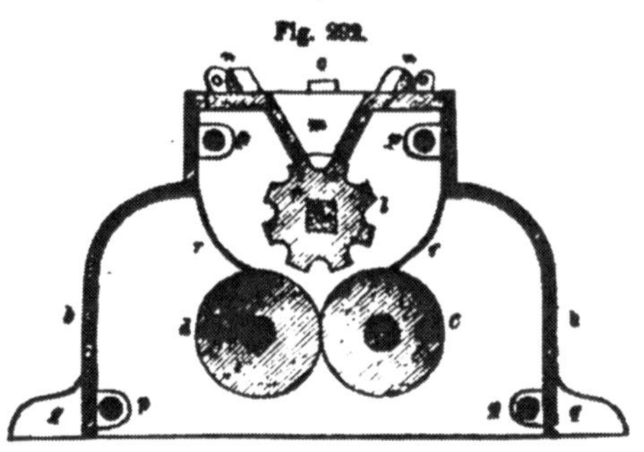

TRANSVERSE SECTION, SHOWING THE RELATION OF THE PRINCIPAL PARTS.

(1390.) This is a very efficient machine for bruising either oats or beans; the adjustment of the plain cylinder to the requisite distance being easily accomplished by the adjusting-screws; and

Fig. 293.

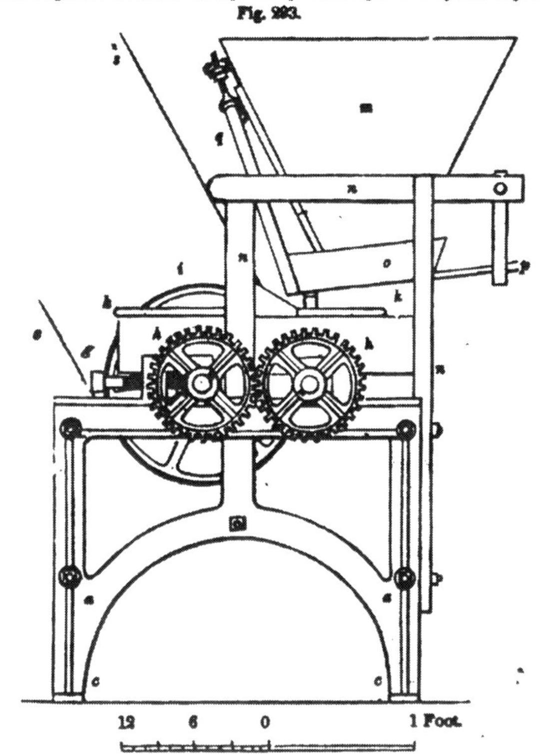

THE ELEVATION OF THE POWER CORN-BRUISER.

to prevent the abrasion of the grooved cylinder, by coming in too close contact with the other, a stopper is applied on each side, to keep the slides from overreaching the due safety distance. From the different velocities of the two cylinders, the grooved one being the fastest, it produces a cutting as well as a bruising action, which renders its effects on the grain more perfect than simple pressure. It can be worked by one man, who will bruise 4 bushels of oats in an hour The price of the machine is £6 10s.

(1391.) Various other forms of this machine are in use, some with both cylinders grooved, others with only one grooved cylinder acting against a grooved plate; in this last state it is much used for bruising beans.

(1392.) Among the varieties of the bruising-machine of the 2d division, I may just notice one that is found very efficient. It consists of two plain-edged wheels or pulleys, as they may be termed, usually about 6 inches broad on the rim or sole; the one ranges from 2½ feet to 4 feet in diameter, and its fellow only half the diameter of the larger. They require to be truly turned on the rim, and work in contact. The smaller one is always driven by the power, and the larger usually by contact with the smaller. The smaller wheel makes, according to its diameter, from 150 to 200 revolutions per minute. Where plain cylinders are employed for bruising, and their surfaces moving with equal velocity, the effect is to press each grain into a flat, hard cake: but when one of the surfaces is left at liberty to move by simple contact, it is found that the effect is different from the above, for the grain passes, bruised indeed, though not into a hard cake, but has apparently undergone a species of tearing, leaving it in a more open and friable state than as described above. This machine, however, does not answer well for bruising beans, for here, again, they come through in the form of a flat cake. If beans, therefore, are used in an establishment where this bruiser is adopted, a separate one, on the principle of fig. 291, is required for the beans alone; that machine though serviceable for a small establishment, being incapable, even with power, to produce the quantity in a reasonable time that would be required in a large one.

(1393.) *Plain Roller Corn-bruisers, for power.*—A very efficient corn-bruiser, adapted for power, is shown in figs. 293, 294, 295. The first being an elevation, the second a plan, and the third a section of the machine; the same letters apply to such corresponding parts as are seen in all the three figures. In fig. 293 *a a* is one of the side-frames, of cast-iron, which are connected together by stretcher-bolts *b*, and the frame so formed is bolted to a floor through the palms at *c c*. On the top bar of the frames there are two strong snugs *d d* cast, sufficient to resist the pressure of the rollers, and are formed also to receive the brass bushes in which the journals of the two rollers are made to run. The two rollers *e* and *f* are respectively 8 and 9½ inches diameter, and are 18 inches in length, fitted with malleable iron shafts 1¾ inches diameter; the roller *f* runs in permanent bearings, but *e* has its bushes movable, for adjustment to the degree of bruising required, and this adjustment is effected by the adjusting-screws *g*. The shaft of each roller carries a wheel *h*, equal in diameter, which is nine inches. The roller *e* has also upon its shaft the driving pulley *i*, which by means of a belt *s s* from any shaft of a threshing-machine or other power having a proper velocity, puts the rollers in motion. The rollers are inclosed in a square wooden case *k k*, in the cover of which a narrow hopper-shaped opening *l* is formed to direct the grain between the rollers. A hopper *m* for receiving the grain is supported on the light wooden frame-work *n n*, which also supports the feeding-shoe *o*, jointed to the frame at *p*, and suspended by the straps *q*, which last is adjustable by a screw at *q* to regulate the quantity of feed. A smooth-edged oblique wheel *r*, fig. 294, is mounted on the shaft of the roller *f*, and by its oscillating revolutions, acting upon a forked arm which descends from the shoe, a vibratory motion is given to the latter, by which a regular and continued supply of the grain is delivered from the hopper to the rollers. After passing the rollers the grain is received into a spout, which either delivers it on the same floor, or through a close spout in the floor below. The velocity of the rollers, which are driven by the belt *s s*, may be 250 revolutions per minute. The dimensions of the frame *a* are 30 inches in length and 24 inches in hight; the width over all being also 24 inches. The price of this machine, as manufactured by James Slight and Co., is £10.—J. S.

Fig. 294.

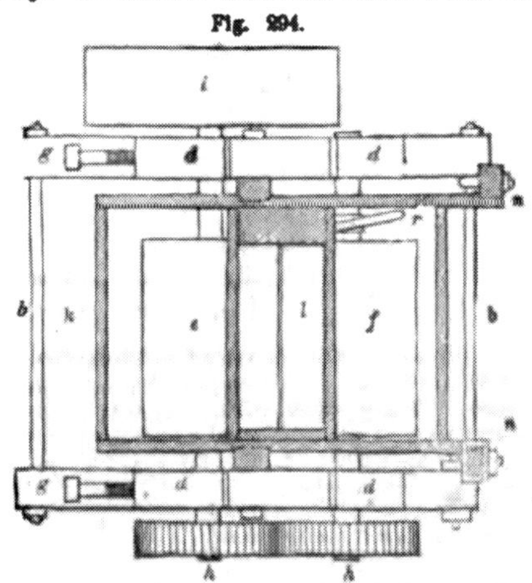

THE PLAN OF THE POWER CORN-BRUISER.

(1394.) The horse is an intelligent animal, and seems to delight in the society of man. It is remarked by those who have much to do with blood-horses, that, when at liberty, and seeing two or more people standing conversing together, they will approach, and seem, as it were, to wish to listen to the conversation. The farm-horse will not do this; but he is quite obedient to call, and distinguishes his name readily from that of his companion, and will not stir when desired to stand until *his own name* is pronounced. He distinguishes the various sorts of work he is put to, and will apply his strength and skill in the best way to effect his purpose, whether in the threshing-mill, the cart, or the plow. He soon acquires a perfect sense of work. I have seen a horse walk very steadily toward a feering-pole, and halt when his head had reached it. He seems also to have a sense of time. I have heard another neigh almost daily about 10 minutes before the time of loosening in the evening, whether in summer or winter. He is capable of distinguishing the tones of the voice, whether spoken in anger or otherwise; and can even distinguish between musical notes. There was a work horse of my own, which, even at his corn, would desist eating and listen attentively, with pricked and moving ears, and steady eyes, the instant he heard the note of low G sounded, and would continue to listen as long as it was sustained; and another, that was similarly affected by a particular high note. The recognition of the sound of the bugle by a trooper, and the excitement occasioned in the hunter when the pack give tongue, are familiar instances of the extraordinary effects of particular sounds on horses.

(1395.) When alluding to the names of horses, I may mention that they should be *short and emphatic*, not exceeding two syllables in length, for longer words are difficult of pronunciation, and inconvenient to utter when quick or sharp action is required of the horse; and a long name is almost always corrupted into a short one. For geldings, Tom, Brisk, Jolly, Tinker, Dragon,

Dobbin, seem very good names; for mares, Peg, Rose, Jess, Molly, Beauty, Mettle, seem as good and as to the name of stallions, they should be somewhat high-sounding, as indicative of greater importance of character, as Lofty, Farmer, Plowboy, Matchem, Diamond, Blaze, Samson, Champion, are names which have all distinguished first rate draught horses.

Fig. 295.

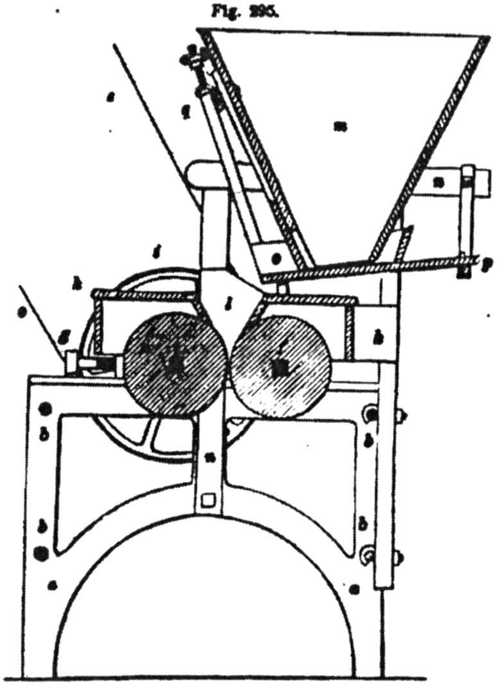

THE SECTION OF THE POWER CORN-BRUISER.

(1396.) This seems a befitting place to say a few words on the farmer's *riding* and *harness* horses. Usually a young lad, or groom, is hired to take charge of these, to go errands and to the post-office, and otherwise make himself serviceable in the house. Sometimes the hedger or shepherd acts the part of groom. My shepherd acted as groom, and his art in grooming is so skillful that many friends remarked to me that they would be glad to see their *professed* grooms turn out a saddle-horse or gig in so good a style as he did. Besides being useful in carrying the farmer to market, or other short distances, a roadster is required to carry him over the farm when it is of large extent, and when the work-people necessarily receive pretty constant attention in the important operations of seed-time and harvest. The harness-horse is useful to a family at all times, as well as to the farmer himself, when he visits his friends; and many farmers now prefer riding to market in a gig or drosky, to horseback; and it must be owned to be the pleasanter mode of the two.

(1397.) I have said that the agricultural pupil should have no horse of his own at first, to tempt him to leave home and neglect his own training. But to know how a riding-horse ought to be kept by a groom, and to be able to correct him when he neglects his duty or performs it in an unsatisfactory manner, I would advise him to undertake the charge of one himself for some time; not merely to superintend its keeping, but to clean it himself, to water and corn it at stated times at morning, noon, and night, and to keep the saddle and bridle in proper order. I groomed a new-broke-in blood filly for four months one winter, and got more insight into its form, temper, and management, and wants, than I could have obtained by observation alone in a much longer time. On coming home at night from visiting a friend, I made it a point with myself to make my charge comfortable for the night before thinking of my own rest.

(1398.) A saddle-horse is treated somewhat differently in the stable from a work-horse. The first thing to be done early in the morning is to shake up the litter nearest the strand with a fork, removing the dung and soiled straw to a court-yard, and sweeping the floor clean. Then give the horse a drink out of the pail which is constantly kept full of water in the stable. The usual practice is to offer the water *immediately* before giving the corn; but I conceive it more conducive to the health of the horse to slake his thirst a while before giving him corn, the water by that time having reached its destined place, and acquired the temperature of the body. Should the horse have to undertake a longer journey than walking about the farm, a stinted allowance of water before starting on the journey is requisite, say to 10 gluts; but if he is to be at home, then he may drink as much as he pleases. He is then groomed by being, in the first place, gently gone over the whole body with the curry-comb, to loosen the mud that may possibly have been left adhering upon the hair from the former night's grooming, and also to raise the scurf from the skin

The whole body should be then wisped down with straw, to clear off all the dust and dirt that the curry-comb may have raised to the surface. The brush follows, to clear the hair of its dust and scurf, the curry-comb being used to clean the brush. Of wisping and brushing, wisping is the more beneficial to the legs, where the hair is short and the tendons and bones are but little covered, because it excites in them warmth, and cleans them sufficiently. Both wisping and brushing should be begun at the head and terminated at the other end of the body, along the lie of the hair, whichsoever way that may be, and which, notwithstanding its different swirls, all tends from the upper to the lower part of the body. Many a groom rests content with the brushing just mentioned; but it does not entirely remove the dust raised to the surface, and therefore a wisping is required to do it. The wisp for this purpose is best made of Russia mat, first wetted, and then beaten to softness, and rolled up somewhat firmly into the form of a wisp sufficient to fill the hand. A wisp of hair-cloth makes the skin clean, but in dry weather it is apt to excite such a degree of electricity in the hair of the horse as to cause it to attract much dust toward it. On the horse being turned round in the stall, his head, neck, counter, and fore-legs, should be well rubbed down with this wisp, and this done, he should again be turned to his former position, and the body, quarters, and hind-legs, then rubbed down; and when all this has been accomplished, the horse may be considered clean. All this grooming implies the bestowal of much more labor than most farmers' riding-horses receive. They are usually scuffled over in the morning with the curry-comb, and then skimmed down with the brush, and with a hasty combing of the mane and tail, the job is considered finished. The mane and tail ought to be carefully combed out, and wetted over at the time of combing with a half-dry water-brush. The sheet should then be thrown over the horse, and fastened (not too tightly) with the roller. On putting on the sheet, it should be thrown more toward the head of the horse than where it is intended to remain, and thence drawn gently *down the hair* with both hands, to its proper position, while standing behind the horse. The litter is then neatly shaken up with a fork, taking care to raise the straw so far up the travis on each side as to form a cushion for the side of the horse to rest against when he lies down. The feed of corn is then given him, and a little hay thrown into the rack; and on the stable-door being shut, he is permitted to enjoy his meal in peace. At mid-day, he should have another drink of water from the pail, the dung removed, the litter shaken up, and another portion of oats given him. At 8 o'clock at night, the sheet should again be taken off, the curry-comb and brush used, and the entire dressing finished again with a wisping of the Russia mat. The sheet is thrown over him as in the morning, the litter shaken up and augmented, water given, and the supper of oats, or a mash, finishes the day's treatment of the saddle-horse.*

(1399.) The treatment just described is most strictly applicable to the horse remaining all day in the stable; but when he is ridden out, a somewhat different procedure is required. When he comes home from a long and dirty ride, the first thing is to get clear of the mud on the belly and legs. A very common practice is to wade the horse through the pond, as the farm-horses are, but such should not be the course pursued with a saddle-horse, because wading through a pond cannot thoroughly clear his legs of mud *to the skin*, he being clean-shanked and smooth-haired, and there still remains the belly to be cleaned by other means than wading. The plan is, being that adapted for winter, to bring the horse into the stable upon the pavement, and, on taking off the saddle and bridle and putting on a halter, scrape all the mud as clean off the belly and legs as can be done with a knife—a blunt table-knife answers the purpose well. Then, with a pailfull of lukewarm water, wash down *the legs*, outside and inside, with a water-brush, and then each foot separately, picking out the mud with the foot-picker; then wash the mud clean from the belly. A scrape with the back of the knife, after the washing will bring out all the superfluous water from among the hair. On going into the stall the horse should be wisped firmly with straw, rubbing the belly first, and then both sides of each leg until they are all thoroughly dry. It is scarcely possible to get the belly dry at once; it should, therefore, get another good wisping with dry clean straw after the legs are dry. On combing out the mane and tail, putting on the sheet, and bedding plentifully with dry straw, the horse will be out of danger, and feel pretty comfortable even for the night; but should he have arrived some time before the evening time for grooming,

---

* In giving these details, and all which follow, we are aware that many of them are too minute, and otherwise ill adapted to American life and management generally; but then the reader must reflect that there are also many patrons of The Farmers' Library who are wealthy men, living in cities, who keep their carriage horses and pet nags for the saddle, with their superfluous grooms and all necessary appliances, to follow out all these directions. And besides, a sensible man never objects to knowing how any department of industry or of pleasure is managed in the country, where such departments have been carried on, as those have in England, to the highest pitch of perfection and refinement. And here we may relate an anecdote of a shrewd Massachusetts man of the world, who was some years since on a tour in England, and on a visit to the Duke of Buccleuch. After showing him many displays of the splendor and luxury of aristocratic living, His Grace took him to his stables, where genius and wealth had combined, to see how much money could be expended, and how much elegance displayed, on such an object. Our Yankee friend perhaps supposing he was expected to be confounded and overcome with astonishment at all he saw, instead of lifting up his eyes and exclaiming as did the Queen of Sheba when she visited Solomon in all his glory, went peering under and about the troughs and stalls; and, on being asked by the Duke if there was anything deficient in his stable appointment; answered "There is but one thing, your Grace, that is wanting to their completeness. I was looking for a *silver pot de chambre*, for each of your Grace's hunters." [Ed. Farm. Lib.

the curry-comb and wisp then applied will remove any moisture or dust that may have been overlooked before.

(1400.) Considerable apprehension is felt in regard to wetting the abdomen of horses, and especially at night, and the apprehension is not ill founded, for if the wet is allowed to remain, even to a small degree, quick evaporation ensues from the excited state of the body consequent on exercise, and rapidly reduces the temperature of the skin. The consequence of this coldness is irritation of the skin, and likely grease on the legs, and this is the danger of wetting the bellies of farm-horses, and of any sort of horse, with *cold* water: for *warm* water cleans the hair and makes it dry sooner, even that on the abdomen, which is generally much longer than that of the legs, but, on the other hand, unless as much labor is bestowed as will dry the skin, and which is usually more than can be expected to be given by ordinary country grooms, it is safer for the horse to remain in a somewhat dirty state, than to run the risk of any inflammation by neglected wet limbs and abdomen. At the same time, if the requisite labor *shall* be bestowed to render the skin completely dry, there is even less risk in wetting the belly than the legs, inasmuch as the legs, in proportion to their magnitude, expose a much larger surface for evaporation, and are not so near the source of animal heat as the body.

(1401.) Saddle-horses receive oats in proportion to the work they have to perform, but the least quantity that is supposed will keep them in such condition as to enable them to do a good day's work at any time, is three half-feeds a day, one in the morning, another at mid-day, and the third at night. When subjected to *daily* exercise, riding-horses require 3 feeds a-day, and an extra allowance for extra work, such as a long journey. A mash once a-week, even when on work daily, is requisite; but when comparatively idle, a part of the mash, whenever prepared for the work-horses, may be administered with much advantage. I am no advocate of a bran-mash to a horse in good health, as it serves only to loosen the bowels without bestowing any nourishment. Boiled barley is far better. A riding-horse should have hay, and not straw, in winter; and he will eat from ½ to ¾ of a stone of 22 lbs. every day.

(1402.) On *cleaning harness* there should be two pairs of girths in use with the saddle, when the horse has much work to do, to allow each pair to be thoroughly cleaned and dried before being again used. The best way to clean girths is first to scrape off the mud with a knife, and then to wash them in *cold* water, and hang them up so as to dry *quickly*. *Warm* water makes them shrink rapidly, and so does long exposure to wet. If there is time, they should be washed in the same day they have been dirtied; but if not, on being scraped at night, they should be washed in the following morning, and hung up in the air to dry, and if the air is damp, let them be hung before the kitchen fire. Girths allowed to dry with the mud on soon become rotten and unsafe. The stirrup leathers should be taken off and sponged clean of the mud, and dried with a cloth. The stirrup-irons and bit should be first washed in water, and then rubbed dry with a cloth immediately after being used. *Fine* sand and water, on a thick woolen rag, clean these irons well, and a dry rub afterward with a cloth makes them bright. Some smear them with oil on setting them past to prevent rust, but oil, on evaporation, leaves a resinous residuum to which dust readily adheres, and is not easily taken off afterward. The curb-chain is best cleaned by washing in clean water, and then rubbed dry and bright by friction between the palms of both hands. The saddle-flaps should be sponged clean of mud, and the seat sponged with a wrung sponge, and rubbed dry with a cloth. Carriage-harness should be sponged clean of mud, kept soft and pliable with fine oil, and, when not japanned, blackened with the best shoe-black. There should be no plating or brass on a farmer's harness; plain iron japanned, or iron covered with leather, forming the neatest, most easily kept, and serviceable mounting. Bright metallic mountings of every kind soon assume the garb of the shabby genteel in the hands of an ordinary rustic groom.

(1403.) In regard to the *diseases* of the horse, if we were to regard in a serious light the list of frightful maladies incident to that animal, which every work on veterinary science contains, we would never purchase a horse; but fortunately for the farmer, his horses are exempt from a large proportion of those maladies, as almost every one relating to the foot, and their consequences, are unknown to them. Nevertheless, many serious and fatal disorders do overtake farm-horses in their usual work, with the symptoms of which you should be so far acquainted as to recognize the nature of the disease; and as you should be able to perform some of the simpler operations to assist the animal in serious cases until the arrival of the veterinary surgeon, a short account of these operations may prove useful. One or more of them, when timely exercised, may have the effect of soon removing the symptoms of less serious complaints. They consist of bleeding, giving physic and drenches, applying fomentations, poultices, injections, and the like.

(1404.) *Bleeding.*—"In the horse and cattle, sheep and dog, bleeding, from its greater facility and rapidity," says Professor Dick, "is best performed in the jugular or neck-vein, though it may also be satisfactorily performed in the *plate* and *saphena* veins, the former coming from the inside of the arm, and running up directly in front of it to the jugular; the latter, or thigh-vein, running across the inside of that limb. Either the fleam or lancet may be used. When blood is to be drawn, the animal is blindfolded on the side to be operated upon, and the head held to the other side; the hair is smoothed along the course of the vein by the moistened finger, the point selected being about 2 inches below the angle of the jaw. The progress of the blood toward the heart is to be obstructed, and the vein thus made sufficiently permanent and tense. A large-bladed fleam and a good-sized lancet are preferable, as the benefit of the operation is much increased by the rapidity with which the blood is drawn. From 6 to 10 pints imperial is a moderate bleeding for the horse and ox, regulated in some degree by the size. From 12 to 16 or even 20 pints is a large one: and sometimes in skillful hands, it is expedient to bleed till fainting is induced, and the animal drops down under the operation. The vessel in which the blood is received should be such that the quantity can be readily ascertained. When this is sufficient, the edges of the wound are to be brought *accurately together*, and kept so by a *small sharp pin* being passed through them, and retained by a little tow. It is of importance, in closing the wound, to see it *quite close*, and that *no hairs or other foreign bodies* interpose. For a time the head should be tied up, and care taken that the horse does not injure the part."

(1405.) The dangers arising from carelessness in blood-letting are not numerous; and "the first of which, though it may alarm the inexperienced, is very trifling. It is a globular swelling, *thrombus*, sometimes as large as the fist, arising immediately around the new-made incision. The filtrating of the blood from the vein into the cellular membrane, which is the cause of the disease, is rarely very copious. Gentle pressure may be used at first, and should be maintained with a well-applied sponge and bandage, kept cool with cold lotion. Occasionally there is *inflammation of the jugular* from bleeding. . . . The cause is usually referred to the use of a foul fleam, or from allowing hairs to interfere with the accurate adjustment of the edges of the wound. The first appearance indicative of the disease is a separation of the cut edges of the integuments, which become red and somewhat inverted. Suppuration soon follows, and the surrounding skin appears tumefied, tight and hard, and the vein itself, above the orifice, feels like a hard cord. After this the swelling of the neck increases, accompanied with extreme tenderness, and now there is constitutional irritation, with tendency to inflammatory fever. . . . In the first stage we must try to relieve by evaporating lotions or by fomentation. If these fail, and as soon as the disease begins to spread in the vein, the appropriate remedy is to touch the spot with the actual cautery, simply to sear the lips of the wound, and apply a blister over it, which may be repeated. Purgatives in full doses must be administered, and the neck, as much as possible, kept steady and upright."

(1406.) *Blistering.*—"Blistering plasters are never applied to horses. An ointment is always used, of which rather more than half is well rubbed into the part to be blistered, while the remainder is thinly and equally spread over the part that has been rubbed. When there is any danger of the ointment running and acting upon places that should not be blistered, they must be covered with a stiff ointment made of hog's lard and beeswax, or kept wet with a little water. . . . The horse's head must be secured in such a way that he cannot reach the blister with his teeth. . . . When the blister has become quite dry the head may be freed. Sometimes it remains itchy and the horse rubs it; in that case he must be tied up again. . . . When the blister is quite dry put some sweet-oil on it, and repeat it every second day. Give time and no work, otherwise the horse may be blemished by the process."

(1407.) *Physicing.*—"Physicing, which, in stable language, is the term used for purging, is employed for improving the condition when in indifferent health, and as a remedy for disease. The medicines chiefly used are—for *horses*, Barbadoes aloes, dose from 3 to 9 drachms; croton *bean*, from 1 scruple to ½ drachm, or *cake*, from ½ drachm to 1 drachm, to which may occasionally be added calomel, from 1 to 1½ drachms. For *cattle*, aloes in doses somewhat larger than for the horse; Epsom salts, or common salt, dose from 1 lb. to 1½ lbs., with some stimulus, as ginger, anise or carraway-seed; also linseed-oil, dose 1 lb., and croton-oil, 15 to 20 drops, or the bean or cake, the same as in the horse. For *dogs*, jalap, dose 1 drachm, combined with 2 grains of calomel; croton-oil, dose 2 drops; bean, 5 grains; and syrup of buckthorn, dose 1 oz. These, it will be observed, are average doses for full-grown animals; in the young and small they may be less, in the large they may require to be greater; but much injury has often been done by too large doses too frequently repeated. To the *horse* physic is usually administered in the form of a bolus or *ball*; to *cattle* by drinking or *drenching*, though for both either way may be employed. A ball is conveniently made of linseed-meal, molasses and the active ingredient, whether purgative, diuretic or cordial; it should be softish, and about the size of a pullet's egg. In administering it the operator stands before the horse, which is unbound and turned with its head out of the stall, with a halter on it. An assistant stands on the left side to steady the horse's head and keep it from rising too high; sometimes he holds the mouth, and grooms generally need such aid. The operator seizes the horse's tongue in his left hand, draws it a little out and to one side, and places his little finger fast upon the under jaw; with his right hand he carries the ball smartly along the roof of the mouth, and leaves it at the root of the tongue; the mouth is closed and the head is held till the ball is seen descending the gullet on the left side. When loth to swallow, a little water may be offered, and it will carry the ball before it. A hot, troublesome horse should be sent at once to a veterinary surgeon. Instruments should, if possible, be avoided, and adding croton farina to the mash often answers the purpose." *Drenches* should be given with caution to either horse or ox; "that no unnecessary force be used, that they be never given by the nostrils, and especially that, if the slightest irritation is occasioned in the windpipe, the animal shall immediately be set at liberty, that, by coughing, he may free himself of the offending matter." "The horse must undergo *preparation for physic*, which is done by gently relaxing the bowels. During the day his food should be restricted to bran-mashes, a ½ peck being sufficient for a feed, and this, with his drink, should be given warm; corn should be withheld and hay restricted. He may have walking and trotting exercise morning and evening. The physic is given on an empty stomach early in the morning; immediately after, a bran-mash is given; that over, the horse goes to exercise for perhaps an hour, and is watered when he returns. The water should be as warm as he will take it; and he should have as much as he pleases throughout the day; bran-mash should be given as often as corn usually is, and better warm than cold; if both are refused bran may be tried, but no corn, and but little hay. Sometimes gentle exercise may be given in the afternoon, and also next day. The physic usually begins to operate next morning, though it rarely takes effect in 12 hours, frequently not for 30. When the physic begins to operate, the horse should stand in the stable till it *sets*, which may be in 12 hours."* The stable should be well littered behind the stall to receive the discharge. "Many practitioners and horse-proprietors," says Mr. Youatt, "have a great objection to the administration of medicines in the form of drinks. . . . There are some medicines, however, which must be given in the form of drink, as in colic. . . . An ox-horn, the larger end being cut slantingly, is the usual and best instrument for administering drinks. The noose of a halter is introduced into the mouth, and then, by means of a stable-fork, the head is elevated by an assistant considerably higher than for the delivery of a ball. The surgeon stands on a pail on the off-side of the horse and draws out the tongue

* Dick's Manual of Veterinary Science.

with the left hand. He then with the right hand introduces the horn gently into the mouth and over the tongue, and, by a dexterous turn of the horn, empties the whole of the drink—not more than about 6 oz.—into the back part of the mouth. The horn is now quickly withdrawn and the tongue loosened, and the greater portion of the fluid will be swallowed. A portion of it, however, will often be obstinately held in the mouth for a long time, and the head must be kept up until the whole is got rid of, which a quick but violent slap on the muzzle will generally compel the horse to do. The art of giving a drink consists in not putting too much in the horn at once, introducing the horn far enough into the mouth, and quickly turning and withdrawing it without bruising or wounding the mouth, the tongue being loosened at the same moment. A bottle is a disgraceful instrument to use, except it be a flat pint bottle with a long and thick neck."* The near-side horn has the most handy twist for administering a drink with the right hand.

(1408.) *Fomentations.*—"Clean water is the best fomentation. It should be as hot as the hand can bear it, yet not hot enough to pain the animal. In fomenting the horse the groom has rarely enough water, and he does not continue the bathing long enough to do any good. If the leg is to be fomented get a pailfull of water as hot as the hand can bear it; put the horse's foot into it, and, with a large sponge, lave the water well above the affected part, and keep it constantly running down the whole limb. Foment for half an hour, and keep the water hot by adding more."

(1409.) *Poultices.*—"Poultices should be formed of those materials which best maintain heat and moisture, and they should be applied as warm as possible and can be safely borne. They are usually made of bran-mash, turnips or oat-meal porridge. Linseed-meal alone makes the best of poultices, and some of it should always be added to the other ingredients. Wet bandages act as poultices."

(1410.) *Lotions.*—"Of *cooling* lotions cold water is the menstruum. It may be made colder by the introduction of a little salt or ice. Sal-ammoniac and vinegar may be added for the same purpose. The object is to reduce heat and promote evaporation. The addition of a little spirits is made with the same object."

(1411.) *The Pulse.*—"Of the *horse* the natural pulse is from 35 to 45 beats in the minute; under fever it rises to 80, 90 and 100. The most convenient spot to examine it is at the edge of the lower jaw, a little before the angle, where the maxillary comes from the neck to be distributed over the face. The pulse is one of the most important indications in all serious disorders."

(1412.) *Injections.*—"Injections, though easily administered by means of the old ox-bladder and pipe, are still more conveniently given with the syringe. For laxative clysters for the horse or cow, from 1 gallon to 12 pints imperial of warm water or gruel, at the temperature of 96° Fahr., with a couple of handfulls of salt or 2 oz. of soft soap, prove useful. Stronger ones may be obtained by adding a few ounces of aloes to the mixture. In cases of diarrhœa or over-purging, the injection should consist of a few pints of warm gruel, to which is added 1 oz. of catechu electuary, or from ½ drachm to 1 drachm of powdered opium. The only art in administering a clyster—where, however, there is often bungling, and even injury by wounding the rectum—is *to avoid frightening the animal*, anointing the pipe well, and *gently* insinuating it *before the fluid is forced up.*"

(1413.) "In general, bran-mashes, carrots, green meat and hay form the sick horse's diet, gruel and tepid water his drink."† Of the diseases themselves, I shall only notice those at present which usually affect *farm*-horses in winter.

(1414.) *Horse-Louse (Trichodectes equi).*—The horse is infested by a louse as well as the ox, and which is represented in fig. 296. Color of the head and thorax bright chesnut, the former very large and somewhat square, the surface with a longitudinal black line toward each side, forming an angle near the middle; antennæ with the third joint longest; abdomen pale, tawny yellow, with fine pubescence, the first eight segments having a dusky transverse band on the upper half, the lateral margins also with a dusky band; legs pale chesnut; length 1 line. Common in the tail-head and neck of the horse, especially when fresh from pasture in autumn. Found also on the ass. A little oil will destroy this animal when first established; but if allowed to remain on for some time, mercurial ointment will be necessary, but in small quantities at a time. The ass, however, has a louse peculiar to itself, the *Hematopinus asini*; of a rusty red; abdomen whitish, tinged with yellow, with a row of dark horny excrescences on each side; head long, with a deep sinuosity behind the antennæ; length 1 to 1½ lines. It frequents the mane and neck, and is common.‡

Fig. 296.

THE HORSE-LOUSE, TRICHODECTES EQUI.

(1415.) *Batts.*—One of the most common complaints among farm-horses is the flatulent colic, gripes or batts. It arises from indigestion, which again is occasioned by various causes, such as hard work immediately after feeding, drinking water largely after a feed of corn, bad state of the food, fast eating, and, in consequence, a paucity of saliva, an overloaded stomach, a sudden change of food from soft to hard and dry, and more likely to occur after eating turnips, potatoes, carrots and grass, than hay and oats, and after peas than barley. The indigestion arises in two forms; the food either undergoing no change, or running rapidly to fermentation. In the former case *acute foot-founder* is apt to arise, and its treatment is purgatives, drenches and injections. In the latter case the symptoms are most alarming. The horse falls down, rolls over, starts up, paws the ground with his fore-foot, strikes his belly with the hind foot, perspiration runs down and agony appears extreme. Relief may be obtained from this dose: Linseed-oil, raw, 1 lb.; oil of turpentine from 2 to 3 oz.; laudanum from 1 to 2 oz., or hartshorn from ½ oz. to 1 oz. The following tincture may be kept in readiness: In 2 lbs. of whisky, digest for 8 days, 2

* Youatt on the Horse, edition of 1842.

† Dick's Manual of Veterinary Science.

‡ Denny's Monographia Anoplurorum Britanniæ

oz. of ginger, 3 oz. of cloves, and then add 4 oz. of sweet spirits of nitre. Half a pint imperial of this tincture is a dose in a quart of warm water. The abdomen should be rubbed, the horse walked slowly about and supplied with a good bed, and with room to roll about. If there is no relief in half an hour a second dose may be given, and ere long, if still required, a third. Farm-horses that have keen appetites and devour their food greedily, and when they have been long in the yoke, are most apt to take this disease.

(1416.) *Inflammation of the Bowels.*—The symptoms of the batts are very similar, at first, to those of inflammation of the bowels, and, if mistaken, serious mistakes may arise, as the treatment of the two complaints is very different. The symptoms may be distinguished thus: In batts, the pulse remains nearly unaltered, whereas in inflammation it is quickened; all the extremities, the ears and feet, feel cold in batts, hot in inflammation. Whenever inflammation is apprehended, blood may be taken; in batts this is not necessary; but under such an apprehension, the assistance of the veterinary surgeon should be obtained as speedily as possible. I have cured many horses of the batts by administering stimulating drinks with a handy cow's horn. I remember of one horse being seized with inflammation of the bowels, on its arrival home from delivering corn at the market-town; and though the usual remedies of bleeding and blistering were resorted to, they proved ineffectual, no doubt from being disproportioned to the exigencies of the case, and the horse sunk in five days in excruciating agony. There was no veterinary surgeon in the district at that time, which was many years ago. Now, however, thanks to the Veterinary College of Edinburgh, through the really practically useful tuition of its indefatigable Principal, Professor Dick, there is not a populous district of the country in which a skillful veterinarian is not settled. To the surgeon, therefore, in a serious case such as this—and, indeed, in all cases of extensive *inflammation*, and especially in the interior of the body—recourse should immediately be had. I say *immediately*, for it is but fair to give the surgeon a *chance* of treating the case correctly from its commencement, and not to impose upon him the task of amending your previous bungling. Inflammation of the *lungs*, as well as inflammation of the *kidneys*, both of which the farm-horse is subject to, should always be treated by the veterinarian; but fortunately, these formidable maladies may, almost with certainty, be evaded with well-timed working, discrimination of work according to the state of the weather, and by good food, supplied with regularity and in due quantity.

(1417.) *Common colds* frequently occur among farm-horses at the commencement of winter, and when not entirely unheeded, but treated with due care, seldom leave serious effects. "A cold requires nothing more but confinement in a moderately warm stable for a few days, with clothing, bran-mashes instead of corn, and a little laxative and diuretic medicine." The evil lies not so much in the complaint itself, as in its ordinary treatment; it is seldom thought seriously of by farmers—"it is *only* a cold," is the usual remark—and, in consequence, the horse goes out every day, feels fatigued, gets wet, becomes worse, and then the lungs not unfrequently become affected, or a chronic discharge is established from one of the nostrils. One season 9 horses out of 12 in one stable were affected, one after another, by a *catarrhal epidemic*, which required bleeding, poulticing, or blistering under the jaw, besides the medical remedies mentioned above. These I was obliged to take charge of myself, there being no veterinarian in the district, and all fortunately recovered. The remaining 3 were slightly affected afterward, and easily brought through; but had the cases been unheeded from the first, very serious loss might have been incurred by death.

(1418.) *Grease.*—"The well-known and unsightly disease called grease," says Professor Dick, "is a morbid secretion from the cutaneous pores of the heels and neighboring parts, of a peculiar greasy, offensive matter, attended with irritation and increased vascular action. It is most frequently seen in coach and cart-horses, but often also in young colts which are badly cared for; and it is most common in the hind-feet, but occurs in all. Its main cause seems to be sudden changes in the condition of the foot from dry to wet, and from heat to cold, greatly augmented, of course, by evaporation." Hence the evil effects of washing the legs at night, without thoroughly drying them afterward. "The first appearance of grease," continues the Professor, "is a dry state of the heels, with heat and itchiness. Swelling succeeds, with a tendency to lameness; the discharge augments in quantity, the hair begins to fall off. . . . . In the early stage the parts should be washed with soap and water, and a solution of sugar of lead and sulphate of zinc applied; this may not be chemically scientific, but we have found it superior to anything else. Even in old and aggravated cases it is very efficacious. . . . . If the horse be strong and full of flesh, laxatives should be given, followed by diuretics; if weak, tonics may be added to these last. The feeding, too, must be varied with the condition: green-meat and carrots should be given, and mashes frequently, as a substitute for corn. During convalescence, exercise should be given, and bandages and pressure hasten the cure." I have no hesitation in saying that it is a disgrace for any steward, and in the want of such a functionary, it is so in the farmer himself, to allow his horses to become greasy. There is a complaint called a *shot of grease*, arising from a different cause from the common grease. "In the horse, plethora," says Professor Dick, "creates a strong disposition to inflammation of the eyes, feet and lungs, and sometimes to an eruption which is called *surfeit*, or the *nettle-rash*. The hair falls off in patches, and the skin is raw and pimpled. There is also a tendency to *grease*, and to what has been designated a *weed* or *shot of grease* in the heavy draught-horse. One of the legs, generally a hind one, suddenly swells, the animal becomes lame; there is pain in the inside of the thigh—increased upon pressure; and fever supervenes. . . . . We have seen it occur chiefly during continued rest after hard work and exposure to weather, in animals which were highly fed. The best treatment is large blood-letting, scarifying the limb, fomenting, and applying hay, straw, or flannel bandages, with purgatives and diuretics. The pressure of a bandage will expedite the reduction of the part to its natural dimensions."

(1419.) *Stomach-staggers.*—"The most prominent symptoms of this disease are the horse's hanging his head, or resting it on the manger, appearing drowsy, and refusing food, the mouth and eyes being tinged with a yellowish color; there is twitching of the muscles of the chest, and the fore-legs appear suddenly to give way, though the horse seldom falls. Inflammation of lungs or bowels, or lockjaw, may supervene. Its cause is long fasting and overwork; but the quality of the food acts as a cause. Its treatment is relieving the stomach and bowels with searching laxatives, such as croton, also aloes and calomel, with ginger. Clysters should also be given, and afterward cordials. Blood-letting from the jugular vein will be attended with advantage. Finally, steady exercise and careful feeding will prevent a recurrence of the disorder."* I had a year-old draught-colt that was affected with this disease. He was a foul-feeding animal, delighting to eat the moistened litter from the stable and byre. He was bled and physiced by a veterinarian, who had established himself in the neighborhood, and the front of his head blistered. He quite recovered, and having been removed from the temptation of foul feeding, he was never again similarly affected. The practice of keeping *he-goats* in the stables of inns, and of those persons who have extensive studs, is supposed, by the common people, to act as a *charm* against the mad staggers; but, as Marshall judiciously observes, the practice may be explained on physiological principles. "The staggers are a nervous disorder," he says, "and as odors, in many cases, operate beneficially on the human nerves, so may the strong scent of the goat have a similar effect on those of the horse. The subject," he adds, "is worthy of inquiry."† And he gives a striking instance of the good effects of the practice.

(1420.) *Thrush and Corns.*—I have said that the feet of the farm-horse are not liable to so many diseases as those of horses subjected to high speed on hard roads. Farm-horses, however, are liable to *thrush* and *corns* in the feet. The former is situate at the hind part of the cleft of the frog, originating principally from continued application of moisture and dirt, and hence it may be most expected to be seen in dirty stables, of which there are not a few in the country. After being thoroughly cleaned out, the hollow may be filled with calomel, which generally cures; or with pledgets of tow dipped in warm tar, or spirit of tar, applied at night, and retained during the day. The general health of the horse should be attended to. *Corns* are usually the consequence of the irregular pressure of the shoe on peculiarly formed hoofs; and are mere bruises, generally produced by the heel of the shoe, and which, from the extravasated blood, assume a reddish or dark color. They usually occur only in the fore-feet; and their site is almost invariably in the inner quarter between the bar and crust, at the heel. The obvious cure is removal of the pressure of the shoe.

(1421.) *Broken wind.*—Besides natural complaints, farm-horses are liable, in the execution of their work, to accidents which may produce serious complaints. Thus overwork, in a peculiar state of condition, may produce *broken wind*, which is the common phrase given to all disorganized affections of the lungs, though the term is defined by veterinarians to be "the rupture of some of the air-cells of the lungs, whereby air-vesicles are produced on the surface, and the expulsion of the air is rendered less direct and easy. It is usually produced by animals being urged to over-exertion when in bad condition, though a horse may become broken-winded in a straw-yard." There are many degrees of broken wind, which receive appellations according to the noise emitted by the horse; and on this account, he is called a piper, trumpeter, whistler, wheezer, roarer, highblower, grunter, and with thick wind, and with broken wind. I had two uncommonly good horses affected in the wind by working much in the *traces* of a four-horse plow, which were employed to rip up old turf dykes intermixed with large stones, and to break up rough ground. These serious effects of such work gave me the hint to relinquish it, and take to the spade, which I soon found did the work much better, and in the end cheaper. The horses got gradually worse under the disease, and at length being unable to maintain their step with the rest, were disposed of as broken-winded horses.

(1422.) *Sprains.*—"A sprain, or strain, is violence inflicted, with extension, often rupture and displacement, upon the soft parts of a joint, including cellular membrane, tendons, ligaments, and all other parts forming the articulation. The dislocation or disruption may be complete, or it may be a mere bruise or stress; and innumerable are the shades of differences between these extremes. Effusion of the fluids is an attendant consequence. Parts of vital importance, as in the neck or back, may be implicated, and the accident be immediately fatal, or wholly irremediable; on the contrary, they may be to that extent only, that, with time and ease, restoration may be accomplished. They constitute a serious class of cases. The marked symptoms are, pain in the injured parts, and inability of motion, sometimes complete. The treatment is at first rest, a regulation of the local action and constitutional disturbance, according to circumstances, by venesection, general and local, the antiphlogistic regimen, fomentation, bandages, and other soothing remedies; and when the sprain is of an older date, counter-irritation, friction, and gentle exercise." Farm-horses are not unfrequently subject to strains, especially in doing work connected with building, draining, and other heavy work; and they are most apt to occur in autumn, when geldings are generally in a weak state. For rough work of this kind, old seasoned horses are best adapted, and such may often be procured for little money at sales of stock.

(1423.) *Saddle-galls.*—When young horses are first put to work, the parts covered by the saddle and collar are apt to become tender, heated, and then inflamed, and if the inflammation is neglected, the parts may break out into sores. Washing with a strong solution of salt in water with tincture of myrrh is a good lotion, while attention should be paid to the packing of both saddle and collar, until they assume the form of the horse intended to wear them. "Tumors, which sometimes result from the pressure of the saddle, go by the name of *warbles*, to which when they ulcerate the name of *sitfasts* is applied, from the callous skin which adheres to the center. Goulard water may be used to disperse the swelling; a digestive ointment will remove the sitfast; and the *sore* should be healed with a solution of sulphate of zinc."

(1424.) *Crib-biting* and *wind-sucking.*—These practices are said to increase the tendency to indigestion and colic, and to lower condition, rendering the horses which practice them unsound. "A crib-biter derives his name from seizing the manger, or some other fixture, with his teeth, arching his neck, and sucking in a quantity of air with a peculiar noise. . . . . Wind-sucking

* Dick's Manual of Veterinary Science.

† Marshall's Rural Economy of Gloucestershire, vol. 2

consists in swallowing air, without fixing the mouth. The horse presses his lip against some hard body, arching his neck, and gathering together his feet." Both vices are said to be prevented by fastening a strap round the neck, studded with one or more sharp points or prickles opposite the lower jaw; but this means will not avail in all cases, for I had a year-old colt, which first began crib-biting in the field, by seizing the gate or any other object he could find. Being prevented using the gate by a few thorns, he pressed his mouth against any object that would resist him, even against the sides or rumps of his companions, and he then began to be a wind-sucker. A strap of the above form was put on, recommended to me by an artillery officer; but though it remained upon the colt for more than a twelvemonth, night and day, and as tight as even to affect his appearance, he continued to crib-bite or wind-suck in spite of it, even to the laceration of his skin by the iron studs. Growing largely to the bone, though very thin, he was taken up to work at the early age of two years, solely with the view of seeing if the yoke would drive him from the practice, but it had no such effect. Whenever he came into the stable he set to with earnestness to bite and suck with the strap on, until he would become puffed up as if to bursting, and preferred sucking wind to eating his corn. At length I was so disgusted with the brute that I sold him to a carrier, to draw a heavy single cart, and got a fair price for him, though sold as a crib-biter.

(1425.) *Dust-ball.*—Millers' horses are most liable to be affected with this disease. It is composed of corn and barley dust, saved in grinding meal, and used as food, and occurs sometimes in the stomach, but more frequently in the intestinal canal. "In an advanced stage no doubt can remain as to the nature of the disorder. The countenance is haggard, the eye distressed, the back up, the belly distended, the respiration becomes hurried, bowels habitually costive, and sometimes the horse will sit like a dog on his haunches. Relief may frequently be afforded. Strong purgatives and large injections must be given, and under their continued action the offending body is sometimes removed." On using barley-dust as food for horses, it would be well to mix it thoroughly with the other prepared ingredients, instead of using it in the dry state."

(1426.) *Worms.*—Farm-horses are sometimes affected with worms. These are of 3 kinds: the round worm, *teres*; the thread-worm, *ascaris*; and the tape-worm, *tænia*. "In the horse the tenia is very rare; in the dog exceedingly common. When the horse is underfed his bowels are full of teres and ascaris; and the appearance of his staring coat, want of flesh and voracious appetite betoken it. They occasion gripes and diarrhœa, but the mischief they produce is not great. The principal habitat of the ascaris is the cæcum, although they are sometimes found in countless multitudes in the colon and rectum. Turpentine is a deadly poison to all these worms; but this medicine, so harmless in man, acts more disagreeably in the lower animals. Hence it must not be given to them pure or in large quantities, but mixed in small proportion with other oils, as linseed, or in a pill; and, with these precautions, it may be found at once safe and efficacious."

(1427.) *Nebulæ or Specks in the Eye.*—Farm-horses are not subject to the more violent diseases of the eye; but, being liable to accidents, the effects of inflammation—nebulæ or specks—do sometimes appear. "The former are superficial, the latter dip more deeply into the substance of the part. Directly in the sphere of vision these, of course, impede it, and cause obscurity of vision. Even here we must proceed gently. These blemishes are the pure consequences of inflammation, and this subdued, their tendency is to disappear. Time and nature will do much, and the duty of the practitioner consists in helping forward the salutary process where necessary, by gently stimulating washes, while irritating powders should be avoided."* With these sensible remarks of Professor Dick I shall conclude what I have to say of the diseases of the farm-horse at this time.

(1428.) The offals of the horse are not of great value. His *hide* is of most value when free of blemishes. It tans well and forms a good leather, which, on being japanned, is chiefly used for covering carriages. I was informed by a friend who settled in Buenos Ayres as a merchant, that he once bought a lot of horses, containing no fewer than 20,000, for the sake of their hides alone, and that some of them would have fetched good prices in England. They were all captured with the *lasso*.

(1429.) *Horse-hair* is used in the manufacture of damask-cloth for sofas and chair-bottoms. The dyeing of it of various beautiful colors and the manufacture of the damask figures have been much improved of late. Horse-hair is also used for making fish-lines, horse-tails for cavalry caps and stuffing for matresses, for which last purpose it is prepared by being wound up hard and baked in an oven.

(1430.) "Hair, of all animal products, is the least liable to spontaneous change. It can be dissolved in water only at a temperature somewhat above 230° Fahr. in Papin's digester, but it appears to be partially decomposed by this heat, since some sulphureted hydrogen is disengaged. By dry distillation hair gives off several sulphureted gases, while the residuum contains sulphate of lime, common salt, much silica, and some oxides of iron and manganese. It is a remarkable fact that fair hair affords magnesia instead of these latter two oxides. Horse-hair yields about 12 per cent. of the phosphate of lime. Hair also yields a bituminous oil, which is black when the hair is black, and yellowish when the hair is red."†

(1431.) "Button-moulds are made of the bones of the horse, ox and sheep. The shavings, saw dust and more minute fragments in making these moulds, are used by the manufacturers of cutlery and iron toys in the operation of case-hardening, so that not the smallest waste takes place."‡ The bones of all these animals, when reduced small, make the valuable manure—bone-dust—now well known to every farmer.

---

* Dick's Manual of Veterinary Science.

† Thomson's Animal Chemistry

‡ Ure's Dictionary of the Arts, arts. *Hair—Buttons.*

END OF VOLUME I. OF BOOK OF THE FARM.

FULL PAGE PLATES

PLATE I.

Fig. 1.

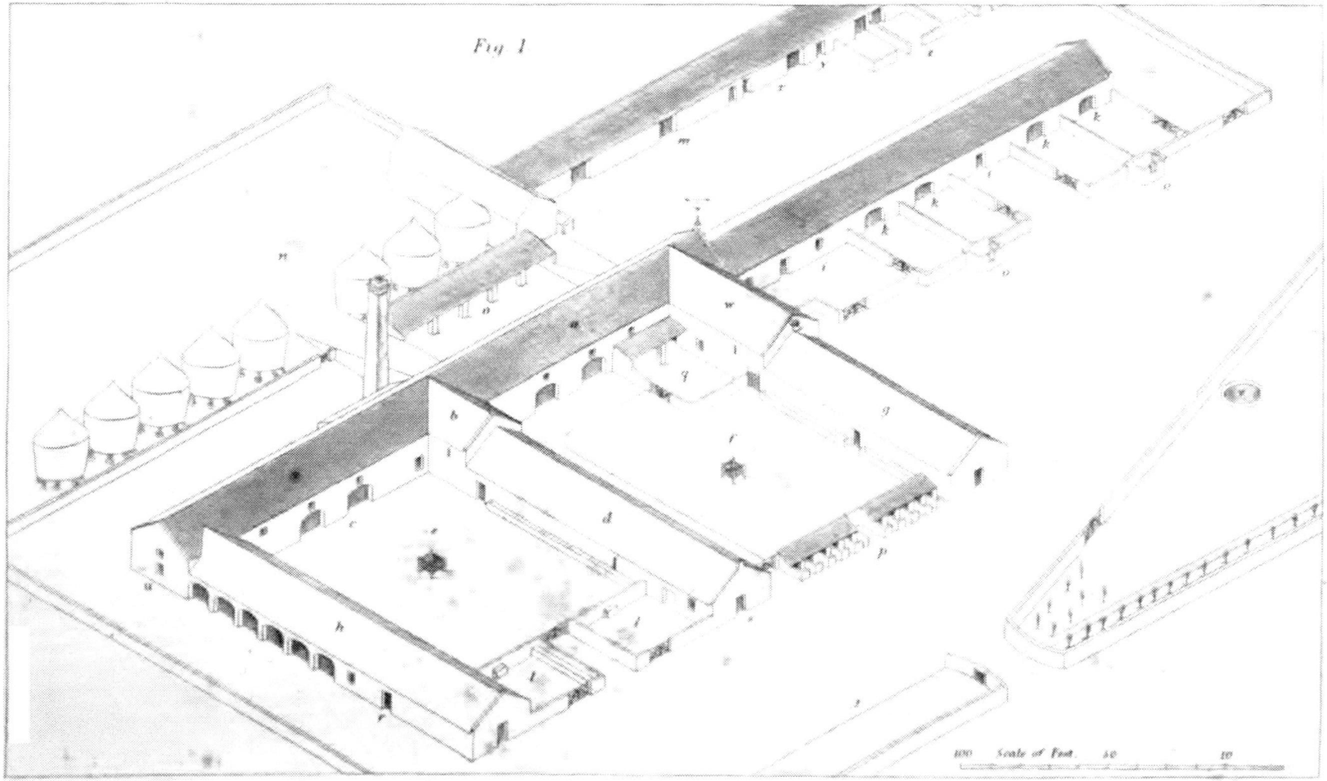

100 Scale of Feet. 50 10

ISOMETRICAL VIEW OF AN EXISTING STEADING

PLATE II.

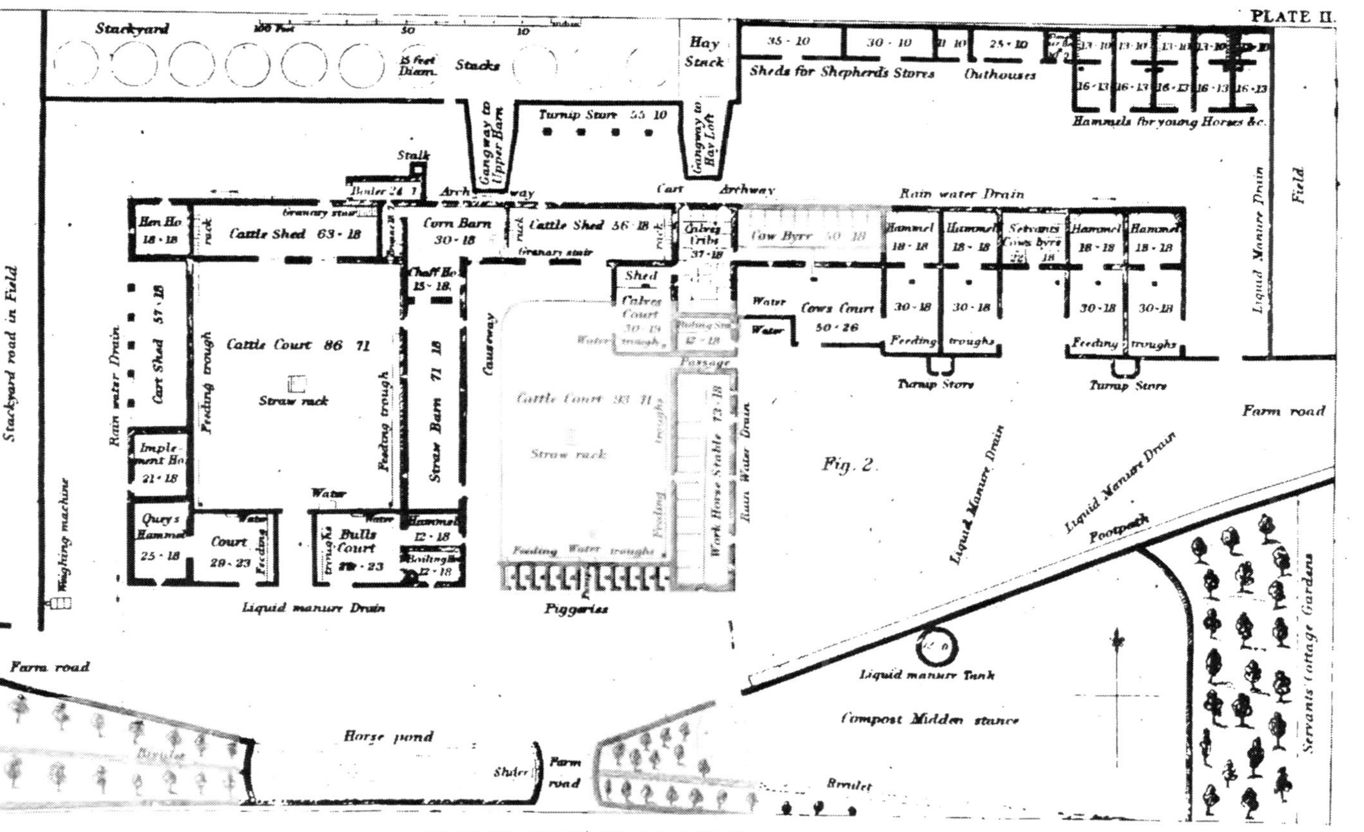

GROUND PLAN OF AN EXISTING STEADING.

PLATE III.

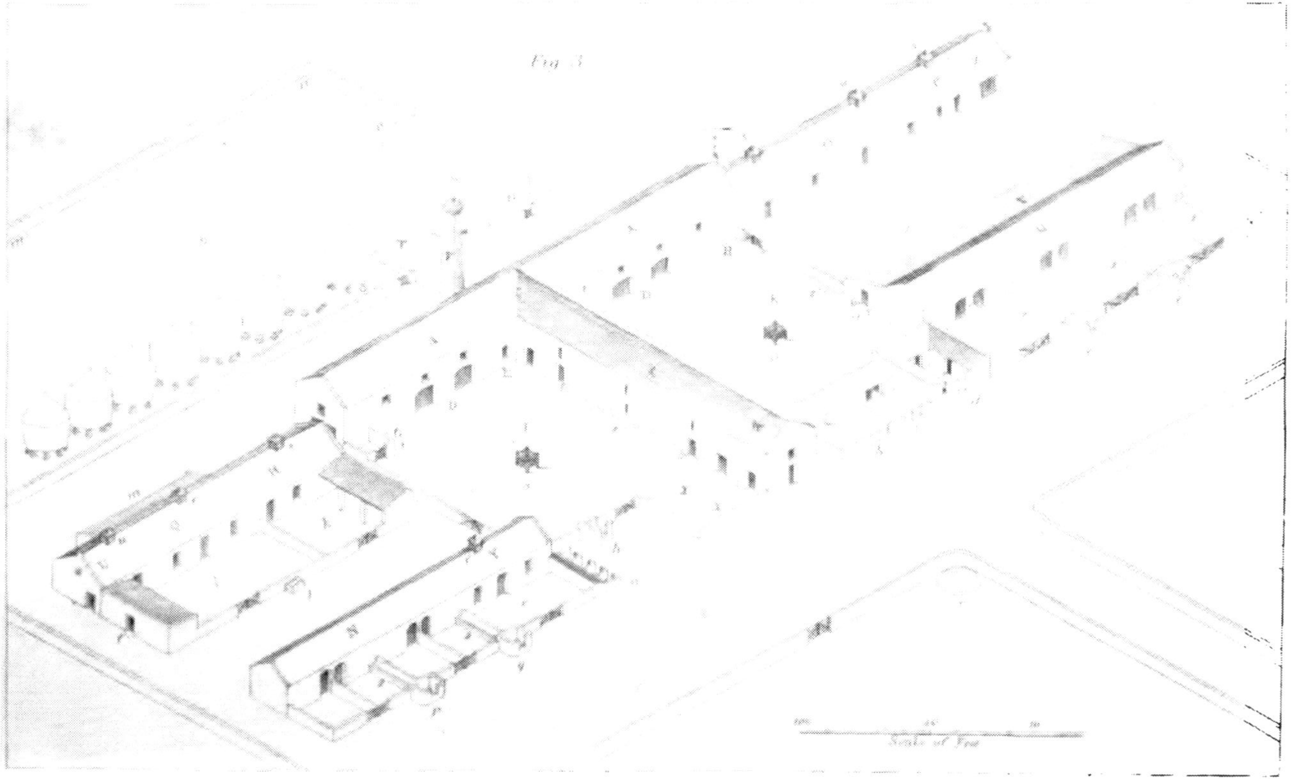

Fig. 3

ISOMETRICAL VIEW OF A PROPOSED STEADING.

Fig. 4.

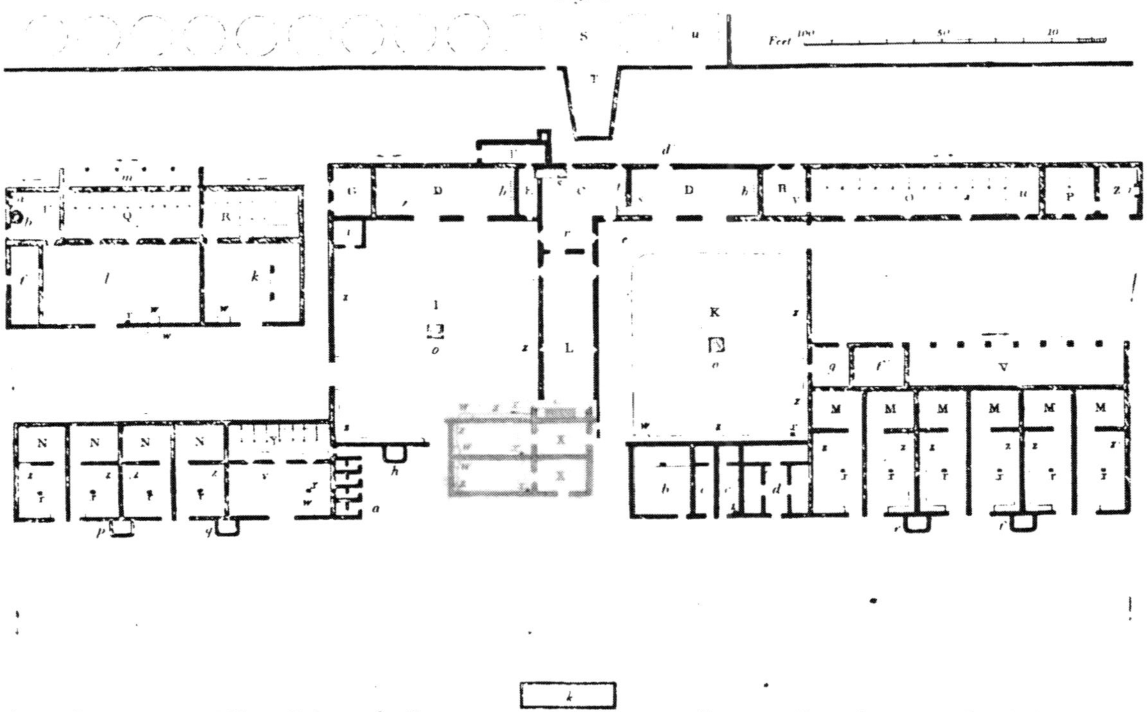

GROUND PLAN OF A PROPOSED STEADING.

MID LOTHIAN OR CURRIE PLOUGH

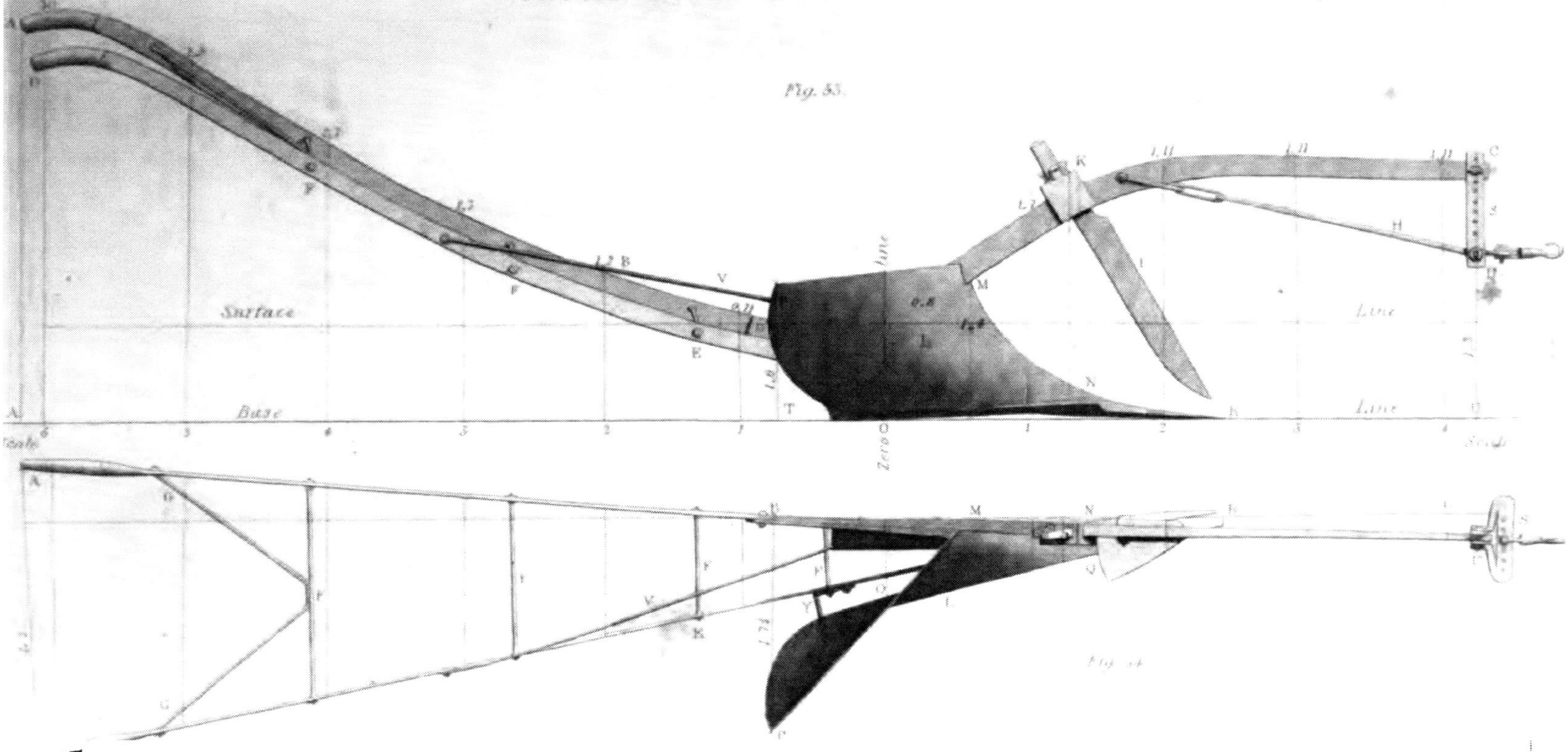

Published by Greeley & McElrath, New York

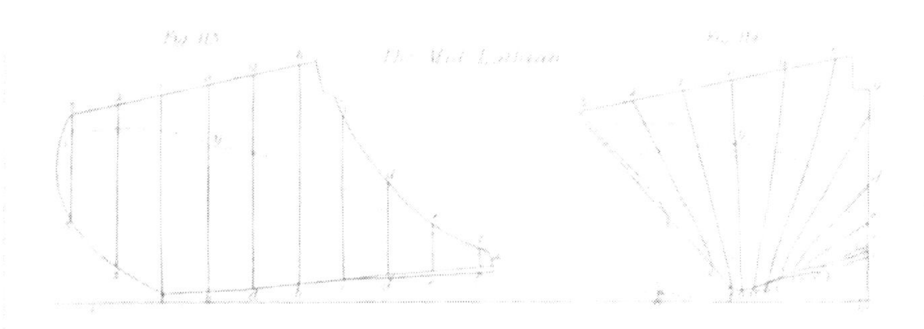
The Mid Lothian

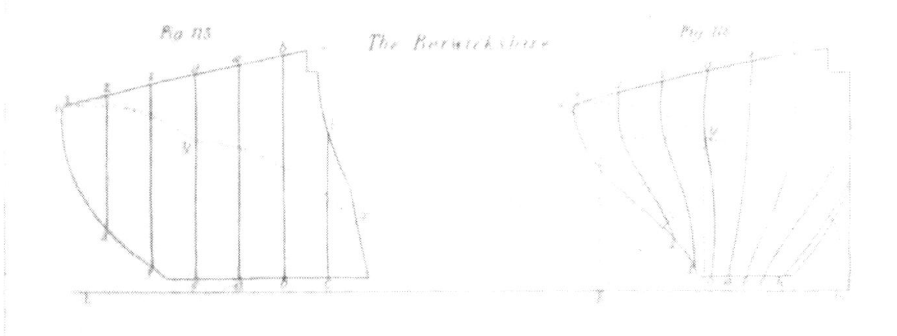
The Berwickshire

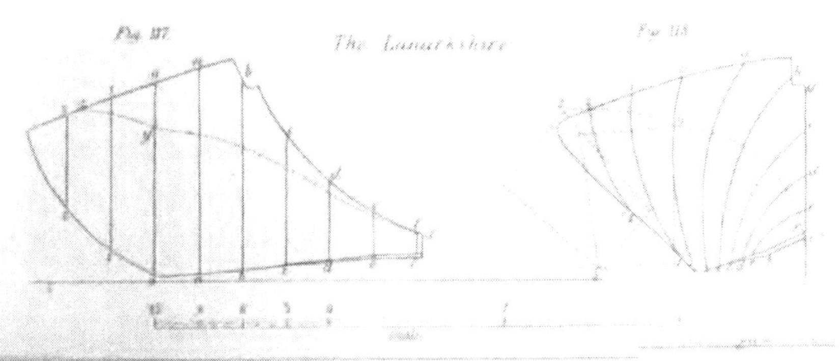
Fig. 117
The Lanarkshire

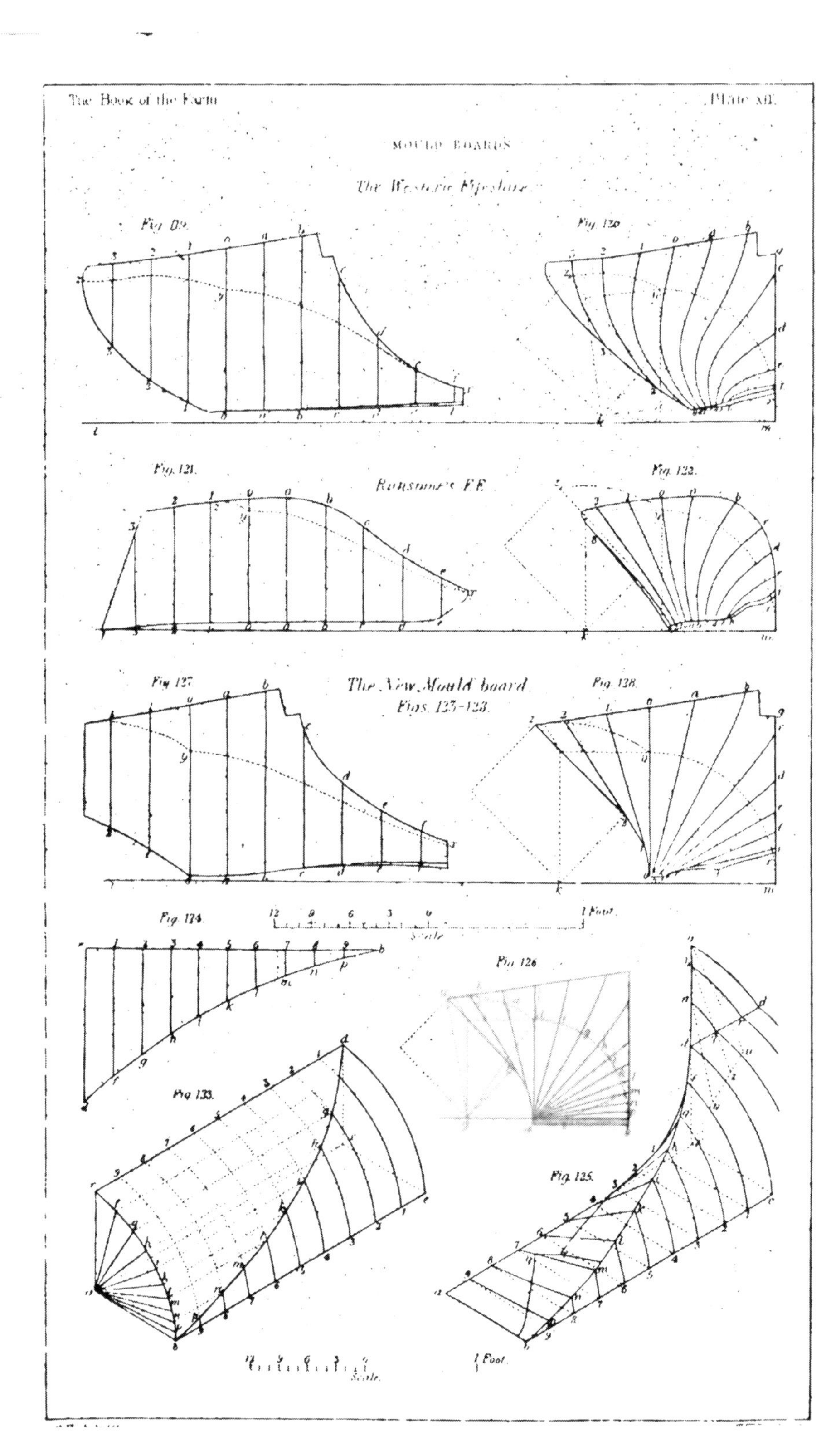

The Book of the Earth
Plate XII.
MOULD BOARDS
Fig. 120.
Fig. 121.
Ransome's F.F.
Fig. 122.
Fig. 127.
The New Mould board
Figs. 123-128.
Fig. 128.
Fig. 124.
Scale
1 Foot.
Fig. 126.
Fig. 133.
Fig. 125.
Scale.
1 Foot.

Also from Benediction Books ...

Wandering Between Two Worlds: Essays on Faith and Art
Anita Mathias
Benediction Books, 2007
152 pages
ISBN: 0955373700

Available from www.amazon.com, www.amazon.co.uk
www.wanderingbetweentwoworlds.com

In these wide-ranging lyrical essays, Anita Mathias writes, in lush, lovely prose, of her naughty Catholic childhood in Jamshedpur, India; her large, eccentric family in Mangalore, a sea-coast town converted by the Portuguese in the sixteenth century; her rebellion and atheism as a teenager in her Himalayan boarding school, run by German missionary nuns, St. Mary's Convent, Nainital; and her abrupt religious conversion after which she entered Mother Teresa's convent in Calcutta as a novice. Later rich, elegant essays explore the dualities of her life as a writer, mother, and Christian in the United States-- Domesticity and Art, Writing and Prayer, and the experience of being "an alien and stranger" as an immigrant in America, sensing the need for roots.

About the Author

Anita Mathias was born in India, has a B.A. and M.A. in English from Somerville College, Oxford University and an M.A. in Creative Writing from the Ohio State University. Her essays have been published in The Washington Post, The London Magazine, The Virginia Quarterly Review, Commonweal, Notre Dame Magazine, America, The Christian Century, Religion Online, The Southwest Review, Contemporary Literary Criticism, New Letters, The Journal, and two of HarperSanFrancisco's The Best Spiritual Writing anthologies. Her nonfiction has won fellowships from The National Endowment for the Arts; The Minnesota State Arts Board; The Jerome Foundation, The Vermont Studio Center; The Virginia Centre for the Creative Arts, and the First Prize for the Best General Interest Article from the Catholic Press Association of the United States and Canada. Anita has taught Creative Writing at the College of William and Mary, and now lives and writes in Oxford, England.

The Greater Trumps, Charles Williams

My Apprenticeship: Volumes I and II, Beatrice Webb

Last and First Men / Star Maker, Olaf Stapledon

Last and First Men, Olaf Stapledon

Darkness and the Light, Olaf Stapledon

The Worst Journey in the World, Apsley Cherry-Garrard

The Schoole of Abuse, Containing a Pleasaunt Invective Against Poets, Pipers, Plaiers, Iesters and Such Like Catepillers of the Commonwelth, Stephen Gosson

Russia in the Shadows, H. G. Wells

Wild Swans at Coole, W. B. Yeats

A hundreth good pointes of husbandrie, Thomas Tusser

The Collected Works of Nathanael West: "The Day of the Locust", "The Dream Life of Balso Snell", "Miss Lonelyhearts", "A Cool Million", Nathanael West

Miss Lonelyhearts & The Day of the Locust, Nathaniel West

The Worst Journey in the World, Apsley Cherry-Garrard

Scott's Last Expedition, V1, R. F. Scott

The Dream of Gerontius, John Henry Newman

The Brother of Daphne, Dornford Yates

The Downfall of Robert Earl of Huntington, Anthony Munday

Clayhanger, Arnold Bennett

The Regent, A Five Towns Story Of Adventure In London , Arnold Bennett

The Card, A Story Of Adventure In The Five Towns , Arnold Bennett

South: The Story of Shackleton's Last Expedition 1914-1917, Sir Ernest Shackketon

Greene's Groatsworth of Wit: Bought With a Million of Repentance, Robert Greene

Beau Sabreur, Percival Christopher Wren

The Hekatompathia, or Passionate Centurie of Love, Thomas Watson

The Art of Rhetoric, Thomas Wilson

Stepping Heavenward, Elizabeth Prentiss

Barker's Delight, or The Art of Angling, Thomas Barker

The Napoleon of Notting Hill, G.K. Chesterton

The Douay-Rheims Bible (The Challoner Revision)

Endimion - The Man in the Moone, John Lyly

Gallathea and Midas, John Lyly,

Mother Bombie, John Lyly

Manners, Custom and Dress During the Middle Ages and During the Renaissance Period, Paul Lacroix

Obedience of a Christian Man, William Tyndale

St. Patrick for Ireland, James Shirley

The Wrongs of Woman; Or Maria/Memoirs of the Author of a Vindication of the Rights of Woman, Mary Wollstonecraft and William Godwin

De Adhaerendo Deo. Of Cleaving to God, Albertus Magnus

Obedience of a Christian Man, William Tyndale

A Trick to Catch the Old One, Thomas Middleton

The Phoenix, Thomas Middleton

A Yorkshire Tragedy, Thomas Middleton (attrib.)

The Princely Pleasures at Kenelworth Castle, George Gascoigne

The Fair Maid of the West. Part I and Part II. Thomas Heywood

Proserpina, Volume I and Volume II. Studies of Wayside Flowers, John Ruskin

Our Fathers Have Told Us. Part I. The Bible of Amiens. John Ruskin

The Poetry of Architecture: Or the Architecture of the Nations of Europe Considered in Its Association with Natural Scenery and National Character, John Ruskin

The Endeavour Journal of Sir Joseph Banks. Sir Joseph Banks

Christ Legends: And Other Stories, Selma Lagerlof; (trans. Velma Swanston Howard)

Chamber Music, James Joyce

Blurt, Master Constable, Thomas Middleton, Thomas Dekker

Since Yesterday, Frederick Lewis Allen

The Scholemaster: Or, Plaine and Perfite Way of Teachyng Children the Latin Tong , Roger Ascham

The Wonderful Year, 1603, Thomas Dekker

Waverley, Sir Walter Scott

Guy Mannering, Sir Walter Scott

Old Mortality, Sir Walter Scott

The Knight of Malta, John Fletcher

The Double Marriage, John Fletcher and Philip Massinger

Space Prison, Tom Godwin

The Home of the Blizzard Being the Story of the Australasian Antarctic Expedition, 1911-1914, Douglas Mawson

Wild-goose Chase , John Fletcher

If You Know Not Me, You Know Nobody. Part I and Part II, Thomas Heywood

The Ragged Trousered Philanthropists, Robert Tressell

The Island of Sheep, John Buchan

Eyes of the Woods, Joseph Altsheler

The Club of Queer Trades, G. K. Chesterton

The Financier, Theodore Dreiser

Something of Myself, Rudyard Kipling

Law of Freedom in a Platform, or True Magistracy Restored, Gerrard Winstanley

Damon and Pithias, Richard Edwards

Dido Queen of Carthage: And, The Massacre at Paris, Christopher Marlowe

Cocoa and Chocolate: Their History from Plantation to Consumer, Arthur Knapp

Lady of Pleasure, James Shirley

The South Pole: An account of the Norwegian Antarctic expedition in the "Fram," 1910-12. Volume 1 and Volume 2, Roald Amundsen

A Yorkshire Tragedy, Thomas Middleton (attrib.)

CPSIA information can be obtained at www.ICGtesting.com
Printed in the USA
BVOW01*0510020414

349502BV00007B/398/P

9 781849 022545